Reaction Green Metrics

Problems, Exercises, and Solutions

Reaction Green Metrics

Problems, Exercises, and Solutions

By

John Andraos

CareerChem
Toronto, ON
Canada
c1000@careerchem.com

CRC Press is an imprint of the
Taylor & Francis Group, an informa business

CRC Press
Taylor & Francis Group
6000 Broken Sound Parkway NW, Suite 300
Boca Raton, FL 33487-2742

Printed on acid-free paper

International Standard Book Number-13: 978-1-138-38895-6 (Hardback)
International Standard Book Number-13: 978-1-138-38894-9 (Paperback)

Visit the Taylor & Francis Web site at
http://www.taylorandfrancis.com

and the CRC Press Web site at
http://www.crcpress.com

Dedication

Dr. Christopher Randall Schmid

(1959–2007)

Inaugural Editor of Organic Process Research & Development

Contents

Preface

An investment in knowledge pays the best interest.

– Benjamin Franklin

Education is what you remember after you have forgotten what you learned in school.

– Albert Einstein

We don't need no education. We don't need no thought control. No dark sarcasm in the classroom. Teachers leave them kids alone.

– Pink Floyd

This book and its sequel volume, *Synthesis Green Metrics: Problems, Exercises, and Solutions*, are the culmination of my 15 years of independent study of the now mature field of green chemistry since 2004. As with all research endeavors, the seed from which this effort emerged is teaching, as was the case when Dmitry Mendeleev came up with the idea of periodicity of the elements or when Gilbert Lewis came up with the idea of drawing dot diagrams for chemical structures to keep track of bonding and non-bonding electrons between atoms. My real education in chemistry began the day I first taught it to others. Up to that point nearly everything I learned from my professors was presented as a set of disjointed facts that needed to be committed to memory for the sole purpose of recall for examination purposes and degree accreditation. So, in order to really understand my subject, I had to relearn it from the beginning. In doing so, two essential tools needed to be mastered. The first was the mapping of the scientific genealogy of the people and ideas that shaped the field. This exercise revealed the connections between those people as a roadmap for career development and it also revealed connections between their ideas, which essentially elucidated the trajectory of important research directions. The second was the fortuitous emergence of green chemistry as a unifying force that linked many scientific disciplines (organic and inorganic chemistries, toxicology, environmental studies, mathematics, computer science, and optimization research) together for a common purpose; namely, to invent improved syntheses as an intellectual endeavor that would have great impact both on preserving and utilizing our planet's finite resources and also on the quality of human life. A great advantage in mastering these skills is quantitative reasoning, which was the key missing ingredient when the 12 Principles of Green Chemistry were announced in 1998 as merely a set of qualitative guiding principles for chemists to think about in the back of their minds. Quantitative reasoning moves that thinking to the front of the mind where decision-making actions are initiated and ultimately carried out.

No accomplishment is made single-handedly. I thank Dr. Floyd H. Dean and his son, Jason, for being intellectual sounding boards on matters of science. I thank Dr. Thuy van Pham for this generous gift of a laptop. I thank Andrei Hent for gathering some of the references from which problems were posed. I thank Hilary LaFoe for her patience as Editor at CRC Press/Taylor & Francis Group throughout this campaign over the last three years. Lastly, I thank my family for their emotional support during the course of my lifelong study of chemistry.

John Andraos
Toronto, Canada

1 Introduction

After 25 years of development, the field of green chemistry has reached critical crossroads on both the education and research fronts. With respect to educating the next generation of chemical scientists, interest in green chemistry is dominated by undergraduate instructors who mainly use it as a means to engage modern students who are active learners and who are seeking ways to make a difference in coming up with innovative solutions to address pressing world problems where science can have a positive impact. Despite this noble cause, green chemistry is still considered a "soft fringe subject" in physical science among research faculty in departments of chemistry as evidenced by the fact that green chemistry is not yet part of the mainstream curriculum for honors degree chemistry programs at leading universities and colleges. In the few places where it does get mentioned in the curriculum, either explicitly or implicitly, it is presented largely in a show-and-tell approach with little quantitative rigor.

In the few dedicated research journals such as *Green Chemistry* (Royal Society), *ChemSusChem* (Wiley-VCH), *Green Chemistry Letters and Reviews* (Taylor & Francis), and *ACS Sustainable Chemical Engineering* (American Chemical Society), which give a voice to cutting-edge research in the field, there are lingering issues. Again, most contributions are presented in a show-and-tell approach with little or no quantitative evidence that backs up the claims of "greenness" made in the research articles. If comparative metrics analysis were to be applied to such pronouncements, it would be readily found that few of them would uphold their claims of "greenness." Up until now, there are no formalized standards put in place on the presentation of so-called green metrics that track any improvements made. Editors have largely left the task to authors' discretion on a voluntary basis for application of metrics in research to justify the implementation of one or more green chemistry principles. Articles focused on metrics themselves, particularly in the areas of material efficiency and life cycle assessment (LCA), are often fraught with "me-too" metrics, where the same ideas are re-branded with new nomenclature depending on the authors, and are presented in a manner without head-to-head comparison of "new" algorithms with prior published algorithms and without framing the discussion in the context of performance, limitations, advantages, and disadvantages among the various methods.

From the point of view of process chemistry, there still exists the narrow-minded view that overall mass balance and inventory control encompasses all that is needed for applying green chemistry metrics when it comes to tracking input material consumption and waste material production. Essentially, metrics such as overall E-factor (mass of waste per unit mass of product) and overall *PMI* (mass of input materials per unit mass of product) are the main concerns—all other material efficiency metrics are considered unduly complicated and thus "over-engineer" the problem of optimization. Of course, this simple approach does not require knowledge of chemistry and only relies on elementary accounting of what goes in versus what comes out of a process. A non-chemistry professional can adequately perform such a task. In doing so, there is little attention paid to making the connection to individual reaction step performances of these metrics and how they relate to the overall metrics for an entire process. When that connection is made, suddenly it becomes apparent that one needs to know the chemistry of each reaction step before waste minimization can begin.

The order of steps undertaken in optimizing a synthesis is also problematic. Solvent reduction and elimination is a top priority since it is obvious that these materials constitute the bulk of input materials used in a chemical process. This is the first line of attack in "greening up" an existing chemical process. However, less attention is paid to inventing new reactions that do not produce by-products from the outset. Coming up with such high atom economy reactions requires imagination. This takes more time and effort. With respect to reducing toxicity and hazard issues, the focus is again on solvents via solvent replacement to arguably more benign ones. The task of synthesis optimization is often carried out in a multi-pronged attack where material metrics, commercial cost and availability, and LCA are optimized simultaneously rather in a structured or layered approach. This leads to a jumble with respect to ranking plans where the inevitable outcome is non-orchestrated optimization. Energy metrics are essentially excluded from experimental procedures and therefore from the discussion of reaction and synthesis optimization.

1.1 WHAT THIS BOOK IS ABOUT

This book addresses the aforementioned problems described in education and research with a focus on material efficiency assessment of individual reactions. In order to successfully navigate the topics presented, the reader is expected to have adequate background knowledge in the following subjects as necessary prerequisites. In chemistry, the reader should have already taken an introductory organic chemistry course, has familiarity with reaction mechanism, has a basic library of named organic reactions committed to memory, and is familiar with drawing and interpreting catalytic cycles. Most importantly, balancing chemical equations is highly emphasized. This skill will be the most challenging to master for a student or a professional, but is a mandatory starting point for solving every example and question posed in this book. No metrics analysis of any kind can begin without first establishing a balanced chemical equation for a given chemical transformation. Time and space are devoted to the following new topics not covered or mentioned in standard undergraduate chemistry courses: (a) drawing schemes for reactions and synthesis plans using the principle of conservation of structural aspect as a key strategy for in-depth understanding and analysis of synthesis strategy and its quantitative parameterization; and (b) connecting balanced chemical equations with their reaction mechanisms. Since this book combines chemistry with quantitative reasoning, there are certain topics in mathematical science that the reader is expected to be fluent in. Specifically, these include basic arithmetic and basic algebra. In order for the reader to take full advantage of the accompanying online suite of spreadsheet algorithms available for Instructors to download by request at https://www.crcpress.com/9781138388949, mastery of Microsoft Excel spreadsheet software and facility with graphing are highly recommended. These spreadsheets and calculators were designed to remove the tedious task of computing metrics by hand.

In addition, extensive databases of key physical and toxicological properties of commonly used industrial chemicals are given in one place so the reader does not have to scour the literature to find them, thereby speeding up the process of conducting calculations and interpreting the results of those calculations.

1.2 STRUCTURE

The topics in this book are presented in a ladder approach where each successive topic follows logically from the previous one. Chapter 2 begins with understanding how to parse the identities and roles of input materials used in a chemical reaction and how to balance chemical equations. Specific attention is paid to redox (reduction-oxidation) reactions because they are the most challenging to balance and yet figure prominently in industrial and process chemistry. Chapter 3 discusses the variety of presentations of experimental procedures appearing in the chemistry literature and how their quality impacts the definiteness of any metrics analysis that is possible. Chapter 4 summarizes the types of chemical reactions that are possible and introduces a standardized nomenclature to describe ring construction strategies. Chapter 5 is an important chapter that is devoted to the principle of conservation of structural aspect and its impact on parameterizing and analyzing synthesis strategy. Standards for the presentation of reaction mechanisms, target bond mapping, and tracking oxidation states in bond forming steps in a synthesis plan are introduced. Chapter 6 introduces metrics that are specific to waste production and input material consumption applied to individual chemical reactions. This chapter focuses on the essential metrics and excludes all superfluous "me-too" metrics that are present in the literature. Most importantly, the selected metrics are connected in a structured and logical hierarchy for immediate comprehension and usage. Chapter 7 is devoted to the concepts of intrinsic greenness and minimum atom economy, and assessing the probability that a given chemical reaction will achieve intrinsic greenness subject to minimum atom economy and reaction yield constraints. An extensive inventory of the library of organic reactions is presented to showcase the present state of the art. Chapter 8 is devoted to presenting all of the Microsoft Excel spreadsheets that have been developed to evaluate material efficiency of individual reactions and synthesis plans in an automated, reliable, and easy-to-use fashion.

1.3 HOW TO USE THE BOOK

For teaching purposes, instructors can use the worked examples in a classroom setting to introduce topics for discussion in lectures and the problems as homework set exercises for deeper thought. It is recommended that the book be used in a one-semester introductory course on green chemistry as a dedicated subject spanning academic and industrial chemistry topics. The style of presentation is brief where definitions of terms are presented first along with an immediate easy example. Each topic is briefly introduced followed by worked examples to illustrate the ideas for immediate understanding and implementation. Advanced problems are posed at the conclusion of each chapter. Consistent with good pedagogy, all examples and problems posed come directly from literature sources, which are cited upfront for both instructors and students to view. Solutions to all problems are available electronically through the publisher's website.

For research purposes, experienced readers may skip topics they are familiar with and focus on those areas they wish to learn more about as needed. The same pedagogical style described above is applicable to experienced chemists for their benefit. Examples and problems of interest to industrial chemists, process chemists, and chemical engineers have been taken from both research journals and patents.

2 Constitution of a Chemical Reaction and Reaction Balancing

2.1 IDENTIFYING COMPONENTS OF ANY CHEMICAL REACTION

2.1.1 Terms, Definitions, and Examples

Additive

A substance added to a reaction mixture that improves reaction performance toward a desired product but whose role is not well defined.

Auxiliary

A substance or material used in the post-processing of a chemical reaction typically in work-up procedures involving washing or extraction and in purification procedures involving chromatography or recrystallization.

By-Product [1, 2]

A product formed in a reaction between reagents as a direct mechanistic consequence of producing the target product assuming a balanced chemical equation that accounts for the production of that target product. By-products and the target product appear on the right-hand side of a balanced chemical equation.

Catalyst

A substance added to a chemical reaction that accelerates the reaction. It participates in the reaction but its structure remains unchanged at the end of the reaction. The consequences of its use are milder reaction conditions such as reduced temperature, shorter reaction times, higher product yield, and higher product selectivity.

Catalyst Loading

A term describing the mass amount of catalyst relative to the mass of substrate (usually the limiting reagent) in a given reaction. The usual unit used is mol%.

Coupled Product(S)

Coupled products arise as a mechanistic consequence of producing the desired target product in a chemical reaction. The term is synonymous with by-products.

Excess Reagent

A reagent used in a chemical reaction at a molar quantity that exceeds that of the limiting reagent as prescribed by the balanced chemical equation. The excess amount also takes into account the accompanying stoichiometric factor associated with that reagent. For example, for a reaction $\nu_A A + \nu_B B \rightarrow \text{products}$, if A is the limiting reagent and x moles of it are used, then the expected stoichiometric number of moles of B needed is $x\nu_B / \nu_A$. However, if y moles of B are used such that $y > x\nu_B / \nu_A$, then we say that reagent B is the excess reagent.

Extraction Solvent

A solvent used in the workup operation in carrying out a chemical reaction. The most common extraction solvents are diethyl ether, petroleum ether, and dichloromethane. The mass of this solvent is counted in the E-aux contribution to E-total.

Feedstock

Feedstock is a general term that refers to any input material used in a synthesis of a target compound in a single reaction or in a synthesis plan composed of several steps. Typically, feedstocks refer to first, second, and third generation high volume industrial commodity chemicals available from the petrochemical industry.

Ligand

In coordination chemistry, a ligand is an ion or molecule (functional group) that binds to a central metal atom to form a coordination complex. In organic synthesis, ligands are associated with structures of organometallic catalysts which contain a central metal ion surrounded by various kinds of functional groups bonded to it.

Limiting Reagent

In a chemical reaction, the reagent that has the least number of moles associated with it corrected for its associated stoichiometric coefficient in a balanced chemical equation.

For a balanced chemical equation given by $\nu_1 S_1 + \nu_2 S_2 \rightarrow P + \nu_3 Q$, where S_1 and S_2 are reagents, P is the target product, and Q is the by-product; and the ν parameters are the respective stoichiometric coefficients, the limiting reagent is one that satisfies the criterion shown in Equation (2.1).

$$\min\left(\frac{\text{moles}_{S_1}}{\nu_1}, \frac{\text{moles}_{S_2}}{\nu_2}\right) = \min\left(\frac{m_{S_1}/(MW)_{S_1}}{\nu_1}, \frac{m_{S_2}/(MW)_{S_2}}{\nu_2}\right) \quad (2.1)$$

where the m parameters refer to the respective masses and the MW parameters refer to the respective molecular weights of reagents. Hence, if

$$\frac{\text{moles}_{S_1}}{\nu_1} < \frac{\text{moles}_{S_2}}{\nu_2}$$

then S_1 is the limiting reagent; otherwise it is the reagent S_2.

Reactant or Reagent

A reactant or reagent is a chemical substance that appears on the left-hand side of a balanced chemical equation. The word *substrate* is also used to designate that reagent or reactant whose structure appears in the highest proportion in the target product of the reaction. Chemists use the three words interchangeably in the literature.

Reaction Solvent

A liquid selected for a conducting a reaction that satisfies the condition that its boiling point matches the desired reaction temperature and that it is able to dissolve the reactants, catalysts, and any other additives. Reaction solvents may or may not act as reagents. Examples of reactions where the reaction solvent is also a reagent are the Fischer esterification of carboxylic acids in alcohols, and the synthesis of acyl chlorides in thionyl chloride.

Side Product [1]

A side product is a product formed in a reaction between reagents, usually undesired, that arises from a competing reaction pathway other than the one that produces the intended target product and its associated by-products.

Side Reaction [1]

A side reaction is a competing, often unwanted, reaction other than the intended reaction between a set of reactants. A side reaction produces side products via a different mechanism than the intended reaction.

Starting Material

In a balanced chemical equation analyzed at the kernel level, the starting materials refer to reagents or reactants. When analyzed at the global level, starting materials also include catalysts, other additives, and reaction solvents.

Target Product

The final desired product in a chemical reaction or synthesis plan.

Waste Material

In a chemical reaction, a waste material is either a reaction by-product, a reaction side product, unreacted reagents, a reaction solvent, a catalyst or other additive, a workup material, or a purification material.

Example 2.1 [3]

Look up the experimental procedure and assign roles to each chemical species used as input material. Write out a balanced chemical equation for the chemical transformation.

Determine which reactant is used in excess. What are the expected waste materials that are produced in this reaction?

Solution

Reactants: tripropargylamine, benzyl azide
Reaction solvents: acetonitrile, water
Catalysts: sodium ascorbate, copper(II)acetate monohydrate

Work-up materials: dichloromethane, 28 wt% ammonium hydroxide solution, saturated sodium chloride solution, magnesium sulfate drying agent
Purification materials: dichloromethane and diethyl ether solvents

Balanced chemical equation:

$$\xrightarrow[\text{sodium ascorbate (cat.)}]{Cu(OAc)_2\ H_2O\ \text{(cat.)}}$$

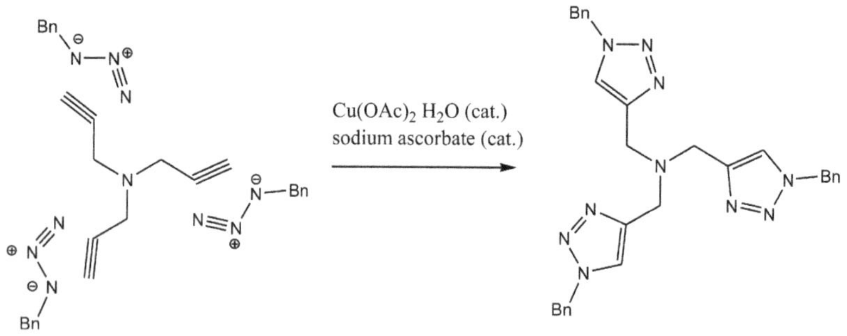

2 g (15.3 mmol) of tripropargylamine
7.34 g (55.2 mmol) of benzyl azide

Since the balanced chemical equation prescribes a molar ratio of 3:1 for azide to alkyne reagents and we find that 55.2/15.3 = 3.61, we conclude that the benzyl azide reagent is used in excess.

There are no by-products produced in this reaction. Hence, the only sources of waste materials are unreacted reagents and auxiliary materials (reaction solvents, catalysts, work-up materials, and purification materials) if they are not reclaimed or recycled.

Example 2.2 [4]

The poisonous alkaloid (–)-cytisine can be extracted from the seeds of the small deciduous tree *Laburnum anagyroides*, which is also known as golden chain due to its very showy large yellow flowers when in bloom. This plant has been used to poison people, as shown in an episode of the murder mystery series *Inspector Lewis* called Wild Justice. Look up the experimental procedure for the extraction of the alkaloid from the seeds and assign roles to each chemical species used as input material. Write out a balanced chemical equation for the chemical transformation. Determine which reactant is used in excess. What are the expected waste materials that are produced in this reaction?

Solution

Reactants:
598 g of *Laburnum anagyroides* seeds
90 mL of 25 wt% ammonium hydroxide solution
Reaction solvents: dichloromethane, methanol
Work-up materials: dichloromethane, 3.3 M HCl solution, 25 wt% ammonium hydroxide solution, magnesium sulfate drying agent

No balanced chemical equation can be written down because no chemical transformation has taken place. Therefore, since there are no formal "reactants" that can be assigned, there is no designated excess reagent. The waste materials are the remains of the seeds after extraction and all the auxiliary solvent materials used including the drying agent.

Note that since no balanced chemical equation is applicable, material efficiency metrics such as reaction yield or atom economy (see Chapter 6) are formally not applicable. However, we may state from the experimental results that 5.25 g of (–)-cytisine are extracted from 598g *L. anagyroides* seeds. We can still determine a process mass intensity as the ratio of masses of all input materials used to collect product, which in this case amounts to 878.

2.1.2 Problems

PROBLEM 2.1 [5]

Tetraphosphorus hexamethylhexaimide is made by reacting phosphorus trichloride with methylamine at –78°C. Look up the experimental procedure and assign roles to each chemical species used as input material. Write out a balanced chemical equation for the chemical transformation. Determine which reactant is used in excess. What are the expected waste materials that are produced in this reaction?

PROBLEM 2.2 [6]

Themes: balancing chemical equations, identifying by-products, reaction mechanisms

Heating 3,4-di-t-butylthiophene-1,1-dioxide with two equivalents of 4-phenyl-1,2,4-triazoline-3,5-dione in refluxing toluene yields a bis-adduct product **1**. In the presence of KOH in methanol this adduct breaks down into 4,5-di-t-butylpyridine (**2**) and other by-products. For the second step, the authors speculated that the reaction proceeds via a bis-hydrazo intermediate that is easily air-oxidized to a bis-azo intermediate.

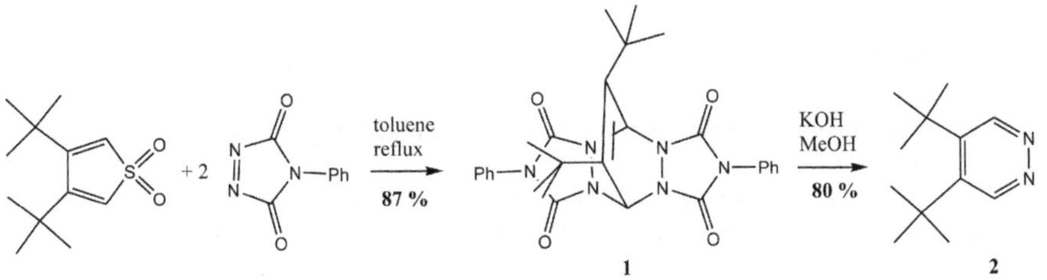

+ 2 — toluene reflux, 87 % → **1** — KOH MeOH, 80 % → **2**

Part 1

Identify the by-products in each reaction and write out mechanisms that rationalize the formation of those products as well as the target product **2**. Write out accompanying balanced chemical equations for each step.

Part 2

Suggest an alternative pathway for the formation of **2** that does not involve air oxidation of intermediates. What are the by-products for such an alternative mechanism? Compare the resulting balanced chemical equation with Part 1.

2.2 BALANCING CHEMICAL EQUATIONS

2.2.1 Terms, Definitions, and Examples

Atom Economy (AE) [7, 8]

In a balanced chemical reaction, atom economy is the molecular weight ratio of the target product to the sum of all reactants as given in Equation (2.2). Essentially, it measures the molecular weight fraction of reactants that end up in the desired product. The concept was introduced in 1991 by Barry Trost at Stanford University.

$$AE = \frac{MW_{\text{product}}}{\sum_j \nu_j \left(MW\right)_{\text{reactant},\, j}} \tag{2.2}$$

where:

ν_j is the stoichiometric coefficient of the jth reactant
MW parameters refer to the respective molecular weights

Balanced Chemical Equation [9]

A chemical equation written with reactants on the left-hand side and all products (target product plus all consequential by-products) on the right-hand side such that the number of each kind of atom appearing on the left is equal to the number of each kind of atom appearing on the right. The concept was introduced by Antoine Lavoisier in 1775 and is a practical consequence of the law of conservation of mass. A balanced chemical equation is the starting point of all metrics analyses.

Molecular Weight

Molecular weight is the mass of 1 mole of a pure chemical substance. Units: g/mol.

Example 2.3

Themes: balancing chemical equations, limiting reagent, percent reaction yield, percent conversion

The following calcination and combustion reactions were carried out by Antoine Lavoisier in his discovery of the law of conservation of mass.

REACTION 1 [10]

61.25 grains of phosphorus when combusted absorbs 69.375 grains of oxygen and produces 114.375 grains of "concrete phosphoric acid" and leaves 16.25 grains of phosphorus unchanged.

100 parts of phosphorus requires 154 parts oxygen for saturation to produce 254 parts of "concrete phosphoric acid"

1 French grain = 0.05306 g

"Concrete phosphoric acid" consists of a compound containing oxygen and phosphorus only.

Write out a balanced chemical equation for the process. Determine the atom economy for the reaction, the reaction yield of reaction with respect to phosphorus, the reaction yield of reaction with respect to oxygen, and the percent conversion of phosphorus.

Solution

Lavoisier's measurements indicated that "concrete phosphoric acid" consists of a compound of O and P made up of a mass ratio of oxygen to phosphorous of about 1.54 to 1. The closest phosphorous oxide that fits this mass ratio is P_2O_5 or its dimer P_4O_{10} (5*16:2*31 = 40:31 = 1.29:1) [11, 12].

$$4\,P + 5\,O_2 \longrightarrow O_{10}P_4$$

4 (31) **5 (32)** **$O_{10}P_4$ Mol. Wt.: 284**

AE = 1, or 100%

	4 P	5 O_2	P_4O_{10}
MW (g/mol)	31	32	284
Mass (g)	61.25*0.05306 = 3.25	69.375*0.05306 = 3.681	114.375*0.05306 = 6.069
Mmol	(3.25/31)*1000 = 104.84	(3.681/32)*1000 = 115.03	(6.069/284)*1000 = 21.37

Limiting reagent is oxygen.
Yield with respect to oxygen = (21.37/115.03)*(5/1) = 0.929 or 92.9%
Yield with respect to phosphorus = (21.37/104.84)*(4/1) = 0.815 or 81.5%
Conversion of phosphorus = (61.25 − 16.25)/61.25 = 0.735 or 73.5%

Supplementary Question

In Lavoisier's actual experiment, the volume of oxygen absorbed was 138.75 cubic inches (French), which he converted to a mass of 69.375 grains at a temperature of 54.5 degrees Fahrenheit and a pressure of 28 inches (French) mercury. Assuming that oxygen is an ideal gas, verify this calculation using the following conversion factors.

28 inches (Fr.) = 29.84 inches (Br.)
1 inch (Fr.) = 1.0659 inch (Br.)
1 inch (Br.) of Hg at 32°F = 0.033421 atm
29.84 inches (Br.) of Hg at 32°F = 29.84*0.033421 = 0.99728 atm
1 inch (Br.) of Hg at 54.5°F = 0.032835 atm

1 French grain = 0.05306 g
F = (9/5)*C + 32
54.5°F = (54.5 − 32)*(5/9) = 12.5°C
29.84 inches (Br.) of Hg at 54.5°F = 29.84*0.032835 = 0.979787 atm
1 cubic inch (Fr.) = $(1.0657)^3$ = 1.21033 cubic inch (Br.)
1 cubic inch (Br.) = 0.01638658 L
138.75*1.21033*0.01638658 = 2.75185 L
P V = *n* R T
R = 0.0820544 L atm/deg mole
n = P V/(R T) = 0.979787*2.75185/(0.0820544*(12.5 + 273.15)) = 0.115032 mol
MW oxygen = 32 g/mol
0.115032 mol oxygen = 0.115032*32 = 3.68104 g
Converting grams to grains: 3.68104/0.05306 = 69.375 grains

REACTION 2 [13]

One pound of phosphorus requires 1 pound, 8 ounces oxygen for its combustion and produces 2 pounds, 8 ounces of "concrete phosphoric acid" (P_4O_{10}). Verify this statement.

1 pound = 9216 grains
1 ounce = 576 grains
1 grain = 0.05306 g

Solution

Input materials:

Mass of phosphorus = 9216*0.05306 = 489.0 g (15.774 moles)
Mass of oxygen = (9216 + 8*576)*0.05306 = 733.5 g (22.922 moles)

Output materials:

Mass of P_4O_{10} = (2*9216 + 8*576)*0.05306 = 1222.5 g (4.305 moles)

Balanced chemical equation:

$$4\,P + 5\,O_2 \longrightarrow P_4O_{10}$$

4 (31) 5 (32)

$O_{10}P_4$
Mol. Wt.: 284

Limiting reagent: phosphorus

Yield = (4.305/15.774)*(4/1) = 1.092, or 109.2% (exceeds 100%)
Theoretical yield of P_4O_{10} = (¼)*15.774*284 = 1119.95 g

The 1222.5 g of "product" collected must have included unreacted phosphorus as well as P_4O_{10}. If we assume a 100% yield, then the mass of unreacted phosphorus must have been 1222.5 − 1119.95 = 102.55 g.

REACTION 3 [14]

Part 1

Four ounces of mercury is heated in air over 12 days yielding 45 grains of red crystals of mercury calx (HgO). At the beginning of the experiment, the volume of air was 50 cubic inches (Fr.) at a standard pressure and temperature. According to Lavoisier, 1 cubic inch of air has a mass of 0.47407 grains at a standard pressure and temperature of 28 inches (Fr.) Hg and 54.5°F, respectively.

From these data, determine the limiting reagent of the reaction, the % conversion of mercury, the % yield of HgO, the % mass decrease of starting air, and the % composition of gases at the end of the experiment.

Part 2

The 45 grains of mercury calx was then heated to produce 41.5 grains of mercury and 7 to 8 cubic inches (Fr.) of oxygen. The volume of gas was determined at a standard pressure and temperature of 28 inches (Fr.) Hg and 54.5°F.

Determine the percent yield of mercury produced and verify the volume of oxygen produced.

1 grain = 0.05306 g
1 ounce = 576 grains
1 inch (Fr.) = 1.0659 inch (Br)

Solution

Part 1

$Hg + ½\, O_2 \rightarrow HgO$

Input materials:

4*576*0.05306 = 122.25 g mercury (0.6095 moles)

Output materials:

45*0.05306 = 2.388 g HgO (0.01102 moles)

If 0.01102 moles of HgO are collected, then 0.5*0.01102 moles of oxygen and 0.01102 moles of mercury must have reacted. Hence, 0.5*0.01102*32 = 0.1763 g oxygen reacted and 0.01102*200.59 = 2.211 g mercury reacted.

Therefore, the mass of mercury that did not react is 122.25 − 2.21 = 120.04 g.

% conversion of mercury = (2.211/122.25)*100 = 1.8%.

The volume of air at the beginning of the experiment was 50 cubic inches at p = 28 inches (Fr.) Hg and T = 54.5°F). According to Lavoisier, 1 cubic inch of air under these conditions has a mass of 0.47407 grains. Therefore 50 cubic inches of air under the same pressure and temperature conditions should have a mass of 50*0.47407 = 23.7035 grains = 23.7035*0.05306 = 1.258 g; 21% of that air was oxygen, amounting to 0.21*1.258=0.264 g. This amounts to 0.264/32=0.00825 moles. When compared with the moles of starting mercury (0.6095), we note that the limiting reagent is oxygen.

- % Yield of HgO with respect to oxygen=(0.01102/0.00825)*(0.5/1)*100=66.8%.

The mass of air at the beginning of the experiment was 1.258 g.

This is composed of 0.78*1.258=0.981 g nitrogen, 0.21*1.258=0.264 g oxygen, and 0.01*1.258=0.013 g trace noble gases.

We know that 0.176 g of oxygen reacted; therefore, 0.264−0.176=0.088 g oxygen remains. The gas at the end of the reaction therefore is composed of 0.981 g nitrogen, 0.088 g oxygen, and 0.013 g trace noble gases, for a total mass of 0.981+0.088+0.013=1.082 g. Therefore the mass reduction of the original air is 1.258−1.082=0.176 g, or (0.176/1.258)*100=14.0%. The composition of the remaining gas at the end of the experiment is 90.7% nitrogen, 8.1% oxygen, and 1.2% trace noble gases.

Note: Lavoisier claimed that the original air had lost one-sixth its mass. These calculations show that it had lost about one-seventh of its original mass. He observed that this remaining air was not respirable, it suffocated animals, and extinguished lit splints, which is consistent with the calculated percent gaseous composition.

Part 2

$HgO \rightarrow Hg + ½ O_2$
Mass of starting HgO=45*0.05306=2.388 g
Mass of Hg produced=41.5*0.05306=2.202 g
Moles of starting HgO=2.388/216.59=0.01103
Moles of Hg produced=2.202/200.59=0.011
Moles of oxygen produced=0.011*0.5=0.00549
Mass of oxygen produced=0.00549*32=0.176 g
% Yield of Hg produced=(0.011/0.01103)*100=99.7%

According to Lavoisier, 1 cubic inch of oxygen at p=28 inches (Fr.) Hg and T=54.5°F has a mass of 0.50506 grains, or 0.50506*0.05306=0.0268 g. Since 0.176 g of oxygen is produced, then this amount must occupy under the same pressure and temperature conditions 0.176/0.0268=6.6 cubic inches. This value is less than Lavoisier's estimate of 7 to 8 cubic inches.

REACTION 4 [15]

100 grains of iron when heated in air increases in mass to 135 grains. At the same time the volume of air was reduced by 70 cubic inches (Fr.) (T=54.5°F and p=28 inches (Fr.) Hg). The mass of oxygen consumed is 0.5 grain per cubic inch of air absorbed.

- 1 grain=0.05306 g
- 1 cubic inch of air has a mass of 0.47407 grains at p=28 inches (Fr.) Hg and T=54.5°F

What kind of iron oxide is formed in this reaction? Determine the percent conversion of iron based on the kind of iron oxide produced.

Solution

Mass of starting Fe = 100*0.05306 = 5.306 g
Mass of product = 135*0.05306 = 7.163 g
Mass of air absorbed = 70*0.47407*0.05306 = 1.761 g
Mass of oxygen consumed = 0.5*70 = 35 grains = 35*0.05306 = 1.857 g
Moles of starting Fe = 5.306/55.85 = 0.0950 moles
Moles of oxygen consumed = 1.857/32 = 0.0580 moles
Assuming product is FeO:
$Fe + ½ O_2 \rightarrow FeO$

Limiting reagent is iron.

Moles of FeO expected = 0.095
Mass of FeO expected = 0.095*71.85 = 6.835 g
The mass of unreacted iron is 7.163 – 6.835 = 0.328 g.
% conversion of iron = 100*(5.306 – 0.328)/5.306 = 93.8%
Assuming product is Fe_2O_3:
$2 Fe + 3/2 O_2 \rightarrow Fe_2O_3$

Limiting reagent is oxygen.

Moles of Fe_2O_3 expected = (2/3)*0.058 = 0.0387
Mass of Fe_2O_3 expected = 0.0387*159.7 = 6.180 g
The mass of unreacted iron is 7.163 – 6.180 = 0.983 g.
% conversion of iron = 100*(5.306 – 0.983)/5.306 = 81.5%
Assuming product is Fe_3O_4:
$3 Fe + 2 O_2 \rightarrow Fe_3O_4$

Limiting reagent is oxygen.

Moles of Fe_3O_4 expected = (1/2)*0.058 = 0.029
Mass of Fe_3O_4 expected = 0.029*231.55 = 6.715 g
The mass of unreacted iron is 7.163 – 6.715 = 0.448 g.
% conversion of iron = 100*(5.306 – 0.448)/5.306 = 91.6%

REACTION 5 [16]

Upon combustion, 1 pound of charcoal absorbs 2 pounds, 9 ounces, 1 gros, and 10 grains of oxygen and yields 3 pounds, 9 ounces, 1 gros, and 10 grains of "acid gas" (carbon dioxide).

1 pound = 9216 grains
1 ounce = 8 gros
1 gros = 72 grains

From these data determine the reaction yield of carbon dioxide.

Solution

$$C + O_2 \longrightarrow CO_2$$

12 32 44

Mass of charcoal = 9216 grains = 9216*0.05306 = 489.001 g
Mass of oxygen absorbed = 2*9216 + 9*576 + 1*72 + 10 = 23698 grains = 23698*0.05306 = 1257.416 g
Mass of carbon dioxide = 3*9216 + 9*576 + 1*72 + 10 = 32914 grains = 32914*0.05306 = 1746.417 g

Mass balance check:

Input: 9216 + 23698 = 32914 grains
Output: 32914 grains
Moles of charcoal = 489.001/12 = 40.750
Moles of oxygen absorbed = 1257.416/32 = 39.294
Moles of carbon dioxide produced = 1746.417/44 = 39.691

The limiting reagent is oxygen.

% Yield of carbon dioxide with respect to oxygen = (39.691/39.293)*100 = 101%.

These experimental data are slightly off, but they suggest the reaction is run to completion in quantitative yield.

REACTION 6 [17]

When 28 grains of charcoal are heated in 85.7 grains of water vapor, 100 grains of "carbonic acid" (carbon dioxide) and 13.7 grains of hydrogen gas are produced.

100 grains of "carbonic acid" consist of 72 grains of oxygen and 28 grains of charcoal.

Therefore, 85.7 grains of water consist of 72 grains of oxygen and 13.7 grains of hydrogen.

1 grain = 0.05306 g

Based on these statements, determine the observed mass ratio of oxygen to carbon in carbon dioxide, and the mass ratio of oxygen to hydrogen in water. Compare with the actual values.

Solution

$C + 2\ H_2O \rightarrow CO_2 + 2\ H_2$
Mass of charcoal = 28*0.05306 = 1.486 g
Mass of water vapor = 85.7*0.05306 = 4.547 g
Mass of carbon dioxide = 100*0.05306 = 5.306 g
Mass of hydrogen = 13.7*0.05306 = 0.727 g
100 grains carbon dioxide = 100*0.05306 = 5.306 g
72 grains oxygen = 72*0.05306 = 3.820 g
28 grains charcoal = 28*0.05306 = 1.486 g

1 mole of carbon dioxide has a mass of 44 g composed of 12 g carbon and 32 g oxygen

Actual mass ratio of oxygen to carbon = 32:12 = **2.67:1**
Observed mass ratio of oxygen to carbon = 72:28 = **2.57:1**
85.7 grains water = 85.7*0.05306 = 4.547 g

72 grains oxygen = 72*0.05306 = 3.820 g
13.7 grains hydrogen = 13.7*0.05306 = 0.727 g

1 mole of water has a mass of 18 g composed of 2 g hydrogen and 16 g oxygen

Actual mass ratio of oxygen to hydrogen = 16:2 = **8:1**
Observed mass ratio of oxygen to hydrogen = 72:13.7 = **5.26:1**

The data for carbon dioxide agrees well, however, those for water are erroneous.

REACTION 7 [18]

When 274 grains of iron are heated in water vapor, 15 grains of hydrogen and 359 grains of iron oxide are produced. The water vapor loses 100 grains.
1 grain = 0.05306 g

SOLUTION

Suppose iron oxide is FeO.

$$\begin{array}{cccc} Fe + & H_2O \longrightarrow & FeO + & H_2 \\ \mathbf{55.85} & \mathbf{18} & \mathbf{71.85} & \mathbf{2} \end{array}$$

Mass of iron at start = 274*0.05306 = 14.538 g
Mass of water that reacts = 100*0.05306 = 5.306 g
Mass of hydrogen produced = 15*0.0536 = 0.804 g
Mass of FeO produced = 359*0.05306 = 19.049 g
Moles of iron at start = 14.538/55.85 = 0.260
Moles of water that reacts = 5.306/18 = 0.295
Moles of hydrogen produced = 0.804/2 = 0.402
Moles of FeO produced = 19.049/71.85 = 0.265
Mole ratio of water that reacts to FeO produced = 0.295/0.265 = 1.11
(Cf. mole ratio of 1:1 according to balanced equation)
Mole ratio of starting iron to FeO produced = 0.260/0.265 = 0.98
(Cf. mole ratio of 1:1 according to balanced equation)
Mole ratio of water that reacts to hydrogen produced = 0.295/0.402 = 0.73
(Cf. mole ratio of 1:1 according to balanced equation)

The data for iron and FeO appears to be fine, however, those for water are faulty.

Suppose iron oxide is Fe_2O_3.

$$\begin{array}{cccc} 2\,Fe + & 3\,H_2O \longrightarrow & Fe_2O_3 + & 3\,H_2 \\ \mathbf{2(55.85)} & \mathbf{3(18)} & \mathbf{159.7} & \mathbf{3(2)} \end{array}$$

Mass of iron at start = 274*0.05306 = 14.538 g
Mass of water that reacts = 100*0.05306 = 5.306 g
Mass of hydrogen produced = 15*0.0536 = 0.804 g
Mass of Fe_2O_3 produced = 359*0.05306 = 19.049 g
Moles of iron at start = 14.538/55.85 = 0.260

Moles of water that reacts = 5.306/18 = 0.295
Moles of hydrogen produced = 0.804/2 = 0.402
Moles of Fe_2O_3 produced = 19.049/159.7 = 0.119
Mole ratio of water that reacts to Fe_2O_3 produced = 0.295/0.119 = 2.48
(Cf. mole ratio of 3:1 according to balanced equation)
Mole ratio of starting iron to Fe_2O_3 produced = 0.260/0.119 = 2.18
(Cf. mole ratio of 2:1 according to balanced equation)
Mole ratio of water that reacts to hydrogen produced = 0.295/0.402 = 0.73
(Cf. mole ratio of 1:1 according to balanced equation)

These data show that the identity of the iron oxide as Fe_2O_3 results in a worse fit than for FeO.

Suppose iron oxide is Fe_3O_4.

$$3\ Fe + 4\ H_2O \longrightarrow Fe_3O_4 + 4\ H_2$$

3(55.85) 4(18) 231.55 4(2)

Mass of Fe_3O_4 produced = 359*0.05306 = 19.049 g
Moles of Fe_3O_4 produced = 19.049/231.55 = 0.0823
Mole ratio of water that reacts to Fe_3O_4 produced = 0.295/0.0823 = 3.58
(Cf. mole ratio of 4:1 according to balanced equation)
Mole ratio of starting iron to Fe_3O_4 produced = 0.260/0.0823 = 3.16
(Cf. mole ratio of 3:1 according to balanced equation)
Mole ratio of water that reacts to hydrogen produced = 0.295/0.402 = 0.73
(Cf. mole ratio of 1:1 according to balanced equation)

These data show that the identity of the iron oxide as Fe_2O_4 results in a worse fit than for FeO.

REACTION 8 [19]

Combustion of 16 ounces of alcohol yields 17 ounces of water.

1 ounce = 576 grains
1 grain = 0.05306 g

Assuming that the combustion reaction is run to completion under stoichiometric conditions, determine the reaction yield.

SOLUTION

$$CH_3CH_2OH + 3\ O_2 \longrightarrow 2\ CO_2 + 3\ H_2O$$

46 3 (32) 2 (44) 3 (18)

Mass of alcohol = 16*576*0.05306 = 489.001 g
Mass of water = 17*576*0.05306 = 519.564 g
Moles of alcohol = 489.001/46 = 10.6305
Moles of water = 519.564/18 = 28.865
% Yield of water with respect to ethanol = (28.865/10.6305)*(1/3)*100 = 90.51%

REACTION 9 [20]

Lavoisier makes the following statement in his book on pages 99 and 102: "One pound of hydrogen absorbs 5 pounds, 10 ounces, 5 gros, and 24 grains of oxygen to form 6 pounds, 10 ounces, 5 gros, and 24 grains of water." He also measured the masses of 1 cubic inch of oxygen and 1 cubic inch of hydrogen at the same temperature and pressure and found that oxygen was 14.29 times heavier than hydrogen. Assuming that the reaction he refers to goes to completion with no by-products, determine from these data what the molecular formula for water is.

1 pound = 9216 grains
1 ounce = 576 grains
1 gros = 72 grains
1 grain = 0.05306 g

SOLUTION

$H_2 + ½ O_2 \rightarrow H_2O$
Mass of hydrogen = 9216*0.05306 = 489.001 g
Mass of oxygen = (5*9216 + 10*576 + 5*72 + 24)*0.05306 = 2771.005 g
Mass of water = (6*9216 + 10*576 + 5*72 + 24)*0.05306 = 3260.006 g
Check: 489.001 + 2771.005 = 3260.006 (mass balance)
According to Lavoisier the mass ratio of oxygen to hydrogen in water is 2771.005/489.001 = 5.67:1
According to Lavoisier, oxygen is 14.29 times heavier than hydrogen.
Molecular formula of water according to Lavoisier: H_nO_m
Let MW of hydrogen be x. Then MW of oxygen is 14.29 x.
m*14.29*x/n*x = 5.67
m*14.29/n = 5.67
m/n = 5.67/14.29 = 0.397
n/m = 2.52
$HO_{2.5}$
H_2O_5

These data show that the measurements made by Lavoisier about the oxygen to hydrogen mass ratio composition of water and the mass ratio of oxygen to hydrogen are both faulty.

REACTION 10 [21]

Mass of tin before reaction = 8 ounces

Mass of air in the sealed vessel containing the tin = 15 and 1/3 grains

Mass of solids in vessel after the reaction = 3.12 grains more than the mass of the original tin

Mass of gas within the sealed vessel = 3.12 grains less than the mass of the original air

1 ounce = 576 grains
1 grain = 0.05306 g

Write out a balanced equation for the oxidation of tin in air.

From the data given determine the masses of reactants and all products. Determine the limiting reagent, the % reaction yield, and the % conversion of tin. What is the composition of the gas in the vessel after the reaction takes place?

SOLUTION

$Sn(s) + O_2(g) \rightarrow SnO_2(s)$
Mass of tin before reaction = 8*576*0.05306 = 244.5 g (2.06 moles)
Mass of air before reaction = (15 + (1/3))*0.05306 = 0.814 g
Mass of oxygen before reaction = 0.21*0.814 = 0.171 g (0.00534 moles)

Limiting reagent is oxygen.

Mass of gas in vessel after reaction = 0.814 – 3.12*0.05306 = 0.648 g
Mass of oxygen that reacted = 3.12*0.05306 = 0.166 g
Mass of SnO_2 + unreacted tin = 244.5 + 3.12*0.05306 = 244.666 g
Number of moles of oxygen that reacted = 0.166/32 = 0.00517
Number of moles of SnO_2 produced = 0.00517
Mass of SnO_2 produced = 0.00517*150.69 = 0.780 g
Mass of unreacted tin = 244.666 – 0.780 = 243.889 g
Yield of SnO_2 with respect to oxygen = (0.00517/0.00534)*100 = 96.9%
Mass of tin that reacted = 244.5 – 243.889 = 0.614 g
Conversion of tin = (0.614/244.5)*100 = 0.25%

Composition of gas in vessel after reaction:
Before reaction:

0.78*0.814 = 0.635 g nitrogen
0.21*0.814 = 0.171 g oxygen
0.01*0.814 = 0.0081 g trace noble gases

After reaction:

0.635 g nitrogen
0.0081 g trace noble gases
0.171 – 0.166 = 0.005 g oxygen
Total = 0.635 + 0.0081 + 0.005 = 0.648 g

Composition:
(0.635/0.648)*100 = 98% nitrogen
(0.0081/0.648)*100 = 1.25% trace noble gases
(0.005/0.648)*100 = 0.77% oxygen

Example 2.4

Themes: balancing chemical equations, atom economy

The following chemical transformations can be balanced in more than one way depending on the choice of by-products. Show these options and determine the associated atom economies.

(a)

OH; NaOH, $(MeO)_2SO_2$; OMe

(b)

OH; NaBr then H_2SO_4; Br

(c)

O, R, R; $NaBH_4$, EtOH; OH, R, R

(d)

O, Cl, $AlCl_3$; O

(e)

O, R, OH; $LiAlH_4$ then H_2O; R, OH

(f)

CN; H_2SO_4, H_2O; COOH

(g)

R, OH; $PhCH_2Br$, K_2CO_3; R, O, Ph

(h)

$Me_3\overset{\oplus}{N}$ $\overset{\ominus}{I}$; Ag_2O

(i)

$$\text{PhCHO} \xrightarrow[]{\text{NaClO}_2,\ \text{Me}_2\text{C=CHMe}} \text{PhCOOH}$$

Solution

(a)

$$2\ \text{PhOH} + \text{MeO–SO}_2\text{–OMe} \xrightarrow[-2\,H_2O\ \ \mathbf{2\,(18)}]{\mathbf{2\,(40)}\ \ 2\,NaOH} 2\ \text{PhOMe} + \text{NaO–SO}_2\text{–ONa}$$

2 (94) **126** **2 (108)** **142**

Balance check:
Reactants: 2*94 + 126 + 2*40 = 394
Products: 2*108 + 142 + 2*18 = 394
AE = 2*108/394 = 0.548

$$\text{PhOH} + \text{MeO–SO}_2\text{–OMe} \xrightarrow[-H_2O\ \ \mathbf{18}]{\mathbf{40}\ \ NaOH} \text{PhOMe} + \text{NaO–SO}_2\text{–OMe}$$

94 **126** **108** **134**

Balance check:
Reactants: 94 + 126 + 40 = 260
Products: 108 + 134 + 18 = 260
AE = 108/260 = 0.415

(b)

$$\text{CH}_3\text{CH}_2\text{CH}_2\text{CH}_2\text{OH} \xrightarrow[-\,NaHSO_4\ \mathbf{(120)},\ -\,H_2O\ \mathbf{(18)}]{NaBr\ \mathbf{(102.9)}\ \text{then}\ H_2SO_4\ \mathbf{(98)}} \text{CH}_3\text{CH}_2\text{CH}_2\text{CH}_2\text{Br}$$

74 **136.9**

Balance check:
Reactants: 74 + 102.9 + 98 = 274.9
Products: 136.9 + 120 + 18 = 274.9
AE = 136.9/274.9 = 0.498

$$2\ \text{CH}_3\text{CH}_2\text{CH}_2\text{CH}_2\text{OH} \xrightarrow[-\,Na_2SO_4\ \mathbf{(142)},\ -\,2\,H_2O\ \mathbf{(36)}]{2\,NaBr\ \mathbf{2(102.9)}\ \text{then}\ H_2SO_4\ \mathbf{(98)}} 2\ \text{CH}_3\text{CH}_2\text{CH}_2\text{CH}_2\text{Br}$$

2 (74) **2 (136.9)**

Balance check:
Reactants: 2*74+2*102.9+98=451.8
Products: 2*136.9+142+2*18=451.8
AE=2*136.9/451.8=0.606

(c)

$$\underset{\mathbf{2\,R+28}}{R_2C{=}O} \xrightarrow[\substack{-\,1/4\ B(OEt)_3\ \mathbf{(36.5)} \\ -\,1/4\ NaOH\ \mathbf{(10)}}]{\substack{1/4\ NaBH_4\ \mathbf{(9.5)} \\ 3/4\ EtOH\ \mathbf{(34.5)} \\ 1/4\ H_2O\ \mathbf{(4.5)}}} \underset{\mathbf{2\,R+30}}{R_2CH{-}OH}$$

Balance check:
Reactants: 2 R+28+9.5+34.5+4.5=2 R+76.5
Products: 2 R+30+36.5+10=2 R+76.5
AE=(2 R+30)/(2 R+76.5)
When R=15, AE(min)=0.563.

$$\underset{\mathbf{2\,R+28}}{R_2C{=}O} \xrightarrow[\substack{-\,B(OEt)_3\ \mathbf{(146)} \\ -\,NaOH\ \mathbf{(40)} \\ -\,3\ H_2\ \mathbf{(6)}}]{\substack{NaBH_4\ \mathbf{(38)} \\ 3\ EtOH\ \mathbf{(138)} \\ H_2O\ \mathbf{(18)}}} \underset{\mathbf{2\,R+30}}{R_2CH{-}OH}$$

Balance check:
Reactants: 2 R+28+38+138+18=2 R+222
Products: 2 R+30+146+40+6=2 R+222
AE=(2 R+30)/(2 R+222)
When R=15, AE(min)=0.238.

$$\underset{\mathbf{2\,R+28}}{R_2C{=}O} \xrightarrow[\substack{-\,Na[B(OEt)_4]\ \mathbf{(214)} \\ -\,3\ H_2\ \mathbf{(6)}}]{\substack{NaBH_4\ \mathbf{(38)} \\ 4\ EtOH\ \mathbf{(184)}}} \underset{\mathbf{2\,R+30}}{R_2CH{-}OH}$$

Balance check:
Reactants: 2 R+28+38+184=2 R+250
Products: 2 R+30+214+6=2 R+250
AE=(2 R+30)/(2 R+250)
When R=15, AE(min)=0.214.

(d)

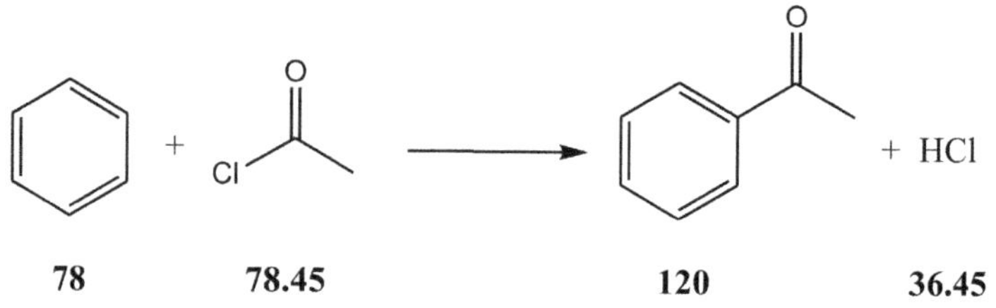

78 **78.45** **120** **36.45**

+ HCl

Balance check:
Reactants: 78 + 78.45 = 156.45
Products: 120 + 36.45 = 156.45
AE = 120/156.45 = 0.767

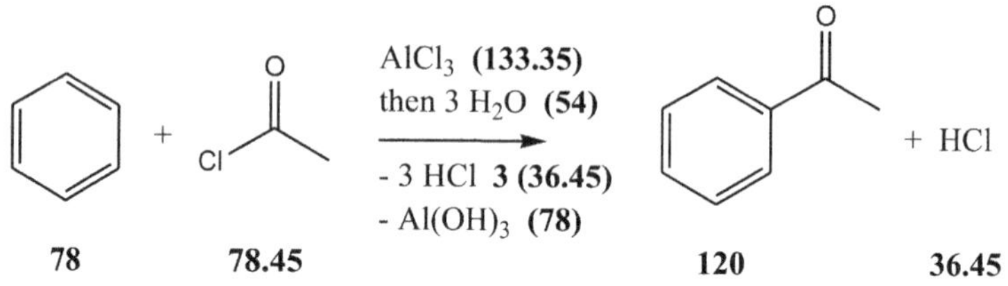

$AlCl_3$ **(133.35)**
then 3 H_2O **(54)**
- 3 HCl **3 (36.45)**
- $Al(OH)_3$ **(78)**

+ HCl

78 **78.45** **120** **36.45**

Balance check:
Reactants: 78 + 78.45 + 133.35 + 54 = 343.8
Products: 120 + 36.45 + 3*36.45 + 78 = 343.8
AE = 120/343.8 = 0.349

(e)

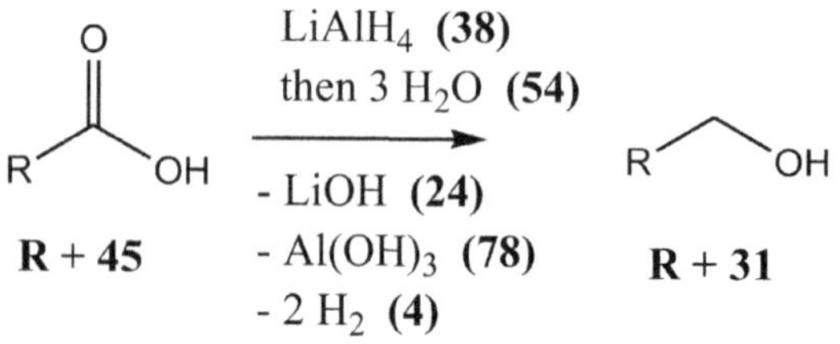

$LiAlH_4$ **(38)**
then 3 H_2O **(54)**
- LiOH **(24)**
- $Al(OH)_3$ **(78)**
- 2 H_2 **(4)**

R + 45 **R + 31**

Balance check:
Reactants: R + 45 + 38 + 54 = R + 137
Products: R + 31 + 24 + 78 + 4 = R + 137
AE = (R + 31)/(R + 137)
When R = 15, AE(min) = 0.303.

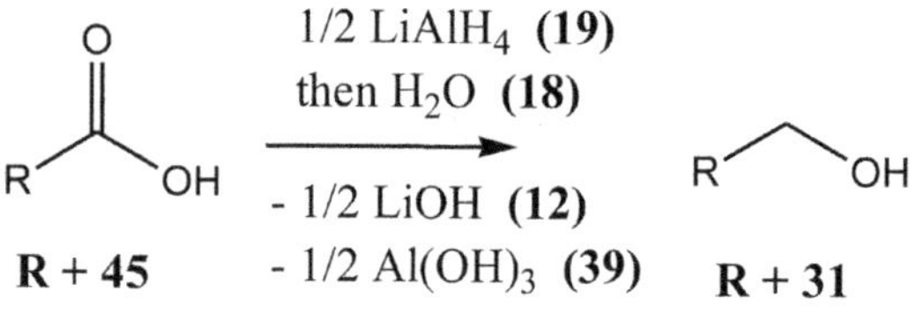

1/2 $LiAlH_4$ **(19)**
then H_2O **(18)**
- 1/2 LiOH **(12)**
- 1/2 $Al(OH)_3$ **(39)**

R + 45 **R + 31**

Balance check:
Reactants: R + 45 + 19 + 18 = R + 82
Products: R + 31 + 12 + 39 = R + 82

AE = (R + 31)/(R + 82)
When R = 15, AE(min) = 0.474

(f)

Benzonitrile (103) + H_2SO_4 **(98)** + 2 H_2O **(36)** → Benzoic acid (122) − $NH_4(HSO_4)$ **(115)**

103 **122**

Balance check:
Reactants: 103 + 98 + 36 = 237
Products: 122 + 115 = 237
AE = 122/237 = 0.515

2 Benzonitrile + H_2SO_4 **(98)** + 4 H_2O **(72)** → 2 Benzoic acid − $(NH_4)_2(SO_4)$ **(132)**

2 (103) **2 (122)**

Balance check:
Reactants: 2*103 + 98 + 72 = 376
Products: 2*122 + 132 = 376
AE = 2*122/376 = 0.649

(g)

RCH_2OH + $PhCH_2Br$ **(170.9)** + K_2CO_3 **(138)** → RCH_2OCH_2Ph − $KHCO_3$ **(100)** − KBr **(118.9)**

R + 31 **R + 121**

Balance check:
Reactants: R + 31 + 170.9 + 138 = R + 339.9
Products: R + 121 + 100 + 118.9 = R + 339.9
AE = (R + 121)/(R + 339.9)
When R = 15, AE(min) = 0.383.

2 RCH_2OH + 2 $PhCH_2Br$ **2(170.9)** + K_2CO_3 **(138)** → 2 RCH_2OCH_2Ph − H_2O **(18)** − CO_2 **(44)** − 2 KBr **2(118.9)**

2 (R + 31) **2 (R + 121)**

Balance check:
Reactants: 2 R + 2*31 + 2*170.9 + 138 = 2 R + 541.8

Products: 2 R + 2*121 + 18 + 44 + 2*118.9 = 2 R + 541.8
AE = (2 R + 242)/(2 R + 541.8)
When R = 15, AE(min) = 0.476.

(h)

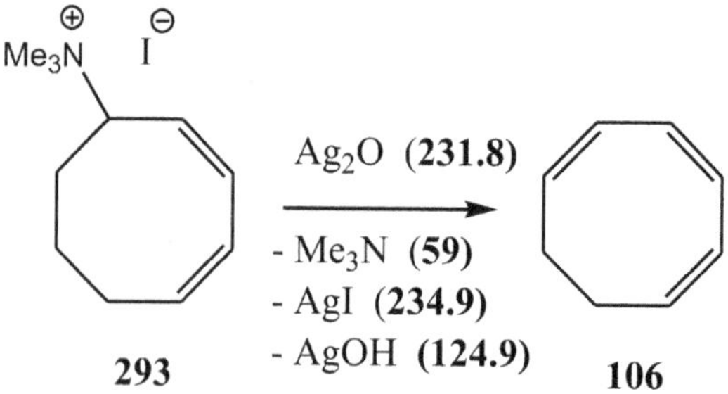

Balance check:
Reactants: 293 + 231.8 = 524.8
Products: 106 + 59 + 234.9 + 124.9 = 524.8
AE = 106/524.8 = 0.202

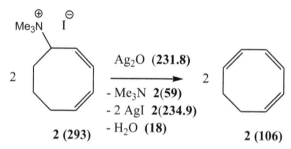

2 $Me_3N^{\oplus}$ $I^{\ominus}$ —— Ag_2O **(231.8)**; - Me_3N **2(59)**; - 2 AgI **2(234.9)**; - H_2O **(18)** ——→ 2

2 (293) **2 (106)**

Balance check:
Reactants: 2*293 + 231.8 = 817.8
Products: 2*106 + 2*59 + 2*234.9 + 18 = 817.8
AE = 2*106/817.8 = 0.259

(i)

PhCHO + (2-methyl-2-butene) —— $NaClO_2$ **(90.45)**; - NaCl **(58.45)** ——→ PhCOOH + (epoxide)

106 **70** **122** **86**

Balance check:
Reactants: 106 + 70 + 90.45 = 266.45
Products: 122 + 86 + 58.45 = 266.45
AE = 122/266.45 = 0.458

PhCHO —— 1/2 $NaClO_2$ **0.5(90.45)**; - 1/2 NaCl **0.5(58.45)** ——→ PhCOOH

106 **122**

Balance check:
Reactants: 106 + 0.5*90.45 = 151.225
Products: 122 + 0.5*58.45 = 151.225
AE = 122/151.225 = 0.807

Example 2.5 [22, 23]

Themes: balancing chemical equations, reaction mechanisms, stereochemistry

The following scheme shows a Walden cycle involving silicon stereogenic centers.

Np = alpha-naphthyl

Part 1

Assign R/S configurations to all structures as appropriate. Balance all chemical equations.

Part 2

For each reaction step, state whether or not it involves inversion or retention of configuration.

Part 3

For each step note the change in sign in optical rotation and the change in R/S configuration. Based on the results of Parts 2 and 3, what are the definitions of retention and inversion of configuration?

Part 4

Explain why the yields of the two products in the lithium borohydride step are not the same.

Solution

Parts 1 and 2

Part 3

Step	R/S configuration change about Si	Sign of optical rotation change	Inversion/retention
1	R → S	+ → −	Retention
2	S → R	− → −	Inversion
3	R → R	− → +	Retention[a]
4	R → R	+ → +	Retention[a]
5	R → R	+ → −	Addition; retention[a]
6	R → R	− → −	Rearrangement
7	R → S	− → −	Retention

[a] Reaction does not occur at Si.

We observe that there is no correlation between retention and changes in R/S configuration. For example, step 1 involves R → S and step 3 involves R → R, yet both are retentions.

We observe that there is no correlation between retention and changes in sign of optical rotation. For example, step 1 involves + → −, step 4 involves + → +, and step 7 involves − → −, yet all three are retentions.

Hence, the definitions of inversion of configuration and retention of configuration cannot include changes in sign of optical rotation or changes in R/S configuration labels. The following definitions apply.

Retention of configuration occurs when the leaving group and incoming group in a substitution reaction occupy the same position relative to the other groups around a stereogenic or non-stereogenic center that undergoes bond cleavage.

Inversion of configuration occurs when the leaving group and incoming group in a substitution reaction occupy opposite positions, 180 degrees to each other, relative to the other groups around a stereogenic or non-stereogenic center that undergoes bond cleavage.

Step	Incoming group	Leaving group	Orientation about Si	Inversion/retention
1	Cl	H	Same	Retention
2	$PhCH_2$	Cl	Opposite	Inversion
3	Br	H	Same	Retention[a]
4	OH	Br	Same	Retention[a]
5	n/a	n/a	Same	Addition; retention[a]
6	n/a	n/a	n/a	Rearrangement
7	H	PhC(Me)(H)O	Same	Retention

[a] Reaction does not occur at Si.

Part 4

Since both products arise as a result of fragmenting the precursor, it is expected that the number of moles of both products will be equal to each other. The observed experimental difference is attributed to losses in the alcohol product during aqueous workup. The alcohol is more soluble in aqueous extraction solvents than the silane, so it is expected that there will be some yield loss in the alcohol product.

2.2.2 Problems

PROBLEM 2.3 [24]

Themes: balancing chemical equations, reaction mechanisms, stereochemistry

The following scheme shows a Walden cycle involving carbon stereogenic centers.

L-(-)-malic acid —PCl_5→ D-(+)-chlorosuccinic acid

D-(+)-chlorosuccinic acid —Ag_2O, H_2O→ D-(+)-malic acid

D-(+)-malic acid —PCl_5→ L-(-)-chlorosuccinic acid

L-(+)-aspartic acid —Cl_2→ L-(-)-chlorosuccinic acid

L-(-)-chlorosuccinic acid —Ag_2O, H_2O→ L-(-)-malic acid

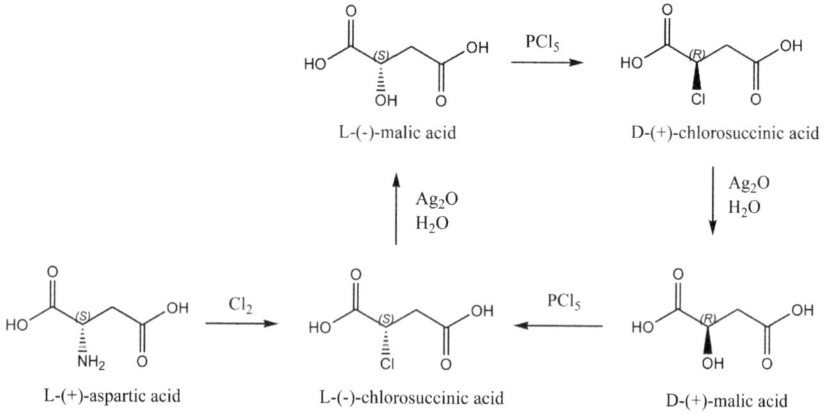

Part 1

Assign R/S configurations to all structures as appropriate. Balance all chemical equations.

Part 2

For each reaction step, state whether or not it involves inversion or retention of configuration.

Part 3

Suggest mechanisms for the chlorination of L-(+)-aspartic acid, the chlorination of L-(-)-malic acid, and the hydroxylation of D-(+)-chlorosuccinic acid.

PROBLEM 2.4 [22, 23]

Themes: balancing chemical equations, reaction mechanisms, stereochemistry

The following scheme shows a Walden cycle involving silicon stereogenic centers.

(+) Ph(Np)(CH$_3$)Si–H —Cl_2 (90 %)→ (-) Ph(Np)(CH$_3$)Si–Cl —Ph_2CHLi (80 %)→ (+) Ph_2CH–Si(CH$_3$)(Ph)(Np) —N-bromosuccinimide, benzoyl peroxide (initiator) (96 %)→ (-) Ph_2CBr–Si(CH$_3$)(Ph)(Np) —CF_3CO_2Ag, H_2O (67 %)→ (+) $Ph_2C(OH)$–Si(CH$_3$)(Ph)(Np) —Na→ [$Ph_2C(ONa)$–Si(CH$_3$)(Ph)(Np)] —HCl (97 %)→ (-) Ph_2CH–O–Si(CH$_3$)(Ph)(Np)

(-) Ph(Np)(CH$_3$)Si–Cl —Ph_2CHOH, $^iPr_2NH$ (13 %)→ (-) Ph_2CH–O–Si(CH$_3$)(Ph)(Np)

(-) Ph_2CH–O–Si(CH$_3$)(Ph)(Np) —$LiAlH_4$ then H_2O (35 %)→ Ph_2CH–OH + (-) H–Si(CH$_3$)(Ph)(Np)

Np = alpha-naphthyl

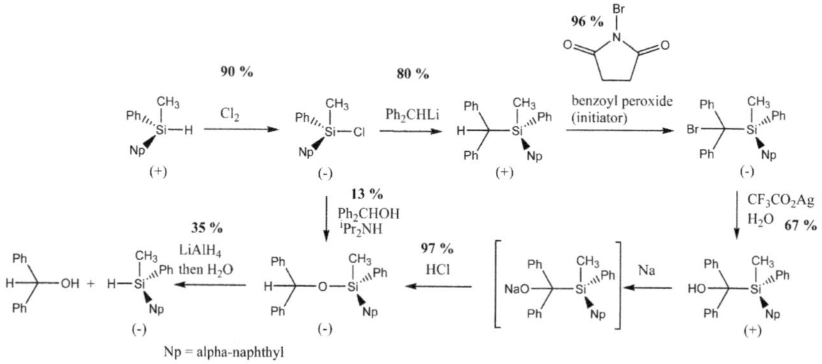

Part 1

Assign R/S configurations to all structures as appropriate. Balance all chemical equations.

Part 2

For each reaction step, state whether or not it involves inversion or retention of configuration.

Part 3

For each step note the change in sign in optical rotation and the change in R/S configuration. Based on the results of Parts 2 and 3 what are the definitions of retention and inversion of configuration?

2.3 REDOX REACTIONS

Redox reactions refer to reduction and oxidation reactions. These types of reactions are very important in the chemical industry and in biological chemistry. When we refer to a reaction as an oxidation reaction, we mean that the substrate of interest has been oxidized. The associated reagent used to carry out the oxidation is called the oxidizing reagent or oxidant. Oxidizing reagents end up being reduced. Similarly, when we refer to a reaction as a reduction reaction, we mean that the substrate of interest has been reduced. The associated reagent used to carry out the reduction is called the reducing reagent or reductant. Reducing agents end up being oxidized. From this we see that redox reactions are composed of two parts, often called half-reactions. One half-reaction refers to oxidation and the other half-reaction refers to reduction. The two half-reactions are always paired together and are called a redox couple.

The mechanisms of redox reactions involve electron transfer steps. From high school chemistry we learn that reduction half-reactions involve a gain of electrons and that oxidation half-reactions involve a loss of electrons. This means that reduction half-reactions have electrons appearing on the left-hand side corresponding to a lowering of the oxidation state of some element in the structure of the reactant. Correspondingly, oxidation half-reactions have electrons appearing on the right-hand side corresponding to a raising of the oxidation state of some element in the structure of the reactant.

Redox reactions therefore involve drastic changes in the electronic properties of atoms. When atoms become oxidized they become more electrophilic since they have more positive charge character and are electron deficient. When atoms become reduced they become more nucleophilic because they have more negative charge character and are electron rich.

Redox reactions may be viewed as a bridge that links other types of reactions as shown in Figure 2.1. Since these types of reactions necessarily produce by-products they generate the most waste compared with other types of reaction classes. They are also the most challenging reactions to elucidate reaction mechanisms and to obtain balanced chemical equations.

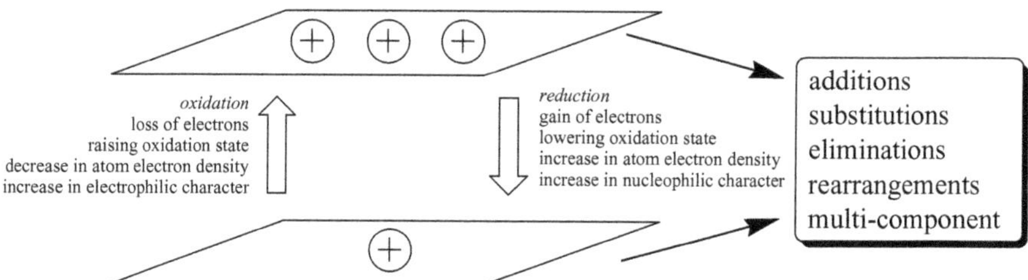

FIGURE 2.1 Relationship between redox reactions and other types of chemical reactions.

2.3.1 Terms, Definitions, and Examples

Additive Oxidation (Reduction) Reaction

Additive oxidation reactions involve a gain of oxygen atoms onto the substrate. Additive reduction reactions involve a gain of hydrogen atoms onto the substrate.

An example of an additive oxidation is the Baeyer–Villiger reaction.

An example of an additive reduction is the Birch reduction.

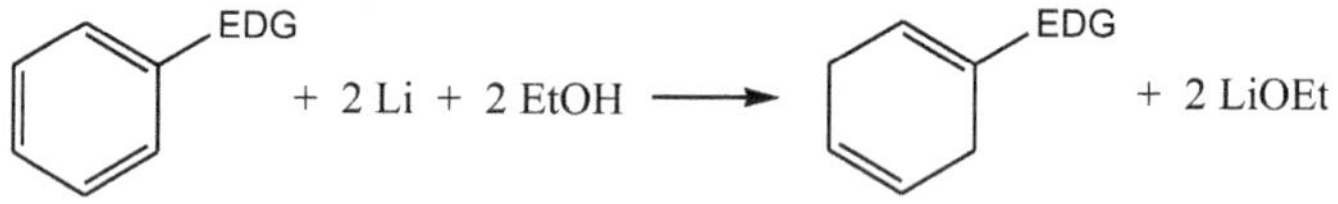

Diophantine Equation

A Diophantine equation is a polynomial equation in two or more independent variables having only integer solutions. It is named in honor of the third-century Greek mathematician Diophantus of Alexandria. For our purposes, cases where multiple sets of stoichiometric coefficients are applicable to balancing a redox reaction are examples of simple linear Diophantine equations of the form $ax+by=c$.

The algorithm to find integer solutions to equations of the form $ax+by=c$ where $a>b$ is as follows.

Step 1:

Find the nearest higher integer when c is divided by a. Find the nearest lower integer when c is divided by b. Then, we have $\left[\frac{c}{a}\right] \le x+y \le \left[\frac{c}{b}\right]$ where [] represents the nearest integer. Let $A=\left[\frac{c}{a}\right]$ and $B=\left[\frac{c}{b}\right]$. Therefore, the set of possible integers for $x+y$ is $x+y=\{A+1, A+2, A+3, \ldots, B-1\}$.

Step 2:

Multiply the $x+y$ possibilities by b so that

$$b(x+y)=\{b(A+1),b(A+2),b(A+3),...,b(B-1)\}.$$

Step 3:
Subtract each member of the set in Step 2 from c.

$$ax+by=c$$

$$-(bx+by)=\{b(A+1),b(A+2),b(A+3),...,b(B-1)\}$$

$$(a-b)x=\{c-b(A+1),c-b(A+2),c-b(A+3),...,c-b(B-1)\}$$

Step 4:
Select elements of set in Step 3 that are multiples of $a-b$.

Step 5:
Substitute those elements found in Step 4 into $ax+by=c$ to find the corresponding set of y values.

Example 2.6

Determine all integer solutions for $7x+2y=41$.

Solution

$$41/7=5.86\rightarrow \text{nearest higher integer is } 6$$

$$41/2=20.5\rightarrow \text{nearest lower integer is } 20$$

$$x+y=\{6,7,8,9,10,11,12,13,14,15,16,17,18,19,20\}$$

$$2(x+y)=\{12,14,16,18,20,22,24,26,28,30,32,34,36,38,40\}$$

$$(7-2)x=\{41-12,41-14,41-16,...,41-40\}$$

$$5x=\{29,27,25,...,1\}$$

Elements of set that are multiples of 5: 25, 20, 15, 10, 5

Therefore, $x=\{5,4,3,2,1\}$. The elements of x that lead to integer values for y are 5, 3, and 1. Therefore, $y=\{3,10,17\}$.

The set of integer solutions is $(x,y)=\{(5,3),(3,10),(1,17)\}$.

PROBLEM 2.5
Determine integer solutions for the following Diophantine equations.
(a) $3x+13y=69$; (b) $6x+19y=35$; (c) $8x+y=27$; (d) $5x+3y=20$.

PROBLEM 2.6
A girl walks into a bicycle and tricycle shop and counts 66 wheels. How many bicycles and tricycles are in the shop?

Half-Reaction

An oxidation half-reaction looks like:

$$X^{\ominus} \longrightarrow X + e$$

A reduction half-reaction looks like:

$$X + e \longrightarrow X^{\ominus}$$

Internal Redox Reaction

A redox reaction that cannot be split into a pair of reduction and oxidation half-reactions, yet oxidation state changes have taken place on atoms on going from reactants to products.

Example 2.7 [25–27]

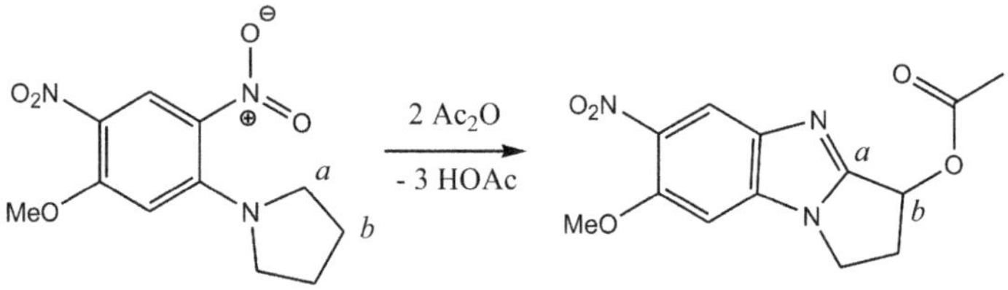

$$\xrightarrow[- 3\ HOAc]{2\ Ac_2O}$$

Atom	Reactant	Product	Change	
N	+3	-3	-6	**Δ = -6**
C_a	-1	+3	+4	**Δ = +6**
C_b	-2	0	+2	

Example 2.8 [28]

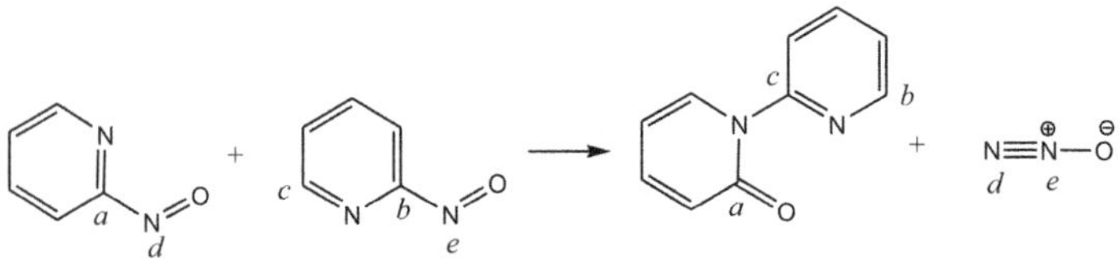

Atom	Reactants	Products	Change	
C_a	+3	+3	0	**Δ = 0**
C_b	+2	0	-2	
C_c	+1	+3	+2	
N_d	+1	0	-1	
N_e	+1	+2	+1	

Redox Couple

A redox couple or pair is the combination of a reduction half-reaction and an oxidation half-reaction.

$$[R] \quad X + n\,e \longrightarrow X^{\ominus n}$$

$$[O] \quad Y^{\ominus m} \longrightarrow Y + m\,e$$

Subtractive Oxidation (Reduction) Reaction

Subtractive oxidation reactions involve a loss of hydrogen atoms from the substrate. Subtractive reduction reactions involve a loss of oxygen atoms from the substrate.

An example of a subtractive oxidation is the Bamford–Stevens reaction of hydrazones.

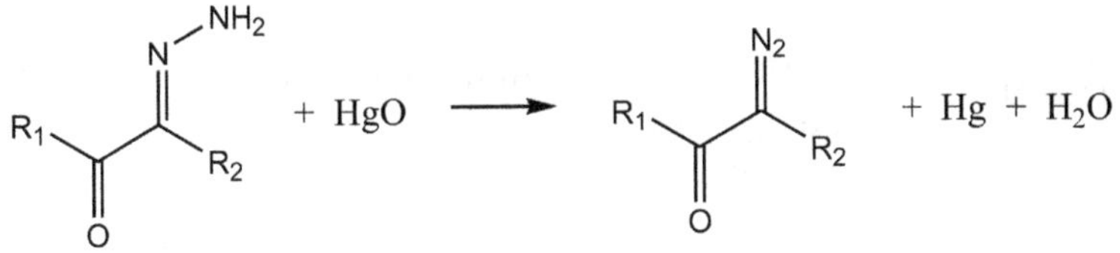

$$R_1C(=O)C(=N\text{-}NH_2)R_2 + HgO \longrightarrow R_1C(=O)C(=N_2)R_2 + Hg + H_2O$$

An example of subtractive reduction is the Corey–Winter reaction.

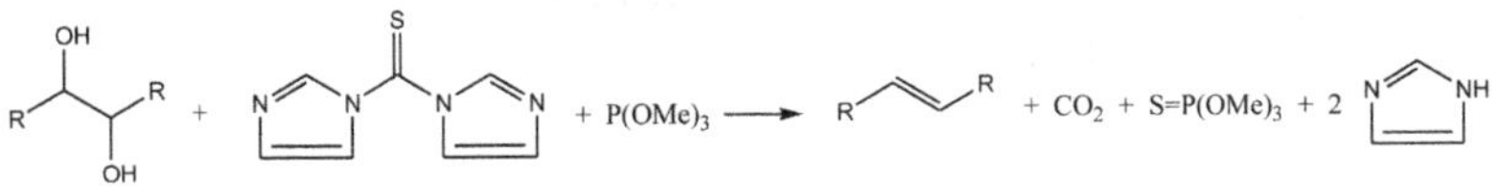

$$RCH(OH)CH(OH)R + (\text{Im})_2C{=}S + P(OMe)_3 \longrightarrow RCH{=}CHR + CO_2 + S{=}P(OMe)_3 + 2\ \text{imidazole}$$

2.3.2 Determining Oxidation States of Atoms in Structures

Example 2.9

Themes: structure drawing, oxidation numbers

Draw chemical structures for the following oxides of nitrogen and assign the oxidation numbers for all nitrogen atoms appearing in each structure.

(a) Nitric acid; (b) nitrous acid; (c) hyponitrous acid; (d) nitric oxide; (e) nitrous oxide; (f) nitrosonium ion; (g) nitronium ion; (h) nitrogen dioxide; (i) nitrogen trioxide; (j) nitrogen tetroxide; (k) nitrogen pentoxide

Solution

(a)

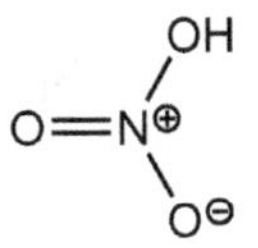

$O{=}N^{\oplus}(OH)O^{\ominus}$ | HNO_3 Mol. Wt.: 63 | nitric acid | +5

(b)

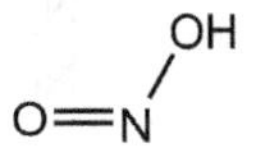

$O{=}N{-}OH$ | HNO_2 Mol. Wt.: 47 | nitrous acid | +3

	Formula	Name	Oxidation state of N
(c)	HNO Mol. Wt.: 31	hyponitrous acid	+1
(d)	NO Mol. Wt.: 30	nitric oxide (colourless gas)	+2
(e)	N_2O Mol. Wt.: 44	nitrous oxide (colourless gas)	N_a +2 N_b 0
(f)	NO^+ Mol. Wt.: 30	nitrosonium ion	+3
(g)	NO_2^+ Mol. Wt.: 46	nitronium ion	+5
(h)	NO_2 Mol. Wt.: 46	nitrogen dioxide (brown gas)	+3
(i)	N_2O_3 Mol. Wt.: 76	nitrogen trioxide (blue liquid at - 100 C)	N_a +2 N_b +4
(j)	N_2O_4 Mol. Wt.: 92	nitrogen tetroxide (colourless gas)	N_a +4 N_b +4
(k)	N_2O_5 Mol. Wt.: 108	nitrogen pentoxide (colourless solid)	N_a +5 N_b +5

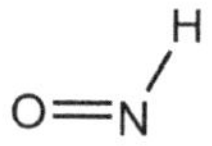

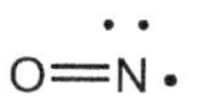

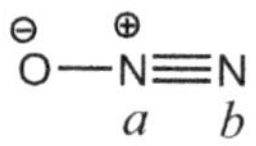

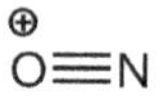

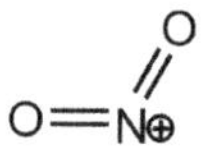

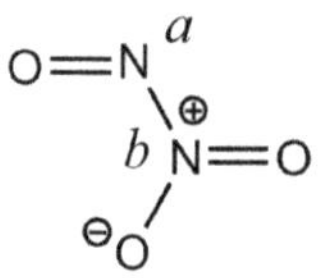

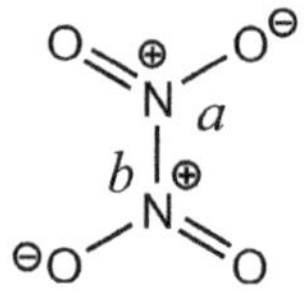

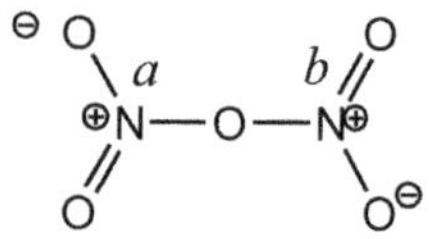

Example 2.10

Themes: nomenclature, oxidation number analysis

A hypervalent iodine reagent is known under the following three names and acronyms in the literature: phenyliododiacetate (PIDA), diacetoxyiodobenzene (DIB), and iodobenzenediacetate (IBDA).

Part 1

Deduce the chemical structure of this reagent.

Part 2

Determine the link between the names and the position of the groups in the chemical structure.

Part 3

Determine the oxidation state of the iodine atom in the molecule. Based on this assignment predict how this molecule will fragment in a redox sense.

Part 4

Determine the oxidation state of the iodine atom in the following hypervalent iodine reagents.

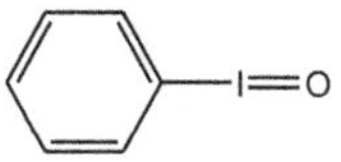

iodosylbenzene

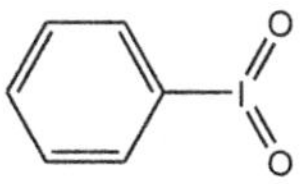

iodylbenzene

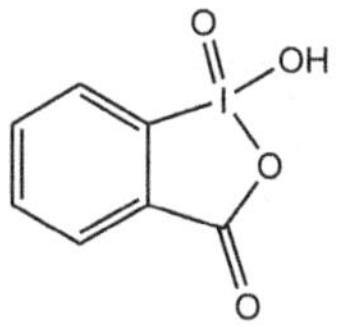

2-iodoxybenzoic acid (IBX)

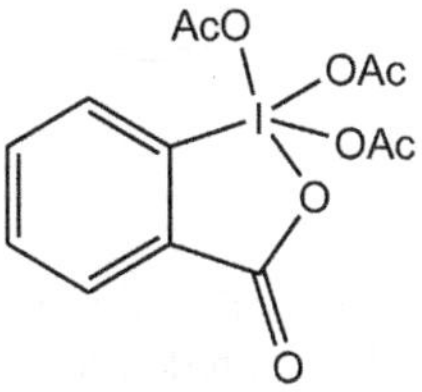

Dess-Martin periodinane (DMP)

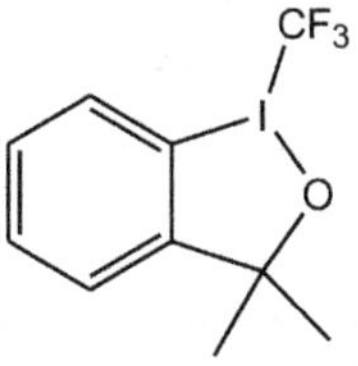

Togni reagent

SOLUTION

Part 1

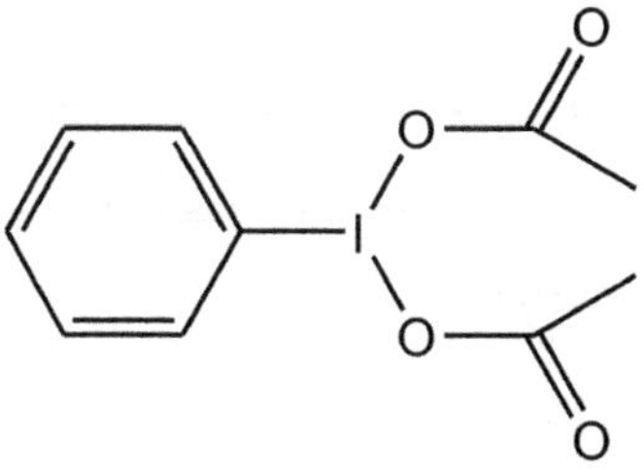

Part 2

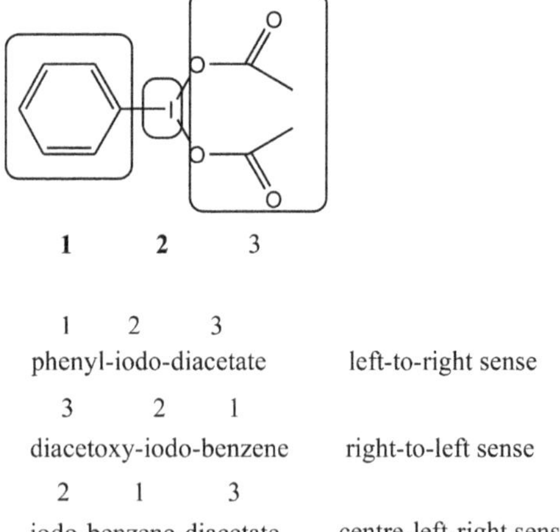

1 2 3 phenyl-iodo-diacetate	left-to-right sense
3 2 1 diacetoxy-iodo-benzene	right-to-left sense
2 1 3 iodo-benzene-diacetate	centre-left-right sense

Part 3

Oxidation number of iodine is + 1.

+1 → +1 $\xrightarrow{e^{\ominus}}$ 0

→ 0 $\xrightarrow{e^{\ominus}}$ -1

In a redox reaction, the hypervalent reagent will be reduced to iodobenzene and diacetate. Therefore, the other partner molecule in a redox reaction will be oxidized. An example reaction is shown below [29].

[R] $\quad$ $Ph\text{-}I(OAc)_2 + 2\,H^+ + 2\,e \longrightarrow Ph\text{-}I + 2\,HOAc$

[O] $\quad$ $2\ Ph{-}C{\equiv}C{-}H \longrightarrow Ph{-}C{\equiv}C{-}C{\equiv}C{-}Ph + 2\,H^+ + 2\,e$

Net reaction:

$2\ Ph{-}C{\equiv}C{-}H + Ph\text{-}I(OAc)_2 \longrightarrow Ph{-}C{\equiv}C{-}C{\equiv}C{-}Ph + Ph\text{-}I + 2\,HOAc$

$2\ Ph{-}C{\equiv}C_a{-}H + Ph\text{-}I(OAc)_2 \longrightarrow Ph{-}C{\equiv}C_a{-}C_a{\equiv}C{-}Ph + Ph\text{-}I + 2\,HOAc$

Atom	Reactant	Product	Change	
C_a	-1	0	+1	$\xrightarrow{X\,2}$ $\Delta = +2$
I	+1	-1	-2	$\Delta = -2$

Part 4

+1 iodosylbenzene

+3 iodylbenzene

+3 2-iodoxybenzoic acid (IBX)

+3 Dess-Martin periodinane (DMP)

-1 Togni reagent

Example 2.11

Themes: balancing chemical equations, oxidation states

In chemical nomenclature, a change of vowel in the name of a compound is indicative of a change of oxidation state. The letter "o" correlates with the highest oxidation state, the letter "i" with the next lower oxidation state, and the letter "e" with the next lower oxidation state. An example is given below for the series: benzenesulfonic acid, benzenesulfinic acid, and benzenesulfenic acid. It appears that there are progressively fewer oxygen atoms bound to the central sulfur atom as one goes from a sulfonic acid to a sulfenic acid.

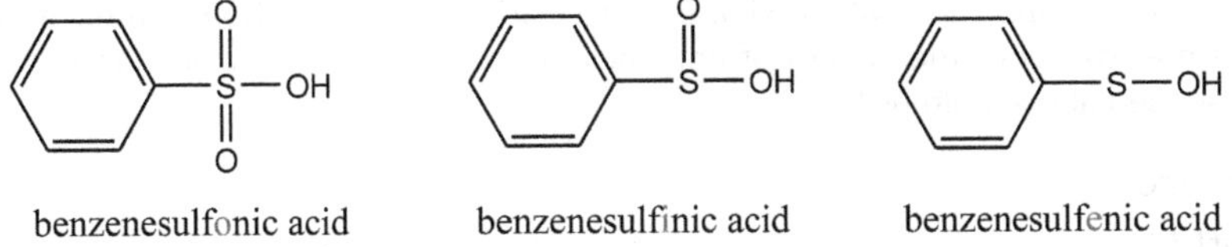

benzenesulfonic acid benzenesulfinic acid benzenesulfenic acid

A similar pattern is observed for phenylphosphonic acid and diphenylphosphinic acid.

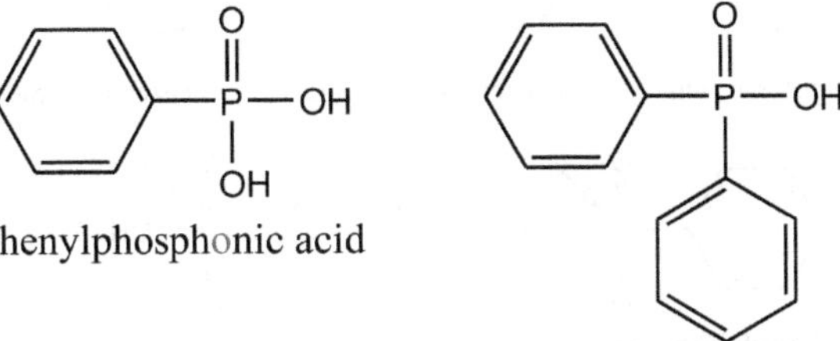

phenylphosphonic acid

diphenylphosphinic acid

Part 1

Determine the oxidation states of the sulfur and phosphorus atoms in the above structures.

Part 2

The structures of phenylboronic acid and diphenylborinic acid are shown below.

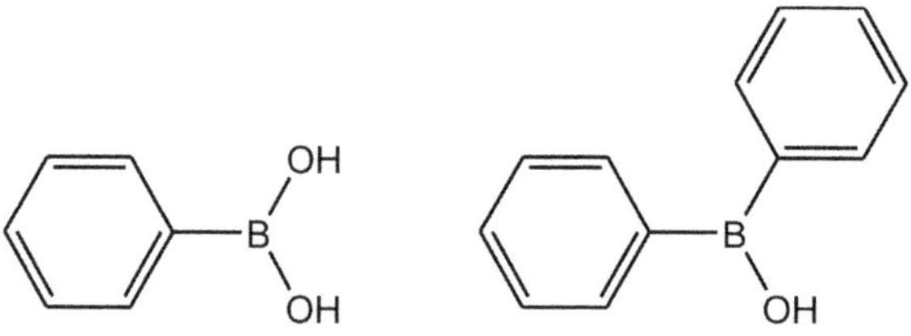

phenylboronic acid   diphenylborinic acid

These compounds are made by the following sequence of reactions.

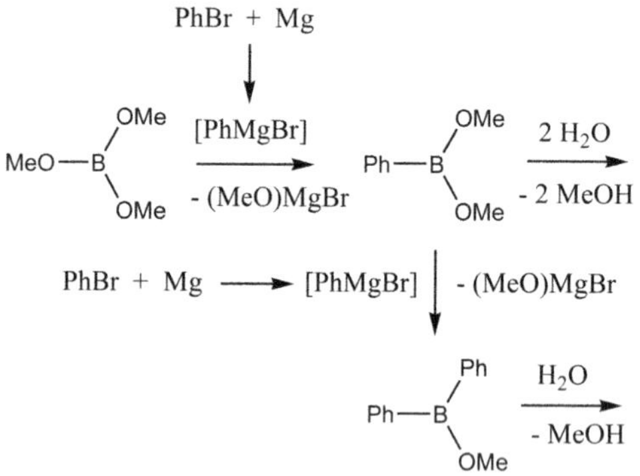

PhBr + Mg → [PhMgBr]; $B(OMe)_3$ + [PhMgBr] → PhB(OMe)$_2$ − (MeO)MgBr; PhB(OMe)$_2$ + 2 H_2O → PhB(OH)$_2$ − 2 MeOH

PhBr + Mg → [PhMgBr]; PhB(OMe)$_2$ + [PhMgBr] → Ph_2B(OMe) − (MeO)MgBr; Ph_2B(OMe) + H_2O → Ph_2B(OH) − MeOH

Based on this information what are the oxidation states of boron in each of these acids? Does the pattern of nomenclature associated with oxidation states described above hold up?

Solution

Part 1

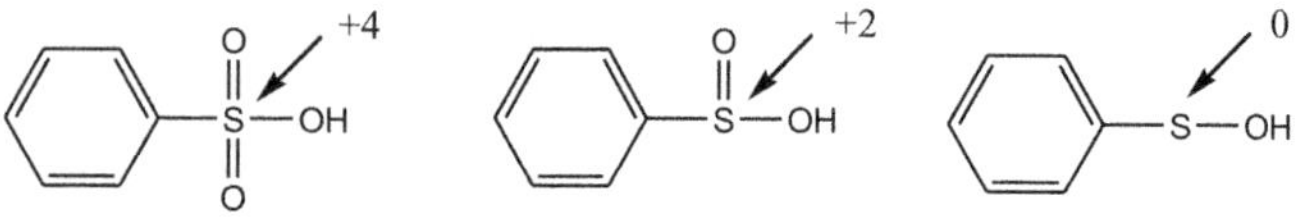

benzenesulfonic acid (+4)   benzenesulfinic acid (+2)   benzenesulfenic acid (0)

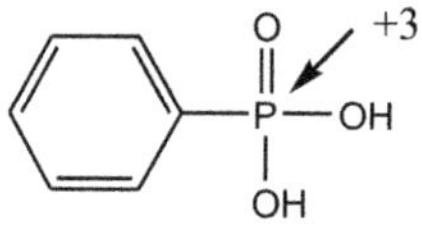

phenylphosphonic acid (+3)   diphenylphosphinic acid (+1)

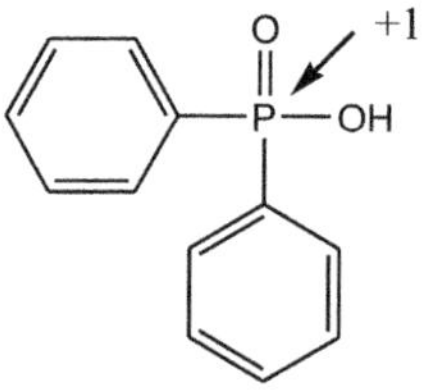

Part 2

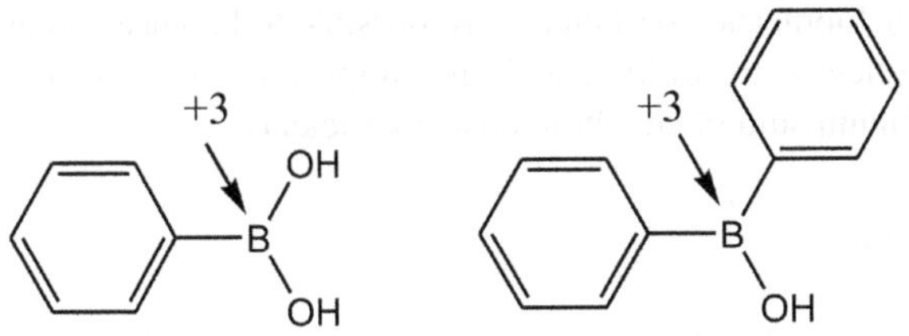

+3 +3

phenylboronic acid diphenylborinic acid

According to the Pauling electronegativity scale, the boron atom (2.04) is more electrophilic than an oxygen (3.44) or carbon (2.55) atom. Hence, the oxidation state of boron is assigned as +3 in both phenylboronic and diphenylborinic acid.

If the boron atom in phenylboronic acid were at a higher oxidation state than in diphenylborinic acid as the nomenclature suggests, then we would expect the following reduction couple.

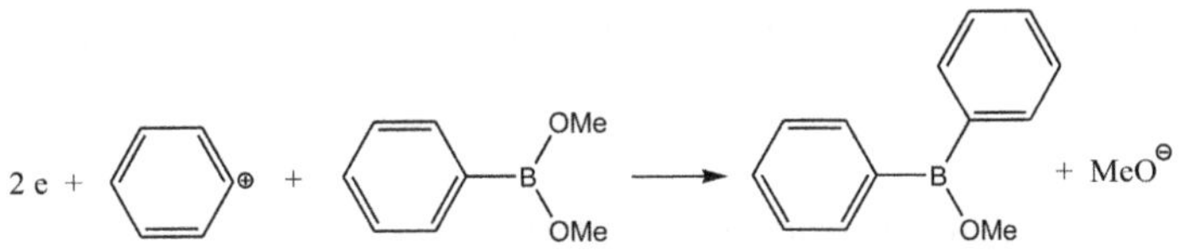

This is inconsistent with how methyl diphenylborinate is synthesized as shown below.

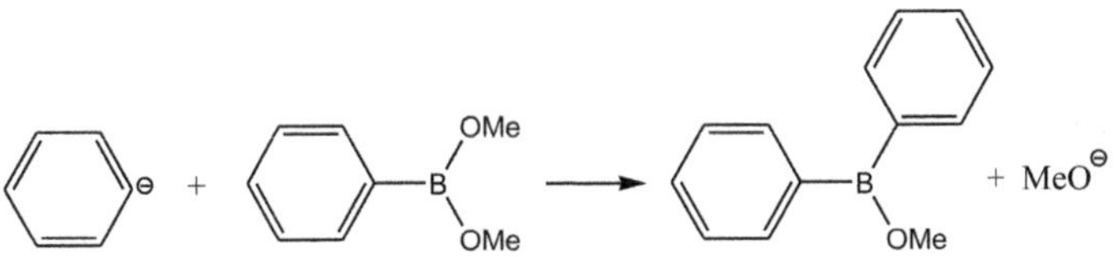

The methoxy group in dimethyl phenylboronate is displaced by a phenyl anion via a Grignard reagent in a simple substitution reaction, which does not involve any redox chemistry in that step. This substitution reaction therefore cannot be decomposed as separate oxidation and reduction couples.

The conclusion is that the names "boronic acid" and "borinic acid" are a misnomer with respect to the original association of the different vowels with different oxidation states.

Boron has only one oxidation state in all its compounds, namely, +3.

2.3.3 Balancing Redox Reactions

In this section, we show examples that illustrate the technique for balancing redox reactions and we make the strong link between balancing chemical equations with reaction mechanism. We also showcase cases of "ambiguous" balancing where more than one set of stoichiometric coefficients can be used to balance a redox reaction. These arise under two cases. This first case occurs when diatomic reactants or products containing one kind of element appear on the left-hand and left-hand sides of the chemical transformation, respectively. The second case arises when there is more than one redox couple occurring simultaneously. Whenever cases arise where multiple sets of stoichiometric coefficients are possible to balance a chemical equation, the set that is selected for linking with the simplest reaction mechanism is the one that has a minimum sum of stoichoimetric coefficients.

Example 2.12

Themes: balancing chemical equations, oxidation states, redox reactions

Borohydride reagents such as lithium aluminum borohydride or sodium borohydride are used to reduce carbonyl groups to alcohols.

Part 1

For the generalized reduction of a ketone to a secondary alcohol, write out its corresponding reduction and oxidation couples.

Part 2

Based on the redox couples, write out the overall balanced chemical equation for the transformation.

Part 3

Identify which atoms are oxidized and which are reduced.

Part 4

Write out a mechanism that is consistent with the overall balanced chemical equation in Part 2.

Solution

Part 1

$$[R]\quad 2\,H^+ + 2\,e + R_1{-}\overset{O}{\overset{\|}{C}}{-}R_2 \longrightarrow R_1{-}\overset{OH}{\underset{H}{C}}{-}R_2$$

$$[O]\quad 3\,H_2O + H{-}\overset{H}{\underset{H}{B^{\ominus}}}{-}H \longrightarrow HO{-}\overset{OH}{B}{-}OH + 7\,H^+ + 8\,e$$

Part 2

Multiplying reduction couple by four and adding result to oxidation couple yields

$$3\,H_2O + H^+ + 4\;R_1{-}\overset{O}{\overset{\|}{C}}{-}R_2 + H{-}\overset{H}{\underset{H}{B^{\ominus}}}{-}H \longrightarrow 4\;R_1{-}\overset{OH}{\underset{H}{C}}{-}R_2 + HO{-}\overset{OH}{B}{-}OH$$

If we use the reagent sodium borohydride, then we need to add one sodium ion to each side.

$$3\,H_2O + H^+ + 4\,R_1C(O)R_2 + NaBH_4 \longrightarrow 4\,R_1CH(OH)R_2 + B(OH)_3 + Na^+$$

Next, we balance off the positively charged ions with hydroxide ions since the reaction is carried out with an aqueous workup. The final balanced chemical equation is given by

$$4\,H_2O + 4\,R_1C(O)R_2 + NaBH_4 \longrightarrow 4\,R_1CH(OH)R_2 + B(OH)_3 + NaOH$$

Part 3

We assume that the R groups have carbon atoms bonded to the carbonyl group in the starting ketone.

$$4\,H_2O + 4\,R_1C(O)R_2 + NaBH_4 \longrightarrow 4\,R_1CH(OH)R_2 + B(OH)_3 + NaOH$$

B: +3 → +3 ($\Delta = 0$); carbonyl C: +2 → 0 ($\Delta = -2$); H (hydride): -1 → +1 ($\Delta = +2$)

We note that the carbonyl carbon atom changes its oxidation state from +2 to 0. Since there are 4 carbonyl carbon atoms, the overall net change is 4*(–2) = –8.

The hydrogen atoms in the borohydride change their oxidation state from –1 to +1 in the alcohol product. Since there are 4 hydride atoms that end up as 4 protic hydrogens, the overall net change is 4*(+2) = +8.

These results are consistent with the redox couple that indicated that the transformation involves eight electrons.

Part 4

$$R_1C(O)R_2 \xrightarrow[-\,BH_3]{BH_4^-} R_1CH(O^-)R_2 \xrightarrow[-\,H^-]{BH_3} R_1CH(OBH_2)R_2$$

$$R_1C(O)R_2 \xrightarrow{H^-} R_1CH(O^-)R_2$$

$$\xrightarrow{-\,H^-} (R_1CHR_2O)_2BH$$

$$R_1C(O)R_2 \xrightarrow{H^-} R_1CH(O^-)R_2 \quad \xrightarrow{-\,H^-}$$

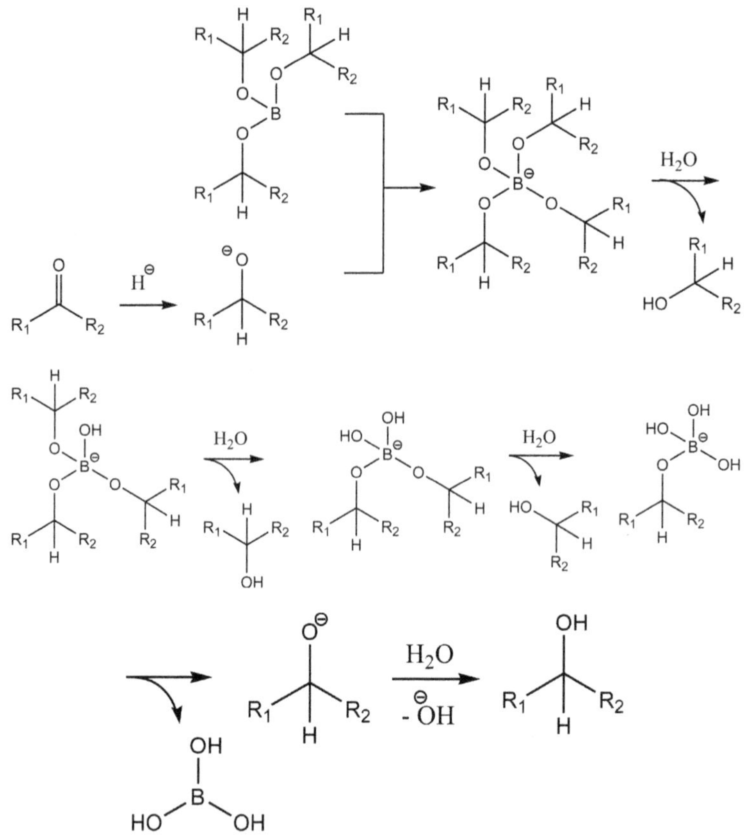

Adding up all the elementary steps in the mechanism leads directly to the overall balanced chemical equation given in Part 2.

Example 2.13 [30]

Themes: balancing chemical equations, atom economy

The chemistry of the roadside alcohol breathalyzer test is based on a redox reaction involving the oxidation of ethanol to acetic acid using potassium dichromate as oxidant and silver nitrate as catalyst.

$$CH_3CH_2OH + K_2Cr_2O_7 + H_2SO_4 \xrightarrow[\text{(cat.)}]{AgNO_3} CH_3COOH + Cr_2(SO_4)_3 + K_2SO_4 + H_2O$$

Part 1

Provide a balanced chemical equation for the transformation.

Part 2

Determine the atom economy for the reaction with respect to acetic acid as product.

Determine the atom economy for the reaction with respect to chromium sulfate as product.

Determine the atom economy for the reaction with respect to potassium sulfate as product.

Determine the atom economy for the reaction with respect to water as product.

SOLUTION

Part 1

$$[O]\quad CH_3CH_2OH + H_2O \longrightarrow CH_3COOH + 4\,H^+ + 4\,e \qquad X\,3$$

$$[R]\quad K_2Cr_2O_7 + 3\,SO_4^{-2} + 14\,H^+ + 6\,e \longrightarrow Cr_2(SO_4)_3 + 2\,K^+ + 7\,H_2O \qquad X\,2$$

Net reaction:

$$3\,CH_3CH_2OH + 3\,H_2O + 2\,K_2Cr_2O_7 + 6\,SO_4^{-2} + 28\,H^+ \longrightarrow$$

$$3\,CH_3COOH + 12\,H^+ + 2\,Cr_2(SO_4)_3 + 4\,K^+ + 14\,H_2O$$

Adding eight sulfate ions to each side yields:

$$3\,CH_3CH_2OH + 3\,H_2O + 2\,K_2Cr_2O_7 + 14\,H_2SO_4 \longrightarrow$$

$$3\,CH_3COOH + 6\,H_2SO_4 + 2\,Cr_2(SO_4)_3 + 2\,K_2SO_4 + 14\,H_2O$$

Cancelling waters and sulfuric acid on both sides yields:

$$3\,CH_3CH_2OH + 2\,K_2Cr_2O_7 + 8\,H_2SO_4 \longrightarrow$$

$$3\,CH_3COOH + 2\,Cr_2(SO_4)_3 + 2\,K_2SO_4 + 11\,H_2O$$

Part 2

$$3\,CH_3CH_2OH + 2\,K_2Cr_2O_7 + 8\,H_2SO_4 \longrightarrow$$

3 (46) 2 (293.992) 8 (98)

$$3\,CH_3COOH + 2\,Cr_2(SO_4)_3 + 2\,K_2SO_4 + 11\,H_2O$$

3 (60) 2 (391.992) 2 (174) 11 (18)

Check balancing:
Reactants: 3*46 + 2*293.992 + 8*98 = 1509.984
Products: 3*60 + 2*391.992 + 2*174 + 11*18 = 1509.984
% atom economy with respect to acetic acid as product
= 100*(3*60)/1509.984 = 11.92%
% atom economy with respect to chromium sulfate as product
= 100*(2*391.992)/1509.984 = 51.92%
% atom economy with respect to potassium sulfate as product
= 100*(2*174)/1509.984 = 23.05%
% atom economy with respect to water as product
= 100*(11*18)/1509.984 = 13.11%

Example 2.14

Themes: balancing chemical equations, redox reactions

By means of redox couple analysis, provide complete balanced equations for the following redox reactions showing stoichiometric coefficients for all reagents and all products.

(a)

$LiAlH_4$

(b)

Zn
HCl

(c)

$FeSO_4$
H_2SO_4

(d)

O_3
Me_2S

(e)

NH_3
Na

(f)

H_2
Lindlar
catalyst

(g)

1. O_3
2. H_2O

(h)

$$\text{PhCH=CH}_2 \xrightarrow[\text{2. NaOH / H}_2\text{O}_2]{\text{1. B}_2\text{H}_6} \text{PhCH}_2\text{CH}_2\text{OH}$$

(i)

$$\text{cyclopentanol} \xrightarrow[\text{H}_2\text{SO}_4]{\text{Na}_2\text{Cr}_2\text{O}_7} \text{cyclopentanone}$$

SOLUTION

(a)

$$[R] \quad 8\,H^+ + 8\,e + \text{MeO(O=)CCH}_2\text{C(=O)OMe} \longrightarrow \text{HOCH}_2\text{CH}_2\text{CH}_2\text{OH} + 2\,MeOH$$

$$[O] \quad LiAlH_4 + 4\,H_2O \longrightarrow LiOH + Al(OH)_3 + 8\,H^+ + 8\,e$$

Adding the half reactions yields

$$\text{MeO(O=)CCH}_2\text{C(=O)OMe} + LiAlH_4 + 4\,H_2O \longrightarrow \text{HOCH}_2\text{CH}_2\text{CH}_2\text{OH} + 2\,MeOH + LiOH + Al(OH)_3$$

(b)

$$[R] \quad 4\,H^+ + 4\,e + \text{cyclohexanone} \longrightarrow \text{cyclohexane} + H_2O$$

$$[O] \quad Zn \longrightarrow Zn^{+2} + 2\,e \qquad \mathbf{X\,2}$$

Adding the half reactions yields

$$2\,Zn + 4\,H^+ + \text{cyclohexanone} \longrightarrow \text{cyclohexane} + H_2O + 2\,Zn^{+2}$$

Adding four chloride ions to each side yields

$$2\,Zn + 4\,HCl + \text{cyclohexanone} \longrightarrow \text{cyclohexane} + H_2O + 2\,ZnCl_2$$

(c)

$$[R] \quad 6\,H^+ + 6\,e + \text{C}_6\text{H}_5\text{–NO}_2 \longrightarrow \text{C}_6\text{H}_5\text{–NH}_2 + 2\,H_2O$$

$$[O] \quad Fe^{+2} \longrightarrow Fe^{+3} + e \qquad \mathbf{X\,6}$$

Adding the half reactions yields

$$6\,H^+ + 6\,Fe^{+2} + \text{C}_6\text{H}_5\text{–NO}_2 \longrightarrow \text{C}_6\text{H}_5\text{–NH}_2 + 2\,H_2O + 6\,Fe^{+3}$$

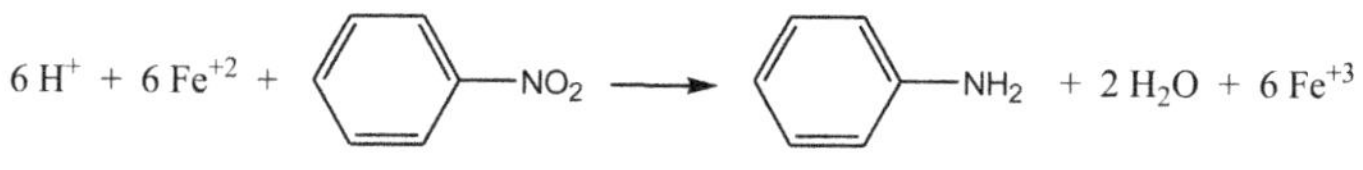

Adding nine sulphate ions to each side yields

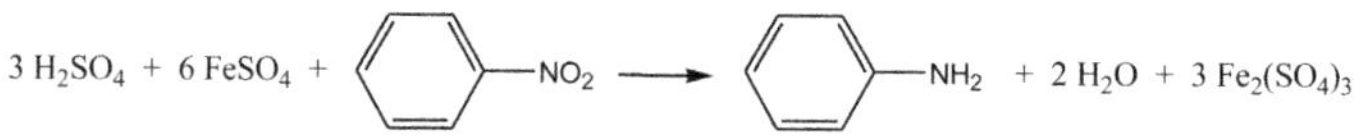

$$3\,H_2SO_4 + 6\,FeSO_4 + \text{C}_6\text{H}_5\text{–NO}_2 \longrightarrow \text{C}_6\text{H}_5\text{–NH}_2 + 2\,H_2O + 3\,Fe_2(SO_4)_3$$

(d)

Ph · O=O⁺–O⁻ → Ph · O O O → Ph · O—O · O

[R] $2\,H^+ + 2\,e +$ Ph · O—O · O $\longrightarrow$ Ph · O · H + O · H + H_2O

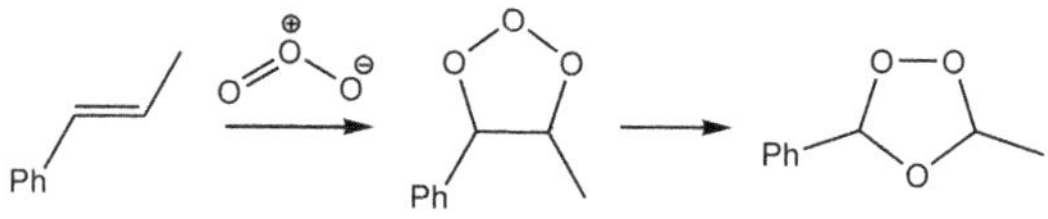

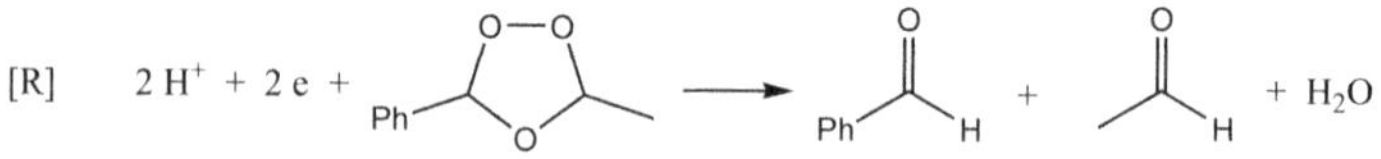

[O] $Me_2S + H_2O \longrightarrow Me_2S{=}O + 2\,H^+ + 2\,e$

Adding the reactions yields

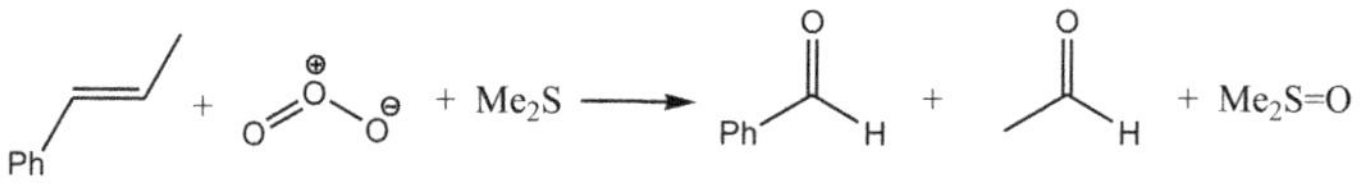

Ph + O=O⁺–O⁻ + $Me_2S \longrightarrow$ Ph · O · H + O · H + $Me_2S{=}O$

(e)

[R] $2\,H^+ + 2\,e +$ [alkyne] $\longrightarrow$ [alkene]

[O] $Na \longrightarrow Na^+ + e$ **X 2**

Adding the half reactions yields

$2\,H^+ + 2\,Na +$ [alkyne] $\longrightarrow$ [alkene] $+ 2\,Na^+$

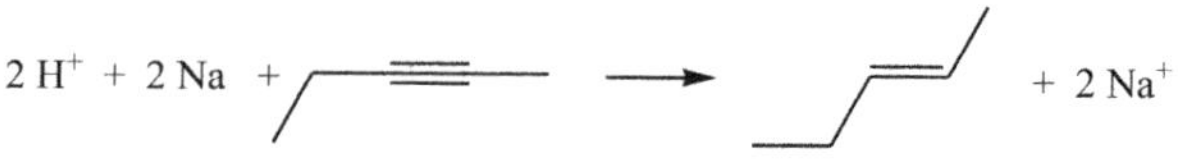

Adding two amide ions to each side yields

$2\,NH_3 + 2\,Na +$ [alkyne] $\longrightarrow$ [alkene] $+ 2\,NaNH_2$

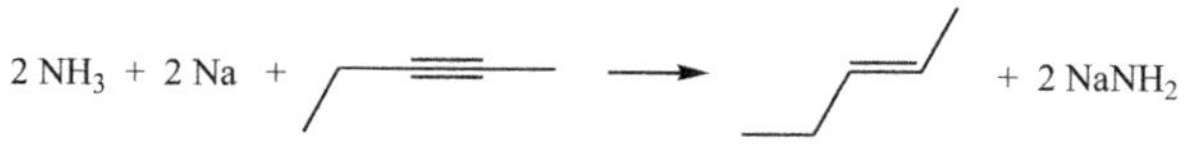

Note that after aqueous workup sodium amide is decomposed to NaOH and ammonia.

(f)

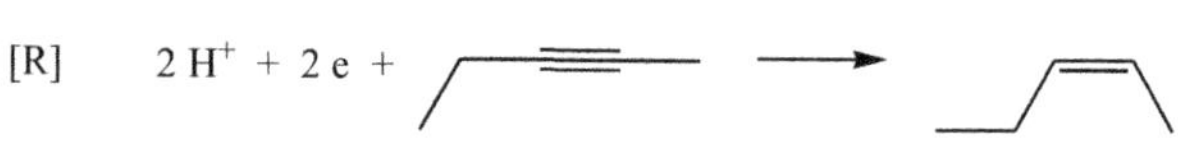

[R] $2\,H^+ + 2\,e +$ [alkyne] $\longrightarrow$ [alkene]

[O] $H_2 \longrightarrow 2\,H^+ + 2\,e$

Adding the half reactions yields

H_2 + [2-pentyne] → [cis-2-pentene]

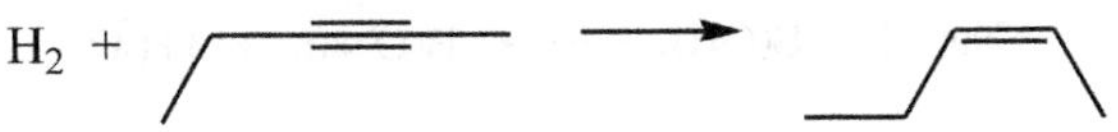

(g)
Mechanism of reaction:

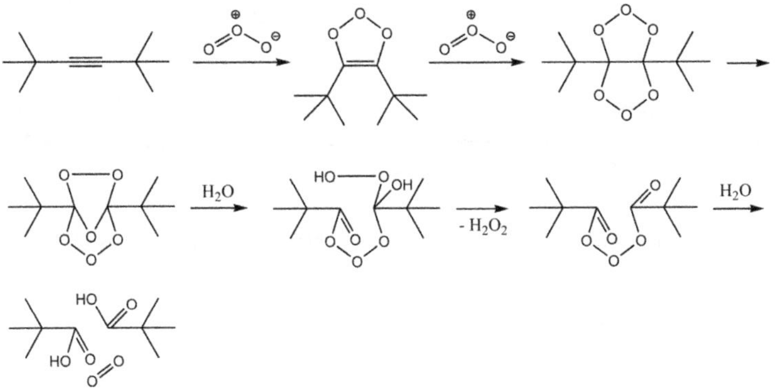

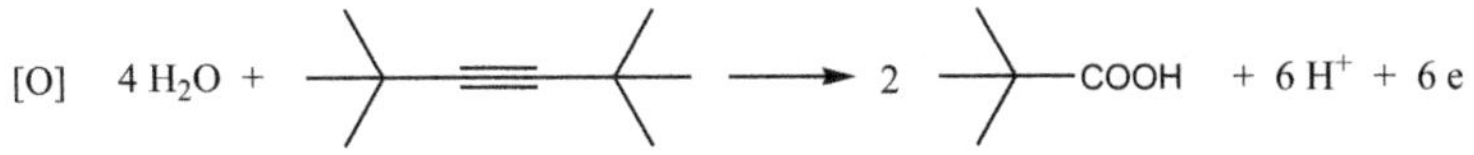

[structural mechanism: alkyne + O_3 → ozonide intermediates; H_2O; $-H_2O_2$; H_2O → carboxylic acids + O_2]

HO O
HO O O=O

[O] $4\,H_2O$ + [di-tert-butylacetylene] → 2 [tert-butyl]–COOH $+ 6\,H^+ + 6\,e$

[R] $2\,O_3 + 6\,H^+ + 6\,e \longrightarrow H_2O_2 + O_2 + 2\,H_2O$

Adding the half reactions yields

$2\,O_3 + 2\,H_2O$ + [di-tert-butylacetylene] → 2 [tert-butyl]–COOH $+ H_2O_2 + O_2$

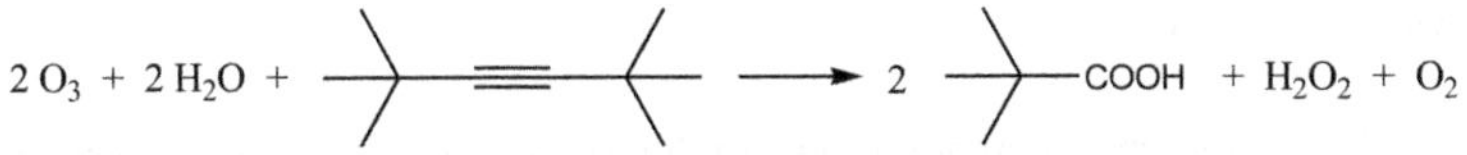

Note that the balanced chemical equation is consistent with the reaction mechanism.

(h)

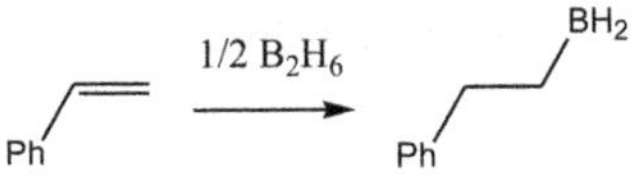

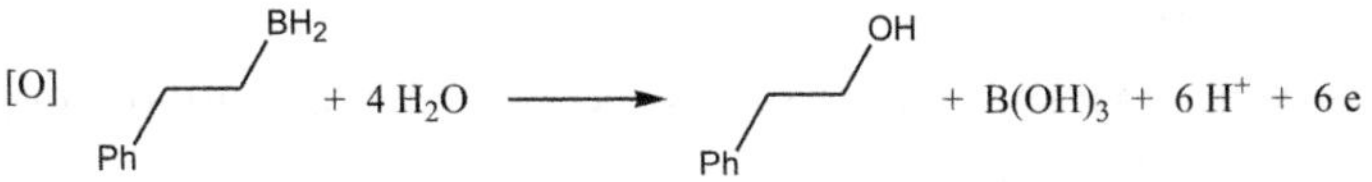

Ph–CH=CH$_2$ $\xrightarrow{1/2\,B_2H_6}$ Ph–CH$_2$CH$_2$–BH$_2$

[O] Ph–CH$_2$CH$_2$–BH$_2$ $+ 4\,H_2O \longrightarrow$ Ph–CH$_2$CH$_2$–OH $+ B(OH)_3 + 6\,H^+ + 6\,e$

[R] $H_2O_2 + 2\,H^+ + 2\,e \longrightarrow 2\,H_2O$ **X 3**

Adding the reactions yields

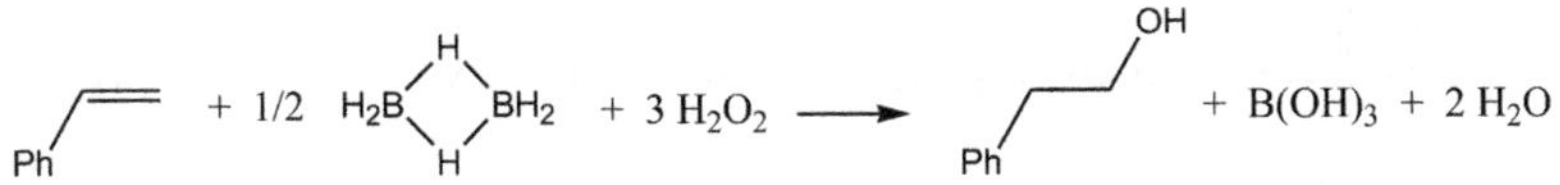

Ph–CH=CH$_2$ + 1/2 $H_2B(\mu\text{-}H)_2BH_2$ $+ 3\,H_2O_2 \longrightarrow$ Ph–CH$_2$CH$_2$–OH $+ B(OH)_3 + 2\,H_2O$

Note that sodium hydroxide neutralizes boric acid according to

$$3\,NaOH + B(OH)_3 \longrightarrow B(ONa)_3 + 3\,H_2O$$

Hence, the overall balanced equation is

$$PhCH{=}CH_2 + 1/2\ H_2B(\mu\text{-}H)_2BH_2 + 3\,H_2O_2 + 3\,NaOH \longrightarrow PhCH_2CH_2OH + B(ONa)_3 + 5\,H_2O$$

(i)

$$[O] \quad C_5H_9\text{–}OH \longrightarrow C_5H_8{=}O + 2\,H^+ + 2\,e \qquad \mathbf{X\,3}$$

$$[R] \quad Na_2Cr_2O_7 + 14\,H^+ + 6\,e \longrightarrow 2\,Na^+ + 2\,Cr^{+3} + 7\,H_2O$$

Adding the half reactions yields

$$Na_2Cr_2O_7 + 8\,H^+ + 3\ C_5H_9\text{–}OH \longrightarrow 3\ C_5H_8{=}O + 2\,Na^+ + 2\,Cr^{+3} + 7\,H_2O$$

Adding four sulphate ions to each side yields

$$Na_2Cr_2O_7 + 4\,H_2SO_4 + 3\ C_5H_9\text{–}OH \longrightarrow 3\ C_5H_8{=}O + Na_2SO_4 + Cr_2(SO_4)_3 + 7\,H_2O$$

Example 2.15 [31]

Themes: balancing chemical equations, multiple ways to balance equations, reaction mechanisms

The following reactions consist of the same reactants and products but are balanced with a different set of stoichiometric coefficients. For each reaction, write out its associated mechanism and check to see if the elementary steps in the mechanism add up to the overall balanced chemical equation.

REACTION #1

$$2\ KO\text{–}Mn(=O)_3 + 3\,H_2SO_4 + 7\,H_2O_2 \longrightarrow K_2SO_4 + 2\,MnSO_4 + 10\,H_2O + 6\,O_2$$

REACTION #2

$$2\ KO\text{–}Mn(=O)_3 + 3\,H_2SO_4 + 3\,H_2O_2 \longrightarrow K_2SO_4 + 2\,MnSO_4 + 6\,H_2O + 4\,O_2$$

REACTION #3

$$2\ KO{-}Mn(=O)_2{=}O + 3\ H_2SO_4 + H_2O_2 \longrightarrow K_2SO_4 + 2\ MnSO_4 + 4\ H_2O + 3\ O_2$$

SOLUTION

Reaction #1

$$KO{-}Mn(=O)_2{=}O \xrightarrow[-\,KHSO_4]{H_2SO_4} HO{-}Mn(=O)_2{=}O \xrightarrow[-\,HSO_4^{\ominus}]{H_2SO_4} H_2O^{\oplus}{-}Mn(=O)_2{=}O \xrightarrow[-\,H_2O]{H\text{-}O\text{-}O\text{-}H} \ldots \rightleftharpoons \ldots \xrightarrow[-\,H_2O,\ -\,O_2]{H\text{-}O\text{-}O\text{-}H} \ldots$$

$$\ldots \rightleftharpoons \ldots \xrightarrow[-\,H_2O,\ -\,O_2]{H\text{-}O\text{-}O\text{-}H} \ldots \rightleftharpoons \ldots \xrightarrow[-\,H_2O,\ -\,O_2]{} Mn^{\oplus}$$

$$H{-}O{-}O{-}H \longrightarrow 2\ HO\cdot$$

$$\left\{\begin{array}{l} Mn^{+} + HO\cdot \longrightarrow Mn^{+2} + HO^{\ominus} \\ HO^{\ominus} + HSO_4^{\ominus} \longrightarrow H_2O + SO_4^{-2} \end{array}\right\} \times 2$$

Adding the elementary steps yields

$$\times 2 \quad KO{-}Mn(=O)_2{=}O + 2\ H_2SO_4 + 3\ H_2O_2 \longrightarrow 4\ H_2O + 3\ O_2 + Mn^{+} + KHSO_4 + HSO_4^{-}$$

$$2\ Mn^{+} + H_2O_2 + 2\ HSO_4^{\ominus} \longrightarrow 2\ Mn^{+2} + 2\ H_2O + 2\ SO_4^{-2}$$

Therefore,

$$2\ KO{-}Mn(=O)_2{=}O + 4\ H_2SO_4 + 7\ H_2O_2 \longrightarrow 2\ KHSO_4 + 2\ MnSO_4 + 10\ H_2O + 6\ O_2$$

The salts on the RHS can be rewritten as separate ions and regrouped as follows.

$$\begin{aligned} 2\ KHSO_4 + 2\ MnSO_4 &= 2\ K^{+} + 2\ HSO_4^{-} + 2\ Mn^{+2} + 2\ SO_4^{-2} \\ &= 2\ K^{+} + 2\ SO_4^{-2} + 2\ H^{+} + 2\ Mn^{+2} + 2\ SO_4^{-2} \\ &= K_2SO_4 + H_2SO_4 + 2\ MnSO_4 \end{aligned}$$

Hence, finally we have

$$2\ KO{-}Mn(=O)_2{=}O + 3\ H_2SO_4 + 7\ H_2O_2 \longrightarrow K_2SO_4 + 2\ MnSO_4 + 10\ H_2O + 6\ O_2$$

Reaction #2

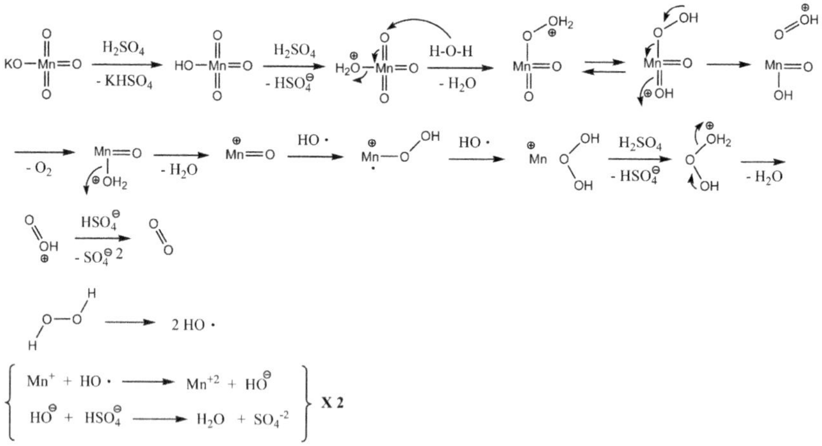

$$O_2 + H_2O_2 \rightarrow 2\,HO\cdot$$

$$\left\{\begin{array}{l} Mn^{+} + HO\cdot \longrightarrow Mn^{+2} + HO^{\ominus} \\ HO^{\ominus} + HSO_4^{\ominus} \longrightarrow H_2O + SO_4^{-2} \end{array}\right\} \times 2$$

Adding the elementary steps yields

$$\times 2 \quad KO{-}MnO_3 + 2\,H_2SO_4 + H_2O_2 \longrightarrow 2\,H_2O + 2\,O_2 + Mn^{+} + KHSO_4 + HSO_4^{-}$$

$$2\,Mn^{+} + H_2O_2 + 2\,HSO_4^{\ominus} \longrightarrow 2\,Mn^{+2} + 2\,H_2O + 2\,SO_4^{-2}$$

Therefore,

$$2\;KO{-}MnO_3 + 4\,H_2SO_4 + 3\,H_2O_2 \longrightarrow 2\,KHSO_4 + 2\,MnSO_4 + 6\,H_2O + 4\,O_2$$

The salts on the RHS can be rewritten as separate ions and regrouped as follows.

$$\begin{aligned} 2\,KHSO_4 + 2\,MnSO_4 &= 2\,K^{+} + 2\,HSO_4^{-} + 2\,Mn^{+2} + 2\,SO_4^{-2} \\ &= 2\,K^{+} + 2\,SO_4^{-2} + 2\,H^{+} + 2\,Mn^{+2} + 2\,SO_4^{-2} \\ &= K_2SO_4 + H_2SO_4 + 2\,MnSO_4 \end{aligned}$$

Hence, finally we have

$$2\;KO{-}MnO_3 + 3\,H_2SO_4 + 3\,H_2O_2 \longrightarrow K_2SO_4 + 2\,MnSO_4 + 6\,H_2O + 4\,O_2$$

Reaction #3

$$HO{-}OH \longrightarrow 2\;\dot{O}H$$

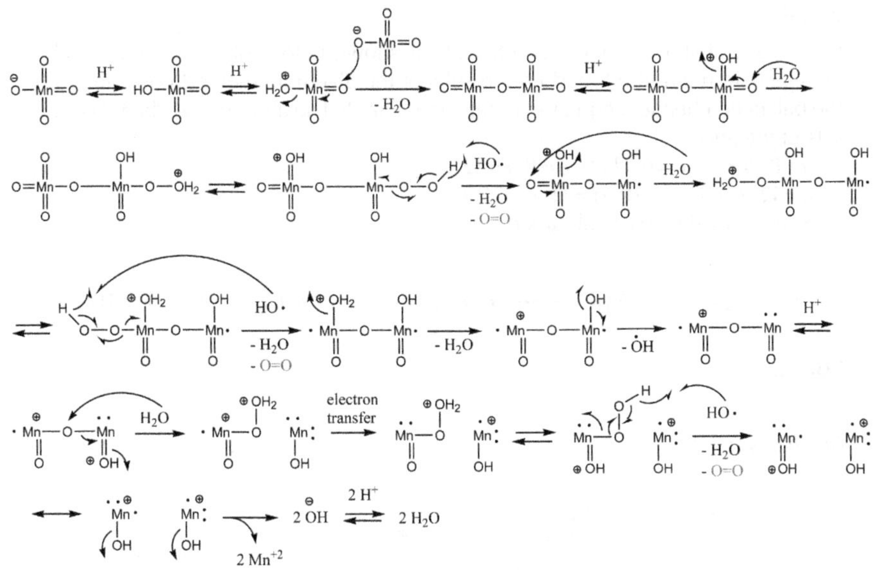

Adding the elementary steps yields

$$2\,MnO_4^- + 6\,H^+ + H_2O_2 \longrightarrow 2\,Mn^{+2} + 3\,O_2 + 4\,H_2O$$

Adding two potassium ions and three sulphate ions to each side yields

$$2\,KMnO_4 + 3\,H_2SO_4 + H_2O_2 \longrightarrow 2\,MnSO_4 + K_2SO_4 + 3\,O_2 + 4\,H_2O$$

Example 2.16 [32]

Themes: balancing chemical equations, redox reactions, reaction mechanism, Diophantine equations

Potassium chlorate reacts with hydrochloric acid to produce chlorine gas and chlorine dioxide according to the following unbalanced transformation.

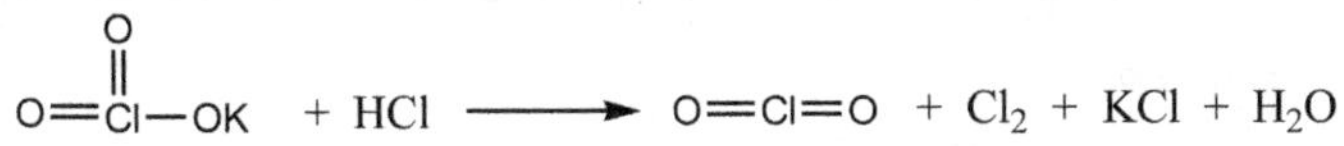

$$\mathrm{O{=}Cl({=}O){-}OK} + HCl \longrightarrow \mathrm{O{=}Cl{=}O} + Cl_2 + KCl + H_2O$$

Part 1

Suggest a mechanism for this transformation.

Part 2

Based on the mechanism obtain the balanced chemical equation.

Part 3

Based on the mechanism write the operative redox couple and verify it by doing an oxidation number analysis.

Part 4

Using general stoichiometric coefficients as shown below, obtain their values by an algebraic method. How do the results of the algebraic method compare with the balanced chemical equation obtained in Part 2? The authors give the following sets of integers:

$a=8, b=24, c=6, d=9, e=8, f=12$

$a=12, b=32, c=10, d=11, e=21, f=16$

Show how they were obtained.

$$a\ KClO_3 + b\ HCl \longrightarrow c\ ClO_2 + d\ Cl_2 + e\ KCl + f\ H_2O$$

Solution

Part 1

Mechanism

$$KClO_3 \xrightarrow[-KCl]{HCl} HClO_3 \xrightarrow{HCl} HOClO_2(OH)Cl \xrightarrow{HCl} [HO-ClO(Cl)-OH_2]^+ Cl^- \xrightarrow{-H_2O}$$

$$[O=Cl(OH)Cl]^+ Cl^- \xrightarrow{-HCl} O_2ClCl \xrightarrow{HCl} Cl-Cl + O=Cl-OH$$

$$O=Cl-OH + KClO_3 \longrightarrow O=Cl-OK + HClO_3$$

$$O=Cl-O^- \ (-K^+) \xrightarrow{-e} O=Cl-O^\bullet \longleftrightarrow O=Cl^\bullet=O$$

$$HClO_3 \longleftrightarrow {}^\bullet O-ClO-OH \xrightarrow{e} {}^-O-ClO-OH \xrightarrow{HCl} HO-ClO-OH\ Cl^- \xrightarrow{HCl} [HO-ClO-OH_2]^+\ Cl^-\ Cl^-$$

$$\xrightarrow{-H_2O} [O=Cl-OH]^+\ Cl^-\ Cl^- \xrightarrow{-HCl} O=Cl^\bullet=O + Cl^\ominus$$

Part 2

Adding the elementary steps leads to

$$2\ KClO_3 + 4\ HCl \longrightarrow 2\ ClO_2 + Cl_2 + 2\ KCl + 2\ H_2O$$

2 (122.45) 4 (36.45) 2 (67.45) 70.9 2 (74.45) 2 (18)

Check balancing:

Reactants: 2*122.45 + 4*36.45 = 390.7

Products: 2*67.45 + 70.9 + 2*74.45 + 2*18 = 390.7

Part 3

Redox Couple

[R] $KClO_3 + 2\,H^+ + e \longrightarrow ClO_2 + K^+ + H_2O$ X 2

[O] $2\,HCl \longrightarrow Cl_2 + 2\,H^+ + 2\,e$

Adding gives

$$2\,KClO_3 + 2\,H^+ + 2\,HCl \longrightarrow 2\,H_2O + Cl_2 + 2\,ClO_2 + 2\,K^+$$

Adding two Cl^- ions on each side gives

$$2\,KClO_3 + 4\,HCl \longrightarrow 2\,H_2O + Cl_2 + 2\,ClO_2 + 2\,KCl$$

Oxidation Number Analysis

$$2\,K\overset{a}{Cl}O_3 + 4\,H\overset{b}{Cl} \longrightarrow 2\,H_2O + \overset{b}{Cl}\text{--}\overset{b}{Cl} + 2\,\overset{a}{Cl}O_2 + 2\,KCl$$

Atom	Reactants	Products	Change	
Cl_a	+5	+4	-1	$\Delta = -2$
Cl_a	+5	+4	-1	
Cl_b	-1	0	+1	X 2 → $\Delta = +2$

The oxidation number difference matches the number of electrons involved.

Part 4

$$a\ O{=}\overset{\overset{O}{\|}}{Cl}{-}OK + b\,HCl \longrightarrow c\ O{=}Cl{=}O + d\,Cl_2 + e\,KCl + f\,H_2O$$

Balancing O atoms: $3\,a = 2\,c + f$
Balancing Cl atoms: $a + b = c + 2\,d + e$
Balancing K atoms: $a = e$
Balancing H atoms: $b = 2\,f$
Therefore,
$3\,a = 2\,c + f$
$a + b = c + 2\,d + a \rightarrow b = c + 2\,d$
$b = 2\,f$
or
$3\,a = 2\,c + 0.5\,b$
$b = c + 2\,d$
$6\,a = 4\,c + b$
$2\,d = -c + b$
Subtracting gives
$6\,a - 2\,d = 5\,c$

We recognize this equation as a Diophantine equation since there exist multiple positive integer solutions. From the balanced chemical equation found in Part 2, we have

$a=2, c=2, d=1$

We can verify that this set of stoichiometric coefficients satisfies the Diophantine equation.

$(6*2) - (2*1) = (5*2)$
$12 - 2 = 10$
$10 = 10$

We may find other sets of stoichiometric coefficients that would satisfy the balancing of the chemical equation by listing all possible redox couples without regard to reaction mechanism as follows.

[R] $KClO_3 + 2\,H^+ + e \longrightarrow ClO_2 + K^+ + H_2O$

[R] $KClO_3 + 6\,H^+ + 6\,e \longrightarrow KCl + 3\,H_2O$

[R] $KClO_3 + 6\,H^+ + 5\,e \longrightarrow 1/2\,Cl_2 + K^+ + 3\,H_2O$

[O] $HCl + 2\,H_2O \longrightarrow ClO_2 + 5\,H^+ + 5\,e$

[O] $HCl \longrightarrow 1/2\,Cl_2 + H^+ + e$

Adding the reduction and oxidation separately gives

[R] $3\,KClO_3 + 14\,H^+ + 12\,e \longrightarrow ClO_2 + 2\,K^+ + KCl + 1/2\,Cl_2 + 7\,H_2O$

[O] $2\,HCl + 2\,H_2O \longrightarrow ClO_2 + 1/2\,Cl_2 + 6\,H^+ + 6\,e$ X 2

Adding the redox couple gives

$$3\,KClO_3 + 2\,H^+ + 4\,HCl \longrightarrow 3\,ClO_2 + 2\,K^+ + KCl + 3/2\,Cl_2 + 3\,H_2O$$

Adding two Cl^- ions on each side gives

$$3\,KClO_3 + 6\,HCl \longrightarrow 3\,ClO_2 + 3/2\,Cl_2 + 3\,KCl + 3\,H_2O$$

3 (122.45) 6 (36.45) 3 (67.45) (3/2) (70.9) 3 (74.45) 3 (18)

Check Balancing

Reactants: $3*122.45 + 6*36.45 = 586.05$
Products: $3*67.45 + 1.5*70.9 + 3*74.45 + 3*18 = 586.05$

From this result we find the following set of integers:

$a=6, c=6, d=3$

This satisfies $6\,a - 2\,d = 5\,c$ since $6*6-2*3=5*6$ or $36-6=30$.

Entry	*a* ($KClO_3$)	*b* (HCl)	*c* (ClO_2)	*d* (Cl_2)	*e* (KCl)	*f* (H_2O)	Comment
1	6	12	6	3	6	6	Based on maximum number of redox couples
2	2	4	2	1	2	2	Based on reaction mechanism (minimum set of coefficients)

The set of integers in entry 2 (based on the reaction mechanism analysis) are found by dividing those in entry 1 (based on the maximum number of redox couples) by three.

The authors give the following sets of integers:

$a=8$, $b=24$, $c=6$, $d=9$, $e=8$, $f=12$
$a=12$, $b=32$, $c=10$, $d=11$, $e=21$, $f=16$

Checking to see if the Diophantine equation $6a - 2d = 5c$ is satisfied:

$6*8 - 2*9 = 5*6$; $48 - 18 = 30$
$6*12 - 2*11 = 5*10$; $72 - 22 = 50$

Solving Diophantine equation

$6a - 2d = 5c$, or
$6a = 5c + 2d$

Set *a* = 1. Then $6 = 5c + 2d$.
Following the Diophantine algorithm, we have the following:

$6/5 < c + d < 3$; $c + d = \{2\}$.
$2(c + d) = \{4\}$.
$(5 - 2)c = \{4\}$.
$3c = \{\text{none}\}$.
No solution.

Set *a* = 2. Then $12 = 5c + 2d$.
$12/5 < c + d < 6$; $c + d = \{3,4,5\}$.
$2(c + d) = \{6,8,10\}$.
$(5 - 2)c = \{6,4,2\}$.
$3c = 6$; $c = 2$.
Therefore, $(c,d) = (2,1)$.

Set *a* = 3. Then $18 = 5c + 2d$.
$18/5 < c + d < 9$; $c + d = \{4,5,6,7,8\}$.
$2(c + d) = \{8,10,12,14,16\}$.
$(5 - 2)c = \{10,8,6,4,2\}$.
$3c = 6$; $c = 2$.
Therefore, $(c,d) = (2,4)$.

Set $a=4$. Then $24=5c+2d$.
$24/5 < c+d < 12$; $c+d=\{5,6,7,8,9,10,11\}$.
$2(c+d)=\{10,12,14,16,18,20,22\}$.
$(5-2)c=\{14,12,10,8,6,4,2\}$.
$3c=\{12,6\}$; $c=\{4,2\}$; $d=\{2,7\}$.
Therefore $(c,d)=(4,2)$, $(2,7)$.

Set $a=5$. Then $30=5c+2d$.
$6 < c+d < 15$; $c+d=\{7,8,9,10,11,12,13,14\}$.
$2(c+d)=\{14,16,18,20,22,24,26,28\}$.
$(5-2)c=\{16,14,12,10,8,6,4,2\}$.
$3c=\{12,6\}$; $c=\{4,2\}$; $d=\{5,10\}$.
Therefore $(c,d)=(4,5)$, $(2,10)$.

Set $a=6$. Then $36=5c+2d$.
$36/5 < c+d < 18$; $c+d=\{8,9,10,11,12,13,14,15,16,17\}$.
$2(c+d)=\{16,18,20,22,24,26,28,30,32,34\}$.
$(5-2)c=\{20,18,16,14,12,10,8,6,4,2\}$.
$3c=\{18,12,6\}$; $c=\{6,4,2\}$; $d=\{3,8,13\}$.
Therefore $(c,d)=(6,3)$, $(4,8)$, $(2,13)$.

Summary

Entry	a ($KClO_3$)	$b=c+2d$ (HCl)	c (ClO_2)	d (Cl_2)	$e=a$ (KCl)	$f=b/2$ (H_2O)	Comment
1	2	4	2	1	2	2	Based on reaction mechanism (minimum set of coefficients)
2	3	10	2	4	3	5	
3	4	8	4	2	4	4	
4	4	16	2	7	4	8	
5	5	14	4	5	5	7	
6	5	22	2	10	5	11	
7	6	12	6	3	6	6	
8	6	20	4	8	6	10	
9	6	28	2	13	6	14	
10	8	24	6	9	8	12	Entry 1 + entry 8
11	12	32	10	11	12	16	Entry 7 + entry 8

2.3.4 Oxidation Number Analysis

In this section, we illustrate using examples how to track the oxidation number changes for atoms involved in a redox transformation and how that relates directly to the number of electrons involved in the associated redox couples.

Example 2.17 [33]

Themes: balancing chemical equations, oxidation numbers

For each transformation, show a balanced chemical equation, identify the target bonds made, assign oxidation numbers to the atoms involved in the bonds made and broken, and determine the oxidation number changes for each atom. Verify that the numerical value of the overall oxidation number change corresponds to the number of electrons involved in the redox couple.

(a)

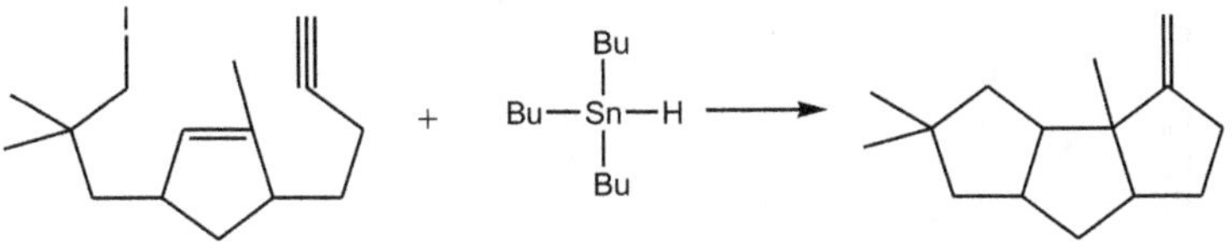

(b)

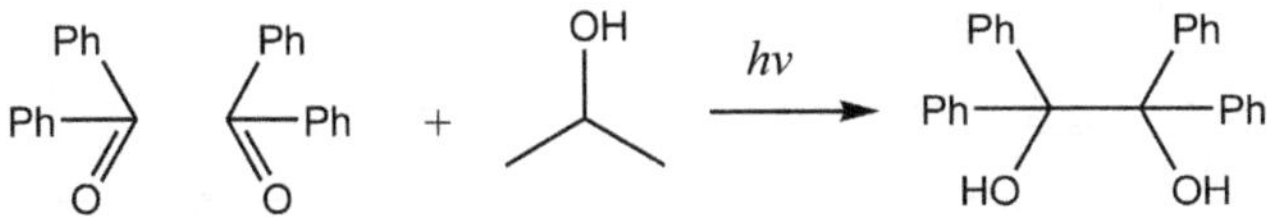

(c)

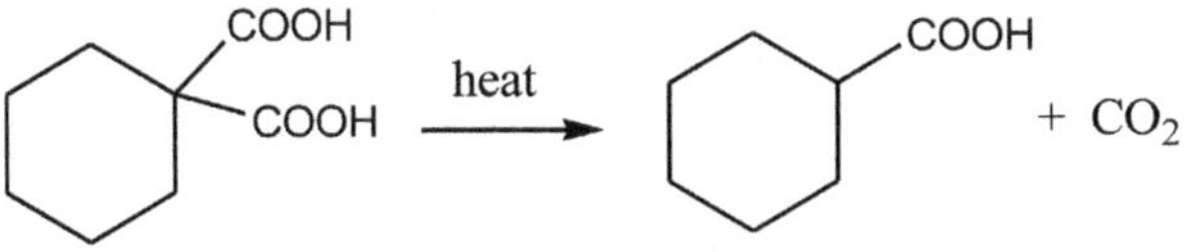

(d)

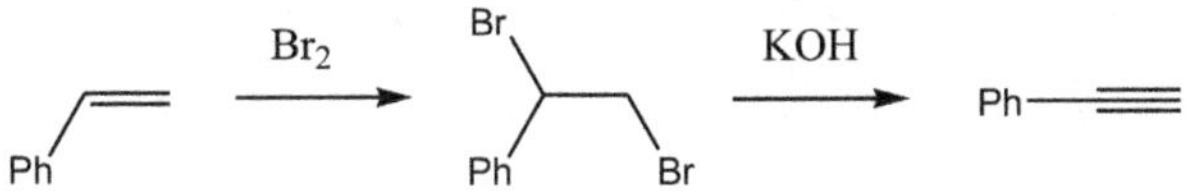

(e)

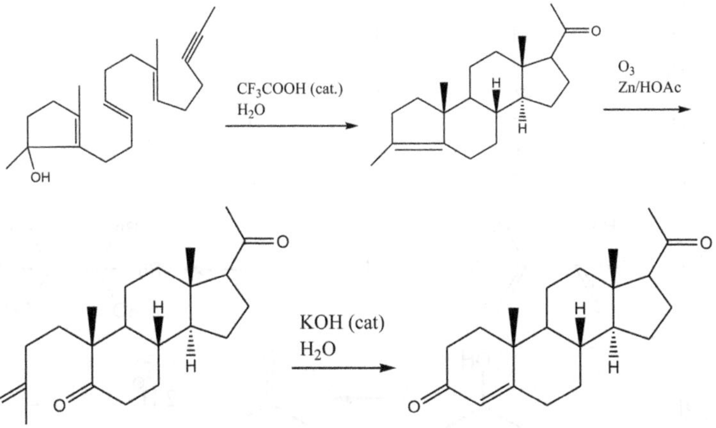

Solution

(a)

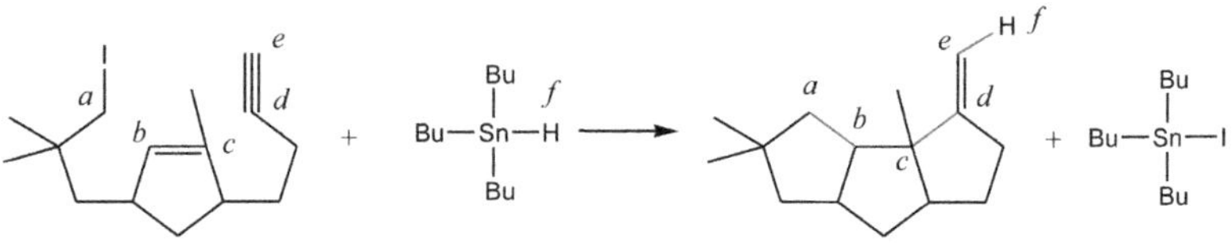

Atom	Reactant	Product	Change	
C_a	-1	-2	-1	
C_b	-1	-1	0	
C_c	0	0	0	$\Delta = -2$
C_d	0	0	0	
C_e	-1	-2	-1	
H_f	-1	+1	+2	$\Delta = +2$

(b)

[R] $+ H^{\oplus} + 2e \longrightarrow$ $+ I^{\ominus}$

[O] $I^{\ominus} +$ Bu₃Sn–H $\longrightarrow$ Bu₃Sn–I $+ H^{\oplus} + 2e$

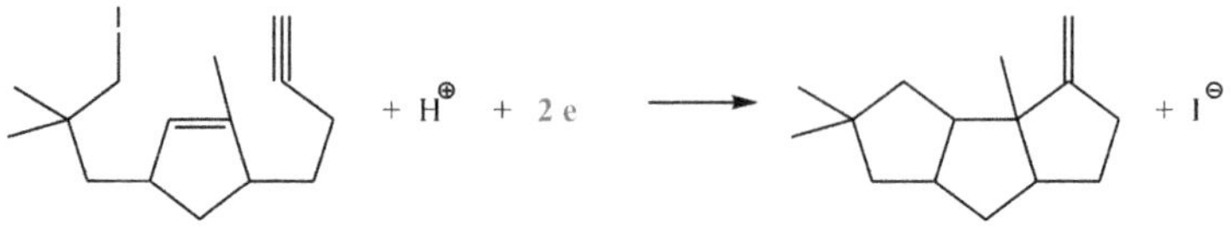

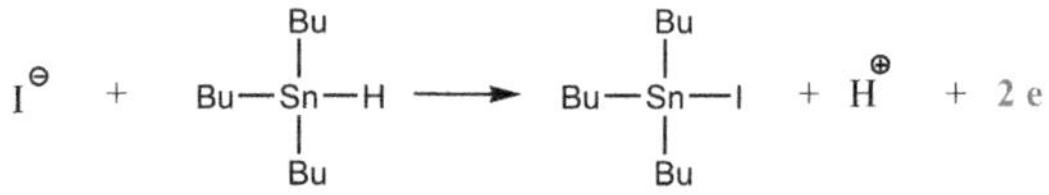

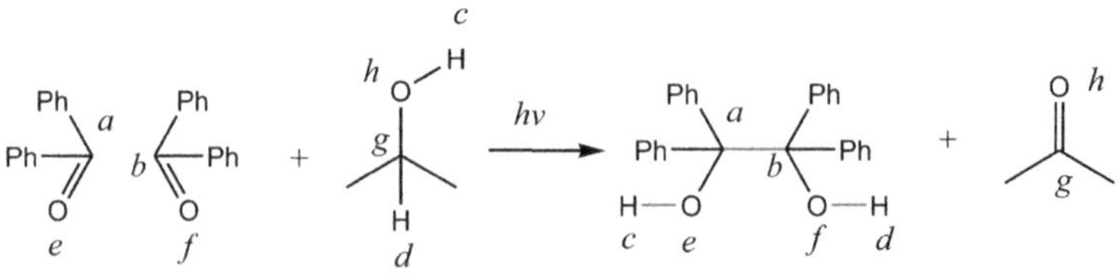

Atom	Reactant	Product	Change	
C_a	+2	+1	-1	$\Delta = -2$
C_b	+2	+1	-1	
H_c	+1	+1	0	
H_d	+1	+1	0	
O_e	-2	-2	0	
O_f	-2	-2	0	
C_g	0	+2	+2	$\Delta = +2$
O_h	-2	-2	0	

[R] $+ 2H^{\oplus} + 2e \longrightarrow$

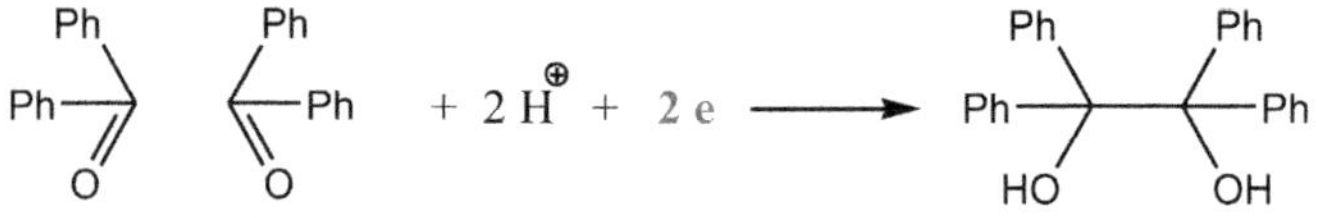

[O] $\longrightarrow$ $+ 2H^{\oplus} + 2e$

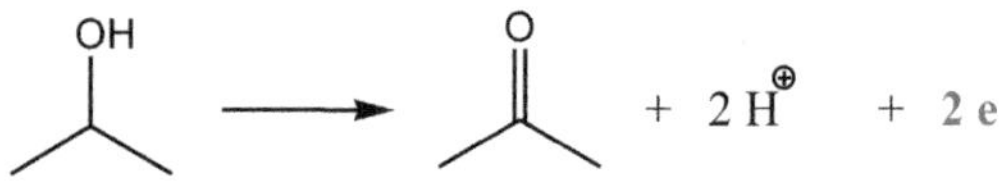

(c)

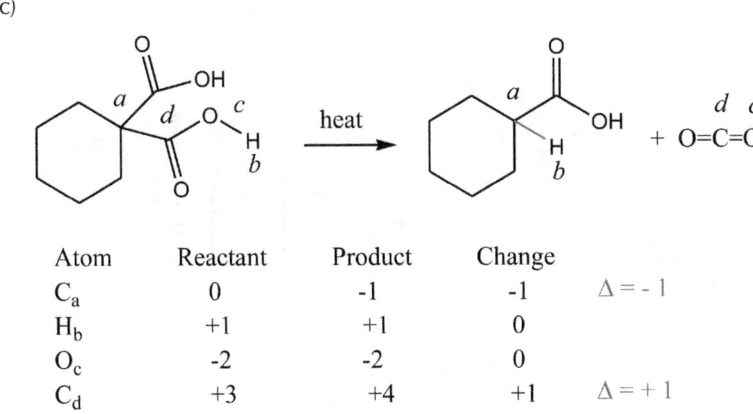

Atom	Reactant	Product	Change	
C_a	0	-1	-1	$\Delta = -1$
H_b	+1	+1	0	
O_c	-2	-2	0	
C_d	+3	+4	+1	$\Delta = +1$

This transformation is an internal redox reaction, so no redox couple can be written.

(d)

Step 1:

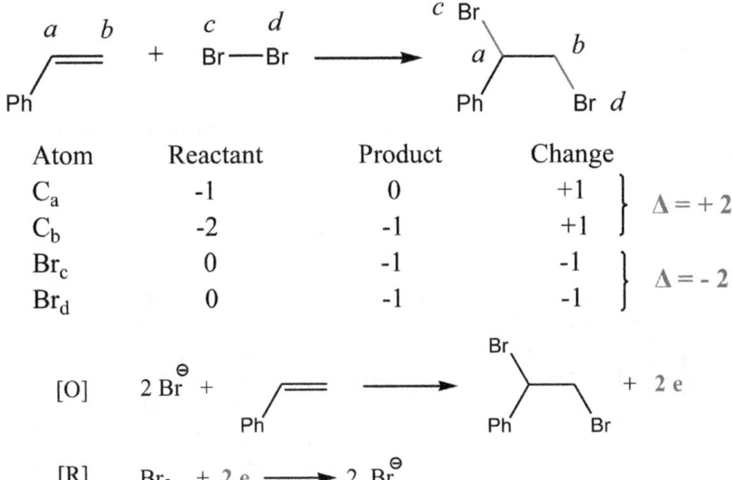

Atom	Reactant	Product	Change	
C_a	-1	0	+1	$\Delta = +2$
C_b	-2	-1	+1	
Br_c	0	-1	-1	$\Delta = -2$
Br_d	0	-1	-1	

[O] $2\ Br^{\ominus}$ + PhCH=CH$_2$ ⟶ PhCHBrCH$_2$Br + 2 e

[R] $Br_2 + 2\ e \longrightarrow 2\ Br^{\ominus}$

Step 2:

PhCHBrCH$_2$Br + 2 KOH ⟶ Ph—C≡CH + 2 KBr + 2 H_2O

Atom	Reactant	Product	Change
C_a	0	0	0
C_b	-1	-1	0

The second step does not involve oxidation number changes.

(e)

Step 1:

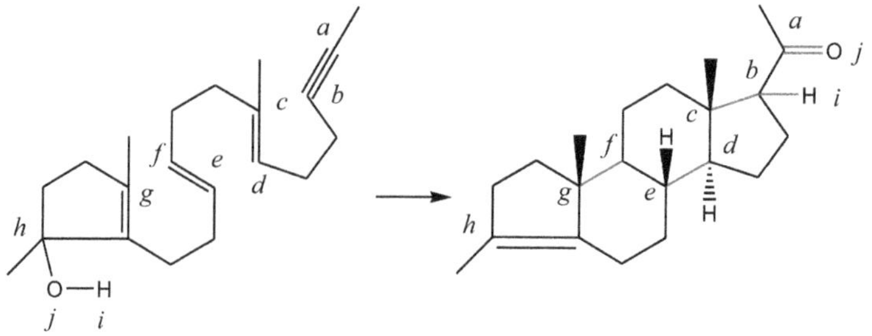

Atom	Reactant	Product	Change	
C_a	0	+2	+2	$\Delta = +2$
C_b	0	-1	-1	$\Delta = -2$
C_c	0	0	0	
C_d	-1	-1	0	
C_e	-1	-1	0	
C_f	-1	-1	0	
C_g	0	0	0	
C_h	+1	0	-1	
H_i	+1	+1	0	
O_j	-2	-2	0	

This transformation is an internal redox reaction, so no redox couple can be written.

Step 2:

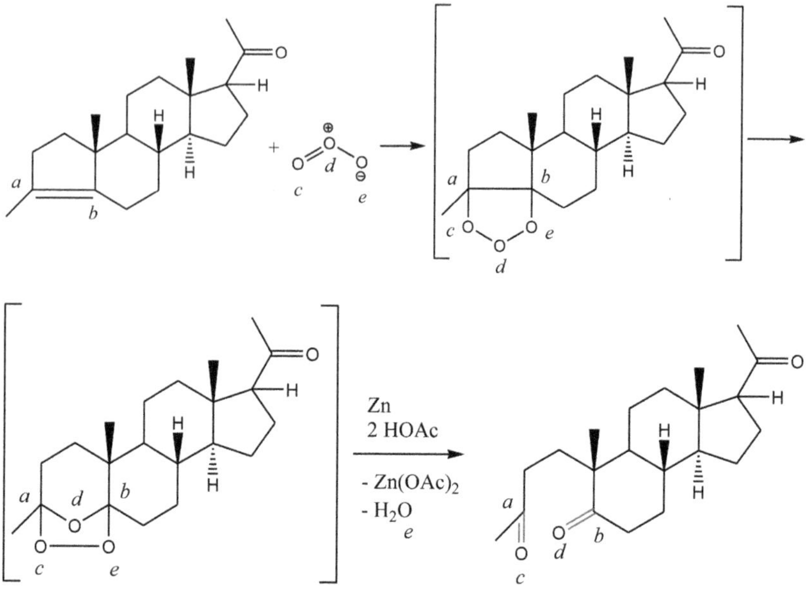

Zn
2 HOAc
- $Zn(OAc)_2$
- H_2O

Atom	Reactant	Product	Change	
C_a	0	+2	+2	$\Delta = -2$
C_b	0	+2	+2	
O_c	0	-2	-2	
O_d	+1	-2	-3	
O_e	-1	-2	-1	
Zn	0	+2	+2	$\Delta = +2$

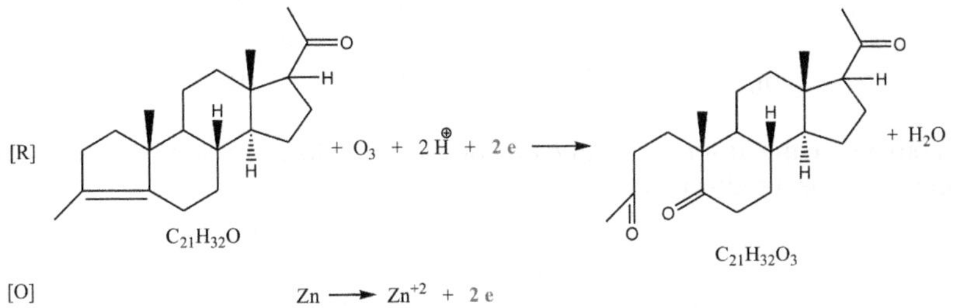

[R] $C_{21}H_{32}O + O_3 + 2H^{\oplus} + 2e \longrightarrow C_{21}H_{32}O_3 + H_2O$

[O] $Zn \longrightarrow Zn^{+2} + 2e$

Step 3:

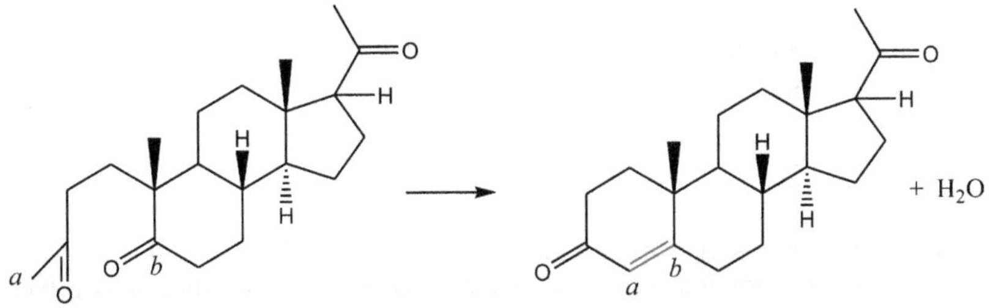

$$\longrightarrow + H_2O$$

Atom	Reactant	Product	Change	
C_a	-3	-1	+2	$\Delta = +2$
C_b	+2	0	-2	$\Delta = -2$

This transformation is an internal redox reaction, so no redox couple can be written.

Example 2.18 [34]

Themes: balancing chemical equations, redox reactions

Decide whether or not the following is a redox reaction using an oxidation number analysis.

$$SO_3^{-2} + S \longrightarrow S_2O_3^{-2}$$

sulfite thiosulfate

Hint: Draw resonance structures for the thiosulfate ion.

Solution

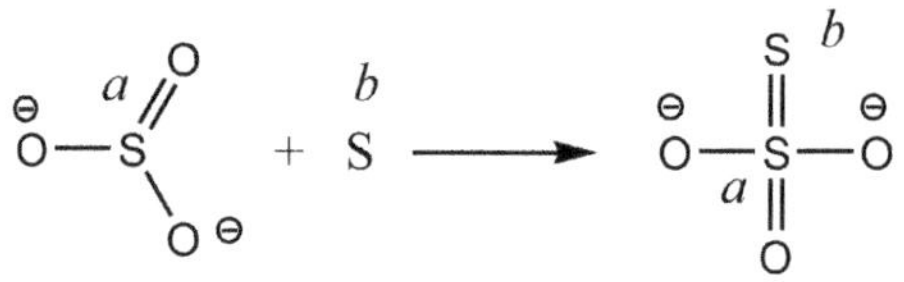

Atom	Reactant	Product	Change
S_a	+4	+4	0
S_b	0	0	0

According to the resonance structure drawn for thiosulfate ion, the reaction appears not be a redox reaction.

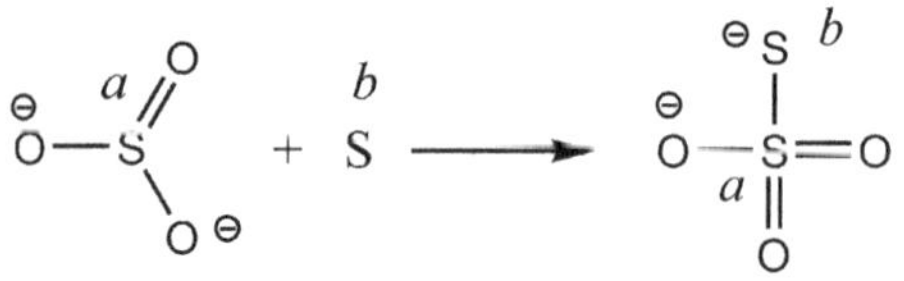

Atom	Reactant	Product	Change
S_a	+4	+5	+1
S_b	0	-1	-1

According to the resonance structure drawn for thiosulfate ion, the reaction appears to be a redox reaction. Note that a redox couple consisting of two half reactions cannot be written for this transformation. Hence, it is an internal redox reaction.

This ambiguous example illustrates that oxidation number assignments to atoms in a structure are dependent on the resonance structures drawn.

Example 2.19 [32]

Themes: balancing chemical equations, redox reactions, reaction mechanism

Given the following unbalanced transformation

$$O{=}N{-}ONa + FeSO_4 + H_2SO_4 \longrightarrow {}^{\bullet}N{=}O + Fe_2(SO_4)_3 + NaHSO_4 + H_2O$$

Part 1

Write the redox couple and determine the appropriate stoichiometric coefficients to balance the equation. Check that the equation is balanced.

Part 2

Do an oxidation number analysis and verify that it agrees with the redox couple.

Part 3

Suggest a mechanism for this transformation.

Solution

Part 1

[R] $2\,H^+ + e + NaNO_2 \longrightarrow NO + Na^+ + H_2O$ X 2

[O] $2\,FeSO_4 + SO_4^{-2} \longrightarrow Fe_2(SO_4)_3 + 2\,e$

Adding the redox couples gives

$$4\,H^+ + 2\,NaNO_2 + 2\,FeSO_4 + SO_4^{-2} \longrightarrow 2\,NO + 2\,Na^+ + 2\,H_2O + Fe_2(SO_4)_3$$

Simplifying LHS gives

$$H_2SO_4 + 2\,H^+ + 2\,NaNO_2 + 2\,FeSO_4 \longrightarrow 2\,NO + 2\,Na^+ + 2\,H_2O + Fe_2(SO_4)_3$$

Adding 2 $[HSO_4]^-$ ions on both sides gives

$$3\,H_2SO_4 + 2\,NaNO_2 + 2\,FeSO_4 \longrightarrow 2\,NO + 2\,NaHSO_4 + 2\,H_2O + Fe_2(SO_4)_3$$

3 (98) 2 (69) 2 (151.85) 2 (30) 2 (120) 2 (18) 399.7

Check Balancing

Reactants: 3*98+2*69+2*151.85=735.7
Products: 2*30+2*120+2*18+399.7=735.7

Part 2

Oxidation Number Analysis

Atom	Reactants	Products	Change	
N	+3	+2	-1	X 2 → Δ = - 2
Fe	+2	+3	+1	X 2 → Δ = + 2

The oxidation number difference matches the number of electrons involved.

Part 3

Mechanism

$$O{=}N{-}O^{\ominus} \xrightarrow{H^{\oplus}} O{=}N{-}OH \xrightarrow{H^{\oplus}} O{=}N{-}OH_2^{\oplus} \xrightarrow[-H_2O]{} {}^{\oplus}O{\equiv}N{:} \longleftrightarrow O{=}N^{\oplus}{:}$$

$$\xrightarrow[-Fe^{+3}]{Fe^{+2}} O{=}\dot{N}{:}$$

Adding the elementary steps gives

$$NO_2^{\ominus} + 2\,H^+ + Fe^{+2} \longrightarrow NO + H_2O + Fe^{+3}$$

Adding Na^+ ions to each side gives

$$NaNO_2 + 2\,H^+ + Fe^{+2} \longrightarrow NO + H_2O + Fe^{+3} + Na^+$$

Adding 2 $[SO_4]^{-2}$ ions to each side gives

$$NaNO_2 + H_2SO_4 + FeSO_4 \longrightarrow NO + H_2O + Fe^{+3} + Na^+ + 2\,SO_4^{-2}$$

Multiplying each stoichiometric coefficient by two gives

$$2\,NaNO_2 + 2\,H_2SO_4 + 2\,FeSO_4 \longrightarrow 2\,NO + 2\,H_2O + 2\,Fe^{+3} + 2\,Na^+ + 4\,SO_4^{-2}$$

Simplifying the RHS gives

$$2\,NaNO_2 + 2\,H_2SO_4 + 2\,FeSO_4 \longrightarrow 2\,NO + 2\,H_2O + Fe_2(SO_4)_3 + Na_2SO_4$$

Adding one H_2SO_4 to each side and rearranging ions on the RHS gives

$$2\,NaNO_2 + 3\,H_2SO_4 + 2\,FeSO_4 \longrightarrow 2\,NO + 2\,H_2O + Fe_2(SO_4)_3 + 2\,NaHSO_4$$

This final equation agrees with the balanced equation found in Part 1.

2.3.5 Problems

PROBLEM 2.7

Themes: structure drawing, oxidation numbers

Draw chemical structures for the following oxides of phosphorus and assign the oxidation numbers for all phosphorus atoms appearing in each structure.

(a) phosphorous pentoxide; (b) phosphorus trioxide; (c) phosphorus tetroxide; (d) P_4O_6; (e) P_4O_{10}; (f) phosphorus oxychloride; (g) sodium pyrophosphate; (h) sodium trimetaphosphate; (i) phosphinic acid; (j) phosphoric acid; (k) phosphorous acid; (l) hypophosphorous acid; (m) trimethyl phosphite; (n) dimethyl methylphosphonate

PROBLEM 2.8

Themes: structure drawing, oxidation numbers

Draw chemical structures for the following oxides of sulfur and assign the oxidation numbers for all sulfur atoms appearing in each structure.

(a) sulfuric acid; (b) sulfurous acid; (c) sulfur dioxide; (d) sulfur trioxide; (e) sodium peroxysulfate; (f) sodium dithionite; (g) sodium dithionate; (h) sulfuryl chloride; (i) thionyl chloride; (j) thiosulfuric acid; (k) pyrosulfuric acid; (l) peroxomonosulfuric acid

PROBLEM 2.9

Themes: balancing chemical equations, redox reactions

A popular demonstration of exothermic reactions performed in chemistry magic shows is the combustion of potassium permanganate in the presence of glycerol. A small depression is made at the top of a mound of ground purple potassium permanganate crystals. A small quantity of glycerol liquid is added to the well created by the depression. At first nothing appears to happen then suddenly a violent reaction takes place with the top of the mound glowing brightly with a flame and emitting a purple vapor as if a volcano is exploding. This demonstration was also popular as a children's activity in chemistry sets sold in the UK and the US in the 1950s.

Part 1

Write out a balanced redox reaction that explains the observations.

Part 2

Show the changes in oxidation numbers of relevant atoms. Do they agree with the results of the redox couple balancing?

Part 3

If we suggest that the fate of permanganate ion is Mn_2O_3, what would be the resultant balanced chemical equation for the redox reaction?

Part 4

Repeat the oxidation number analysis on Part 3.

PROBLEM 2.10 [35]

Themes: balancing chemical equations, sacrificial reagents

2-Bromoethylbenzene and 2-nitroethylbenzene can be converted to phenylacetic acid in the presence of sodium nitrite and acetic acid in DMSO solvent. Sodium nitrite acts as a sacrificial reagent in each transformation.

$NaNO_2$, HOAc, DMSO

2-nitroethylbenzene → phenylacetic acid ← 2-bromoethylbenzene

Write out mechanisms that account for the role of sodium nitrite as a sacrificial reagent. For each mechanism add up the elementary steps to obtain the corresponding overall balanced chemical equations. Which reaction requires a higher stoichiometric amount of sodium nitrite?

PROBLEM 2.11 [36]

Themes: balancing chemical equations, redox reactions, by-products, side products

L-phenylalanine undergoes selective oxidative decarboxylation to phenylacetonitrile via vanadium chloroperoxidase (VCPO) according to the reaction shown below. Another product of the reaction is phenylacetaldehyde.

VCPO, H_2O_2, NaBr (cat.), citrate buffer, pH 5.6

L-phenylalanine (100 % conversion) → phenylacetonitrile + phenylacetaldehyde (3 : 1)

Part 1

Write out the redox couple pertaining to the production of phenylacetonitrile.
Write out the redox couple pertaining to the production of phenylacetaldehyde.

Part 2

Write out a balanced chemical equation for the production of phenylacetonitrile.
Write out a balanced chemical equation for the production of phenylacetaldehyde.

Part 3

Determine the atom economies and kernel reaction mass efficiencies of both reactions.

PROBLEM 2.12 [37]

Themes: balancing chemical equations, reaction mechanism, redox reactions, oxidation number analysis

Naphthalene is oxidized to phthalonic acid with potassium permanganate.

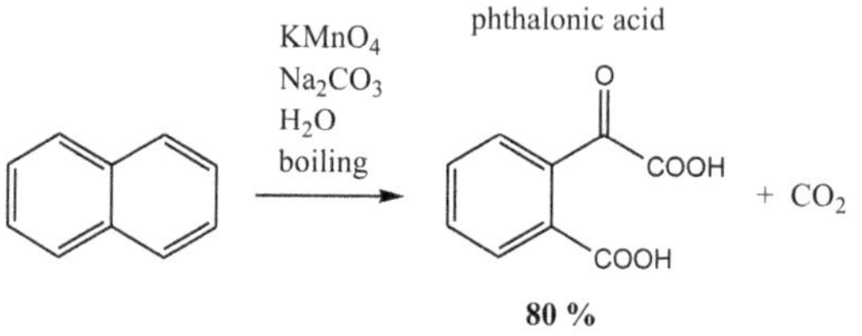
phthalonic acid
KMnO4
Na2CO3
H2O
boiling
O
COOH
COOH
+ CO2
80 %

The following intermediates were ruled out during the course of the oxidation:

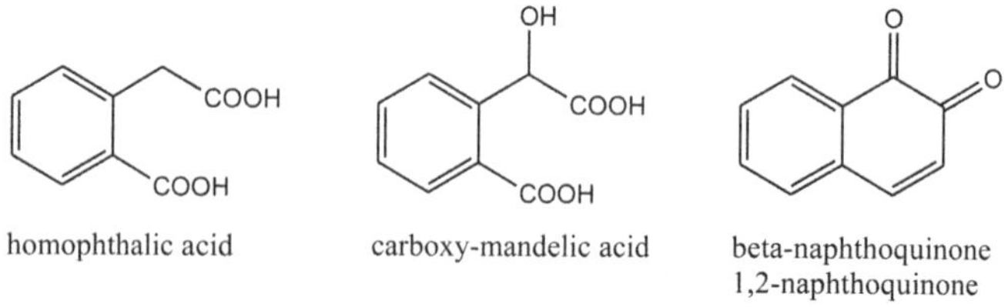
COOH
COOH
homophthalic acid
OH
COOH
COOH
carboxy-mandelic acid
O
O
beta-naphthoquinone
1,2-naphthoquinone

The following intermediates were suspected as plausible during the course of the oxidation:

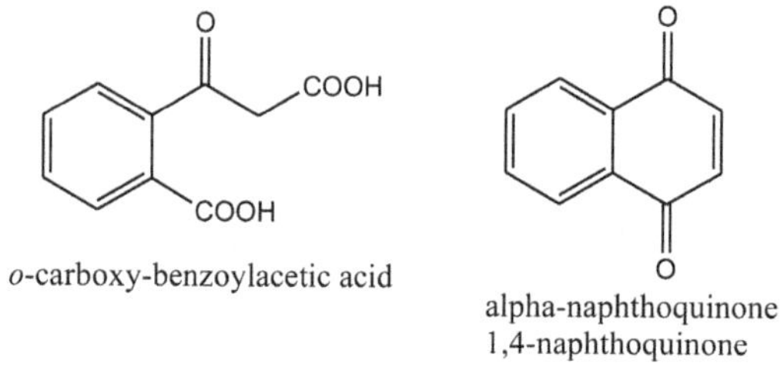
O
COOH
COOH
o-carboxy-benzoylacetic acid
O
O
alpha-naphthoquinone
1,4-naphthoquinone

Part 1

Based on the evidence given suggest a reaction mechanism for this transformation. Is the mechanism consistent with the plausible intermediates suggested by the authors?

Part 2

Write a balanced chemical equation for the reaction.

Part 3

Write the redox couple for the reaction.

Part 4

Show the oxidation number analysis that is consistent with the redox couple.

PROBLEM 2.13 [38]

Themes: balancing chemical equations, redox reactions

The cyanidation process of gold extraction from crushed raw gold ore is shown schematically below.

$$\text{Raw Au ore} \xrightarrow[\text{air}]{0.1\ \%\ \text{NaCN (aq)}} [Au(CN)_2]^{\ominus} \xrightarrow{Zn} Au^0$$

Part 1

Write out balanced chemical equations for each step in the process.

Part 2

Excess cyanide in step 1 is treated by exposure to light in air and excess zinc metal in step 2 is treated with sulfuric acid. What reactions are taking place in each of these treatments? What are the fates of cyanide ion and zinc?

Part 3 [39, 40]

An alternative process that avoids the use of cyanide involves treating the raw gold ore with *aqua regia* (3 parts conc. HCl: 1 part conc. HNO_3). The strong acid solution dissolves the gold and results in an aqueous solution of chloroauric acid ($HAuCl_4$). Another way to obtain chloroauric acid is to treat the raw gold ore with hydrochloric acid and chlorine gas. The chloroauric acid is then treated with either hydrazine hydrochloride, sulfur dioxide, oxalic acid, ferrous sulfate, sodium metabisulfite, or formic acid to precipitate the purified gold. Write out balanced chemical equations for all transformations described.

Part 4 [41–46]

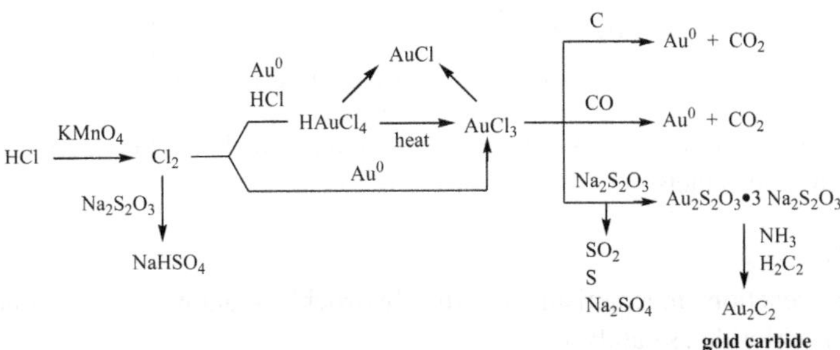

For each step in the reaction network shown, write out balanced chemical equations and identify reaction by-products.

PROBLEM 2.14

Themes: balancing chemical equations, redox reactions

Part 1

Cinnabar, HgS, is the primary ore from which liquid mercury is extracted via any one of the following thermal reactions:

1. heating in air
2. heating with lime
3. heating with lime in air
4. heating with iron
5. heating with hematite.

Write out balanced chemical equations for each of these transformations showing all reaction by-products.

Part 2 [47–50]

Half of the world's mined mercury ore is destined to manufacture mercury(II) chloride, which is used as a catalyst to make vinyl chloride from acetylene and hydrogen chloride. China is the leading manufacturer of vinyl chloride by this method since it also has the largest reserves of coal from which acetylene is obtained. As a consequence of this process, China is also the highest mercury polluter. Below is a reaction network for the production of mercury(II) chloride via various routes all starting from elemental mercury.

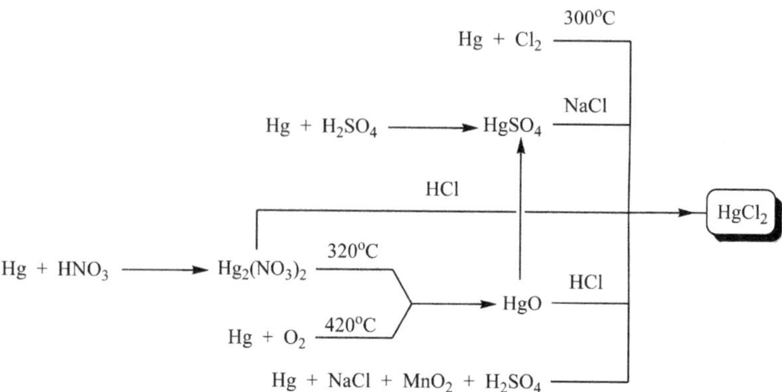

For each reaction in the network write out a balanced chemical equation identifying reaction by-products.

Part 3 [51]

Suggest a reaction mechanism for the hydrochlorination of acetylene using mercury(II) chloride as catalyst.

PROBLEM 2.15

Themes: balancing chemical equations, redox reactions, reaction mechanism, Diophantine equations

Chromium dioxide is a ferromagnetic material used in the manufacture of recording tape cassettes. It can be made by either of the following routes.

Route #1 [52]

$$Cl{-}Cr({=}O)_2{-}Cl \xrightarrow[O_2]{360^{o}C} O{=}Cr{=}O$$

chromyl chloride

Route #2 [53]

$$O{=}Cr{-}O{-}Cr{=}O \xrightarrow[3000\ atm]{360^{o}C} O{=}Cr{=}O$$

chromic oxide
chromium sesquioxide

Route #3 (hydrothermal process) [54–56]

$$CrO_3 + Cr_2O_3 \xrightarrow{800\ K,\ 200\ MPa,\ H_2O} CrO_2 + O_2$$

For each route, suggest a reaction mechanism and provide a balanced chemical equation.

PROBLEM 2.16

Themes: balancing chemical equations, redox reactions, Diophantine equations

Methane can be incompletely combusted to various ratios of carbon dioxide and carbon monoxide according to the general chemical equation given below.

$$a\,CH_4 + b\,O_2 \longrightarrow c\,CO_2 + d\,CO + e\,H_2O$$

Part 1

Determine the possible positive integer values of the stoichiometric coefficients.

Part 2

Determine the general Diophantine equation for the incomplete combustion of a general linear hydrocarbon C_nH_{2n+2}.

PROBLEM 2.17

Themes: balancing chemical equations, redox reactions, oxidation number analysis

Phthalocyanine can be made from the four starting materials shown below.

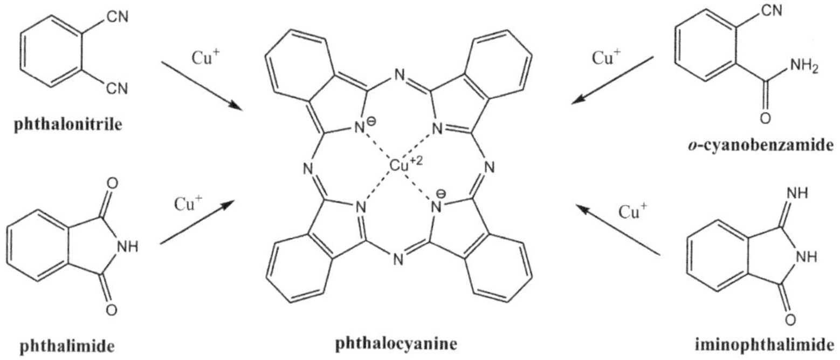

phthalonitrile

o-cyanobenzamide

phthalimide

phthalocyanine

iminophthalimide

Part 1

For each starting material, write out a balanced chemical equation that leads to the phthalocyanine product.

Part 2

Do an oxidation number analysis to pinpoint which atoms are reduced and oxidized.

PROBLEM 2.18

Themes: balancing chemical equations, reaction mechanisms, redox reactions

An example Clemmensen reduction is shown below.

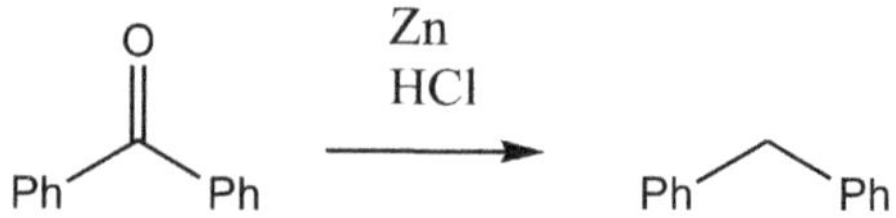

Part 1

Write the redox couple and determine the appropriate stoichiometric coefficients to balance the equation. Check that the equation is balanced.

Part 2

Do an oxidation number analysis and verify that it agrees with the redox couple.

Part 3

Suggest a mechanism for this transformation.

PROBLEM 2.19 [57]

Themes: balancing chemical equations, reaction mechanisms, redox reactions

For the transformation shown below, the fate of the nitrogen atom in the oxime was not described in the reference.

$$Ph_2C{=}N{-}OH \xrightarrow[\text{microwaves}]{(NH_4)_2S_2O_8,\ \text{silica}} Ph_2C{=}O$$

Part 1

Write the redox couple and determine the appropriate stoichiometric coefficients to balance the equation suggesting a possible fate for the nitrogen atom. Check that the equation is balanced.

Part 2

Do an oxidation number analysis and verify that it agrees with the redox couple.

Part 3

Suggest a mechanism for this transformation.

PROBLEM 2.20 [58, 59]

Themes: balancing chemical equations, redox reactions

Two procedures have been developed to produce deuterated chloroform which is the most used solvent for running NMR spectra. They differ in the production of deuterated sodium hydroxide.

Procedure 1

$$Na + D_2O \longrightarrow NaOD + D_2$$

$$NaOD + HC(O)CCl_3 \longrightarrow HC(O)ONa + CDCl_3$$

Procedure 2

$$Na_2O_2 + D_2O \longrightarrow NaOD + O_2$$

$$NaOD + HC(O)CCl_3 \longrightarrow HC(O)ONa + CDCl_3$$

Part 1

For each procedure balance the first chemical equation for the production of deuterated sodium hydroxide.

Part 2

Write out redox couples for the production of deuterated sodium hydroxide.

Part 3

Suggest mechanisms for the production of deuterated sodium hydroxide.

PROBLEM 2.21

Themes: balancing chemical equations, redox reactions, Diophantine equations

Two unbalanced chemical reactions used in rocket fuel are shown below.

Reaction A

$$H_2O_2 \xrightarrow[]{\text{MnO}_2 \text{ (cat.)}} H_2O + O_2$$

Reaction B

$$Al + NH_4ClO_4 \longrightarrow Al_2O_3 + AlCl_3 + H_2O + N_2$$

Part 1

Write out the corresponding redox couples and balance both equations.

Part 2

Write out a catalytic cycle for Reaction A that is consistent with its balanced chemical equation.

Part 3

Using general stoichiometric coefficients as shown below for Reaction B obtain their values by an algebraic method. How do the results of the algebraic method compare with the balanced chemical equation obtained in Part 1?

$$a\,Al + b\,NH_4ClO_4 \longrightarrow c\,Al_2O_3 + d\,AlCl_3 + e\,H_2O + f\,N_2$$

PROBLEM 2.22

Themes: balancing chemical equations, redox reactions

The chemistry of airbags in automobiles involves rapid production of nitrogen gas in two stages. The first stage is a fast electric discharge applied to sodium azide. This produces sodium as well as nitrogen gas. The sodium produced in the first stage then reacts in a second stage with potassium nitrate to produce more nitrogen and potassium and sodium oxides. These oxides are finally neutralized in a third stage by silica to produce potassium and sodium silicates.

Stage 1

$$NaN_3 \longrightarrow Na + N_2$$

Stage 2

$$Na + KNO_3 \longrightarrow K_2O + Na_2O + N_2$$

Stage 3

$$K_2O + Na_2O + SiO_2 \longrightarrow K_2SiO_3 + Na_2SiO_3$$

Part 1

For each stage, write out appropriate redox couples as necessary and provide balanced chemical equations for all three stages.

Part 2

Write out an overall balanced equation for the entire three-stage process and determine the mole ratio of nitrogen to sodium azide.

PROBLEM 2.23 [60, 61]

Themes: balancing chemical equations, mechanism, redox reactions, oxidation number analysis, atom economy

A reaction step in the synthesis of estrone is shown below.

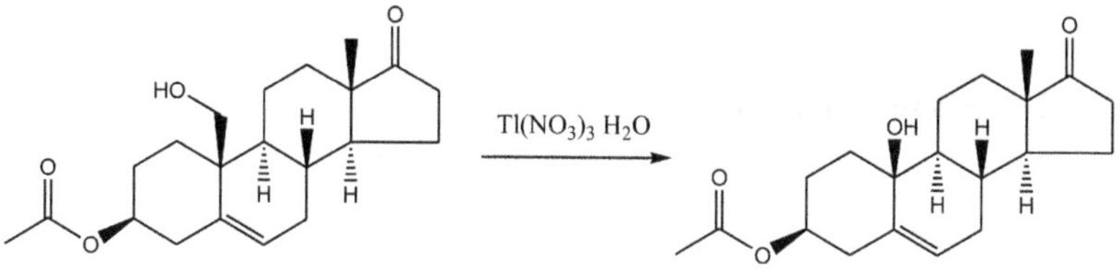

$Tl(NO_3)_3$ H_2O

Part 1

Write a mechanism for the transformation.

Part 2

Balance the chemical equation showing all by-products. Determine the atom economy.

Part 3

Show the redox couple and do an oxidation number analysis.

PROBLEM 2.24

Themees: balancing chemical equations, mechanism, catalytic cycles, atom economy, stoichiometric versus catalytic conditions, overall process mass intensity, E-factor profiles

Method A (stoichiometric conditions) [62]
Method B (catalytic conditions, Upjohn procedure) [63]
Method C (biomimetic) [64, 65]
Method D (catalytic conditions, polymer supported) [66]

The *cis*-dihydroxylation of olefins with osmium tetroxide can be carried out either under stoichiometric amounts of oxidant or under catalytic conditions. Method A shows an example under the former case and Methods B, C, and D under the latter. In Method B, N-methylmorpholine oxide (NMO) is used to regenerate the osmium tetroxide catalyst. In Method C, a biomimetic procedure is employed that involves a triple redox cascade where hydrogen peroxide is the terminal oxidant. As in Method B,

NMO regenerates osmium tetroxide and the resulting N-methylmorpholine reduces a flavin hydroperoxide to a flavinium hydroxide which in turn reduces hydrogen peroxide to water. In Method D, the osmium tetroxide is microencapsulated in polystyrene and can be easily retrieved and recycled.

Method A (Stoichiometric)

OsO_4
pyridine

KOH

Method B (Upjohn Procedure)

OsO_4
(cat.)
H_2O

Method C (Biomimetic Procedure)

OsO_4 (cat.)
N-methylmorpholine (cat.)
flavin analogue (cat.)

+ H_2O_2

Method D

Method D

OsO_4 on
polystyrene (cat.)
H_2O

82 117 116 101

Part 1

For each method, determine the overall balanced chemical equation and its atom economy.

Part 2

For each method, write out a reaction mechanism that is consistent with its corresponding balanced chemical equation. For method B write out the catalytic cycle.

Part 3

Write out the redox couples involved in each method.

Part 4

In Methods B and C, sodium hydrosulfite (sodium dithionite) is used in the work-up. Explain its purpose. Show a mechanism for the reaction involved.

PROBLEM 2.25

Themes: balancing chemical equations, redox reactions

Tellurophene is an analog of furan, thiophene, and selenophene. It is made according to the following routes. Route 1 is a direct synthesis. Route 2 generates *in situ* the same starting materials as Route 1. The third route is a three-step sequence that traps tellurophene as a dichloro derivative so it can be better separated from the by-product dibutyltelluride.

Route 1 [67]

+ Na_2Te —MeOH, 69 %→ Te

Route 2 [68]

Cl Cl —KOH, H_2O→ [ ] ; Te —$NaBH_4$→ [Na_2Te]

[ ] + [Na_2Te] —32 %→ Te

Route 3 [69]

Cl Cl —NaOH, pyridine→ [ ] —Bu-Te-Te-Bu, $NaBH_4$→ 93 % Te Bu —Bu-Te-Te-Bu, $NaBH_4$→

70 % Bu Te Te Bu —BuLi then H_2O, - Bu_2Te→ [Te] —SO_2Cl_2→ 61 % Te Cl Cl

Part 1

For each route, provide balanced chemical equations.

Part 2

Write out mechanisms that support the balanced equations determined in Part 1.

Part 3

For each route, perform an oxidation number analysis to track changes in oxidation states of key atoms in the transformation.

PROBLEM 2.26 [70–72]

Themes: balancing chemical equations, redox reactions

The major component of the dyestuff mauveine is made by reacting a mixture of aniline and three geometric isomers of toluidine according to the reaction shown below. The correct structure of mauveine was determined in 1994, 138 years after its accidental discovery by William Perkin Jr. in 1856.

$K_2Cr_2O_7$, H_2SO_4

$Cr_2O_4^{\ominus 2}$

Part 1

Highlight the target synthesis bonds in the product structure depicting the constituent anilines.

Part 2

Balance the redox reaction and write out a balanced chemical equation for the transformation.

PROBLEM 2.27 [73]

Themes: balancing chemical equations, redox reactions, reaction mechanism, Diophantine equations

Oxygen difluoride reacts with hydroxide ion and with hydrochloric acid to produce oxygen gas and chlorine gas, respectively, according to the reactions shown below.

$$\text{(i)}\quad F{-}O{-}F \xrightarrow{\;^{\ominus}OH\;} O{=}O + F^{\ominus} + H_2O$$

$$\text{(ii)}\quad F{-}O{-}F \xrightarrow{\;HCl\;} Cl{-}Cl + HF + H_2O$$

Part 1

Suggest mechanisms for these transformations.

Part 2

Based on the mechanisms obtain the balanced chemical equations.

Part 3

Based on the mechanism write the operative redox couples and verify them by doing an oxidation number analysis.

Part 4

Using general stoichiometric coefficients as shown below, obtain their values by an algebraic method. How do the results of the algebraic method compare with the balanced chemical equations obtained in Part 2?

$$\text{(i)}\quad a\ OF_2 + b\ {}^{\ominus}OH \longrightarrow c\ O_2 + d\ F^{\ominus} + e\ H_2O$$

$$\text{(ii)}\quad a\ OF_2 + b\ HCl \longrightarrow c\ Cl_2 + d\ HF + e\ H_2O$$

REFERENCES

1. Andraos, J. Application of green metrics analysis to chemical reactions and synthesis plans, in green chemistry metrics. In David C. Constable, Alexei Lapkin (Eds.), *Green Chemistry Metrics*, Blackwell Scientific: Oxford, 2008, p. 84.
2. Watson, W. *Org. Process Res. Dev.* 2012, *16*, 1877.
3. Hein, J.E.; Krasnova, L.B.; Iwasaki, M.; Fokin, V.V. *Org. Synth.* 2011, *88*, 238.
4. Dixon, A.J.; McGrath, M.J.; O'Brien, P. *Org. Synth.* 2006, *83*, 141.
5. Holmes, R.R.; Forstner, J.A. *Inorganic Syntheses* 1966, *8*, 63.
6. Nakayama, J.; Hirashima, A. *Heterocycles* 1989, *29*, 1241.
7. Trost, B.M. *Science* 1991, *254*, 1471.
8. Trost, B.M. *Angew. Chem. Int. Ed.* 1995, *34*, 259.
9. Lavoisier, A. Mémoire sur la nature du principe qui se combine avec les métaux pendant leur calcination, et qui en augmente le poids. *Mèmoires de l'Académie Royale des Sciences (1775)* 1778, 520.

10. Lavoisier, A. *The Elements: In a New Systematic Order, Containing All the Modern Discoveries* (Robert Kerr, transl. 1790), Dover Publications: New York, 1965, pp. 55–56.
11. Partington, J.R. *A Short History of Chemistry*, McMillan & Co.: London, 1957, p. 130.
12. Keiter, R.L.; Gamage, C.P. *J. Chem. Educ.* 2001, *78*, 908.
13. Lavoisier, A. *The Elements: In a New Systematic Order, Containing All the Modern Discoveries* (Robert Kerr, transl. 1790), Dover Publications: New York, 1965, p. 100.
14. Lavoisier, A. *The Elements: In a New Systematic Order, Containing All the Modern Discoveries* (Robert Kerr, transl. 1790), Dover Publications: New York, 1965, pp. 31–36.
15. Lavoisier, A. *The Elements: In a New Systematic Order, Containing All the Modern Discoveries* (Robert Kerr, transl. 1790), Dover Publications: New York, 1965, p. 44.
16. Lavoisier, A. *The Elements: In a New Systematic Order, Containing All the Modern Discoveries* (Robert Kerr, transl. 1790), Dover Publications: New York, 1965, p. 64, 101.
17. Lavoisier, A. *The Elements: In a New Systematic Order, Containing All the Modern Discoveries* (Robert Kerr, transl. 1790), Dover Publications: New York, 1965, p. 80.
18. Lavoisier, A. *The Elements: In a New Systematic Order, Containing All the Modern Discoveries* (Robert Kerr, transl. 1790), Dover Publications: New York, 1965, pp. 88–89.
19. Lavoisier, A. *The Elements: In a New Systematic Order, Containing All the Modern Discoveries* (Robert Kerr, transl. 1790), Dover Publications: New York, 1965, p. 96.
20. Lavoisier, A. *The Elements: In a New Systematic Order, Containing All the Modern Discoveries* (Robert Kerr, transl. 1790), Dover Publications: New York, 1965, p. 99, 102.
21. Lavoisier, A. Memoir on the calcination of tin in closed vessels and on the cause of the gain in weight which this metal acquires in the operation. *Mémoires de l'Académie Royale des Sciences (1774)*, 1777, 351.
22. Brook, A.G.; Warner, C.M.; Limburg, W.W. *Can. J. Chem.* 1967, *45*, 1231.
23. Sommer, L.H.; Frye, C.L. *J. Am. Chem. Soc.* 1959, *81*, 1013.
24. Walden, P. *Chem. Ber.* 1896, *29*, 133.
25. Zhou, R.; Skibo, E.B. *J. Org. Chem.* 1996, *39*, 4321.
26. Skibo, E.B.; Islam, I.; Schulz, W.G.; Zhou, R.; Bess, L.; Boruah, R. *Synlett* 1996, 297.
27. Grantham, R.K.; Meth-Cohn, O. *J. Chem. Soc. C* 1969, 70.
28. Taylor, E.C.; Harrison, K.A.; Rampal, J.B. *J. Org. Chem.* 1986, *51*, 101.
29. Hassner, A.; Namboothiri, I. *Organic Synthesis Based on Name Reactions*, 3rd ed., Elsevier: Amsterdam, 2012, p. 500.
30. Labianca, D.A. *J. Chem. Educ.* 1990, *67*, 259.
31. Lockwood, K.L. *J. Chem. Educ.* 1968, *45*, 731.
32. Balasubramanian, K. *J. Math. Chem.* 2001, *30*, 219.
33. Shibley, I.A. Jr.; Amaral, K.E.; Aurentz, D.J.; McCaully, R.J. *J. Chem. Educ.* 2010, *87*, 1351.
34. Karen, P.; McArdle, P.; Takats, J. *Pure Appl. Chem.* 2014, *86*, 1017.
35. Matt, C.; Wagner, A.; Mioskowski, C. *J. Org. Chem.* 1997, *62*, 234.
36. But, A.; Le Nôtre, J.; Scott, E.L.; Wever, R.; Sanders, J.P.M. *ChemSusChem* 2012, *5*, 1199.
37. Daly, R.A. *J. Phys. Chem.* 1907, *11*, 93.
38. Renner, H.; Schlamp, G.; Hollmann, D.; Luschow, H.M.; Tews, P.; Rothaut, J.; Dermann, K.; Knodler, A.; Hecht, C.; Schlott, M.; Drieselmann, R.; Peter, C.; Schiele, R. *Ullmann's Encyclopedia of Industrial Chemistry*, Wiley-VCH: Weinheim, 2012, vol. 17, pp. 93–143.
39. Mooiman, M.B.; Simpson, L. *Gold Ore Processing—Project Development and Operations*, M.D. Adams (Ed.), Elsevier: Amsterdam, 2016, pp. 595–615.

40. Kyriakakis, G. *Gold Ore Processing—Project Development and Operations*, M.D. Adams (Ed.), Elsevier: Amsterdam, 2016, pp. 857–870.
41. Chloroauric acid: King, S.R.; Massicot, J.; McDonagh, A.M. *Metals* 2015, *5*, 1454.
42. Reduction of gold(III) chloride: Avery, D. *J. Soc. Chem. Ind.* 1908, *27*, 255.
43. Gold carbide: Mathews, J.A.; Watters, L.L. *J. Am. Chem. Soc.* 1900, *22*, 108.
44. Gold(I) chloride: Biltz, W.; Wein, W. *Z. Anorg. Allg. Chem.* 1925, *148*, 192.
45. Gold(III) chloride: Fischer, W.; Biltz, W. *Z. Anorg. Allg. Chem.* 1928, *176*, 81.
46. Bauer, G. *Handbook of Preparative Inorganic Chemistry*, Academic Press: New York, 1973, pp. 1055–1057.
47. Peplow, M. *ACS Cent. Sci.* 2017, *3*, 261.
48. Liu, X.; Conte, M.; Elias, D.; Lu, L.; Morgan, D.J.; Freakley, S.J.; Johnston, P.; Kiely, C.J.; Hutchings, G.J. *Catal. Sci. Technol.* 2016, *6*, 5144.
49. Johnston, P.; Carthey, N.; Hutchings, G.J. *J. Am. Chem. Soc.* 2015, *137*, 14548.
50. Simon, M.; Jonk, P.; Wuhl-Couturier, G.; Halbach, S. *Ullmann's Encyclopedia of Industrial Chemistry*, Wiley-VCH: Weinheim: 2012, Volume 22, pp. 559–594.
51. Barton, D.H.R. *J. Soc. Chem. Ind.* 1950, *69*, 75.
52. Cox, N.L. US 3078147 (duPont, 1963).
53. Cox, N.L. US 3278263 (duPont, 1966).
54. Cox, N.L.; Hicks, W.T. GB 1155508 (duPont, 1969).
55. Crandall, T.G. US 4747974 (duPont, 1988).
56. Anger, G.; Halstenberg, J.; Hochgeschwender, K.; Scherhag, C.; Korallus, U.; Knopf, H.; Schmidt, P.; Ohlinger, M. Chromium compounds. In *Ullmann's Encyclopedia of Industrial Chemistry*, Wiley-VCH: Weinheim, 2012, Vol. 9, pp. 157–191.
57. Chen, F.; Liu, A.; Yan, Q.; Liu, M.; Zhang, D.; Shao, L. *Synth. Commun.* 1999, *29*, 1049.
58. Breuer, F.W. *J. Am. Chem. Soc.* 1935, *57*, 2236.
59. Kluger, R. *J. Org. Chem.* 1964, *29*, 2045.
60. Kocovsky, P.; Baines, R.S. *Tetrahedron Lett.* 1993, *34*, 6139.
61. Kocovsky, P.; Baines, R.S. *J. Org. Chem.* 1994, *59*, 5439.
62. Criegee, R.; Marchand, B.; Wannowius, H. *Ann. Chem.* 1942, *550*, 99.
63. Van Rheenen, V.; Cha, D.Y.; Hartley, W.M. *Org. Synth. Coll.* 1988, *6*, 342.
64. Bergstadt, K.; Jonsson, S.Y.; Bäckvall, J.E. *J. Am. Chem. Soc.* 1999, *121*, 10424.
65. Jonsson, S.Y.; Färnegardh, K.; Bäckvall, J.E. *J. Am. Chem. Soc.* 2001, *123*, 1365.
66. Nagayama, S.; Endo, M.; Kobayashi, S. *J. Org. Chem.* 1998, *63*, 6094.
67. Mack, W. *Angew. Chem.* 1966, *78*, 940.
68. Sweat, D.P.; Stephens, C.E. *J. Organometallic Chem.* 2008, *693*, 2463.
69. Dabdoub, M.J.; Dabdoub, V.B.; Guerrero, P.G. Jr.; Silveira, C.C. *Tetrahedron* 1997, *53*, 4199.
70. Brown, T.M.; Cooksey, C.J.; Dronsfield, A.T. *Educ. Chem.* **May** 2000, 75.
71. Meth-Cohn, O.; Smith, M. *J. Chem. Soc. Perkin Trans. 1* 1994, 5.
72. Perkin, W.J. GB 1984 (1856).
73. Cotton, A.F.; Wilkinson, G. *Advanced Inorganic Chemistry*, 2nd ed., Interscience Publishers: New York, 1966, pp. 369–371.

3 Experimental Description of Chemical Reactions

3.1 STRUCTURE OF LITERATURE EXPERIMENTAL PROCEDURES

Key features of a well-documented experimental procedure include:

1. statement of grams and moles of all reactants used;
2. statement of grams and moles of all catalysts and ligands used;
3. statement of volumes of all reaction solvents used;
4. statement of reaction temperature, reaction pressure (if different from 1 atm), and reaction time;
5. for the work-up procedure, statement of the volumes of all aqueous and organic extraction solvents, and masses of decolorizing charcoal, filtering agents, and drying agents;
6. for the purification procedure, statement of the volumes of recrystallization solvents and chromatographic solvents, and mass of silica gel or alumina packing material;
7. documentation of any color changes, gas evolution, precipitation, or heat evolution that may occur during the course of the reaction;
8. documentation of all cautionary and safety notes in carrying out the reaction;
9. statement of the size of the reaction vessel and the number of ports (necks);
10. statement of any specialized glassware that is required (e.g., dropping funnel, Schlenk tube, gas manifold, glove box);
11. statement of the mass, % purity, and % yield of product collected;
12. if the product is a chiral molecule, then some kind of optical characterization is required such as optical rotation (concentration, solvent specified, temperature), and enantiomeric (diastereomeric) excess or ratio;
13. if the product is a solid, a melting point should be recorded;
14. if the product is a liquid (oil), a distillation boiling point should be recorded;
15. itemization of spectroscopic data: ^{1}H and ^{13}C NMR along with structural assignments; IR peaks along with characteristic functional group absorptions; mass spectrum MW fragments and molecular ion MW;
16. if microwave irradiation is used as a heat source, then the power (Watts), reaction time, reaction temperature, and energy efficiency of the apparatus should be stated; and
17. if continuous flow apparatus is used, then the flow rates, dimensions of tubing and reaction chamber, residence time, and process time should all be stated.

A standard example template of a properly written generic experimental procedure is as follows:

> Compound A (___ g, ___ moles) was dissolved in ___ mL of solvent S1 in a ___ mL two-necked round-bottomed flask. The reaction mixture was cooled to *T* degrees C with a __________ bath. To this reaction mixture was added dropwise with stirring a solution of compound B (___ g, ___ moles) dissolved in ___ mL of solvent S2. After ___ h the reaction was warmed to room temperature and quenched with the slow addition of ___ mL water. The aqueous layer was separated and extracted with ___ mL diethyl ether. The organic layer was washed successively with (___ X ___ mL) saturated sodium bicarbonate solution, and (___ X ___ mL) saturated brine solution. The combined organic solutions were dried with ___ g sodium sulfate. After rotary evaporation of the solvents, the crude residue was taken up in ___ mL (__:__ hexane:ethyl acetate) and flash chromatographed on a column containing ___ g silica gel and eluted with ___ mL (__:__ hexane:ethyl acetate). The fractions containing the product were evaporated to afford ___ g of product P (___ % yield).

The only sources for well-written experimental procedures in the whole of the chemistry literature are *Organic Syntheses* (est. 1921) and *Inorganic Syntheses* (est. 1939). Even these well-respected sources are not error-free, as noted by a recent exhaustive survey of procedures in *Inorganic Syntheses* that contain structural and various errors with respect to balancing chemical equations [1]. The rest of the chemistry literature should be viewed with cautionary skepticism.

3.2 QUALITY OF LITERATURE EXPERIMENTAL PROCEDURES

It is well known that the quality of experimental procedures varies widely across the chemistry literature. However, it is a myth to suggest that journals of the highest quality that have "rigorous peer review standards" in place have the highest quality experimental procedures. One can find both poor- and high-quality descriptions of experimental accounts in high impact and "substandard" research journals with equal frequency.

The problem of irreproducibility in chemical science [2–9] and its impact on the integrity of green chemistry [10–20] has been documented in the literature. An example of irreproducibility that has been caught by Sir John Cornforth is the synthesis of the alkaloid aconitine [21–23]. An extreme example of faked data that passed the sieve of "expert peer review" and caused turmoil in the process chemistry literature is the enantiomeric resolution of an intermediate in the synthesis of the selective serotonin reuptake inhibitor (SSRI) anti-depressant citalopram [24–28].

Poorly laid out experimental procedures contain the following phrases that should be considered warning flags of increased likelihood of irreproducibility: "typical procedure," or "general procedure" for a range of substrate structures (one size fits all approach). Other footprints are specification of moles of reactants and not the corresponding masses, and specification of % yield of product and not the corresponding mass. Literature papers that are classified as "communications" or "letters" are notorious for non-disclosure of experimental details under the cloak of brevity.

The modern trend of scuttling experimental procedures in the electronic supporting information (SI) of a published full paper encourages devaluation of experimental procedures and their reproducibility [9]. In today's world of tsunamis of junk or passable science making it through the sieve of "expert peer review," reviewers customarily do not take the time to read the SI as part of the standard peer review process. The only way mistakes or fraud will be caught and exposed is if someone (usually a graduate student or post-doctoral fellow) actually follows a given literature procedure for their own research work and makes the unpleasant discovery that a prep in fact does not work. One hopes that these occurrences do not also cause personal harm or injury to those workers, or negatively impact their careers.

For the purpose of conducting any kind of metrics analysis as described in this book, experimental procedures are often lacking any one or more of the following items: masses of drying agents used (magnesium or sodium sulfate), masses of silica gel or alumina used in flash chromatography, masses (volumes) of chromatographic solvents used as eluents, masses of catalysts used, masses (volumes) of reaction solvents, masses (volumes) of wash solvents, masses (volumes) of recrystallization solvents, masses (volumes) of extraction solvents, masses of decolorizing charcoal, masses of Celite, and masses of gaseous reagents. Missing quantities preclude proper assessment of material efficiency according to the determination of true E-factor and process mass intensity (*PMI*) values. Given this scenario, only lower limit estimates of these quantities will be obtained. This makes reliable and fair ranking of the material performance of one procedure against another impossible. Needless to say, this translates into a serious problem in the field of green chemistry where advertising procedures as green, often called "green washing," is commonplace. There is some sign in the most recent literature that editors of journals are considering taking concrete enforceable steps to address this growing problem [29].

Examples of well laid out experimental procedures in the green chemistry and general chemistry literature follow.

Example 3.1 [30]

Themes: balancing chemical equations, experimental procedures

O O Ph Ph + Ph $\xrightarrow[CH_2Cl_2]{AuCl_3/AgOTf}$ O O Ph Ph Ph

EXPERIMENTAL PROCEDURE:

An oven-dried, 300-mL, 3-necked, round-bottomed flask equipped with a nitrogen inlet adapter, a rubber septum, a stopcock, and a Teflon-coated magnetic stir bar is cooled to room temperature under a flow of nitrogen. While temporarily removing the septum, the flask is charged with gold(III) chloride ($AuCl_3$) (0.303 g, 1 mmol, 5 mol% Au) and silver trifluoromethanesulfonate (AgOTf) (0.77 g, 3 mmol, 15 mol%) (Note 1) under N_2. Dichloromethane (150 mL) (Note 2) is added, and the catalyst mixture is stirred at room temperature for 1.5–2 h, resulting in a brown suspension (Note 3). Dibenzoylmethane (4.48 g, 20 mmol) (Note 4) is added to the reaction flask over 2 min while temporarily removing the septum and the color of the reaction mixture turns green. Styrene (3.45 mL, 30 mmol) (Notes 5, 6) is diluted to 25 mL with CH_2Cl_2 and added dropwise to the reaction mixture by syringe pump through the septum over 6–7 h at room temperature (Note 7). The mixture is then stirred for another 30 min (Note 8). A deep-green solution is obtained (Note 9). The solution is filtered through approximately 40 mL of silica gel in a 60-mL glass Büchner filter funnel to remove the catalyst (Note 10). The silica gel is washed with ether (6×50 mL), and the filtrate is concentrated by rotary evaporation at room temperature. The residue, a clear, orange-brown oil, is then purified by flash column chromatography (Note 11) to provide the crude product as a pale yellow solid (6.44 g, 98%) (Note 12). Further purification is achieved by recrystallization. The crude product is dissolved in 30 mL of hot EtOAc and diluted with 90 mL of hexanes to the point of cloudiness. The mixture is allowed to cool slowly to room temperature overnight and the resulting solids are collected by filtration, rinsing with 10 mL of hexanes. The solids are air-dried, then placed in a vacuum oven for 4 h to provide analytically pure product 1,3-diphenyl-2-(1-phenyl-ethyl)-propane-1,3-dione as a white solid (4.16 g, 12.7 mmol, 63%) (Notes 13, 14).

Note 11:

The oil was diluted with 50 mL of CH_2Cl_2, and 13 g of SiO_2 (Baker, 40 µm, flash chromatography grade) was added. The resulting slurry was concentrated on the rotary evaporator to a tan powder, which was charged to the top of a column prepared from flash silica gel slurried in 2% EtOAc-hexanes (column dimensions 3.5 cm diameter×18 cm height). Elution with 2% EtOAc-hexanes (1.5 L, removed unreacted dibenzoylmethane) followed by 10% EtOAc-hexanes (6 L) afforded the product in fractions 19–74 (75 mL fractions were collected). The product containing fractions were combined and concentrated to provide the crude product.

Example 3.2 [31]

Themes: balancing chemical equations, experimental procedures

$Ru^{III}Cl_3\ 3\ H_2O$ + [cyclopentadiene] —PPh_3 / EtOH→ CpRu^{II}(Cl)(PPh_3)$_2$ —≡—Ph / MeOH→ [CpRu^{IV}(=C=CHPh)(PPh_3)$_2$]$^{\oplus}$ $Cl^{\ominus}$ —Na→ CpRu^{II}(C≡CPh)(PPh_3)$_2$

EXPERIMENTAL PROCEDURES:

Step 1

The reaction is carried out in a 500 mL two-necked round-bottomed flask equipped with a magnetic stir bar, dropping funnel, and a reflux condenser topped with a nitrogen bypass. The apparatus is purged with nitrogen and charged with triphenylphosphine (8.4 g, 32 mmol) and degassed ethanol (300 mL). The triphenylphosphine is dissolved in the ethanol by heating to reflux. Hydrated ruthenium trichloride (2.1 g, 8 mmol) is dissolved in boiling ethanol (40 mL), and then allowed to cool. Freshly distilled cyclopentadiene (4 mL, 50 mmol) is added to the ruthenium trichloride solution, and the mixture is transferred to the dropping funnel. The dark brown solution is then added to the refluxing triphenylphosphine solution over 10 min. The reaction mixture is dark brown in color, which after 1 h lightens to a dark red-orange with an orange precipitate. The solution is now safe to expose to air. The solution is cooled to room temperature to complete the precipitation. The solid is collected by filtration, washed with ethanol (4 × 10 mL) and hexanes (4 × 10 mL), and dried in vacuum. Yield: 5.14 g (89%).

Step 2

$(\eta^5\text{-}C_5H_5)RuCl(PPh_3)_2$ (1.45 g, 2 mmol) is suspended in methanol (100 mL) in a 200 mL two-necked flask equipped with a reflux condenser and a nitrogen bypass. Phenylacetylene (0.33 mL, 3 mmol) is added dropwise to the suspension, which is kept strictly under nitrogen, and then the mixture is heated to reflux temperature for 30 min. $(\eta^5\text{-}C_5H_5)RuCl(PPh_3)_2$ gradually dissolves to form a dark red solution. After the solution has cooled, sodium metal (0.2 g, 9 mmol) is added in small pieces whereupon a yellow crystalline precipitate forms as the sodium dissolves. The solid is filtered from the mother liquor, washed with methanol (4 × 3 mL) and hexanes (4 × 3 mL), and dried in vacuum. Yield: 1.37 g (86%).

Write out balanced equations for each transformation.

Solution

$Ru^{III}Cl_3$ 3 H_2O + [cyclopentadiene] → (2.5 PPh_3; - 2 HCl; - 2.5 H_2O; - 0.5 O=PPh_3) $(\eta^5\text{-}C_5H_5)Ru^{II}Cl(PPh_3)_2$ → (Ph-C≡CH) $[(\eta^5\text{-}C_5H_5)Ru^{IV}(=C=CH\text{-}Ph)(PPh_3)_2]^{\oplus}$ $Cl^{\ominus}$

→ (Na, MeOH; - NaCl; - H_2) $(\eta^5\text{-}C_5H_5)Ru^{II}(C\equiv C\text{-}Ph)(PPh_3)_2$

Examples of less reliable experimental procedures in the green chemistry and general chemistry literature follow.

Example 3.3 [32]

Themes: balancing chemical equations, experimental procedures, ring construction strategy 2-amino-2-chromenes have been synthesized under base catalyzed solvent-free conditions from either 1- or 2-naphthols and arylidenemalononitriles as shown below.

An experimental procedure is also given.

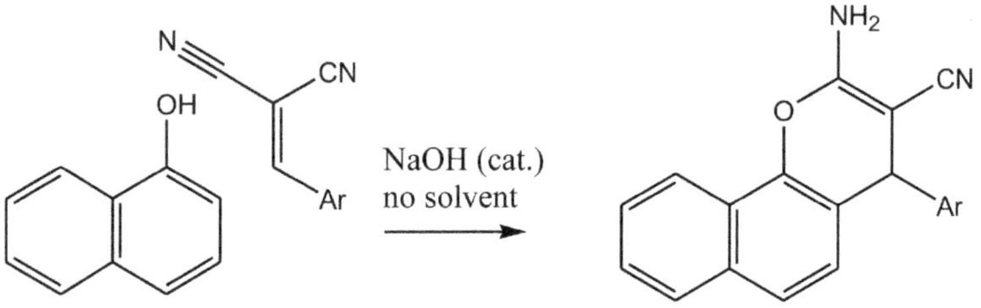

NaOH (cat.)
no solvent

1-naphthol

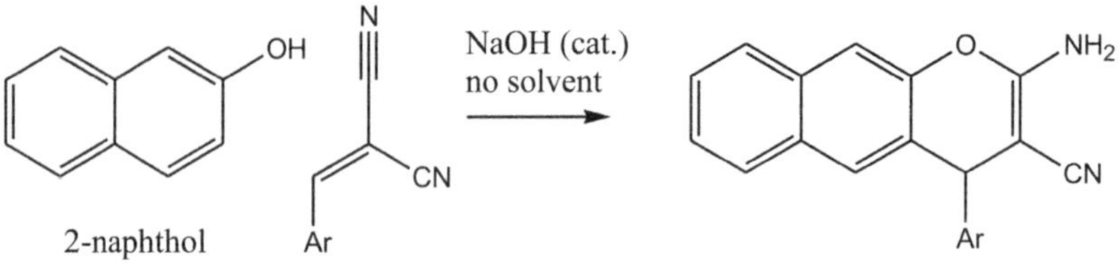

NaOH (cat.)
no solvent

2-naphthol

General procedure for preparation of substituted 2-amino-2-chromenes:

1-naphthol or 2-naphthol (10 mmol), arylidenemalononitrile (10 mmol), sodium hydroxide (1 mmol) were stirred and heated at 100°C for 1/2 h, cooled to room temperature, and washed with cold water; the crude solid was purified by crystallization from methanol.

Part 1

For each reaction shown, write out a balanced chemical equation and determine its atom economy.

Part 2

What is the ring construction strategy used?

Part 3

Explain why the experimental procedure is poorly written with respect to determining material efficiency metrics.

Solution

Part 1

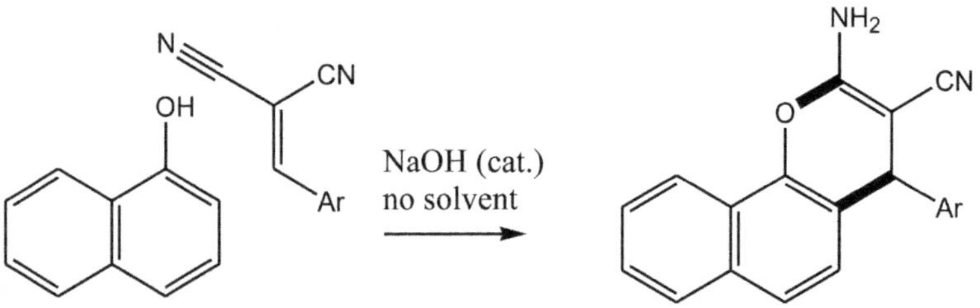

NaOH (cat.)
no solvent

144 **Ar + 77** **Ar + 221**

NaOH (cat.)
no solvent

144 **Ar + 77** **Ar + 221**

Both reactions have 100% atom economy.

Part 2

Ring construction strategy: [3 + 3]

Part 3

The general procedure is lacking in the following details:

Volume of water used in washing

Volume of methanol used in recrystallization

No specified product yield information is given

From a reproducibility standpoint, the generic reaction time of 30 min for a variety of substrates cannot be taken as true.

Example 3.4 [33]

Themes: balancing chemical equations, material efficiency metrics

BuLi
THF

From the experimental procedure, state any assumptions and other manipulations that would be necessary in order to carry out a materials efficiency metrics calculation.

Solution

Assumptions:

10 g magnesium sulfate drying agent.

Note: 55 mL of 1.6 M *n*BuLi solution in hexane is used. The density of this solution is 0.68 g/mL. The mass of *n*BuLi used is 0.055*1.6*64 = 5.632 g. The mass of solution used is 55*0.68 = 37.4 g. Therefore, the mass of hexane solvent used is 37.4 − 5.632 = 21.77 g.

Example 3.5 [34]

Themes: balancing chemical equations, material efficiency metrics

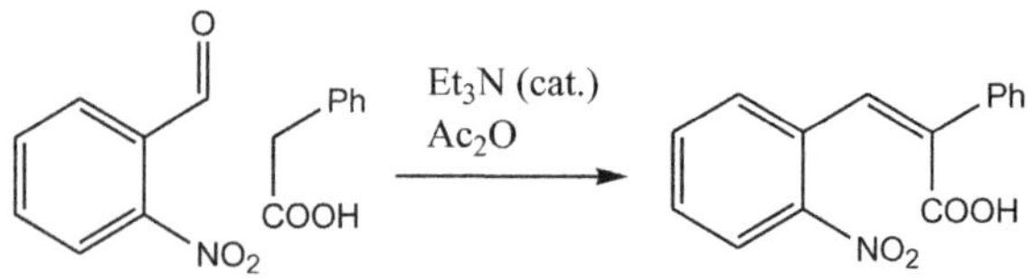

From the experimental procedure state any assumptions and other manipulations that would be necessary in order to carry out a materials efficiency metrics calculation.

Solution

Assumptions:

60 mL water washing

Example 3.6 [35]

Themes: balancing chemical equations, material efficiency metrics, experimental procedures

EXPERIMENTAL PROCEDURE:

To a mixture of aldehyde (2 mmol) in 5 M LPDE (lithium perchlorate/diethylether) (4 ml) was added hydroxylamine (2.2 mmol) at room temperature. The mixture was stirred for 2 min and trimethylsilylcyanide (2.2 mmol) was added. The mixture was stirred for 15 min then water was added and the product was extracted with CH_2Cl_2. The organic phase was collected, dried (Na_2SO_4) and evaporated to afford the crude product. The product was purified by flash chromatography (hexane–ethyl acetate). Yield: 97%.

Part 1

Write out the balanced chemical equation and determine its atom economy. Show the target bonds made in the product structure.

Part 2

From the experimental procedure, state any assumptions and other manipulations that would be necessary in order to carry out a materials efficiency metrics calculation.

Solution

Part 1

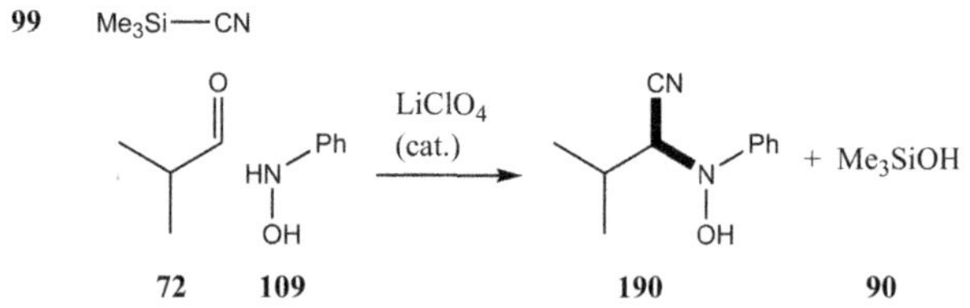

99 $Me_3Si—CN$

72 109 → ($LiClO_4$ (cat.)) 190 + Me_3SiOH 90

Reactants: 99 + 72 + 109 = 280
Products: 190 + 90 = 280
AE = 190/280 = 0.679, or 67.9%

Part 2

Assumptions:

10 mL water, 10 mL dichloromethane, and 2 g sodium sulfate in workup

Example 3.7 [36]

EXPERIMENTAL PROCEDURE:

Equimolar amounts of neat reactants, aldehyde, malononitrile, and resorcinol were taken in an Erlenmeyer flask, and 10 ml saturated solution of K_2CO_3 in water was added to it. The reaction mixture was subjected to MWI (microwave irradiation) for a specific time at low power (560 W). The progress of the reaction was monitored by TLC examination at an interval of every 30 s. On completion of reaction, the reaction mixture was cooled and was triturated with 2–3 ml of ice-cold water to get the solid product, leaving behind K_2CO_3 dissolved in water. The product obtained was filtered, washed with cold water, dried, and recrystallized from ethanol. Yield: 91% (R=Ph).

Write out a balanced chemical equation for the transformation and critique the quality of this experimental procedure from the point of view of conducting a materials efficiency metrics analysis.

SOLUTION

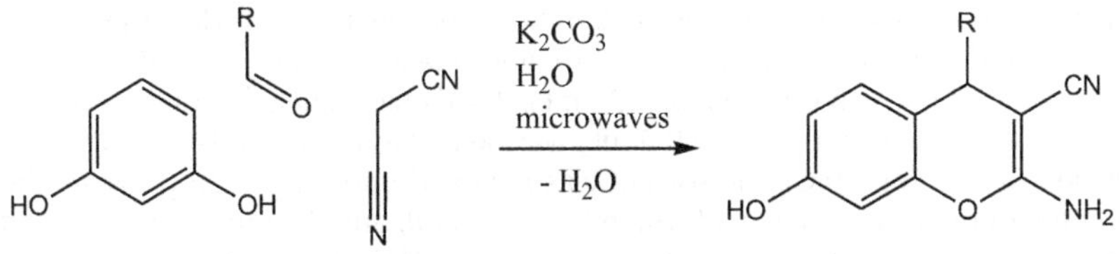

R
CN
K_2CO_3
H_2O
microwaves
- H_2O
HO OH
N
R
CN
HO O NH_2

The scale of the reaction is not specified. Volumes of cold water and ethanol are not specified.

Example 3.8 [37]

An improved three-step synthesis of the anti-obesity drug rimonabant was recently reported by Dr. Reddy's Laboratories.

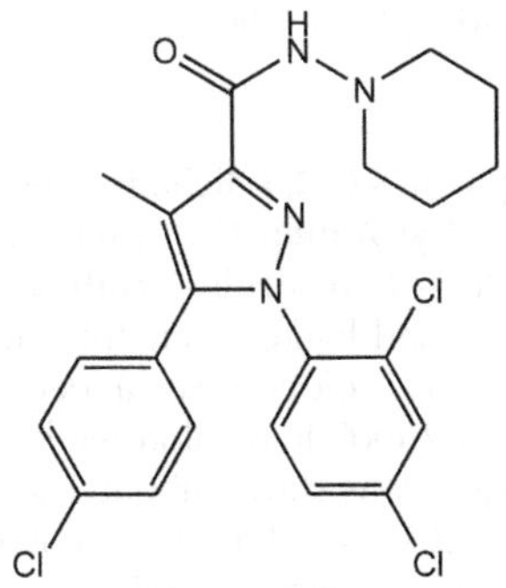

O H N N N N Cl Cl Cl

rimonabant

Carefully read the experimental procedure given by the authors. Draw out the synthesis plan showing balanced chemical equations for each step. Include the molecular weights of the starting materials, reagents, and intermediates. Are there any inconsistencies in the protocol?

EXPERIMENTAL PROCEDURE:

Step 1

Preparation of Lithium Salt of Ethyl 4-(4-Chlorophenyl)-3-methyl-4-oxydo-2-oxobuten-3-oate (3). A mixture of methylcyclohexane (300 mL) and lithium hexamethyl disilazane (300 mL, 1.07 mol) was charged into a round bottom flask under nitrogen atmosphere and cooled to 15–25°C. To this solution was added a solution of chloropropiophenone (**2**, 50.0 g, 1 mol) in methyl cyclohexane (125 mL) over a period of 30–45 min, and the mixture was stirred for 2.5 h. To this mixture, diethyl oxalate (47.8 g, 1.1 mol) was added over a period of 30–45 min and was stirred for 17 h. On completion of reaction (TLC), the solid material was filtered and washed with methyl cyclohexane (100 mL). The wet solid was taken up with methylcyclohexane (250 mL) and charged into a round bottom flask, and the mixture was stirred about 45 min, filtered, and washed with methyl cyclohexane (100 mL), followed by being dried under vacuum for about 3 h to afford lithium salt of 4-(4-chloro phenyl)-3-methyl-2,4-dioxo ethyl butyrate (**3**, 45.1 g, 55.2%).

Step 2

Preparation of 5-(4-Chlorophenyl)-1-(2,4-dichlorophenyl)-4-methyl-1 *H*-pyrazole-3-carboxylic Acid (6). A mixture of 4-(4-chlorophenyl)-3-methyl-2, 4-dioxo-ethyl butyrate (**3**, 8.0 g, 1 mol), ethanol (200 mL), 2,4-dichlorophenyl hydrazine (6.2 g, 1 mol), and 50% sulfuric acid (80.0 mL) was heated to reflux for 4–6 h. On completion of reaction (TLC), the solvent was removed under reduced pressure, the second lot of 50% sulfuric acid (160 mL) was added, and the mixture was heated to reflux for 6–8 h. On completion of reaction (TLC), the mixture was cooled to 25–35°C, and the reaction mass was poured into ice-water (200 mL), stirred about 15 min, filtered, washed with water (80 mL), and dried under vacuum for about 2 h. A mixture of the wet solid and water (220 mL) was stirred about 10 min at ambient temperature, adjusted to pH 10–12 with caustic lye (2.2 mL), and washed with petroleum ether. The aqueous layer was separated, adjusted to pH 2 with 12 N hydrochloric acid (2.0 mL), and stirred about 15 min at 25–35°C. The Solid material was filtered, washed with water (1000 mL), and dried at 35°C to constant weight to afford 5-(4-chlorophenyl)-1-(2,4-dichloro phenyl)-4-methyl-1 *H*-pyrazole-3-carboxylic acid (**6**, 7.7 g, 70%).

Step 3

Preparation of 5-(4-Chloro phenyl)-1-(2,4-dichlorophenyl)-4-methyl-1 *H*-pyrazole-3-carboxylic Acid Piperdin-1-yl Amide (1). A mixture of 5-(4-chlorophenyl)-1 -(2,4-dichlorophenyl)-4-methyl-1*H*-pyrazole-3-carboxylic acid (**4**, 45.0 g, 1 mol) in dichloromethane (450 mL) and hydroxybenztriazole (HOBt, 18.7 g, 1.2 mol) was stirred about 10 min. To this solution was added a solution of dicyclohexyl carbodiimide (DCC, 29.3 g, 1.2 mol) in dichloromethane (900 mL) over a period of 15–30 min, and the mixture was stirred about 1.5 h at ambient temperature. On completion of reaction (TLC), the solution was cooled to 0–5°C, stirred about 45 min, then the dicyclohexyl urea was filtered and washed with dichloromethane (135 mL), and the solvent of the filtrate was removed under reduced pressure.

A mixture of the above crude, dichloromethane (900 mL), and potassium carbonate (15.4 g, 1 mol) was stirred for min, a solution of 1-amino piperidine (11.8 g, 1.0 mol) in dichloromethane (450 mL) was added over a period of 15–30 min, and the mixture was stirred for 45–60 min at ambient temperature. On completion of reaction (TLC), the reaction mixture was cooled to 0–5°C and stirred for 45 min, and precipitated hydroxy benztriazole was filtered, washed with dichloromethane (135 mL), and the solvent of the filtrate was removed under reduced pressure. To the above crude was added petroleum ether (450 mL), and the mixture was stirred for 45 min at 50°C, filtered, washed with petroleum ether (225 mL), and dried in air to constant weight.

A mixture of the above solid in acetone (225 mL) was heated to reflux, 12 N hydrochloric acid (10 mL) was slowly added, and the mixture was stirred for 30 min, cooled to 0–5°C, and stirred for 45 min. Finally, the precipitated solid was filtered and washed with acetone (45 mL). The wet solid was charged in methanol (225 mL) and adjusted to pH 10–12 with caustic lye (8.0 mL), and DM water (450 mL) was added. After the solution was stirred for 45 min at ambient temperature, the solid material was filtered, washed with DM water (450 mL), and dried at 50°C under high vacuum to constant weight to afford 5-(4-chlorophenyl)-1-(2,4-dichlorophenyl)-4-methyl-1*H*-pyrazole-3-carboxylic acid piperdin-1-yl amide (**1**, 39.6 g, 72%).

Solution

$LiN(SiMe_3)_2$
146
- $HN(SiMe_3)_2$
- EtOH
168.45
55 %
274.45
compound **3**
H_2SO_4
- $LiHSO_4$
176.9
- H_2O
- EtOH

69 %
381.35
compound **6**
100
Cy-N=C=N-Cy 206
135
- Cy-NH-CO-NH-Cy
- 1-hydroxybenzotriazole
72 %
463.35
compound **1**
rimonabant

Inconsistencies found in experimental protocol:

Step 1:

1 A volume of 300 mL is quoted for lithium hexamethyl disilazane instead of the expected mass for a solid. This material is made *in situ* by reacting hexamethyl disilazane with *n*-BuLi in an ether solvent. Hence, there is a missing step in the protocol.
2. The authors quote 50 g (1 mol) for chloropropiophenone yet its molecular weight is 168.45 g/mol.
3. The authors quote 47.8 g (1.1 mol) for diethyl oxalate yet its molecular weight is 146 g/mol.
4. The authors quote that they collected 45.1 g of compound **3** as the product which translates to 45.1/274.45 = 0.164 mol. This number is far less than the 1 mol scale indicated by the starting materials used.

Step 2:

1. The authors quote 8.0 g (1 mol) for compound **3** yet its molecular weight is 274.45 g/mol.
2. The authors quote 6.2 g (1 mol) for 2,4-dichlorophenyl hydrazine yet its molecular weight is 176.9 g/mol.
3. The authors quote that they collected 7.7 g of compound **6** as the product which translates to 7.7/381.35 = 0.0202 mol. This number is far less than the 1 mol scale indicated by the starting materials used.

Step 3:

1. The starting material should be compound **6**, not compound **4**. Hence, the authors mislabeled their compounds.
2. The authors quote 45.0 g (1 mol) for compound **6** yet its molecular weight is 381.35 g/mol.
3. The authors quote 18.7 g (1.2 mol) for 1-hydroxybenzotriazole yet its molecular weight is 135 g/mol.
4. The authors quote 29.3 g (1.2 mol) for DCC yet its molecular weight is 206 g/mol.
5. The authors quote 11.8 g (1 mol) for 1-aminopiperidine yet its molecular weight is 100 g/mol.
6. The authors quote that they collected 39.6 g of rimonabant (compound **1**) as the product which translates to 39.6/463.35 = 0.0855 mol. This number is far less than the 1 mol scale indicated by the starting materials used.

In summary, every quoted mass of reactant used or product collected does not correspond to the number of moles. There are two possible conclusions that could be drawn from this protocol. Either it is a complete fabrication or the authors mistook "mol" to mean "equivalent." If we are to believe the masses of reactants are true, then for step 1, the limiting reagent is chloropropiophenone (0.297 mol) and the % yield of product compound **3** is 55.2%, which agrees with the authors value of 55%. Similarly, for step 2, the limiting reagent is compound **3** (0.0291 mol) and the % yield of product compound **6** is 69.4%, which agrees with the authors value of 69%. Similarly, for step 3, the limiting reagent is 1-aminopiperidine (0.118 mol) and the % yield of product compound **1** is 72.5%, which agrees with the authors value of 72%. It appears from this evidence that the authors likely mistook "mol" to mean "equivalent." Nevertheless, the reviewers of this paper should have picked up on the discrepancy between masses and moles that occurred throughout if they actually read the experimental section that was given as part of the manuscript and not relegated as supporting information.

3.3 PROBLEMS

PROBLEM 3.1

For each experimental procedure given below, write out a balanced chemical equation for the transformation and critique the quality of the details given from the point of view of conducting a materials efficiency metrics analysis. All of these literature procedures were advertised as fulfilling green chemistry principles.

1. [38]
   Meldrum's acid (10 mmol), ethyl acetoacetate or methyl acetoacetate (10 mmol), the corresponding aromatic aldehyde (10 mmol), ammonium acetate (30 mmol), and 10 mL of acetic acid were added to a 50 mL Pyrex flask. The mixture was continuously irradiated at room temperature for 15–20 min with a sonic horn at 18 kHz. The reactions were followed by thin layer chromatography (TLC) using hexane:chloroform:ethyl acetate (3:2:1) as eluent. After completion of the reaction, the mixture was poured into ice-water. The solid that precipitated was collected by filtration. Further purification was accomplished by recrystallization from ethanol. Yield: 85%.
2. [39]
   A mixture of aniline (1 mmol), dimedone (2 mmol), formaldehyde (3 mmol, 37%–41% aqueous solution), and a catalytic amount of STA (silica tungstic acid) (10 mol %) in dichloromethane (5 mL) was stirred at room temperature for 4 h. The progress of the reaction was monitored by TLC. After the completion of the reaction, STA was removed from the reaction mixture by filtration. The solvent was removed under reduced pressure. The crude product was then purified directly by crystallization from ethanol. Yield: 93%.
3. [40]
   A mixture of resorcinol (0.110 g, 1 mmol), malononitrile (0.066 g, 1 mmol), aldehyde (1 mmol), and NaBr (0.05 g, 0.5 mmol) in propanol (25 mL) was stirred and electrolyzed in an undivided cell equipped with an iron cathode (5 cm^2) and a magnesium anode (5 cm^2) at room temperature under a constant current density of 10 mA/cm^2 ($I = 50$ mA). After completion of the reaction (monitored by TLC, ethyl acetate/*n*-hexane 1:1), the solvent was evaporated under reduced pressure, water (20 mL) was added to the reaction mixture and the resulting solid was separated by filtration. The crude product was recrystallized from ethanol to afford pure product. Yield: 83%.
4. [41]
   A mixture of substituted (triethoxymethyl)arene (1 mmol), 1-methyl-1*H*-indol-2-ol (1 mmol) and cyanoacetamide (1.1 mmol), and polymeric catalyst

(0.2 mmol) in CH_3CN was stirred for 4 h. The reaction was monitored by TLC, and after completion of the reaction, the catalyst was simply recovered by filtration and washed by water. The solvent was removed from filtrate by distillation and crude product was purified by column chromatography on silica gel eluting by petroleum ether-EtOAc (5:1). Yield: 90%.

5. [42]
   A solution of alkyne (1 mmol) in one mL of dichloromethane was added slowly to a mixture of Cu(I)-zeolite (30 mg), sulfonyl azide (1 mmol), nitrone (1 mmol) and Et_3N (1.2 mmol) taken in 2 mL of dichloromethane under N_2 atmosphere. After stirring at room temperature for 3 h, the mixture was diluted with ethyl acetate (5 mL). After removing the catalyst by filtration, followed by solvent evaporation under reduced pressure, the resulting crude product was finally purified by column chromatography on silica gel (60–120 mesh) with ethyl acetate and petroleum ether as eluting solvent to give the desired product. The recovered catalyst was thoroughly washed with ethyl acetate and used it for next run. Yield: 85%.
6. [43]
   A mixture of benzaldehyde (1 mol) ethylcyanoacetate (1 mol), β-naphthol (1 mol) and $TiCl_4$ (10 mol %) was stirred for 5 min at room temperature (monitored by TLC). The mixture was extracted with ethyl acetate and was washed with water. The organic layer was dried over Na_2SO_4 and concentrated to give crude product, which is purified by column (2:8 Ethyl acetate: Hexane). Yield: 95%.
7. [44]
   To a magnetically stirred of equivalent amount of neat adduct diketene (1 mmol) and aldehyde (1 mmol) and alcohol in excess amount of equivalent (4 mL) in SSA (silica sulphuric acid, 10 mol%) for 5 min at reflux condition, urea (1.2 mmol) was added, and then heated in neat state with stirring in refluxing condition, for an appropriate time (TLC monitoring). After that, hot ethanol (20 mL) was added and the mixture was filtered to remove the catalyst. The reaction was cooled to room temperature and the solid was washed with cooled water, petroleum ether/ether. All the products were previously reported and were characterized by comparing physical data with those. Spectral data for selected products: methyl 6-methyl-2 -oxo-4-phenyl-1,2,3,4-tetrahydropyrimidine-5-carboxylate. Yield: 90%.
8. [45]
   A mixture of 2-bromobenzaldehyde (1.5 mmol), primary amine (1.8 mmol), sodium azide (2 mmol) and catalytic amounts of Cu(II)–hydrotalcite (20 mg) in dimethylsulfoxide (DMSO) (1.5 mL) were stirred at 120°C temperature for 6 h. After completion of the reaction, as indicated by TLC, the reaction mixture was centrifuged to separate the catalyst. The reaction mixture was poured into EtOAc (30 mL) and washed with distilled water (3×20 mL), and the combined layers were dried over anhydrous Na_2SO_4, concentrated in vacuum and purified by column chromatography on 60–120 mesh silica gel, using ethyl acetate and hexane as the eluent (1:9) to afford the pure 2*H*-indazoles. Yield: 91%.

REFERENCES

1. Andraos, J. *J. Chem. Educ.* 2016, *93*, 1330.
2. Cornforth, J. *Chem. Brit.* 1975, *11*, 432.
3. Cornforth, J.W. *Austr. J. Chem.* 1993, *46*, 157.
4. Wernerova, M., Hudlicky, T. *Synlett* **2010**, 2701.
5. Carlson, R., Hudlicky, T. *Helv. Chim. Acta.* 2012, *95*, 2052.
6. Kovac, P. *Carbohydrate Chemistry: Proven Synthetic Methods*, vol. 1. CRC Press: Boca Raton, 2012, p. xix.
7. Danheiser, R.L. *Org. Synth.* 2011, *88*, 1.
8. Laird, T. *Org. Proc. Res. Dev.* 2011, *15*, 305.
9. Laird, T. *Org. Proc. Res. Dev.* 2011, *15*, 729.
10. Tundo, P.; Andraos, J. in *Green Syntheses,* vol. *1*, CRC Press: Boca Raton, 2014, pp. 1–12.
11. Laird, T. *Org. Proc. Res. Dev.* 2012, *16*, 1.
12. Tucker, J.L. *Org. Process Res. Dev.* 2006, *10*, 315.
13. Tucker, J.L. *Org. Process Res. Dev.* 2010, *14*, 328.
14. Carey, J.S., Laffan, D., Thomson, C., Williams, M.T. *Org. Biomol. Chem.* 2006, 4, 2337.
15. Constable, D.J.C., Dunn, P.J., Hayler, J.D., Humphrey, G.R., Leazer, J.L., Linderman, R.J., Lorenz, K., Manley, J., Pearlman, B.A., Wells, A., Zaks, A., Zhang, T.Y. *Green Chem.* 2007, *9*, 411.
16. Warhurst, M. *Green Chem.* 2002, *4*, G20.
17. Diehlmann, A., Kreisel, G. *Green Chem.* 2002, *4*, G15.
18. Watson, W.J.W. *Green Chem.* 2012, *14*, 251.
19. Welton, T. *Green Chem.* 2011, *13*, 225.
20. Bennett, G.D. Green chemistry as an expression of environmental ethics. In Sharma S.K.; Mudhoo A. (Eds.), *Green Chemistry for Environmental Sustainability.* CRC Press: Boca Raton, 2010, p. 116.
21. Chatterjee, S. *Tetrahedron Lett.* 1979, *20*, 3249.
22. Cornforth, J. *Tetrahedron Lett.* 1980, *21*, 709.
23. Cornforth, J.; Pengelly, T. *Tetrahedron Lett.* 1982, *23*, 2213.
24. Elati, C.R.; Kolla, N.; Vankawala, P.J.; Gangula, S.; Chalamala, S.; Sundaram, V.; Bhattacharya, A.; Vurimidi, H.; Mathad, V.T. *Org. Process Res. Dev.* 2007, *11*, 289.
25. Dancer, R.J.; de Diego, H.L. *Org. Process Res. Dev.* 2009, *13*, 23.
26. Elati, C.R.; Kolla, N.; Mathad, V.T. *Org. Process Res. Dev.* 2009, *13*, 34.
27. Dancer, R.J.; de Diego, H.L. *Org. Process Res. Dev.* 2009, *13*, 38.
28. Lowe, D. Your Paper is a Sack of Raving Nonsense. Thank You. (http://blogs.sciencemag.org/pipeline/archives/2009/02/27/your_paper_is_a_sack_of_raving_nonsense_thank_you; February 27, 2009)
29. Allen, D.T.; Carrier, D.J.; Gong, J.; Hwang, B.J.; Licence, P.; Moores, A.; Pradeep, T.; Sels, B.; Subramaniam, B.; Tam, M.K.C.; Zhang, L.; Williams, R.M. *ACS Sust. Chem. Eng.* 2018, *6*, 1.
30. Li, C.J.; Yao, X. *Org. Synth.* 2007, *84*, 222.
31. Lemke, F.R. *Inorg. Synth.* 2014, *36*, 233.
32. Zhang, A.Q.; Zhang, M.; Chen, H.H.; Chen, J.; Chen, H.Y. *Synth. Commun.* 2007, *37*, 231.
33. Meth-Cohn, O. *Org. Synth. Coll.* 1993, *8*, 350.
34. DeTar, D.F. *Org. Synth. Coll.* 1963, *4*, 730.
35. Heydari, A.; Larijani, H.; Emani, J.; Karami, B. *Tetrahedron Lett.* 2000, *41*, 2471.
36. Kidwai, M.; Saxena, S.; Khan, M.K.R.; Thukral, S.S. *Bioorg. Med. Chem. Lett.* 2005, *15*, 4295.

37. Kotagiri, V.K.; Suthrapu, S.; Reddy, J.M.; Rao, C.P.; Bollugoddhu, V.; Bhattacharya, A.; Bandichhor, R. *Org. Process Res. Dev.* 2007, *11*, 910.
38. Ruiz, E.; Rodriguez, H.; Coro, J.; Salfran, E.; Suarez, M.; Martinez-Alvarez, R.; Martin, N. *Ultrasonics Sonochem.* 2011, *18*, 32.
39. Atar, A.B.; Jeong, Y.T. *Tetrahedron Lett.* 2013, *54*, 1302.
40. Makarem, S.; Mohammadi, A.A.; Fakhari, A.R. *Tetrahedron Lett.* 2008, *49*, 7194.
41. Zohuri, G.H.; Damavandi, S.; Sandaroos, R. *Res. Chem. Intermed.* 2013, *39*, 2115.
42. Namitharan, K.; Pitchumani, K. *Org. Lett.* 2011, *13*, 5728.
43. Kumar, B.S.; Srinivasulu, N.; Udupi, R.H.; Rajitha, B.; Reddy, T.; Reddy, P.N.; Kumar, P.S. *J. Heterocyclic Chem.* 2006, *43*, 1691.
44. Rostamnia, S.; Lamei, K. *Chin. Chem. Lett.* 2012, *23*, 930.
45. Prasad, A.N.; Srinivas, R.; Reddy, B.M. *Catal. Sci. Technol.* 2013, 3, 654.

4 Chemical Reaction Classifications

4.1 IDENTIFYING TYPES OF REACTIONS

There are five major classes of chemical reaction types encountered in organic chemistry: additions, eliminations, rearrangements, redox reactions, and substitutions. Every organic reaction that has ever been discovered or will be discovered in the future can be slotted in at least one of these categories. Generally, most of the named organic reactions that form the workhorse toolbox for organic chemists are categorized uniquely into one of these classes. Modern reactions have attributes that have features of more than one of these classes. Each of these reaction classes may be depicted using a visual algebraic approach using LEGO®-like cartoons to illustrate what is occurring.

4.1.1 Terms, Definitions, and Examples

Additions

Addition reactions involve coupling of two or more components in an acyclic (linear) or a cyclic sense. The visual algebraic operation is addition.

acyclic

cyclic

Friedel-Crafts acylation

$AlCl_3$ (cat.)

+ HCl

Biginelli reaction [3 + 2 + 1]

Lewis acid catalyst

+ 2 H_2O

Eliminations (Fragmentations)

Eliminations or fragmentations involve the splitting apart of a substrate or an intermediate. The visual algebraic operation is subtraction.

Hofmann degradation

$R_1R_2C(NH_2)$ + 3 $\mathbf{CH_3I}$ → (- 2 HI) [$R_1R_2C{-}N^{\oplus}Me_3$ $I^{\ominus}$] → (**NaOH**; - NaI, - H_2O, - Me_3N) $R_1R_2C{=}CH_2$

structures shown in bold text are sacrificial

Multi-Component

Multi-component reactions are a special sub-class of addition reactions involving at least three substrate structures combining together to produce a product. The order of addition of those substrates may or may not matter depending on the reaction mechanism. This is the most powerful method of constructing ring-containing compounds with the least number of reaction steps.

Rearrangements

Rearrangement reactions involve a reshuffling of atom connectivity of the substrate or a reaction intermediate. The visual algebraic operation is reshuffling. Rearrangements of substrates always have 100% atom economy. Reactions involving rearrangements of transient intermediates always have atom economies less than 100% since by-products always arise.

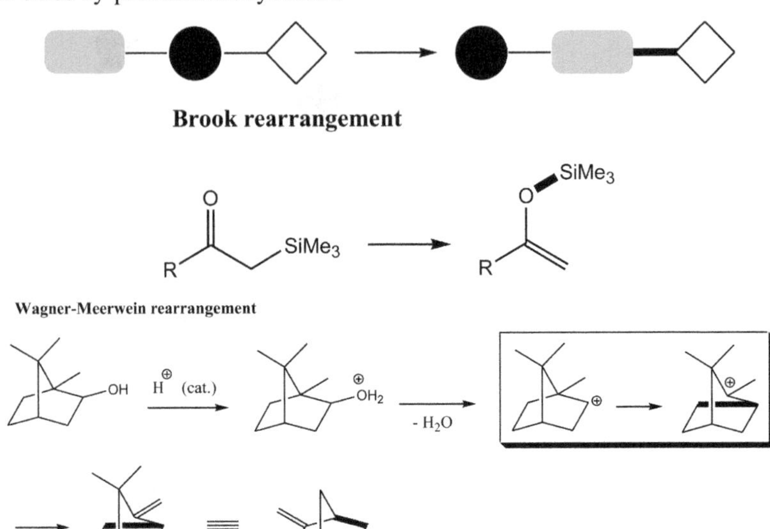

Redox Reactions

Redox reactions involve either reduction or oxidation of the substrate where the oxidation states of key atoms in the substrate decrease or increase, respectively. There are two sub-classes of redox reactions: additive or subtractive. For additive oxidations, oxygen atoms are added to the substrate; whereas, for subtractive oxidations, hydrogen atoms are subtracted from the substrate. For additive reductions, hydrogen atoms are added to the substrate; whereas for subtractive reductions, oxygen atoms are subtracted from the substrate. Redox reactions have the lowest atom economies since significant production of by-products arise. They also have the most opportunities for recycling those by-products back to reagents via other redox-type reactions. This class of reaction is discussed in more depth in Chapter 2.

Substitutions

Substitution reactions involve an exchange of functional groups. The visual algebraic operation is exchange or metathesis. A substitution reaction where the leaving group has a greater molecular weight than the incoming group will generally have a low atom economy. The opposite scenario leads generally to a higher atom economy.

Aromatic nitration

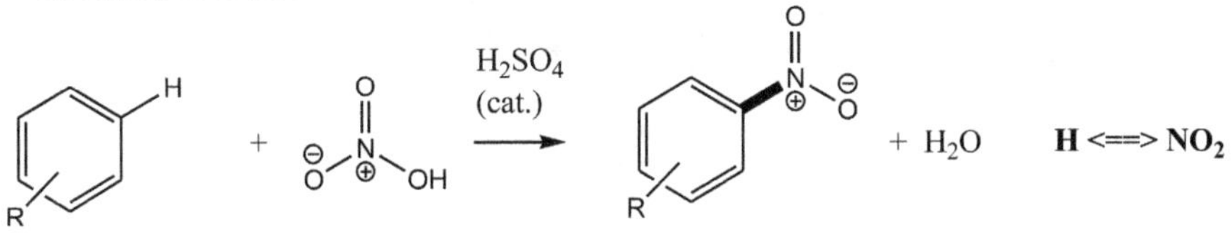

H_2SO_4 (cat.) + H_2O **H <==> NO_2**

Schotten-Baumann reaction

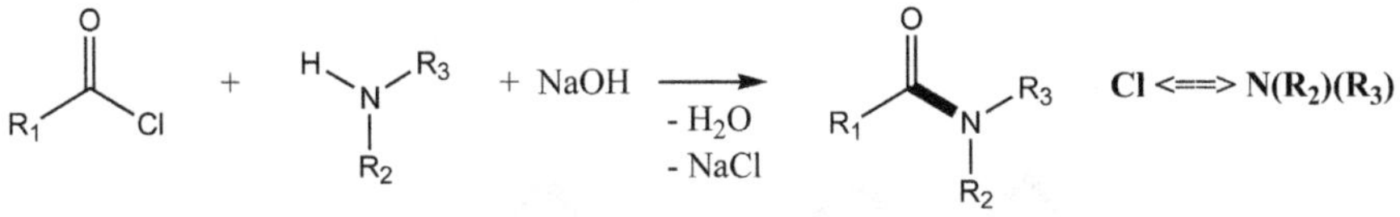

+ NaOH - H_2O - NaCl **Cl <==> $N(R_2)(R_3)$**

Example 4.1

Themes: classification of reaction types

The following cartoons depict various reaction motifs encountered in organic chemical reactions that are characterized as *aufbau* (building up or synthesis) reactions or *abbau* (breaking down or degradation) reactions. Categorize each motif into one or more of the following reaction types: addition, fragmentation, substitution, and rearrangement. Highlight the target bonds broken and made in each transformation. *Hint*: View the polygons as building blocks as in a LEGO set and the lines connecting the polygons as chemical bonds. Which reaction types are expected to have high atom economies? How about low atom economies?

(a)

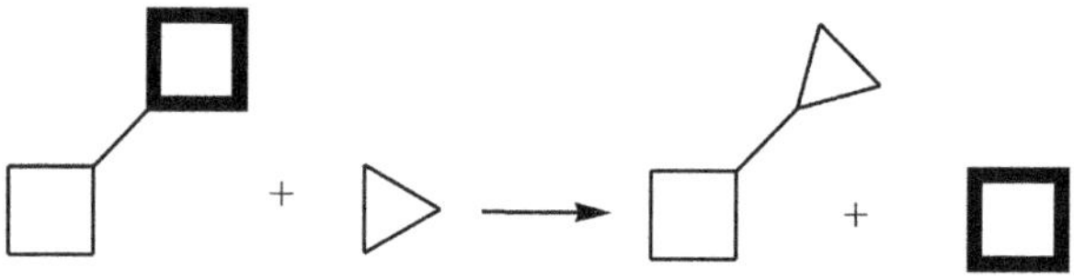

+ → +

(b)

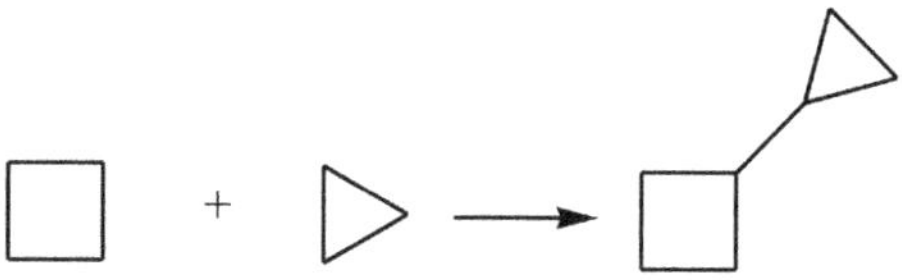

+ →

(c)

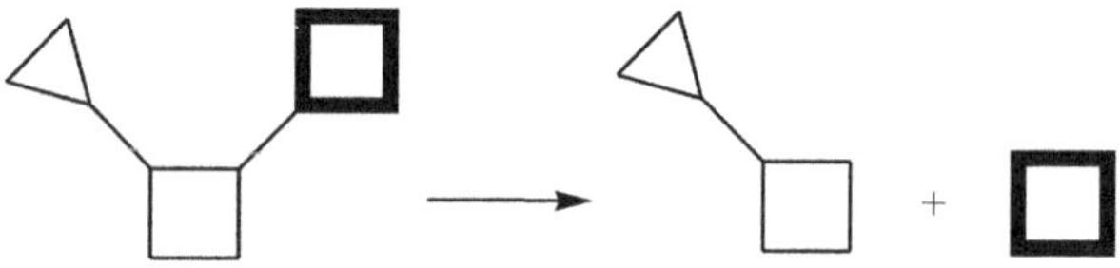

→ +

(d)

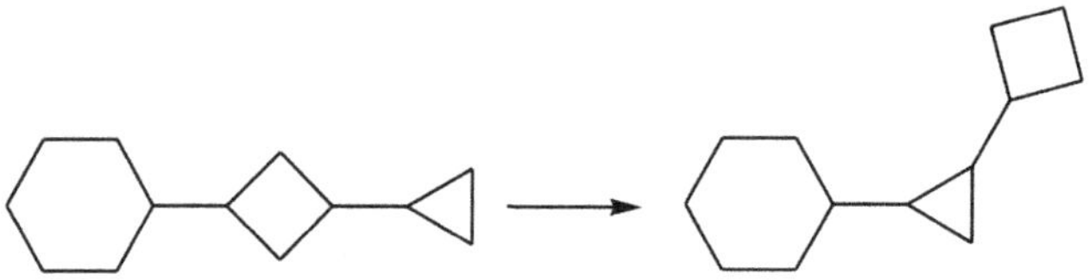

→

(e)

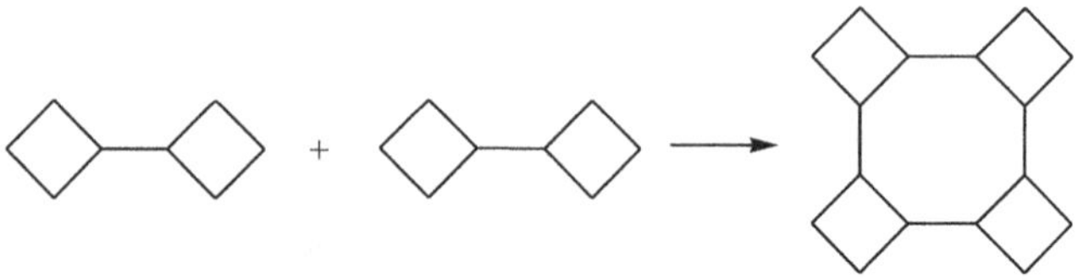

+ →

(f)

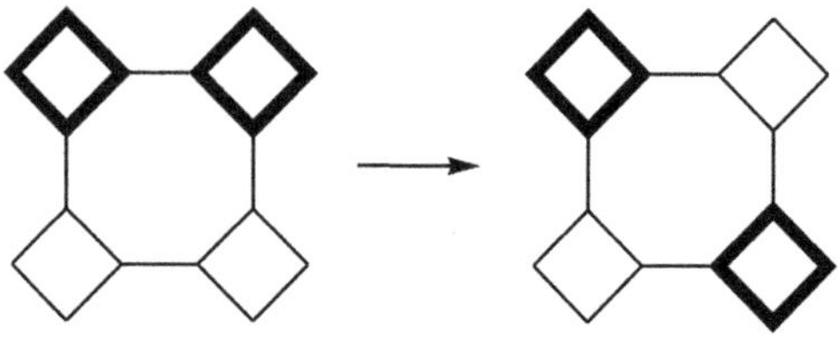

→

(g)

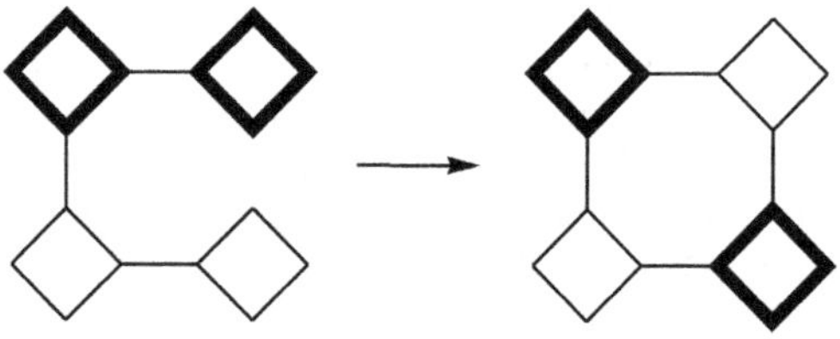

→

(h)

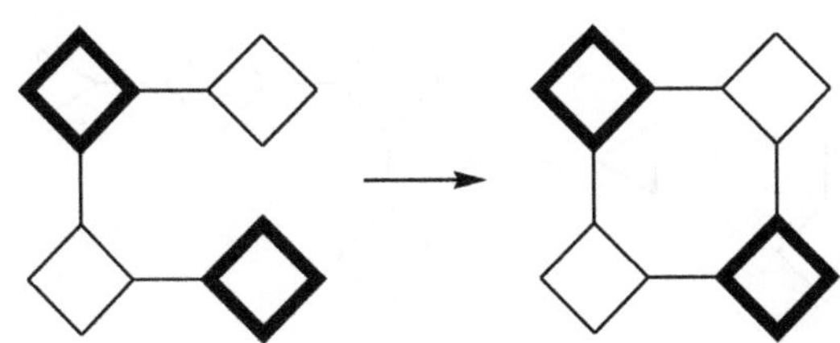

(i)

(j)

(k)

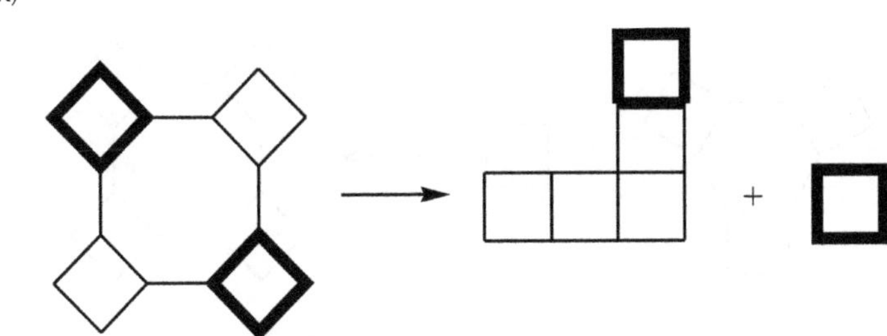

(l)

(m)

(n)

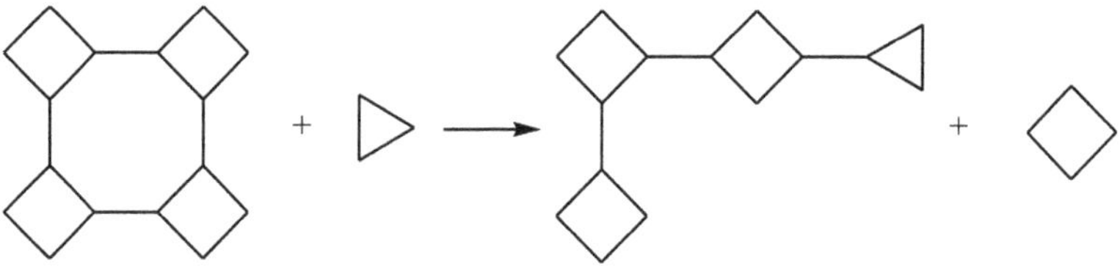

+ → +

(o)

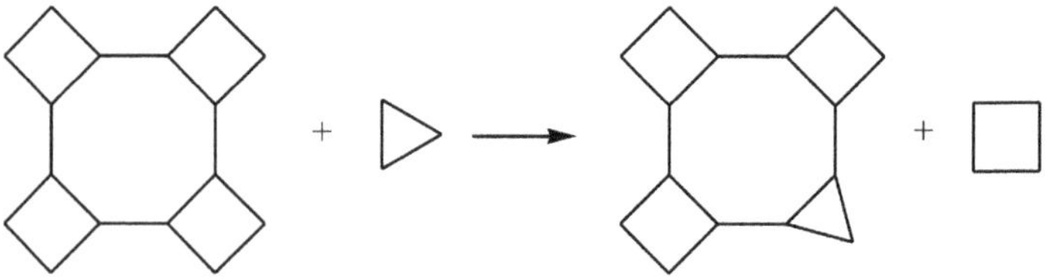

+ → +

(p)

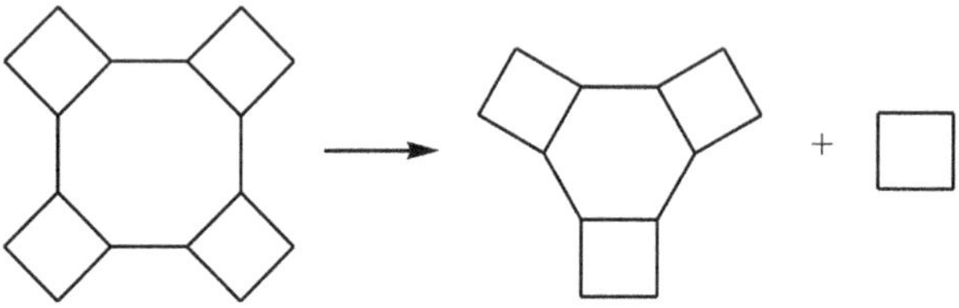

→ +

(q)

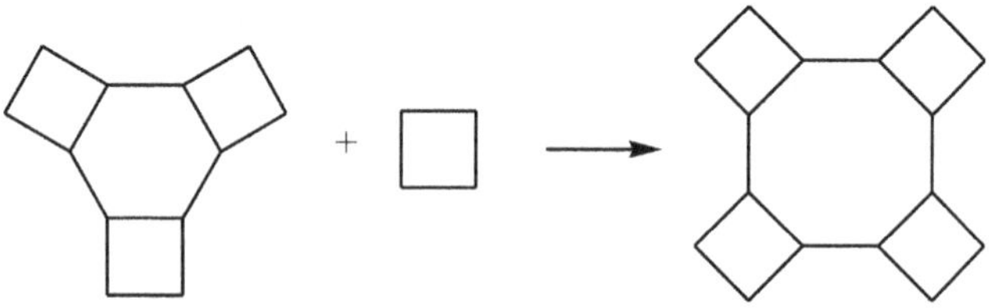

+ →

(r)

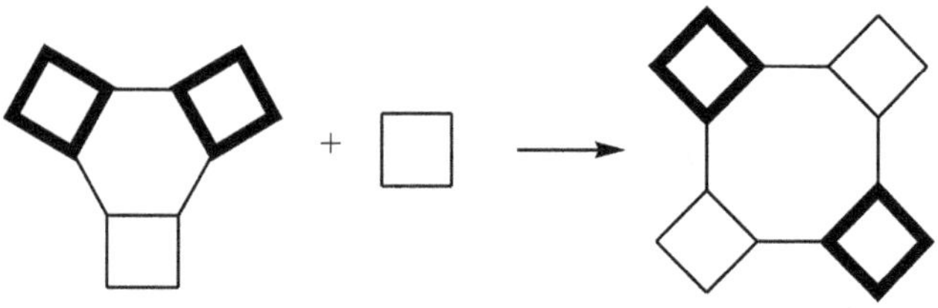

+ →

(s)

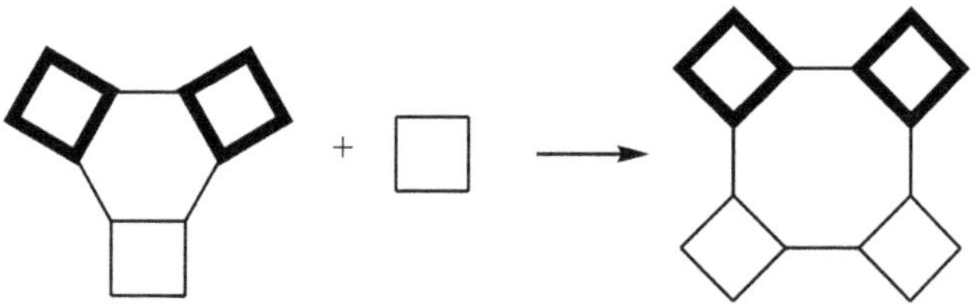

+ →

(t)

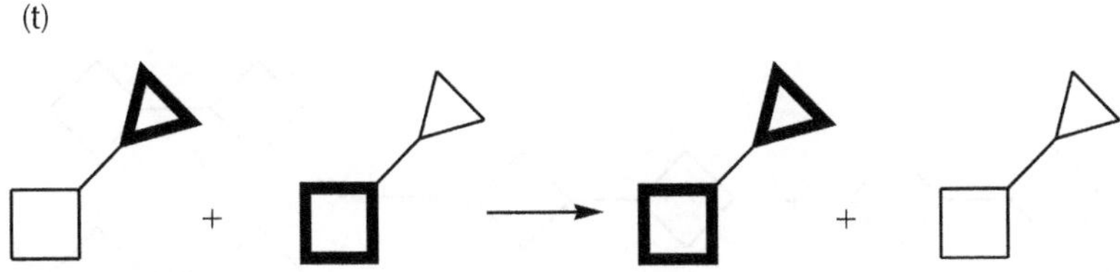

Solution

(a)

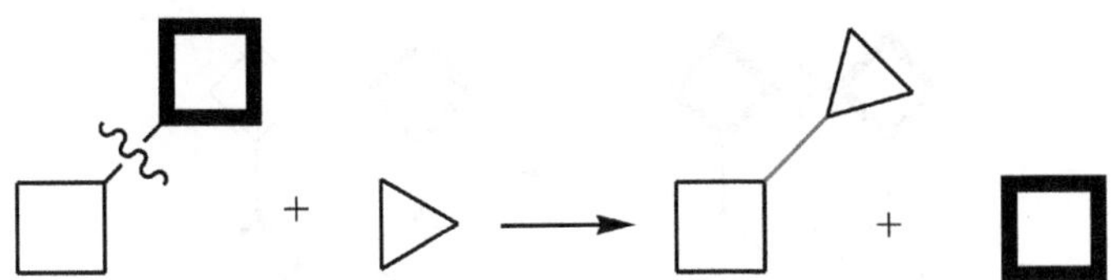

Reaction Type: Substitution

(b)

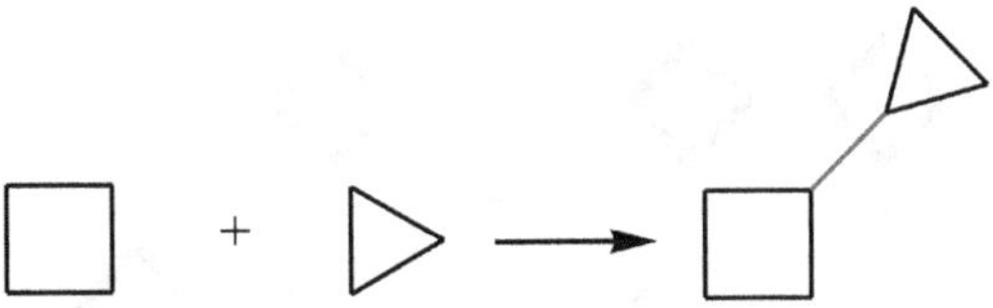

Reaction Type: Addition

(c)

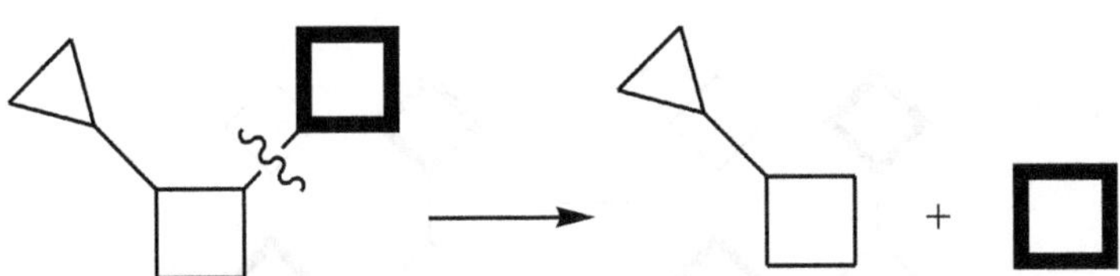

Reaction Type: Fragmentation

(d)

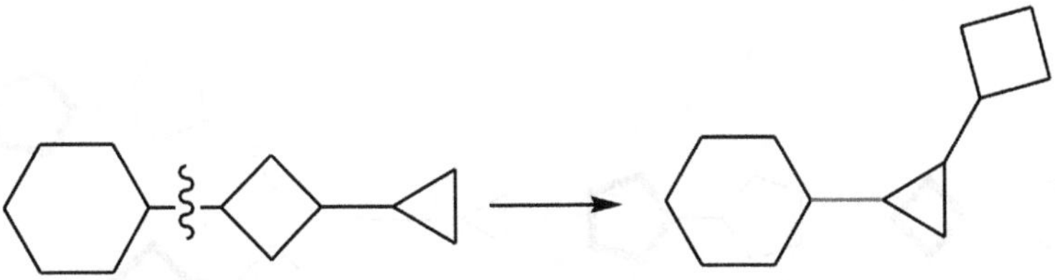

Reaction Type: Rearrangement

(e)

Reaction Type: Addition, Specifically Cycloaddition

(f)

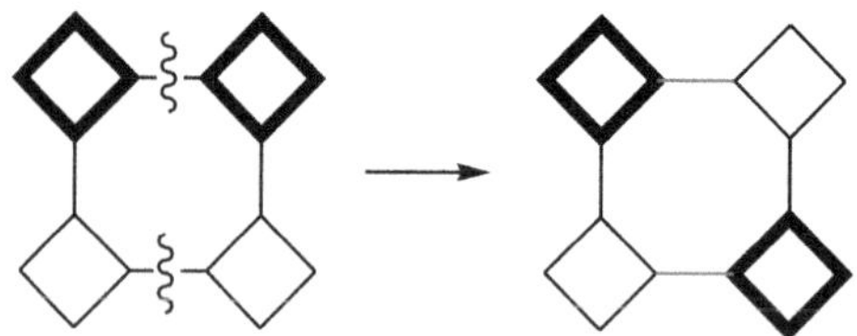

Reaction Type: Rearrangement

(g)

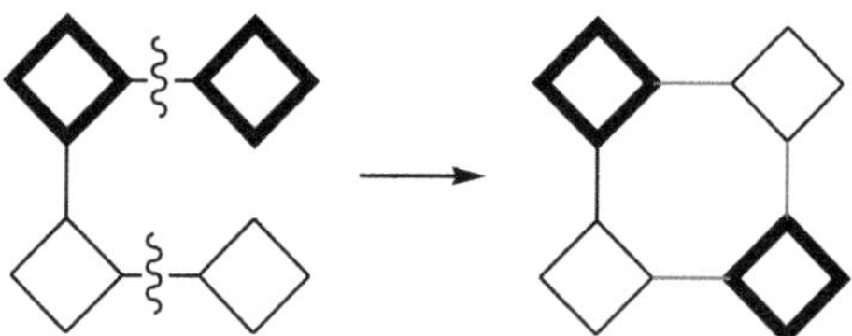

Reaction Type: Rearrangement

(h)

Reaction Type: Unimolecular Cyclization

(i)

Reaction Type: Addition, Multi-Component Addition

(j)

Reaction Type: Addition with Fragmentation

(k)

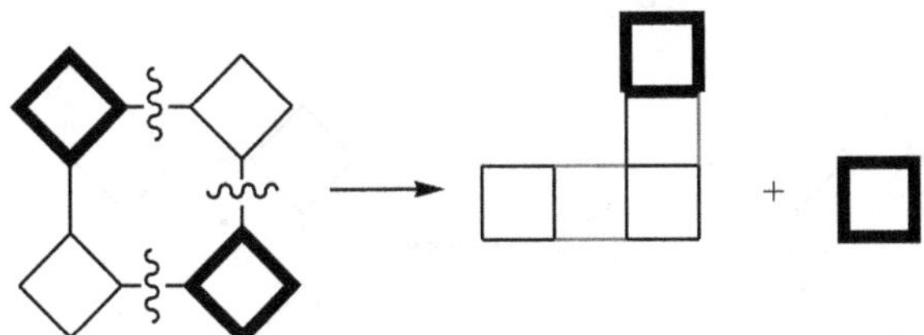

or

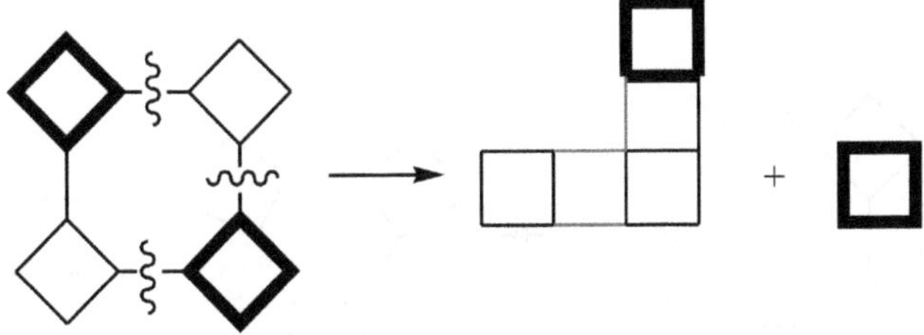

Reaction Type: Mixed Fragmentation and Addition

(l)

Reaction Type: Tandem Addition (at Different Reaction Centers)

(m)

Reaction Type: Substitution (Tandem Addition Fragmentation at Same Reaction Center)

(n)

Reaction Type: Mixed Fragmentation and Addition

(o)

Reaction Type: Mixed Fragmentation and Addition; or Substitution

(p)

Reaction Type: Fragmentation and Reattachment; Extrusion; Ring Contraction

(q)

Reaction Type: Tandem Fragmentation Cycloaddition; Ring Expansion

(r)

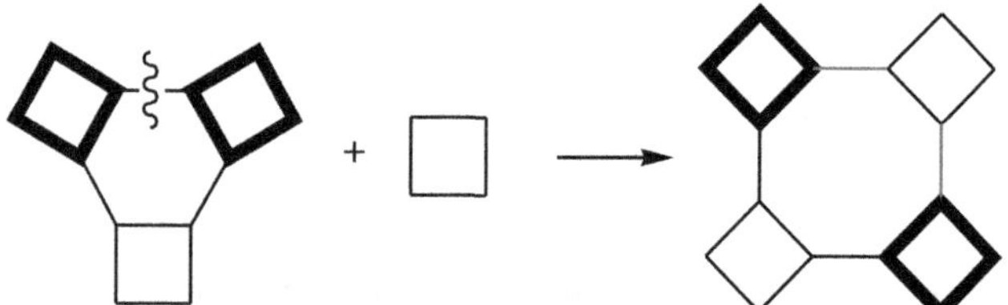

Reaction Type: Tandem Fragmentation Cycloaddition; Ring Expansion

(s)

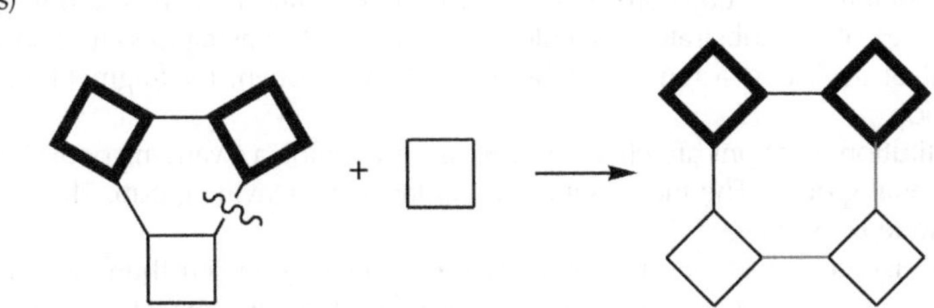

Reaction Type: Tandem Fragmentation Cycloaddition; Ring Expansion

(t)

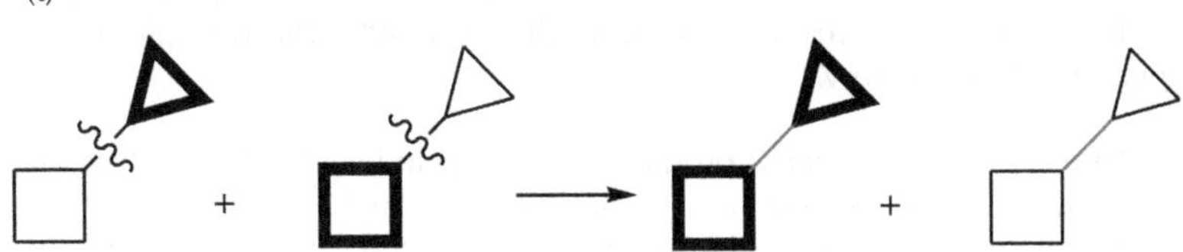

Reaction Type: Tandem Fragmentation Substitution; Metathesis

The reaction types that are expected to have high atom economies are additions, leaving no by-products and rearrangements of substrate structures. Substitutions where the incoming group is heavier than the leaving group are the next highest atom economical reactions. Reaction types having low atom economies are fragmentations yielding significant by-products and substitutions where the incoming group is lighter than the leaving group.

Example 4.2

Themes: reaction types

All known organic reactions may be categorized into one or more of the following categories depicted in the chart below.

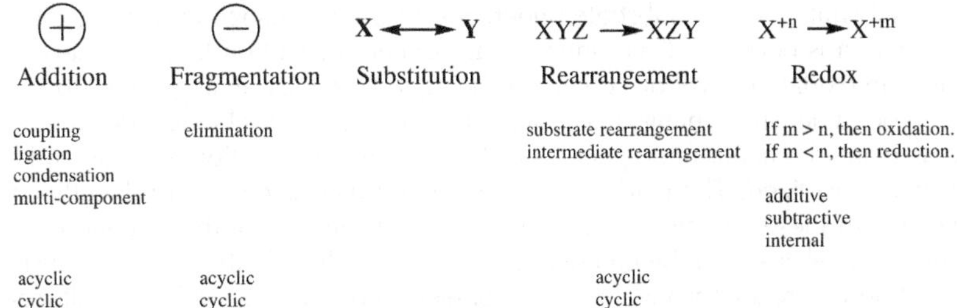

Addition reactions are depicted with an operational plus sign indicating that at least two molecules are added together to produce a product which retains some or all of the atoms found in the starting two substrates. Synonyms for addition reactions are couplings, ligations, condensations, and multi-component reactions. Condensations are uniquely characterized as addition reactions that also produce water or alcoholic by-products. Multi-component reactions are addition reactions involving at least three substrate molecules.

Fragmentation reactions are depicted with an operational minus sign indicating that part of the substrate molecule is subtracted off leaving behind a smaller fragment of itself and a remainder by-product. A synonym for fragmentation is elimination.

Substitution reactions are characterized as undergoing a swapping or exchange of functional groups. The incoming group replaces the leaving group. The operational mode is exchange.

Rearrangement reactions are characterized as undergoing a reshuffling or reorganization of connectivity operation between the atoms in the starting substrate structure or in an intermediate structure along the reaction pathway. The operational mode is reshuffling of atom connectivity.

Additions, fragmentations, and rearrangement reaction categories may take place in linear (acyclic) manifolds or in cyclic manifolds. The possible manifold changes are shown below.

- Linear substrate → linear product (no change in length, chain extension via addition, or chain truncation via fragmentation)
- Linear substrate → cyclic product (designated as ring closure or cyclization)
- Cyclic substrate → cyclic product (no change in ring size, ring expansion, or ring contraction)
- Cyclic substrate → linear product (designated as ring opening)

All four reaction categories may operate under one of two oxidation level regimes. If there is no change in the oxidation level for all key atoms involved in forming new bonds, then the reaction is redox neutral. If key atoms involved in forming new bonds undergo a change in oxidation level, then the reaction is a redox reaction. If such a reaction can be written as a redox couple, that is, as the sum of an oxidation half-reaction and a reduction half-reaction, then the overall reaction is a formal redox reaction. If the substrate of interest undergoes an increase in oxidation level, then the reaction is said to be an oxidation reaction. If the substrate of interest undergoes a decrease in oxidation level, then the reaction is said to be a reduction. In either case, the associated reagents will also undergo a change in oxidation level equal in magnitude but opposite in direction to that of the substrate. So, if a substrate undergoes a positive change in oxidation level of $+x$, i.e., it is being oxidized, then the associated reagent undergoes a negative change in oxidation level of $-x$, i.e., it is being reduced. By analogy, if a substrate undergoes a negative change in oxidation level of $-x$, i.e., it is being reduced, then the associated reagent undergoes a positive change in oxidation level of $+x$, i.e., it is being oxidized. The oxidation and reduction half reactions are directly correlated with these oxidation level changes. The magnitude of the oxidation level changes correlates with the number of electrons involved in the electron transfer. Reduction half reactions involve a gain of electrons whereas oxidation half reactions involve a loss of electrons. If the substrate of interest loses oxygen atoms it is a subtractive reduction; if the substrate of interest loses hydrogen atoms it is a subtractive oxidation; if the substrate of interest gains oxygen atoms it is an additive oxidation; and if the substrate of interest gains hydrogen atoms it is an additive reduction. If the reaction cannot be decomposed into a combination of half reactions with separable positive and negative changes in electron count, then the reaction is an internal redox reaction with no net gain or loss of oxidation level. Such reactions do not involve reagents whose atoms change in oxidation level.

Atoms undergoing oxidation level changes also undergo concomitant changes in philicity. Therefore, an atom increasing in oxidation level becomes more electrophilic (it is losing electron density) and an atom decreasing in oxidation level becomes more nucleophilic (it is gaining electron density).

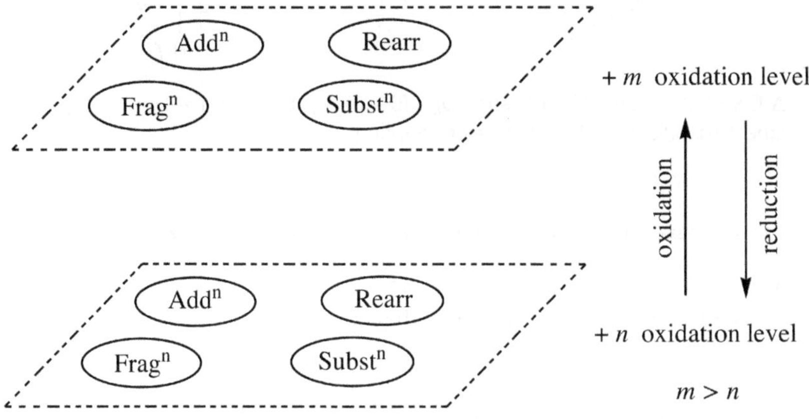

Part 1

From the named organic reaction database give examples of the following:

- An addition reaction that operates on a linear manifold and whose atoms involved in making new synthesis bonds do not undergo oxidation level changes.
- An addition reaction that operates on a linear manifold and whose atoms involved in making new synthesis bonds undergo oxidation level changes.
- An addition reaction that operates on a cyclic manifold and whose atoms involved in making new synthesis bonds do not undergo oxidation level changes.
- An addition reaction that operates on a cyclic manifold and whose atoms involved in making new synthesis bonds undergo oxidation level changes.

Solution

Linear manifold, $\Delta Ox = 0$: Baylis-Hillman reaction

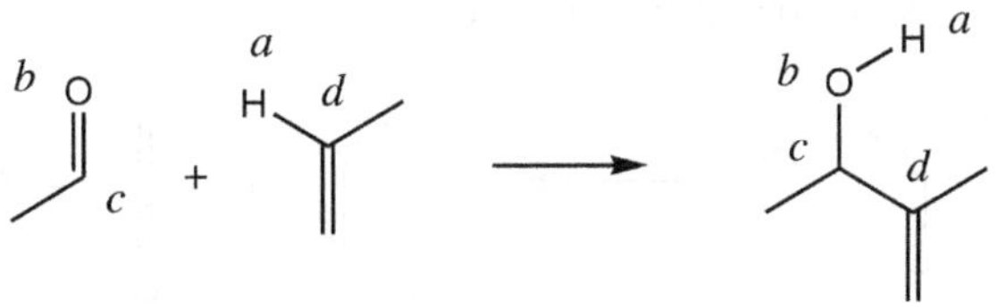

$\Delta\ Ox = (1-1)_a + (-2-(-2))_b + (0-1)_c + (0-(-1))_d = -1 + 1 = 0$

Cyclic manifold, $\Delta Ox = 0$: Bergmann cyclization

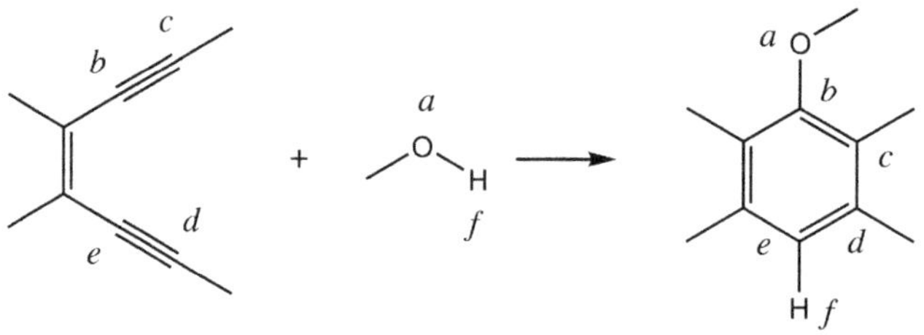

Δ Ox = $(-2-(-2))_a + (1-0)_b + (0-0)_c + (0-0)_d + (-1-0)_e + (1-1)_f = 1-1=0$

Linear manifold, $\Delta Ox \neq 0$: Glaser coupling

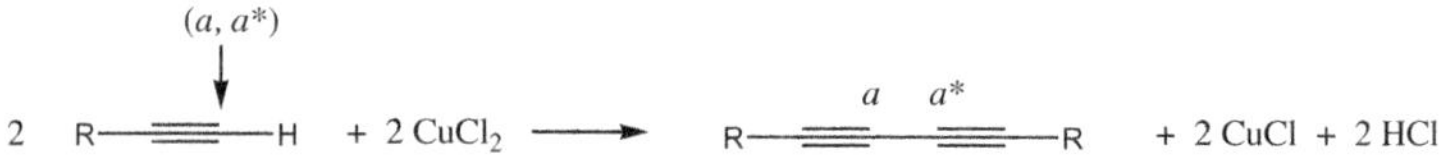

$$2\ R{-}{\equiv}{-}H + 2\,CuCl_2 \longrightarrow R{-}{\equiv}{-}{\equiv}{-}R + 2\,CuCl + 2\,HCl$$

Δ Ox = $(0-(-1))_a + (0-(-1))_{a^*} = 1+1 = +2$

Cyclic manifold, $\Delta Ox \neq 0$: Kulinkovich reaction

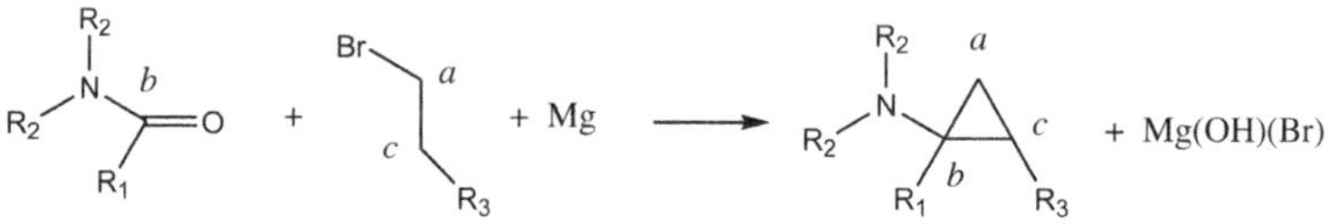

Δ Ox = $(-2-(-1))_a + (1-3)_b + (-1-(-2))_c = -1-2+1 = -2$

Part 2

- A fragmentation reaction that operates on a linear manifold and whose atoms involved in making new synthesis bonds do not undergo oxidation level changes.
- A fragmentation reaction that operates on a linear manifold and whose atoms involved in making new synthesis bonds undergo oxidation level changes.
- A fragmentation reaction that operates on a cyclic manifold and whose atoms involved in making new synthesis bonds do not undergo oxidation level changes.
- A fragmentation reaction that operates on a cyclic manifold and whose atoms involved in making new synthesis bonds undergo oxidation level changes.

Solution

Linear manifold, $\Delta Ox = 0$: Chugaev reaction

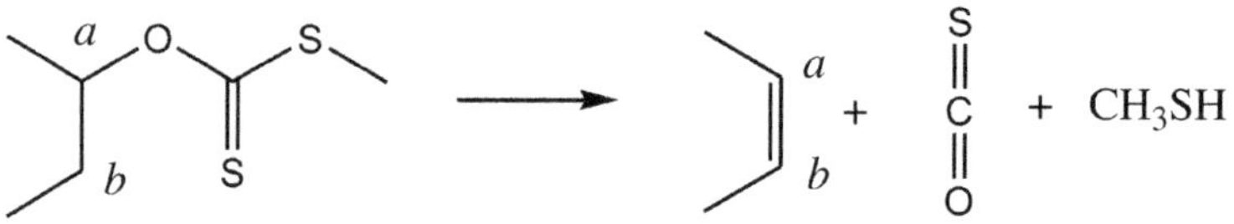

Δ Ox = $(-1-0)_a + (-1-(-2))_b = -1+1 = 0$

Cyclic manifold, $\Delta Ox = 0$: Grob fragmentation

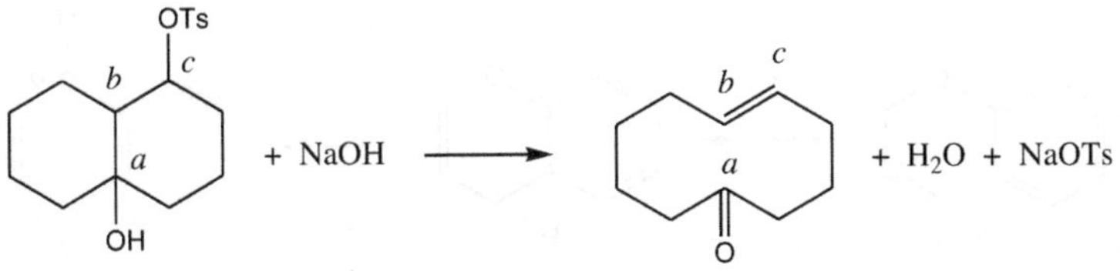

$\Delta\ Ox = (2-1)_a + (-1-(-1))_b + (-1-0)_c = 1-1 = 0$

Linear manifold, $\Delta Ox \neq 0$: Kochi reaction

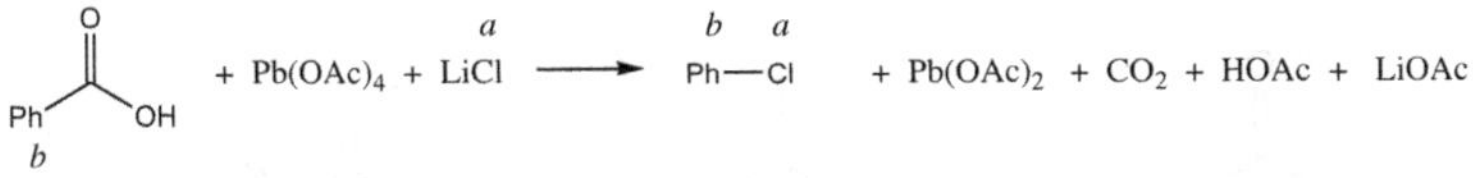

$\Delta\ Ox = (-1-(-1))_a + (1-0)_b = +1$

Cyclic manifold, $\Delta Ox \neq 0$: Tiffeneau–Demjanov reaction

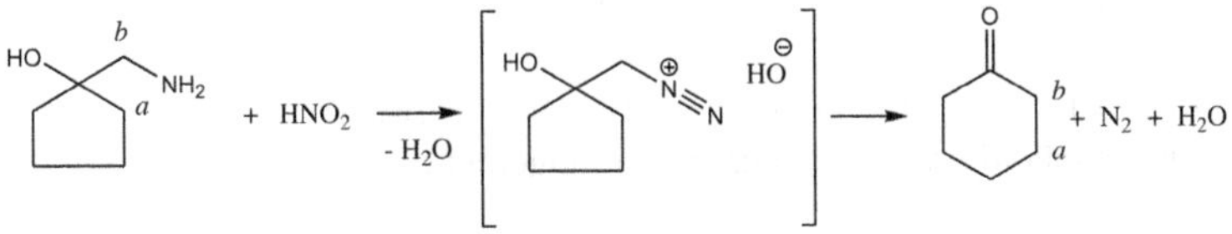

$\Delta\ Ox = (-2-(-2))_a + (-2-(-1))_b = -1$

Part 3

- A rearrangement reaction that operates on a linear manifold and whose atoms involved in making new synthesis bonds do not undergo oxidation level changes.
- A rearrangement reaction that operates on a linear manifold and whose atoms involved in making new synthesis bonds undergo oxidation level changes.
- A rearrangement reaction that operates on a cyclic manifold and whose atoms involved in making new synthesis bonds do not undergo oxidation level changes.
- A rearrangement reaction that operates on a cyclic manifold and whose atoms involved in making new synthesis bonds undergo oxidation level changes.

Solution

Linear manifold, $\Delta Ox = 0$: Cope rearrangement

$\Delta\ Ox = (-2-(-2))_a + (-2-(-2))_b = 0$

Linear manifold, $\Delta Ox \neq 0$: Claisen rearrangement

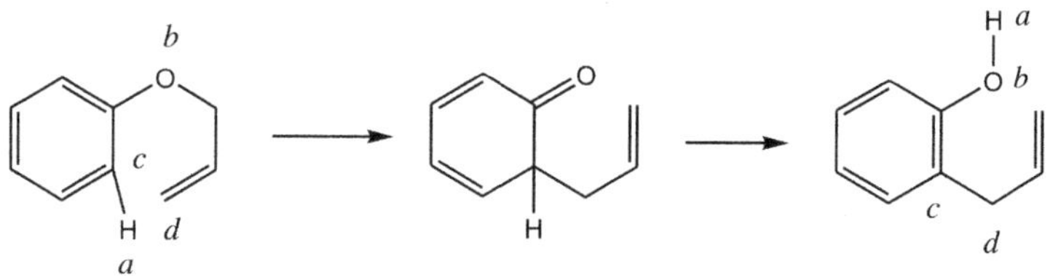

$$\Delta\ Ox=(1-1)_a+(-2-(-2))_b+(0-(-1))_c+(-2-(-2))_d=+1$$

Cyclic manifold, $\Delta Ox = 0$: di-π-methane rearrangement

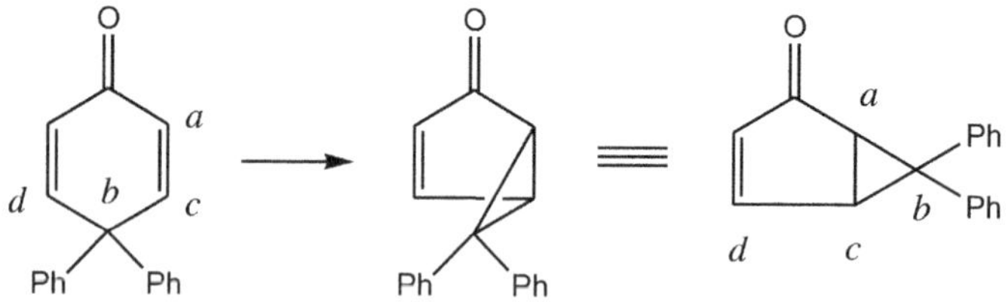

$$\Delta\ Ox=(-1-(-1))_a+(0-0)_b+(-1-(-1))_c+(-1-(-1))_d=0$$

Cyclic manifold, $\Delta Ox \neq 0$: Perkin rearrangement

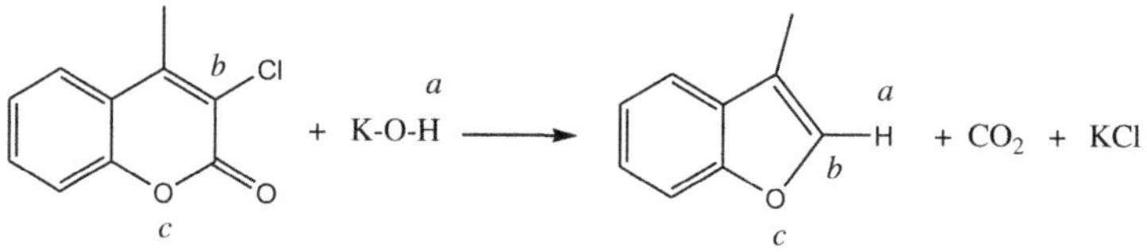

+ K-O-H → + CO_2 + KCl

$$\Delta\ Ox=(1-1)_a+(0-1)_b+(-2-(-2))_c=-1$$

Part 4

- A substitution reaction whose atoms involved in making new synthesis bonds do not undergo oxidation level changes.
- A substitution reaction whose atoms involved in making new synthesis bonds undergo oxidation level changes.

Solution

$\Delta Ox = 0$, aromatic nitration:

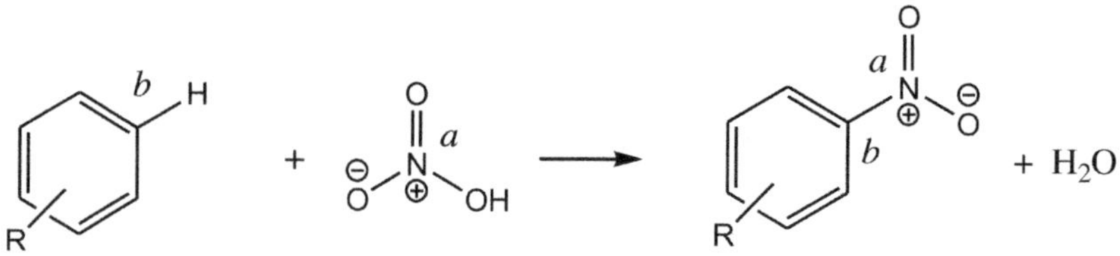

+ H_2O

$$\Delta\ Ox=(3-5)_a+(1-(-1))_b=-2+2=0$$

$\Delta Ox \neq 0$, Lapworth reaction:

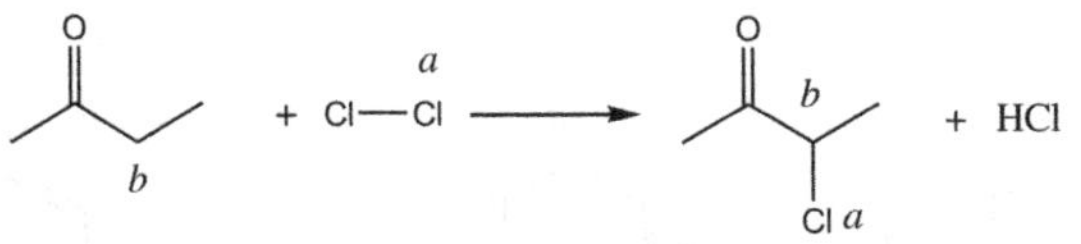

$$\Delta\ \mathrm{Ox} = (-1-0)_a + (0-(-2))_b = -1+2 = +1$$

Part 5

Tandem reactions are composed of sequential building block reactions. Considering the four main theme categories of available types of organic reactions, enumerate all possible pairwise tandem reaction combinations. For each combination suggest an example. For each example label the atoms in both the product and in the starting material structures that are involved in bond making and bond breaking processes. Determine the net change in oxidation number for the transformation based on the labeled atoms.

Solution

(i) No change in philicity

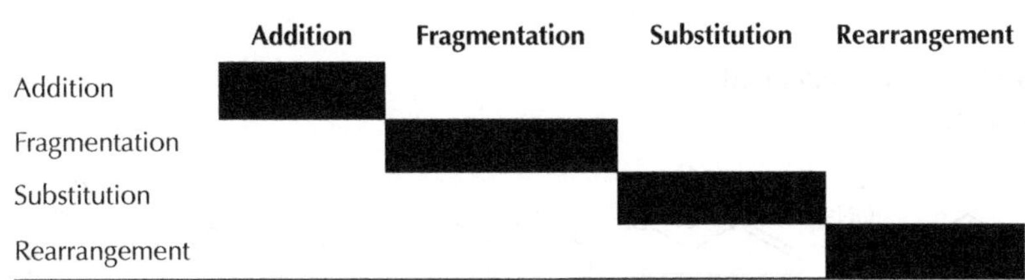

	Addition	Fragmentation	Substitution	Rearrangement
Addition				
Fragmentation				
Substitution				
Rearrangement				

Addition-Fragmentation

Example 1

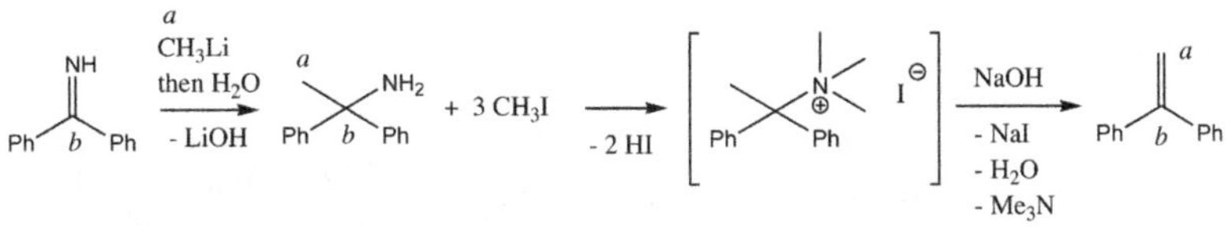

$$\Delta\ \mathrm{Ox} = (-2-(-4))_a + (0-2)_b = 2-2 = 0$$

Example 2

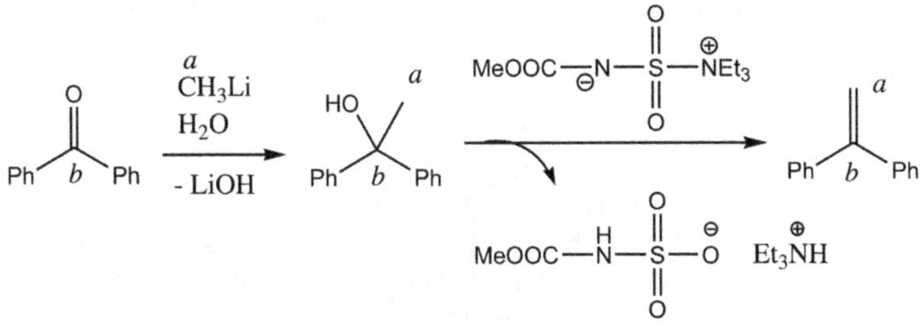

$$\Delta\ \mathrm{Ox} = (-2-(-4))_a + (0-2)_b = 2-2 = 0$$

Example 3

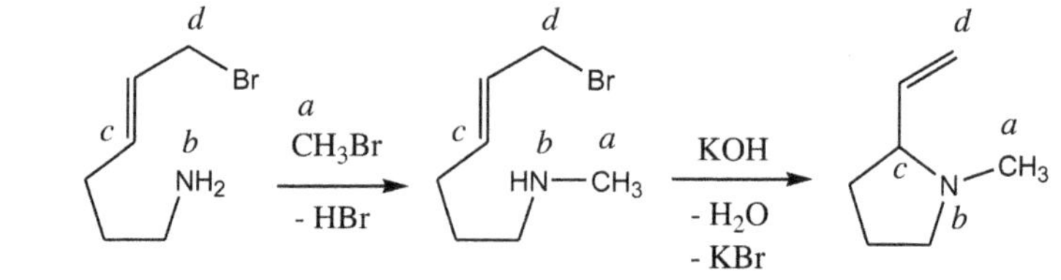

Δ Ox = $(-2-(-2))_a+(-3-(-3))_b+(0-(-1))_c+(-2-(-1))_d=1-1=0$

Addition-Substitution

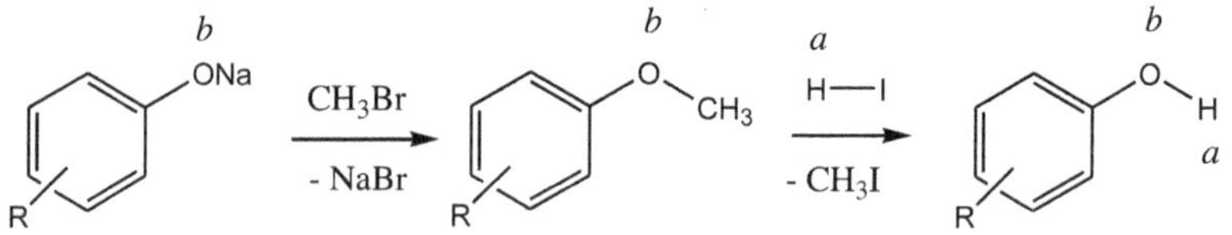

Δ Ox = $(1-1)_a+(-2-(-2))_b=0$

Addition-Rearrangement

Example 1

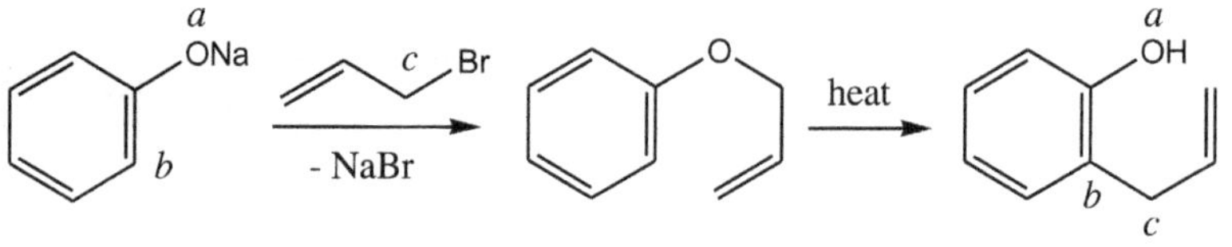

Δ Ox = $(-2-(-2))_a+(0-(-1))_b+(-2-(-1))_c=1-1=0$

Example 2

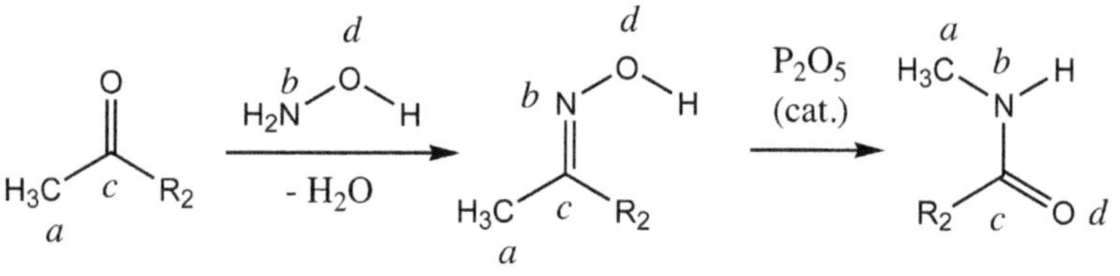

Δ Ox = $(-2-(-3))_a+(-3-(-1))_b+(3-2)_c+(-2-(-2))_d=1-2+1=0$

Example 3

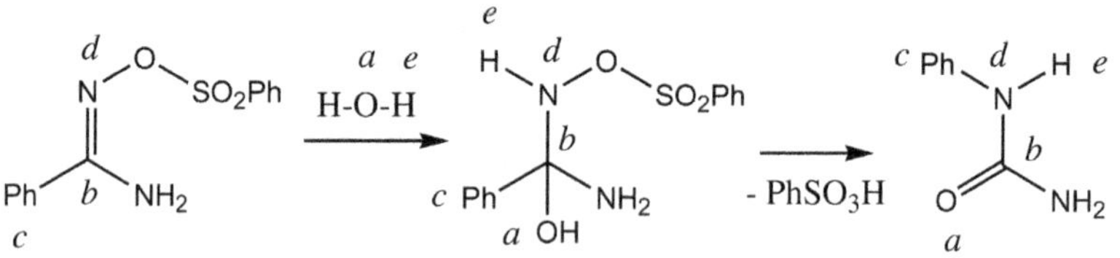

Δ Ox = $(-2-(-2))_a+(4-3)_b+(1-0)_c+(-3-(-1))_d+(1-1)_e=1+1-2=0$

Example 4

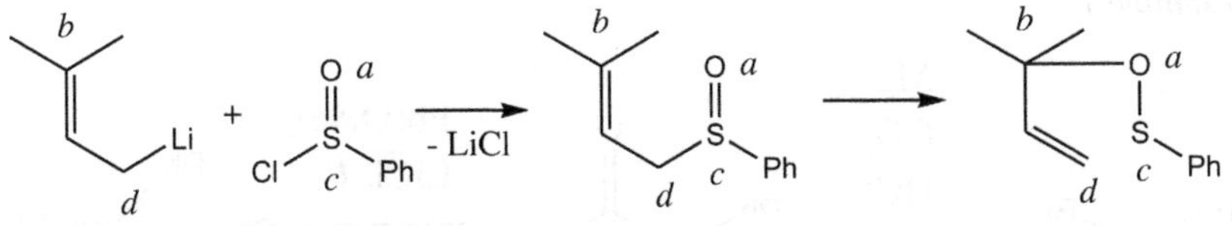

$$\Delta\,\mathrm{Ox} = (-2-(-2))_a + (1-0)_b + (0-2)_c + (-2-(-3))_d = 1-2+1 = 0$$

Example 5

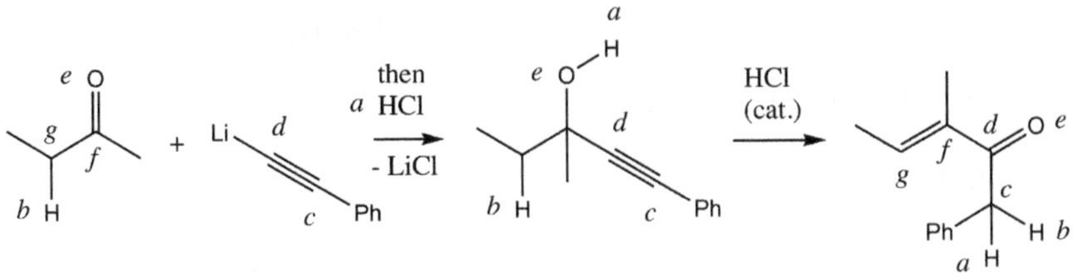

$$\Delta\,\mathrm{Ox} = (1-1)_a + (1-1)_b + (-2-0)_c + (2-(-1))_d + (-2-(-2))_e + (0-2)_f + (-1-(-2))_g$$
$$= -2+3-2+1 = 0$$

Example 6

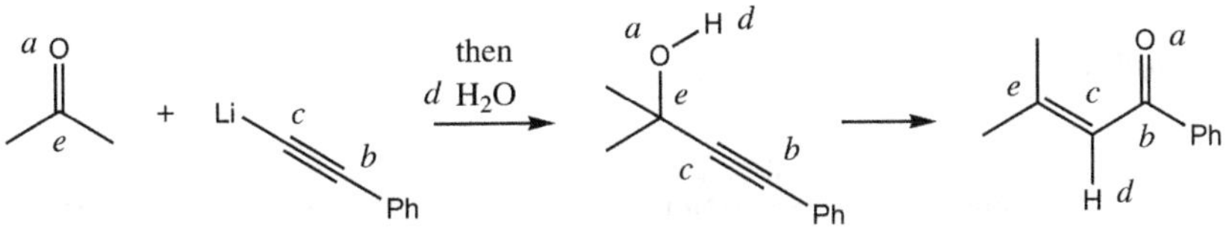

$$\Delta\,\mathrm{Ox} = (-2-(-2))_a + (2-0)_b + (-1-(-1))_c + (1-1)_d + (0-2)_e = 2-2 = 0$$

Example 7

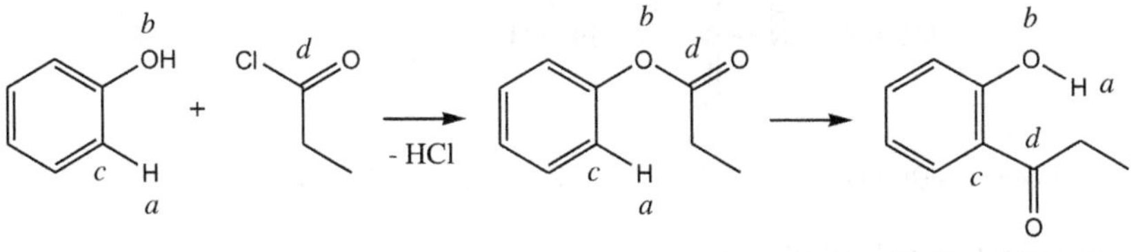

$$\Delta\,\mathrm{Ox} = (1-1)_a + (-2-(-2))_b + (0-(-1))_c + (2-3)_d = 1-1 = 0$$

Example 8

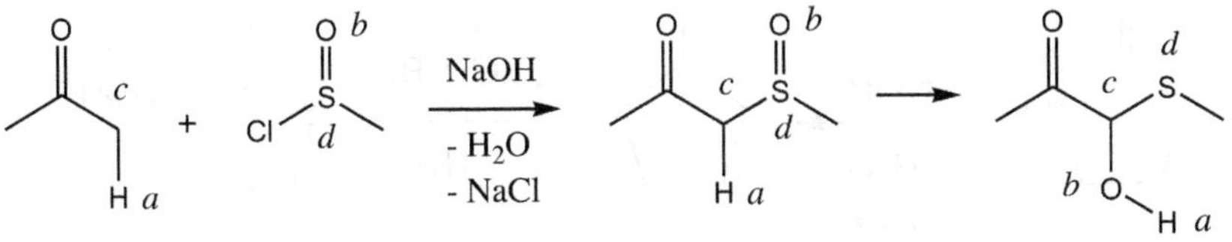

$$\Delta\,\mathrm{Ox} = (1-1)_a + (-2-(-2))_b + (1-(-3))_c + (-2-2)_d = 4-4 = 0$$

Fragmentation-Addition

Example 1

Ph Br, *a* → (Mg, CO_2, HCl; $- Mg(Br)(Cl)$) → Ph–CH_2–COOH, *b* → ($Pb(OAc)_4$, LiCl *b*; $- Pb(OAc)_2$, $- CO_2$, - HOAc, - LiOAc) → Ph–CH_2–Cl, *a b*

$$\Delta\ Ox = (-1-(-1))_a + (-1-(-1))_b = 0$$

Example 2

c, *b*: $Ph_2C(CH_3)NH_2$ + 3 CH_3I → (- 2 HI) → [$Ph_2C(CH_3)N^+Me_3$ I^-] → (NaOH; - NaI, - H_2O, - Me_3N) → $Ph_2C{=}CH_2$ → (*a* *a** H—H, Pd/C (cat.)) → Ph_2CH–CH_2–H (*a*, *b*, *c*, *a**)

$$\Delta\ Ox = (1-0)_a + (1-0)_{a*} + (-1-1)_b + (-3-(-3))_c = 1 + 1 - 2 = 0$$

Example 3

HO–CPh_2–CH_3 (*b*, *c*) → ($MeOOC{-}N^{-}{-}SO_2{-}N^{+}Et_3$; → $MeOOC{-}NH{-}SO_2{-}O^{-}$ Et_3NH^{+}) → $Ph_2C{=}CH_2$ → (*a* *a** H—H, Pd/C (cat.)) → Ph_2CH–CH_2–H (*a*, *b*, *c*, *a**)

$$\Delta\ Ox = (1-0)_a + (1-0)_{a*} + (-1-1)_b + (-3-(-3))_c = 1 + 1 - 2 = 0$$

Fragmentation-Substitution

Br–CH_2(*d*)–CH=CH(*c*)–$(CH_2)_4$–NH_2(*b*) → (KOH; - H_2O, - KBr) → 2-vinylpyrrolidine (N–H) → (*a* CH_3Br; - HBr) → N-methyl-2-vinylpyrrolidine (*a* CH_3, *b* N, *c*, *d*)

$$\Delta\ Ox = (-2-(-2))_a + (-3-(-3))_b + (0-(-1))_c + (-2-(-1))_d = 1 - 1 = 0$$

Fragmentation-Rearrangement

Example 1

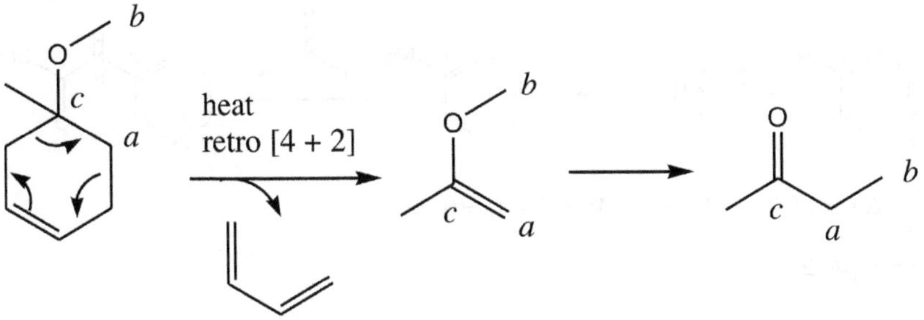

heat
retro [4 + 2]

Δ Ox = (–2 – (–2))$_a$ + (–3 – (–2))$_b$ + (2 – 1)$_c$ = –1 + 1 = 0

Example 2

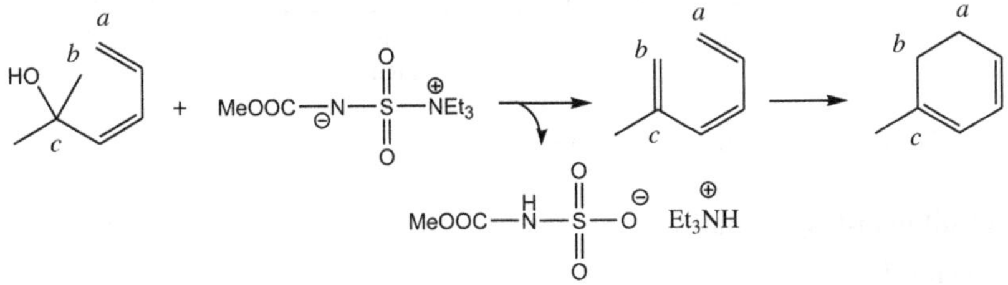

HO + $MeOOC-\overset{\ominus}{N}-SO_2-\overset{\oplus}{N}Et_3$ → $MeOOC-NH-SO_2-O^{\ominus}$ $Et_3\overset{\oplus}{N}H$

Δ Ox = (–2 – (–2))$_a$ + (–2 – (–3))$_b$ + (0 – 1)$_c$ = 1 – 1 = 0

Example 3

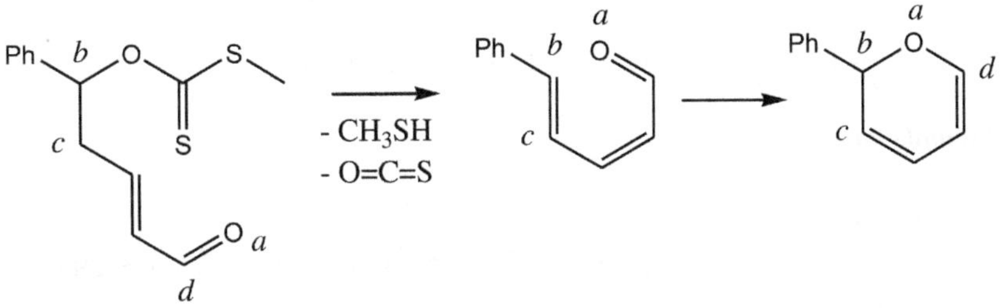

- CH_3SH
- O=C=S

Δ Ox = (–2 – (–2))$_a$ + (0 – 0)$_b$ + (–1 – (–2))$_c$ + (0 – 1)$_d$ = 1 – 1 = 0

Example 4

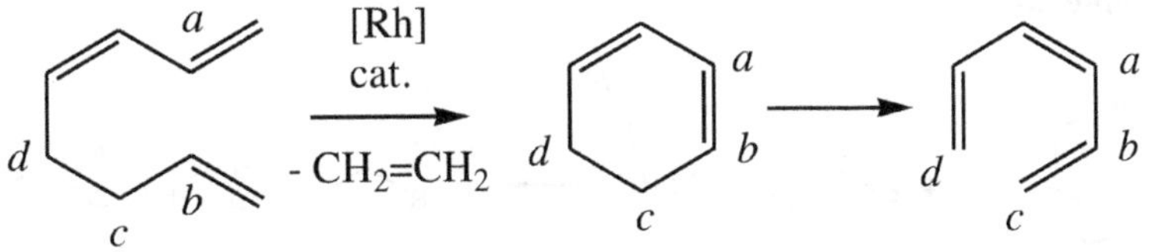

[Rh]
cat.
- CH_2=CH_2

Δ Ox = (–1 – (–1))$_a$ + (–1 – (–1))$_b$ + (–2 – (–2))$_c$ + (–2 – (–2))$_d$ = 0

Substitution-Addition

Example 1

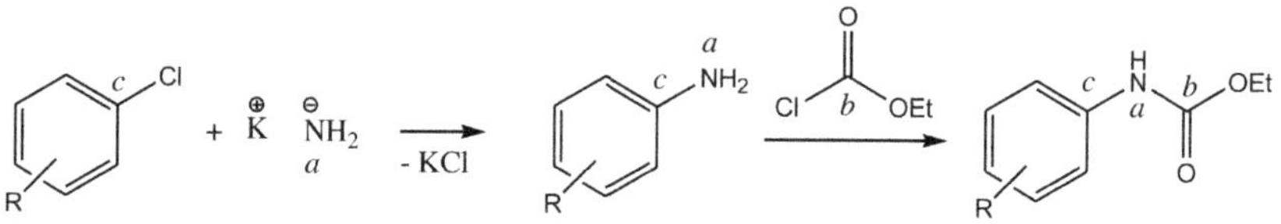

$$\Delta\ Ox = (-3-(-3))_a + (4-4)_b + (1-1)_c = 0$$

Example 2

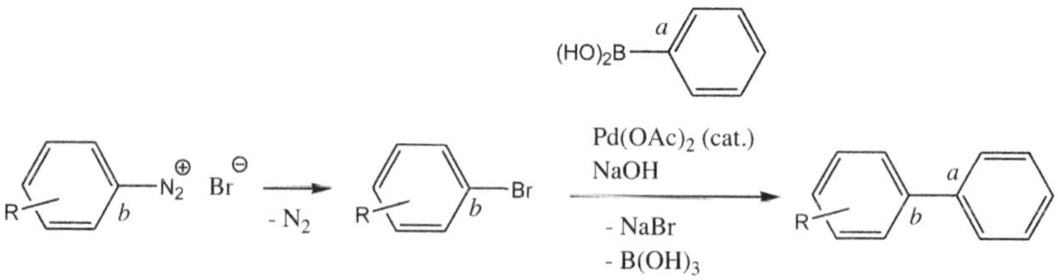

$$\Delta\ Ox = (0-(-1))_a + (0-1)_b = 1-1 = 0$$

Substitution-Fragmentation

Example 1

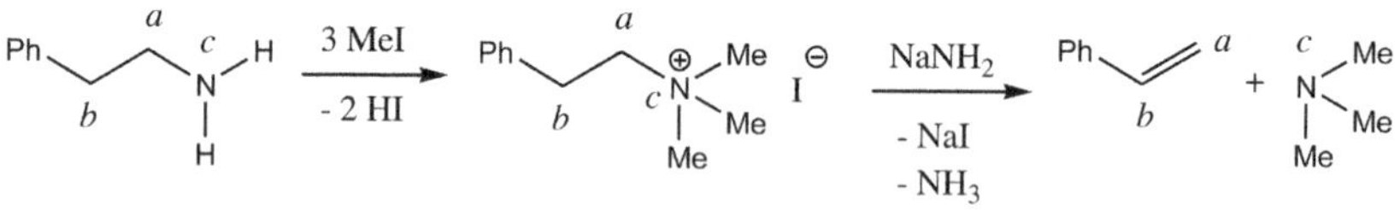

$$\Delta\ Ox = (-2-(-1))_a + (-1-(-2))_b + (-3-(-3))_c = -1+1 = 0$$

Example 2a

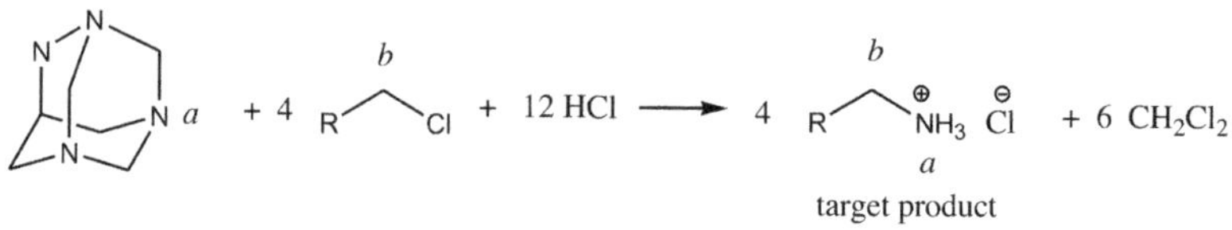

$$\Delta\ Ox = (-3-(-3))_a + (-1-(-1))_b = 0$$

Example 2b

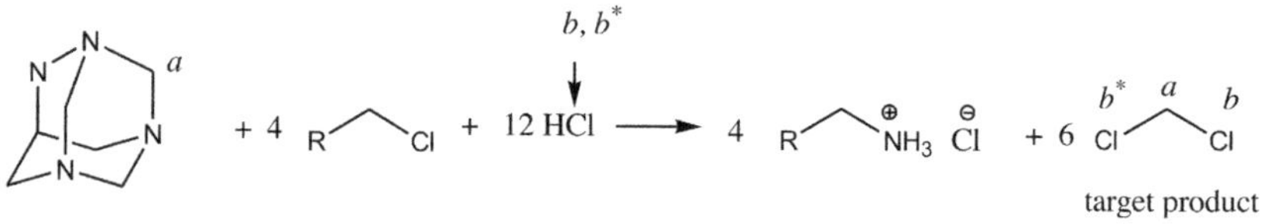

$$\Delta\ Ox = (0-0)_a + (-1-(-1))_b + (-1-(-1))_{b^*} = 0$$

Substitution-Rearrangement

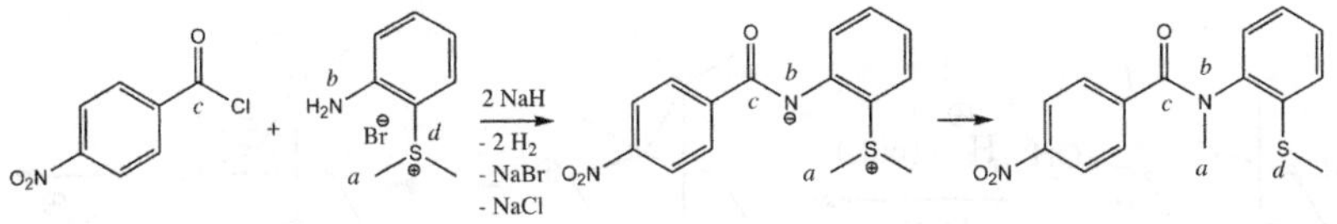

Δ Ox = $(-2-(-2))_a + (-3-(-3))_b + (3-3)_c + (-2-(-2))_d = 0$

Rearrangement-Addition

Example 1

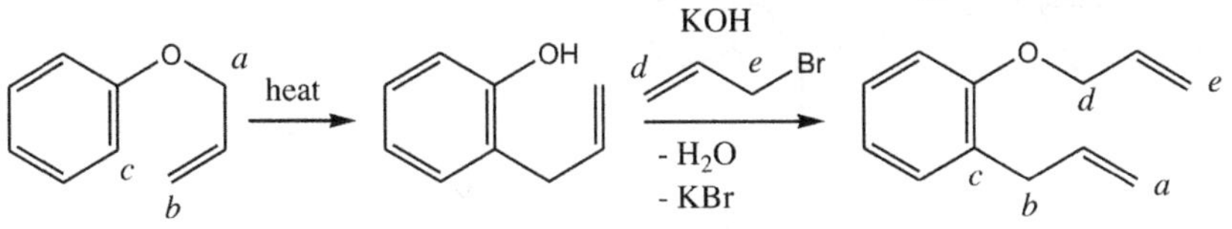

Δ Ox = $(-2-(-1))_a + (-2-(-2))_b + (0-(-1))_c + (-1-(-2))_d + (-2-(-1))_e$
$= -1 + 1 + 1 - 1 = 0$

Example 2

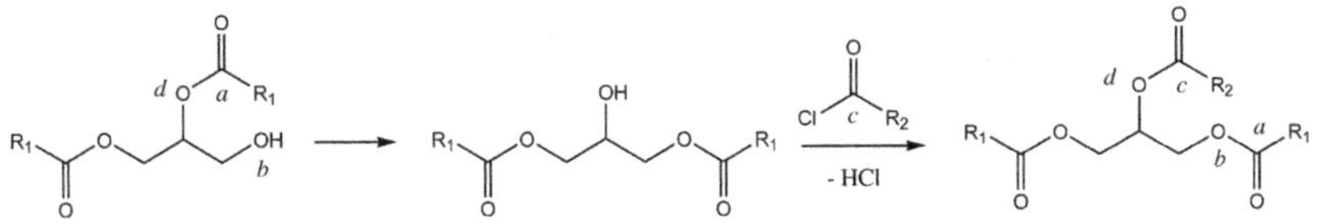

Δ Ox = $(3-3)_a + (-2-(-2))_b + (3-3)_c + (-2-(-2))_d = 0$

Example 3

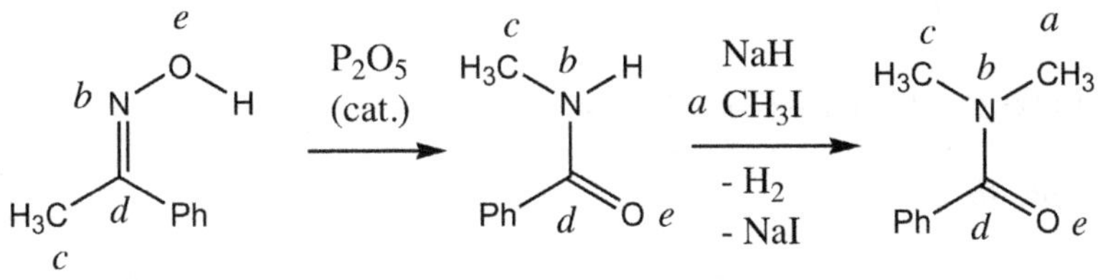

Δ Ox = $(-2-(-2))_a + (-3-(-1))_b + (-2-(-3))_c + (3-(2))_d + (-2-(-2))_e = -2 + 1 + 1 = 0$

Rearrangement-Fragmentation

Example 1

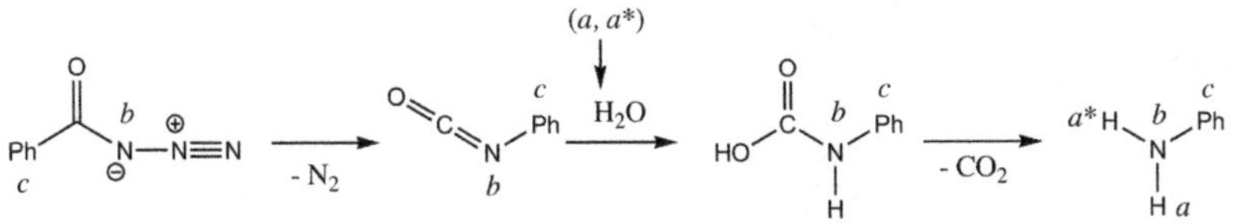

Δ Ox = $(1-1)_a + (1-1)_{a*} + (-3-(-2))_b + (1-0)_c = -1 + 1 = 0$

Example 2

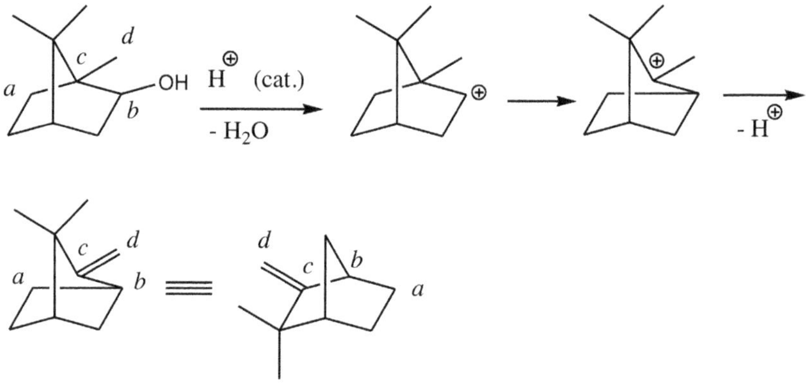

$$\Delta \text{ Ox} = (-2-(-2))_a + (-1-0)_b + (0-0)_c + (-2-(-3))_d = -1+1 = 0$$

Example 3a

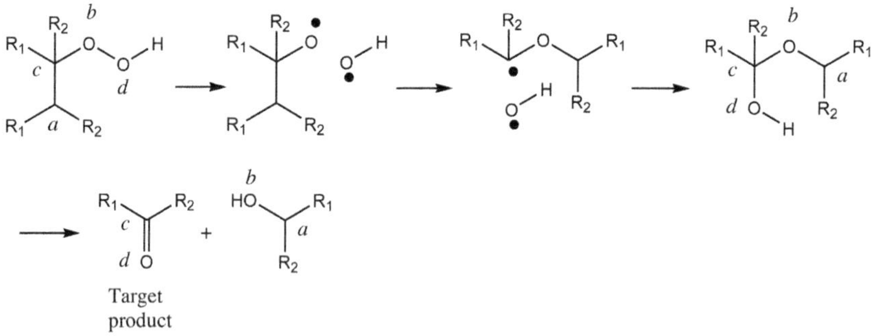

Target product

$$\Delta \text{ Ox} = (2-1)_c + (-2-(-1))_d = 1-1 = 0$$

Example 3b

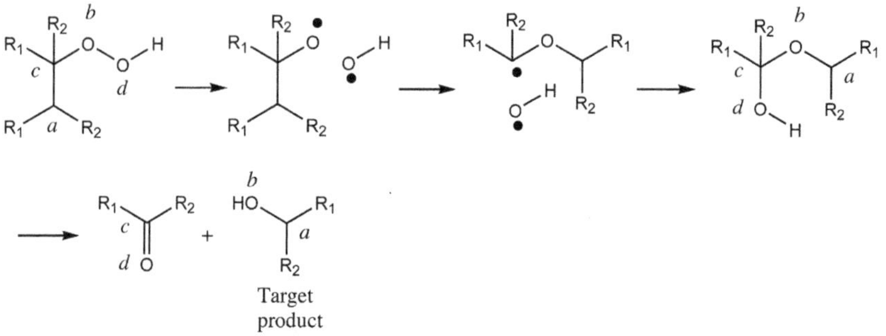

Target product

$$\Delta \text{ Ox} = (0-(-1))_a + (-2-(-1))_b = 1-1 = 0$$

Example 4a

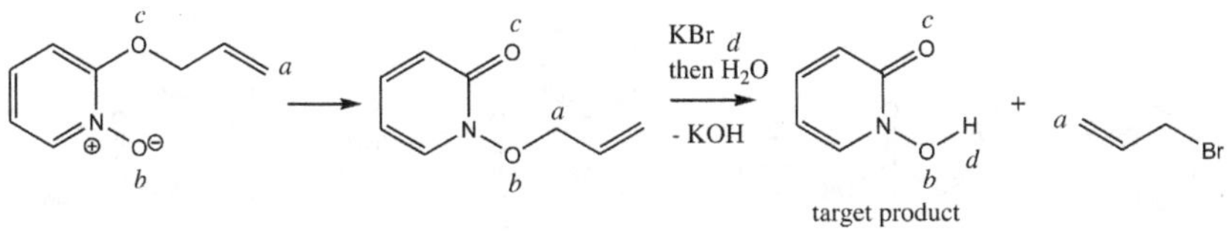

KBr *d* then H_2O - KOH

target product

$$\Delta\ \mathrm{Ox} = (-2-(-2))_b + (-2-(-2))_c + (1-1)_d = 0$$

Example 4b

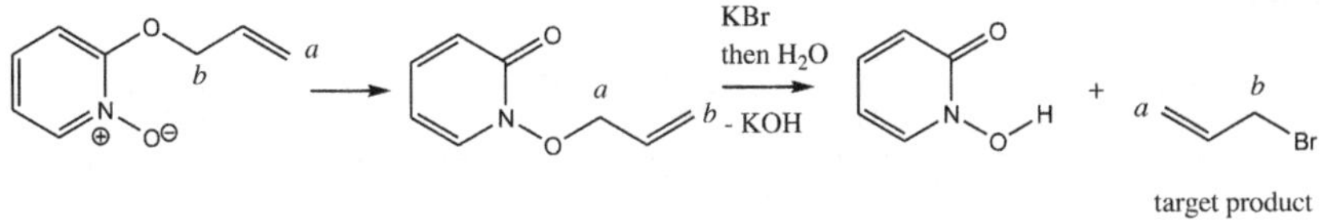

KBr then H_2O - KOH

target product

$$\Delta\ \mathrm{Ox} = (-2-(-2))_a + (-1-(-1))_b = 0$$

Rearrangement-Substitution

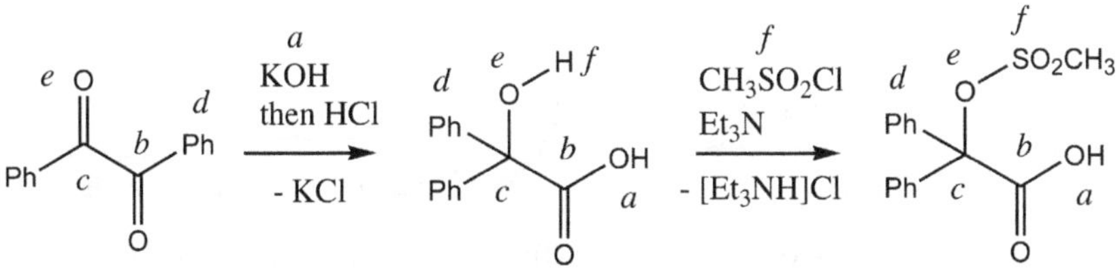

a KOH then HCl - KCl

f CH_3SO_2Cl Et_3N - $[Et_3NH]Cl$

$$\Delta\ \mathrm{Ox} = (-2-(-2))_a + (3-2)_b + (1-2)_c + (0-0)_d + (-2-(-2))_e + (4-4)_f = 1-1 = 0$$

(ii) Positive change in philicity (oxidation)

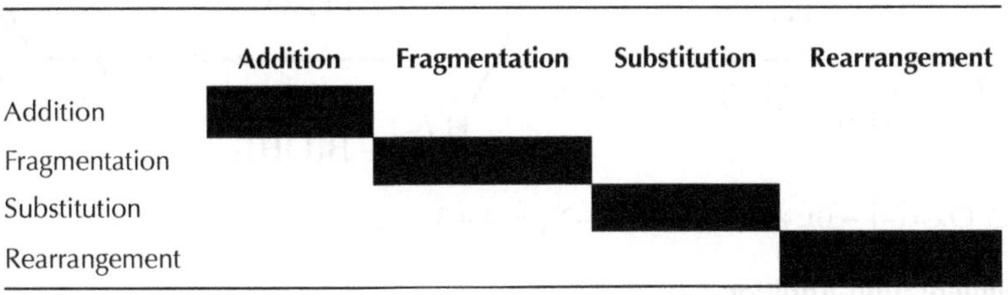

	Addition	Fragmentation	Substitution	Rearrangement
Addition				
Fragmentation				
Substitution				
Rearrangement				

Addition-Fragmentation

- H_2O - NH_3

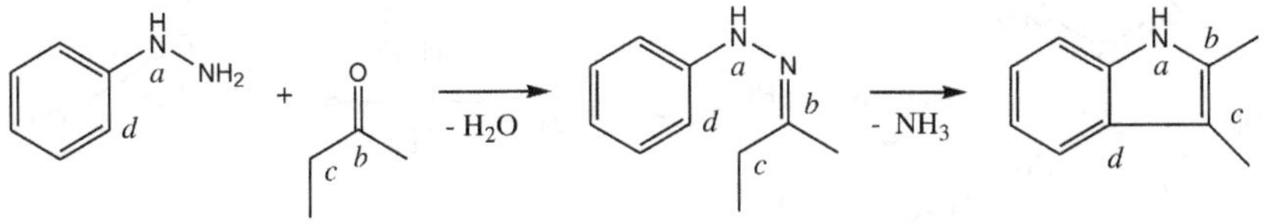

$$\Delta\ \mathrm{Ox} = (-3-(-2))_a + (1-2)_b + (0-(-2))_c + (0-(-1))_d = -1-1+2+1 = +1$$

Addition-Substitution

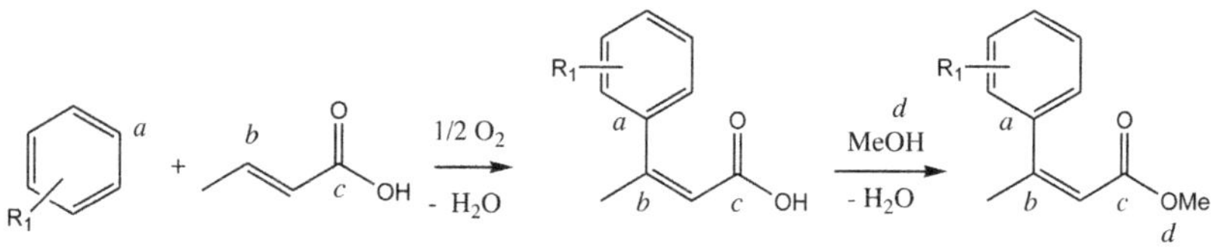

Δ Ox = $(0-(-1))_a + (0-(-1))_b + (3-3)_c + (-2-(-2))_d = 1+1 = +2$

Addition-Rearrangement

Example 1

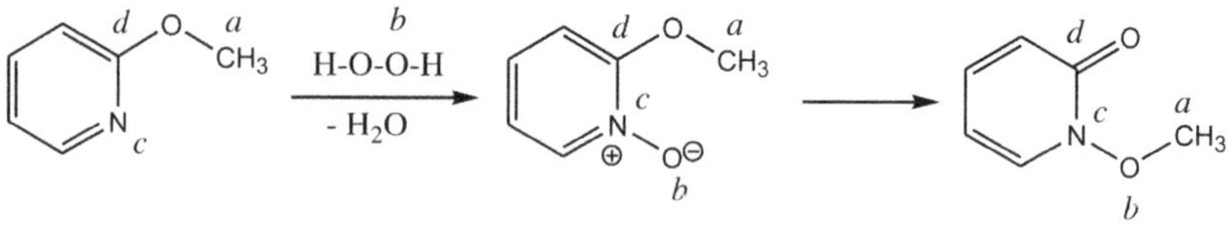

Δ Ox = $(-2-(-2))_a + (-2-(-1))_b + (-1-(-3))_c + (3-2)_d = -1+2+1 = +2$

Example 2

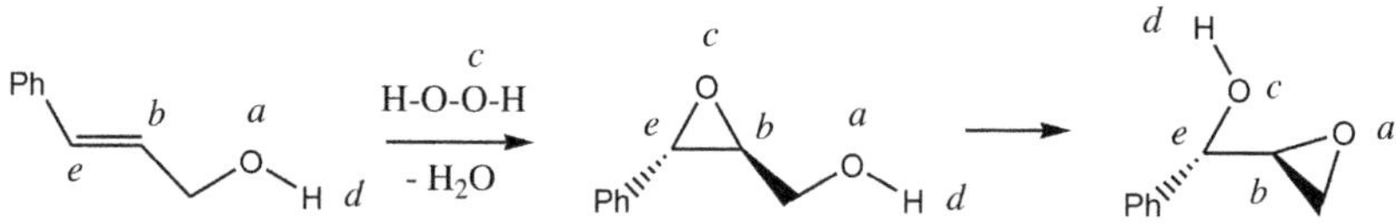

Δ Ox = $(-2-(-2))_a + (0-(-1))_b + (-2-(-1))_c + (1-1)_d + (0-(-1))_e = 1-1+1 = +1$

Example 3

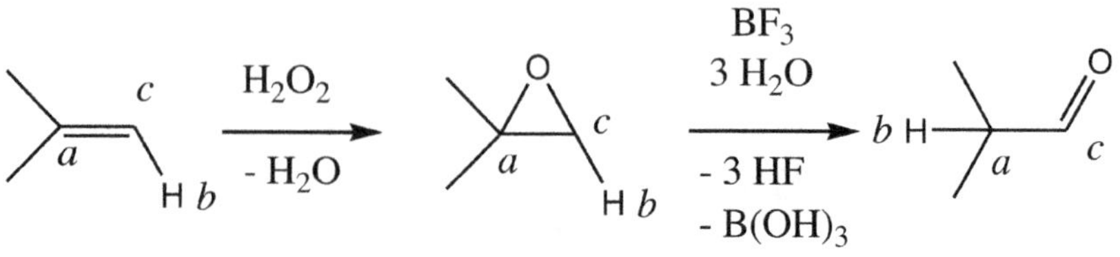

Δ Ox = $(-1-0)_a + (1-1)_b + (1-(-2))_c = -1+3 = +2$

Fragmentation-Addition

Example 1

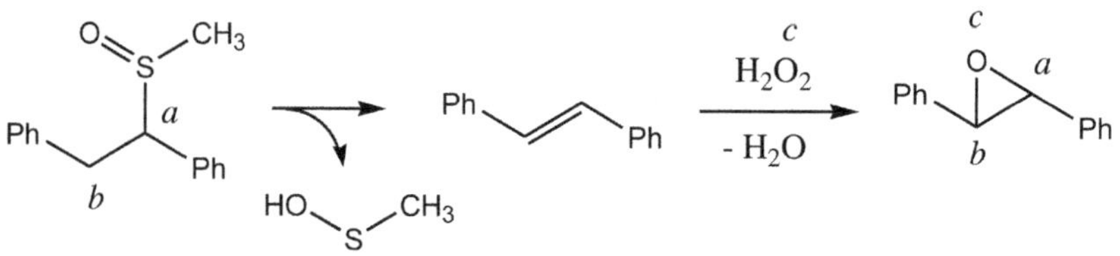

Δ Ox = $(0-0)_a + (0-(-2))_b + (-2-(-1))_c = 2-1 = +1$

Example 2

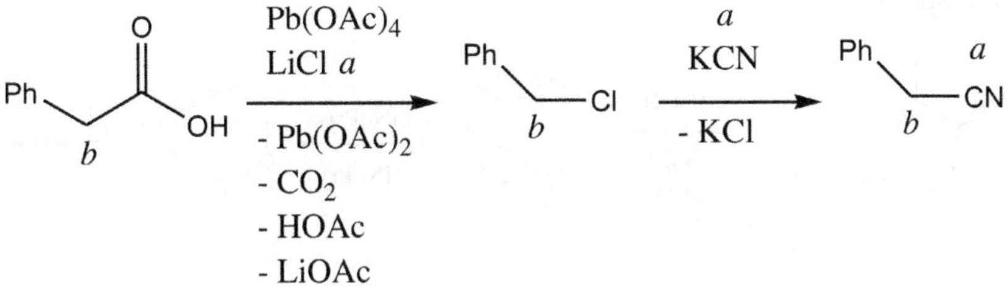

Ph–CH$_2$(*b*)–C(=O)OH → [$Pb(OAc)_4$, LiCl *a*; − $Pb(OAc)_2$, − CO_2, − HOAc, − LiOAc] → Ph–CH$_2$(*b*)–Cl → [*a* KCN; − KCl] → Ph–CH$_2$(*b*)–CN(*a*)

$$\Delta \text{Ox} = (3-2)_a + (-2-(-2))_b = +1$$

Example 3

Ph–C(*c*)(=O)–CH$_3$ → [*b* 4 Na-O-H, 3 Br_2; − 3 NaX, − $CHBr_3$, − 3 H_2O] → Ph–C(*c*)(=O)–O$^{\ominus}$(*b*) $Na^{\oplus}$ → [Cl–CH$_2$(*a*)–Ph; − NaCl] → Ph–C(*c*)(=O)–O(*b*)–CH$_2$(*a*)–Ph

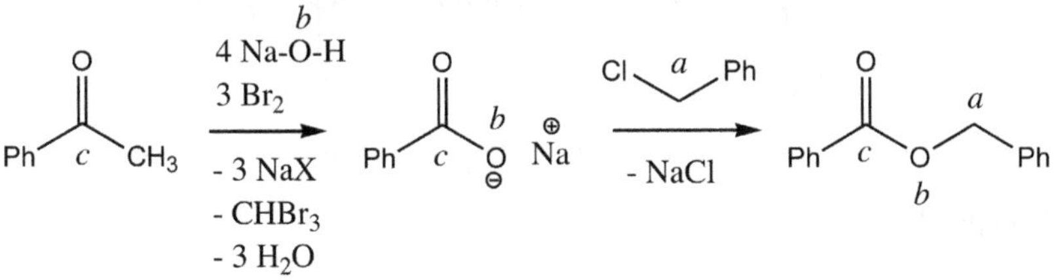

$$\Delta \text{Ox} = (-1-(-1))_a + (-2-(-2))_b + (3-2)_c = +1$$

Fragmentation-Substitution

Example 1

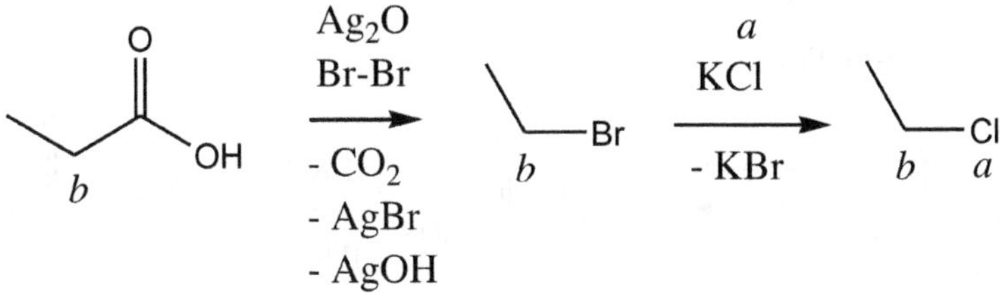

CH$_3$–CH$_2$(*b*)–C(=O)OH → [Ag_2O, Br-Br; − CO_2, − AgBr, − AgOH] → CH$_3$–CH$_2$(*b*)–Br → [*a* KCl; − KBr] → CH$_3$–CH$_2$(*b*)–Cl(*a*)

$$\Delta \text{Ox} = (-1-(-1))_a + (-1-(-2))_b = +1$$

Example 2

CH$_3$–CH$_2$(*b*)–C(=O)OH → [Ag_2O, Br-Br; − CO_2, − AgBr, − AgOH] → CH$_3$–CH$_2$(*b*)–Br → [*a* Bu_3SnH; − Bu_3SnBr] → CH$_3$–CH$_2$(*b*)–H(*a*)

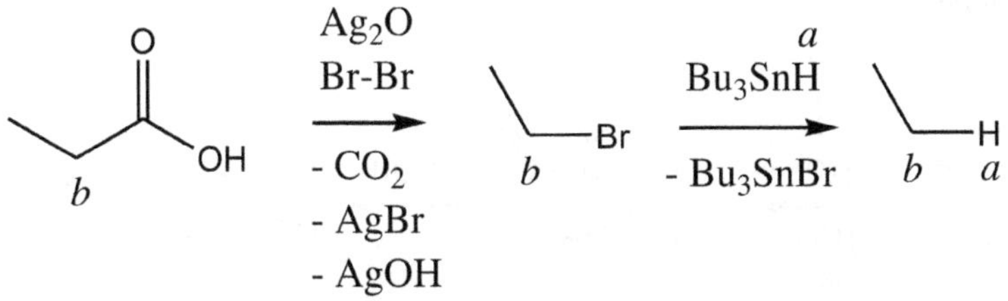

$$\Delta \text{Ox} = (1-(-1))_a + (-3-(-2))_b = 2 - 1 = +1$$

Fragmentation-Rearrangement

Example 1

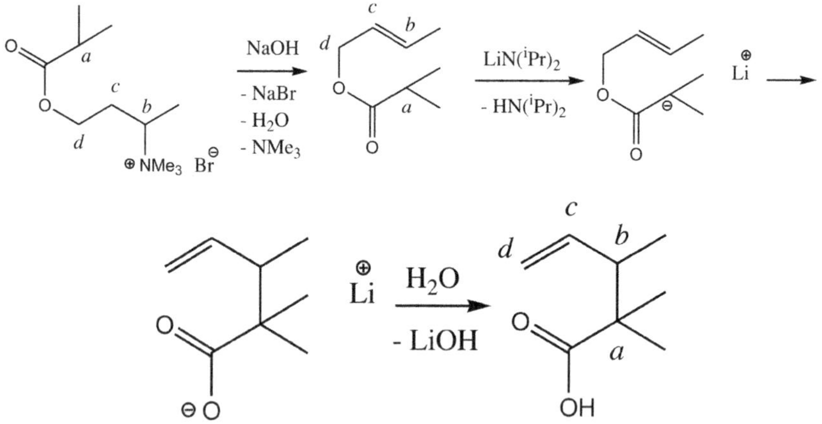

NaOH, - NaBr, - H_2O, - NMe_3; $LiN(^iPr)_2$, - $HN(^iPr)_2$; H_2O, - LiOH

$$\Delta\ Ox = (0-(-1))_a + (-1-(-1))_b + (-1-(-2))_c + (-2-(-1))_d = 1+1-1 = +1$$

Example 2

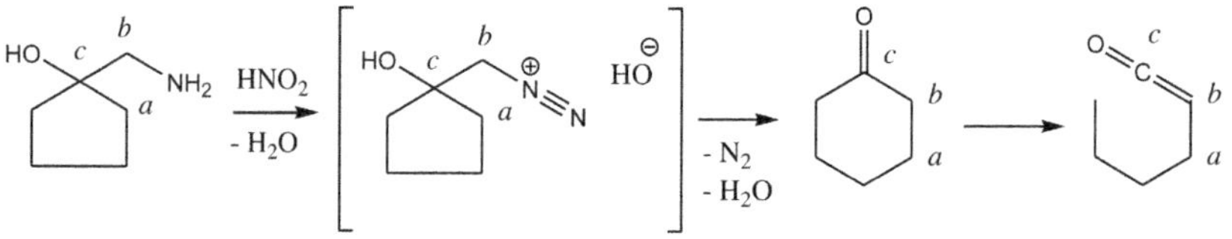

HNO_2, - H_2O; - N_2, - H_2O

$$\Delta\ Ox = (-2-(-2))_a + (-1-(-1))_b + (2-1)_c = +1$$

Substitution-Addition

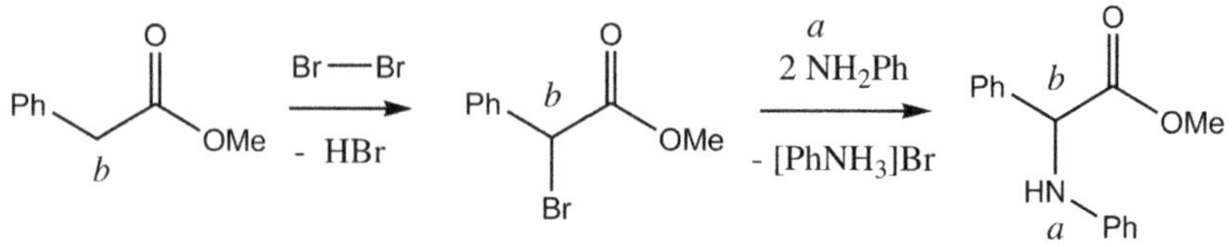

Br—Br, - HBr; 2 NH_2Ph, - $[PhNH_3]Br$

$$\Delta\ Ox = (-3-(-3))_a + (0-(-2))_b = +2$$

Substitution-Fragmentation

Example 1

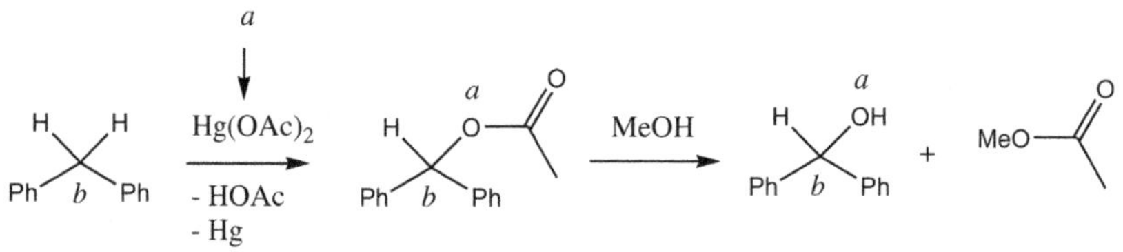

$Hg(OAc)_2$, - HOAc, - Hg; MeOH

$$\Delta\ Ox = (-2-(-2))_a + (0-(-2))_b = +2$$

Example 2

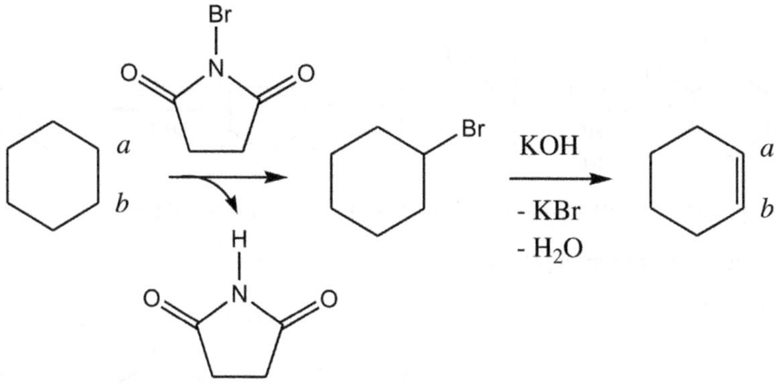

Δ Ox = $(-1-(-2))_a+(-1-(-2))_b=1+1=+2$

Example 3

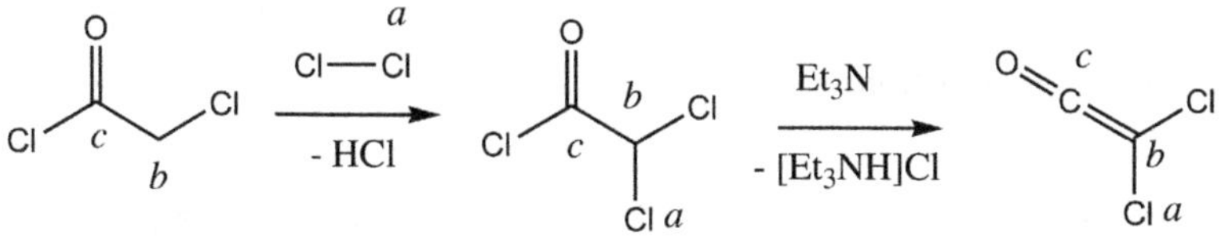

Δ Ox = $(-1-0)_a+(2-(-1))_b+(2-3)_c=-1+3-1=+1$

Substitution-Rearrangement

Example 1

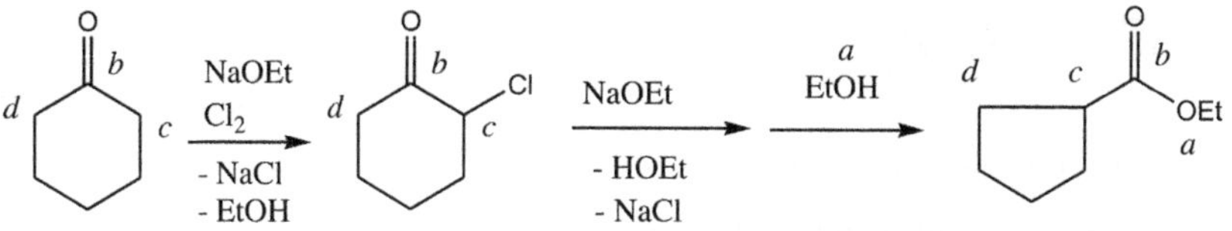

Δ Ox = $(-2-(-2))_a+(3-2)_b+(-1-(-2))_c+(-2-(-2))_d=1+1=+2$

Example 2

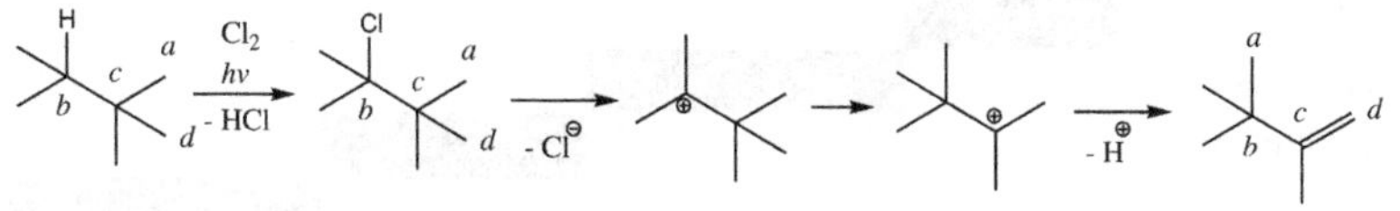

Δ Ox = $(-3-(-3))_a+(0-(-1))_b+(0-0)_c+(-2-(-3))_d=1+1=+2$

Example 3

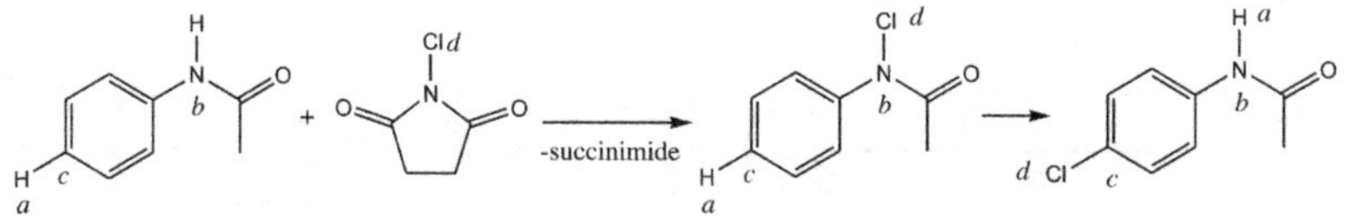

Δ Ox = $(1-1)_a+(-3-(-3))_b+(1-(-1))_c+(-1-(-1))_d=+2$

Rearrangement-Addition

KO^tBu; $- HO^tBu$; $- KCl$ → $- SO_2$ (d) → H_2O_2 (c); $- H_2O$

$\Delta\ Ox = (0 - 1)_a + (0 - (-1))_b + (-2 - (-1))_c + (4 - 2)_d = -1 + 1 - 1 + 2 = +1$

Rearrangement-Fragmentation

→ 4 Na-O-H (e); 3 Cl_2; - 3 NaCl; - $CHCl_3$; - 3 H_2O

$\Delta\ Ox = (-2 - (-2))_a + (3 - 0)_b + (-1 - 0)_c + (1 - 1)_d + (-2 - (-2))_e = 3 - 1 = +2$

Rearrangement-Substitution

→ $NaNO_2$; 2 HCl (a); - 2 NaCl; - 2 H_2O → - N_2

$\Delta\ Ox = (-1 - (-1))_a + (1 - 1)_b + (1 - (-1))_c + (-2 - (-2))_d = +2$

(iii) Negative change in philicity (reduction)

	Addition	Fragmentation	Substitution	Rearrangement
Addition	■			
Fragmentation		■		
Substitution			■	
Rearrangement				■

Addition-Fragmentation

Example 1

- HCl → Na_2CO_3; - N_2; - $NaHCO_3$; - Me-C_6H_4-SO_2Na

$\Delta\ Ox = (1 - 1)_a + (1 - 3)_b = -2$

Example 2

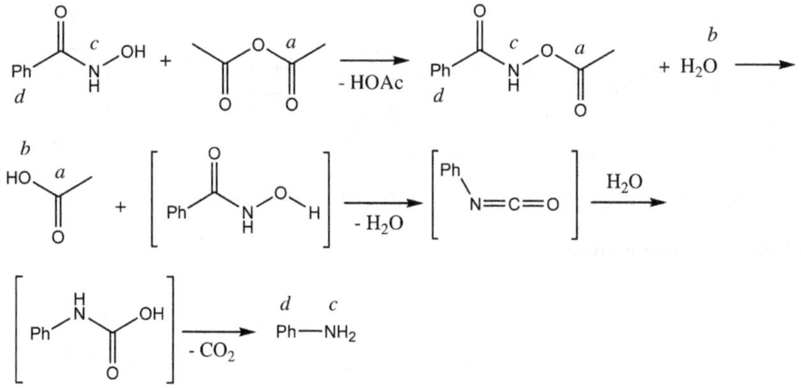

$$\Delta\,Ox = (-3 - (-1))_a + (1 - 0)_d = -2 + 1 = -1$$

Example 3

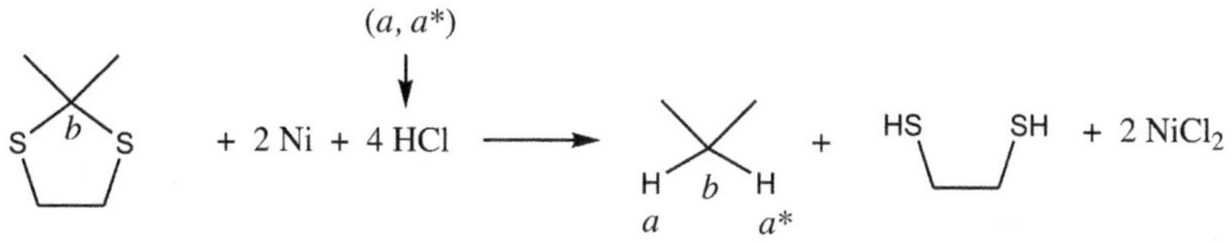

$$\Delta\,Ox = (1 - 1)_a + (1 - 1)_{a*} + (-2 - 2)_b = -4$$

Example 4

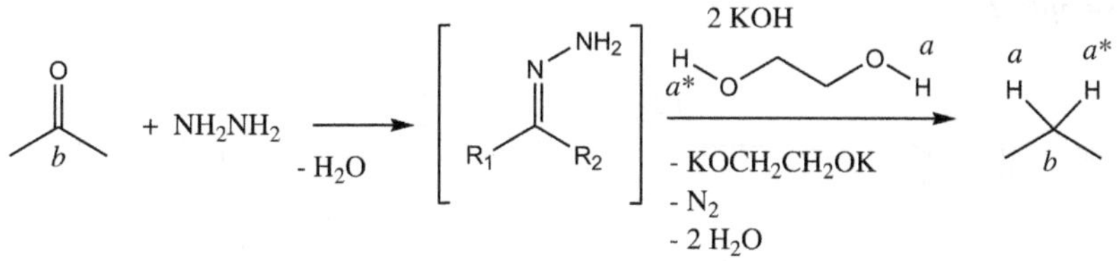

$$\Delta\,Ox = (1 - 1)_a + (1 - 1)_{a*} + (-2 - 2)_b = -4$$

Addition-Substitution

Example 1

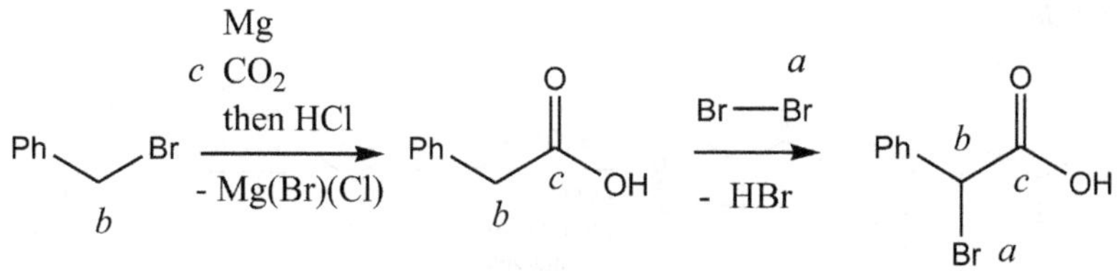

$$\Delta\,Ox = (-1 - 0)_a + (0 - (-1))_b + (3 - 4)_c = -1 + 1 - 1 = -1$$

Example 2

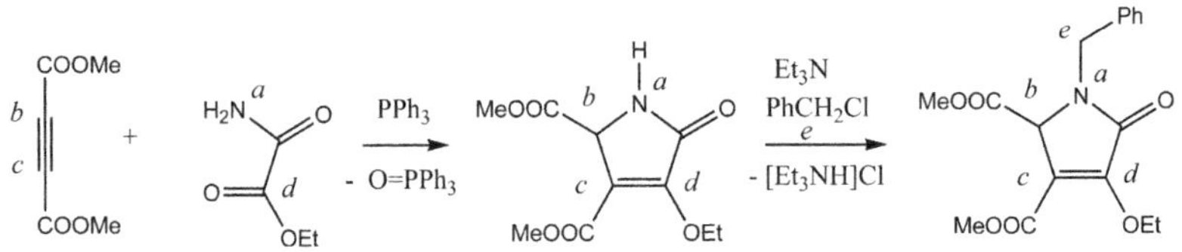

Δ Ox = $(-3-(-3))_a + (0-0)_b + (0-0)_c + (0-3)_d + (-1-(-1))_e = -3$

Addition-Rearrangement

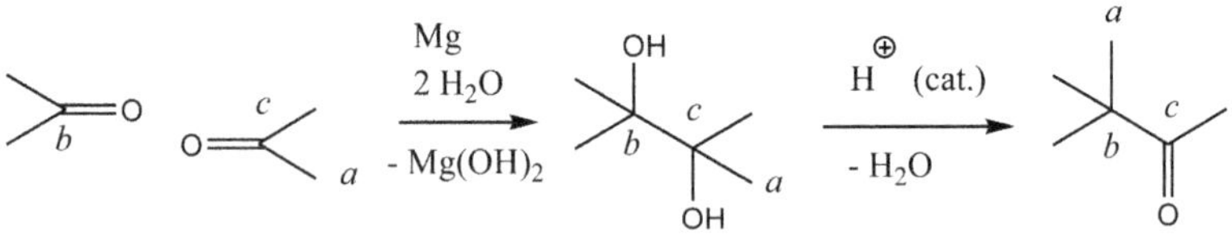

Δ Ox = $(-3-(-3))_a + (0-2)_b + (2-2)_c = -2$

Fragmentation-Addition

Example 1

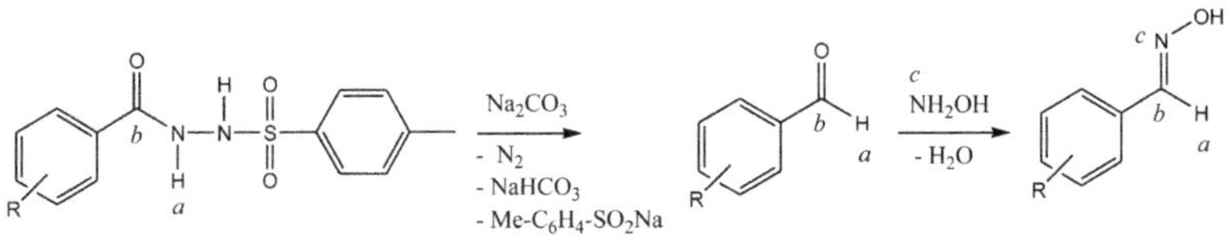

Δ Ox = $(1-1)_a + (1-3)_b + (-1-(-1))_c = -2$

Example 2

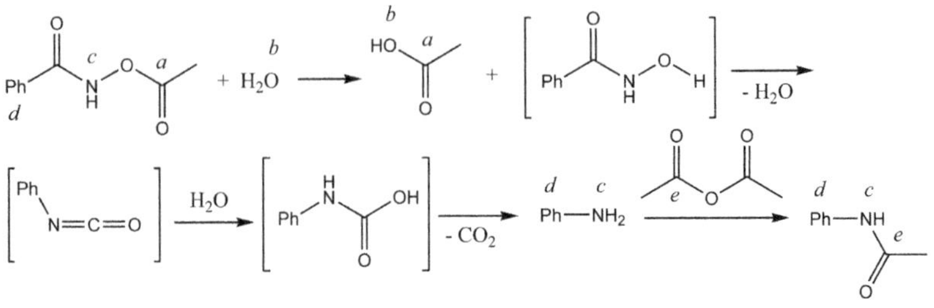

Δ Ox = $(-3-(-1))_c + (1-0)_d + (3-3)_e = -2 + 1 = -1$

Example 3

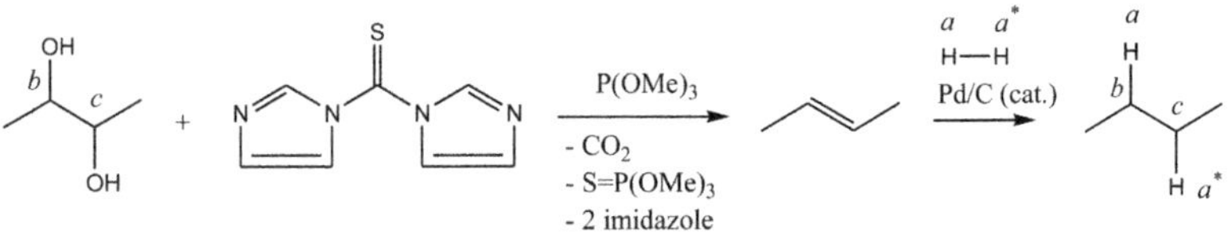

Δ Ox = $(1-0)_a + (1-0)_{a*} + (-2-0)_b + (-2-0)_c = 1 + 1 - 2 - 2 = -2$

Example 4

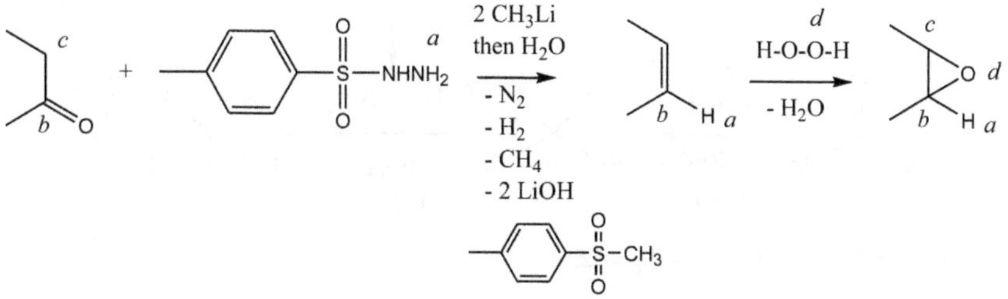

$$\Delta\ Ox = (1-1)_a + (0-2)_b + (0-(-2))_c + (-2-(-1))_d = -2+2-1 = -1$$

Example 5

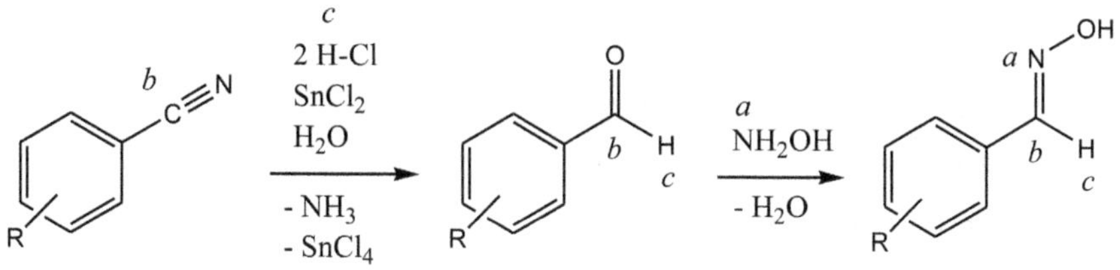

$$\Delta\ Ox = (-1-(-1))_a + (1-3)_b + (1-1)_c = -2$$

Fragmentation-Substitution

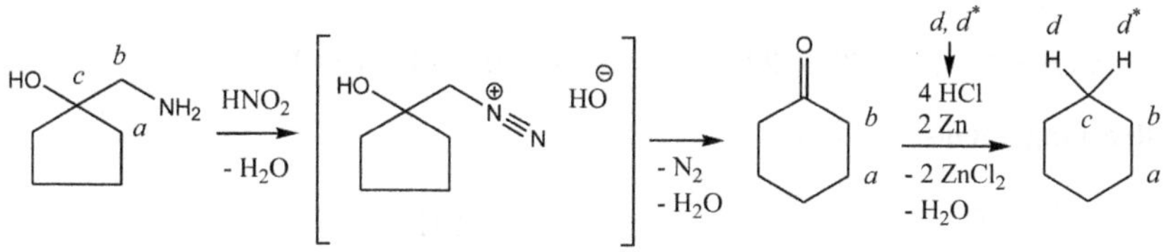

$$\Delta\ Ox = (-2-(-2))_a + (-2-(-1))_b + (-2-1)_c + (1-1)_d + (1-1)_{d^*} = -1-3 = -4$$

Fragmentation-Rearrangement

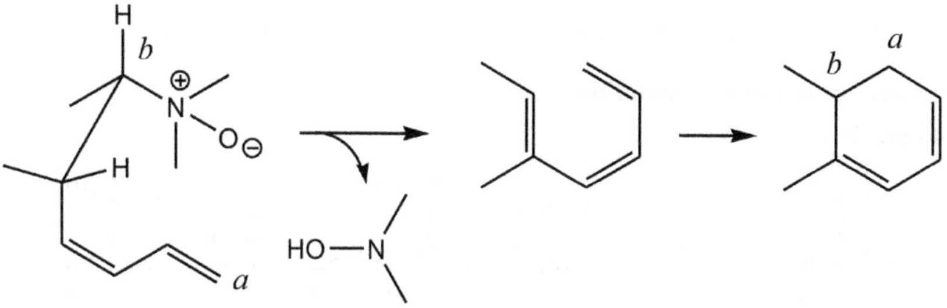

$$\Delta\ Ox = (-2-(-2))_a + (-1-0)_b = -1$$

Substitution-Addition

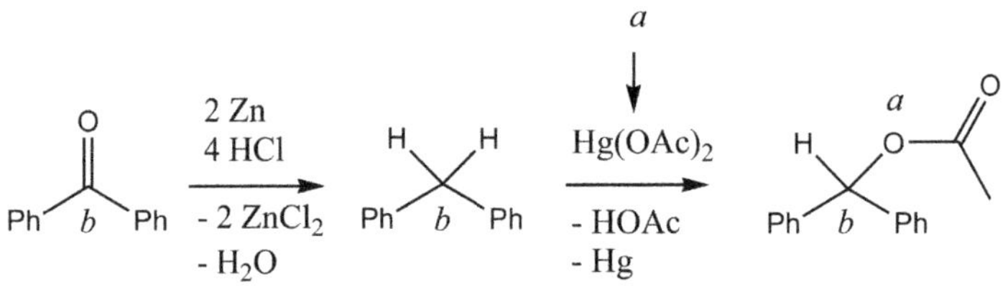

Δ Ox = $(-2-(-2))_a+(0-2)_b=-2$

Substitution-Fragmentation

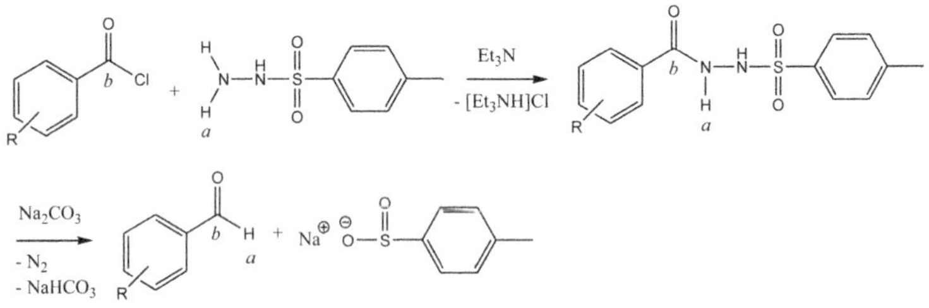

Δ Ox = $(1-1)_a+(1-3)_b=-2$

Substitution-Rearrangement

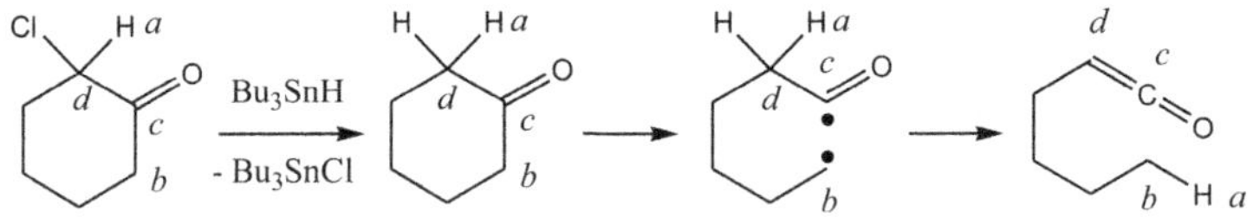

Δ Ox = $(1-1)_a+(-3-(-2))_b+(2-2)_c+(-1-0)_d=-1-1=-2$

Rearrangement-Addition

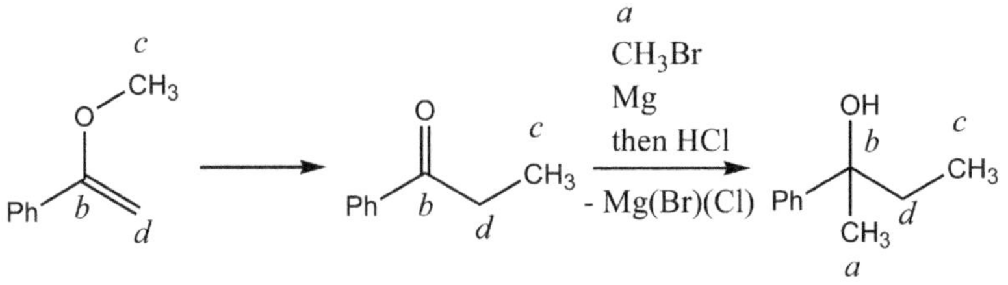

Δ Ox = $(-3-(-2))_a+(1-1)_b+(-3-(-2))_c+(-2-(-2))_d=-1-1=-2$

Rearrangement-Fragmentation

Example 1

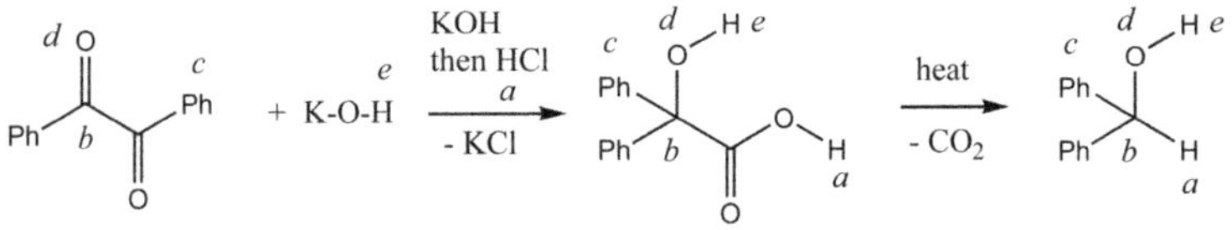

Δ Ox = $(1-1)_a+(0-2)_b+(0-0)_c+(-2-(-2))_d+(1-1)_f=-2$

Example 2

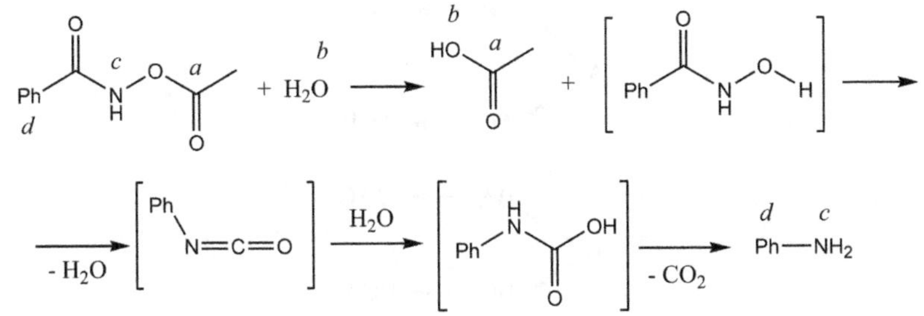

$$\Delta\, Ox = (-3 - (-1))_c + (1 - 0)_d = -2 + 1 = -1$$

Rearrangement-Substitution

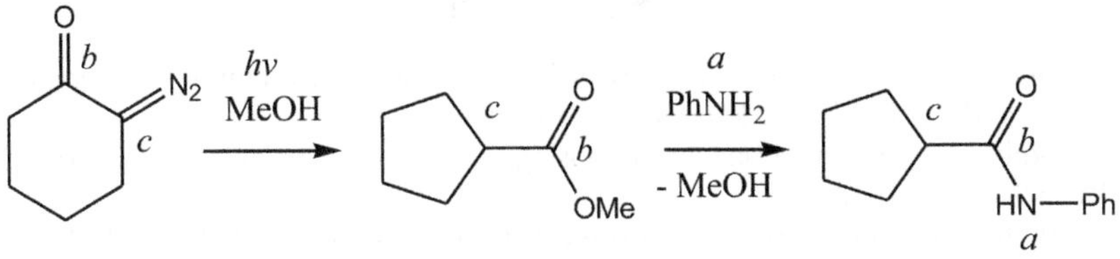

$$\Delta\, Ox = (-3 - (-3))_a + (3 - 2)_b + (-1 - 2)_c = 1 - 3 = -2$$

Example 4.3

Themes: multi-component reactions, order of addition of reagents

Multi-component reactions consist of at least three reagents coming together to form a product. Sometimes the order of addition of the reagents leads to different product outcomes. In other cases, adding all the reagents at once leads to one unique product.

Part 1

For a general three-component coupling reaction, how many ways can the reagents be added in a sequential fashion?

Part 2

For a general four-component coupling reaction, how many ways can the reagents be added in a sequential fashion?

Part 3

Generalize the results for an *N*-component coupling reaction.

Solution

Part 1

A general three-component coupling reaction is shown below involving reagents A, B, and C.

$$A + B + C \rightarrow P$$

Possible permutations:

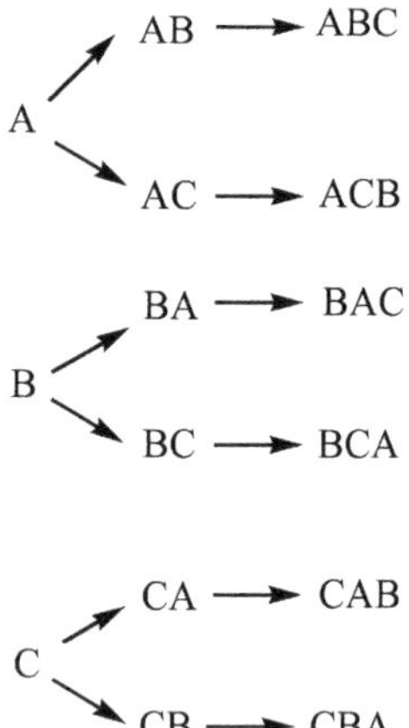

There are three possible ways to order the sequential addition of reagents, which yield six possible ordered linear sequences.

Number of combinations for choosing 2 reagents out of 3 = C(3, 2) = 3!/(2!1!) = 3*2/(2*1*1) = 3

Number of combinations for choosing 1 reagent out of 1 = C(1, 1) = 1!/(1!0!) = 1

Total number of combinations = 3*1 = 3

Number of ways of ordering 3 items = 3*2*1 = 6

Part 2

A general four-component coupling reaction is shown below involving reagents A, B, C, and D.

A + B + C + D → P

Possible permutations:

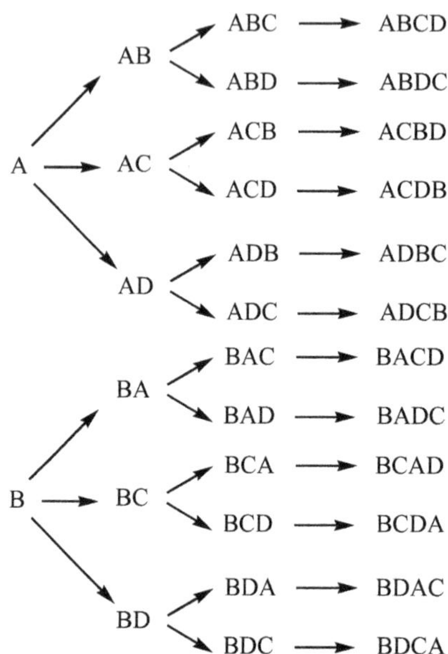

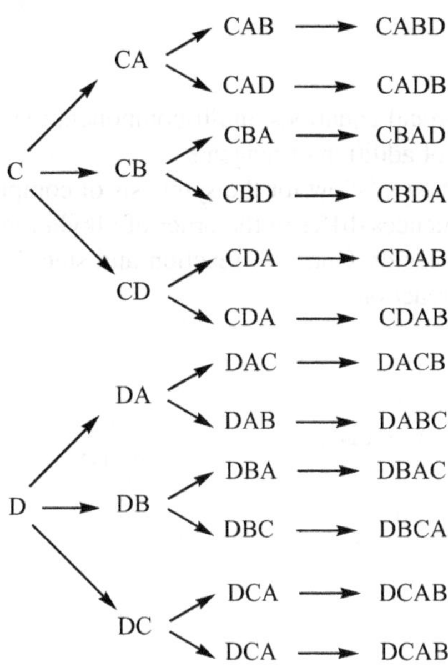

There are 12 possible ways to order the sequential addition of reagents, which yield 24 possible ordered linear sequences.

Number of combinations for choosing 2 reagents out of 4 = $C(4, 2) = 4!/(2!2!) = 4*3*2/(2*1*2*1) = 6$

Number of combinations for choosing 1 reagent out of 2 = $C(2, 1) = 2!/(1!1!) = 2$

Number of combinations for choosing 1 reagent out of 1 = $C(1, 1) = 1!/(1!0!) = 1$

Total number of combinations = 6*2*1 = 12

Number of ways of ordering 3 items = 4*3*2*1 = 24

Part 3

For an N-component reaction:

Number of combinations for choosing 2 reagents out of $N = C(N, 2)$

Number of combinations for choosing 1 reagent out of $N-2 = C(N-2, 1)$

Number of combinations for choosing 1 reagent out of $N-3 = C(N-3, 1)$

Number of combinations for choosing 1 reagent out of $N-4 = C(N-4, 1)$

…

Number of combinations for choosing 1 reagent out of $1 = C(1, 1)$

Total number of combinations

$$C(N,2)\prod_{j=1}^{N-2} C(j,1) = \frac{N!}{2!(N-2)!}\left[1*2*3*\ldots*(N-2)\right] = \frac{N!(N-2)!}{2!(N-2)!} = \frac{N!}{2}$$

Number of ways of ordering N items = $N(N-1)(N-2)\ldots1 = N!$

Therefore, for an N-component reaction there are $N!/2$ possible ways to order the sequential addition of reagents, which yield $N!$ possible ordered linear sequences.

4.1.2 Problems

PROBLEM 4.1 [1]

Themes: balancing chemical equations, multi-component reactions, conservation of structural aspect, order of addition of reagents

Two sequences are shown below for the synthesis of complex isomeric heterocyclic structures. The sequences differ in the order of alkylating reagents used in step 1. Step 2 is a Thorpe–Ziegler–Guarashi reaction and step 3 is a tandem Michael–hetero-Thorpe–Ziegler reaction.

Sequence 1

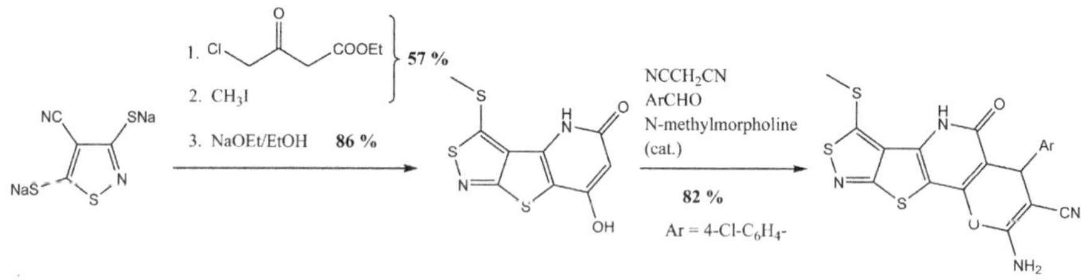

1. Cl–CH$_2$C(O)CH$_2$–COOEt
2. CH_3I
(57 %)
3. NaOEt/EtOH 86 %

$NCCH_2CN$, ArCHO, N-methylmorpholine (cat.), 82 %, Ar = 4-Cl-C_6H_4-

Sequence 2

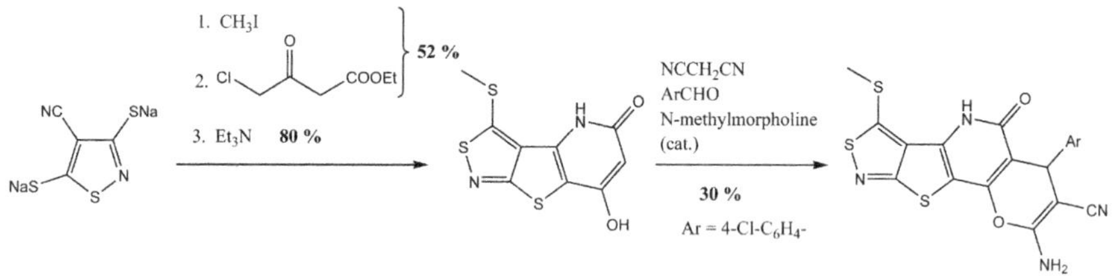

1. CH_3I
2. Cl–CH$_2$C(O)CH$_2$–COOEt
(52 %)
3. Et_3N 80 %

$NCCH_2CN$, ArCHO, N-methylmorpholine (cat.), 30 %, Ar = 4-Cl-C_6H_4-

Part 1

Rewrite each sequence using the principle of conservation of structural aspect with respect to the product highlighting the target bonds made in each step. From the target bond mapping in the product structure, determine the overall ring construction strategy employed.

Part 2

Suggest why the site of alkylation shown below is the more reactive one in the starting material. The authors did not discuss this important point in the paper.

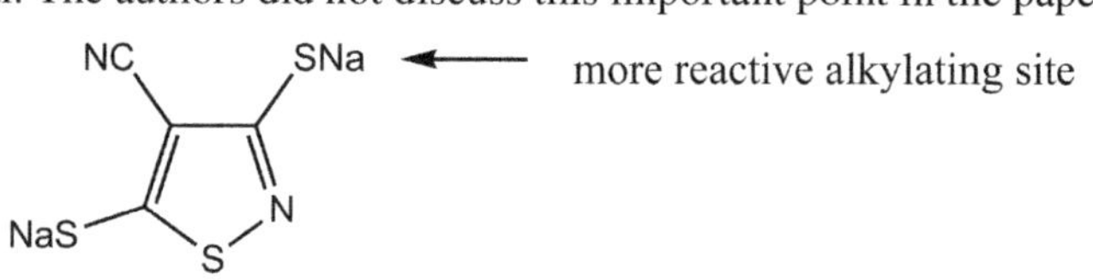

NC, SNa ⟵ more reactive alkylating site; NaS

PROBLEM 4.2 [2]

Themes: balancing chemical equations, multi-component reactions, order of addition of reagents

A four-component reaction involving 2-formylbenzoic acid, malononitrile, an isocyanide, and an amine leads to two different product outcomes depending on the order of addition of reagents as shown below.

Reaction 1

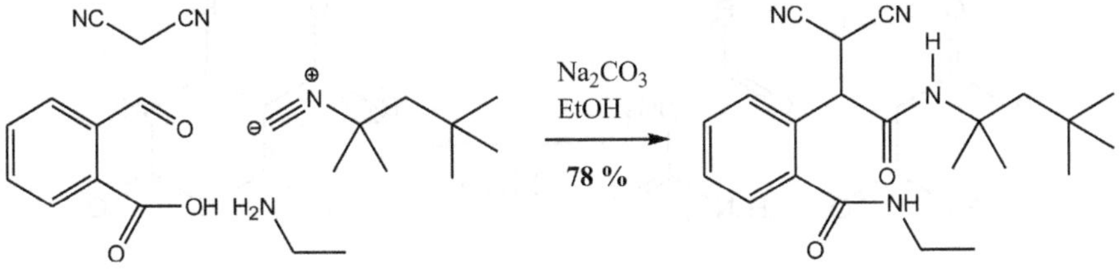

NC CN; OH; O; O; H_2N; ⊖ ⊕ N; Na_2CO_3 EtOH → 78 %; NC CN H N; O; NH; O

Order of addition:

2-formylbenzoic acid + malononitrile + sodium carbonate
then isocyanide
then ethylamine

Reaction 2

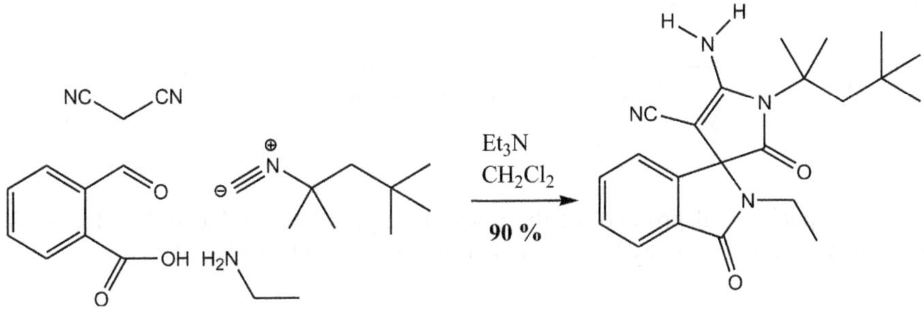

NC CN; O; OH; O; H_2N; ⊖ ⊕ N; Et_3N CH_2Cl_2 → 90 %; H H N; NC; N; O; N; O

Order of addition:

2-formylbenzoic acid + malononitrile + isocyanide + triethylamine
then ethylamine

Part 1

For each reaction, highlight the target bonds made in the product structure, provide balanced chemical equations, and determine the atom economy.

Part 2

Provide a mechanism for reaction 1.

Part 3

In reaction 2, an adduct was isolated after the reaction of 2-formylbenzoic acid, malononitrile, and isocyanide. Show the structure of this adduct before ethylamine is added. Show a mechanism to justify your answer.

PROBLEM 4.3 [3]

Themes: balancing chemical equations, multi-component reactions, order of addition of reagents

The Biginelli reaction produces 3,4-dihydro-1H-pyrimidin-2-one rings via coupling of aldehydes, urea, and β-dicarbonyl compounds.

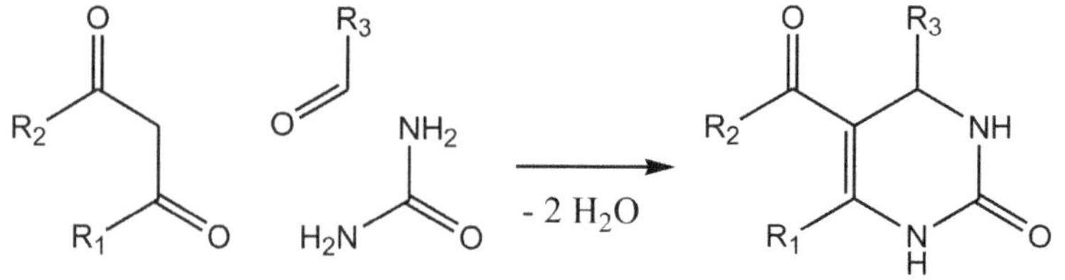

$$\text{R}_2\text{C(=O)CH}_2\text{C(=O)R}_1 + \text{R}_3\text{CHO} + \text{H}_2\text{NC(=O)NH}_2 \xrightarrow{-\,2\,H_2O} \text{product}$$

Show mechanistically that the product formed is independent of the order of addition of reagents.

PROBLEM 4.4 [4]

Themes: balancing chemical equations, multi-component reactions, order of addition of reagents

2-Aryl-3-aminobenzofurans were synthesized according to the 3-component coupling reaction shown below.

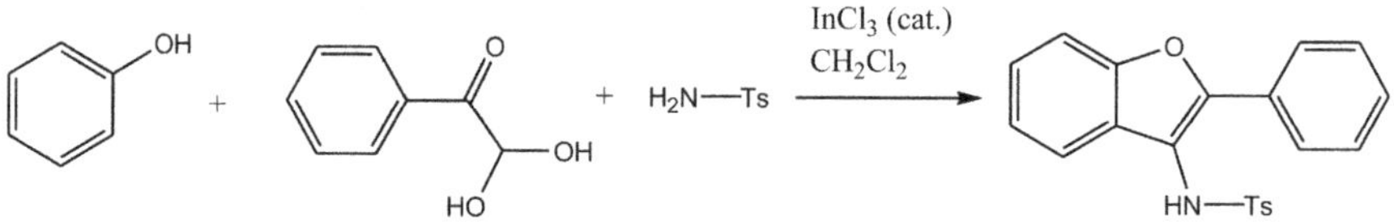

OH + HO, OH, O + H_2N—Ts → ($InCl_3$ (cat.), CH_2Cl_2) O, HN—Ts

The experimental protocol of the reaction required that toluene-4-sulfonamide and phenylglyoxal monohydrate be added first before adding phenol.

Part 1

Provide a balanced equation for the reaction and determine its atom economy. Show the target bond mapping in the product structure.

Part 2

Explain mechanistically the strict order of reactants.

Part 3

What products would be expected if phenylglyoxal monohydrate and phenol were added first before toluene-4-sulfonamide?

4.2 NOMENCLATURE FOR RING CONSTRUCTION STRATEGIES

4.2.1 Terms, Definitions, and Examples

Monocycle

A compound containing one ring in its structure.

Bicycle

A compound containing two rings in its structure. These rings may be separated, fused, or in a *spiro* arrangement.

Ring Construction Mapping

Ring construction mapping refers to the tracking of the target bonds made in a ring-containing structure via a single reaction or a synthesis plan over several steps. The target bond map shows the visual representation of the bonds made in the target product structure. The precise description of the mapping has the following general notational representations:

1. for monocycles: $[n+m]$
2. for bicycles with separated rings: $[n+m]_A+[u+v]_B$
3. for bicycles with fused rings: $[(n+m)_A+(u+v)_B]$
4. for bicycles with spiro rings: *spiro*-$[(n+m)_A+(u+v)_B]$
5. for tricycles with separated rings: $[n+m]_A+[u+v]_B+[a+b]_C$
6. for tricycles with fused rings: $[(n+m)_A+(u+v)_B+(a+b)_C]$

In all cases, the integers refer to the number of atoms involved in each fragment. If two integers appear inside parentheses or brackets, then this means that the ring is formed from two separate fragments. The sum of the integers inside a set of parentheses is equal to the resulting ring size made as a result of adding the fragments together.

Figure 4.1 shows some examples of these representations.

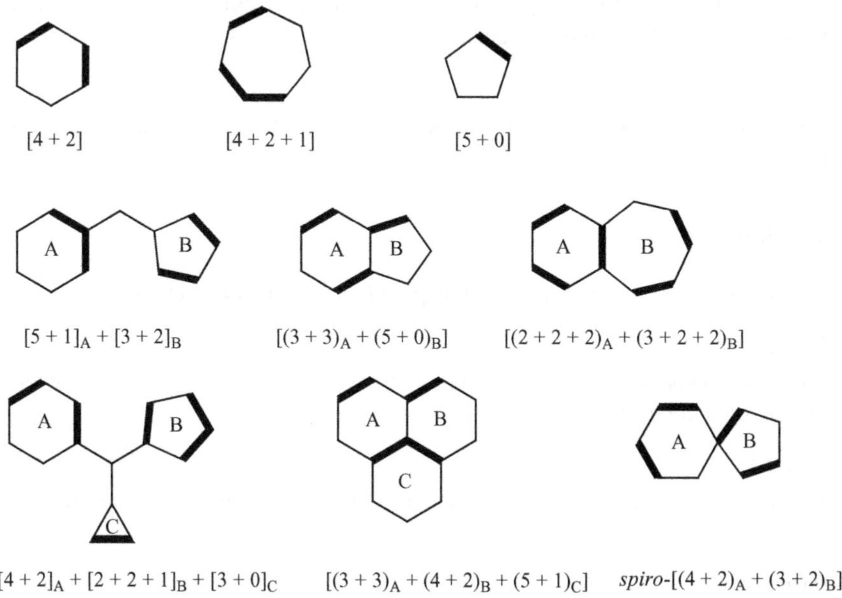

FIGURE 4.1 Examples of ring construction mappings and their notational representations.

Ring Construction Strategy

Ring construction strategy refers to the sequence of steps in constructing a target ring-containing structure. For a single reaction, the ring construction strategy is identical to the ring construction mapping. The notation takes the form shown by the examples given below.

Example A

$$\xrightarrow{[n+m]_A} A_p \xrightarrow{[u+v]_B} B_q \xrightarrow{[a+b]_C} C_r$$

The notation is read as follows: ring A is made in step p via an $[n+m]$ cycloaddition, ring B is made in step q via a $[u+v]$ cycloaddition, and ring C is made in step r via an $[a+b]$ cycloaddition.

Example B

$$\xrightarrow{\left[(n+m)_A+(u+v)_B\right]} [4.3.0]_p \xrightarrow{[a+b]_C} C_q \xrightarrow{[c+d]_D} D_r$$

The notation is read as follows: [4.3.0] bicyclic ring is made in step p via simultaneous $[n+m]$ and $[u+v]$ cycloadditions for rings A and B, respectively; ring C is made in step q via an $[a+b]$ cycloaddition; and ring D is made in step r via a $[c+d]$ cycloaddition.

Ring Partitioning [5]

Ring partitioning refers to the dissection of a ring into its potential individual building blocks. The combinatorial partitioning of monocyclic rings parallels exactly the combinatorial partitioning of integers.

Tricycle

A compound containing three rings in its structure. These rings may be separated or fused.

Example 4.4

Themes: ring construction strategy, ring partitioning and enumeration

Part 1

Enumerate all possible two-component partition strategies for constructing a general six-membered ring.

Part 2

Enumerate all possible two-component partition strategies for constructing cyclohexanone.

Part 3

Enumerate all possible two-component partition strategies for constructing 1,2-cyclohexadione.

Part 4

Enumerate all possible two-component partition strategies for constructing 1,3-cyclohexadione.

Part 5

Enumerate all possible two-component partition strategies for constructing 1,4-cyclohexadione.

Solution

Part 1

Possible two-partitions for a six-membered ring: [5 + 1], [4 + 2], and [3 + 3].

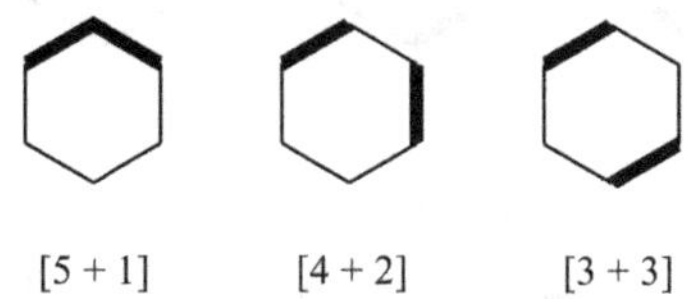

Part 2

[5 + 1] partitions of cyclohexanone:

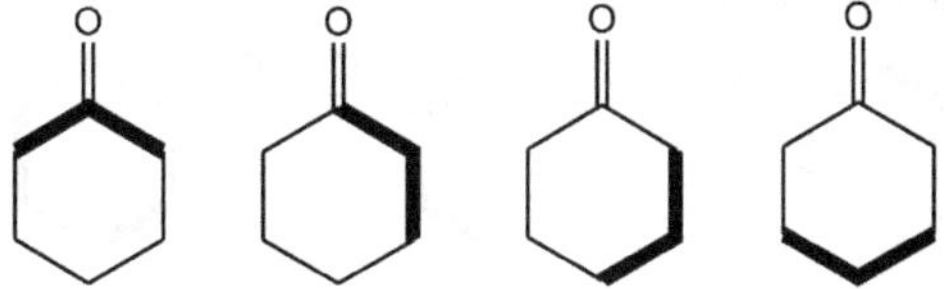

[4 + 2] partitions of cyclohexanone:

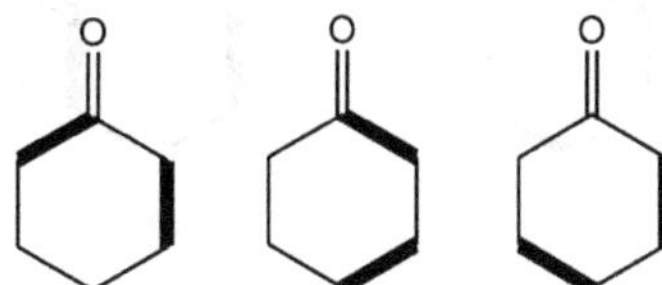

[3 + 3] partitions of cyclohexanone:

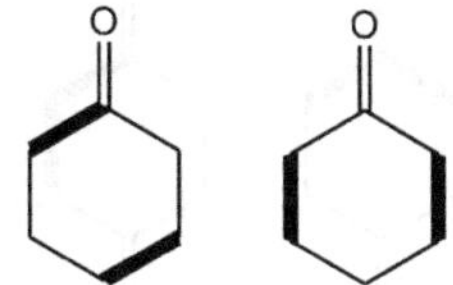

Part 3

[5 + 1] partitions of 1,2-cyclohexadione:

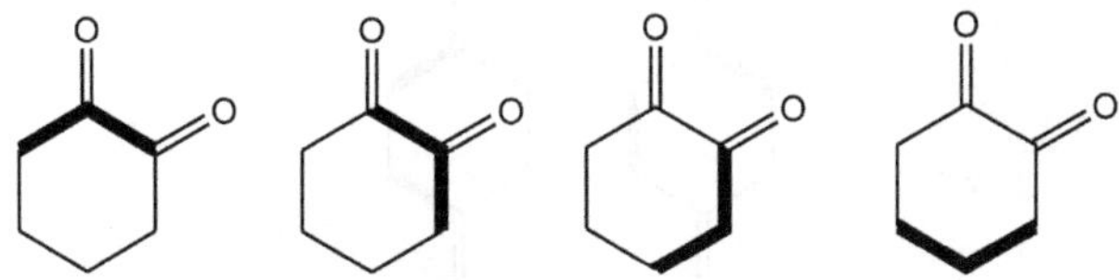

[4 + 2] partitions of 1,2-cyclohexadione:

[3 + 3] partitions of 1,2-cyclohexadione:

Part 4

[5 + 1] partitions of 1,3-cyclohexadione:

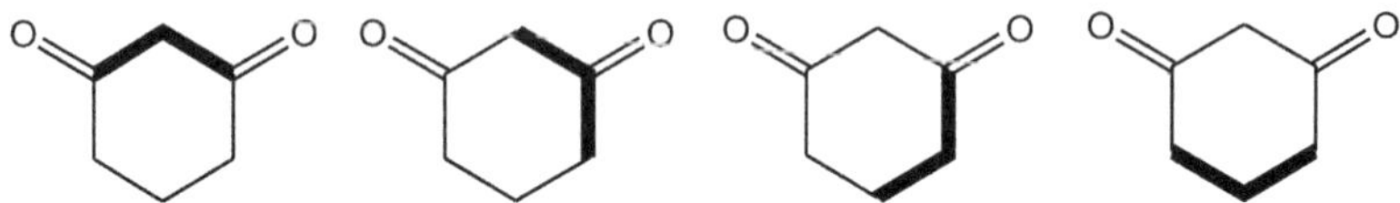

[4 + 2] partitions of 1,3-cyclohexadione:

[3 + 3] partitions of 1,3-cyclohexadione:

Part 5

[5 + 1] partitions of 1,4-cyclohexadione:

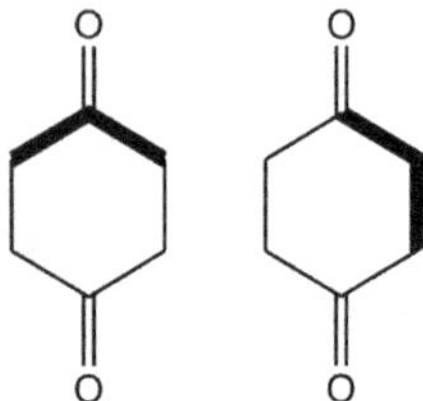

[4 + 2] partitions of 1,4-cyclohexadione:

[3 + 3] partitions of 1,4-cyclohexadione:

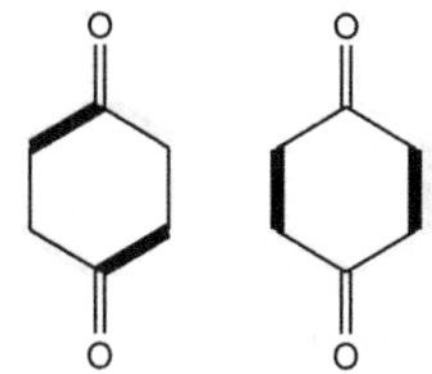

Example 4.5

Themes: ring construction strategy, ring partitioning and enumeration

Part 1

Enumerate all possible three-component partition strategies for constructing a general six-membered ring.

Part 2

Enumerate all possible three-component partition strategies for constructing cyclohexanone.

Part 3

Enumerate all possible three-component partition strategies for constructing 1,2-cyclohexadione.

Part 4

Enumerate all possible three-component partition strategies for constructing 1,3-cyclohexadione.

Part 5

Enumerate all possible three-component partition strategies for constructing 1,4-cyclohexadione.

Solution

Part 1

Possible three-partitions for a six-membered ring: [4 + 1 + 1], [3 + 2 + 1], and [2 + 2 + 2].

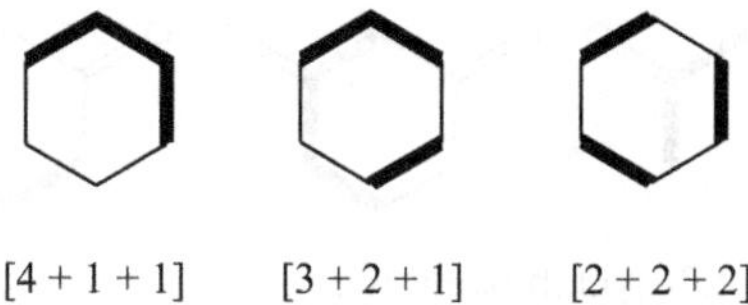

Part 2

[4 + 1 + 1] partitions of cyclohexanone:

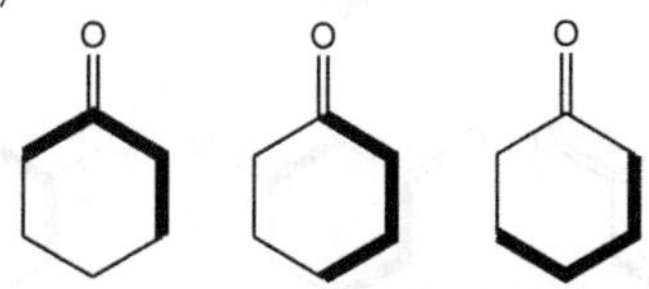

[3 + 2 + 1] partitions of cyclohexanone:

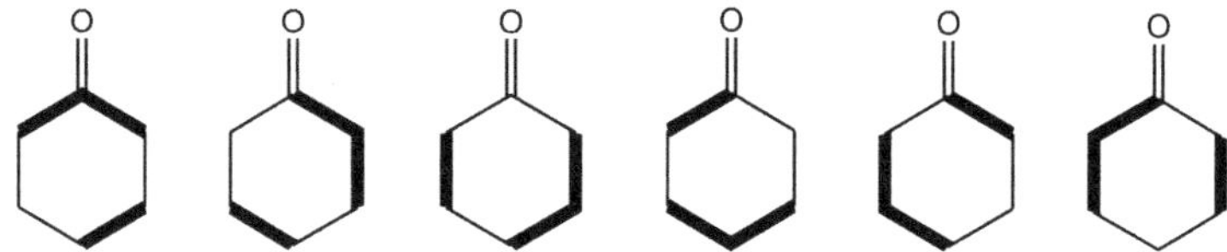

[2 + 2 + 2] partitions of cyclohexanone:

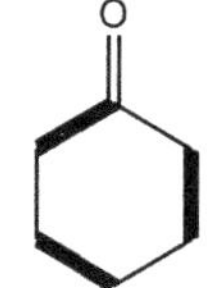

Part 3

[4 + 1 + 1] partitions of 1,2-cyclohexadione:

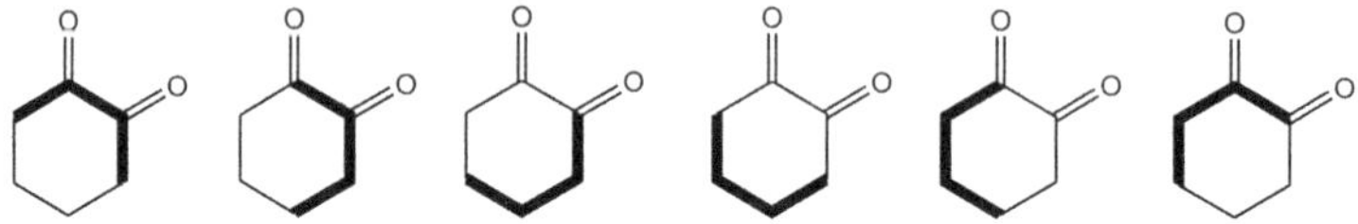

[3 + 2 + 1] partitions of 1,2-cyclohexadione:

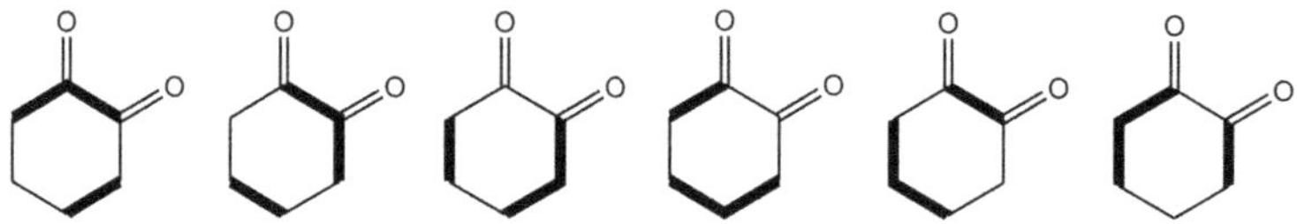

[2 + 2 + 2] partitions of 1,2-cyclohexadione:

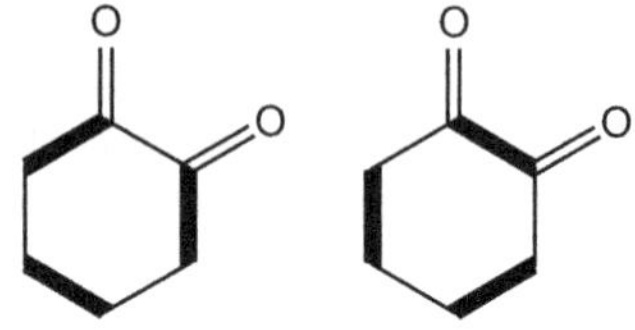

Part 4

[4 + 1 + 1] partitions of 1,3-cyclohexadione:

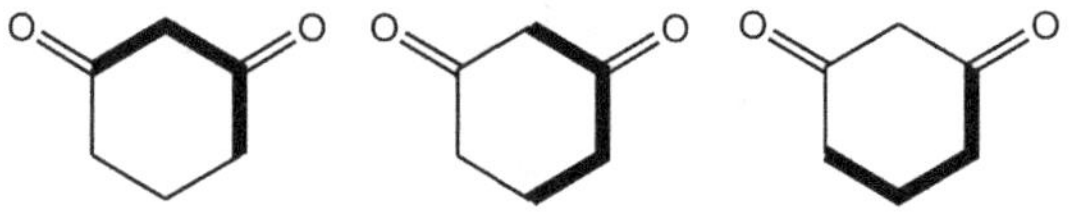

[3 + 2 + 1] partitions of 1,3-cyclohexadione:

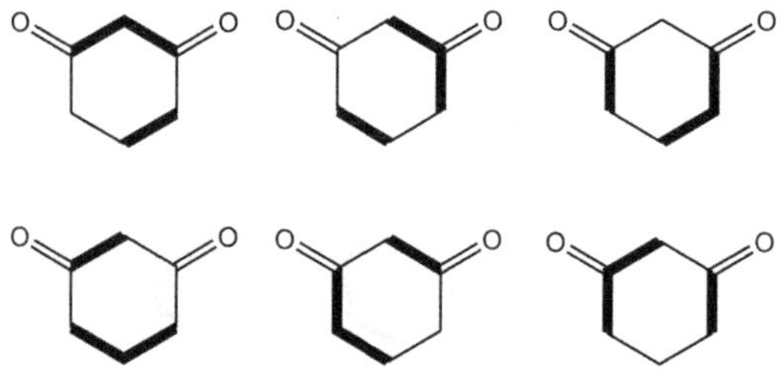

[2 + 2 + 2] partitions of 1,3-cyclohexadione:

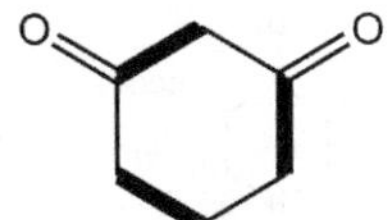

Part 5

[4 + 1 + 1] partitions of 1,4-cyclohexadione:

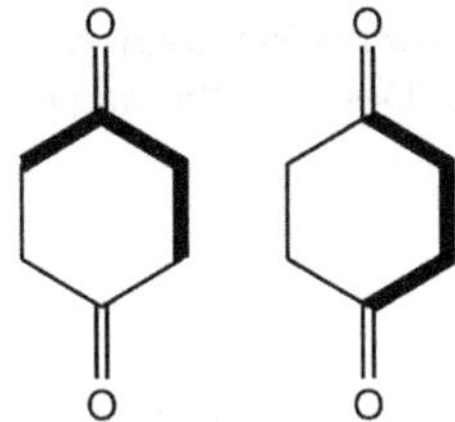

[3 + 2 + 1] partitions of 1,4-cyclohexadione:

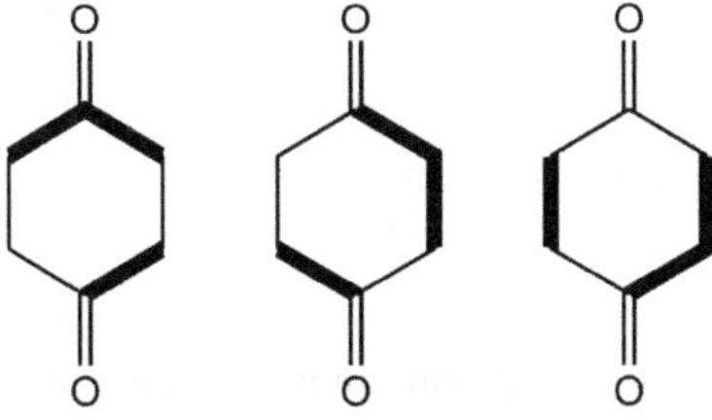

[2 + 2 + 2] partitions of 1,4-cyclohexadione:

Example 4.6 [6]

Themes: balancing chemical equations, ring construction strategies and partitioning, target bond mapping

4-Piperidone is the starting material for the synthesis of fentanyl, a controversial opioid painkiller that has significant addictive properties. It can be made by the four-component Petrenko–Kritschenko condensation reaction from formaldehyde, ammonia, and dimethyl 3-oxo-pentanedioate. This reaction can also be described as a double Mannich reaction.

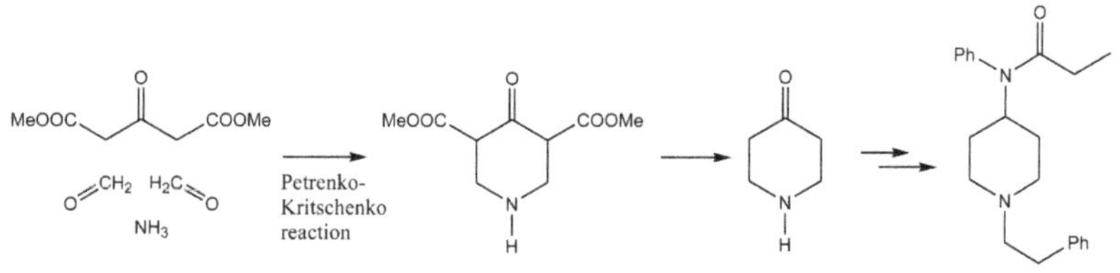

Part 1

Show the target bond mapping for the Petrenko–Kritschenko reaction and give the expected reaction by-products. Describe the ring construction strategy.

Solution

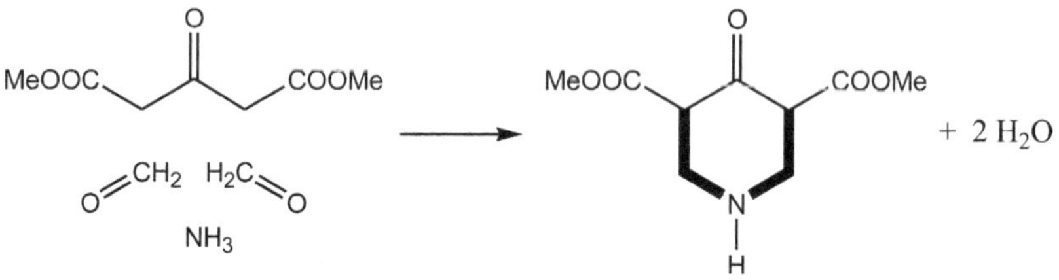

Ring construction strategy: [3 + 1 + 1 + 1]

Part 2

Show all possible four-component combinations to make 4-piperidone.

Solution

[3 + 1 + 1 + 1] partitions:

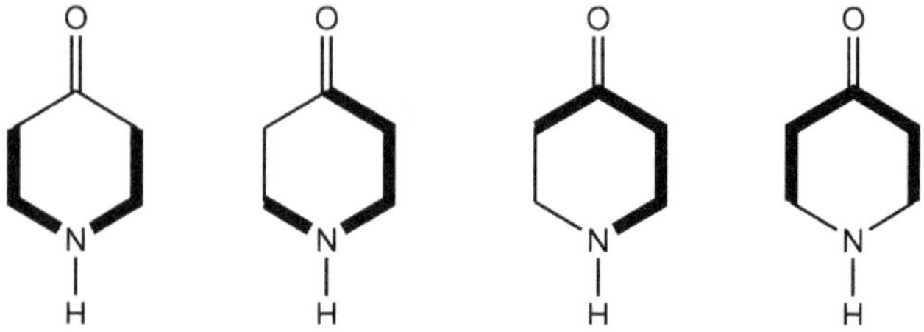

[2 + 2 + 1 + 1] partitions:

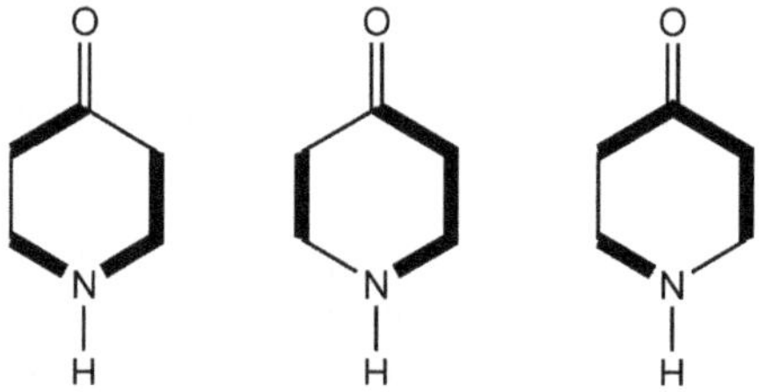

[2 + 1 + 2 + 1] partitions:

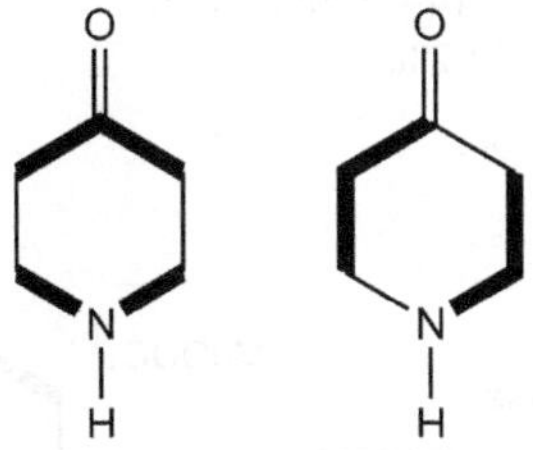

Part 3

From the list of possible four-component partitions found in Part 2, select those that align with the symmetry of 4-piperidone. Based on these selections suggest an alternative four-component strategy to make the 4-piperidone derivative in Part 1. Show the balanced chemical equation and compare its atom economy with that for the Petrenko–Kritschenko reaction.

Solution

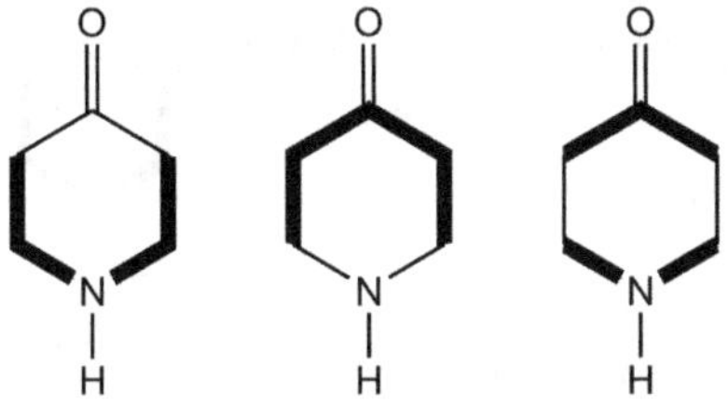

Petrenko-Kritschenko

Alternative strategy #1:

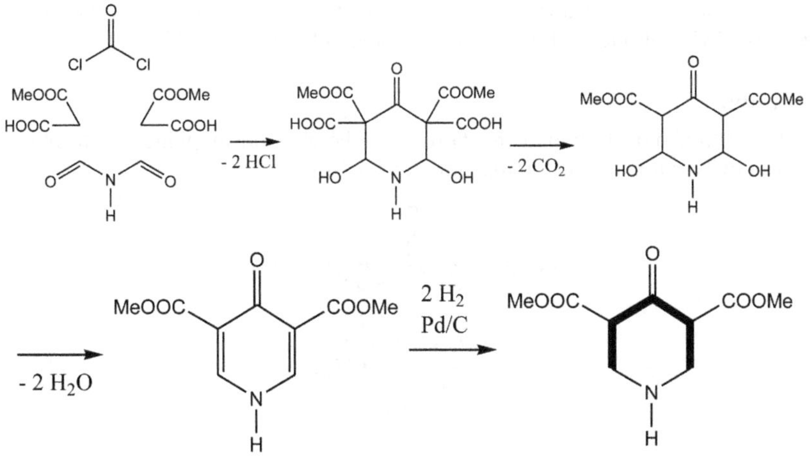

- 2 HCl

- 2 CO_2

- 2 H_2O

2 H_2 Pd/C

This sequence involves two separate reactions. The first reaction is the [3 + 1 + 1 + 1] cycloaddition and the second is the reduction.

Reactants: 98.9 + 2*118 + 73 + 2*2 = 411.9
Products: 215 + 2*36.45 + 2*44 + 2*18 = 411.9
AE = 215/411.9 = 0.522, or 52.2%
Alternative strategy #2:

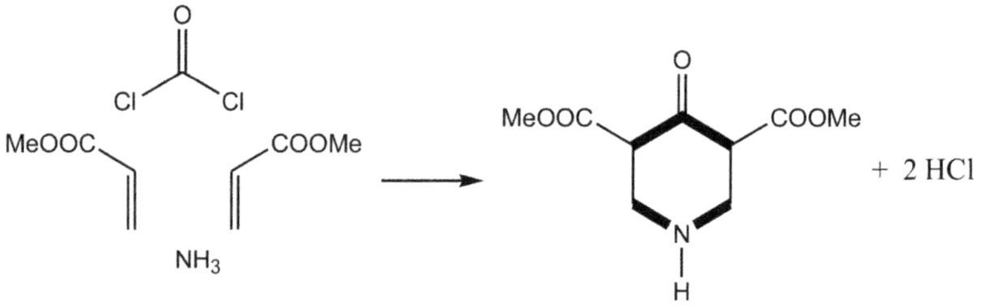

Reactants: 17 + 2*86 + 98.9 = 287.9
Products: 215 + 2*36.45 = 287.9
AE = 215/287.9 = 0.747, or 74.7%
Petrenko–Kritschenko strategy:

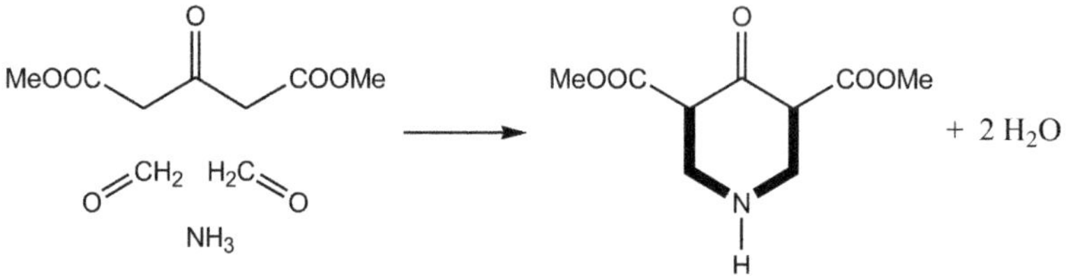

Reactants: 174 + 2*30 + 17 = 251
Products: 215 + 2*18 = 251
AE = 215/251 = 0.857, or 85.7%

Part 4

Comment if the order of addition of reagents matters for the Petrenko–Kritschenko reaction. What about for the alternative strategies found in Part 3?

Solution

In the Petrenko–Kritschenko reaction, the following nucleophilic–electrophilic connectivity is implied by the reaction mechanism.

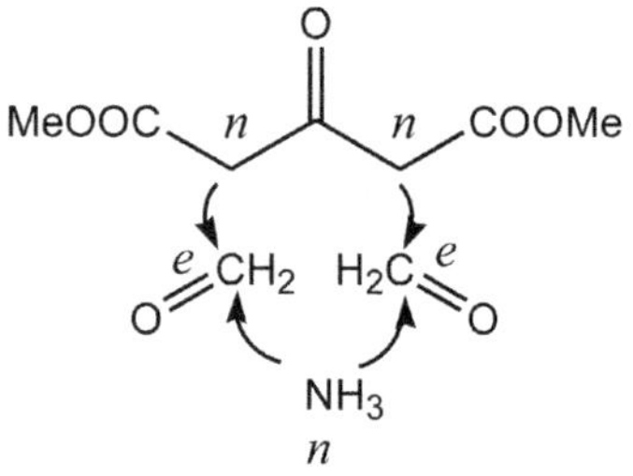

n = nucleophilic center; *e* = electrophilic center

The order of addition of reagents in the Petrenko–Kritschenko is not critical. Ammonia is not expected to add to the 3-oxo position of dimethyl

3-oxopentanedioate since that compound will have a significant enol content and so the 3-oxo position is not a true carbonyl group. Moreover, the carbonyl group of formaldehyde is more reactive and so ammonia is expected to add there.

MeOOC COOMe ⇌ MeOOC COOMe

In the case of the [2+1+2+1] strategy #2 shown below, order of addition does matter since ammonia can react with phosgene to produce urea. Hence, in order to avoid this side reaction ammonia is first added to two equivalents of methyl acrylate. The resulting adduct is then reacted with phosgene using a non-nucleophilic base.

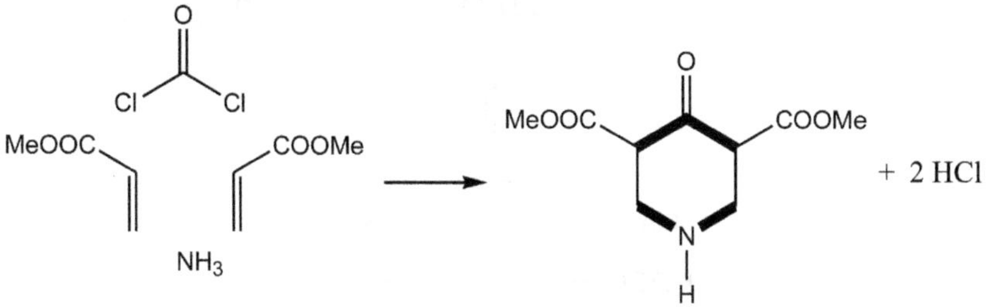

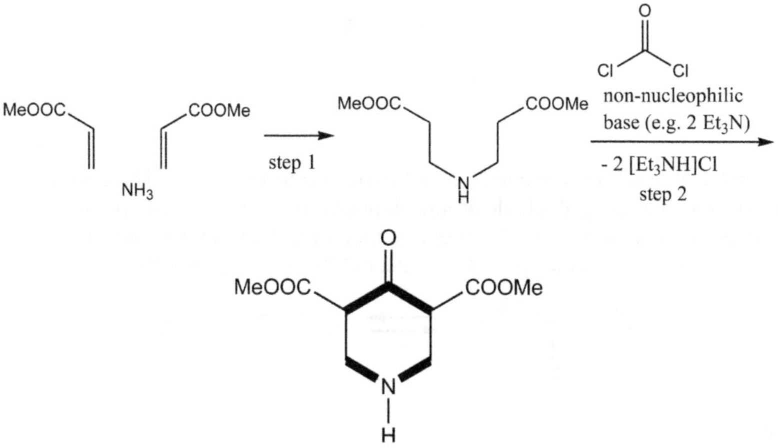

Cl Cl MeOOC COOMe NH_3 → MeOOC COOMe + 2 HCl

MeOOC COOMe NH_3 → step 1 MeOOC COOMe → non-nucleophilic base (e.g. 2 Et_3N) - 2 $[Et_3NH]Cl$ step 2

MeOOC COOMe

The intended nucleophilic–electrophilic combination is as follows.

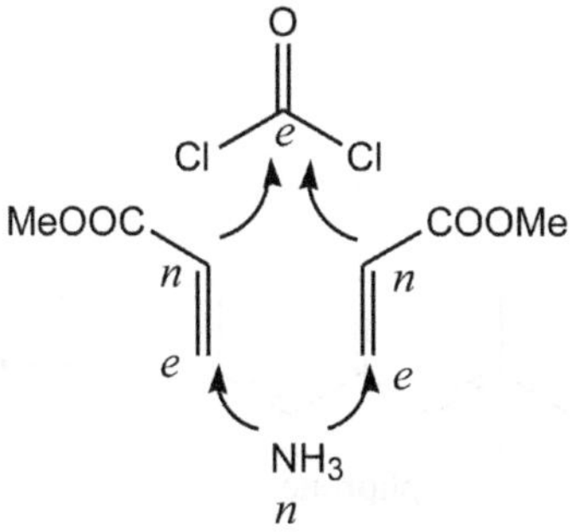

Cl *e* Cl MeOOC COOMe *n* *n* *e* *e* NH_3 *n*

For the [3+1+1+1] strategy #1 shown below the intended nucleophilic–electrophilic combination is as follows.

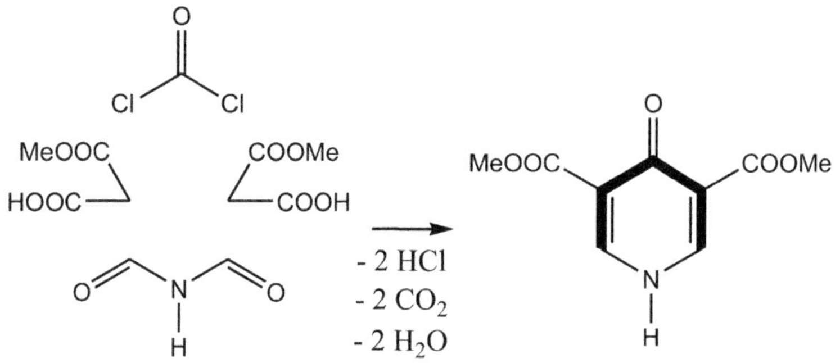

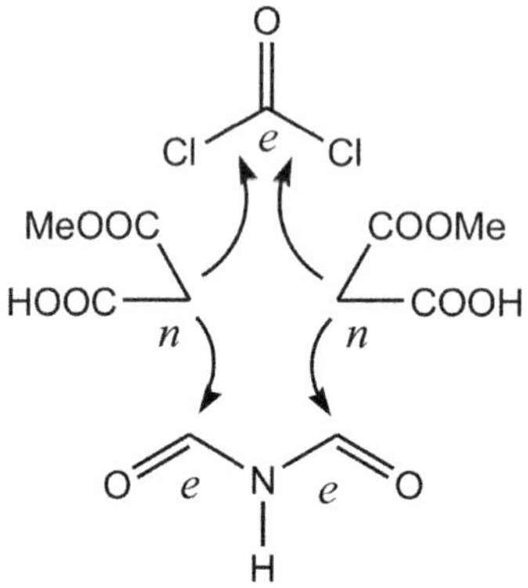

- 2 HCl
- 2 CO_2
- 2 H_2O

Order of addition of reagents is not expected to be critical. The two flanking electron withdrawing aldehyde groups diminish the nucleophilicity of the nitrogen atom of diformylammonia. This suggests that a reaction between diformylammonia and phosgene via nitrogen atom nucleophilic attack is unlikely.

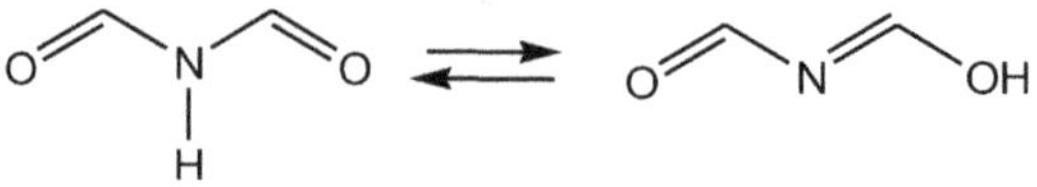

Part 5

For the following reactions to prepare 4-piperidone derivatives, show the target bond mapping and describe the ring construction strategy.

REACTION #1 [7]

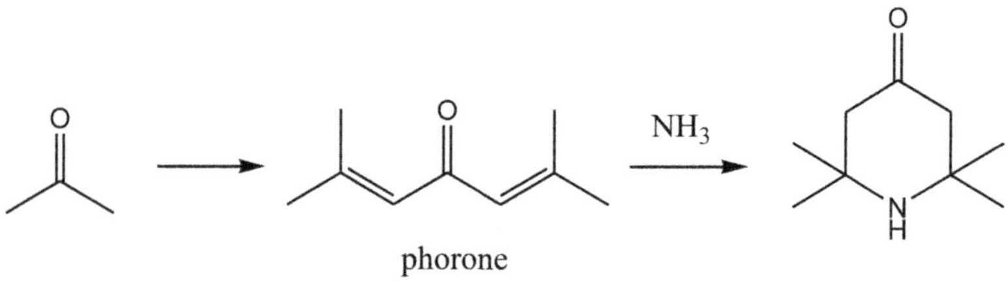

NH_3

phorone

REACTION #2 [8, 9]

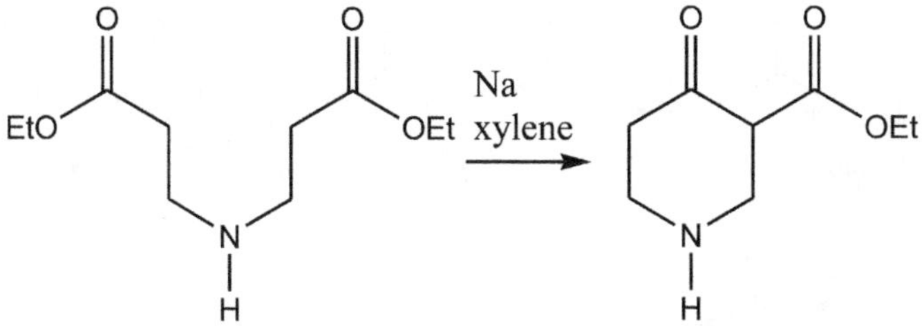

Na
xylene

REACTION #3 [10]

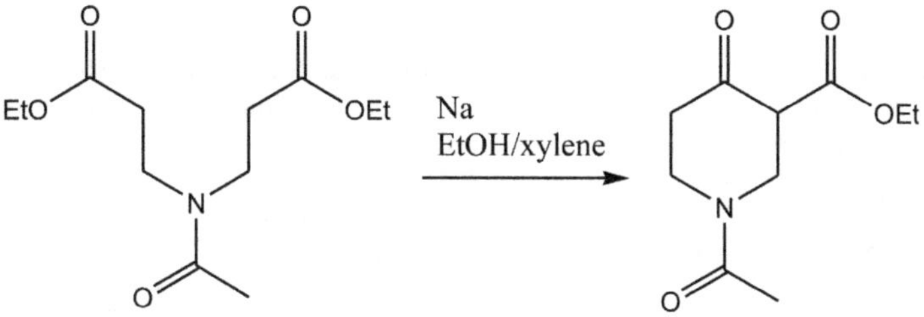

Na
EtOH/xylene

SOLUTION

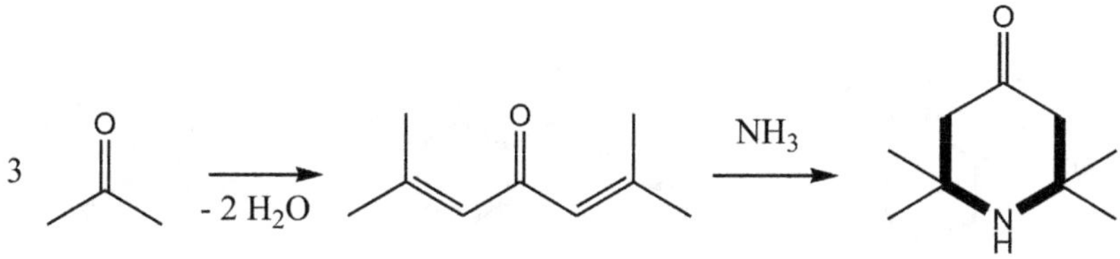

3 → $- 2\ H_2O$ → NH_3 →

[3 + 1 + 1 + 1]

This strategy is the same as the Petrenko–Kritschenko strategy.

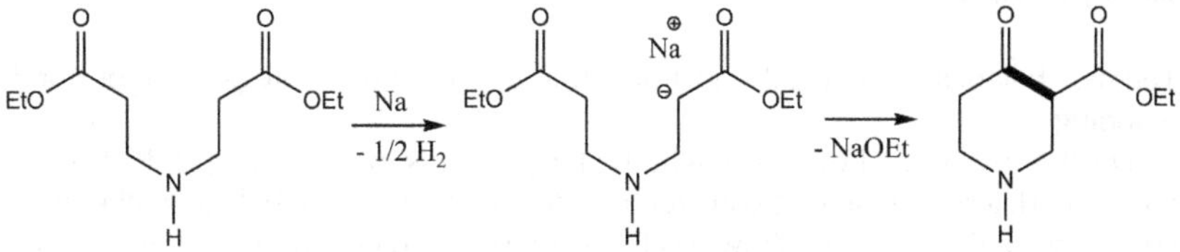

Na
$- 1/2\ H_2$
- NaOEt

[6 + 0]

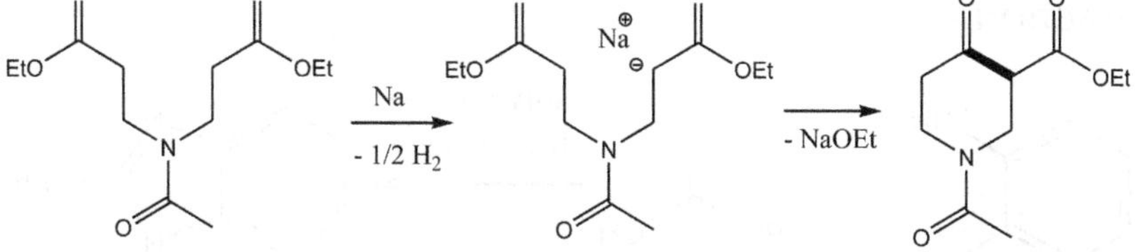

Na
$- 1/2\ H_2$
- NaOEt

[6 + 0]

Both of these strategies are equivalent and suggest a [3 + 2 + 1] cycloaddition as follows:

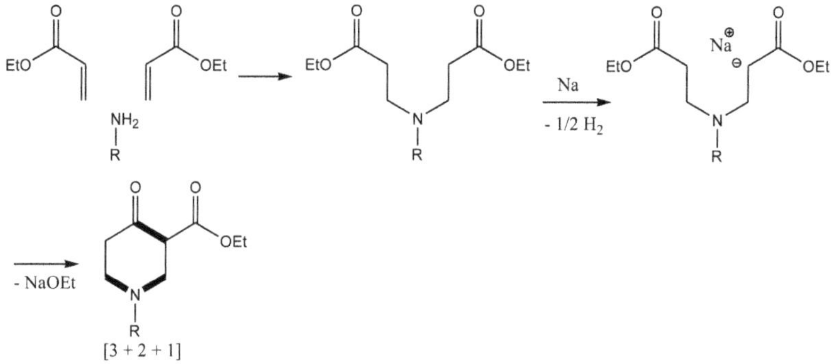

Note that the first adduct is the same as that found in alternative strategy #2 in Part 3. In the above case ring closure is carried out without the need of another carbonyl containing component such as phosgene.

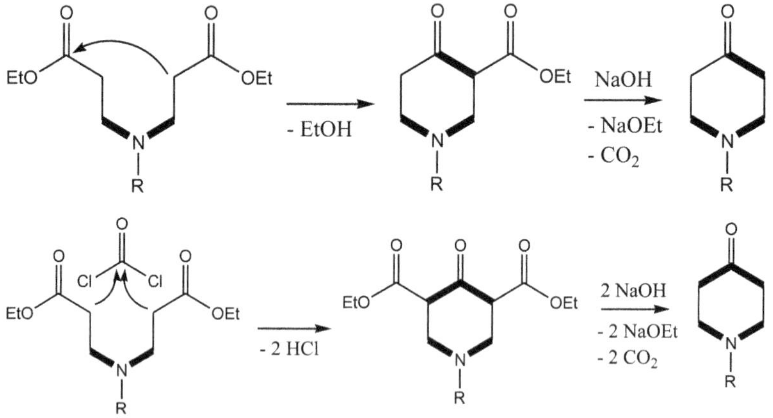

Example 4.7 [11]

Themes: balancing chemical equations, ring construction strategies, target bond mapping

For the reactions shown highlight the target bonds made in the product structures and determine the ring construction strategies. Write out balanced chemical equations for the reactions showing all by-products and determine their respective atom economies.

REACTION 1

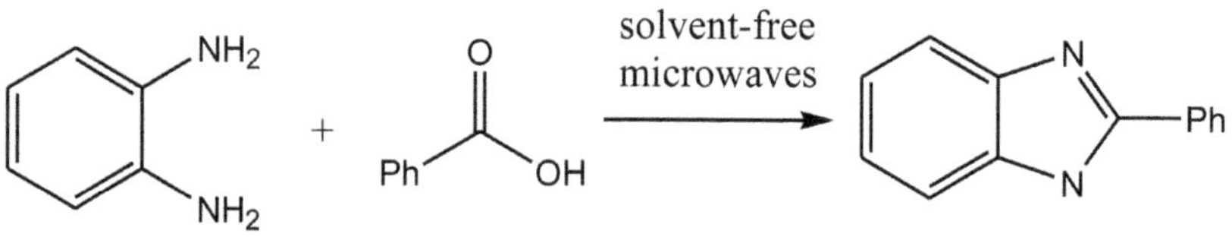

REACTION 2

Ph(C=O)–(C=O)Ph + PhCHO + NH_4OAc —(HOAc, microwaves)→ 2,4,5-triphenylimidazole

SOLUTION

Reaction 1

o-phenylenediamine + PhCOOH → 2-phenylbenzimidazole + $2\ H_2O$

108 **122** **194** **2 (18)**

- Reactants: 108 + 122 = 230
- Products: 194 + 2*18 = 230
- AE = 194/230 = 0.843, or 84.3%
- Ring construction strategy: [4 + 1]

Reaction 2

Ph(C=O)–(C=O)Ph + PhCHO + $2\ NH_4OAc$ —($-3\ H_2O$)→ 2,4,5-triphenylimidazole + $2\ HOAc$

210 **106** **2 (77)** **296** **2 (60)**

- Reactants: 210 + 106 + 2*77 = 470
- Products: 296 + 2*60 + 3*18 = 470
- AE = 296/470 = 0.630, or 63.0%
- Ring construction strategy: [2 + 1 + 1 + 1]

Example 4.8 [12]

Themes: balancing chemical equations, catalytic cycle, reaction mechanism, oxidation states, ring construction strategies

In the synthesis of the natural product (+)-shiromool, the following reaction was employed. It was described as a Pd-catalyzed umpolung allylation reaction utilizing 1.5 equivalents of diethylzinc.

$PdCl_2(PPh_3)_2$
Et_2Zn
K_2CO_3
N,N-dimethylacetamide

42 %

(+)-shiromool

Part 1

Write out a catalytic cycle for the reaction explaining the umpolung nature of the cyclization.

Part 2

Write out a balanced chemical equation for the transformation.

Part 3

Show the target bonds made in the product structure and describe the ring construction strategy.

Solution

Part 1

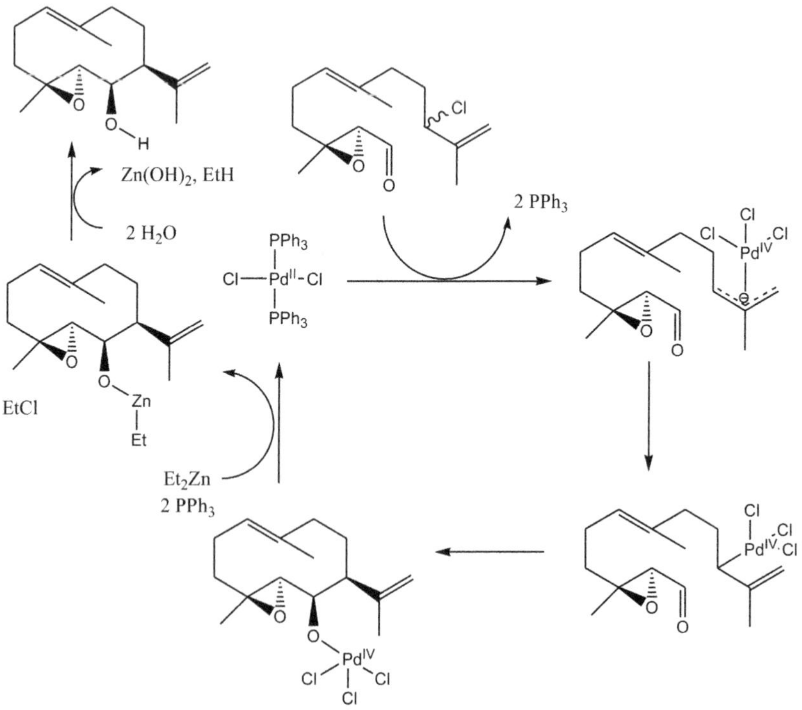

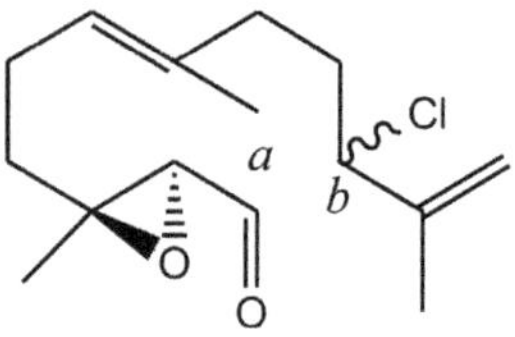

The carbon centers at C_a and C_b are both electrophilic. In order to create a bond between them, the C_b center is converted to a nucleophilic one via oxidative addition by the Pd(II) catalyst. The reversal of an electrophilic center to a nucleophilic one is the umpolung aspect of the reaction. Hence, C_b is reduced from an oxidation state of 0 to –2, while Pd is oxidized from +2 to +4. When the C_a-C_b bond is made, C_a changes from +1 in the substrate to 0 in the product, and similarly C_b changes from 0 to –1. Hence, both carbon atoms are reduced in the reaction. This is compensated by the oxidation of one of the carbon atoms in the ethyl groups of diethylzinc to ethyl chloride (–3 to –1).

A full oxidation number analysis for the reaction is shown below.

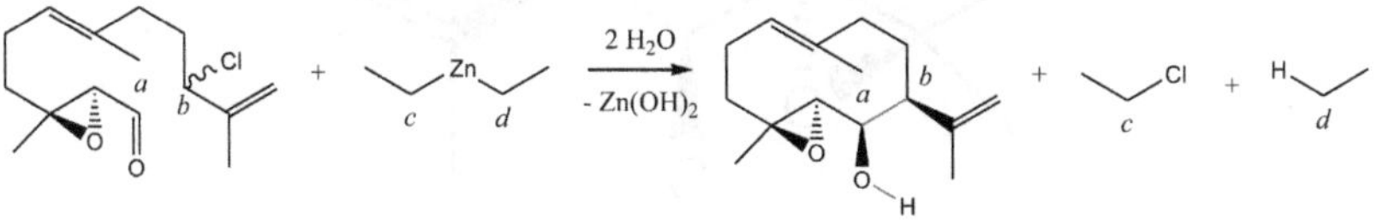

$$\text{substrate} + Et_2Zn \xrightarrow[- Zn(OH)_2]{2\,H_2O} \text{product} + EtCl + EtH$$

Atom	Reactant	Product	Change	
C_a	+1	0	- 1	$\Delta = -2$
C_b	0	- 1	- 1	
C_c	- 3	- 1	+ 2	$\Delta = +2$
C_d	- 3	- 3	0	

The redox couple for the reaction is given below.

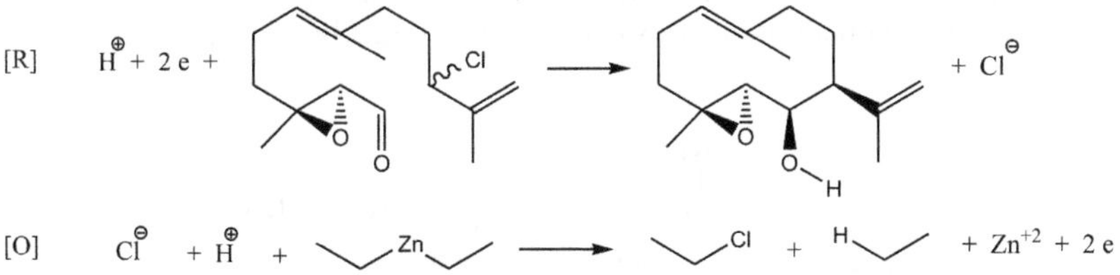

[R] $H^{\oplus} + 2e$ + substrate → product + $Cl^{\ominus}$

[O] $Cl^{\ominus} + H^{\oplus}$ + Et_2Zn → EtCl + EtH + $Zn^{+2} + 2e$

Net reaction:

$2H^{\oplus}$ + Et_2Zn + substrate → product + EtCl + EtH + Zn^{+2}

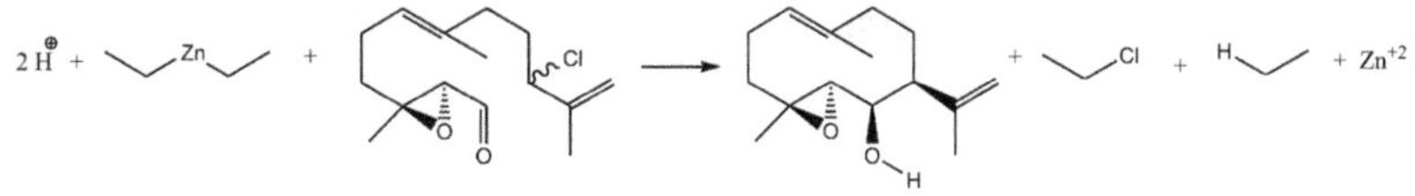

Adding two hydroxide ions to each side yields

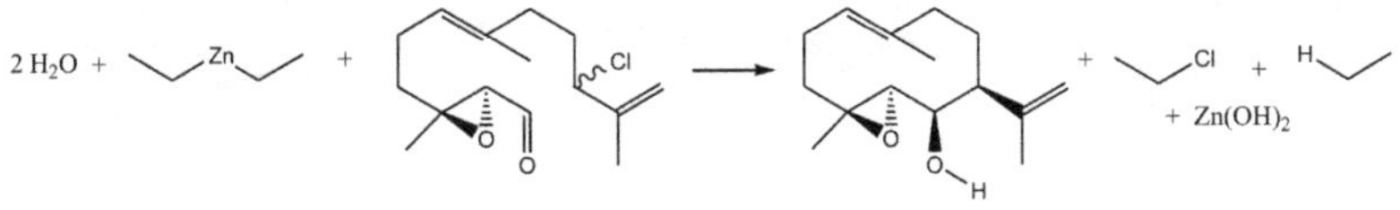

$2H_2O$ + Et_2Zn + substrate → product + EtCl + EtH + $Zn(OH)_2$

Part 2

$$\text{substrate} + Et_2Zn \xrightarrow[- Zn(OH)_2]{2\,H_2O} \text{product} + EtCl + EtH$$

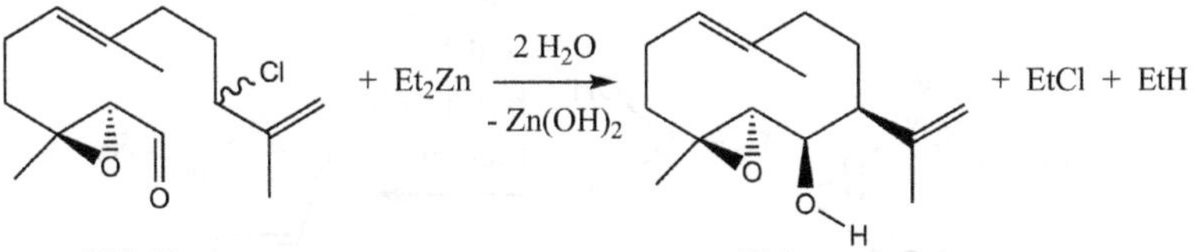

270.45 **236**

Reactants: 270.45 + 123.38 + 2*18 = 429.83
Products: 236 + 30 + 64.45 + 99.38 = 429.83

Part 3

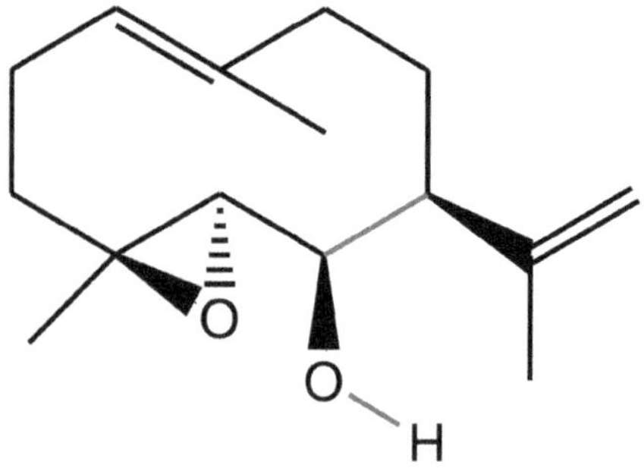

Ring construction strategy: [10 + 0]

Example 4.9

Themes: balancing chemical equations, combined MCR and redox reactions, minimum atom economies, product structure target bond maps

For the following combined multi-component redox reactions, provide balanced chemical equations, a redox couple, and a target bond map.

(a) [13]

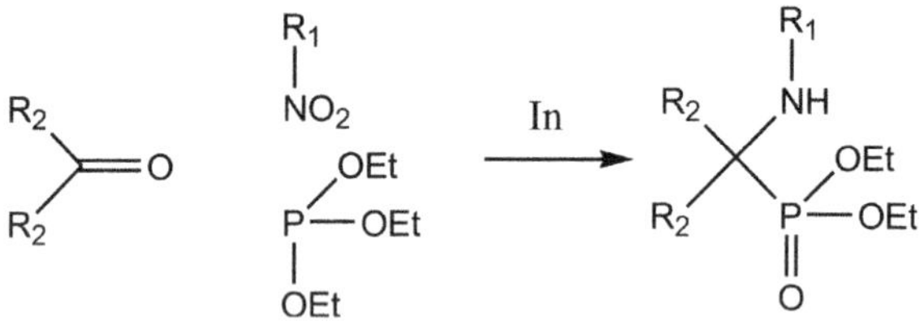

(b) [14]

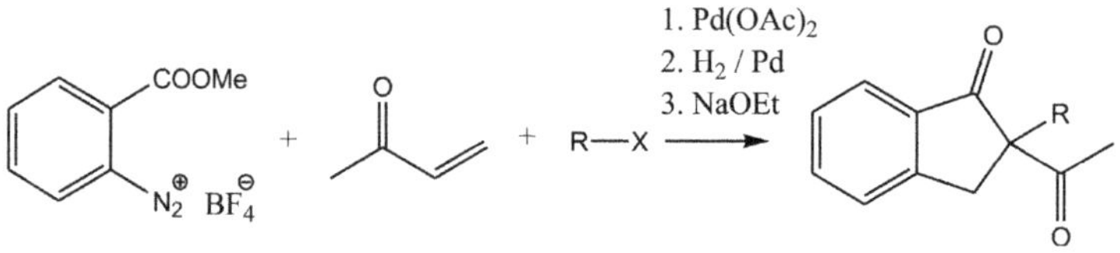

1. $Pd(OAc)_2$
2. H_2 / Pd
3. NaOEt

(c) [15]

Cu/Al_2O_3 (cat.)
ball milling
then H_2O

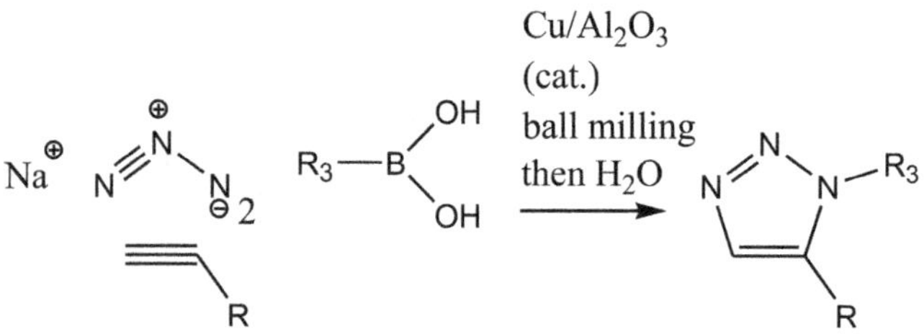

(d) [16]

AgX

(e) [17]

$PdCl_2$ (cat.)
CuX_2
1 atm O_2
H_2O

(f) [18]

H_2O

Solution

(a)

[R] R_1-NO_2 + $R_2C(=O)R_2$ + $P(OEt)_3$ + 6 H^+ + 6 e ⟶ $R_2C(NHR_1)(P(O)(OEt)_2)R_2$ + EtOH + 2 H_2O

[O] In ⟶ In^{+3} + 3 e X 2

2 In
4 H_2O
- 2 $In(OH)_3$
- EtOH

(b)

[R] + + R—X + 2 H^+ + 2 e ⟶ + N_2 + HBF_4 + HX + MeOH

[O] $H_2 \longrightarrow 2\,H^+ + 2\,e$

1. $Pd(OAc)_2$
2. H_2 / Pd
3. NaOEt

- HBF_4
- N_2
- MeOH
- NaX
- EtOH

(c)

[O] H_2O + ⟶ + $B(OH)_3$ + Na^+ + H^+ + 2 e

[R] $O_2 + 4\,H^+ + 4\,e \longrightarrow 2\,H_2O$ X 1/2

1/2 O_2
H_2O

- $B(OH)_3$
- NaOH

(d)

[O] ⟶ + H_2O + H^+ + 2 e

[R] $Ag^+ + e \longrightarrow Ag$ X 2

2 AgX

- 2 Ag
- HBr

(e)

[O] H_2O + $X^{\ominus}$ + ... → ... + EtOH + H^+ + 2 e

[R] Cu^{+2} + e → Cu^+ X 2

2 CuX_2, H_2O; - 2 CuX, - EtOH, - HX

(f)

[R] 12 H^+ + 12 e + ... → ... + 2 NH_3

[O] ... → CO_2 + NH_4^+ + H^+ + 2 e X 6

6 ... + ... → (6 H_2O; - 6 CO_2, - 2 NH_3, - 6 NH_4OH)

Example 4.10

Themes: balancing chemical equations, ring construction strategies, product structure target bond maps, minimum atom economy.

Pyrazoles have been made by various three-component reaction strategies. A number of examples are listed below which show how this 5-membered ring can be made by [2+2+1] and [3+1+1] couplings.

(a) [19]

Cu(OAc)$_2$

(b) [20, 21]

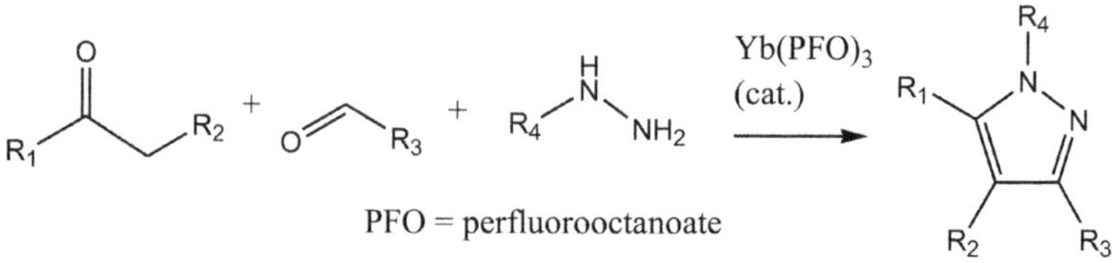

$Yb(PFO)_3$ (cat.)

PFO = perfluorooctanoate

(c) [22]

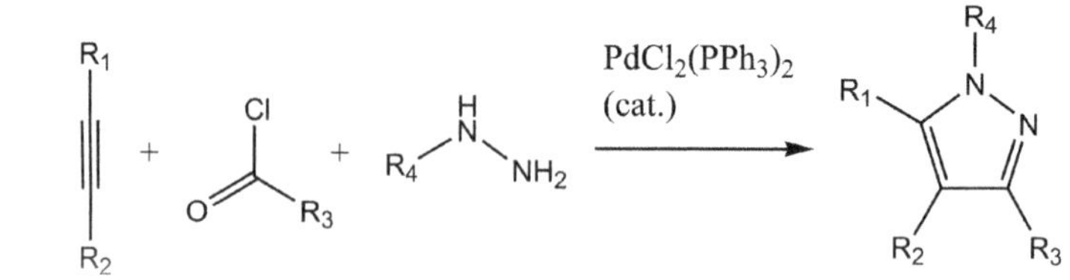

$PdCl_2(PPh_3)_2$ (cat.)

(d) [23]

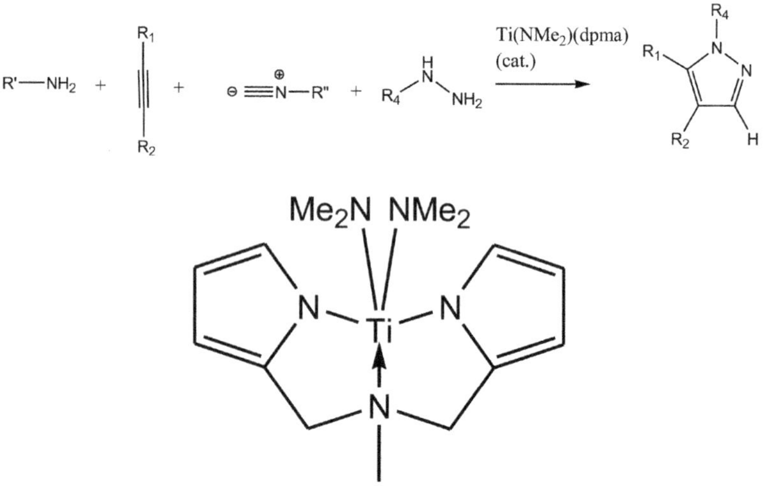

$Ti(NMe_2)(dpma)$ (cat.)

$Ti(NMe_2)(dpma)$

(e) [24]

$LiN(SiMe_3)_2$

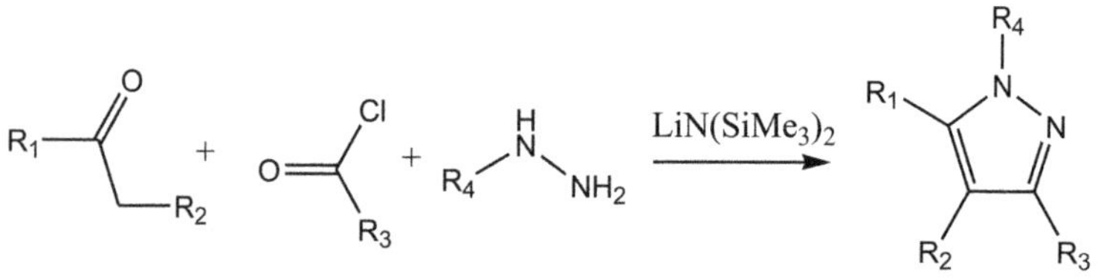

(f) [25]

$$R_1C{\equiv}CH + R_4NHNH_2 + X{-}R_3 \xrightarrow[Cs_2CO_3]{Mo(CO)_6} \text{pyrazole}$$

(g) [26]

$$R{-}\overset{\oplus}{N}{\equiv}\overset{\ominus}{C} + MeOOC{-}C{\equiv}C{-}COOMe + Ar{-}NH{-}NH{-}C(=O){-}NH{-}Ph \longrightarrow \text{pyrazole}$$

(h) [27]

$$EtOC(=O)H + CH_3NHNH_2 + EtOOC{-}CH_2{-}C(=O){-}R \longrightarrow \text{pyrazole}$$

(i) [28]

$$(MeO)_2P(=O){-}C(=N_2){-}C(=O)CH_3 + R_1CHO + R_2CH_2CN \xrightarrow[MeOH]{KOH} \text{pyrazole}$$

Part 1

For each example, provide a balanced chemical equation, highlight which bonds are the synthesis target bonds, and determine the associated minimum atom economy.

Part 2

For a generally substituted pyrazole ring, write out all possible [2 + 2 + 1] and [3 + 1 + 1] target bond maps. Are there any combinations that were not considered in the examples in Part 1? If so, suggest three-component reactions that would follow those missing ring construction strategies. For each of these suggested reactions determine the minimum AE.

Solution

Part 1

(a)

Reaction scheme: ketone + R_4NH_2 → (− H_2O) → [imine] → [enamine + nitrile $N{\equiv}C{-}R_3$] → $Cu(OAc)_2$ (− Cu^0, − 2 HOAc) → pyrazole (R_1, R_2, R_3, R_4)

AE(min) = 0.393

Glorius (2010)

(b)

Reaction scheme: ketone + R_4NHNH_2 + aldehyde R_3CHO → (− H_2O) → hydrazone → (− H_2O) → → $Yb(PFO)_3$ (cat.), 1/2 O_2 (− H_2O) → pyrazole

PFO = perfluorooctanoate

AE(min) = 0.64

Shen (2008)

(c)

Reaction scheme: alkyne $R_1C{\equiv}CR_2$ + R_4NHNH_2 + acid chloride R_3COCl → Et_3N (− $[Et_3NH]Cl$) → hydrazide → $PdCl_2(PPh_3)_2$ (cat.) (− H_2O) → pyrazole

AE (min) = 0.444

Müller (2008)

(d)

Reaction scheme: $R'{-}NH_2$ + alkyne $R_1C{\equiv}CR_2$ + isonitrile $R''N{\equiv}C$ → $Ti(NMe_2)(dpma)$ (cat.) → intermediate + R_4NHNH_2 → (− $R'NH_2$, − $R''NH_2$) → pyrazole (R_1, R_2, H, R_4)

AE(min) = 0.640

Odom (2009)

$Ti(NMe_2)(dpma)$

(e)

LiN(SiMe$_3$)$_2$

- HN(SiMe$_3$)$_2$

- LiCl

- 2 H_2O

AE(min) = 0.341

Heller (2006)

(f)

$Mo(CO)_6$

Cs_2CO_3

- $Mo(CO)_5$

- H_2O

- CsX

- $CsHCO_3$

AE(min) = 0.102

Stonehouse (2008)

(g)

- $PhNH_2$

- CO

AE(min) = 0.705

Adib (2008)

(h)

- 2 H_2O

- EtOH

AE(min) = 0.672

Raw (2009)

(i)

KOH

MeOH

- 2 H_2O

- KCN

- CH_3COOMe

AE(min) = 0.538

Mohanan (2010)

Part 2

All possible target bond maps for constructing a pyrazole ring via [2+2+1] and [3+1+1] strategies:

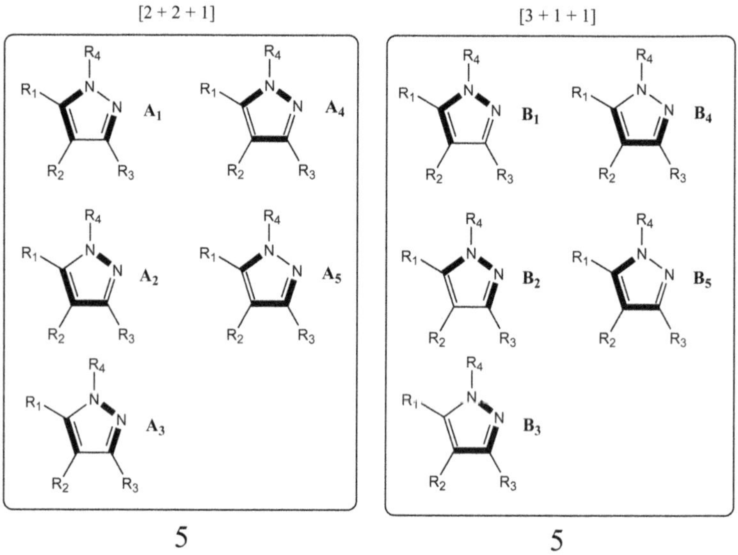

[2 + 2 + 1] [3 + 1 + 1]

R4 R1 N N A1 R2 R3 — R4 R1 N N A4 R2 R3 — R4 R1 N N B1 R2 R3 — R4 R1 N N B4 R2 R3

R4 R1 N N A2 R2 R3 — R4 R1 N N A5 R2 R3 — R4 R1 N N B2 R2 R3 — R4 R1 N N B5 R2 R3

R4 R1 N N A3 R2 R3 — R4 R1 N N B3 R2 R3

5 5

Target bond maps not considered by examples given in Part 1:

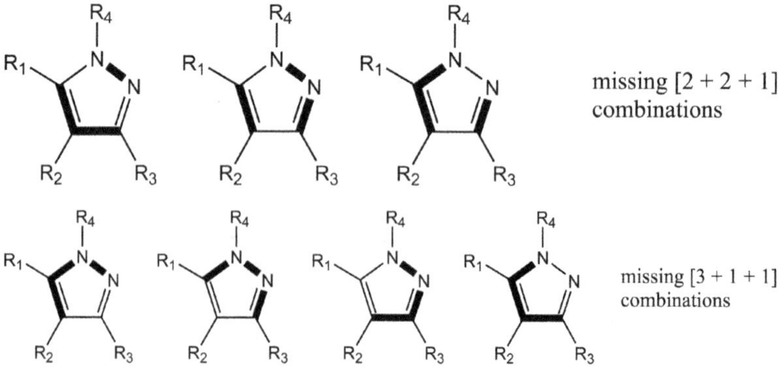

missing [2 + 2 + 1] combinations

missing [3 + 1 + 1] combinations

Below are listed some hypothetical three-component reactions that may lead to these missing combinations along with their minimum atom economies. There are several answers to this part.

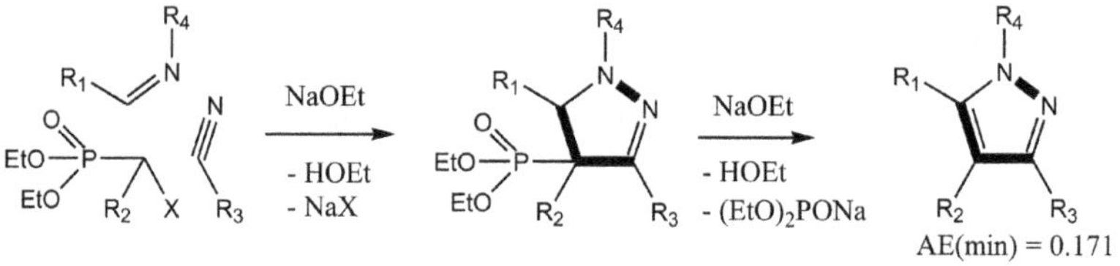

NaOEt, - HOEt, - NaX; NaOEt, - HOEt, - $(EtO)_2PONa$

AE(min) = 0.171

NH_3, $- H_2O$; $- H_2O$

AE(min) = 0.640

NaOEt, - HOEt; NaOEt, - NaX, - HOEt; - NaOH

AE(min) = 0.254

$- N_2$; $- H_2O$; [O], - 2 H

AE(min) < 0.582

$- N_2$; $- H_2O$; $- H_2O$

AE(min) = 0.500

$- N_2$, - HCl; NaOMe, - HOMe, - NaCl; $- H_2O$

AE(min) = 0.273

2 NaOMe, - 2 HOMe, - NaOMe; - NaOMe

AE(min) = 0.177

Example 4.11 [29]

Themes: balancing chemical equations, ring construction strategies, target bond mapping, mechanisms, synthesis tree diagram, materials efficiency metrics

By analogy with the [4+2] Diels–Alder cycloaddition, the hexadehydro-Diels–Alder reaction (HDDA) also proceeds via [4+2] cycloaddition as follows.

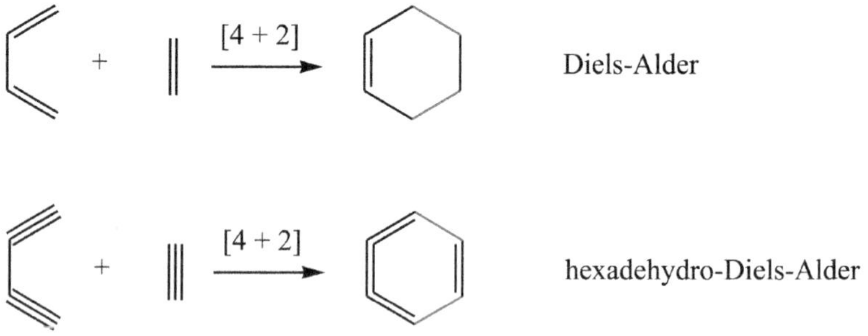

[4 + 2] Diels-Alder

[4 + 2] hexadehydro-Diels-Alder

The adduct of the HDDA reaction is a benzyne intermediate.

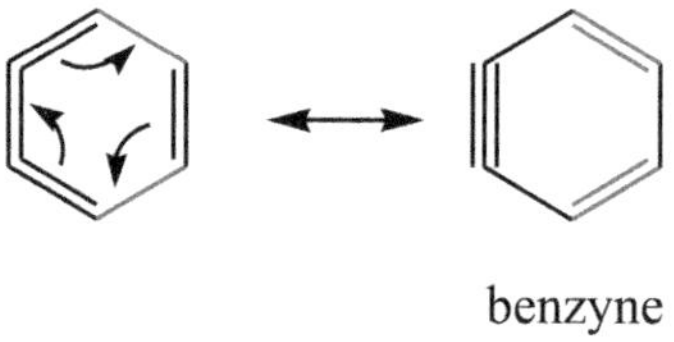

benzyne

Part 1

Explain the origin of the prefix name "hexadehydro" in describing this transformation.

By analogy, draw out schemes for the didehydro- and tetradehydro-Diels–Alder reactions.

Part 2

The reaction was discovered serendipitously after an attempted oxidation of the substrate shown with manganese dioxide. Show the expected oxidation product and how it cyclized to the observed product via the HDDA reaction. Highlight the target bonds made and describe the ring construction strategy.

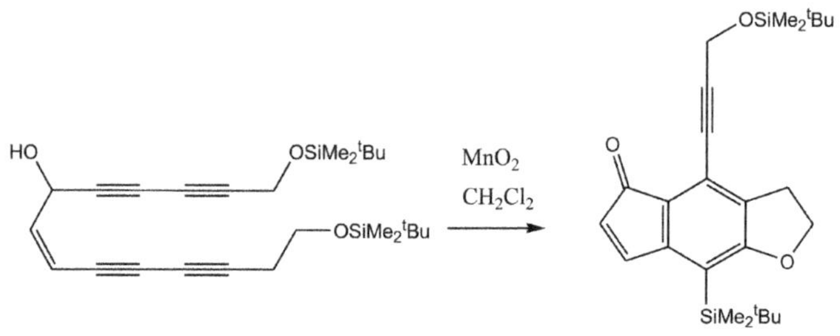

HO OSiMe$_2$^{t}Bu OSiMe$_2$^{t}Bu MnO_2 CH_2Cl_2 O OSiMe$_2$^{t}Bu O SiMe$_2$^{t}Bu

Part 3

The HDDA reaction was applied to the synthesis of blue emitting arylalkynyl naphthalene derivatives as shown by the synthesis plan given below [30].

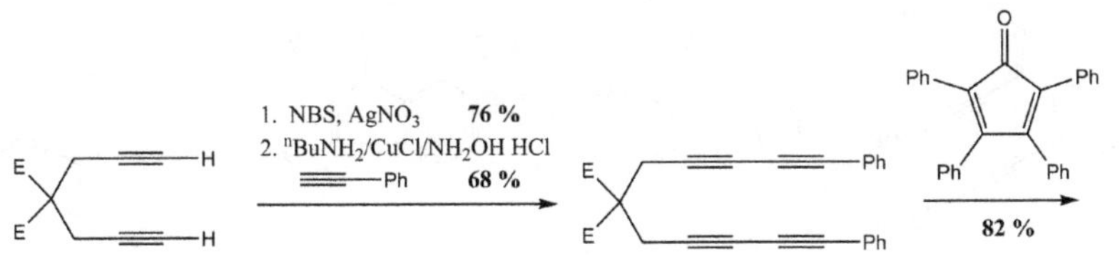

1. NBS, $AgNO_3$ 76 %
2. $^{n}BuNH_2/CuCl/NH_2OH$ HCl
≡—Ph 68 %

82 %

E = COOMe

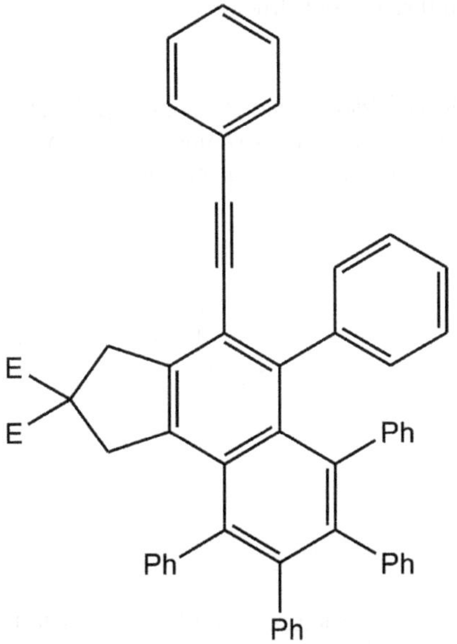

Show balanced equations for each step and determine the respective atom economies. Show the target bond map for the product structure and describe the ring construction strategy.

Part 4

Show how the fluorenone ring system could be constructed using the HDDA reaction.

Part 5

Show how the following alkaloids could be constructed using the HDDA reaction.

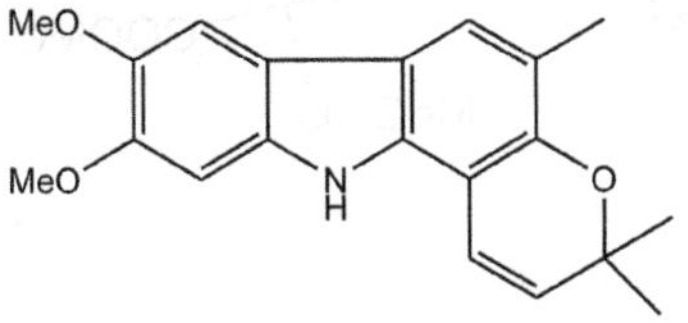

koenidine 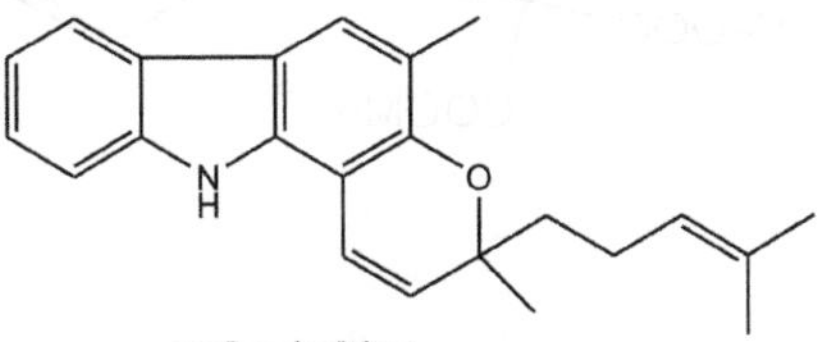 **mahanimbine**

Part 6

The following 3-component coupling reaction was developed using the HDDA strategy.

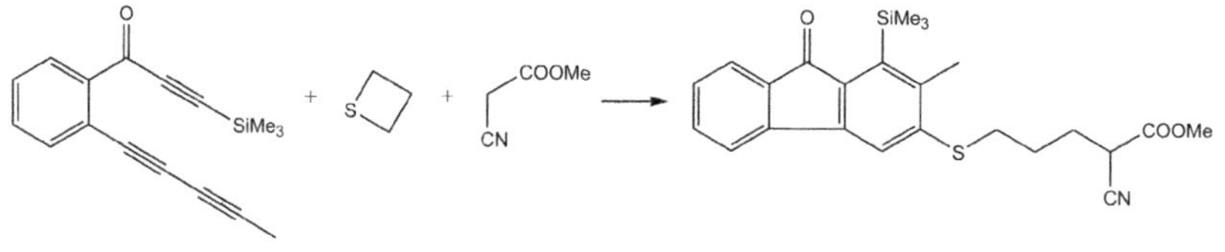

Show a mechanism for how this transformation takes place and indicate the target bonds made in the product structure.

Part 7

The pentadehydro-Diels–Alder reaction has been applied to the synthesis of fused pyridine derivatives as shown by the example below. Show a mechanism for the transformation and explain the logic of calling it a pentadehydro-Diels–Alder reaction.

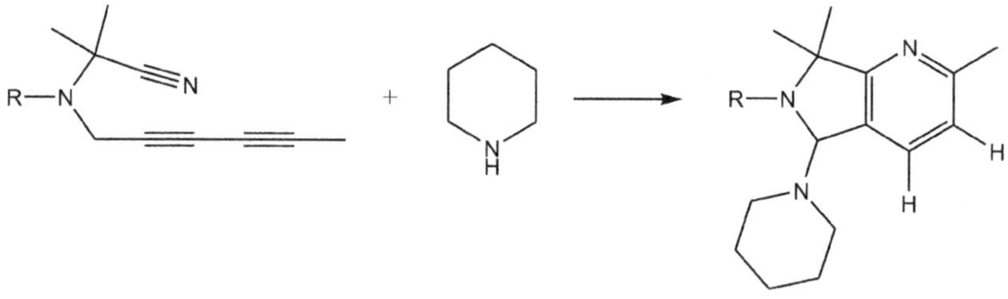

Part 8

Suggest how the following structure could be constructed from perylene using the HDDA strategy. Show the target bond mapping in the product structure and describe the ring construction strategy.

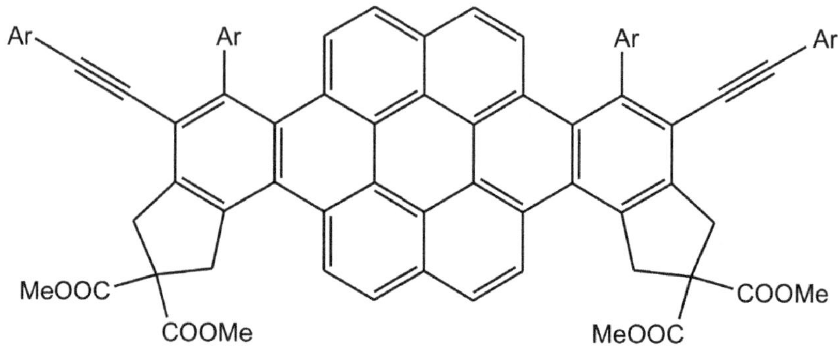

Solution

Part 1

C_4H_6 + C_2H_4 —[4 + 2]→ C_6H_{10} Diels-Alder

C_4H_2 + C_2H_2 —[4 + 2]→ C_6H_4 hexadehydro-Diels-Alder

On comparing C_6H_{10} with C_6H_4, we see that they differ in six hydrogen atoms, hence there is a loss of six hydrogen atoms from the parent Diels–Alder reaction.

C_4H_6 + C_2H_2 —[4 + 2]→ C_6H_8 didehydro-Diels-Alder (deficit of 10 - 8 = 2 H atoms wrt Diels-Alder)

C_4H_4 + C_2H_2 —[4 + 2]→ C_6H_6 tetradehydro-Diels-Alder (deficit of 10 - 6 = 4 H atoms wrt Diels-Alder)

Note: The adduct of the tetradehydro-Diels–Alder reaction can tautomerize to benzene via a [1,3]H shift.

H H

Part 2

$OSiMe_2{}^tBu$ HO O $SiMe_2{}^tBu$ —MnO_2, - $Mn(OH)_2$→ $OSiMe_2{}^tBu$ O O $SiMe_2{}^tBu$ —[4 + 2]→ $OSiMe_2{}^tBu$ O O $SiMe_2{}^tBu$ ↔ $OSiMe_2{}^tBu$ O O $SiMe_2{}^tBu$

expected product

Ring construction mapping: $[(5+0)_A+(4+2)_B+(5+0)_C]$

Part 3

Step 1

Reactants: 208 + 2*177.9 = 563.8
Products: 365.8 + 2*99 = 563.8
AE = 365.8/563.8 = 0.649, or 64.9%

Step 2

Reactants: 365.8 + 2*102 + 2*73 = 715.8
Products: 408 + 2*153.9 = 715.8
AE = 408/715.8 = 0.570, or 57.0%

Step 3

Reactants: 408 + 384 = 792
Products: 764 + 28 = 792
AE = 764/792 = 0.965, or 96.5%

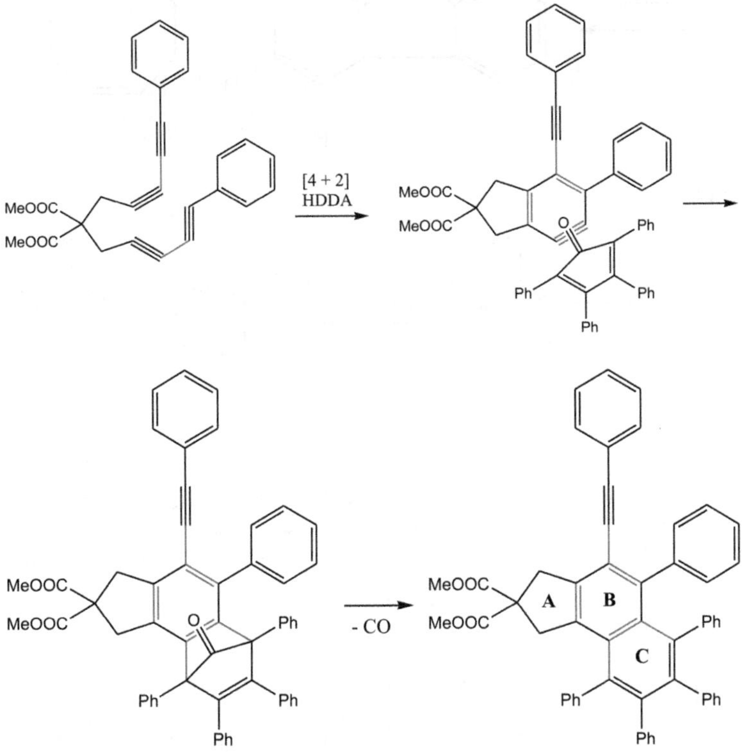

Ring construction mapping:

$$\left[(5+0)_A+(2+2+2)_B+(4+1+1)_C\right]$$

Ring construction strategy:

$$\xrightarrow{[(5+0)_A+(4+2)_B]}(A+B)_1\xrightarrow{[4+2]}C_2$$

Part 4 [31]

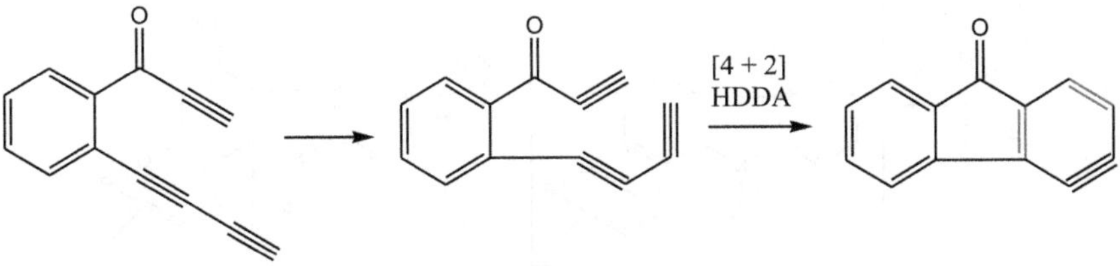

O

H

H

+

Part 5 [32]

SiMe3

MeO

MeO

N

Ts

[4 + 2]
HDDA

SiMe3

MeO

MeO

N

Ts

O

[2 + 2]

SiMe3

MeO

MeO

N

Ts

O

retro
[2 + 2]

SiMe3

MeO

MeO

N

Ts

O

SiMe3

MeO

MeO

N

Ts

O

2 [$^nBu_4N$]F
then 2 H_2O

- Me_3SiF
- TsF
- 2 [$^nBu_4N$]OH

H

MeO

MeO

N

H

O

koenidine

SiMe3

N

Ts

[4 + 2]
HDDA

SiMe3

N

Ts

O R

[2 + 2]

SiMe3

N

Ts

O

R

retro
[2 + 2]

SiMe3

N

Ts

O

R

[3,3]

$2 [^nBu_4N]F$ then $2 H_2O$

$- Me_3SiF$
- TsF
$- 2 [^nBu_4N]OH$

mahanimbine

R = H_2C

Part 6 [33]

[4 + 2] HDDA

Part 7 [34]

[1,3]H

[4 + 2]

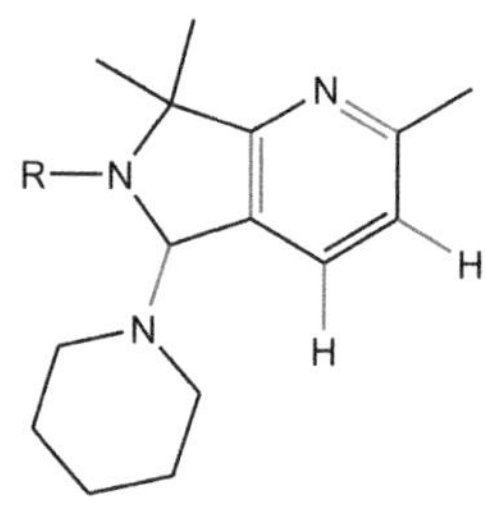

R—N N H H

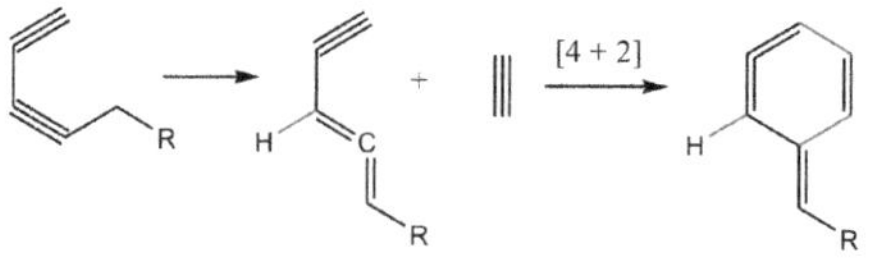

[4 + 2]

H C R H R

pentadehydro-Diels-Alder
(deficit of 10 - 5 = 5 H atoms wrt Diels-Alder)

RC_5H_3 RC_5H_3 RC_2H_2 RC_7H_5

Part 8 [35]

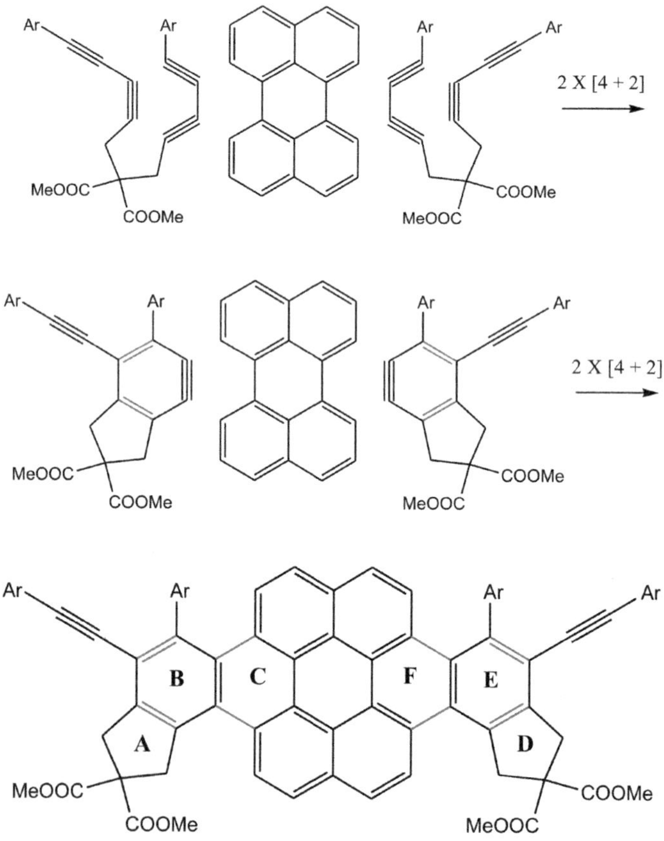

Ar Ar Ar Ar MeOOC COOMe COOMe MeOOC

2 X [4 + 2]

Ar Ar Ar Ar MeOOC COOMe COOMe MeOOC

2 X [4 + 2]

Ar Ar Ar Ar A B C F E D MeOOC COOMe COOMe MeOOC

Ring construction mapping:

$$\left[(5+0)_A+(4+2)_B+(4+2)_C\right]+\left[(5+0)_D+(4+2)_E+(4+2)_F\right]$$

Ring construction strategy:

$$\xrightarrow{[(5+0)_A+(4+2)_B]+[(5+0)_D+(4+2)_E]}(A+B)_1,(D+E)_1\xrightarrow{[4+2]_C+[4+2]_F}(C+F)_2$$

Example 4.12 [36]

Themes: balancing chemical equations, ring construction strategies, target bond mapping, overall kernel, overall AE, overall yield

The Weiss–Cook condensation involves reacting glyoxal derivatives with 3-ketoglutarates under aqueous mild basic conditions to give [3.3.0] bicyclic ring frameworks.

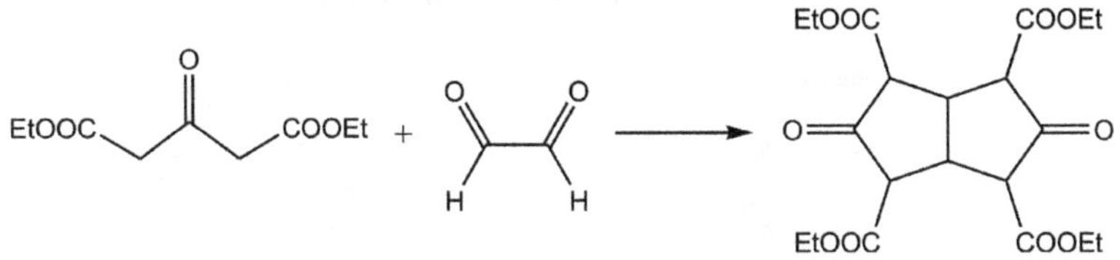

EtOOC COOEt O EtOOC COOEt + O O H H → EtOOC COOEt O O EtOOC COOEt

Write out a balanced chemical equation for the transformation and determine the atom economy. Draw the target bond map and describe the ring construction strategy.

Solution

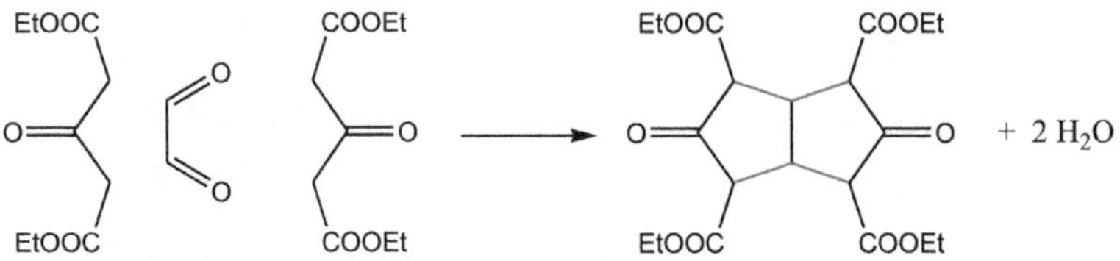

EtOOC O EtOOC O O COOEt O COOEt → EtOOC COOEt O O EtOOC COOEt + $2\ H_2O$

202 **58** **202** **426** **2 (18)**

- Reactants: 202 + 58 + 202 = 462
- Products: 426 + 2*18 = 462
- AE = 426/462 = 0.922, or 92.2%
- Ring construction strategy: $[(3+2)+(3+2)]_{[3.3.0]}$

Example 4.13

The Weiss–Cook condensation has been used as a synthesis strategy to construct polyquinane and polyquinene ring systems such as the ones shown below. The core bicyclic ring system common to all rings is tricyclo[3.3.3.0 [1, 5]]undecane or bicyclo[3.3.0]octane.

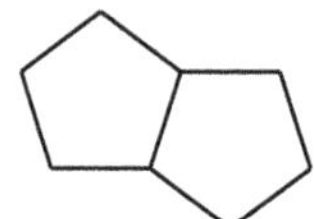

bicyclo[3.3.0]octane

tricyclo[3.3.3.$0^{1,5}$]undecane

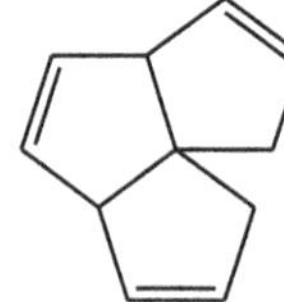

tricyclo[6.3.0.$0^{1,5}$]undeca-[3,6,9]-triene

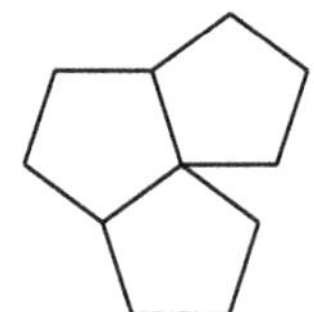

tricyclo[6.3.0.$0^{1,5}$]undecane

tricyclo[5.2.1.$0^{4,10}$]deca-2,5,8-triene

triquinacene

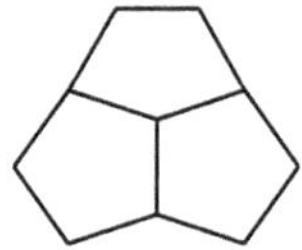

tricyclo[5.2.1.$0^{4,10}$]decane

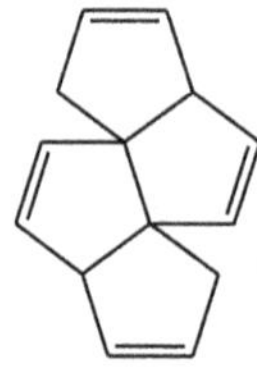

tetracyclo[6.6.0.$0^{1,5}$.$0^{8,12}$]tetradeca-3,6,10,13-tetraene

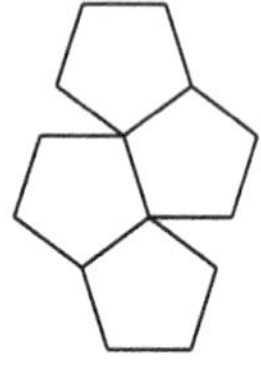

tetracyclo[6.6.0.$0^{1,5}$.$0^{8,12}$]tetradecane

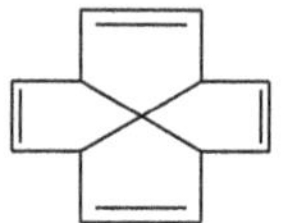

tetracyclo[5.5.1.$0^{4,13}$.$0^{10,13}$]trideca-2,5,8,11-tetraene

staurane-2,5,8,11-tetraene

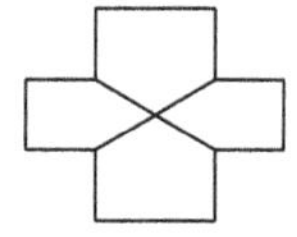

tetracyclo[5.5.1.$0^{4,13}$.$0^{10,13}$]tridecane

staurane

For each of these ring systems find the tricyclo[3.3.3.0[1, 5]]undecane motif.

SOLUTION

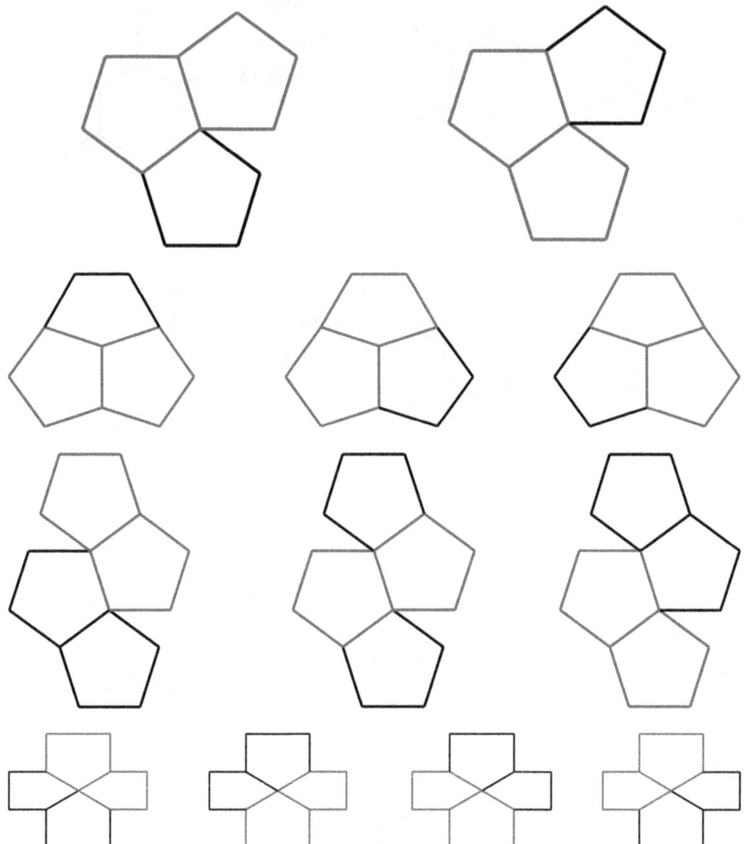

4.2.2 PROBLEMS

PROBLEM 4.5

Themes: balancing chemical equations, ring construction strategy

For the following reactions to prepare cyclohexadiones and cyclohexanone, show the ring construction strategy used by highlighting the target bonds in the product structure. For each reaction show balanced chemical equations.

(a) [37]

EtO O O OEt

NaOEt/EtOH
then H_2SO_4

O COOEt EtOOC O

(b) [38]

KO^{t}Bu
THF
then H_2O

(c) [39]

KO^{t}Bu
pentane
then H_2O

(d) [40]

NaOEt
then HCl
then H_2O

(e) [41]

nBuLi
I_2
heat
nBuLi
I_2
HCl
H_2O

(f) [42]

Ca^{+2}
heat

(g) [43]

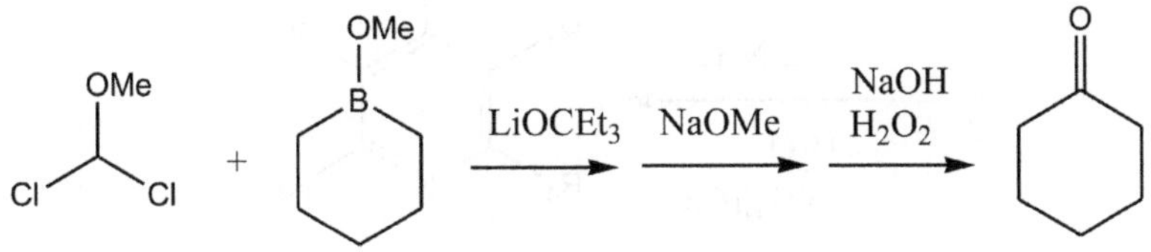

(h) [44]

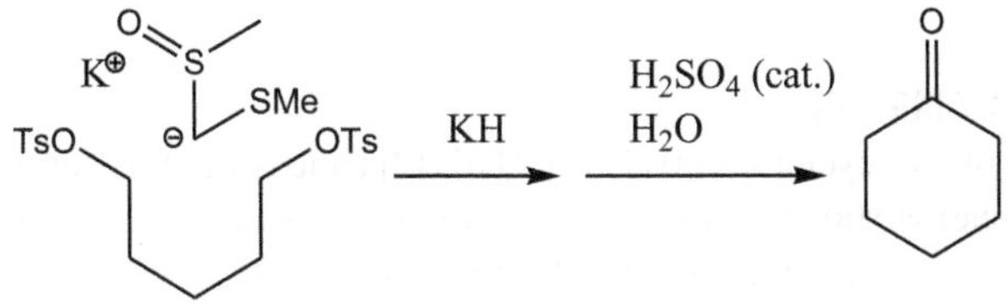

(i) [45]

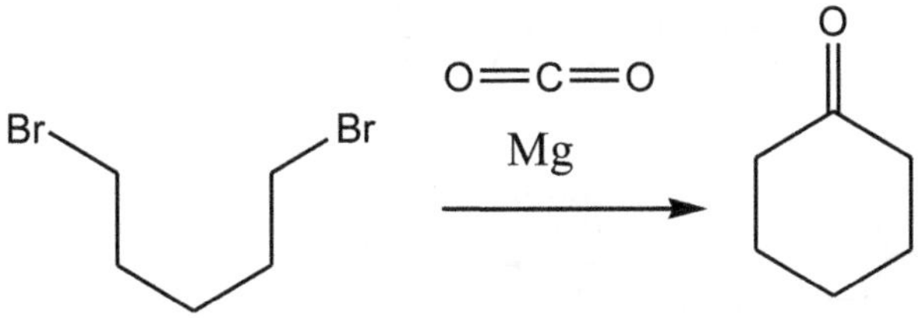

(j) [46]

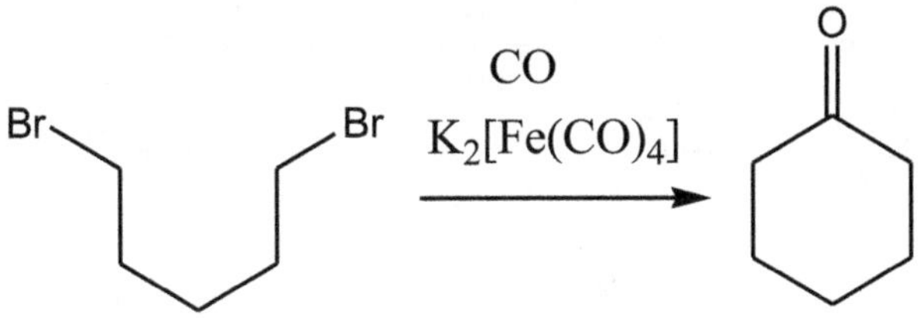

PROBLEM 4.6 [47, 48]

Themes: balancing chemical equations, multi-component reaction, dynamic kinetic resolution

The general form of the Balaban–Nenitzescu–Praill synthesis of substituted pyridines via pyrylium salts is given below.

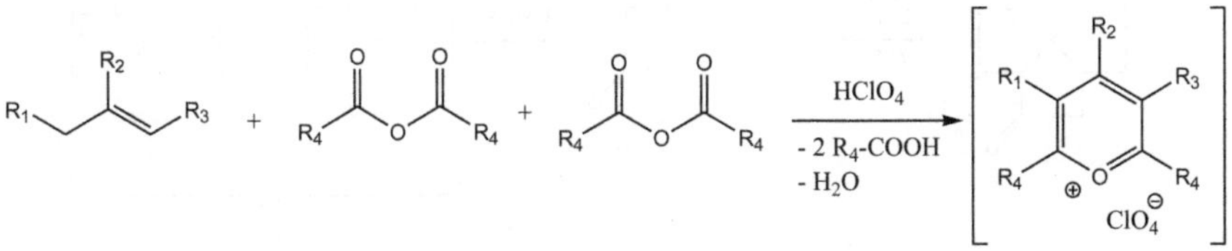

$\xrightarrow[\text{- H}_2\text{O, - [NH}_4\text{][ClO}_4\text{]}]{2\ NH_3}$ (pyridine with substituents R_1, R_2, R_3, R_4, R_4)

Suggest how this reaction could be applied in a dynamic kinetic resolution sense for the synthesis of both enantiomers of 2-(1-methyl-pyrrolidin-2-yl)-pyridine.

PROBLEM 4.7 [49, 50]

The synthesis of tetracyclo[5.5.1.0 [4, 13].0 [10, 13] trideca-2,5,8,11-tetraene (staurane-2,5,8,11-tetraene) is shown below. Rewrite the scheme showing balanced chemical equations for each step. Highlight the target bonds made in the final product structure and indicate the ring construction strategy.

$SOCl_2$, 85 %; CH_2N_2, 87 %; PPh_3, 93 %; $NaNO_2$, HCl, 60 %

H_2O, pH 8.3, 90 %; HCl/HOAc, H_2O, 90 %

OsO_4, 92 %; CrO_3, H_2SO_4, 70 %; naphthalene-1-sulfonic acid (cat.), 70 %

B_2H_6, THF, 92 %; heat, HMPA; 4 : 1, 61 %

PROBLEM 4.8 [51, 52]

The synthesis of tricyclo[5.2.1.0 [4, 10]deca-2,5,8-triene is shown below. Rewrite the scheme showing balanced chemical equations for each step. Highlight the target bonds made in the final product structure and indicate the ring construction strategy.

H_2SO_4; 72 %

K_2CO_3, $NaHCO_3$, H_2O/MeOH; 93 %

CH_2N_2; 95 %

KH

HCl/HOAc, heat; 90 %, 2 steps

OsO_4 (cat.), $NaIO_4$; 90 %

HCl, THF; 92 %

B_2H_6, THF; 92 %

HMPA, heat; 80 %

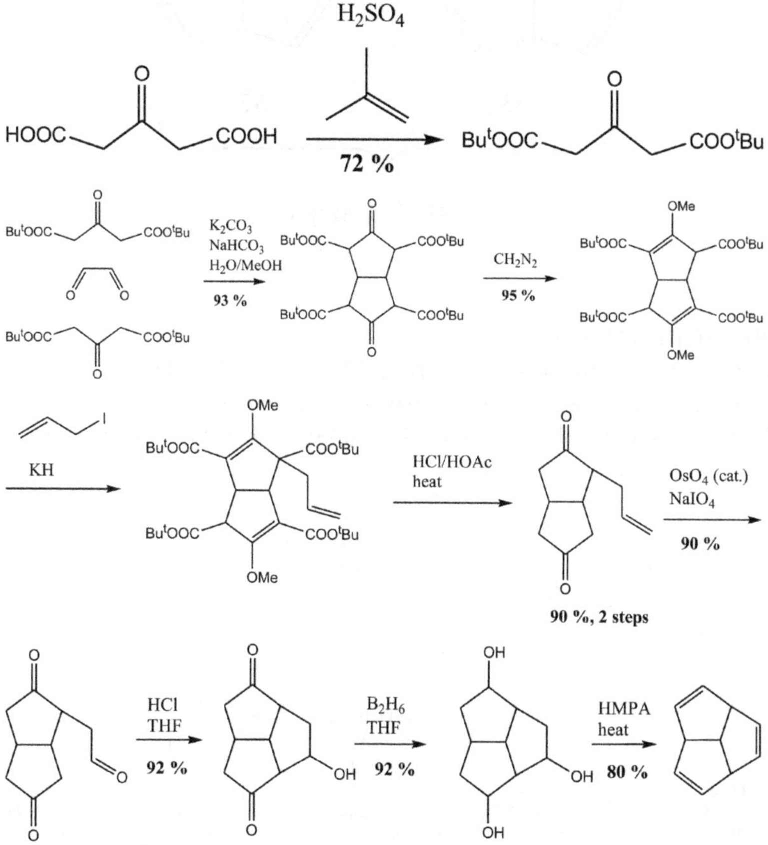

PROBLEM 4.9 [53–55]

The synthesis of tricyclo[6.3.0.0[1, 5]]undeca-[3,6,9]-triene is shown below. Rewrite the scheme showing balanced chemical equations for each step. Highlight the target bonds made in the final product structure and indicate the ring construction strategy.

$NaHCO_3$, H_2O; 80 %

HOAc, HCl; 70 %

naphthalene-1-sulfonic acid (cat.), benzene/dioxane; 60 %

BH_3 THF then MeOH

HMPA, heat

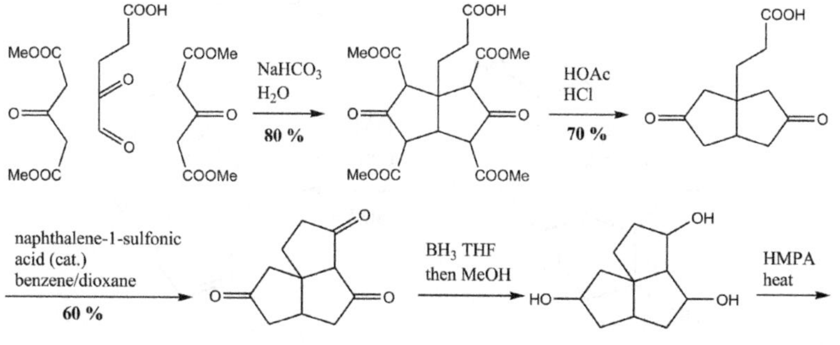

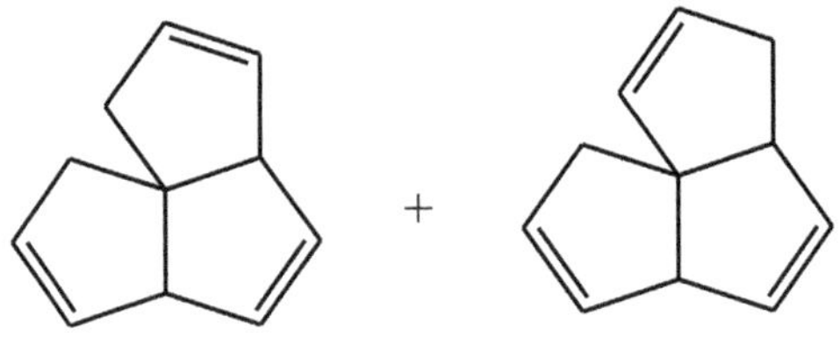

+

43 : 57

60 %, 2 steps

PROBLEM 4.10 [56, 57]

The synthesis of tetracyclo[6.6.0.0[1, 5].0[8, 12]]tetradeca-3,6,10,13-tetraene is shown below. Rewrite the scheme showing balanced chemical equations for each step. Highlight the target bonds made in the final product structure and indicate the ring construction strategy.

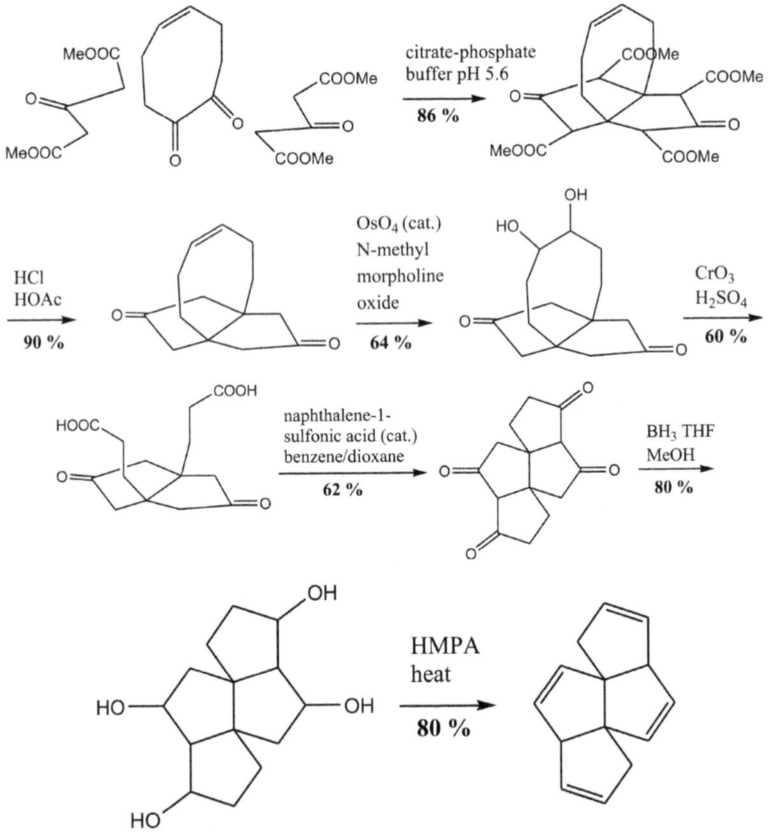

MeOOC, O, MeOOC, O, O, COOMe, O, COOMe

citrate-phosphate buffer pH 5.6, 86 %

COOMe, COOMe, O, O, MeOOC, COOMe

HCl, HOAc, 90 %

O, O

OsO_4 (cat.), N-methyl morpholine oxide, 64 %

OH, HO, O, O

CrO_3, H_2SO_4, 60 %

COOH, HOOC, O, O

naphthalene-1-sulfonic acid (cat.), benzene/dioxane, 62 %

O, O, O, O

BH_3 THF, MeOH, 80 %

OH, HO, OH, HO

HMPA, heat, 80 %

PROBLEM 4.11 [58–60]

Themes: ring construction strategy, reaction mechanism, target bond map

Morellin is a caged natural product found in the seeds of tropical trees and shrubs of the genus *Garcinia*.

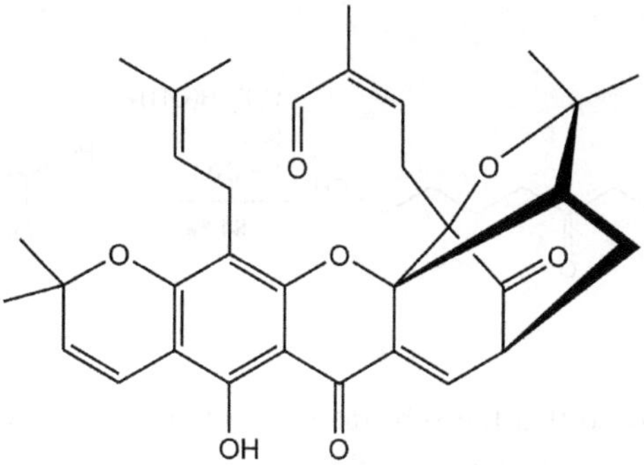
O
O
O
O
O
OH
O

The following late stage sequence was utilized in its total synthesis. Deduce the structure of the intermediate and highlight the target bonds in the final structure. *Hint*: Maintain the same structural aspect from structure to structure to better trace the bond connectivities in each step.

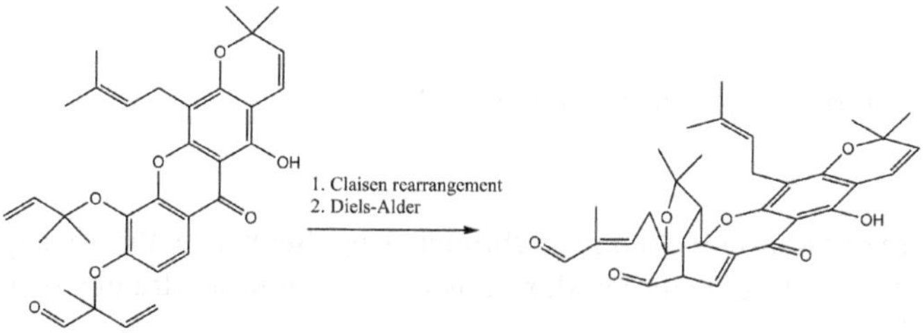
1. Claisen rearrangement
2. Diels-Alder
OH
OH

PROBLEM 4.12 [61]

Themes: balancing chemical equations, catalytic cycles, ring construction strategies, target bond mapping

Boronic acids are used to catalyze carbo- and heterocyclization reactions of allyl alcohols as shown in the example below.

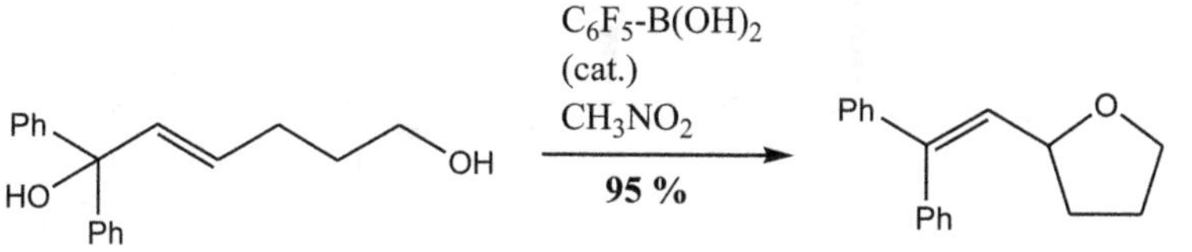
C6F5-B(OH)2
(cat.)
CH3NO2
95 %
Ph
HO
Ph
OH
Ph
Ph
O

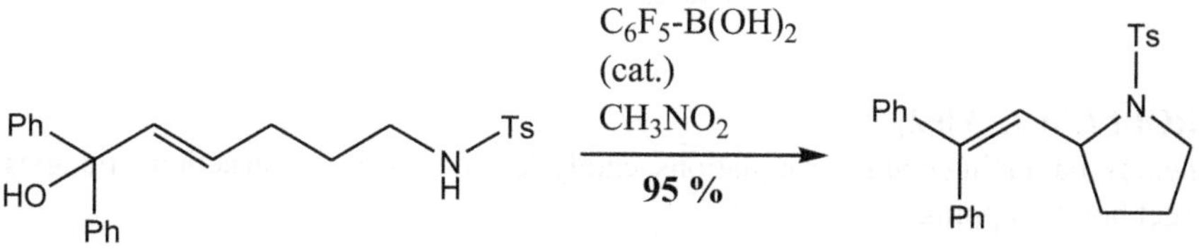
C6F5-B(OH)2
(cat.)
CH3NO2
95 %
Ph
HO
Ph
N
H
Ts
Ph
Ph
Ts
N

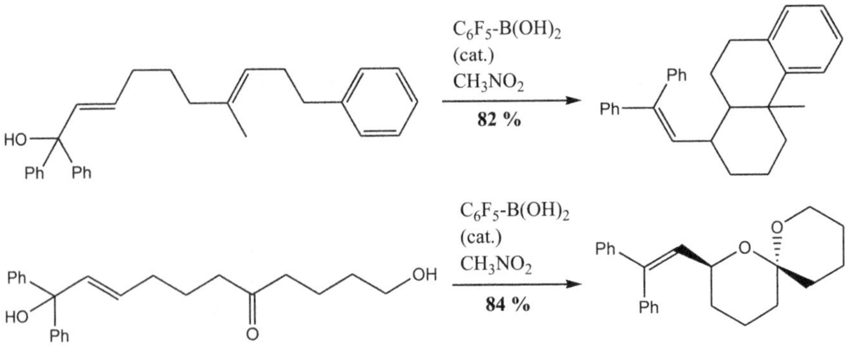

C_6F_5-B(OH)$_2$ (cat.) CH_3NO_2 82 %

C_6F_5-B(OH)$_2$ (cat.) CH_3NO_2 84 %

Part 1

Redraw the reactions so that the substrates have the same structural aspect as the products. Show balanced chemical equations.

Part 2

Based on the drawings in Part 1, highlight the target bonds made in each reaction and deduce the ring construction strategies.

Part 3

Write out a catalytic cycle for the first example.

Part 4

The structure of vitamin E contains a chroman ring. Show how this moiety could be made from the starting materials shown using the synthesis strategy demonstrated in this problem.

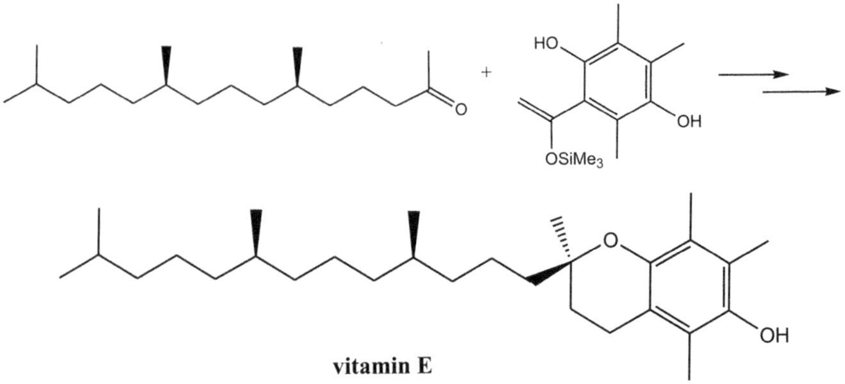

vitamin E

PROBLEM 4.13 [62]

Themes: balancing chemical equations, catalytic cycles, ring construction strategies, target bond mapping

Boronic acids are used to catalyze [3+2] cycloaddition reactions to produce triazole, isoxazole, and isoxazolidine ring containing molecules as shown in the example below.

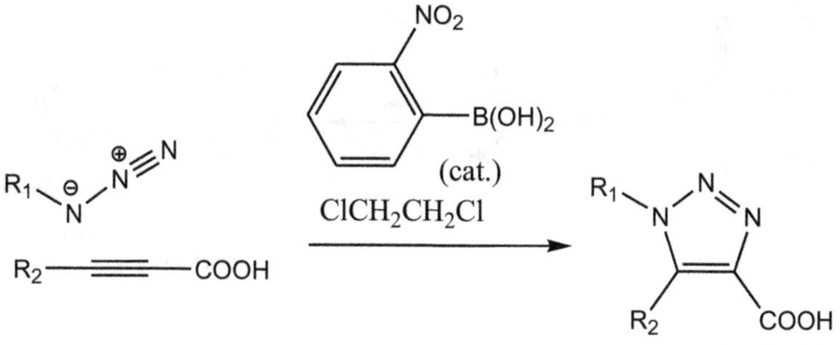
NO2
B(OH)2
(cat.)
ClCH2CH2Cl
R1
N
N
N
R2
COOH

1,2,3-triazoles

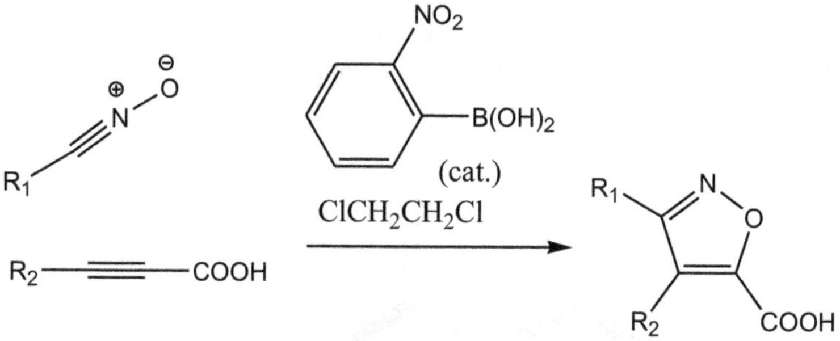
NO2
B(OH)2
(cat.)
ClCH2CH2Cl
R1
N
O
R2
COOH

isoxazoles

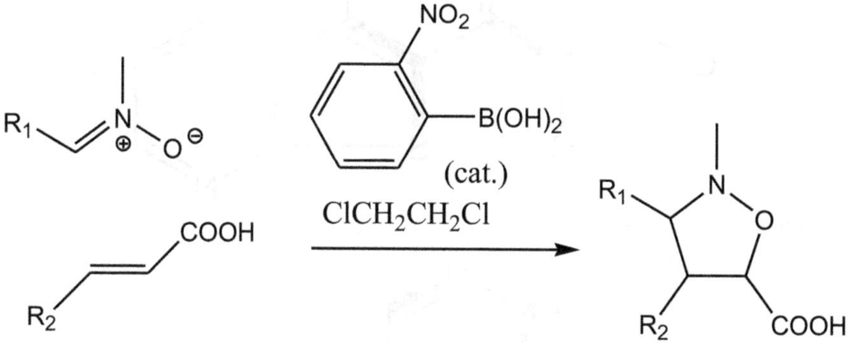
NO2
B(OH)2
(cat.)
ClCH2CH2Cl
R1
N
O
R2
COOH

isoxazolidines

Part 1

Highlight the target bonds made in each reaction.

Part 2

Write out a catalytic cycle for the first example.

Part 3

Propose a synthesis of the anticonvulsant rufinamide from the starting materials shown using the synthesis strategy demonstrated in this problem.

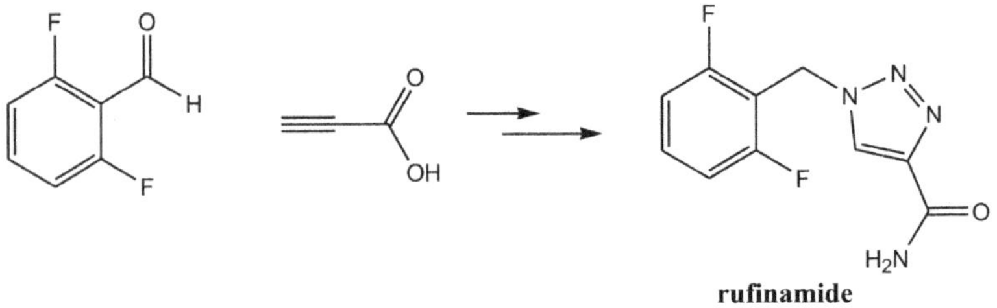

rufinamide

PROBLEM 4.14

Themes: balancing chemical equations, ring construction strategies, product structure target bond maps

For the following product structure target bond maps, suggest the starting material components that are required for the intended multi-component reactions. Balance all chemical equations as necessary.

(a) [63]

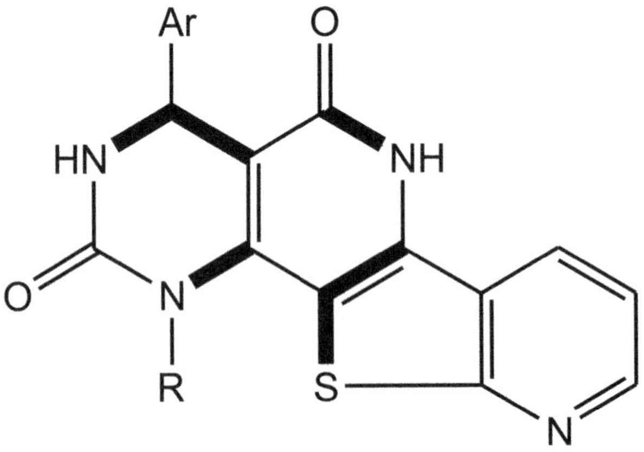

(b) [64]

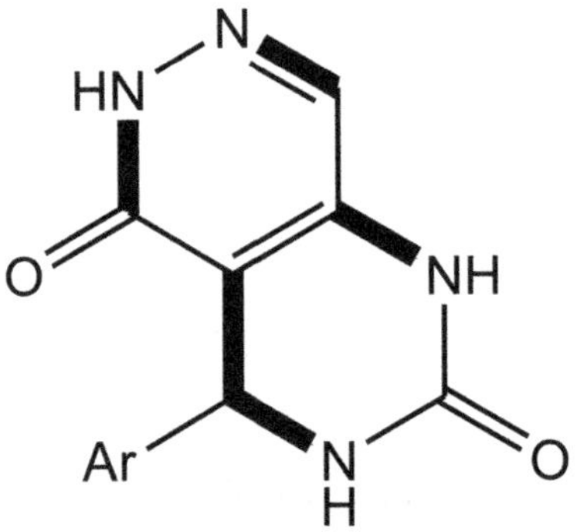

(c) [65]

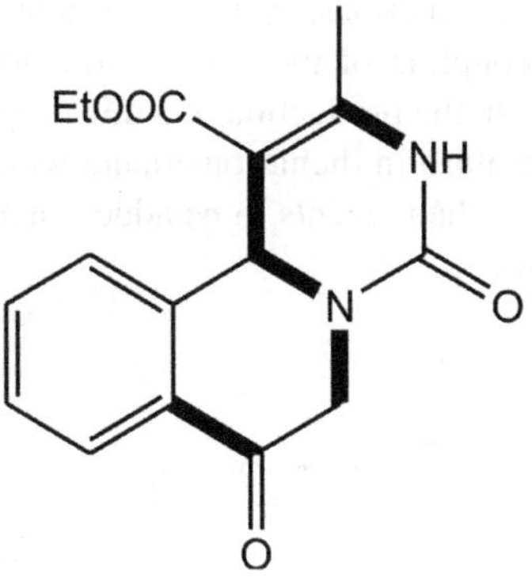

EtOOC NH N O O

(d) [66]

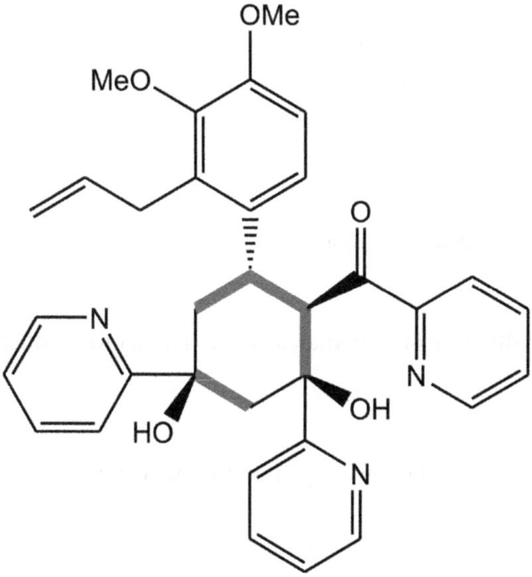

OMe MeO N O N HO OH N

(e) [67]

O R N N R O NH N H O

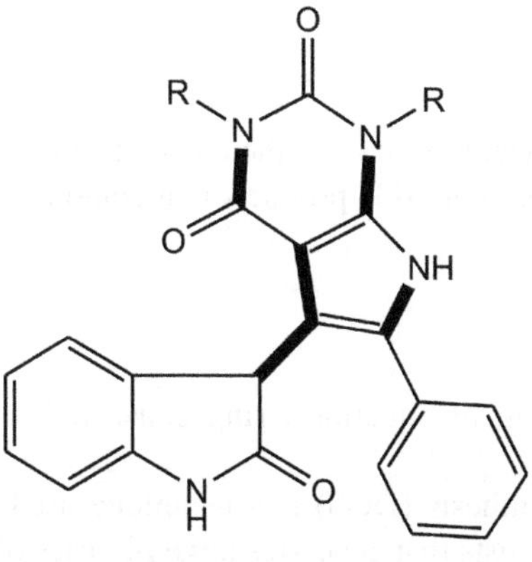

PROBLEM 4.15 [68]

Themes: balancing chemical equations, multi-component reactions, atom economy

A novel 5-component coupling of two amines, diketene, malononitrile, and an aromatic aldehyde resulted in the production of a dihydropyridine. The reaction was carried out at room temperature in dichloromethane with no added catalysts. This assembly, however, required the reagents to be added in a specific order as implied by the scheme shown below.

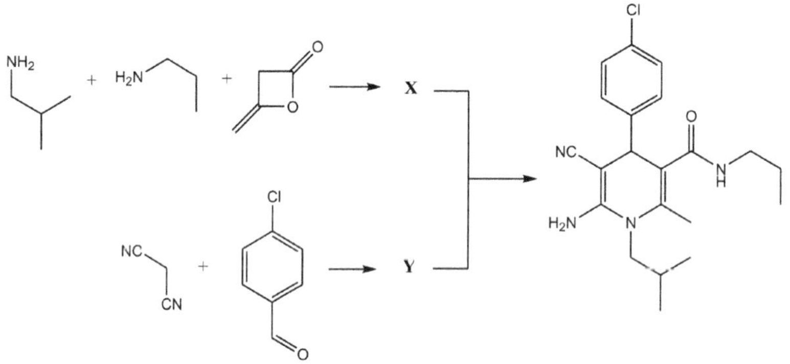

Part 1

Deduce the structures of intermediates X and Y.

Part 2

Show mechanistically how intermediates X and Y react to product the dihydropyridine product.

Part 3

Show an overall balanced chemical equation for the reaction and determine its atom economy.

Part 4

Highlight the target bonds made in the final product and describe the ring construction strategy.

Part 5

The authors obtained a yield of 82% for the reaction. Based on mechanistic considerations, suggest other possible side products that could arise from intermediates X and Y.

PROBLEM 4.16 [69]

Themes: balancing chemical equations, ring construction strategies, target bond mapping

Diversity-oriented synthesis (DOS) is a technique used by synthetic medicinal and pharmaceutical chemists that generates large libraries of compounds in the lead discovery process. In this strategy a scaffold structure of constant structure having pre-selected properties is decorated with a diverse array of groups that can be varied until a finalized small set of lead compounds is obtained after an automated high-throughput screening process against a given assay.

An example synthesis sequence having DOS potential is shown below.

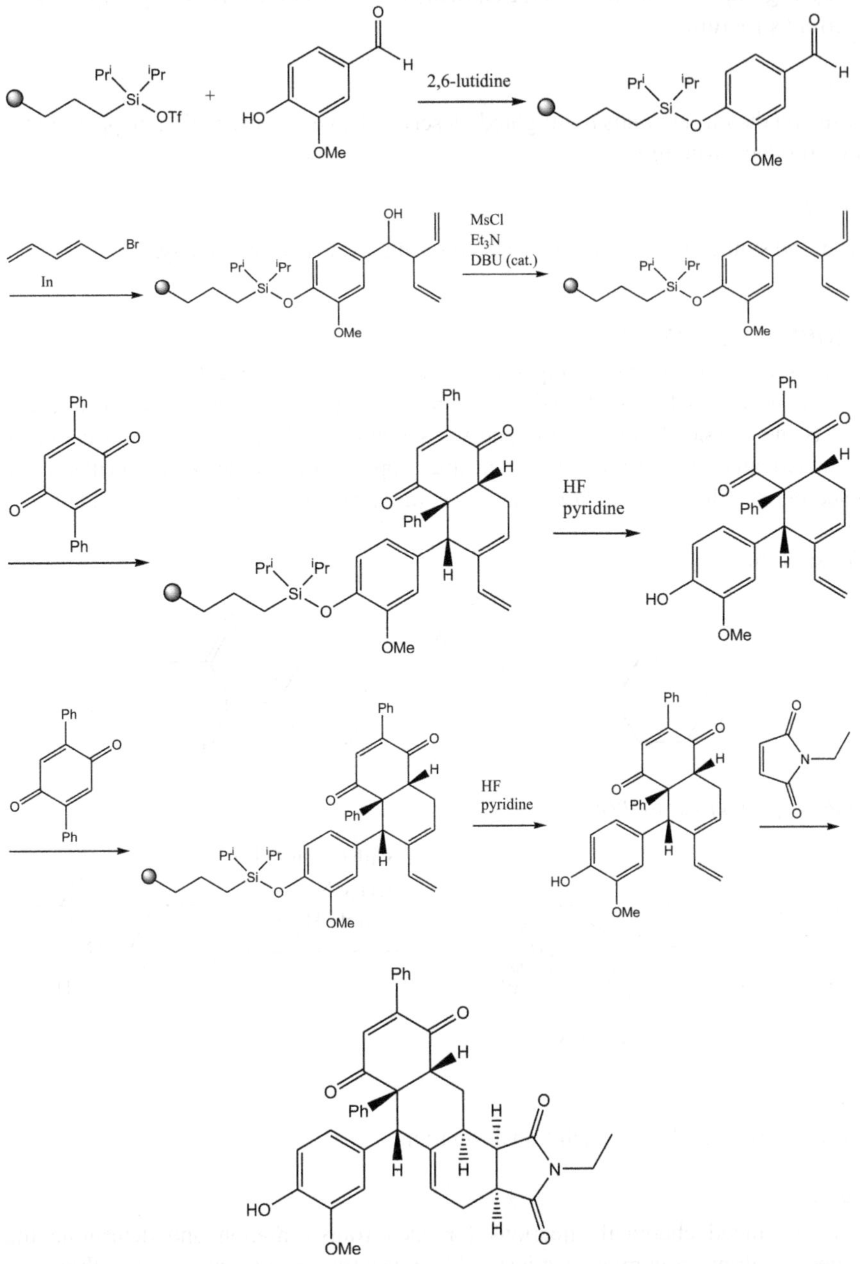

Part 1

Rewrite the scheme showing all by-products in each step and highlight the target bonds made along the way.

Part 2

Using R groups rewrite the scheme showing the points of diversity of groups and the scaffold structure.

Part 3

Based on the target bonds highlighted, describe the ring construction mapping in the final product structure.

Part 4

From the synthesis sequence, describe the ring construction strategy.

PROBLEM 4.17 [70–72]

Themes: balancing chemical equations, reaction comparison, catalytic cycles

Two strategies for synthesizing *N*-cyclohexyl-acetamide are shown below. The first is the classic Ritter reaction and the second is a Ritter-type CH amination of an unactivated sp^3 carbon center that uses copper(II)bromide, zinc(II)triflate, and Selectfluor®. The authors claim that two equivalents of Selectfluor® are needed.

Classic Ritter

OH H_2SO_4 CH_3CN H_2O O N H

Ritter-Type CH Amination

+ F—N⊕ ⊖PF_6 N⊕ Cl ⊖PF_6 2 equiv. 1. $CuBr_2$, $Zn(OTf)_2$ CH_3CN 2. $NaOH/H_2O$ O N H

Part 1

Write catalytic cycles for each transformation.

Part 2

Write balanced chemical equations for each transformation and determine the respective atom economies. Is it justifiable to use two equivalents of Selectfluor®?

Part 3

In a separate experiment Selectfluor® (F-TEDA-BF_4) reacted with copper(II)bromide in acetonitrile to give a white solid which was verified by X-ray crystallography to have the structure H-TEDA-BF_4.

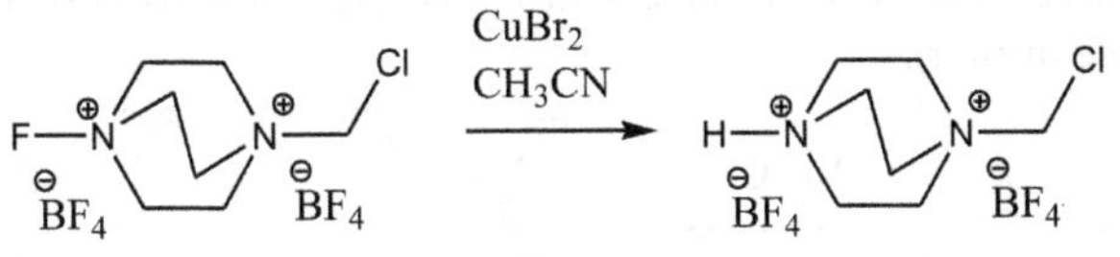

$CuBr_2$
CH_3CN

F-TEDA-BF_4

H-TEDA-BF_4
white solid
structure verified by X-ray crystallography

If this reaction is operative in parallel with the amination reaction, how does this change the results of Part 2?

PROBLEM 4.18 [73]

Themes: ring construction strategy, product structure target bond maps

The Alder–Rickert reaction is a thermal [4+2] tandem Diels–Alder retro-Diels–Alder sequence between a 1,3-cyclohexadiene and an acetylene that yields highly substituted aromatic ring products. The intermediate is a bicyclo[2.2.2]octadiene framework. An example is shown below.

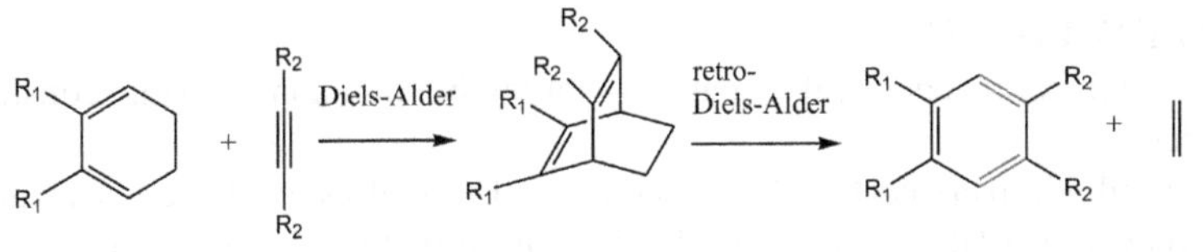

Diels-Alder

retro-Diels-Alder

By analogy show the expected products from the following substrates. In each case, identify what the by-products are.

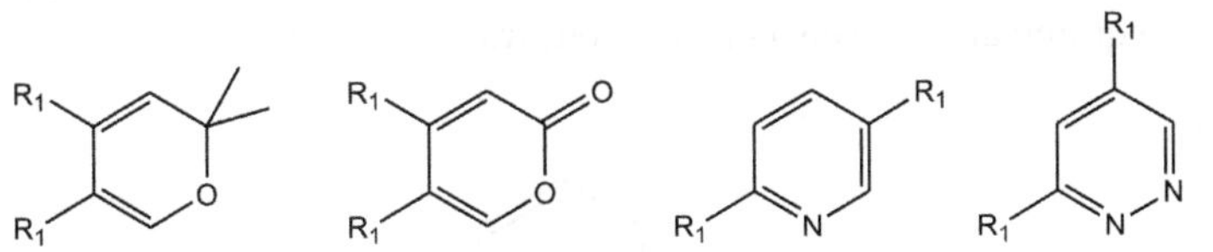

(a) (b) (c) (d)

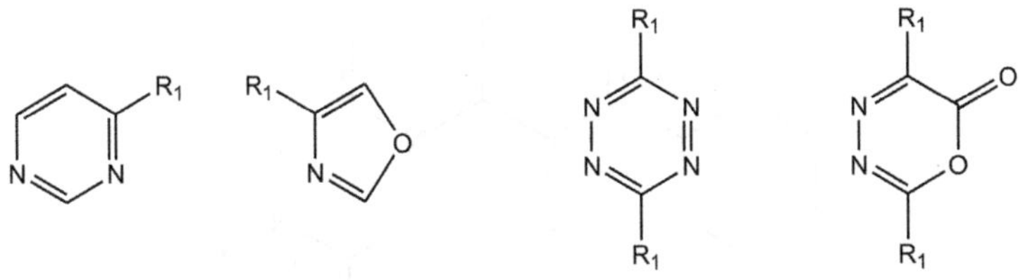

(e) (f) (g) (h)

PROBLEM 4.19 [74]

Themes: balancing chemical equations, reaction mechanism, ring construction strategies, ring partitioning

The Claisen isoxazole synthesis methodology involves condensing β-ketoesters with hydroxylamine. Two ring products are possible: isoxazol-3-ol (major) and 4*H*-isoxazol-5-one (minor).

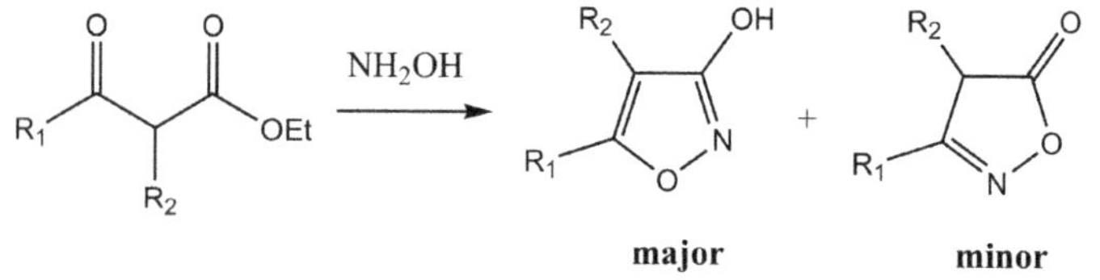

NH_2OH

major **minor**

Part 1

Write out mechanisms that account for the formation of these ring structures.

Part 2

Write out balanced chemical equations that lead to each of these ring structures and determine their minimum atom economies.

Part 3

Show all possible two-partitions of the isoxazole ring and determine which partition corresponds to the Claisen methodology. Based on the other combinations conjecture other two-component reactions that can lead to this product.

PROBLEM 4.20 [75]

Themes: balancing chemical equations, target bond mapping, ring construction strategies

Fused dihydrotriazolo[1,5-a]pyrazinones and triazolobenzodiazepines were synthesized by an Ugi/alkyne–azide cycloaddition synthetic sequence. This sequence is ordered so that the resulting adduct of the Ugi condensation is isolated first before carrying out the click reaction. For the following products shown, highlight the target bonds made in the product structures and deduce the structures of the starting materials required and the ring construction strategies employed.

Example 1

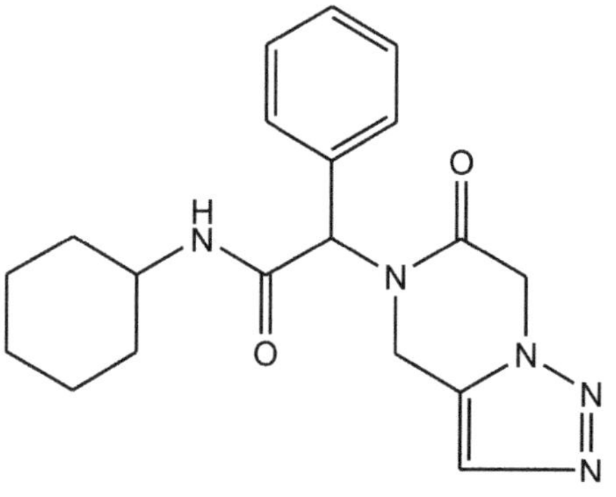

Example 2

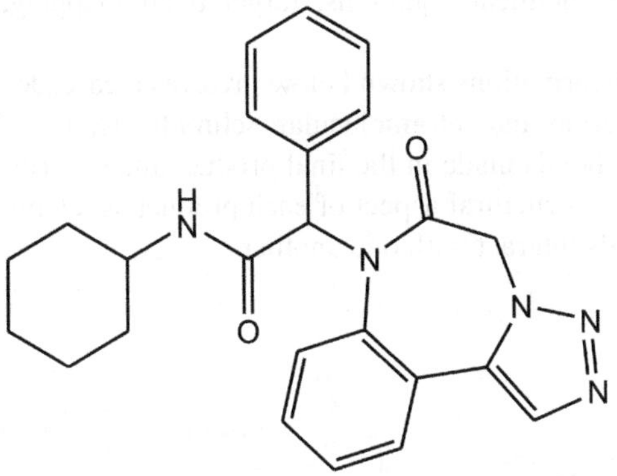

Example 3

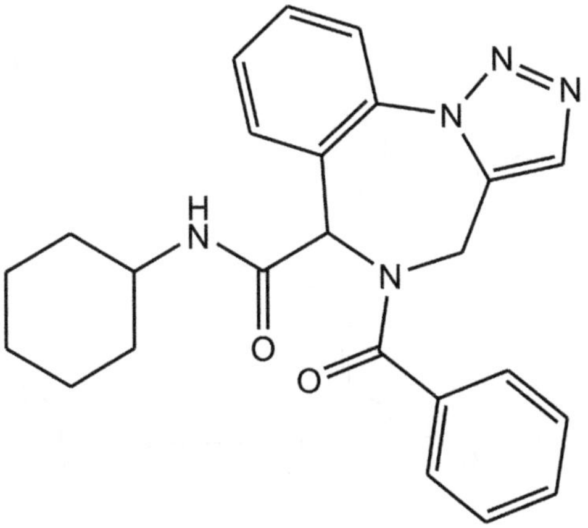

Example 4

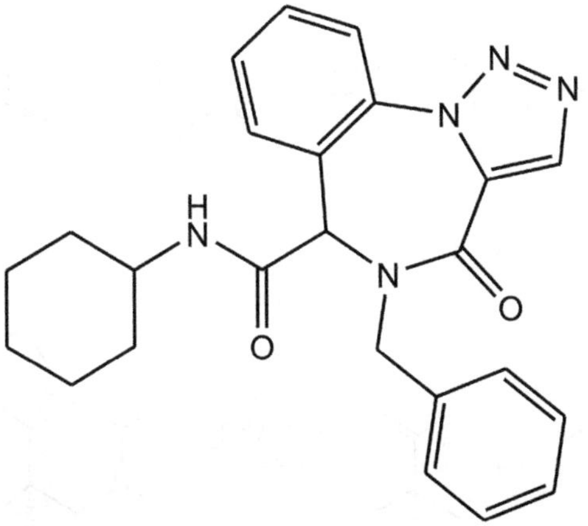

PROBLEM 4.21 [76]

Themes: balancing chemical equations, target bond mapping, ring construction strategies

Each of the transformations shown below involves a cascade sequence involving Diels–Alder cyclization and intramolecular Schmidt insertion. For each example, highlight the target bonds made in the final product and describe the ring construction strategy. Use the structural aspect of each product as a template to deduce how the starting materials interact with one another.

Example 1

N_3 + → (CH_3AlCl_2) N, O

Example 2

$OSiMe_3$, N_3 + → ($SnCl_4$) N, O, O

Example 3

$OSiMe_3$, N_3 + → ($SnCl_4$) N, O, O

Example 4

$OSiMe_3$, N_3 + O, H → (CH_3AlCl_2) O, NH, O + O, H, N, O

major minor

Example 5

N_3, O + → (CH_3AlCl_2) N, O, H

REFERENCES

1. Shestapalov, A.M.; Laionova, N.A.; Federov, A.E.; Rodinovskaya, L.A.; Mortikov, V.Y.; Zubarev, A.A.; Bushmarinov, I.S. *ACS Comb. Sci.* 2013, *15*, 541.
2. Soleimai, E.; Zainali, M.; Ghasemi, N.; Notash, B. *Tetrahedron* 2013, *69*, 9832.
3. Biginelli, P. *Chem. Ber.* 1891, *24*, 1317.
4. Chen, C.X.; Liu, L.; Yang, D.P.; Wang, D.; Chen, Y.J. *Synlett* 2005, 2047.
5. Andraos, J. *Beilstein J. Org. Chem.* 2016, *12*, 2420.
6. Petrenko–Kritschenko, P.; Zoneff, N. *Chem. Ber.* 1906, *39*, 1358.
7. Guareschi, J. *Chem. Ber.* 1895, *28*, 160 referate [abstract].
8. Kuettel, G.M.; McElvain, S.M. *J. Am. Chem. Soc.* 1931, *53*, 2692.
9. Ruzicka, L.; Fornasir, V. *Helv. Chim. Acta* 1920, *3*, 806.
10. Dickerman, S.C.; Lindwall, H.G. *J. Org. Chem.* 1949, *14*, 530.
11. Zhao, N.; Wang, Y.L.; Wang, J.Y. *J. Chin. Chem. Soc.* 2005, *52*, 535.
12. Foo, K.; Usui, I.; Götz, D.C.G.; Werner, E.W.; Holte, D.; Baran, P.S. *Angew. Chem. Int. Ed.* 2012, *51*, 11491.
13. Das, B.; Satyalakshmi, G.; Suneel, K.; Damodar, K. *J. Org. Chem.* 2009, *74*, 8400.
14. Nassar-Hardy, L.; Fabre, S.; Amer, A.M.; Fouquet, E.; Felpin, F.X. *Tetrahedron Lett.* 2012, *53*, 338.
15. Mukherjee, N.; Ahammed, S.; Bhadra, S.; Ranu, B.C. *Green Chem* 2013, *15*, 389.
16. Jayakumar, J.; Parthasarathy, K.; Cheng, C.H. *Angew. Chem. Int. Ed.* 2012, *51*, 197.
17. Huang, L.; Wang, Q.; Liu, X.; Jiang, H. *Angew. Chem. Int. Ed.* 2012, *51*, 5696.
18. Shares, J.; Yehl, J.; Kowalsick, A.; Byers, P.; Haaf, M.P. *Tetrahedron Lett* 2012, *53*, 4426.
19. Neumann, J.J.; Suri, M.; Glorius, F. *Angew. Chem. Int. Ed.* 2010, *49*, 7790.
20. Shen, L.; Cao, S.; Liu, N.J.; Wu, J.J.; Zhu, L.J.; Qian, X.H. *Synlett* 2008, 1341.
21. Shen, L.; Zhang, J.; Cao, S.; Yu, J.; Liu, N.; Wu, J.; Qian, X. *Synlett* 2008, 3058.
22. Willy, B.; Müller, T.T.J. *Eur. J. Org. Chem.* 2008, 4157.
23. Majumder, S.; Gipson, K.R.; Staples, R.J.; Odom, A.L. *Adv. Synth. Catal.* 2009, *351*, 2013.
24. Heller, S.T.; Natarajan, S.R. *Org. Lett.* 2006, *8*, 2675.
25. Stonehouse, J.P.; Chekmarev, D.S.; Ivanova, N.V.; Lang, S.; Pairandeau, G.; Smith, N.; Stocks, M.J.; Sviridov, S.I.; Utkina, L.M. *Synlett* 2008, 100.
26. Adib, M.; Mohammadi, B.; Bijanzadeh, H.R. *Synlett* 2008, 3180.
27. Raw, S.A.; Turner, A.T. *Tetrahedron Lett.* 2009, *50*, 696.
28. Mohanan, K.; Martin, A.R.; Toupet, L.; Smietana, M.; Vasseur, J.J. *Angew. Chem. Int. Ed.* 2010, *49*, 3196.
29. Hoye, T.R.; Baire, B.; Niu, D.; Willoughby, P.H.; Woods, B.P. *Nature* 2012, *490*, 208.
30. Xu, F.; Hershey, K.W.; Holmes, R.J.; Hoye, T.R. *J. Am. Chem. Soc.* 2016, *138*, 12739.
31. Niu, D.; Willoughby, P.H.; Woods, B.P.; Baire, B.; Hoye, T.R. *Nature* 2013, *501*, 531.
32. Wang, T.; Hoye, T.R. *J. Am. Chem. Soc.* 2016, *138*, 13870.
33. Chen, J.; Palani, V.; Hoye, T.R. *J. Am. Chem. Soc.* 2016, *138*, 4318.
34. Wang, T.; Naredla, R.R.; Thompson, S.K.; Hoye, T.R. *Nature* 2016, *532*, 484.
35. Xu, F.; Xiao, X.; Hoye, T.R. *Org. Lett.* 2016, *18*, 5636.
36. Bertz, S. H.; Cook, J. M.; Gawish, A.; Weiss, U. *Org. Synth.* 1986, *64*, 27.
37. Nielsen, A.T.; Carpenter, W.R. *Org. Synth. Coll.* 1973, *5*, 288.
38. Ishikawa, T.; Kadoya, R.; Arai, M.; Takahashi, H.; Kaisi, Y.; Mizuta, T.; Yoshikai, K.; Saito, S. *J. Org. Chem.* 2001, *66*, 8000.
39. van den Berg, E. M. M.; Jansen, F. J. H. M.; de Goede, A. T. J. W.; Baldew, A. U.; Lugtenburg, J. *Rec. Trav. Chim.* 1990, *109*, 287.
40. Cui, B.; Wang, R.H.; Chen, L.Z.; Yan, J.; Han, G.F. *Synth. Commun.* 2011, *41*, 1064.
41. Shatzmiller, S.; Lidor, R. *Synthesis* 1983, 580.

42. Feist, F. *Chem. Ber.* 1895, *28*, 738.
43. Brown, H.C.; Phadke, A.S.; Rangaishenvi, M.V. *J. Am. Chem. Soc.* 1988, *110*, 6263–6264.
44. Ogura, K.; Yamashita, M.; Suzuki, M.; Furukawa, S.; Tsuchinashi, G. *Bull. Chem. Soc. Jpn.* 1984, *57*, 1637.
45. Grignard, V.; Vignon, L. *Compt. Rend.* 1907, *144*, 1359.
46. Yamashita, M.; Uchida, M.; Tashika, H.; Suemitsu, R. *Bull. Chem. Soc. Jpn.* 1989, *62*, 2728.
47. Balaban, A.T.; Nenitzescu, C.D. *Ann. Chem.* 1959, *625*, 74.
48. Praill, P.F.G.; Whitear, A.L. *Proc. Chem. Soc.* 1961, 312.
49. Venkatachalam, M.; Deshpande, M.N.; Jawdosiuk, M.; Wehrli, S.; Cook, J.M.; Weiss, U. *Tetrahedron* 1986, *42*, 1597.
50. Deshpande, M.N.; Jawdosiuk, M.; Kubiak, G.; Venkatachalam, M.; Weiss, U.; Cook, J.M. *J. Am. Chem. Soc.* 1985, *107*, 4786.
51. Bertz, S.H.; Lannoye, G.; Cook, J.M. *Tetrahedron Lett.* 1985, *26*, 4695.
52. Gupta, A.K.; Lannoye, G.S.; Kubiak, G.; Schkeryaatz, J.; Wehrli, S.; Cook, J.M. *J. Am. Chem. Soc.* 1989, *111*, 2169.
53. Oehldrich, J.; Cook, J.M.; Weiss, U. *Tetrahedron Lett.* 1976, *17*, 4549.
54. Venkatachalam, M.; Kubiak, G.; Cook, J.M.; Weiss, U. *Tetrahedron Lett.* 1985, *26*, 4863.
55. Venkatachalam, M.; Deshpande, M.N.; Jawdosiuk, M.; Wehrli, S.; Cook, J.M.; Weiss, U. *Tetrahedron* 1986, *42*, 1597.
56. Mitschka, R.; Oehldrich, J.; Takahashi, K.; Cook, J.M.; Weiss, U. *Tetrahedron* 1981, *17*, 4521.
57. Venkatachalam, M.; Deshpande, M.N.; Jawdosiuk, M.; Wehrli, S.; Cook, J.M.; Weiss, U. *Tetrahedron* 1986, *42*, 1597.
58. Quillinan, A.J.; Scheinmann, F. *Chem. Commun.* 1971, 966.
59. Tisdale, E.J.; Slobodov, I.; Theodorakis, E.A. *Proc. Natl. Acad. Sci. USA* 2004, *101*, 12030.
60. Karanjgaonkar, C.G.; Nair, P.M.; Venkataraman, K. *Tetrahedron Lett.* 1966, 7, 687.
61. Zheng, H.; Ghanbari, S.; Nakamura, S.; Hall, D.G. *Angew. Chem. Int. Ed.* 2012, *51*, 6187.
62. Zheng, H.; McDonald, R.; Hall, D.G. *Chem. Eur. J.* 2010, *16*, 5454.
63. Lebedyeva, I.O.; Dotsenko, V.V.; Turovtsev, V.V.; Krivokolysko, S.G.; Povstyanoy, V.M.; Povstyanoy, M.V. *Tetrahedron* 2012, *68*, 9729.
64. Patra, G.C.; Bhunia, S.C.; Roy, M.K.; Pal, S.C. *Helv. Chim. Acta.* 2013, *96*, 130.
65. Singh, J.; Mahajan, N.; Sharma, R.L.; Razdan, T.K. *J. Heterocyclic Chem.* **2011**, *48*, 1398.
66. Chang, M.Y.; Wu, M.H. *Tetrahedron* 2012, *68*, 9616.
67. Rad-Moghadam, K.; Azimi, S.C. *Tetrahedron* 2012, *68*, 9706.
68. Rezvanian, A.; Heravi, M.M.; Shaabani, Z.; Tajbakhsh, M. *Tetrahedron* 2017, *73*, 2009.
69. Kwon, O.; Park, S.B.; Schreiber, S.L. *J. Am. Chem. Soc.* 2002, *124*, 13402.
70. Ritter, J. J.; Minieri, P. P. *J. Am. Chem. Soc.* 1948, *70*, 4045.
71. Ritter, J. J.; Kalish, J. *J. Am. Chem. Soc.* 1948, *70*, 4048.
72. Michaudel, Q.; Thevenet, D.; Baran, P.S. *J. Am. Chem. Soc.* 2012, *134*, 2547.
73. Alder, K.; Rickert, H.F. *Ann. Chem.* 1936, *524*, 180.
74. Lie, J.J. *Name Reactions: A Collection of Detailed Mechanisms and Synthetic Applications*, 5th ed., Springer: Basel, 2014, p. 138.
75. Akritopoulou-Zanze, I.; Gracias, V.; Djuric, S.W. *Tetrahedron Lett.* 2004, *45*, 8439.
76. Zeng, Y.; Reddy, D.S.; Hirt, E.; Aubé, J. *Org. Lett.* 2004, *6*, 4993.

5 Drawing Chemical Structures

In this chapter, we address a topic that is never discussed in the undergraduate or graduate curriculum or in chemistry textbooks; namely, *how* to draw proper chemical structures of reactants and products in any chemical equation so that the equation can be visually read without need for unnecessary mental gymnastics such as reflections, rotations, and translations of chemical structures. As we have seen in Chapter 2, chemists abandoned the fundamental properties of an "equation" from a numerical sense when they replaced the equals sign ("=") with a reaction arrow ("→") in writing out chemical transformations. Since that time, generations of chemists have been taught from the outset that the most important concern for a chemist in transcribing a chemical transformation and communicating that information to other chemists is to draw out correct structures for the starting substrate of interest and the resulting product along with annotating the reaction conditions (reagents, solvents, catalysts, reaction temperature, and reaction time) above the reaction arrow. In this practice, structures of by-products are completely ignored. Also, ignored is the relative aspect of *how* the substrate and product structures appear on the page. Modern chemists who wish to practice green chemistry principles have rediscovered the notion of balancing equations based on element count and charge balance applied to the right- and left-hand sides. This became mandatory in order to determine material efficiency since by-products needed to be specifically identified. Here, we add one more layer to the display of a balanced chemical equation. Since the constituents of a chemical equation are drawings or pictures of chemical structures, the equation also has the component of a *visual equation* as well as a numerical one. This means that when a visual equation is read from left to right, that is, from the reactant side to the product side, it must be made obvious as far as possible to show which parts of the starting structure remain unchanged and which are different as a consequence of the transformation when comparing the reactant structures with the product structures. In Section 5.1, we introduce the principle of conservation of structural aspect as a guiding principle in depicting how chemical equations are to be displayed. This idea has significant advantages when tracking the target bonds made from step to step throughout a synthesis plan until the final product structure is reached. We also take the time in Section 5.2 to refresh and reinforce key ideas concerning writing reaction mechanisms such as the connection between elementary steps and the final balanced chemical equation, careful usage of the curly arrow notation in tracking electron flow, translating mechanism steps written in linear format to a cyclic format and vice versa, and depicting electrochemically based reaction mechanisms. Target bond mapping discussed in Section 5.3 is at once made obvious by, and is greatly facilitated from, the principle of conservation of structural aspect.

Once the target bonds are identified, the oxidation states of the atoms involved in each bond that is made can be tracked as well, and this leads to the determination of a useful parameter called the hypsicity index (*HI*), which is discussed in Section 5.4.

5.1 PRINCIPLE OF CONSERVATION OF STRUCTURAL ASPECT

The principle of conservation of structural aspect states that the picture of the structure of a reaction product appearing in a balanced chemical equation has the same aspect as the picture of the structure of its progenitor substrate. Or, conversely, the aspect of the picture of the key substrate is consistent with that of the reaction product of interest so that atom-by-atom and bond-by-bond comparisons between these structures can be readily made. In the research journal literature and in teaching textbooks, there is no attention or care paid by authors to follow this very simple idea. In order to see the absurdity of the current practice of not paying attention to the structural aspect between reactants and products, one can imagine if this practice was also adopted in the presentation of mathematical symbolism when equations are written out. Imagine for example that the number six with the symbol "6" was suddenly written upside down with the symbol "9" in the same manner how a ring-containing reactant may be unnecessarily written upside down in a chemical equation relative to how that structural moiety appears in the product structure. In this case, a six suddenly becomes a nine with a completely different meaning, which results in something not adding up in a simple arithmetic expression! We contend that it is precisely the failure of adopting this practice of consistent display of structural aspect that causes untold anguish among students in understanding and seeing what is going on when they look at a drawing of a chemical transformation involving chemical structures. Adhering to the principle greatly simplifies and quickens the pace of understanding among students, improves their memory retention of reactions learned, and most importantly, makes the experience much more enjoyable. It also has a significant consequence for researchers when it comes to "reading" the table of contents graphics. Chemical equations that are written out following this principle can be read and understood much faster than ones that are not. This translates into more papers being scanned and read with less eye fatigue. If a research author writes out his or her chemical transformation in a way that requires the reader to play mental visual gymnastics in order to follow what the transformation is about, it will risk being glossed over and deemed "unimportant." An even worse practice that is common in the literature is to change the structural aspects of intermediates along the way in a scheme showing a total synthesis of a particular target product. Sometimes the aspects appear alternating! This is a very easy way to lose the thread of tracking the strategies employed. The method used to depict a synthesis scheme that employs a consistent structural aspect throughout is to choose a structural aspect for the target product and work backwards step-by-step following the structures of the intermediates drawn with the same consistent aspect all the way back to the starting substrate. When read in the forward direction such a scheme is very easy to read visually, thus greatly facilitating target bond mapping and elucidating any ring construction strategies employed. When the principle of conservation of structural aspect is adopted for synthesis plans it enhances their aesthetic presentation as well as improving and speeding up the process of understanding what synthesis strategies are in the plan. Given the tsunami of literature nowadays from traditional established journals and innumerable online electronic-only journals in a digital format, the research community would do well to pay more care and heed this advice and adopt this principle as a routine practice in communicating chemical science to their peers and to the next generation of scientists.

For any given substance, there are multiple ways of drawing its chemical structure. Drawings are often linked to the kind of typesetting technology that is available to publishers. In the nineteenth century, the structure for cyclohexane was drawn as a ring with CH_2 units linked by lines depicting chemical bonds in contrast to the structure of benzene, which was drawn at the time as a hexagon. Nowadays, a drawing of a hexagon implies the structure of cyclohexane. The problem in the nineteenth century was that double bonds could not be depicted in a ring structure, hence the reason for drawing cyclohexane with CH_2 units specified. Examples 5.1 and 5.2 show several ways to draw the structures of cyclohexane and benzene. This insight has a very important consequence. If one wants to read the chemistry literature during this period one needs to read it with the eye of that time period. This is particularly true if one wants to read the large body of work published in the nineteenth century on the chemistry of dyestuffs, for example. Apart from technological difficulties there is another important insight about drawings of chemical structures. Although there may be more than one way to draw a chemical structure for a given substance, each of those drawings must be carefully examined to select those that are well suited to illustrate and communicate a concept. In Example 5.3, not all drawings shown for the [2.2.1] bicyclic ring frame are equally suitable to depict chemical reactions. Context is therefore important in selecting how a structure is to be displayed. In Example 5.4, the concept of inversion of configuration in S_N2 reactions can be illustrated by any one of the drawings shown, however, only a handful actually communicate that concept in an effective visual manner. Yet, one may ask themselves how many of the other less advantageous drawings are used to convey the concept in textbooks or are even used as visual "tricks" to foil students on examinations! Examples 5.5 and 5.6 show drawings of chemical reactions appearing in a paper about green chemistry that illustrate the problems of not following the principle of conservation of structural aspect. These examples nicely show how to redraw them so that they are more visually understandable. Example 5.7 is particularly important because the structural aspect of the product does not correspond to the structural aspect of the reactant, and in doing so, the drawing obscures which methyl group is undergoing the 1,2-methyl migration in the rearrangement reaction. The mismatch of aspects between the initial and final structures as drawn by the author of the undergraduate textbook is a nice demonstration of the impediment of understanding of what is going on in this reaction, and ultimately of bad pedagogy. As drawn, it artificially creates a scenario in the mind of a student that such reactions are "hard" to follow and understand. In contrast, the two possible solutions shown readily clarify the situation. Example 5.8 illustrates the principle in creating a target bond map for the synthesis of a simple pharmaceutical, antipyrine, that was once used as an undergraduate laboratory experiment exercise.

5.1.1 Examples

Example 5.1

Show different ways to draw the structure of cyclohexane.

Solution

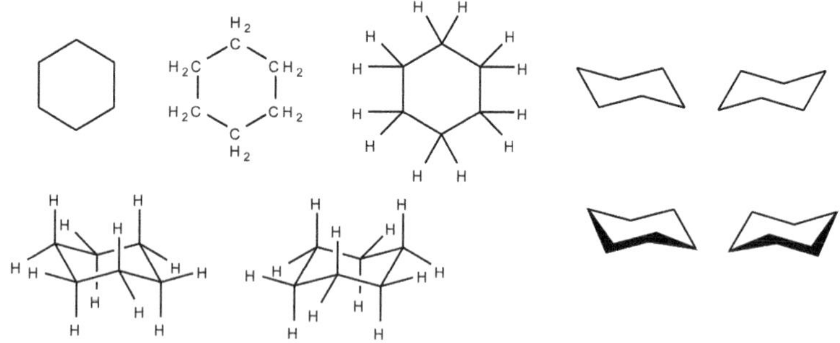

Example 5.2

Show different ways to draw the structure of benzene.

Solution

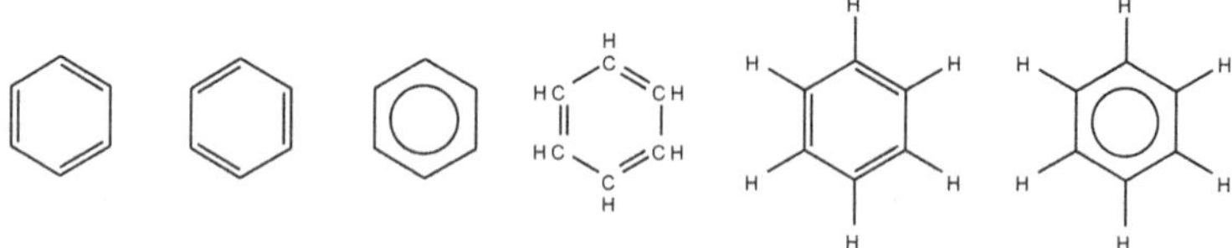

Example 5.3

Show different ways to depict bicyclic [2.2.1] ring frames.

Solution

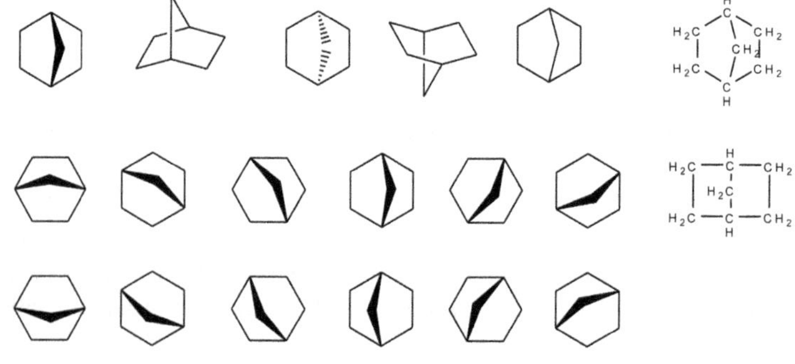

Example 5.4

Which diagrams illustrate better the concept of inversion of configuration by an S_N2 reaction?

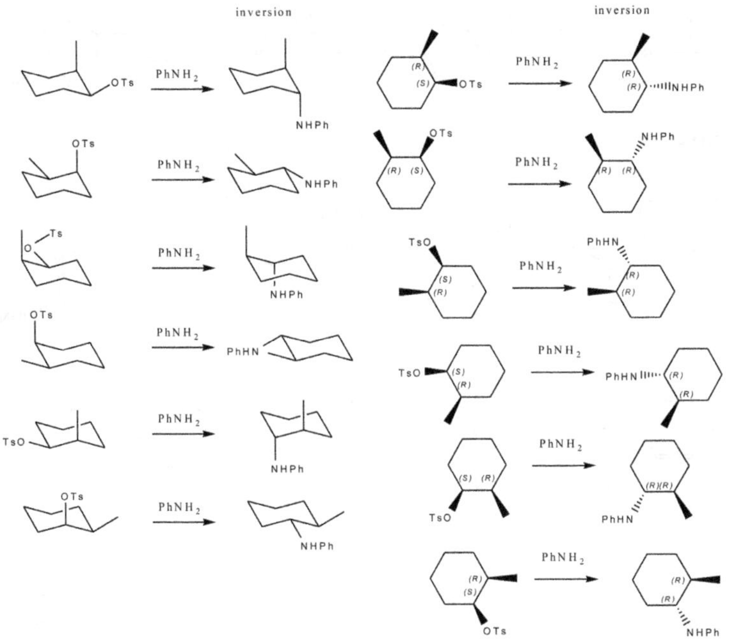

Solution

Diagrams on the right are better at illustrating the concept than the chair conformations.The best chair conformations to use are 1st, 2nd, 4th, 5th, and 6th entries.

Example 5.5 [1]

Redraw the reaction below using the principle of structural aspect. Identify the target bonds made.

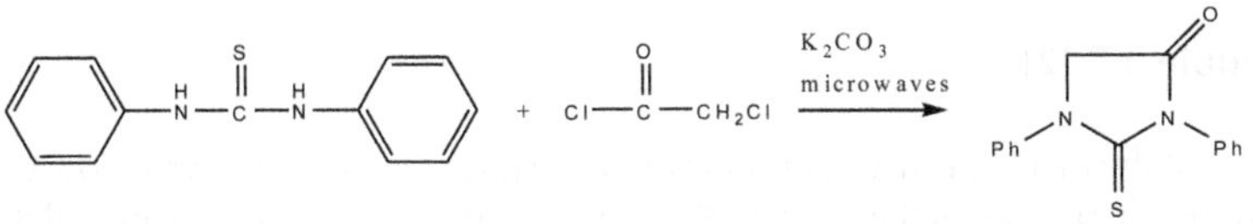

Solution

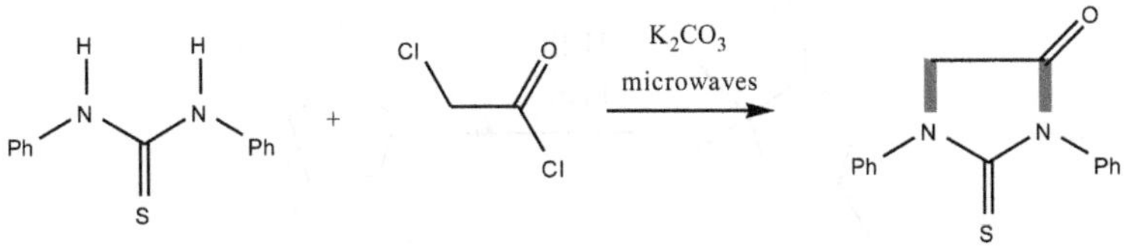

Example 5.6 [1]

Redraw the reaction below using the principle of structural aspect. Identify the target bonds made.

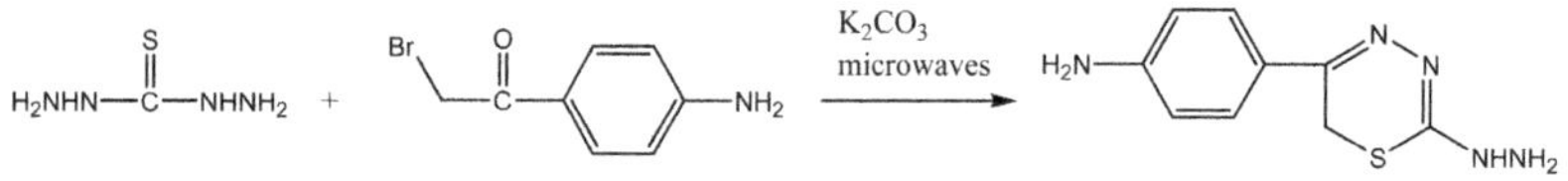

Solution

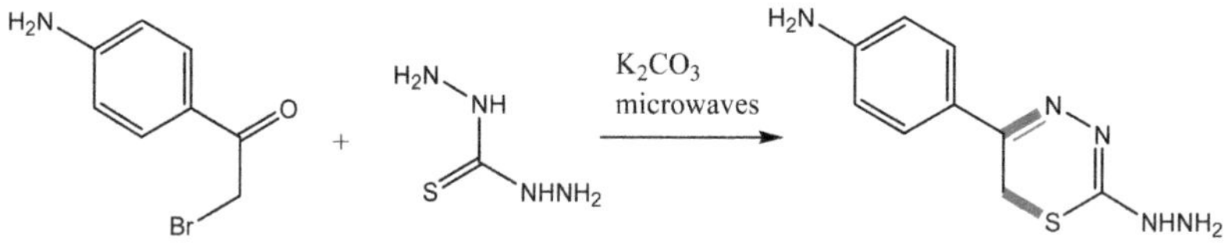

or

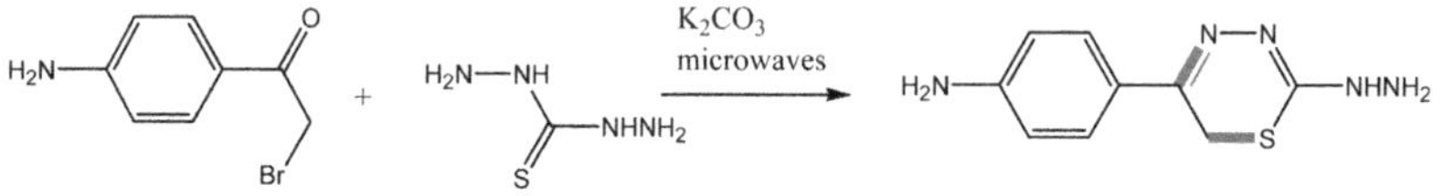

or

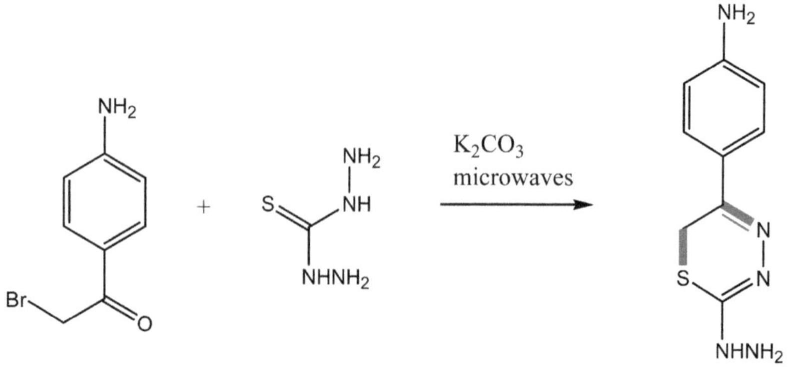

Example 5.7 [2]

Write out the mechanism for the acid catalyzed rearrangement shown below using the principle of structural aspect applied to the starting substrate structure. Repeat the exercise, applying the principle to the final product structure. The drawing shown below is exactly how it appeared in Solomon's book.

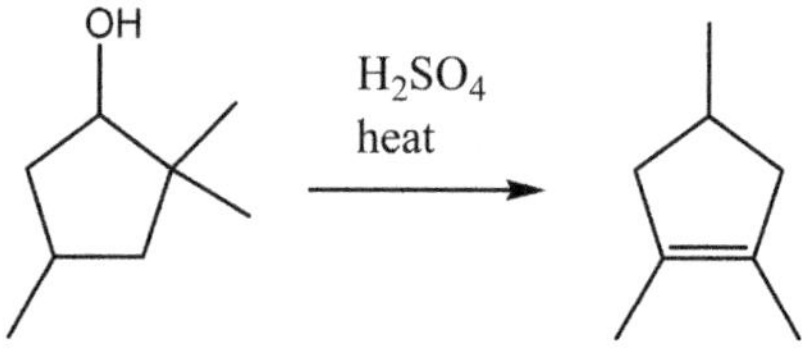

Solution

Aspect with respect to starting substrate

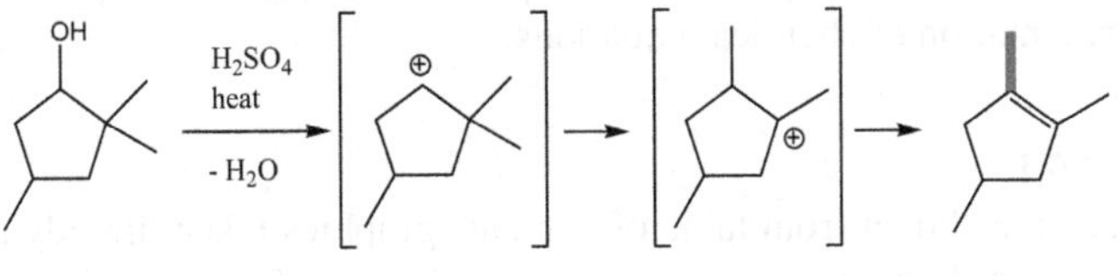

OH

H_2SO_4
heat
$- H_2O$

Aspect with respect to product

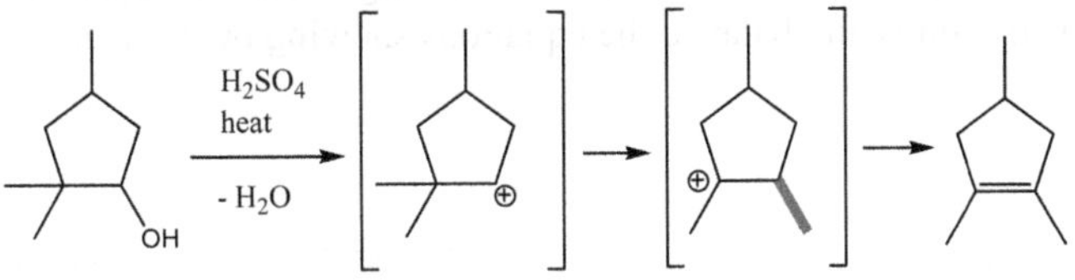

H_2SO_4
heat
$- H_2O$

OH

Example 5.8 [3, 4]

Redraw the synthesis scheme shown below for the preparation of antipyrine using the principle of conservation of structural aspect. Show balanced chemical equations in each step and determine what the ring construction mapping is in the final product structure.

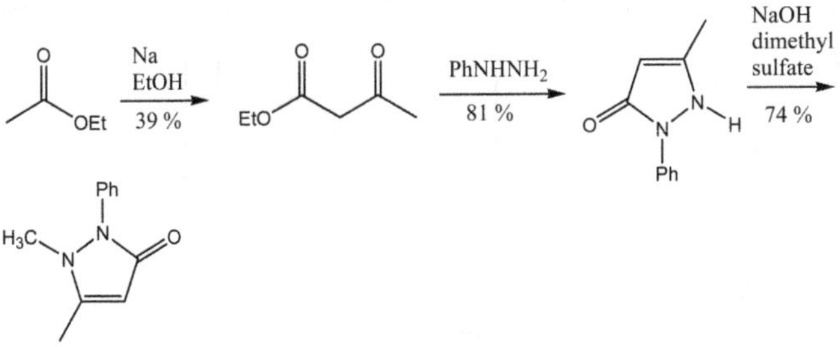

O OEt — Na, EtOH, 39 % → EtO O O — $PhNHNH_2$, 81 % → O N N H Ph — NaOH, dimethyl sulfate, 74 % →

H_3C N N Ph O

antipyrine

Solution

O CH_3 OEt — Na, $- 1/2\ H_2$ → [O $CH_2^{\ominus}$ $Na^{\oplus}$ OEt] — EtO O, $-$ NaOEt → **39 %** O O OEt — $PhNHNH_2$, $- H_2O$, $-$ EtOH →

81 % O N N H Ph — NaOH, $(MeO)_2SO_2$, $- H_2O$, $- (MeO)SO_2(ONa)$ → **74 %** O N N CH_3 Ph

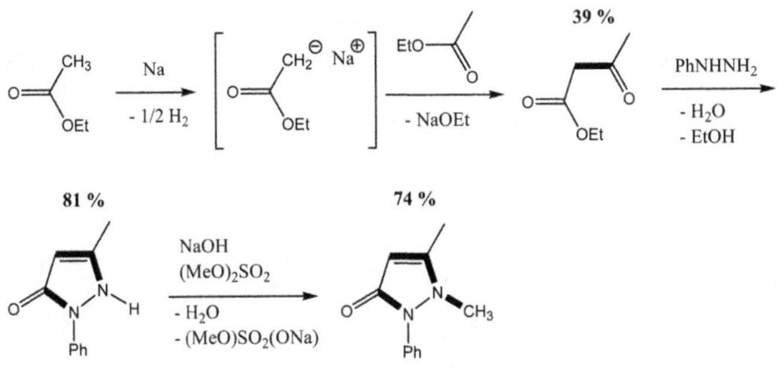

antipyrine

Ring construction mapping: [2 + 2 + 1].

5.1.2 Problems

In this section we pose problems taken from the literature where the principle of conservation of structural aspect can be implemented to greatly improve the quality of visual presentation of chemical equations.

PROBLEM 5.1

For each reaction shown from table of contents graphics taken directly as depicted in the literature, redraw the reaction using the principle of conservation of structural aspect applied to the product structures shown. Highlight the target bonds made in the product structures and balance the equations showing by-products.

(a) [5]

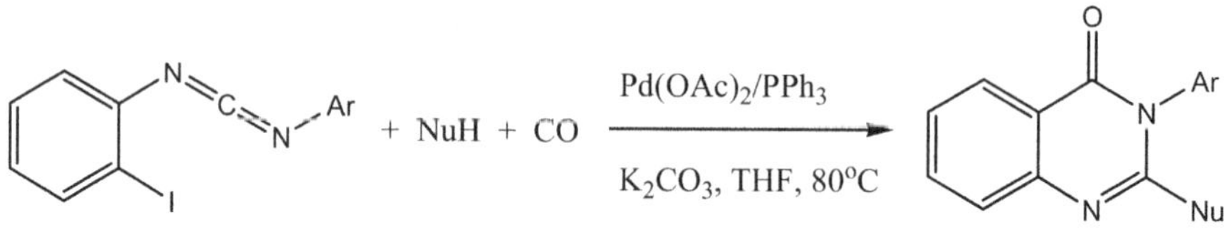

+ NuH + CO

$Pd(OAc)_2/PPh_3$

K_2CO_3, THF, 80°C

(b) [6]

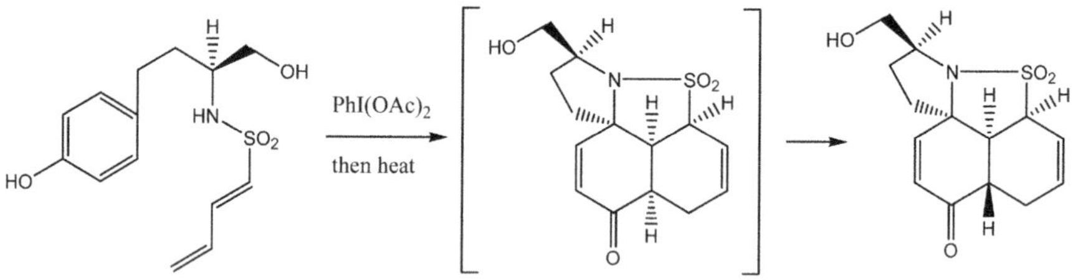

$PhI(OAc)_2$

then heat

(c) [7]

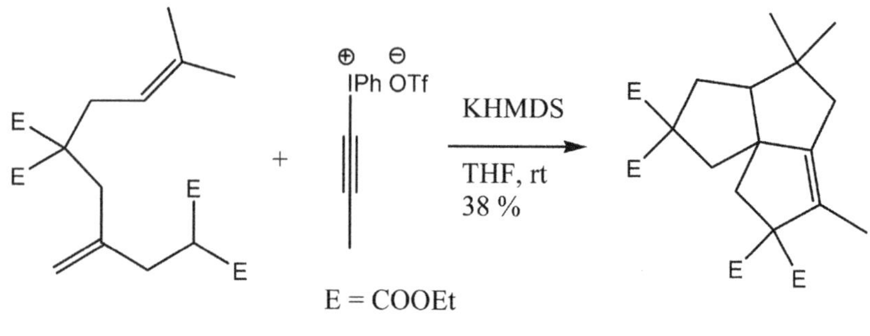

KHMDS

THF, rt
38 %

E = COOEt

(d) [8]

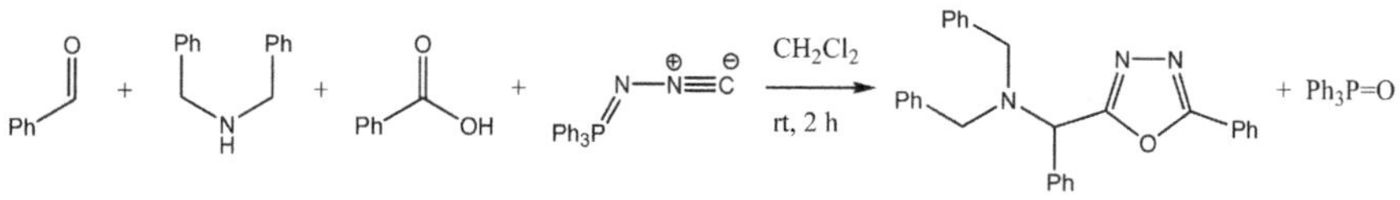

CH_2Cl_2

rt, 2 h

+ $Ph_3P{=}O$

(e) [9]

Ar, H, HN, Boc, Me, O

Me_2AlCl (1.5 eq.)

CH_2Cl_2
- 78°C --> - 20°C
88 %

NHBoc, Me, O, H, Ar, H, H

Boc, N, Me, HO, H, Ar, H, H

(f) [10]

R_2, R_1, CN, OH + OH, O

Et_3N, MeCN

20-25°C, 48 h

CN, O, O, R_1 R_2 + H_2O

(g) [11]

I, R_2, N=C=N, R_1, I + R^3NH_2

$Pd(OAc)_2/PPh_3$

K_2CO_3, THF
CO (500 psi),
120°C

O, R_2, R_1, N, N, N, O, R_3

(h) [12]

N H, CHO + N_3, EWG, R

Cs_2CO_3, DMF

rt, 24 h

R, EWG, N, N

(i) [13]

Ar, Ar, NH, SO_2R

$Cu(OTf)_2$ (20 mol%)
(R,R)-Ph-box (25 mol%)

K_2CO_3, $PhCF_3$
120°C, 24 h

X, Ar, N, SO_2R

(j) [14]

PROBLEM 5.2 [15]

Themes: balancing chemical equations, catalytic cycle, ring construction strategy, target bond mapping

A tandem Nazarov–Diels–Alder reaction mediated by boronic acid catalysts was developed.

Rewrite the scheme maintaining a consistent structural aspect throughout. Highlight the target bonds made in the product and describe the ring construction strategy.

PROBLEM 5.3 [16]

Themes: conservation of structural aspect

Rewrite the following scheme for the synthesis of (+)-lycoricidine using the principle of conservation of structural orientation and determine the ring construction mapping in the product structure. Show balanced equations in each reaction step.

1. PhSH, *hv*, tol. 27°C, 97 %

2. SmI_2, THF, 86 %
3. TFA, 90 %

OH OH OH NH O O O

(+)-lycoricidine

(a) $NaIO_4$, CH_2Cl_2; (b) CBr_4, PPh_3, NEt_3, 80% over two steps; (c) L-selectride, Et_2O, - 78°C; (d) HCl-H_2NOBn, pyridine, 90 % over two steps; (e) ^nBuLi, Et_2O, - 90°C, 93 %

5.2 WRITING A REACTION MECHANISM

5.2.1 Elementary Steps

For any chemical reaction, a reaction mechanism is a series of sequential steps, often called elementary steps, that describe the sequence of bond making and bond breaking events that occur between the initial reactant structures and the final product structures. For the purpose of quantifying material efficiency for any chemical reaction, the key insight that connects the reaction mechanism and elementary steps is the following statement. *Since each elementary step is itself a balanced chemical equation, when elementary steps comprising a reaction mechanism are added up, one obtains the fully balanced chemical equation (BCE) for the given reaction.* As we saw in Chapter 2, the balanced chemical equation describes the number and identity of each kind of reactant and the number and identity of each kind of product produced in the reaction. All material efficiency metrics, which will be discussed in detail in Chapter 6, originate from such a balanced chemical equation. The general logic of the constitution of a reaction mechanism and its connection to the balanced chemical equation is summarized in Figure 5.1. For clarity in determining the sum of elementary steps, the reaction mechanism is written out as a linear series of separate elementary steps in a sequential fashion.

From Figure 5.1 we note that for an N-step mechanism there are $N-1$ intermediates and that the by-products, Q_i, are formed as a direct *mechanistic consequence* of producing the desired target product P. By-products are therefore distinguishable from side products since the latter arise from separate parallel reactions that undergo different mechanism pathways.

5.2.2 Curly Arrow Notation

In writing out a sensible reaction mechanism that is consistent with experimental observations, there are certain ground rules when it comes to how such a mechanism is to be written out. The curly arrow notation introduced by Robinson in 1922 [17] is a bookkeeping device that helps a chemist track the electron flow between nucleophiles elementary steps:

- step 1: $S_1 + \ldots \longrightarrow I_1 + \ldots$
- step 2: $I_1 + S_2 + \ldots \longrightarrow I_2 + \ldots$
- Step 3: $I_2 + \ldots \longrightarrow I_3 + \ldots$
- ....
- Step N: $I_{N-1} + \ldots \longrightarrow P + \ldots$

SUM: $\nu_1 S_1 + \nu_2 S_2 + \ldots \longrightarrow \nu_P P + \nu_{q1} Q_1 + \nu_{q2} Q_2 + \ldots$ BCE

ν_i = stoichiometric coefficient pertaining to chemical species i
S_i = reactant i
P = final target product
Q_i = by-product i

FIGURE 5.1 Conceptual connection between reaction mechanism and balanced chemical equation applicable to any chemical reaction.

and electrophiles so that all bond making and bond breaking events from structure to structure until the structure of the final target is reached. Unfortunately, in undergraduate instruction, explicit rules about how to use and interpret the curly arrow notation are not taught properly or not at all. Here we take the time to make sure the reader is fully aware so that they get the essential information in simple terms along with illustrative examples to reinforce the concepts.

For two-electron movements we have the following rules.

Rule #1: Arrow heads always point toward electrophilic centers.
Rule #2: Arrow tails always point toward nucleophilic centers.
Rule #3: Electron flow goes from a nucleophilic source to an electrophilic sink.
Rule #4: A series of arrows are always unidirectional. This means that an arrow head is always followed by an arrow tail, so that two arrow heads or two arrow tails never meet.

Figure 5.2 shows all possible two-electron flow combinations between an electron source and an electron sink.

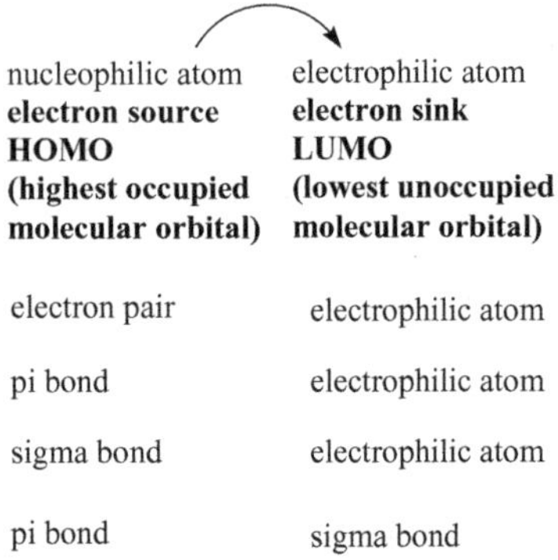

FIGURE 5.2 Curly arrow notation for two-electron flow combinations between various electron sources and electron sinks.

Example 5.9

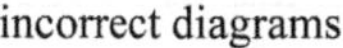
incorrect diagrams

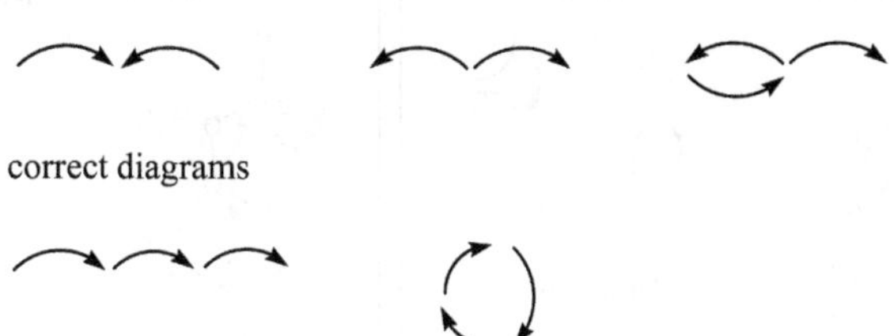
correct diagrams

Example 5.10 [18]

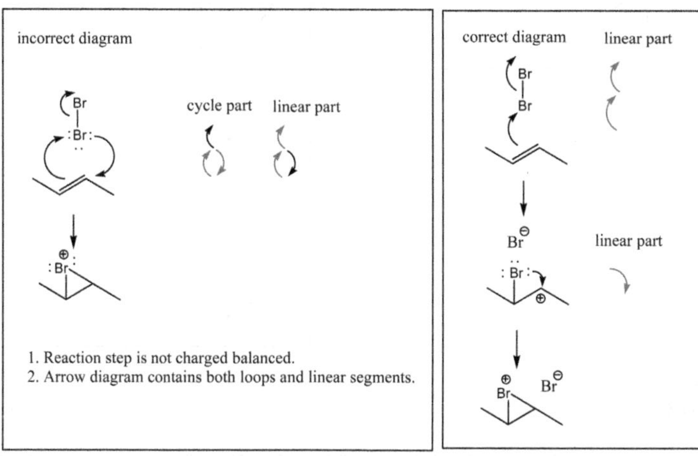
incorrect diagram
cycle part
linear part
Br
:Br:
:Br
1. Reaction step is not charged balanced.
2. Arrow diagram contains both loops and linear segments.
correct diagram
linear part
Br
Br
Br
linear part
: Br :
Br
Br

Example 5.11 [18]

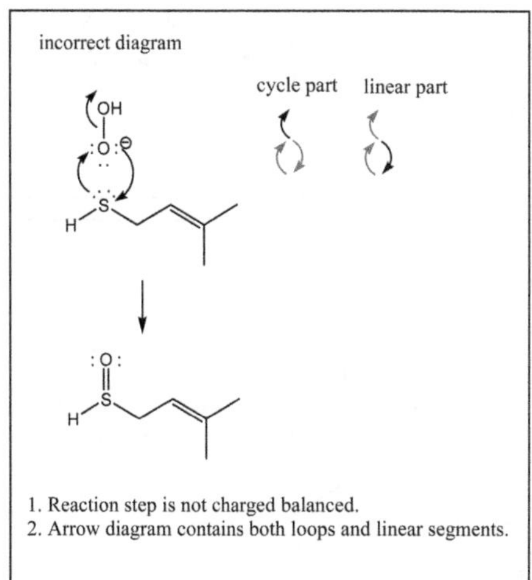
incorrect diagram
cycle part
linear part
OH
:O:
S
H
:O:
S
H
1. Reaction step is not charged balanced.
2. Arrow diagram contains both loops and linear segments.

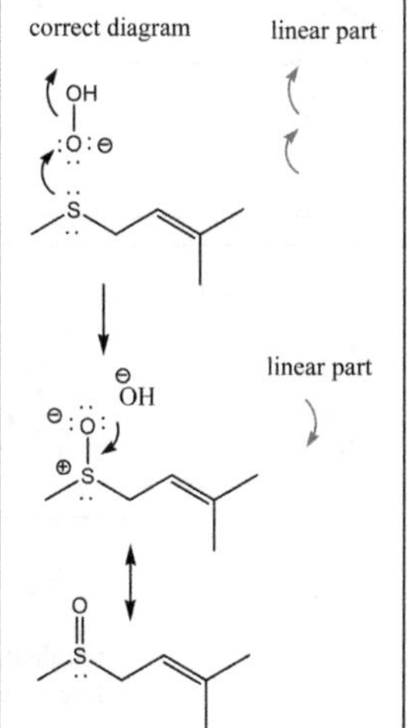
correct diagram
linear part
OH
:O:
S
linear part
OH
:O:
S
O
S

Example 5.12 [18]

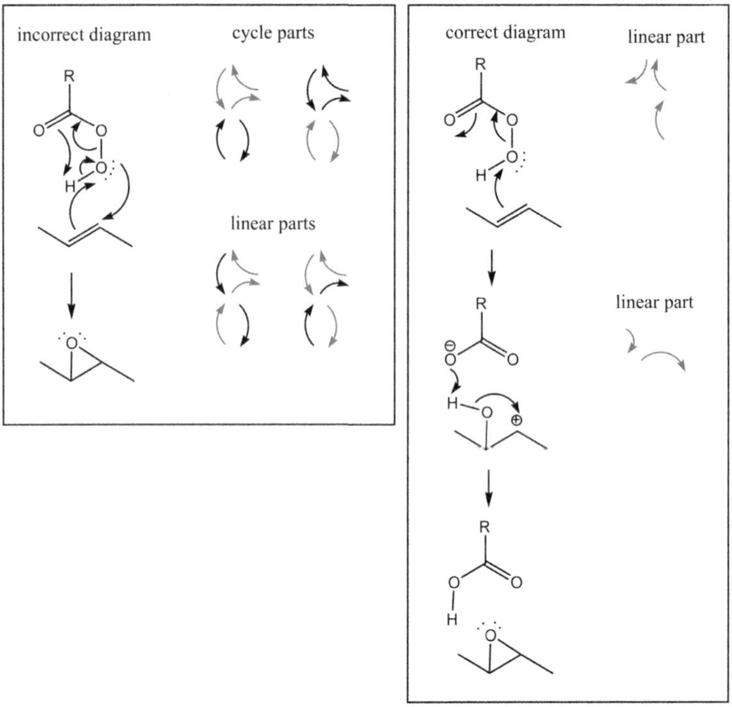

Example 5.13

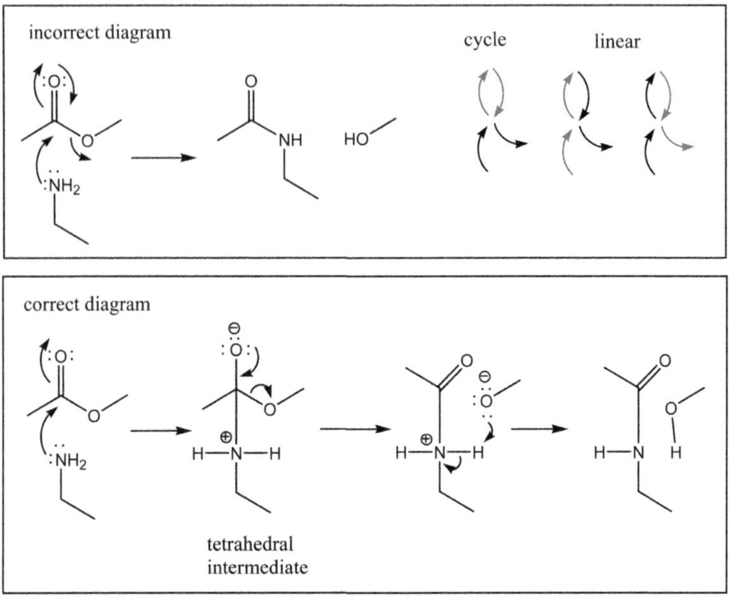

Example 5.14

This is not a cycle since head of arrow *a* is not going toward tail of arrow *b*. Therefore, diagram is correct.

It is possible to draw an alternative stepwise diagram to depict this transformation more clearly as shown below.

For one-electron movements applicable to homolytic bond cleavages and reformations in radical-type mechanisms, we have the following rules.

1. Rule #1: Curly arrows have half a head to depict one-electron transfer steps.
2. Rule #2: Two half arrow heads coming together depict bond formation.
3. Rule #3: Two arrow tails emanating from a bond depict homolytic bond cleavage.
4. Rule #4: A series of arrows are not unidirectional.

Example 5.15

incorrect diagrams

correct diagrams

Example 5.16

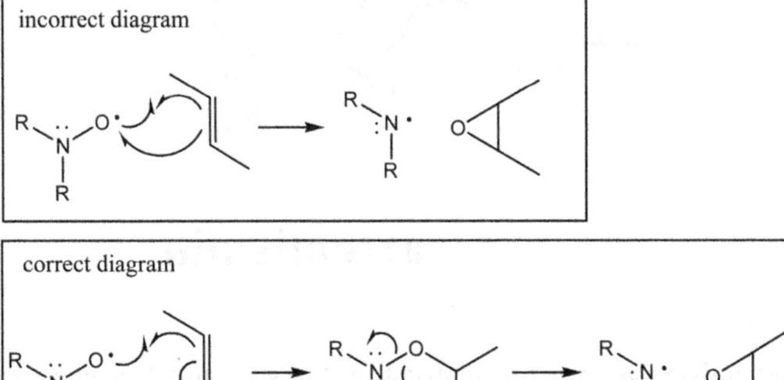

Example 5.17

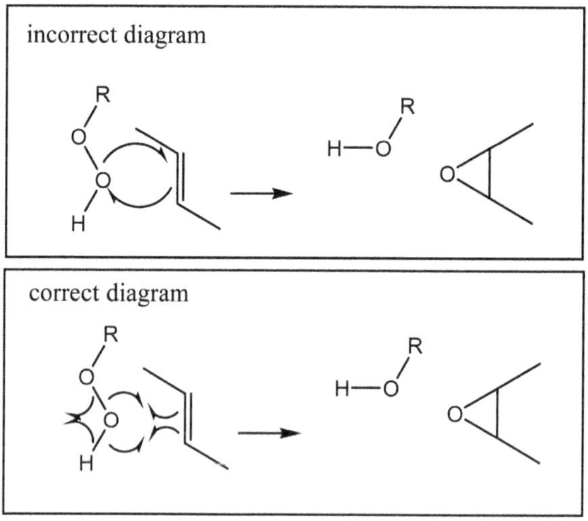

incorrect diagram

correct diagram

PROBLEM 5.4 [19]

Themes: balancing chemical equations, reaction mechanism, redox reactions

In a synthesis of the anti-malarial natural product artemisinin, the following redox step was employed.

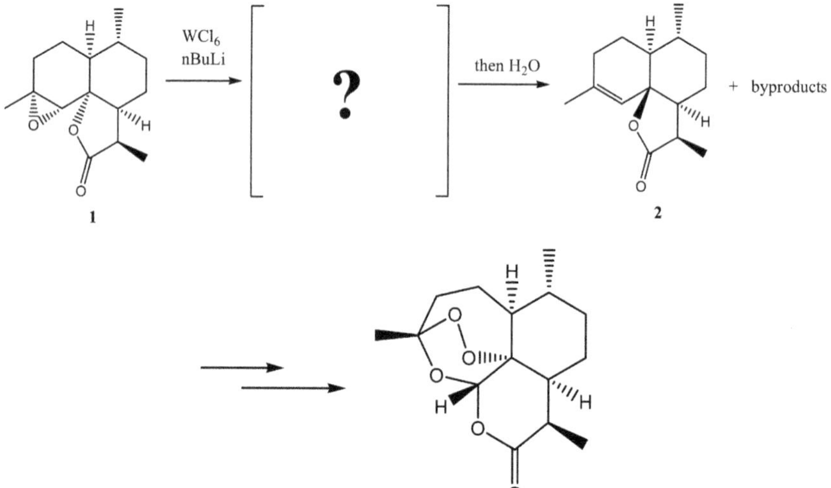

WCl_6
nBuLi

?

then H_2O

+ byproducts

1

2

artemisinin

Part 1

Write out a mechanism that transforms structure **1** to structure **2**. Suggest a structure for the missing intermediate and explain how the epoxide moiety is opened. What is the redox couple for the reaction?

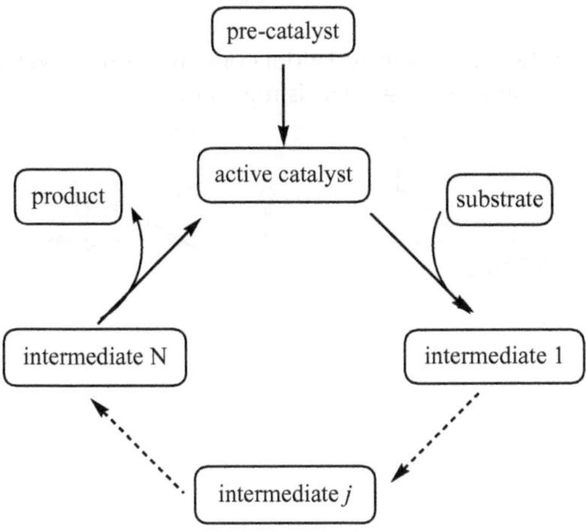

FIGURE 5.3 General template depicting a catalytic cycle or single Tolman loop.

Part 2

From the mechanism write out the balanced chemical equation.

Part 3

Show an oxidation number analysis for the transformation.

5.2.3 Catalytic Cycles

Catalytic cycles can be drawn using the general template shown in Figure 5.3. The notation of catalytic cycles was invented by inorganic chemist Chad Tolman, who worked at duPont Central Research and Development (Wilmington, Delaware). In his honor, catalytic cycles are also known as Tolman loops [20, 21].

Example 5.18

Part 1

The mechanism for the acid catalyzed aldol condensation is given below. Rewrite it as a catalytic cycle showing that H^+ is regenerated.

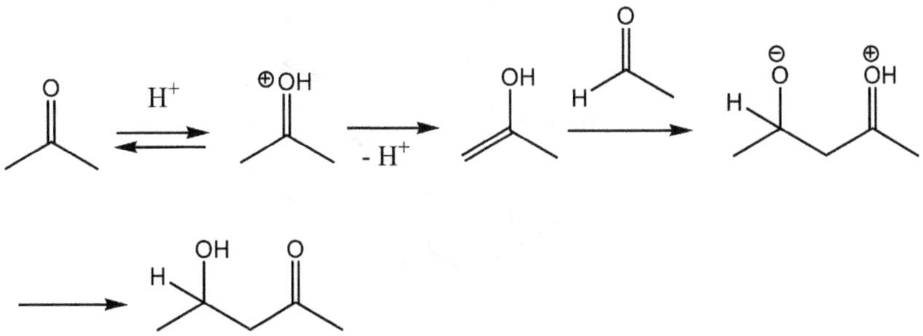

Part 2

The mechanism for the base catalyzed aldol condensation is given below. Rewrite it as a catalytic cycle showing that OH⁻ is regenerated.

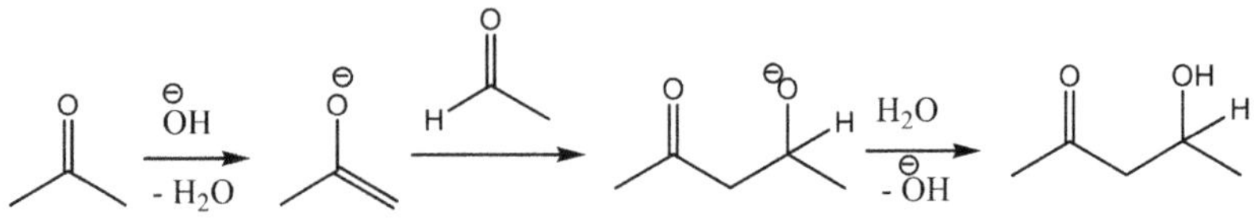

Solution

Part 1

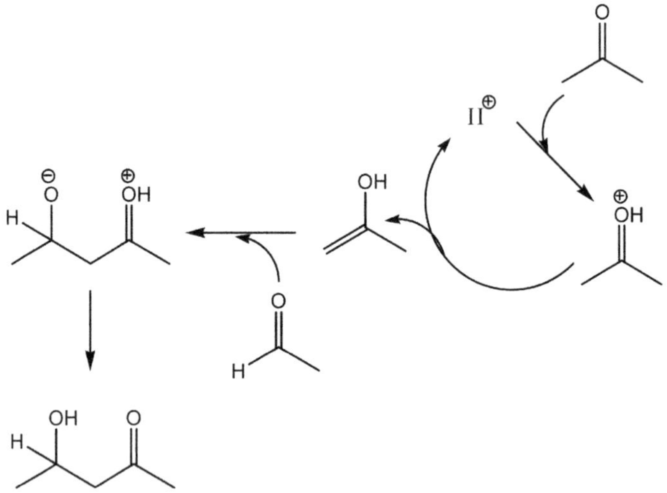

Part 2

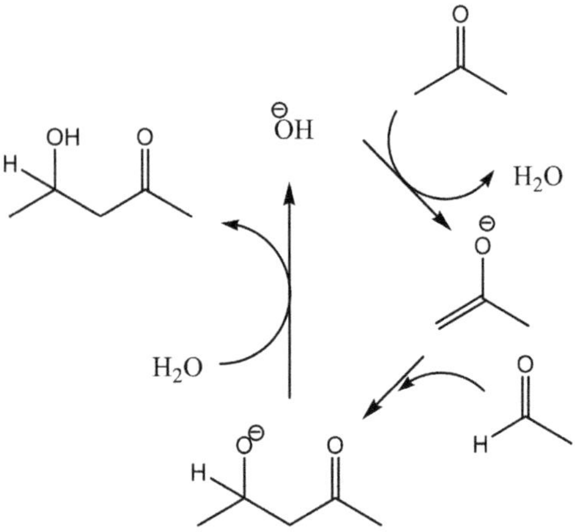

Example 5.19

Part 1

The mechanism for the acid catalyzed ketonization of enols is given below. Rewrite it as a catalytic cycle showing that H⁺ is regenerated.

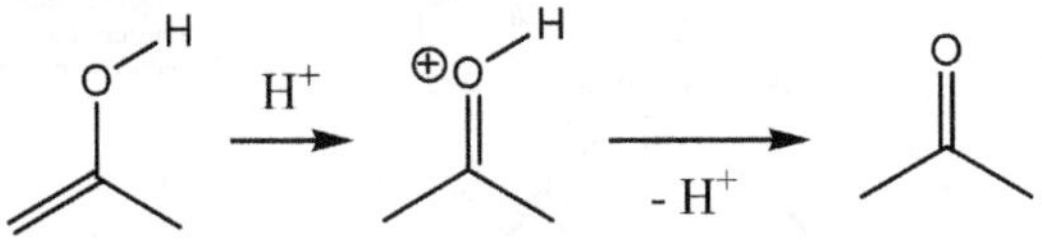

H^+

$- H^+$

Part 2

The mechanism for the base catalyzed ketonization of enols is given below. Rewrite it as a catalytic cycle showing that OH⁻ is regenerated.

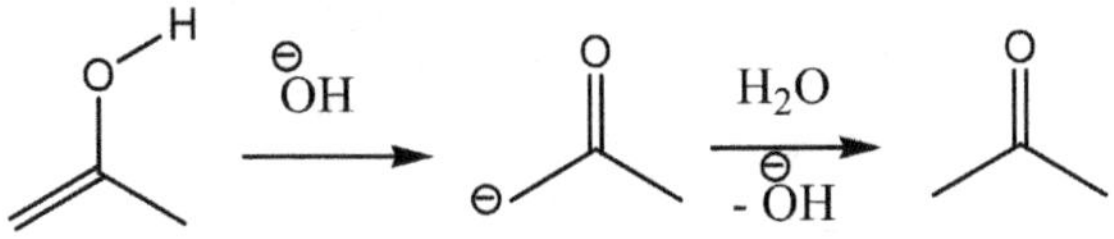

$\overset{\ominus}{O}H$

H_2O

$- \overset{\ominus}{O}H$

Solution

Part 1

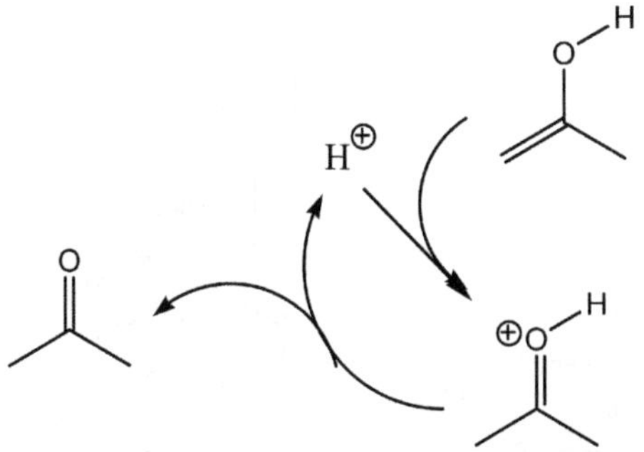

$H^{\oplus}$

Part 2

$\overset{\ominus}{O}H$

H_2O

H_2O

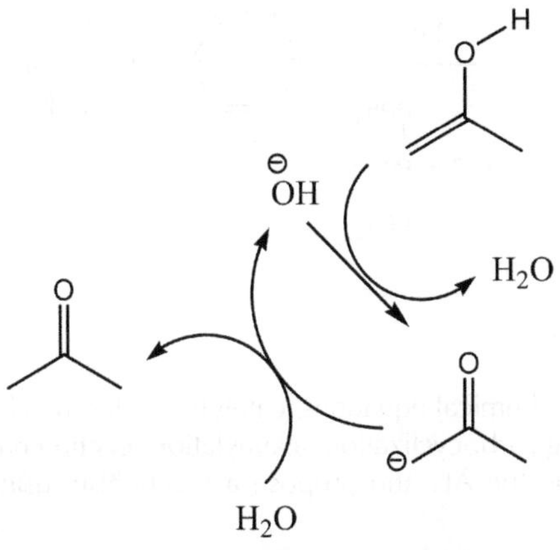

Example 5.20

The mechanism for the Suzuki coupling using Pd(0) complexes is given below. Rewrite it in linear format.

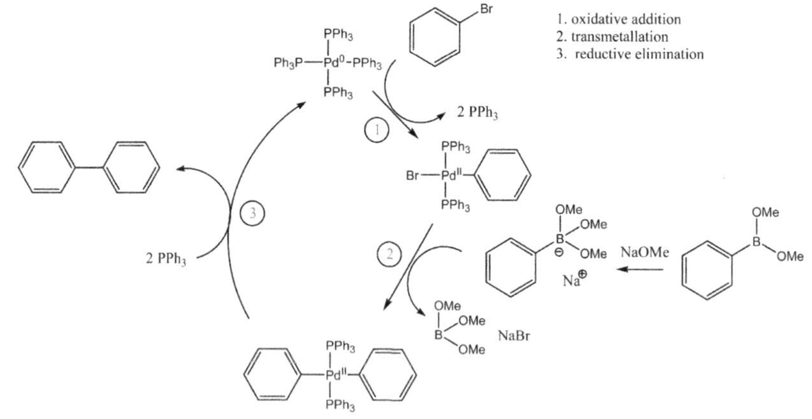
1. oxidative addition
2. transmetallation
3. reductive elimination
2 PPh3
NaOMe
NaBr
2 PPh3

Solution

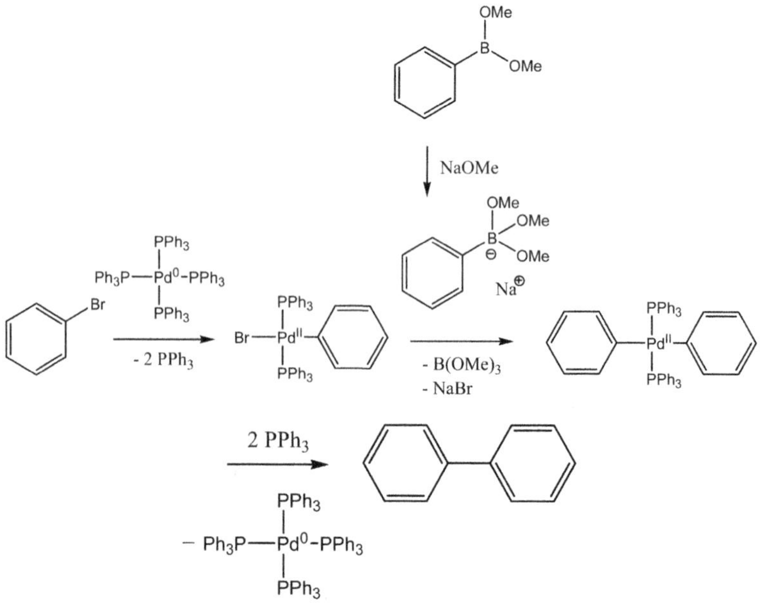
NaOMe
- 2 PPh3
- B(OMe)3
- NaBr
2 PPh3

Example 5.21 [22]

Themes: balancing chemical equations, catalytic cycles, mechanism

For the following carbocyclization–alkynylation reaction balance the chemical equation, determine the AE, and propose a mechanism using a catalytic cycle notation.

Ph—≡ + MeOOC COOMe → MeOOC COOMe Ph

$Pd(OAc)_2$ (cat.)
Co(salophen) (cat.)
benzoquinone (cat.)
O_2

Solution

MeOOC COOMe 236 + 1/2 O_2 → MeOOC COOMe + H_2O

Ph 102 1/2 (32) Ph 336 18

Reactants: 236 + 102 + 0.5*32 = 354
Products: 336 + 18 = 354

1/2 O_2
2 HOAc
H_2O
Pd^0 → $Pd(OAc)_2$

MeOOC COOMe Ph

MeOOC COOMe Pd^{II} Ph

MeOOC COOMe Pd^{II} AcO OAc

BQ HOAc

HOAc + BQ

Ph—≡

MeOOC COOMe Pd^{II} AcO BQ

MeOOC COOMe AcO Pd^{II} BQ

BQ = benzoquinone

Example 5.22 [23–27]

Themes: balancing chemical equations, catalytic cycles, redox reactions, reaction mechanism

The Bart reaction converts aryl diazonium salts to aryl arsonates via catalysis with copper(I)chloride.

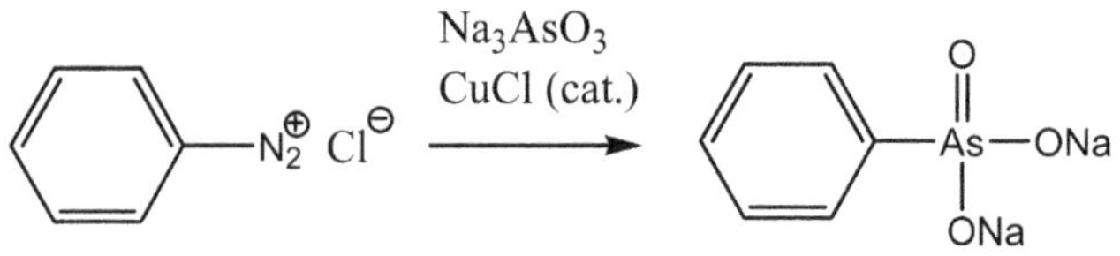

$$C_6H_5N_2^{\oplus}Cl^{\ominus} \xrightarrow[\text{CuCl (cat.)}]{Na_3AsO_3} C_6H_5As(=O)(ONa)_2$$

Part 1

Write out a catalytic cycle for the reaction that depicts the mechanism of the reaction.

Identify reaction by-products.

Part 2

Write out a balanced chemical equation for the process.

Part 3

Identify any oxidation number changes in the atoms from reactants to products. Can a redox couple be written for the reaction?

Solution

Part 1

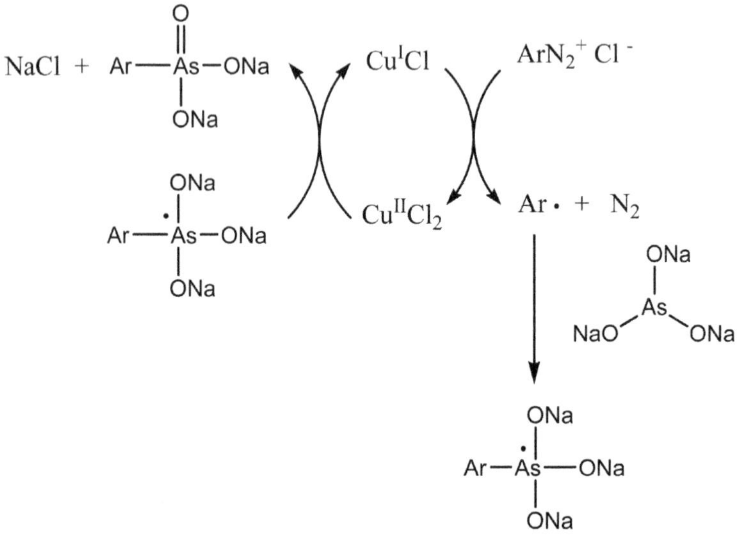

NaCl + Ar—As(=O)(ONa)—ONa

$Cu^{I}Cl$ $ArN_2^+\,Cl^-$

$Ar\cdot + N_2$

$Cu^{II}Cl_2$

$As(ONa)_3$

Ar—•As(ONa)₃

Part 2

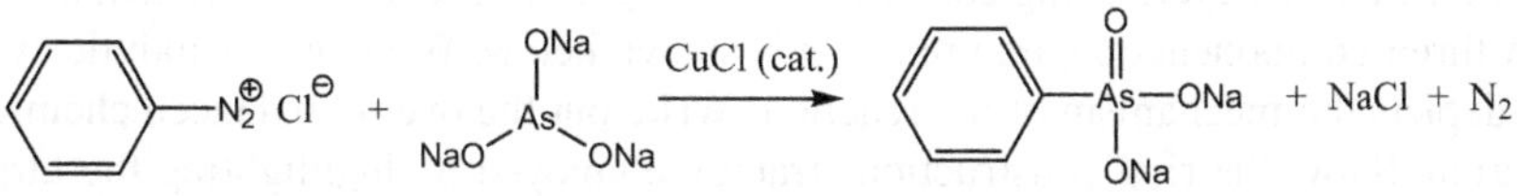

CuCl (cat.)

+ NaCl + N_2

140.45 **191.92** **245.92** **58.45** **28**

Reactants: 140.45 + 191.92 = 332.37
Products: 245.92 + 58.45 + 28 = 332.37

Part 3

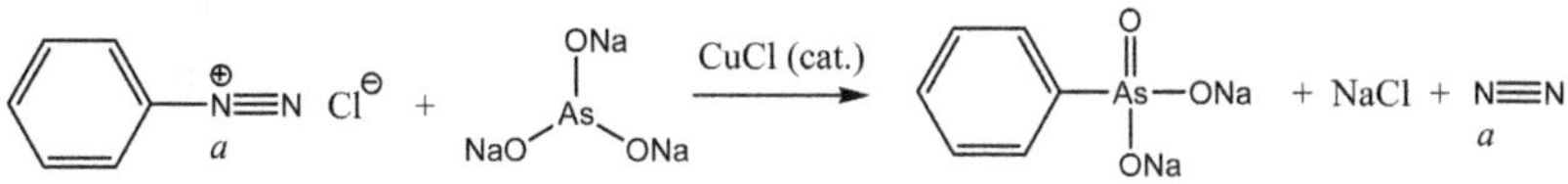

CuCl (cat.)

+ NaCl + N≡N

Atom	Reactants	Products	Change
N_a	0	0	0
As	+3	+3	0

There are no oxidation changes in any of the key elements. Therefore, there is no redox couple to consider.

PROBLEM 5.5

The mechanism for the Lewis acid catalyzed Friedel–Crafts acylation is given below. Rewrite it as a catalytic cycle showing that the Lewis acid is regenerated.

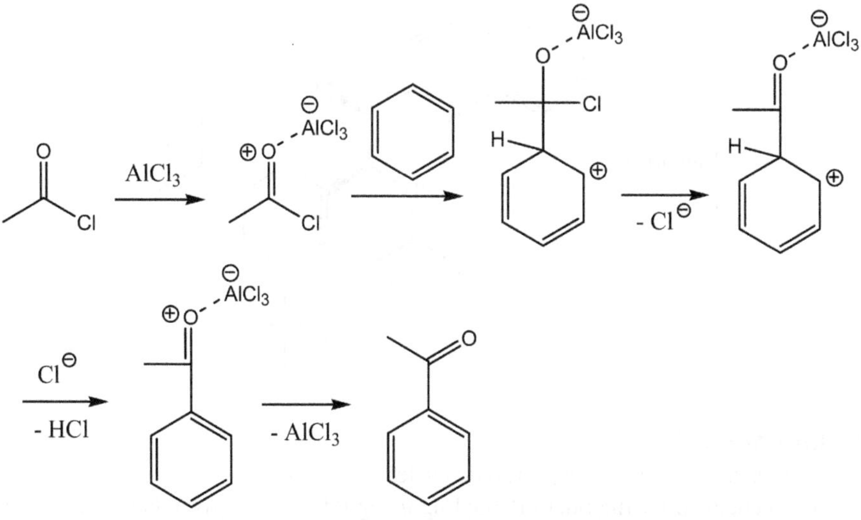

$AlCl_3$

- $Cl^{\ominus}$

$Cl^{\ominus}$

- HCl

- $AlCl_3$

PROBLEM 5.6 [28]

Themes: catalytic cycles, ring construction strategy, multi-component reaction

A three-component coupling reaction is shown below. Write out a catalytic cycle that depicts the mechanism of the reaction. Write out the overall balanced chemical equation. Show the ring construction strategy employed by highlighting the target bonds formed in the product structure.

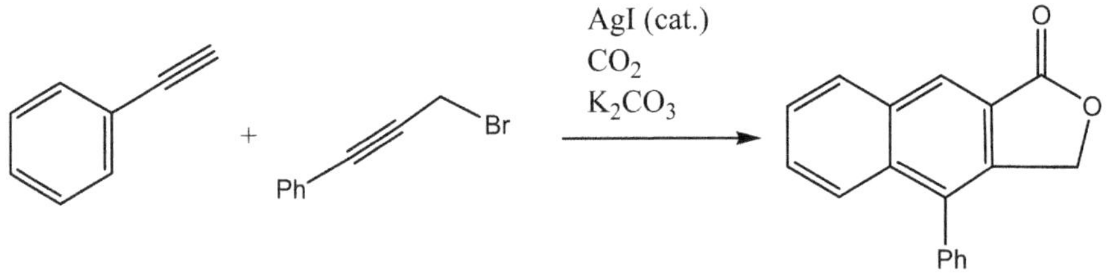

AgI (cat.)
CO_2
K_2CO_3

Br

Ph

O

O

Ph

PROBLEM 5.7 [29]

Themes: balancing chemical equations, catalytic cycles, mechanism

The following stereoselective methylene C-H activation reaction took 15 years to develop.

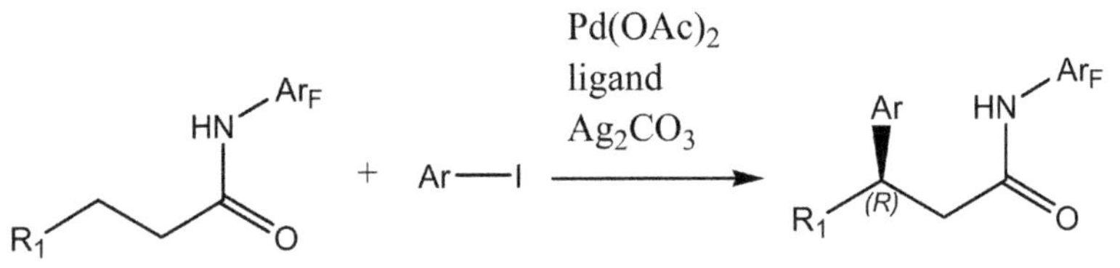

$Pd(OAc)_2$
ligand
Ag_2CO_3

HN
Ar_F
R_1
O
Ar—I
Ar
(R)

Ligand =

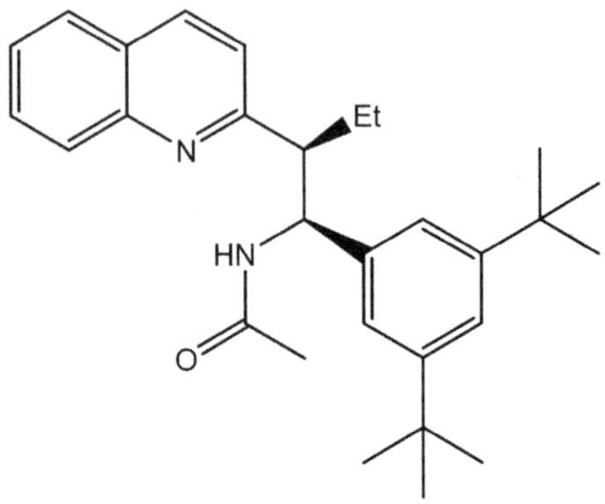

N
Et
HN
O

PROBLEM 5.8

Themes: reaction mechanisms, catalytic cycles

The mechanism for the Suzuki coupling using Pd^{+2} salts is given below. Rewrite it in linear format.

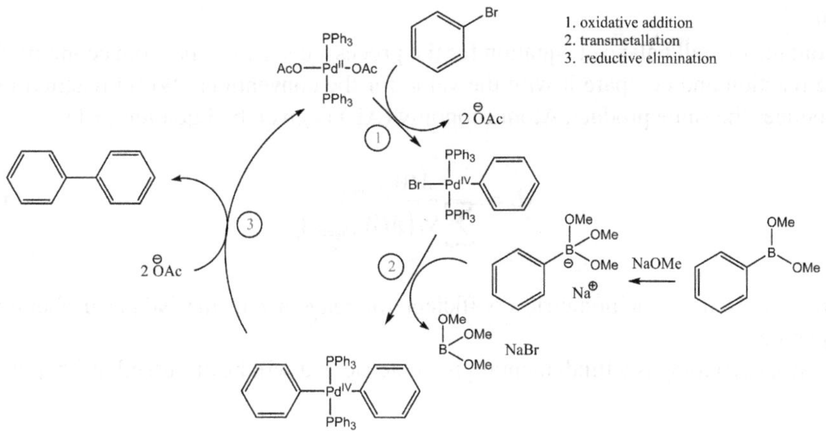

1. oxidative addition
2. transmetallation
3. reductive elimination

PROBLEM 5.9 [30]

Themes: catalytic cycles; reaction mechanism

Draw the catalytic cycle for the reaction shown below and highlight the target bonds in the product structure.

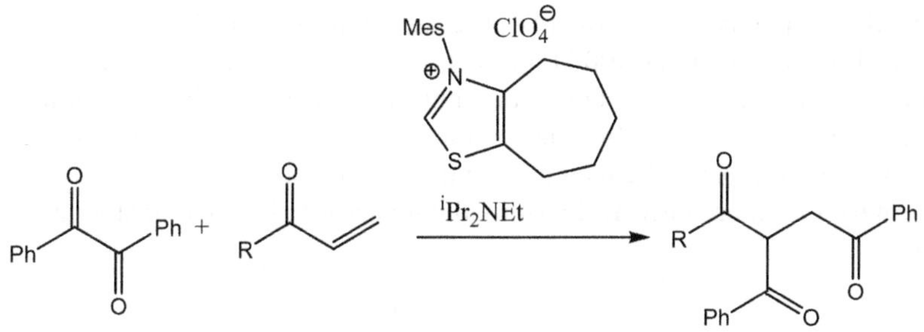

Mes $ClO_4^{\ominus}$

$^{i}Pr_2NEt$

PROBLEM 5.10 [31]

Themes: balancing chemical equations, catalytic cycle, reaction mechanism, atom economy

Part 1

Draw the catalytic cycle for the Wittig reaction shown below. Balance all reaction steps.

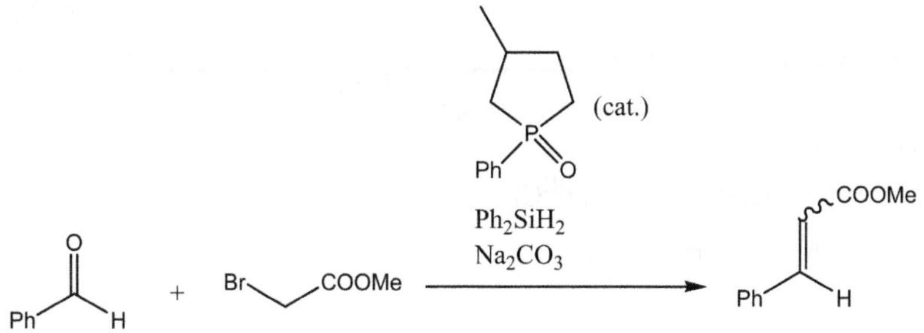

(cat.)

Ph_2SiH_2

Na_2CO_3

Part 2

From the overall balanced equation for the process, determine the atom economy for the reaction and compare it with the value for the conventional Wittig reaction that produces the same product. Atom economy (AE) is given by Equation (5.1).

$$AE = \frac{MW_{\text{product}}}{\sum_i \nu_i \left(MW_{\text{reagent}}\right)_i}, \tag{5.1}$$

where ν_i is the stoichiometric coefficient for reagent i in the balanced chemical equation.

Atom economy is a fundamental green metric that will be discussed in Chapter 6.

PROBLEM 5.11 [32–36]

Themes: balancing chemical equations, catalytic cycles, experimental procedures

Half of the world's mined mercury ore is destined to manufacture mercury(II) chloride, which is used as a catalyst to make vinyl chloride from acetylene and hydrogen chloride. China is the leading manufacturer of vinyl chloride by this method since it also has the largest reserves of coal from which acetylene is obtained. As a consequence of this process, China is also the highest mercury polluter. It is estimated that 1000 tonnes of mercury per annum is needed to manufacture the catalyst and that 600 g of catalyst escapes into the environment per ton of vinyl chloride made. In order to mitigate this situation, recent research has been directed to replace the mercury catalyst with more benign alternatives. One such catalyst is chloroauric acid adsorbed onto activated carbon or silicon carbide support.

$$\text{H—C}\equiv\text{C—H} + \text{HCl} \xrightarrow[\text{T = 185°C}]{\text{HAuCl}_4\text{/ activated C}} \text{CH}_2\text{=CHCl}$$

The authors suggested that the catalyst is a combination of gold(I) and gold(III); however, they did not write out a catalytic cycle for the reaction. Based on computational studies, they wrote out the following mechanisms.

$$\text{H—C}\equiv\text{C—H} + \text{Au}^{\text{III}}\text{Cl}_3 \longrightarrow [\text{Au}^{\text{III}}\text{Cl}_3(\eta^2\text{-HC}\equiv\text{CH})] \longrightarrow [\text{Cl}_2\text{Au}^{\text{III}}\text{-CH=CH-Cl}] \xrightarrow{\text{HCl}} \text{CH}_2\text{=CHCl} + \text{Au}^{\text{III}}\text{Cl}_3$$

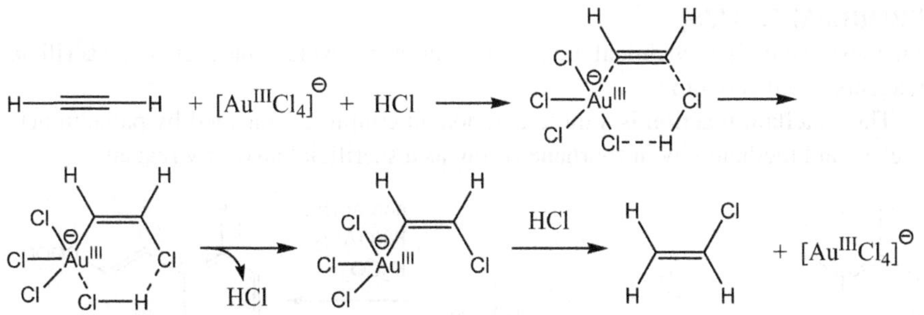

Write out catalytic cycles for the reaction using gold(III) and gold(I) species. Suggest how the gold(I) species arises.

Explain the observed enhanced conversion of acetylene when ceric sulfate is added to the catalyst.

PROBLEM 5.12 [37]

Themes: balancing chemical equations, catalytic cycle, reaction mechanism

A double catalytic cycle is shown below for the stereoselective epoxide formation from sulfur ylides and aldehydes.

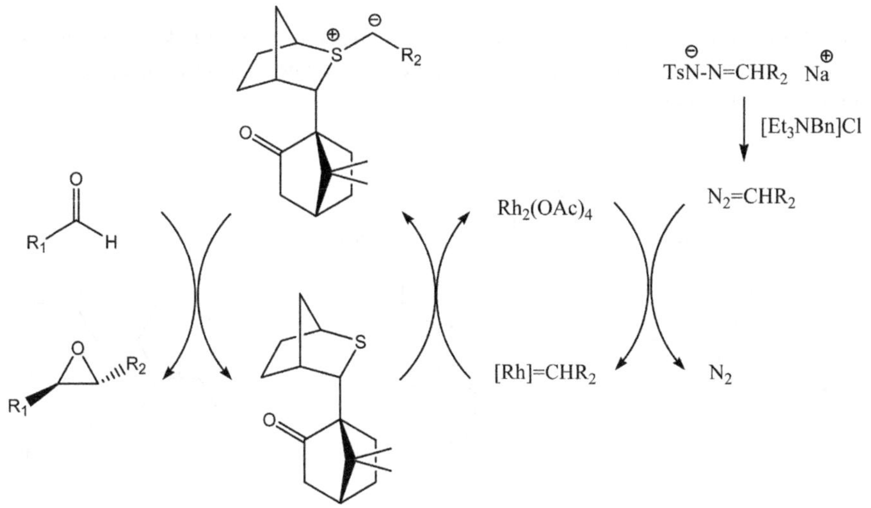

Part 1

From the catalytic cycles shown, write out the balanced kernel chemical reaction.

Part 2

Highlight the target bonds made in the epoxide product.

Part 3

Write out a step-by-step mechanism for the reaction.

PROBLEM 5.13 [38]

Themes: balancing chemical equations, catalytic cycles, mechanism, sacrificial reagents, atom economy

The Catellani reaction is a three-component coupling catalyzed by palladium(II) acetate and mediated by norbornene acting as a sacrificial auxiliary reagent.

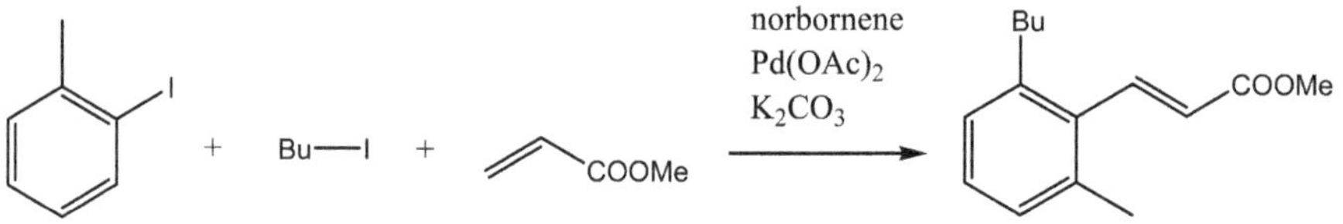

norbornene
$Pd(OAc)_2$
K_2CO_3

Bu—I + COOMe → Bu, COOMe

Part 1

Write a catalytic cycle that shows the roles of norbornene, palladium(II)acetate, and potassium carbonate.

Part 2

Based on the catalytic cycle write out an overall balanced equation for the transformation and determine its atom economy.

Part 3

Show how the following products could be constructed using this kind of reaction. For each case write out a balanced chemical equation and highlight the target bonds in the product structure.

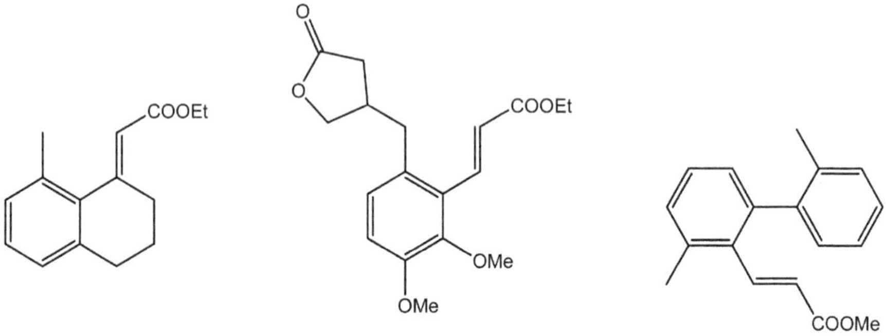

COOEt O O COOEt OMe OMe COOMe

5.2.4 Electrochemical Reactions

Electrochemical reactions may be depicted using a diagram as shown in Figure 5.4. The reduction half-reaction takes place at the cathode (negative electrode) and the oxidation half-reaction takes place at the anode (positive electrode). Note that the sign convention used by physicists for the anode and cathode are exactly the reverse of the convention used by chemists. Cations are positively charged chemical species that are attracted to the negatively charged cathode. Anions are negatively charged chemical species that are attracted to the positively charged anode.

FIGURE 5.4 Diagram depicting half-reactions taking place at cathode and anode electrodes.

Example 5.23

Themes: balancing chemical equations, redox reactions, electrochemical reactions, cathode–anode diagrams

For each of the following electrochemical transformations complete the cathode–anode diagrams and write the balanced chemical equations.

(a) [39]

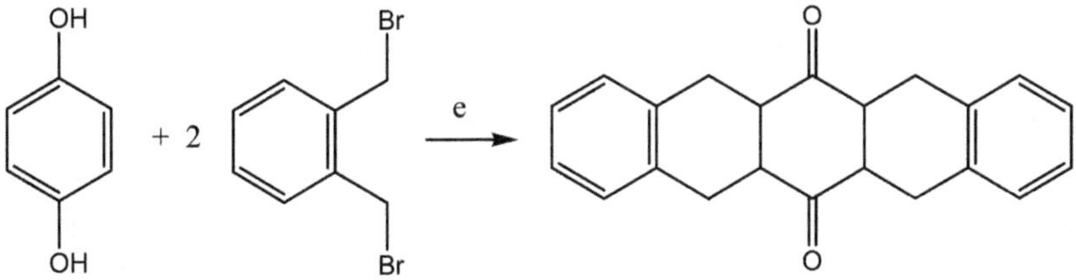

(b) [40]

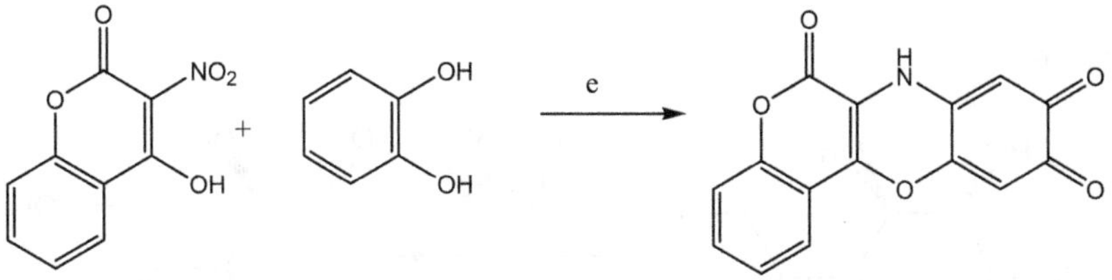

(c) [41]

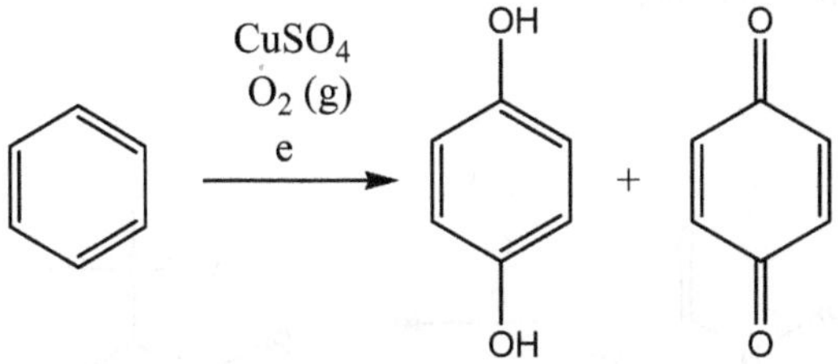

(d) [42]

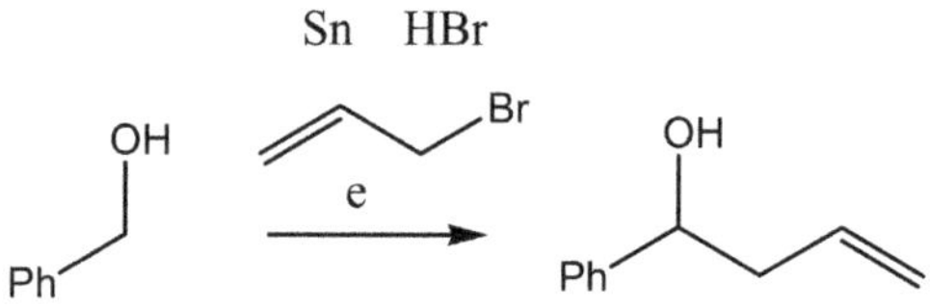

Sn HBr

Br

OH OH

e

Ph Ph

(e) [43]

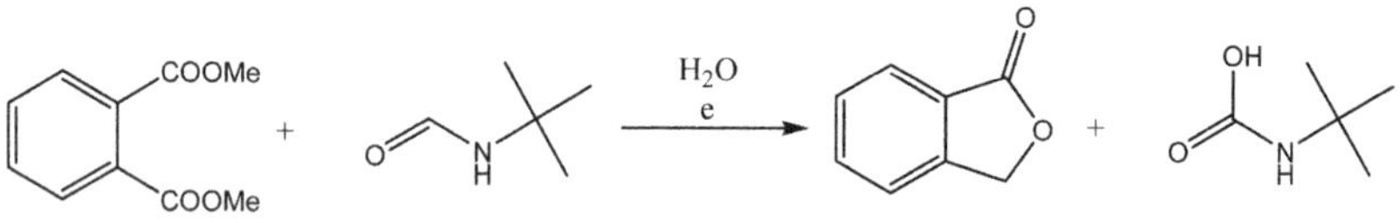

COOMe + O N H + H_2O e → O O + OH O N H

COOMe

(f) [44]

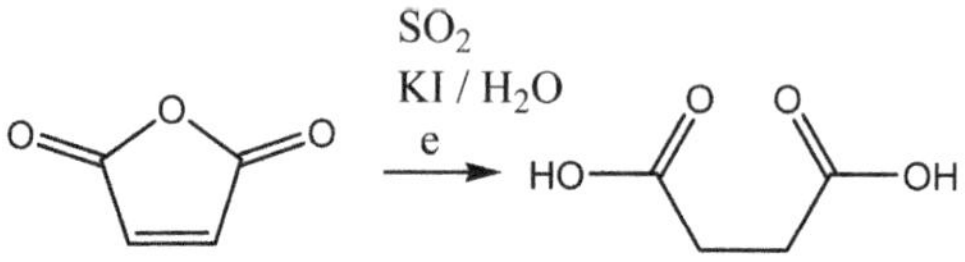

SO_2

KI / H_2O

O O O O O

e

→ HO OH

(g) [45]

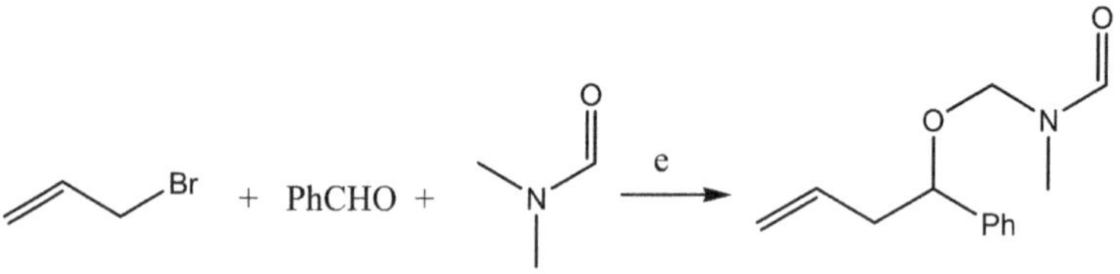

Br + PhCHO + O N e → O N O Ph

(h) [46]

HO OH HOAc O OH

O_2 / Fe^{+2}

OH e OH

NaO

O HO OH HO OH

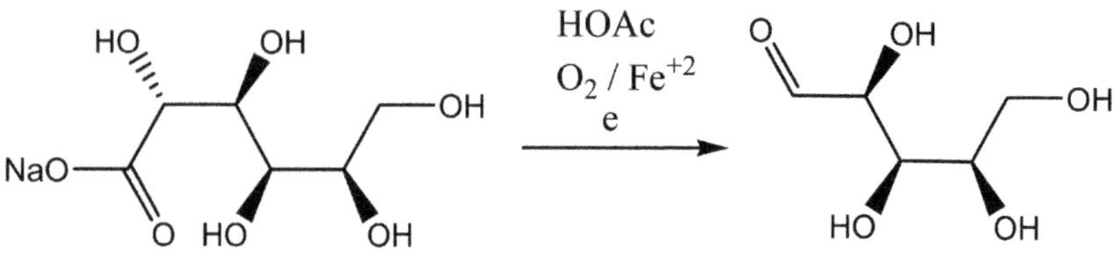

sodium gluconate D-arabinose

(i) [47]

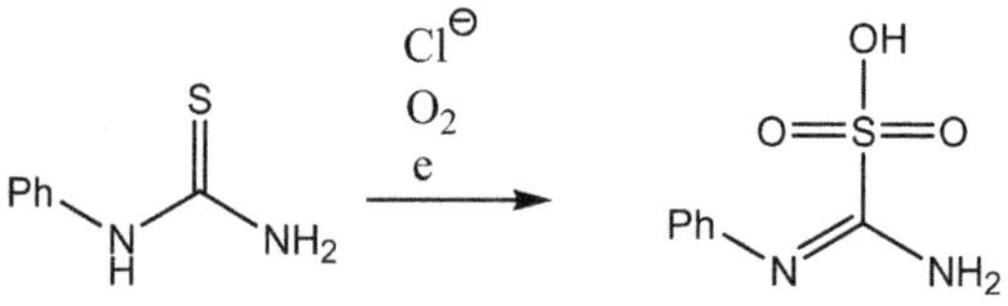

$Cl^{\ominus}$

S O_2 OH

O=S=O

Ph N H NH2 e Ph N NH2

(j) [48]

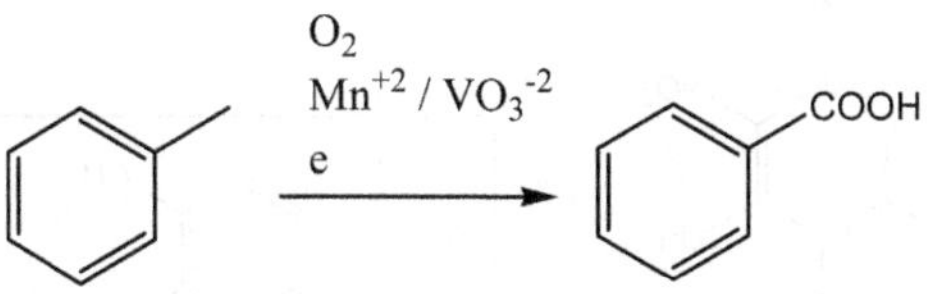

O_2
Mn^{+2} / VO_3^{-2}
e

COOH

Solution

(a)

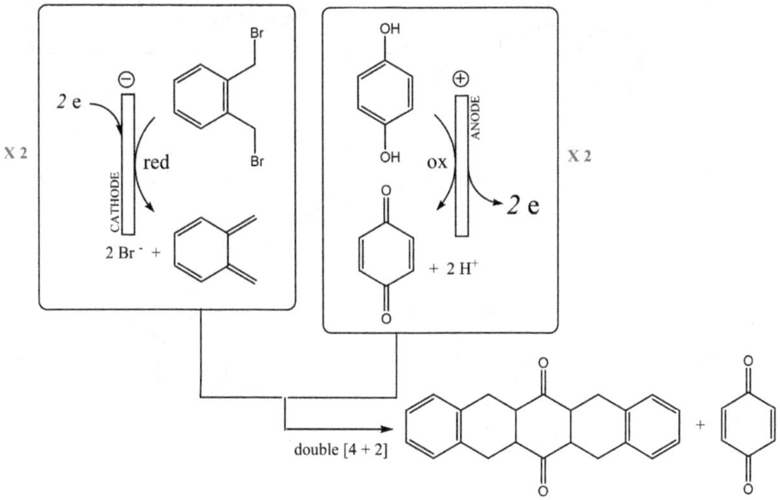

Br
2 e
X 2
red
CATHODE
Br
2 Br⁻ +
OH
ANODE
OH ox
X 2
O
2 e
+ 2 H⁺
O
O
double [4 + 2]
O
O
+
O

Br

CATHODE 2 e⁻ + → + 2 Br⁻

Br

OH O

ANODE → + 2 H⁺ + 2 e⁻

OH O

OH Br O O

2 + 2 → +
- 4 HBr

OH Br O O

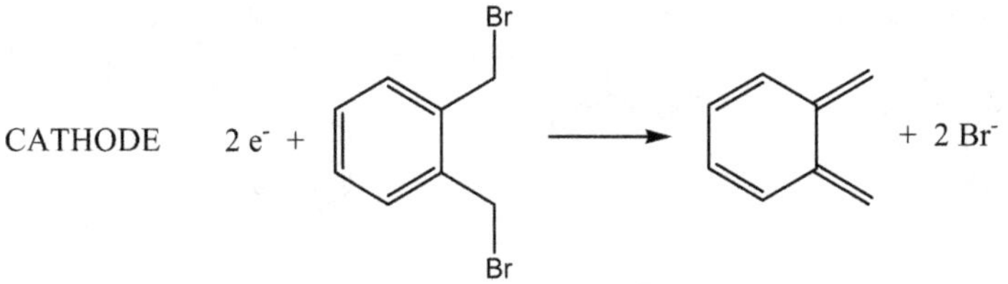

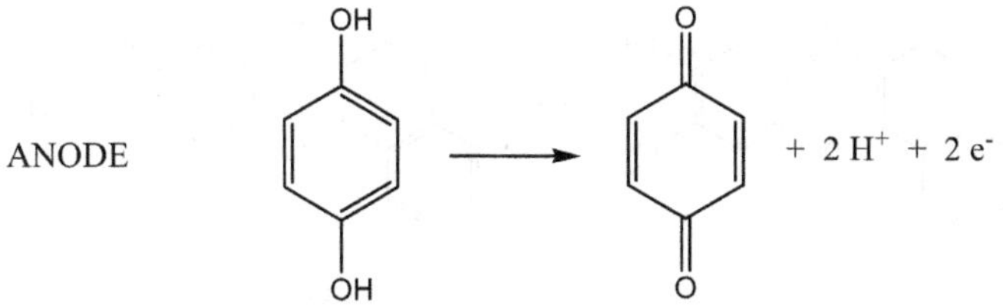

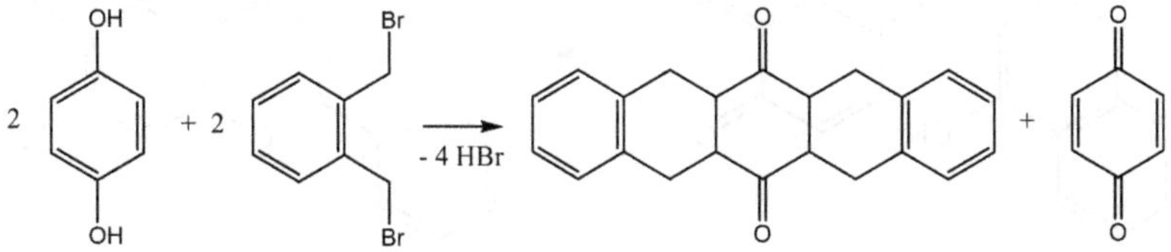

(b)

NO2
OH
6 e
red
CATHODE
NH2
OH
+ 2 H_2O
OH
OH
ANODE
ox
2 e
2 H^+ +
X3
ANODE
OH
OH
+ 2 H^+ + 2 e^-
CATHODE 6 H^+ + 6 e^- +
NO2
OH
NH2
OH
+ 2 H_2O
NH2
OH
+
H
N
OH
OH
OH
ANODE
H
N
OH
OH
OH
H
N
OH
+ 2 H^+ + 2 e
H
N
OH
H
N
OH
OH

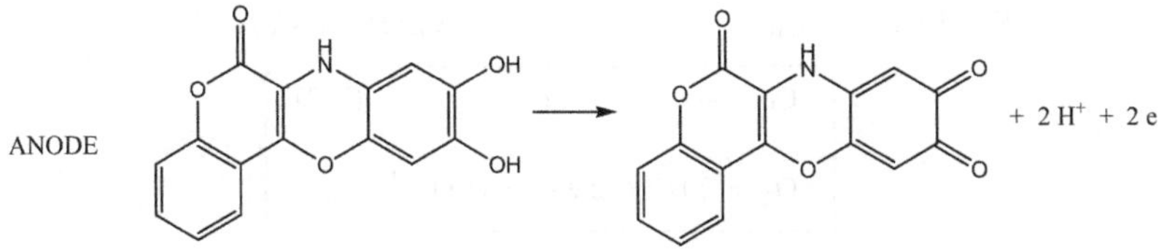

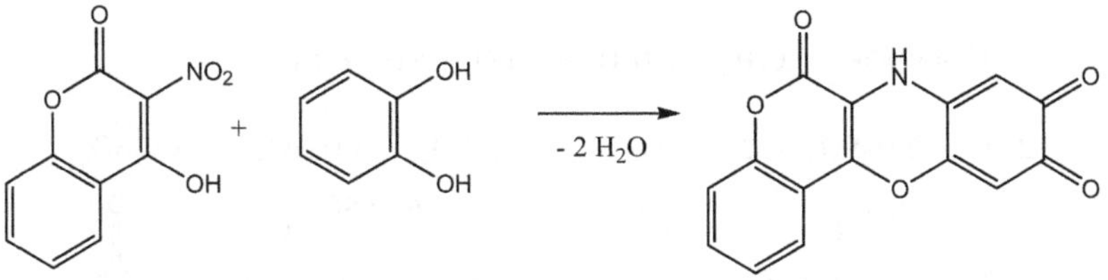

ANODE

$+ 2 H^+ + 2 e$

$- 2 H_2O$

(c)

6 e, CATHODE, red, $3 Cu^{+2}$, $6 Cu^+$, $3 H_2O_2$, $3 O_2 + 6 H^+$, $6 H_2O$, ANODE, ox, 6 e

4 e, CATHODE, red, $2 Cu^{+2}$, $4 Cu^+$, $2 H_2O_2$, $2 O_2 + 4 H^+$, $2 H_2O$, ANODE, ox, 4 e

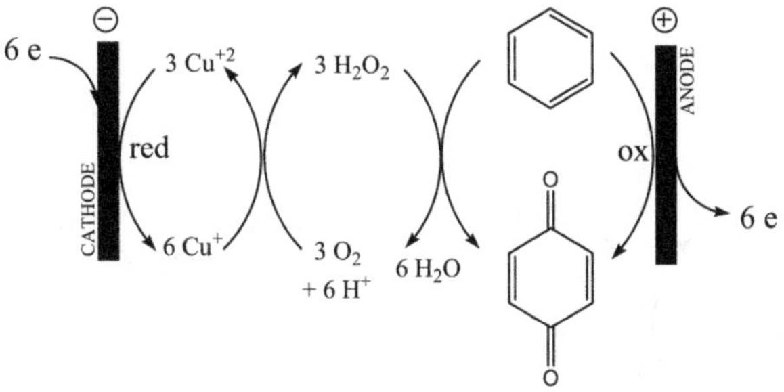

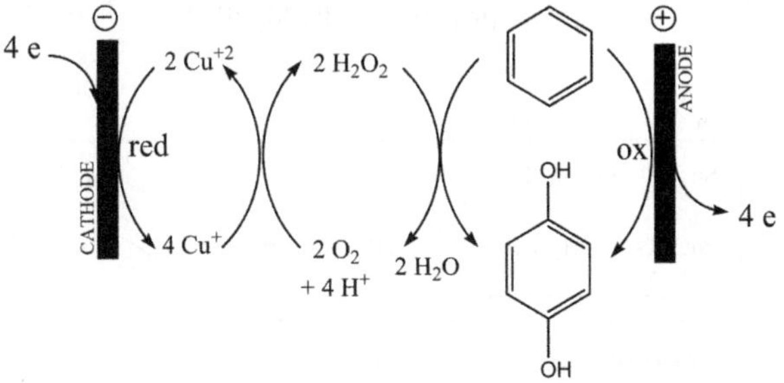

CATHODE $Cu^{+2} + e \Longrightarrow Cu^+$ (X 2) (X 2)

$Cu^+ \Longrightarrow Cu^{+2} + e$ (X 2)

$O_2 + 2 H^+ + 2 e \Longrightarrow H_2O_2$

} (X 2)

$H_2O_2 + 2 H^+ + 2 e \Longrightarrow 2 H_2O$ (X 2)

ANODE $C_6H_6 + 2 H_2O \Longrightarrow HO\text{-}C_6H_4\text{-}OH + 4 H^+ + 4 e$

CATHODE $Cu^{+2} + e \Longrightarrow Cu^{+}$ (X 2) (X 2)(X 2)

$Cu^{+} \Longrightarrow Cu^{+2} + e$ (X 2)

$O_2 + 2\,H^{+} + 2\,e \Longrightarrow H_2O_2$

(X 3)

$H_2O_2 + 2\,H^{+} + 2\,e \Longrightarrow 2\,H_2O$ (X 3)

ANODE $C_6H_6 + 2\,H_2O \Longrightarrow BQ + 6\,H^{+} + 6\,e$

$2\,O_2 + 2\,H_2SO_4 + 2\,Cu_2SO_4$, $-4\,CuSO_4$; benzene $\xrightarrow[-2\,H_2O]{[2\,H_2O_2]}$ hydroquinone (OH, OH)

$3\,O_2 + 3\,H_2SO_4 + 3\,Cu_2SO_4$, $-6\,CuSO_4$; benzene $\xrightarrow[-4\,H_2O]{[3\,H_2O_2]}$ benzoquinone (O, O)

(d)

2 e; CATHODE; red; Sn^{+2}; Sn^{0}; PhCH(OH)CH$_2$CH=CH$_2$; PhCHO, $CH_2{=}CHCH_2Br$; PhCHO; $PhCH_2OH$; $2\,Br^{-}$; Br_2; ox; ANODE; 2 e

CATHODE $Sn^{+2} + 2\,e \Longrightarrow Sn$

$Sn \Longrightarrow Sn^{+2} + 2\,e$

$PhCHO + CH_2{=}CHCH_2Br + H^{+} + 2\,e \longrightarrow$ PhCH(OH)CH$_2$CH=CH$_2$ $+ Br^{-}$

$PhCH_2OH \Longrightarrow PhCHO + 2\,H^{+} + 2\,e$

$Br_2 + 2\,e \Longrightarrow 2\,Br^{-}$

ANODE $2\,Br^{-} \Longrightarrow Br_2 + 2\,e$

$PhCH_2OH \xrightarrow[-2\,HBr]{Br_2} PhCHO \xrightarrow[-SnBr_2]{Sn,\ HBr,\ CH_2=CHCH_2Br}$ PhCH(OH)CH$_2$CH=CH$_2$

(e)

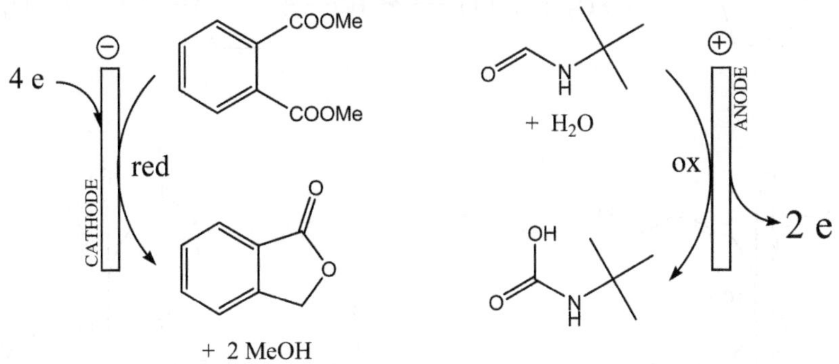

4 e CATHODE red COOMe COOMe O O + 2 MeOH

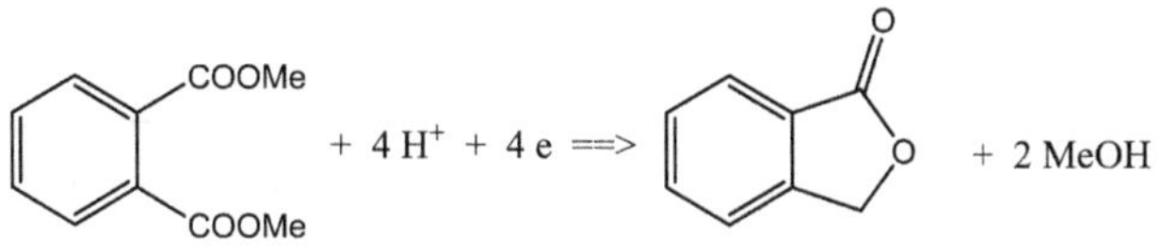

O N H + H_2O ANODE ox 2 e OH O N H

CATHODE COOMe COOMe + 4 H^+ + 4 e ==> O O + 2 MeOH

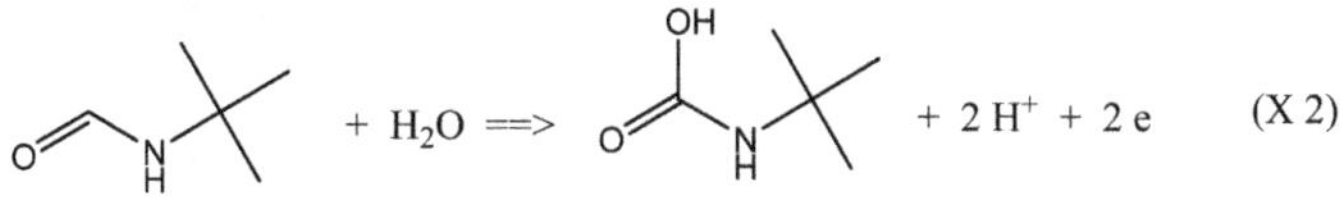

ANODE O N H + H_2O ==> OH O N H + 2 H^+ + 2 e (X 2)

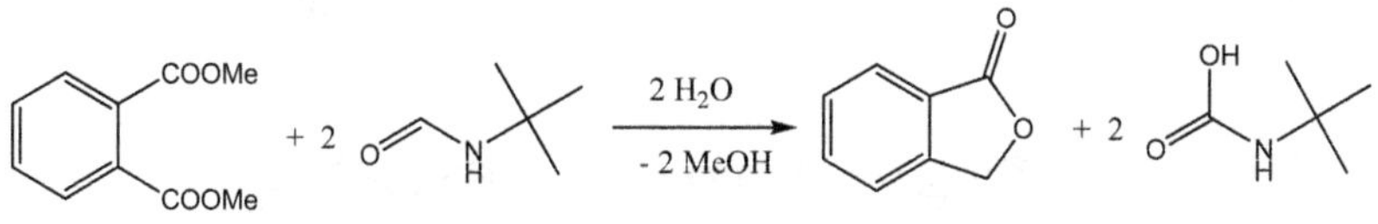

COOMe COOMe + 2 O N H 2 H_2O / - 2 MeOH O O + 2 OH O N H

(f)

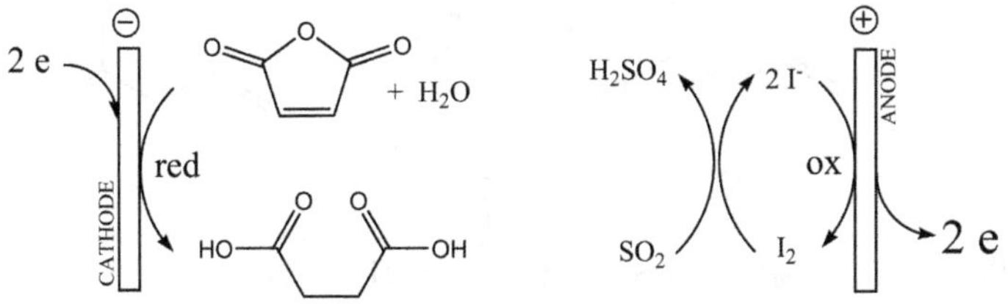

2 e CATHODE red O O O + H_2O HO O O OH

H_2SO_4 2 I^- ANODE ox 2 e SO_2 I_2

CATHODE O O O + H_2O + 2 H^+ + 2 e ==> HO O O OH

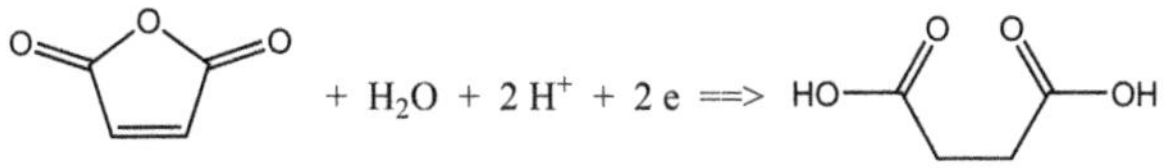

SO_2 + 2 H_2O ==> H_2SO_4 + 2 H^+ + 2 e

I_2 + 2 e ==> 2 I^-

ANODE 2 I^- ==> I_2 + 2 e

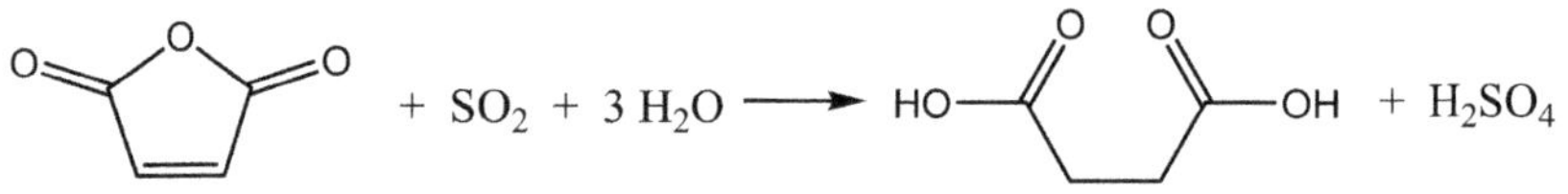

$+ SO_2 + 3\,H_2O \longrightarrow$ HO—C(=O)CH$_2$CH$_2$C(=O)—OH $+ H_2SO_4$

(g)

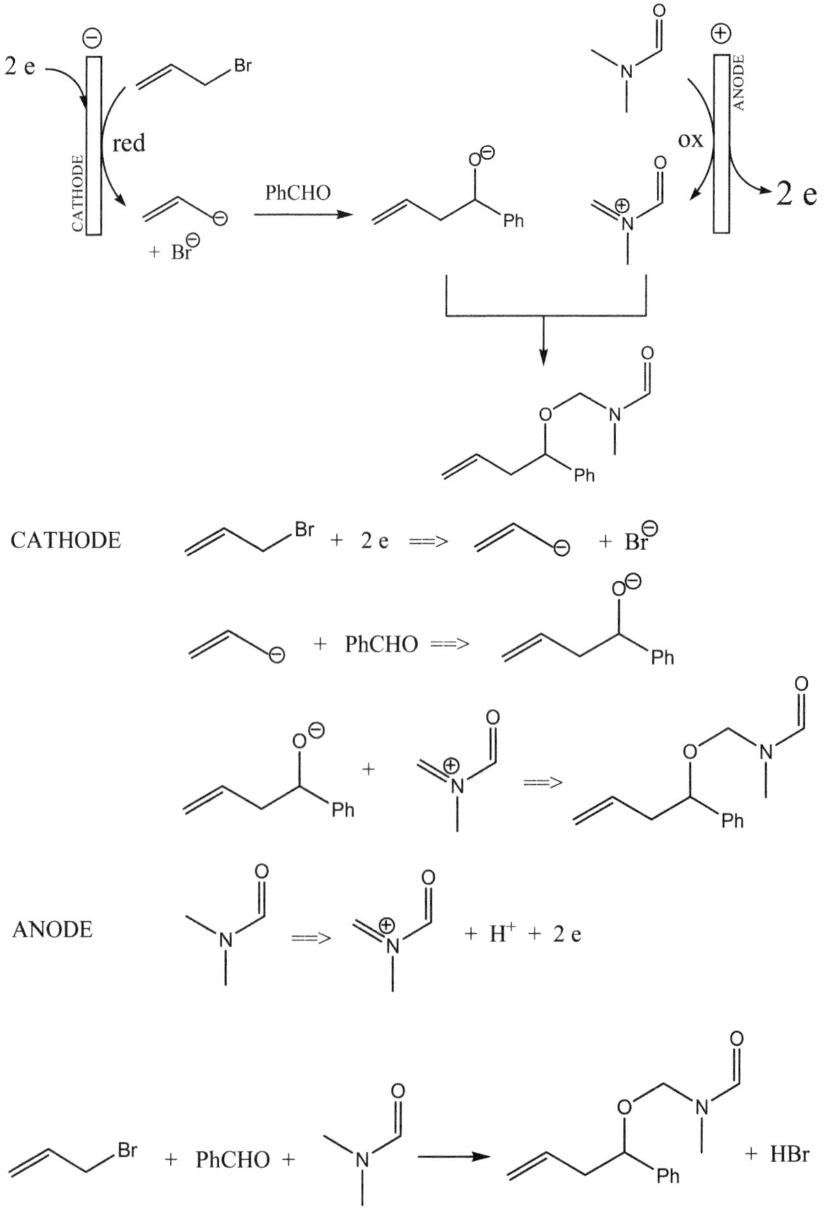

2 e
CATHODE
red
Br
PhCHO
$O^{\ominus}$
Ph
$+ Br^{\ominus}$
O
N
ANODE
ox
2 e
O
Ph

CATHODE + 2 e ==> + $Br^{\ominus}$

+ PhCHO ==>

+ ==>

ANODE ==> + H^+ + 2 e

+ PhCHO + ⟶ + HBr

(h)

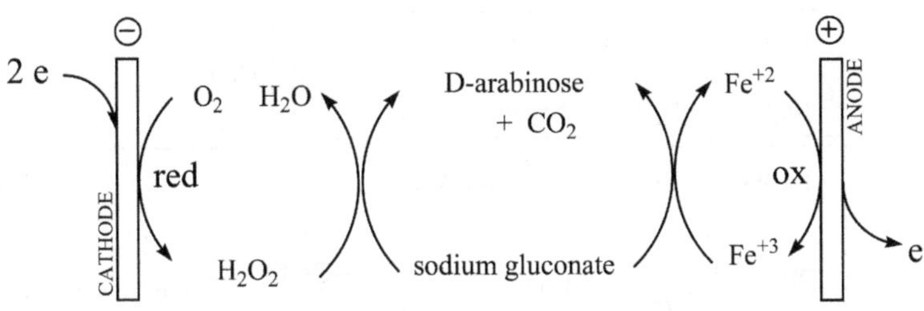

CATHODE $O_2 + 2\,H^+ + 2\,e \Longrightarrow H_2O_2$

$H_2O_2 + 2\,H^+ + 2\,e \Longrightarrow 2\,H_2O$

Na gluconate $\Longrightarrow$ D-arabinose $+ CO_2 + Na^+ + H^+ + 2\,e$

Na gluconate $\Longrightarrow$ D-arabinose $+ CO_2 + Na^+ + H^+ + 2\,e$

$Fe^{+3} + e \Longrightarrow Fe^{+2}$ (X 2)

ANODE $Fe^{+2} \Longrightarrow Fe^{+3} + e$ (X 2)

2 NaO— HO OH —OH O HO OH → 2 HOAc, O_2, $-\,2\,CO_2$, $-\,2$ NaOAc, $-\,2\,H_2O$ → 2 O OH —OH HO OH

sodium gluconate D-arabinose

(i)

2 e CATHODE red O_2 H_2O H_2O_2 NaO-W(=O)(=O)(-O-O)-ONa NaO-W(=O)(=O)-ONa Ph-NH-C(=S)-NH$_2$ Ph-N=C(-S(=O)(=O)OH)-NH$_2$ Cl_2 2 Cl^- ox ANODE 2 e

CATHODE $O_2 + 2\,H^+ + 2\,e \Longrightarrow H_2O_2$ (X3)

$H_2O_2 + 2\,H^+ + 2\,e \Longrightarrow 2\,H_2O$

$Na_2WO_4 + H_2O \Longrightarrow Na_2WO_5 + 2\,H^+ + 2\,e$ } (X3)

$Na_2WO_5 + 2\,H^+ + 2\,e \Longrightarrow Na_2WO_4 + H_2O$ (X3)

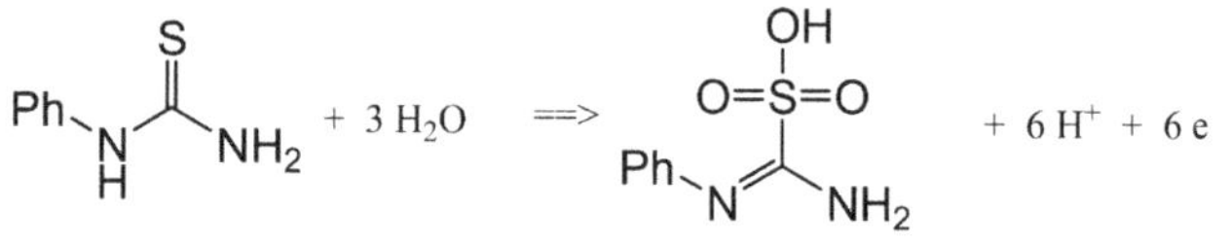

S
Ph
N
H
NH2
+ 3 H2O
==>
OH
O=S=O
Ph
N
NH2
+ 6 H+ + 6 e

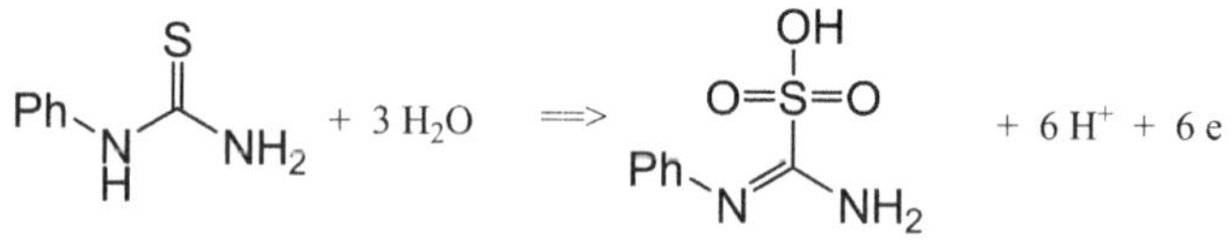

S
Ph
N
H
NH2
+ 3 H2O
==>
OH
O=S=O
Ph
N
NH2
+ 6 H+ + 6 e

$Cl_2 + 2\,e \Longrightarrow 2\,Cl^-$ (X 3)

ANODE $2\,Cl^- \Longrightarrow Cl_2 + 2\,e$ (X 3)

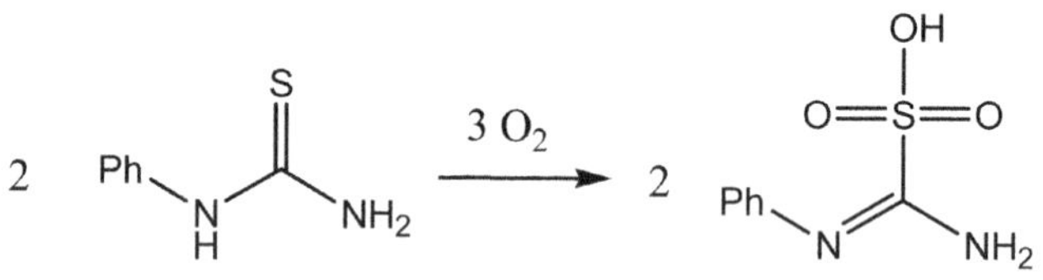

2
S
Ph
N
H
NH2
3 O2
2
OH
O=S=O
Ph
N
NH2

(j)

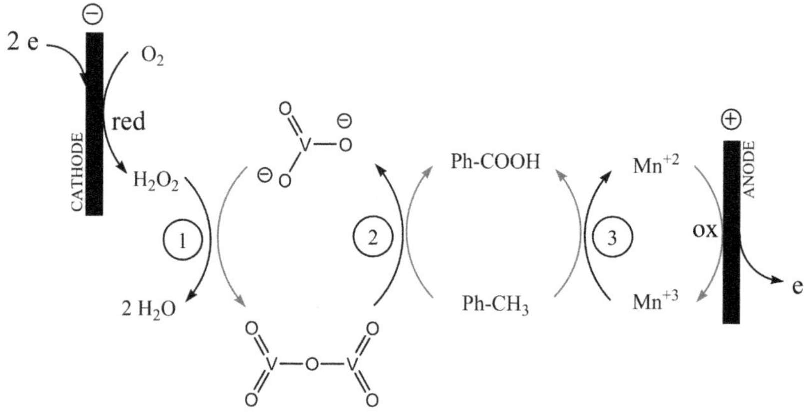

2 e
O2
red
CATHODE
H2O2
1
2 H2O
2
Ph-COOH
Ph-CH3
3
Mn+2
Mn+3
ox
ANODE
e

CATHODE $O_2 + 2H^+ + 2e \Longrightarrow H_2O_2$ (X 3)

(1)
$H_2O_2 + 2H^+ + 2e \Longrightarrow 2H_2O$
$2VO_3^{-2} + 2H^+ \Longrightarrow V_2O_5 + H_2O + 2e$
(X 3)

(2)
$V_2O_5 + H_2O + 2e \Longrightarrow 2VO_3^{-2} + 2H^+$ (X 3)
$PhCH_3 + 2H_2O \Longrightarrow PhCOOH + 6H^+ + 6e$

(3)
$Mn^{+3} + e \Longrightarrow Mn^{+2}$ (X 6)
$PhCH_3 + 2H_2O \Longrightarrow PhCOOH + 6H^+ + 6e$

ANODE $Mn^{+2} \Longrightarrow Mn^{+3} + e$ (X 6)

$$2\ PhCH_3 \xrightarrow[-2H_2O]{3O_2} PhCOOH$$

5.3 TARGET BOND MAPPING

Concession Step (Sacrificial Step) [49–51]

A reaction step in a synthesis plan that does not result in the formation of a target bond that appears in the final target structure of the plan. It is typically a redox correction step, a functional group protection step, a functional group deprotection step, or a substitution correction step. Well-strategized synthesis plans have few concession steps.

Construction Step (Target Bond Forming Step) [49–51]

A reaction step in a synthesis plan that results in the formation of a target bond that appears in the final target structure of the plan. Well-strategized synthesis plans have a high proportion of construction steps, often containing construction steps that produce more than one target bond in the same reaction.

Ideality [50–52]

Applied to a synthesis plan ideality is the fraction of reaction steps in a synthesis that are not concession or sacrificial steps.

Tanimoto Similarity Index [53]

The Tanimoto index is a pairwise similarity score that compares the number of common and uncommon highlighted target forming bonds between two target bond maps for the same chemical structure. Each target bond map corresponds to a given synthesis plan and highlights the set of target forming bonds made. It is used to compare multiple synthesis strategies employed to make a given target molecule. It is useful in discerning unique strategies in a set of strategies and thus in categorizing strategies into groups. This is particularly useful in grouping ring construction strategies that can be made by various assemblages. Strategic novelty is characterized by a low similarity score.

The Tanimoto index, *T*, is given by Equation (5.2).

$$T = \frac{b(A \cap B)}{b(A \cup B)} = \frac{b(A \cap B)}{b(A) + b(B) - b(A \cap B)} \tag{5.2}$$

where:

$b(A \cap B)$ is the number of common highlighted target forming bonds found in both structures *A* and *B*

$b(A)$ is the number of highlighted target forming bonds found only in structure A

$b(B)$ is the number of highlighted target forming bonds found only in structure B.

If $T = 1$, then both target bond maps A and B highlight the same set of highlighted target forming bonds, and hence it can be concluded that the strategies employed to make the target molecule are very similar. If $T = 0$, then there are no common target forming bonds found in target bond maps *A* and *B*, and hence it can be concluded that the strategies employed to make the target molecule are very different. The Tanimoto index does not take into account the order of reaction steps, so two synthesis plans resulting in identical target bond maps but using reactions in different orders will be counted as having the same Tanimoto index.

Example 5.24 [54–56]

A computer-assisted program called SYNSUP was used to propose three routes to 4,4′-diisopropylcyclohexanone, which is an intermediate for the synthesis of the anti-arthritic compound atiprimod. All routes begin from the common starting material 4-heptanone.

Route A

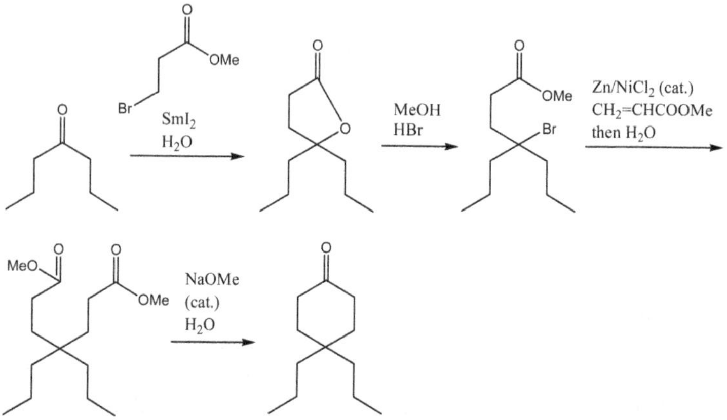

SmI_2, H_2O → MeOH, HBr → Zn/$NiCl_2$ (cat.), CH_2=CHCOOMe then H_2O → NaOMe (cat.), H_2O

Route B

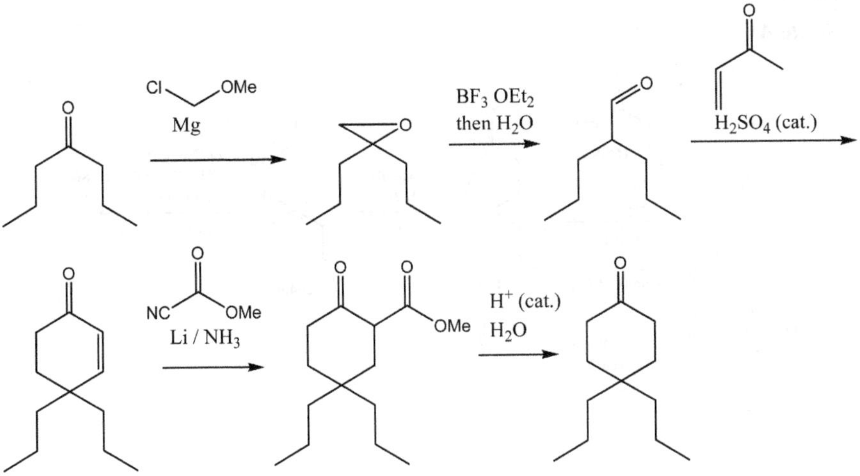

$Cl\!\!-\!\!CH_2\!\!-\!\!OMe$, Mg; $BF_3\,OEt_2$ then H_2O; methyl vinyl ketone, H_2SO_4 (cat.); NC–CO–OMe, Li / NH_3; H^+ (cat.) H_2O

Route C

NC–CH₂–COOMe, base (cat.); H_2SO_4 H_2O; $SOCl_2$; $CH_2{=}N_2$; heat MeOH; NaOMe (cat.) then H_2O

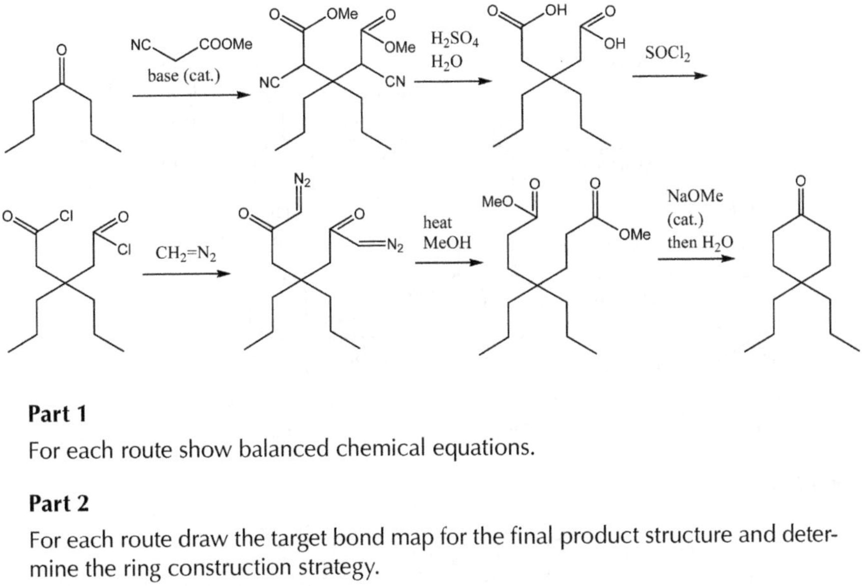

Part 1

For each route show balanced chemical equations.

Part 2

For each route draw the target bond map for the final product structure and determine the ring construction strategy.

Part 3

From the product target bond maps determine the Tanimoto similarity index for all pairwise map comparisons. Which pair of synthesis strategies is most similar?

Solution

Part 1

Route A

O
OMe
Br
166.9
O
SmI_2
H_2O
- MeOH
- I_2
- Sm(OH)(Br)
114
O
O
170
MeOH
HBr
- H_2O
O
OMe
Br
264.9
$Zn/NiCl_2$ (cat.)
CH_2=CHCOOMe **86**
then H_2O
- Zn(Br)(OH)

MeO
O
O
OMe
272
NaOMe
(cat.)
H_2O
- 2 MeOH
- CO_2
O
182

Route B

80.45
Cl OMe
Mg
O
- Mg(Cl)(OMe)
114
O
128
BF_3 OEt_2
then 3 H_2O
- $B(OH)_3$
- 3 HF
- Et_2O
O
128
O
70
H_2SO_4 (cat.)
- H_2O

O
180
O
85
NC OMe
2 Li / NH_3
- LiCN
- $LiNH_3$
O O
OMe
240
H^+ (cat.)
H_2O
- MeOH
- CO_2
O
182

Route C

99
2 NC COOMe
base (cat.)
- H_2O
O
114
O OMe O
OMe
NC CN
294
H_2SO_4
6 H_2O
- 2 MeOH
- 2 CO_2
- $(NH_4)_2SO_4$
O OH O
OH
216
2 $SOCl_2$
- 2 SO_2
- 2 HCl

O Cl O
Cl
252.9
2 CH_2=N_2
- 2 HCl
N_2
O O
N_2
264
heat
2 MeOH
- 2 N_2
MeO
O O
OMe
272
NaOMe
(cat.)
then H_2O
- 2 MeOH
- CO_2
O
182

Part 2

Product structure target bond maps:

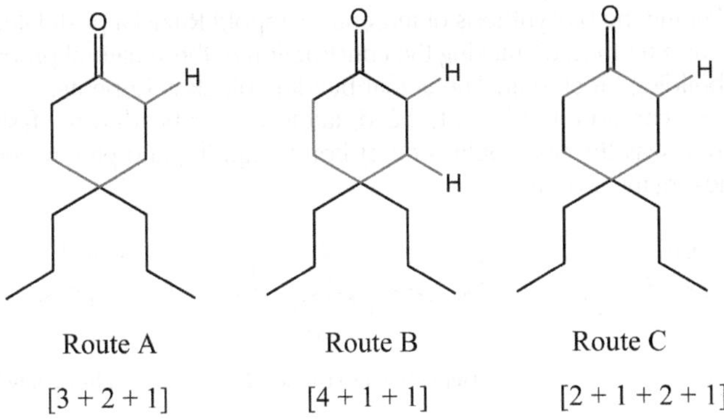

Route A [3 + 2 + 1]

Route B [4 + 1 + 1]

Route C [2 + 1 + 2 + 1]

Part 3

A versus B: $b(A \cap B) = 3$, $b(A) + b(B) = 4 + 5 = 9$ T = 3/(9 – 3) = 0.5

A versus C: $b(A \cap C) = 4$, $b(A) + b(C) = 4 + 5 = 9$ T = 4/(9 – 4) = 0.8

B versus C: $b(B \cap C) = 3$, $b(B) + b(C) = 3 + 5 = 8$ T = 3/(8 – 3) = 0.6

Routes	**A**	**B**	**C**
A		0.5	0.8
B			0.6
C			

Therefore, routes A and C are the most similar in synthesis strategy.

Target Bond Map

A drawing of a chemical structure that highlights the target bonds made according to the synthesis plan followed to make it.

Target Bond Profile

A bar graph or histogram that plots the number of target bonds made in each reaction step. This diagram is useful in categorizing which steps are target bond forming (construction) steps and which steps are sacrificial (concession) steps in a synthesis plan.

Example 5.25 [57]

Themes: target bond mapping

In explaining the biosynthesis of terpenes, Leopold Ruzicka postulated the isoprene rule as a method of tracking the construction of these natural products from a simple building block unit. The actual building block is isopentenyl pyrophosphate, which can add in a head-to-head, tail-to-tail, or head-to-tail fashion. The isoprene rule was the first application of bond mapping and pattern recognition used in bio-organic synthesis.

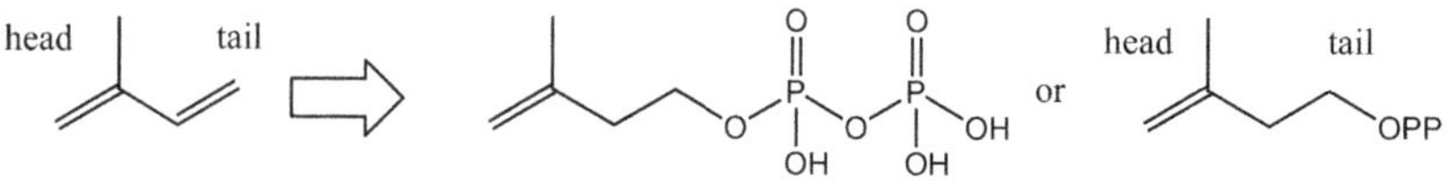

head tail ⇨ O O P P O O OH OH OH or head tail OPP

isoprene **isopentenyl pyrophosphate** **abbreviated structure**

Isopentenyl pyrophosphate is in equilibrium with dimethylallyl pyrophosphate.

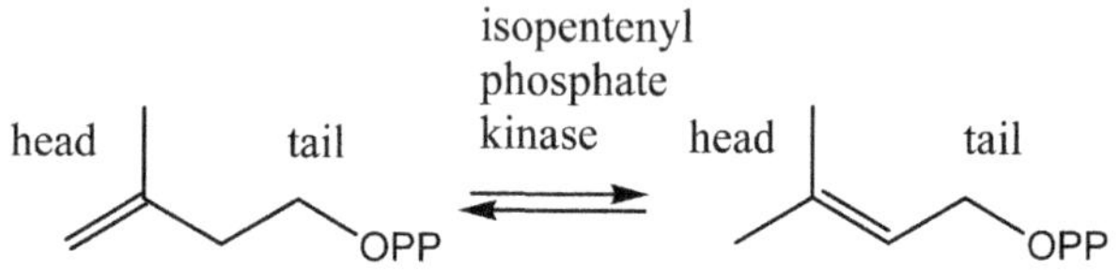

head tail OPP ⇌ (isopentenyl phosphate kinase) head tail OPP

isopentenyl PP **dimethylallyl PP**

The biosynthetic pathway for the production of isopentenyl pyrophosphate from acetyl-CoA via mevalonic acid is shown below.

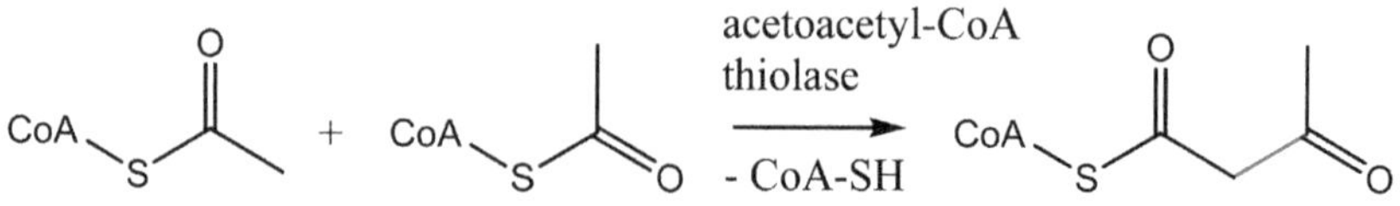

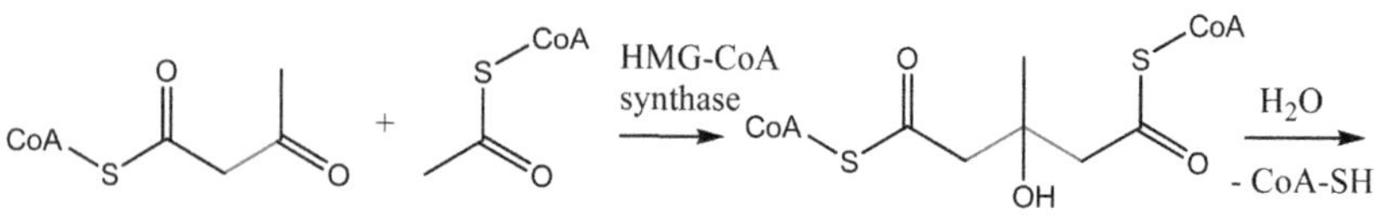

CoA–S–C(O)CH3 + CoA–S–C(O)CH3 → (acetoacetyl-CoA thiolase; - CoA-SH) CoA–S–C(O)CH2C(O)CH3

CoA–S–C(O)CH2C(O)CH3 + CoA–S–C(O)CH3 → (HMG-CoA synthase) HMG-CoA (OH) → (H_2O; - CoA-SH)

HMG = 3-hydroxy-3-methyl-glutaryl

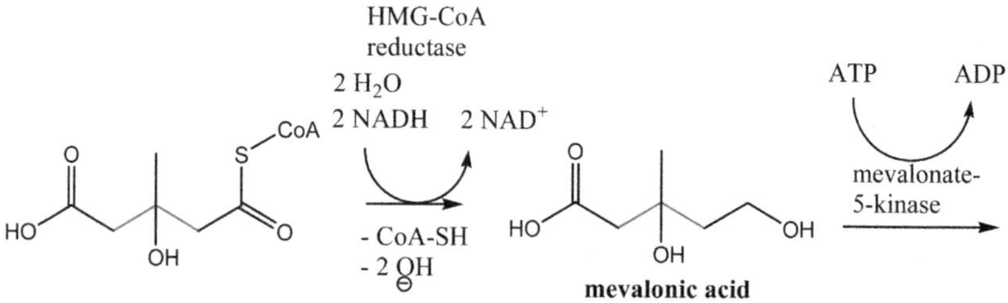

HO–C(O)–CH2–C(OH)(CH3)–CH2–C(O)–S–CoA → (HMG-CoA reductase; 2 H_2O; 2 NADH → 2 NAD^+; - CoA-SH; - 2 OH^-) HO–C(O)–CH2–C(OH)(CH3)–CH2CH2–OH → (ATP → ADP; mevalonate-5-kinase)

mevalonic acid

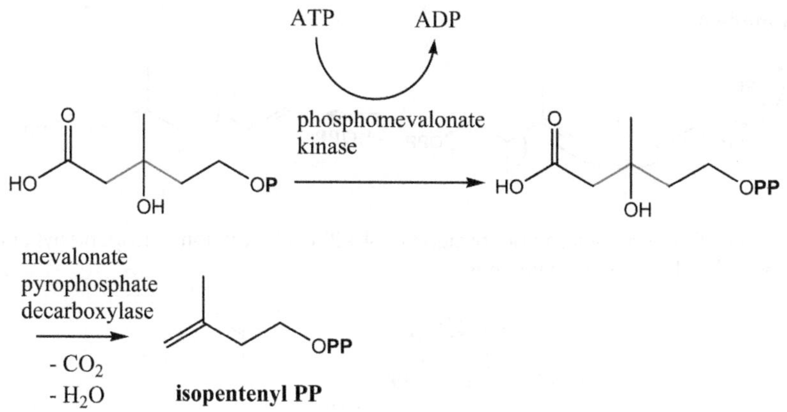

ATP ADP

phosphomevalonate kinase

HO OP OH → HO OPP OH

mevalonate pyrophosphate decarboxylase

- CO_2

- H_2O

OPP

isopentenyl PP

The C5 isopentenyl and C5 dimethylallyl pyrophosphate units can link up in head-to-tail and head-to-head fashions to create linear C10 dimers called monoterpenes.

Head-to-tail linkage:

Example 1

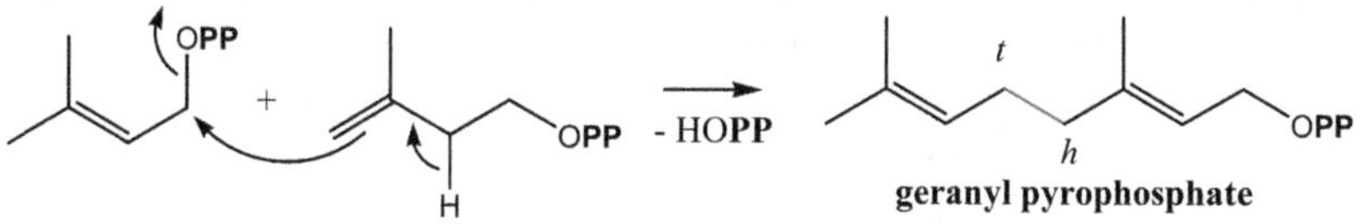

OPP + OPP H - HOPP

t OPP *h*

geranyl pyrophosphate

Example 2

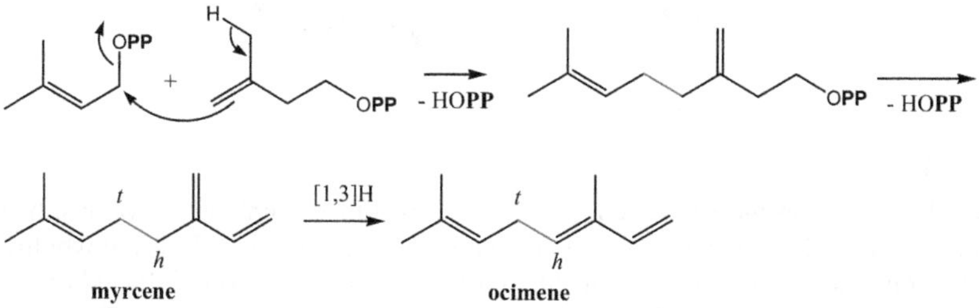

OPP H + OPP - HOPP

OPP - HOPP

t *h* **myrcene** [1,3]H *t* *h* **ocimene**

Head-to-head linkage:

Example 3

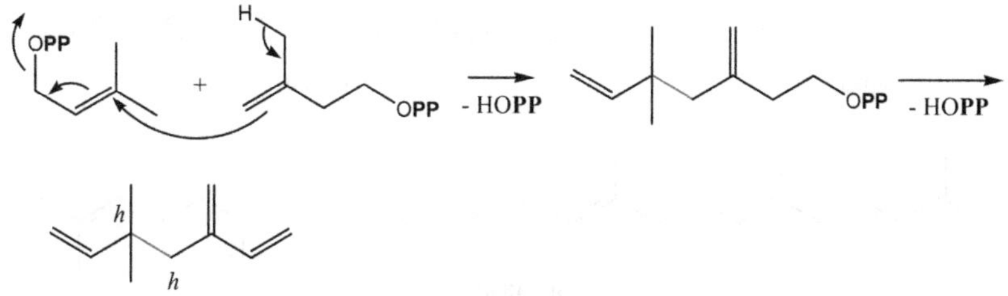

OPP H + OPP - HOPP

OPP - HOPP

h *h*

Example 4

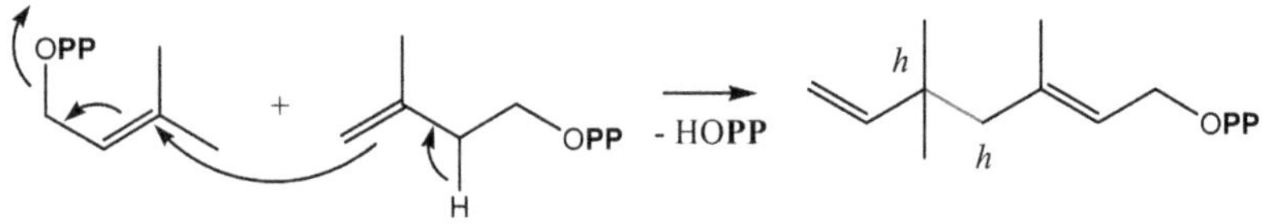

OPP
+
OPP
- HOPP
H
h
h
OPP

Cyclic C10 terpenes can be formed via [4+2] cycloaddition of isopentenyl and dimethylallyl pyrophosphate units.

H
PPO
OPP
- 2 HOPP
H

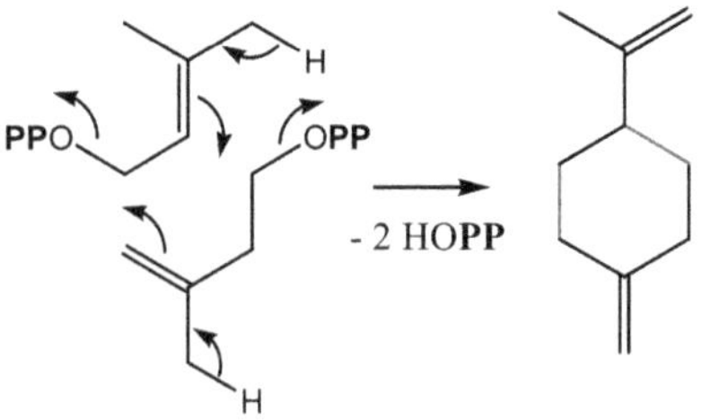

Ruzicka was likely inspired by the discovery of the [4+2] Diels–Alder cycloaddition reaction in 1928 when he formulated the isoprene rule to explain the biosynthesis of C10 cyclic monoterpenes. The above [4+2] example can be translated in Diels–Alder form as shown below.

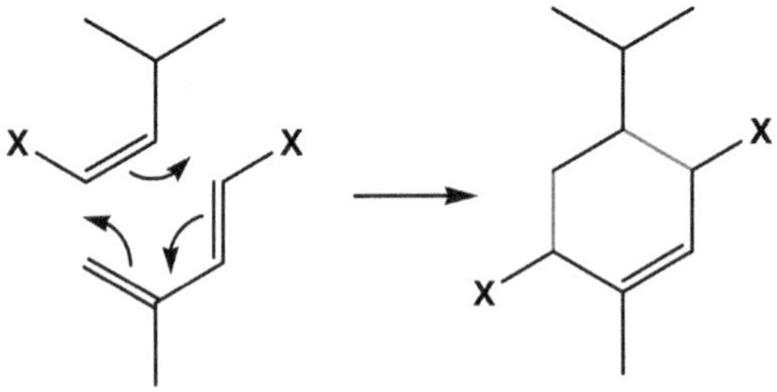

X
X
X
X

Part 1

For the following terpenes parse the structures into isopentenyl pyrophosphate and dimethylallyl pyrophosphate units in a jigsaw puzzle sense. Indicate whether the connecting bonds are head-to-head, head-to-tail, or tail-to-tail. Based on the C5 isoprene unit, classify the terpenes as mono- (C10 = 2 × C5 = 1 × C10), sesqui- (C15 = 3 × C5 = 1.5 × C10), di-(C20 = 4 × C5 = 2 × C10), ses-(C25 = 5 × C5 = 2.5 × C10), tri- (C30 = 6 × C5 = 3 × C10), or tetraterpenes (C40 = 8 × C5 = 4 × C10).

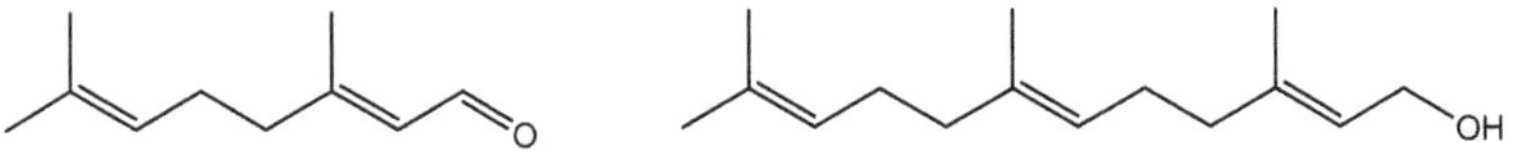

O
citral

OH
farnesol

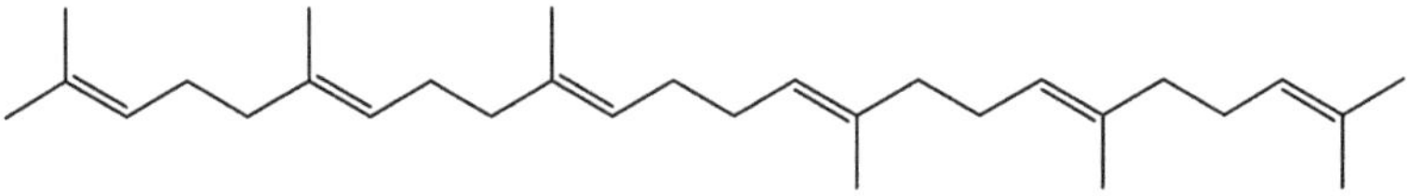

squalene

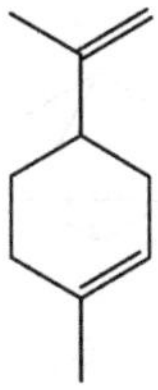

OH

OH

limonene 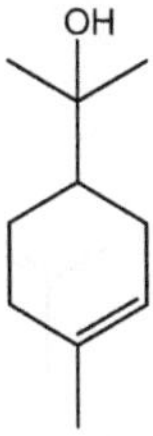 **α-terpineol** 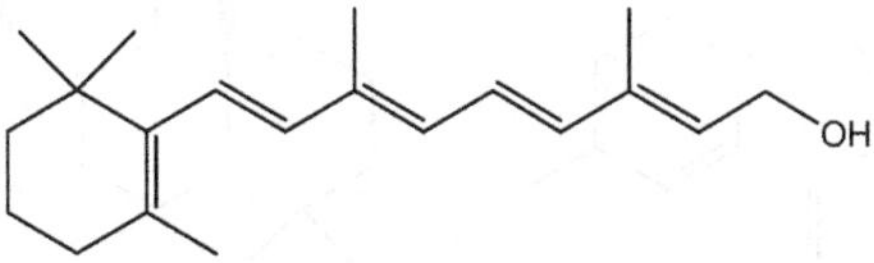 **retinol (vitamin A)**

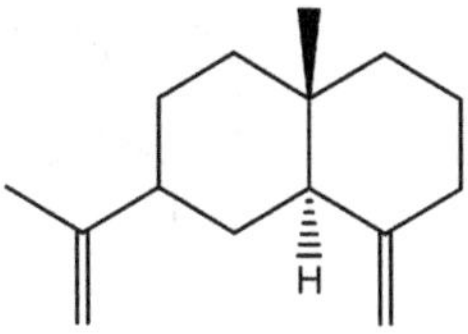

H

OH

α-selinene 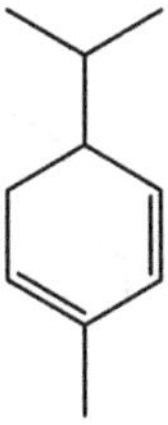 **phellandrene** 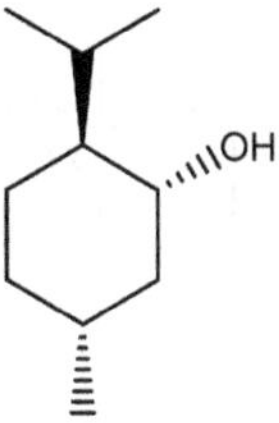 **(-)-menthol**

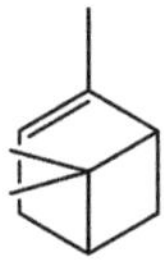

OH

OH

α-pinene 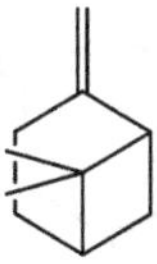 **β-pinene** 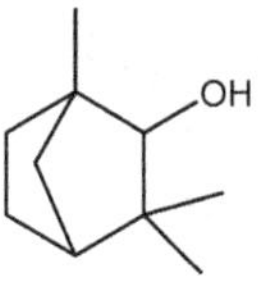 **fenchol** 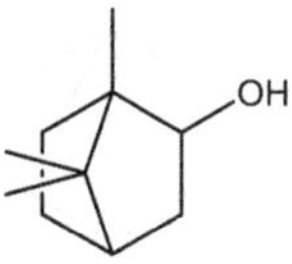 **borneol** **Δ^4-carene**

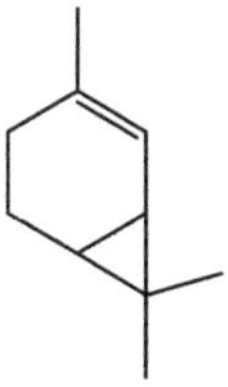

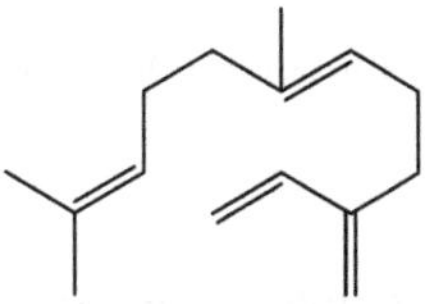

OH

farnesene 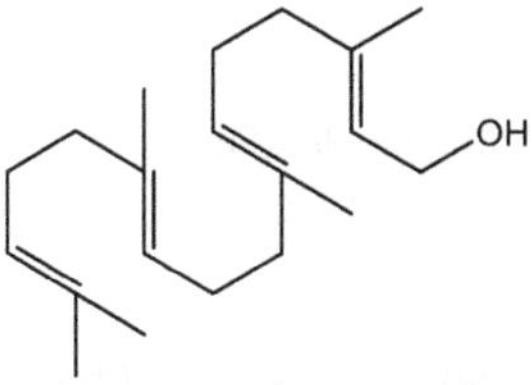 **geranylgeraniol** 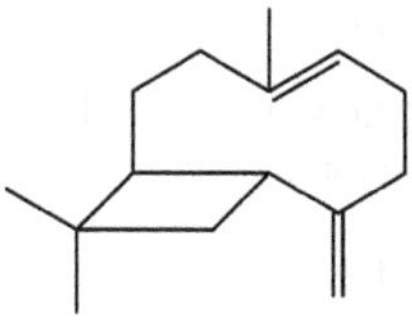 **β-caryophyllene**

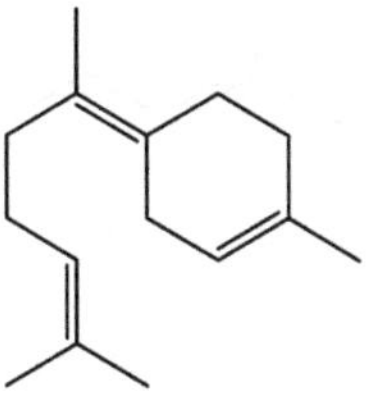

COOH

bisabolene 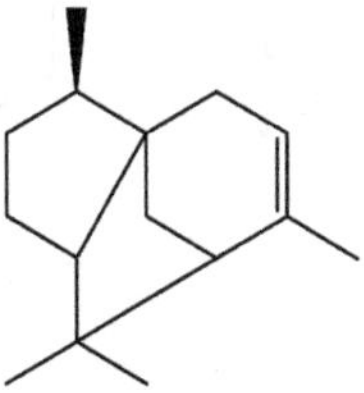 **cedrene** 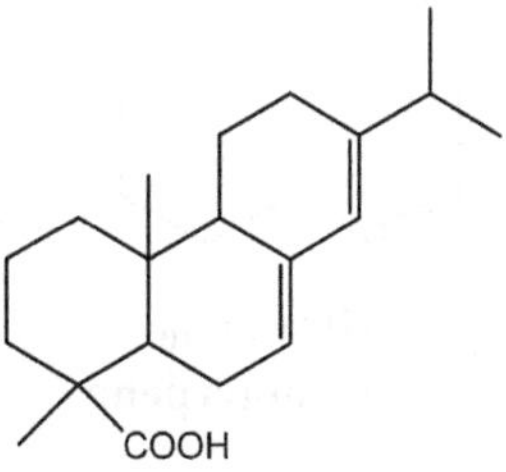 **abietic acid**

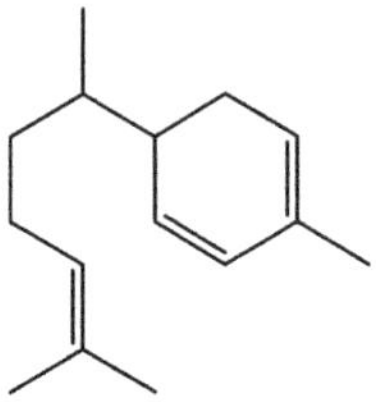

zingiberene 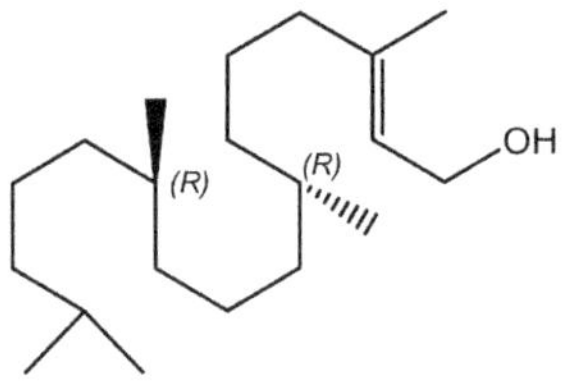 phytol 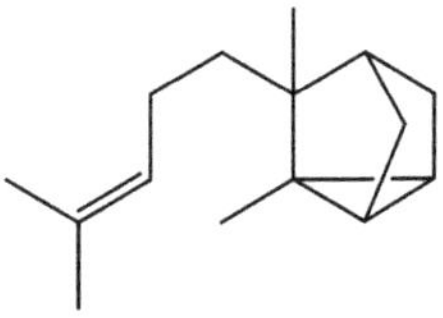 α-santalene

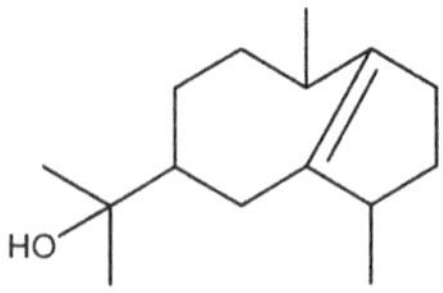

guaiol 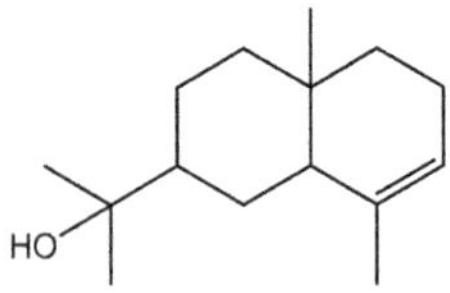 α-eudesmol 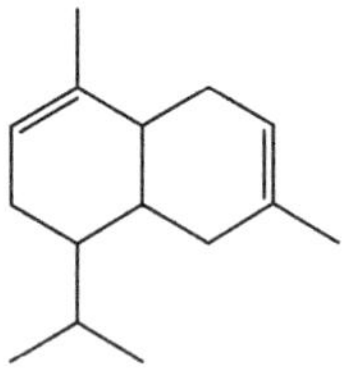 cadinene

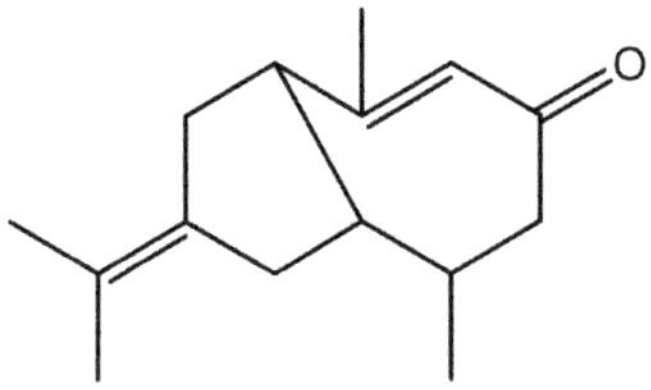

vetivone 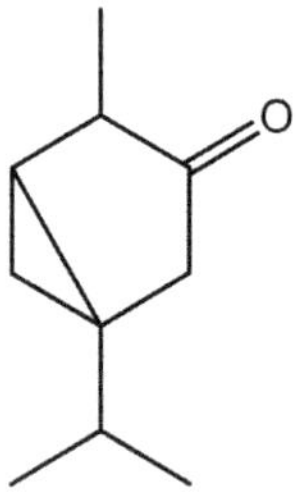 thujone

Part 2

Explain why a tail-to-tail linkage leading to a linear C10 monoterpene is not possible.

Part 3

Explain why cyclic C10 monoterpenes cannot be constructed from two isopentenyl pyrophosphate units, or two dimethylallyl pyrophosphate units.

Solution

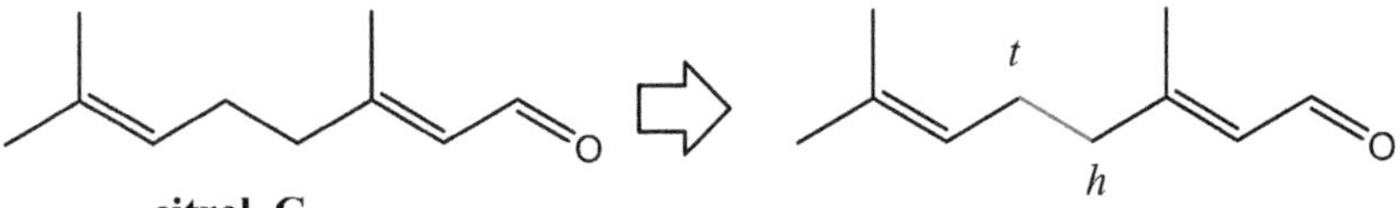

citral, C_{10} monoterpene

farnesol, C_{15}
sesquiterpene

squalene, C_{30}
triterpene

retinol (vitamin A), C_{20}
diterpene

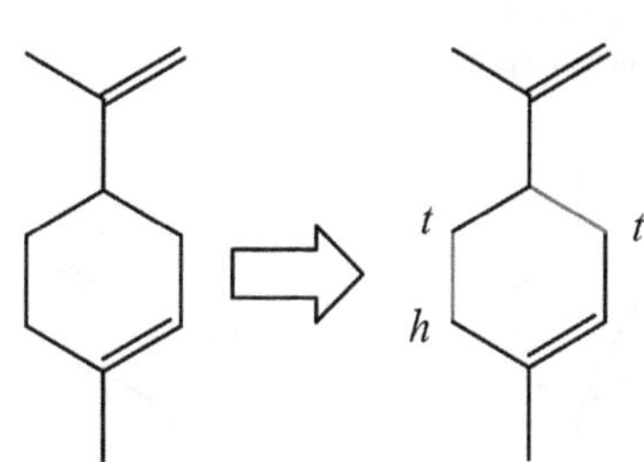

limonene, C_{10}
monoterpene

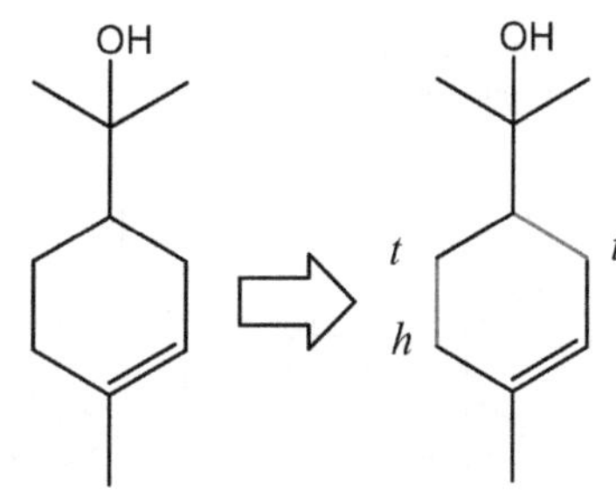

α-terpineol, C_{10}
monoterpene

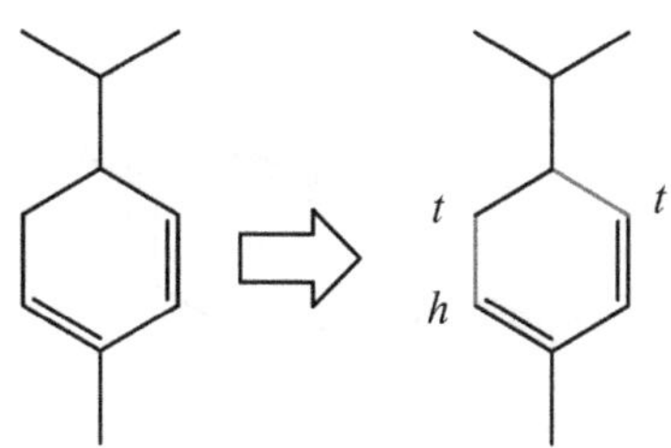

phellandrene, C_{10}
monoterpene

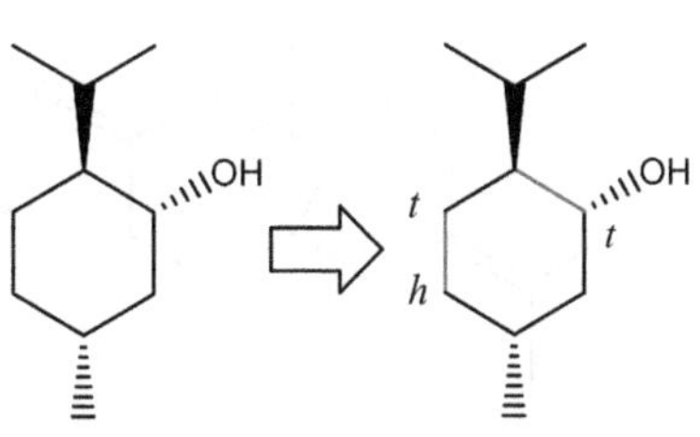

(-)-menthol, C_{10}
monoterpene

α-selinene, C_{15}
sesquiterpene

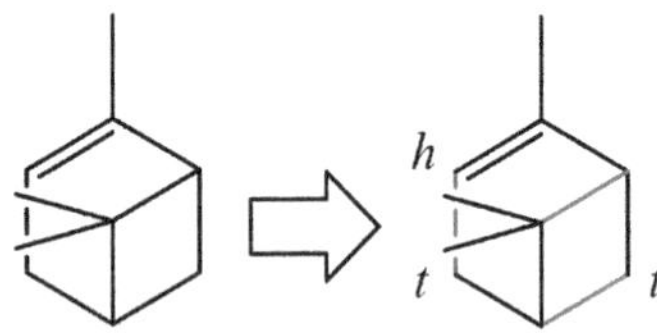

α-pinene, C_{10}
monoterpene

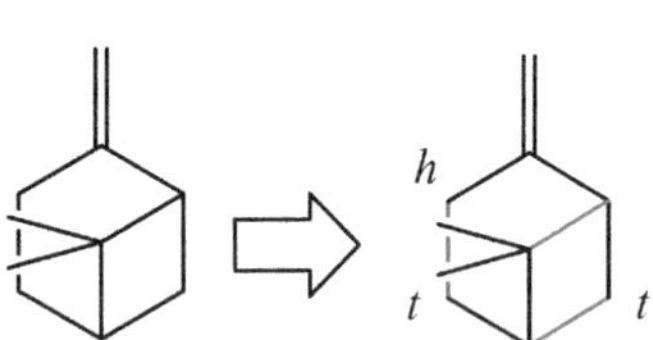

β-pinene, C_{10}
monoterpene

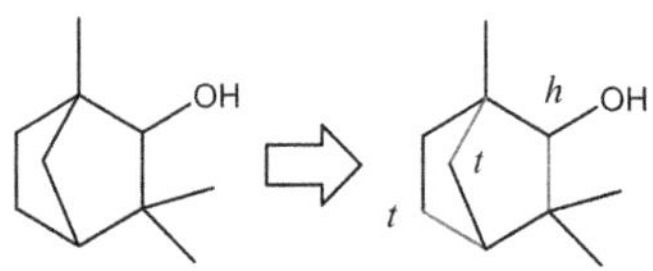

fenchol, C_{10}
monoterpene

borneol, C_{10}
monoterpene

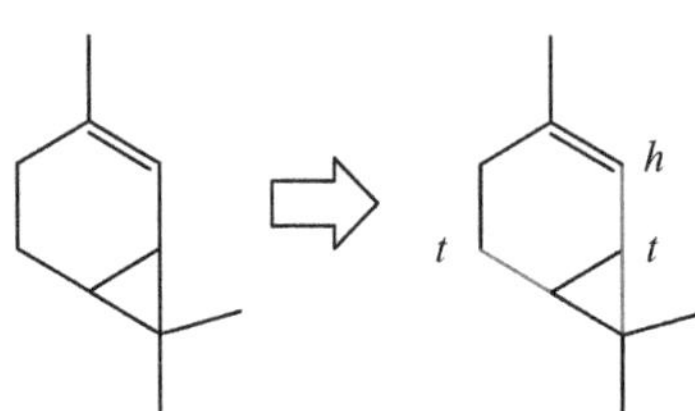

Δ^4-carene, C_{10}
monoterpene

thujone, C_{10}
monoterpene

vetivone, C_{15}
sesquiterpene

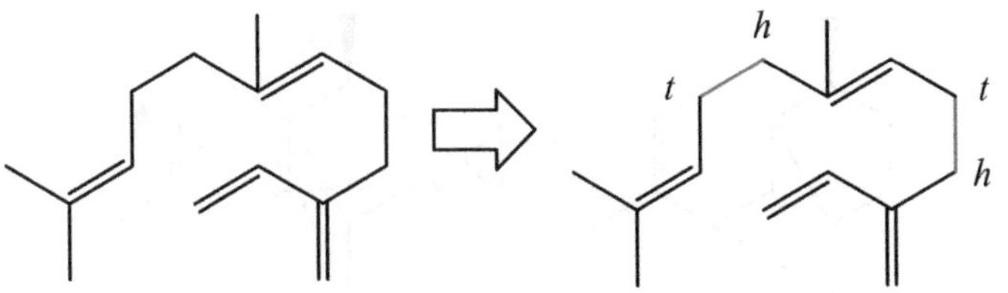

h
t t
h

farnesene, C_{15}
sesquiterpene

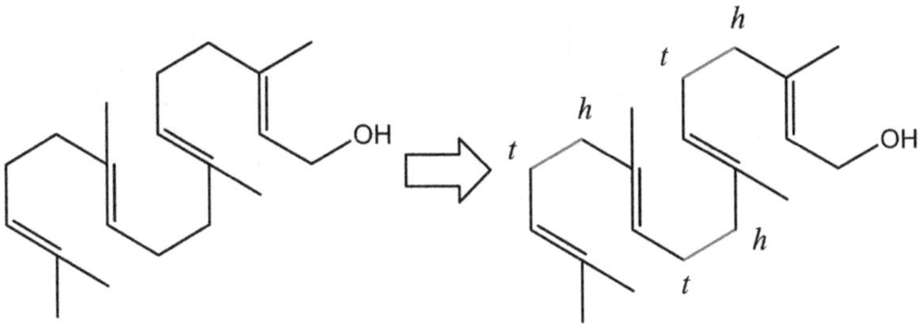

h
t
h
OH
t
OH
h
t

geranylgeraniol, C_{20}
diterpenes

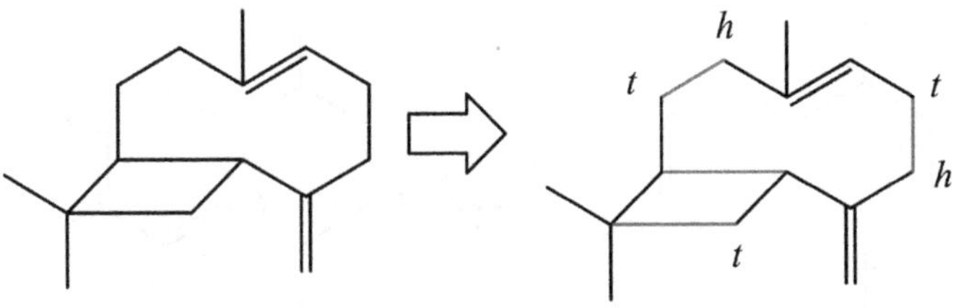

h
t t
h
t

β-caryophyllene, C_{15}
sesquiterpene

t
h h
t t

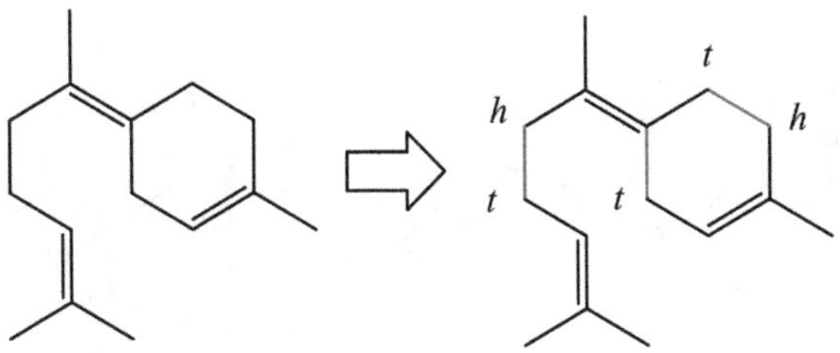

bisabolene, C_{15}
sesquiterpene

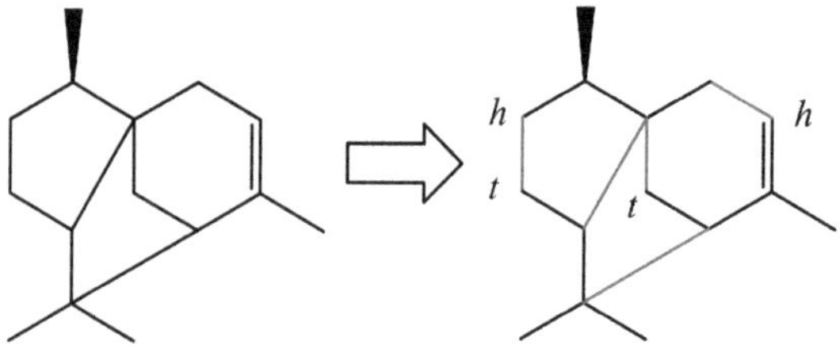

h h t t

cedrene, C_{15} sesquiterpene

t t h h t h h t COOH COOH

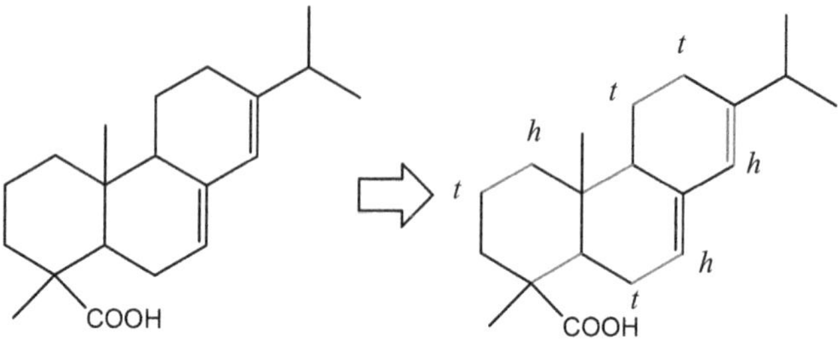

abietic acid, C_{20} diterpene

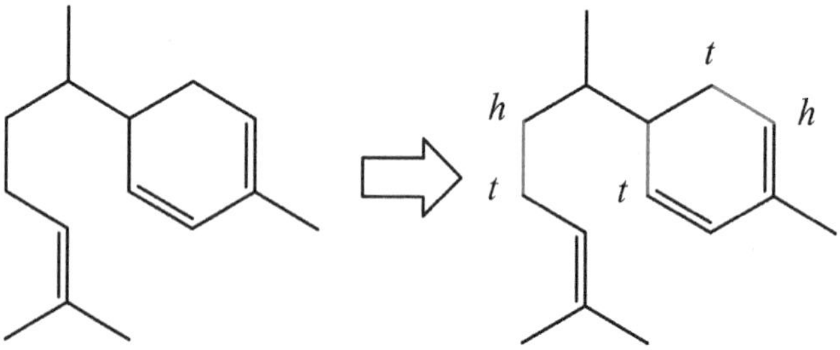

t h h t t

zingiberene, C_{15} sesquiterpene

h t h t (R) (R) OH (R) (R) OH h t

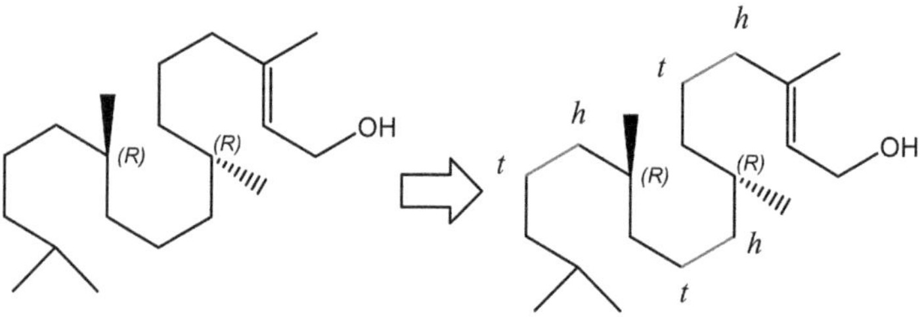

phytol, C_{20} diterpene

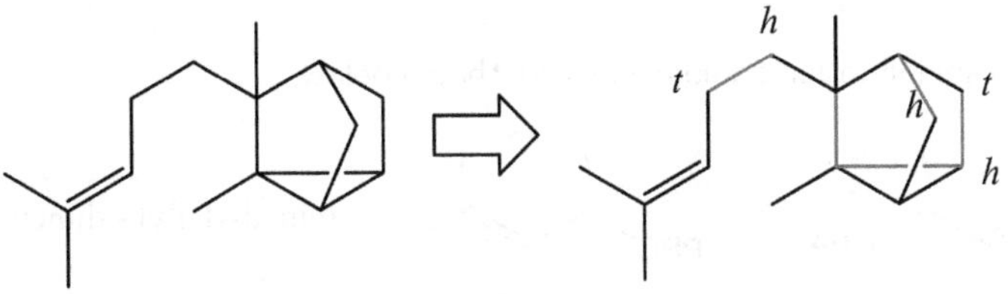

h t t h h

α-santalene, C_{15} sesquiterpene

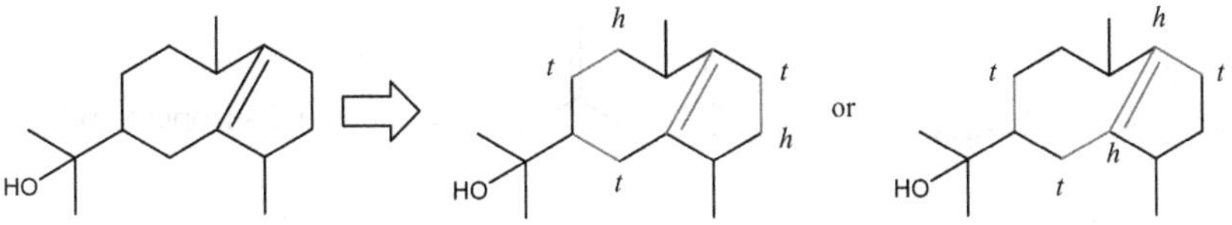

h t t h or h t t h t HO HO HO

guaiol, C_{15} sesquiterpene

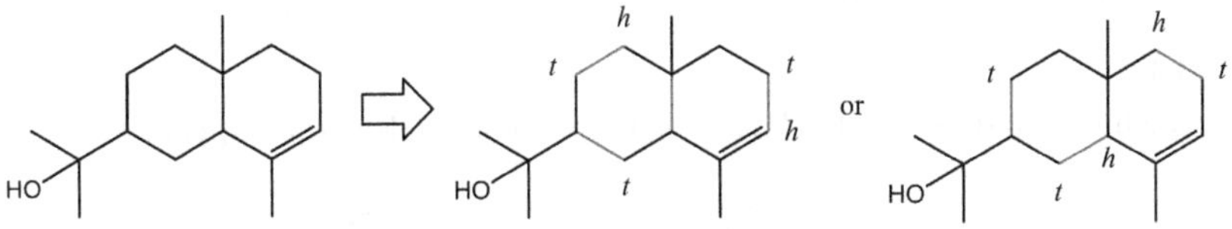

h t t h or h t t h t HO HO HO

α-eudesmol, C_{15} sesquiterpene

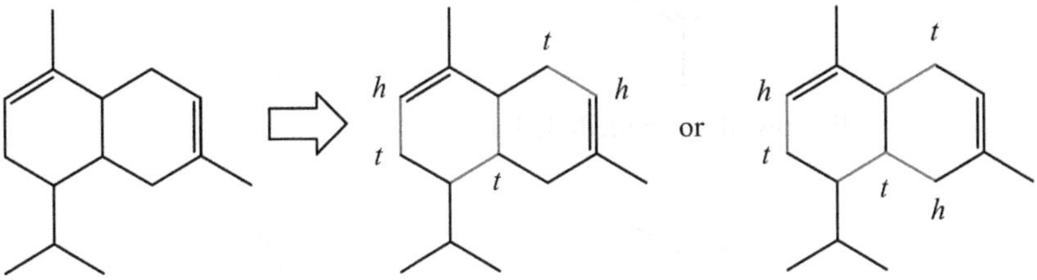

t h h t t or t h t t h

cadinene, C_{15} sesquiterpene

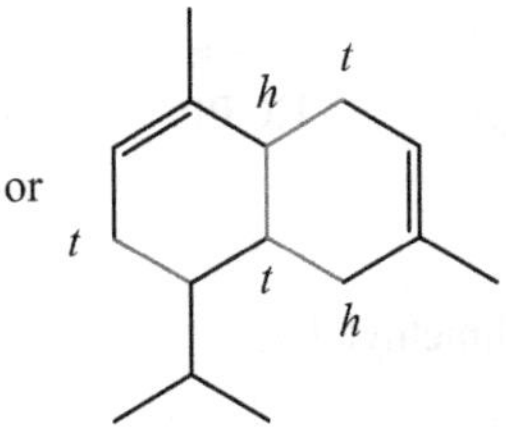

or t h t t h

Part 2

The relevant tail-to-tail combinations are shown below.

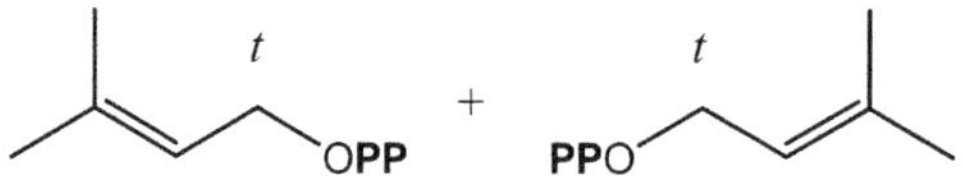

dimethylallyl - dimethylallyl

isopentenyl - dimethylallyl

isopentenyl - isopentenyl

In each case, there is a suggestion of two carbon atoms with leaving groups attached coming together. This is not possible since to make a bond between isoprene units there needs to be a nucleophilic (electron source) end attacking an electrophilic (electron sink) end. Hence, the pairwise carbon atoms marked with a "*t*" cannot be connected.

Part 3

A dimethylallyl–dimethylallyl linkage leads to a linear C10 monoterpene.

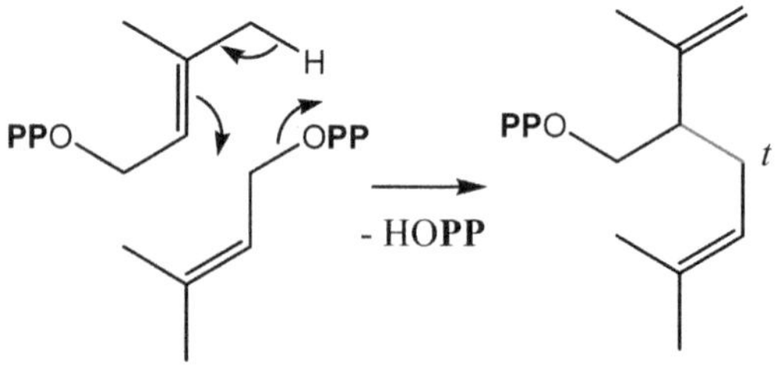

- HO**PP**

dimethylallyl - dimethylallyl

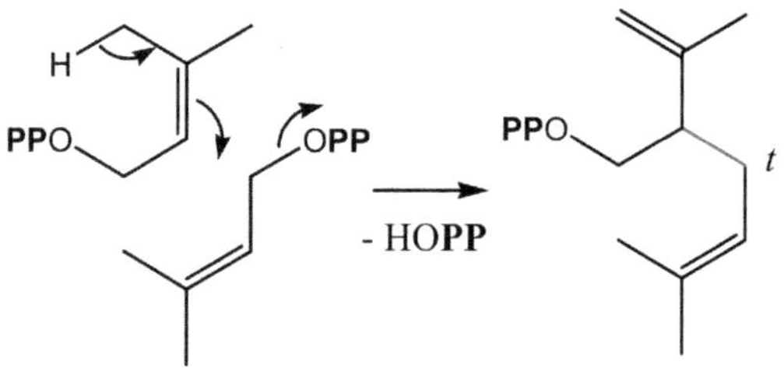

- HO**PP**

dimethylallyl - dimethylallyl

An isopentenyl–isopentenyl linkage leads to a linear C10 monoterpene.

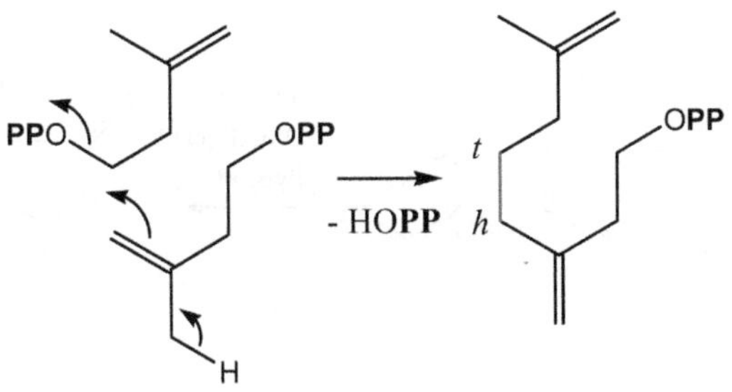

isopentenyl - isopentenyl

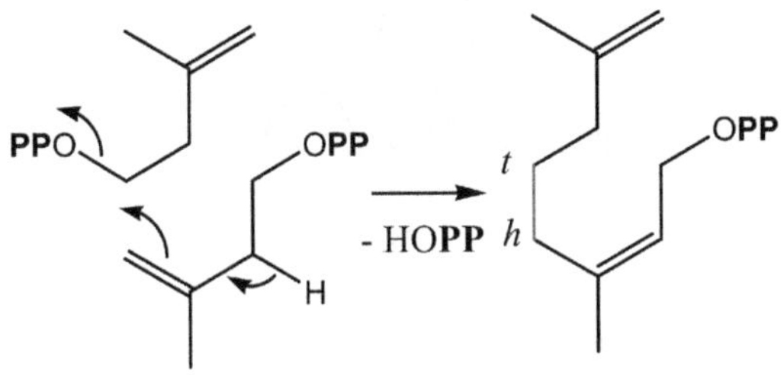

isopentenyl - isopentenyl

Example 5.26 [58–62]

Themes: balancing chemical reactions, target bond mapping, ring construction strategy

Each reaction shown below is an example of transforming a set of starting materials into a complex structure in a single step. For each case provide balanced chemical equations, show the target bond mapping in the product structure, and describe the ring construction strategy.

Hint: Begin with the structural aspect of the product and disassemble it to the starting materials drawn in the same aspect as the product.

Example 1

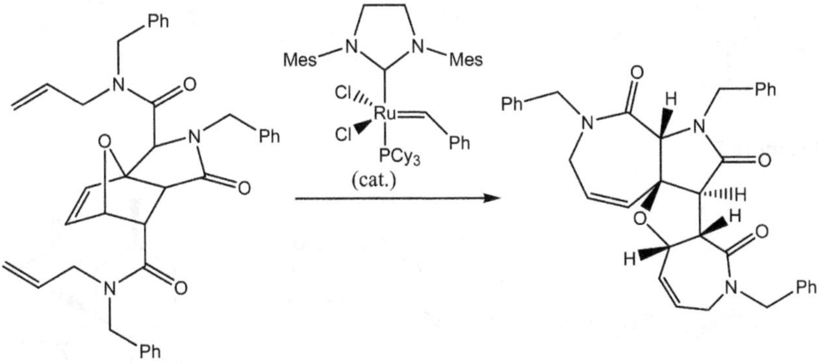

Example 2

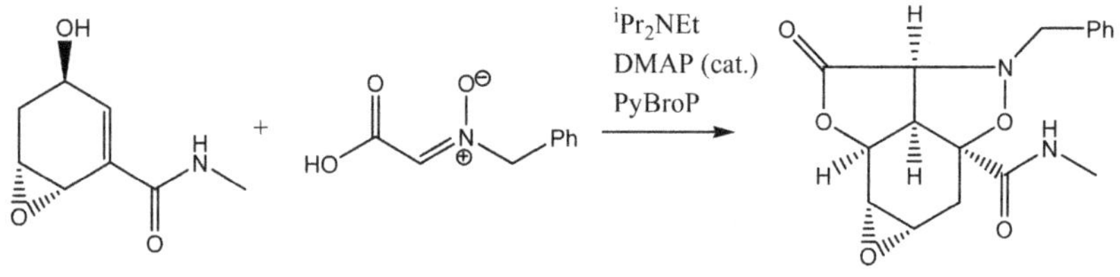

$^{i}Pr_2NEt$
DMAP (cat.)
PyBroP

PyBroP = bromotris(pyrrolidino)phosphonium hexafluorophosphate

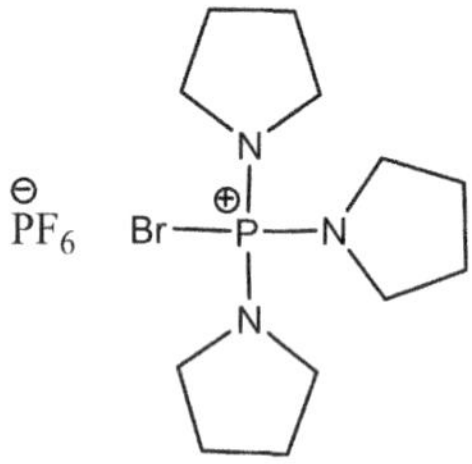

Example 3

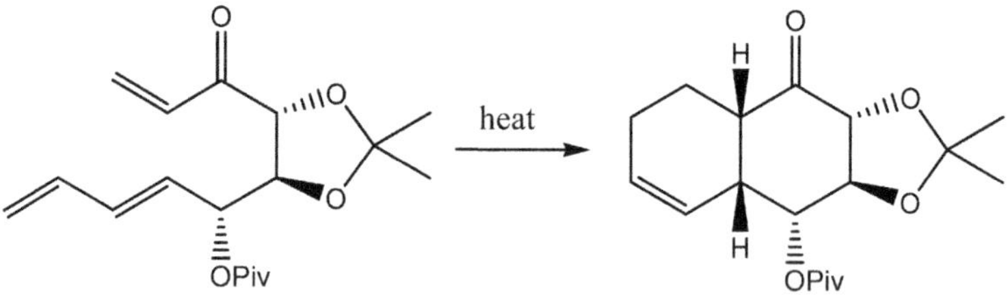

heat

Example 4

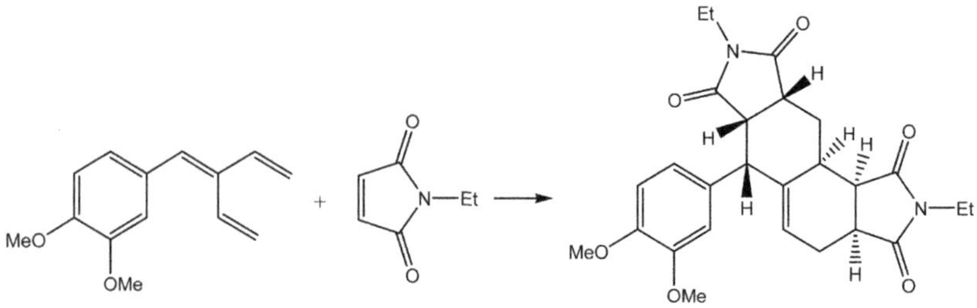

Example 5

μW
CH_3CN

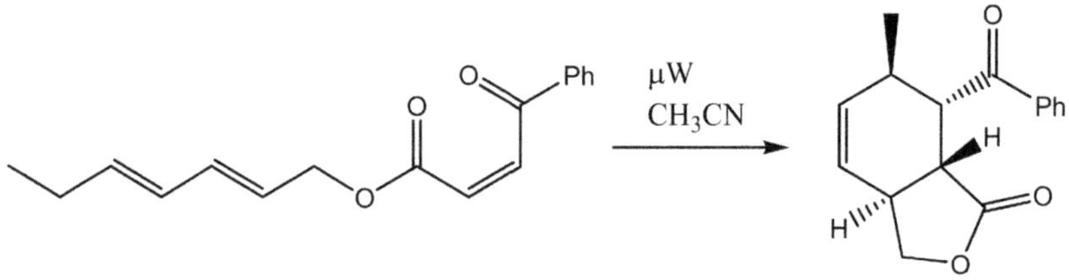

Example 6

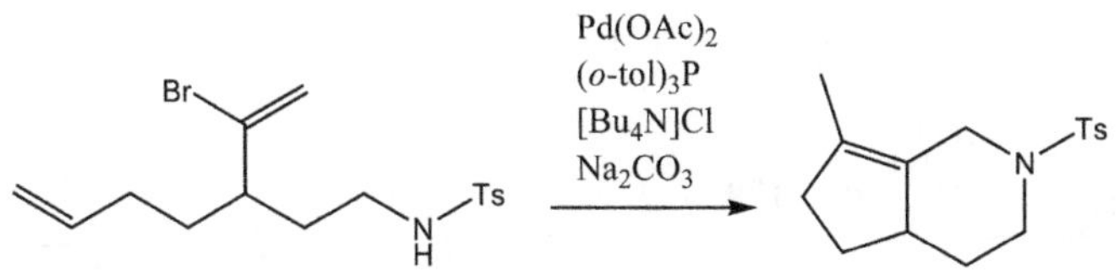

Pd(OAc)$_2$
(o-tol)$_3$P
[Bu$_4$N]Cl
Na$_2$CO$_3$

Example 7

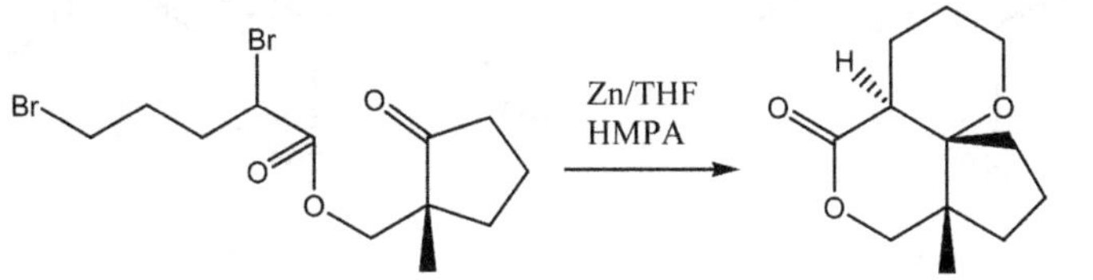

Zn/THF
HMPA

Example 8

1. NH$_3$
2. HOAc

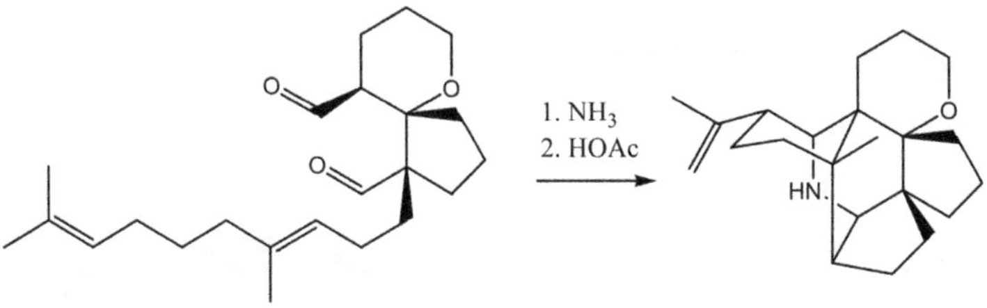

Example 9

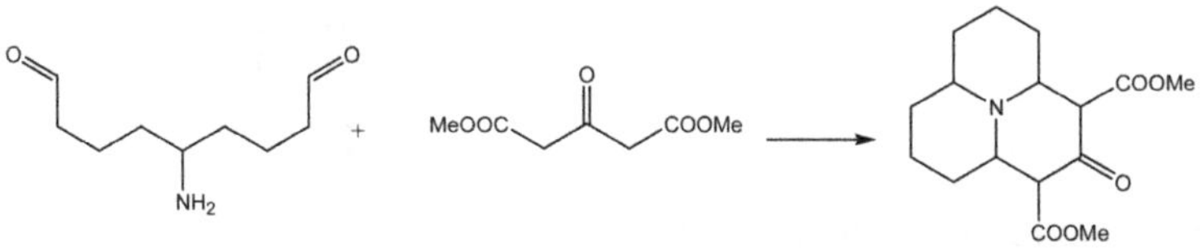

Solution

Example 1

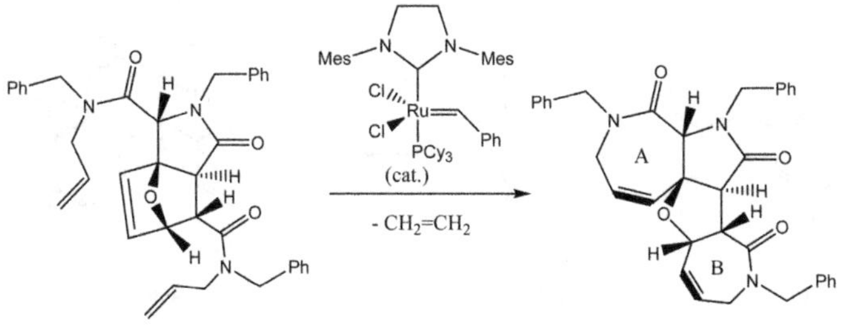

(cat.)

- CH$_2$=CH$_2$

587

559

Reactants: 587
Products: 559 + 28 = 587
Ring construction strategy: $[7+0]_A + [7+0]_B$

Example 2

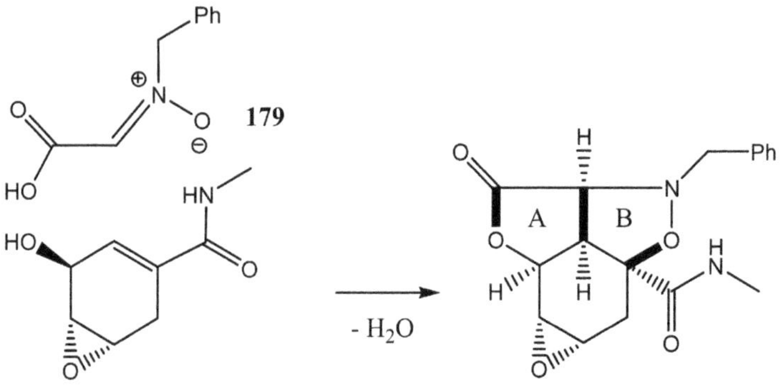

179

169

- H_2O

330

Reactants: 169 + 179 = 348
Products: 330 + 18 = 348
Ring construction strategy: $[(3+2)_A + (3+2)_B]$

Example 3

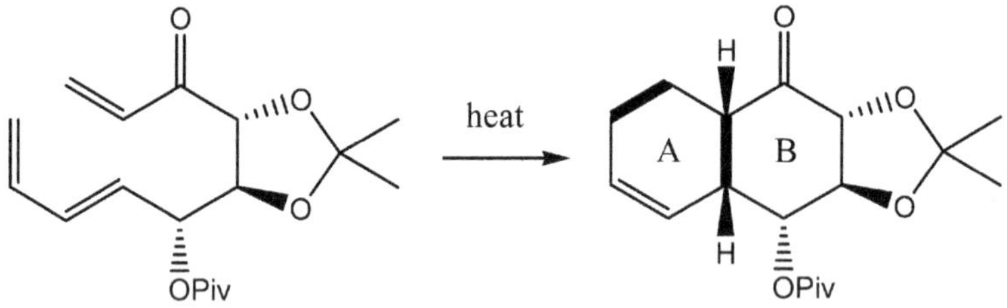

heat

322

322

Ring construction strategy: $[(4+2)_A + (6+0)_B]$

Example 4

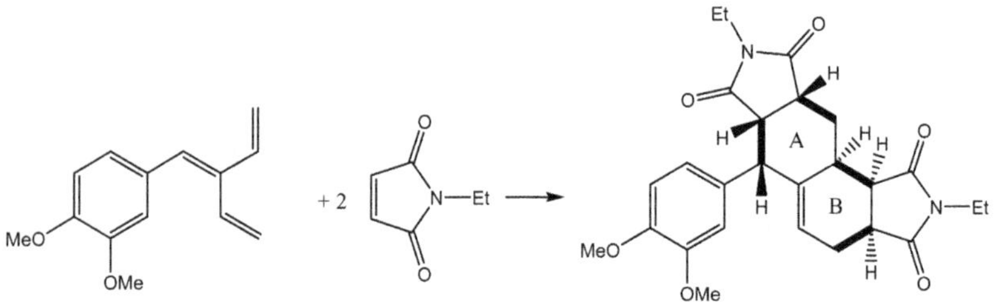

+ 2

216

2 (125)

466

Reactants: 216 + 2*125 = 466
Products: 466
Ring construction strategy: $[(4+2)_A + (4+2)_B]$

Example 5

O Ph O O → O Ph A H H B O O

256 **256**

Ring construction strategy: $[(4+2)_A+(5+0)_B]$

Example 6

Br HN Ts Na_2CO_3 → - NaBr - $NaHCO_3$ Ts N A B

371.9 **291**

Reactants: 371.9 + 106 = 477.9
Products: 291 + 84 + 102.9 = 477.9
Ring construction strategy: $[(5+0)_A+(6+0)_B]$

Example 7

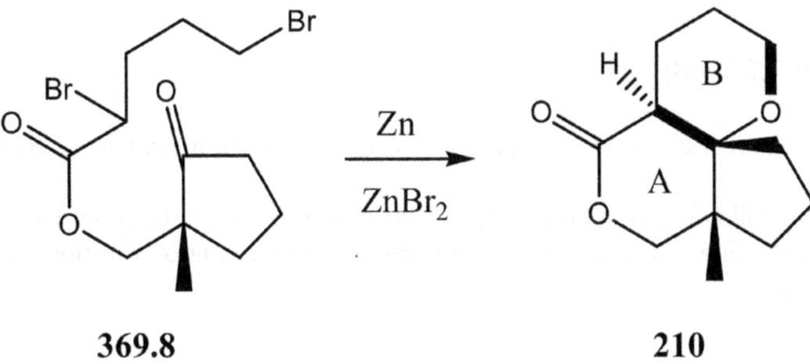

Reactants: 369.8 + 65.38 = 435.18
Products: 210 + 225.18 = 435.18
Ring construction strategy: $[(6+0)_A+(4+2)_B]$

Example 8

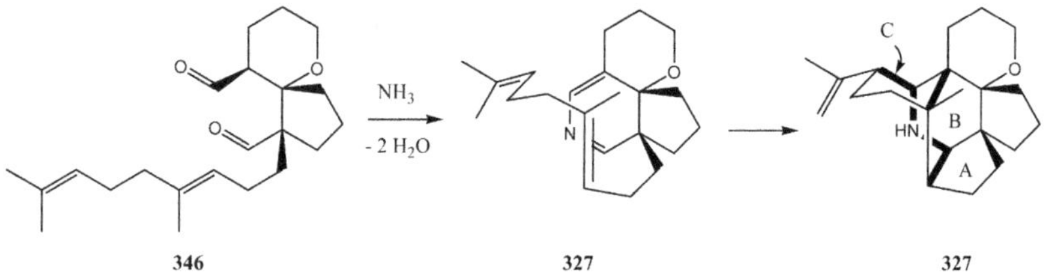

346 327 327

Reactants: 346 + 17 = 363
Products: 327 + 2*18 = 363
Ring construction strategy: $[(5+0)_A + (5+1)B + (4+2)_C]$

Example 9

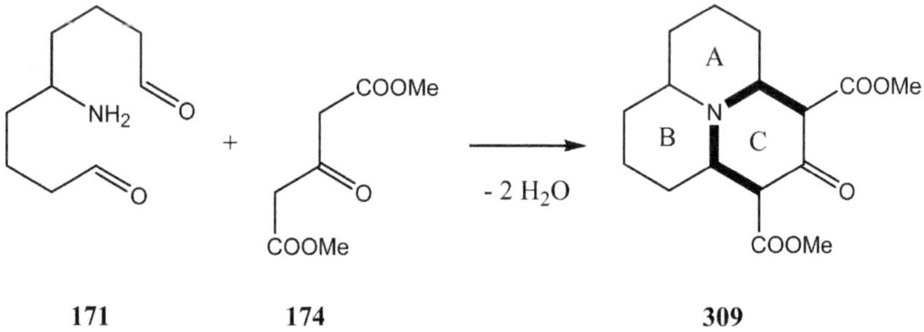

171 174 309

Reactants: 171 + 174 = 345
Products: 309 + 2*18 = 345
Ring construction strategy: $[(6+0)_A + (6+0)_B + (3+1+1+1)_C]$

Example 5.27 [63]

Themes: balancing chemical equations, ring construction strategy, target bond mapping

For the following reaction identify the reaction by-products and the target bonds made. Suggest a mechanism for the reaction. Describe the ring construction strategy.

NH_4OAc, 220°C, 80 %

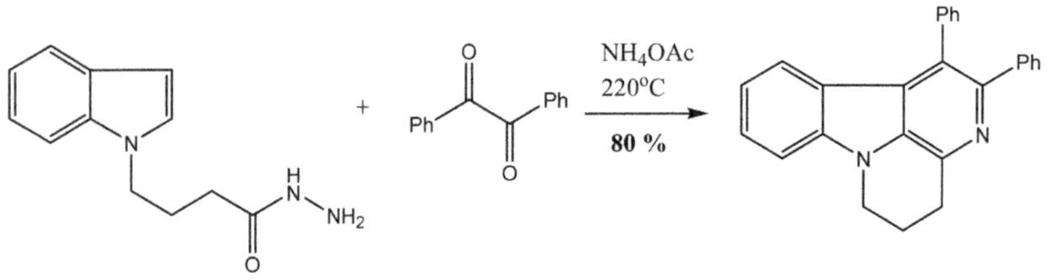

Solution

Mechanism

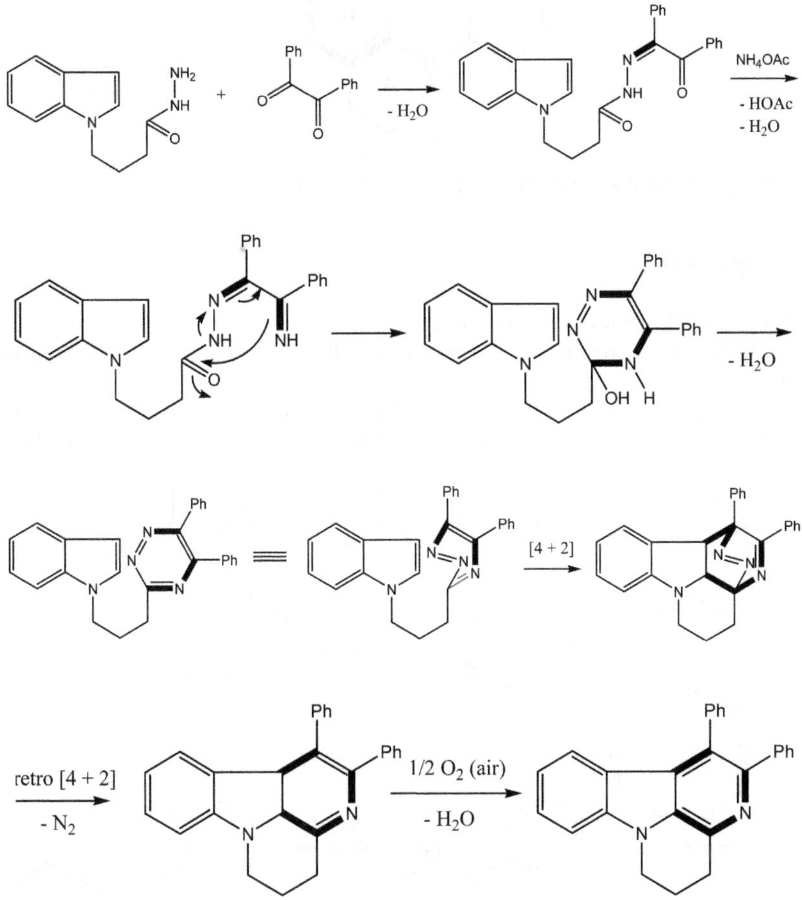

Net reaction:

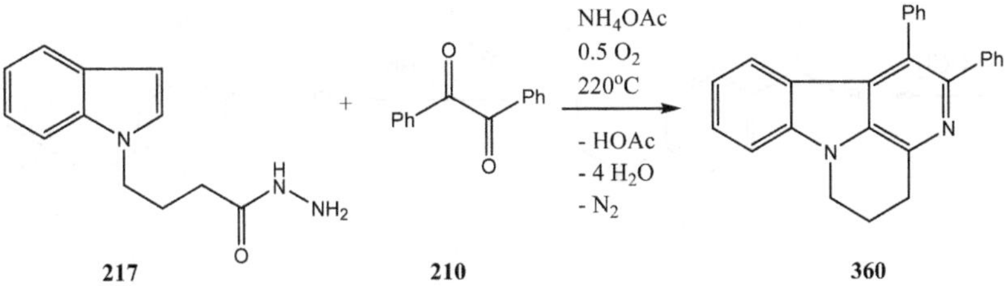

Reactants: 217 + 210 + 77 + 0.5*32 = 520

Products: 360 + 60 + 4*18 + 28 = 520

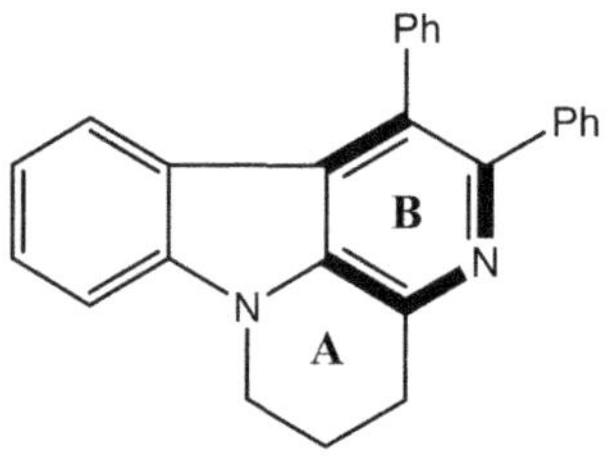

Ph
Ph
B
N
N
A

Ring construction strategy: $[6+0]_A + [2+2+1+1]_B$

Example 5.28

Themes: product structure target bond map

Captodiame is a sedative and tranquilizer that is produced by a convergent synthesis.

Working backwards in the synthesis plan, determine the product structure target bond map. Draw the target bond profile for the plan and determine its ideality.

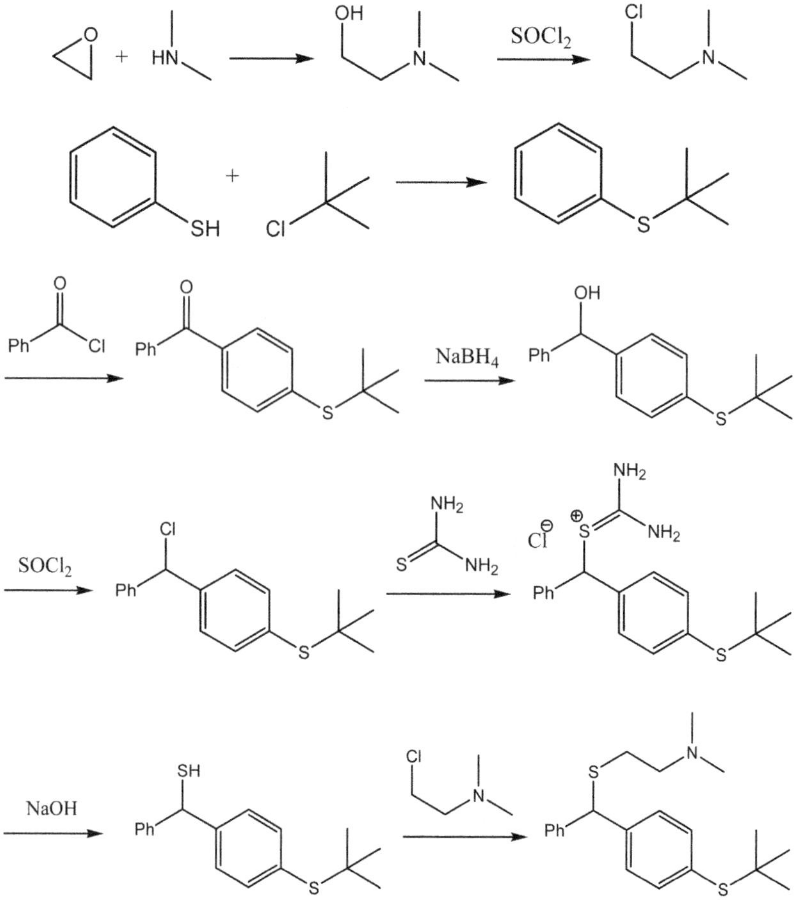

O + HN → OH N —$SOCl_2$→ Cl N

SH + Cl → S

Ph(C=O)Cl → Ph(C=O)…S —$NaBH_4$→ Ph OH … S

—$SOCl_2$→ Ph Cl … S —(S=C(NH_2)NH_2)→ $Cl^{\ominus}$ $\oplus$S=C(NH_2)NH_2 Ph … S

—NaOH→ Ph SH … S —(Cl…N)→ Ph S…N … S

Solution

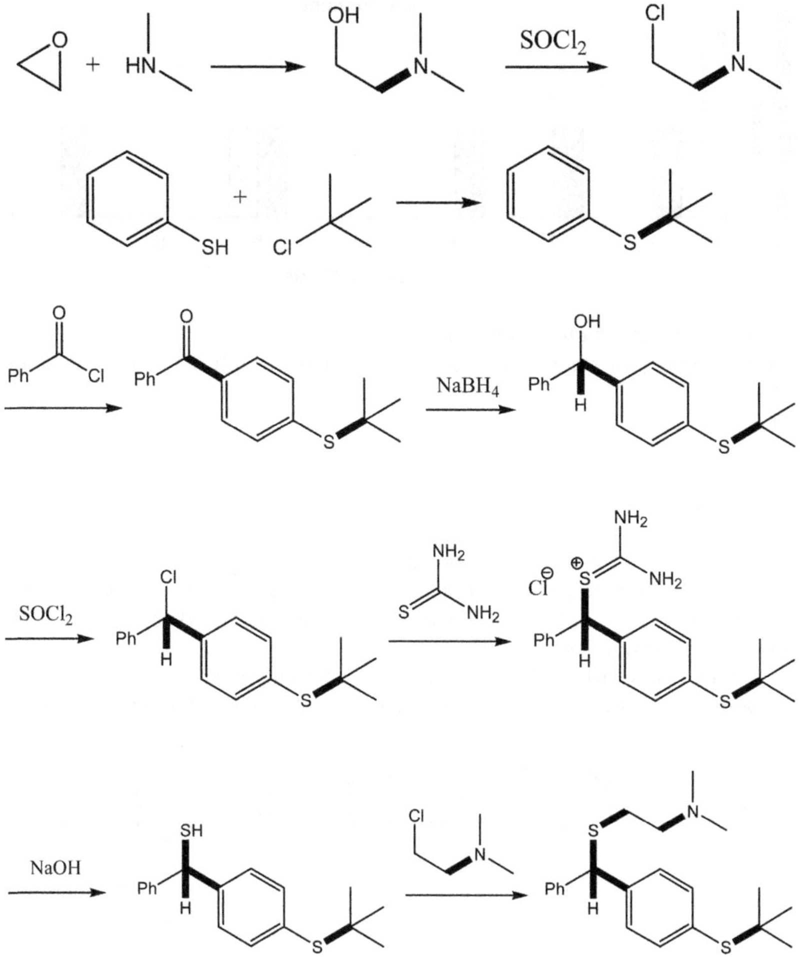

Product structure target bond map:

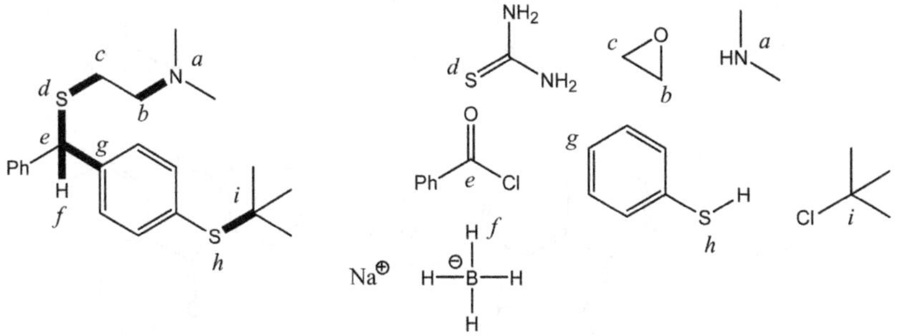

Target bond profile:

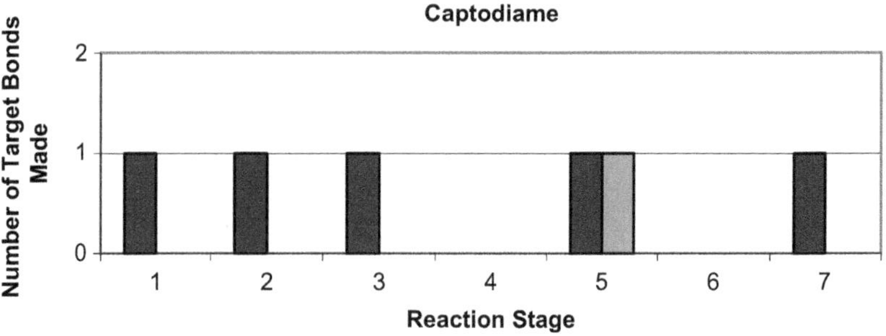

Number of construction steps = 6

Total number of steps = 7 + 2 = 9 (convergent plan with a branch containing 2 steps)

Ideality = 6/9

Example 5.29 [64–67]

Themes: product structure target bond map

Rohypnol (flunitrazepam) is a controversial sedative because it has been misused to spike people's drinks and hence it is known as the "date rape" drug. Its synthesis is shown below.

Working backwards in the synthesis plan, determine the product structure target bond map. Draw a target bond profile for the plan and determine its ideality.

F Br_2 *hν* F Br NaCN F N H_2SO_4 H_2O F O OH $SOCl_2$

F O Cl PhH $AlCl_3$ F O Ph HN NH_2 F Ph N NH

H_2SO_4 F NH Ph O_3 Me_2S F O NH O Ph

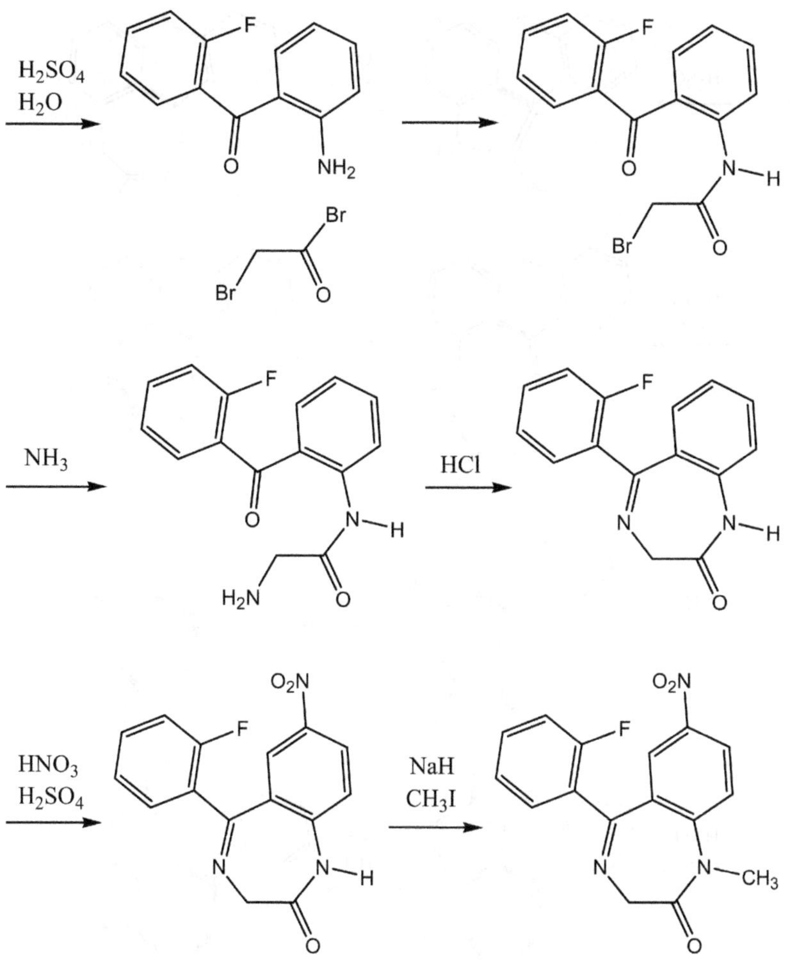
H2SO4
H2O
F
O
NH2
Br
Br
O
N–H
HCl
NH3
H2N
HNO3
H2SO4
O2N
NaH
CH3I
N–CH3
N

Solution

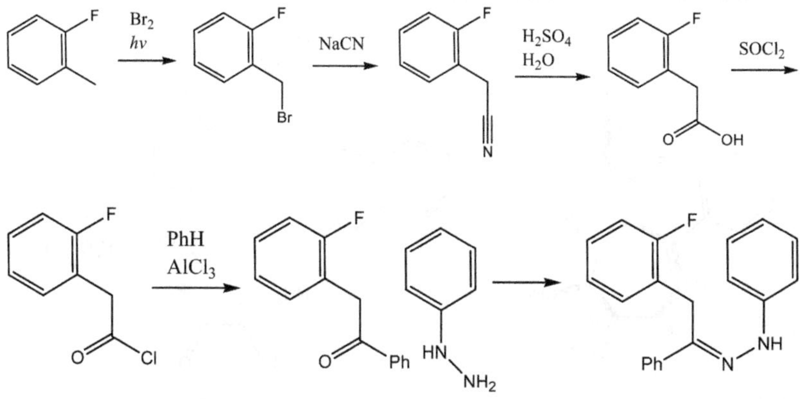
F
Br2
hv
Br
NaCN
N
H2SO4
H2O
OH
O
SOCl2
Cl
PhH
AlCl3
Ph
HN
NH2
NH

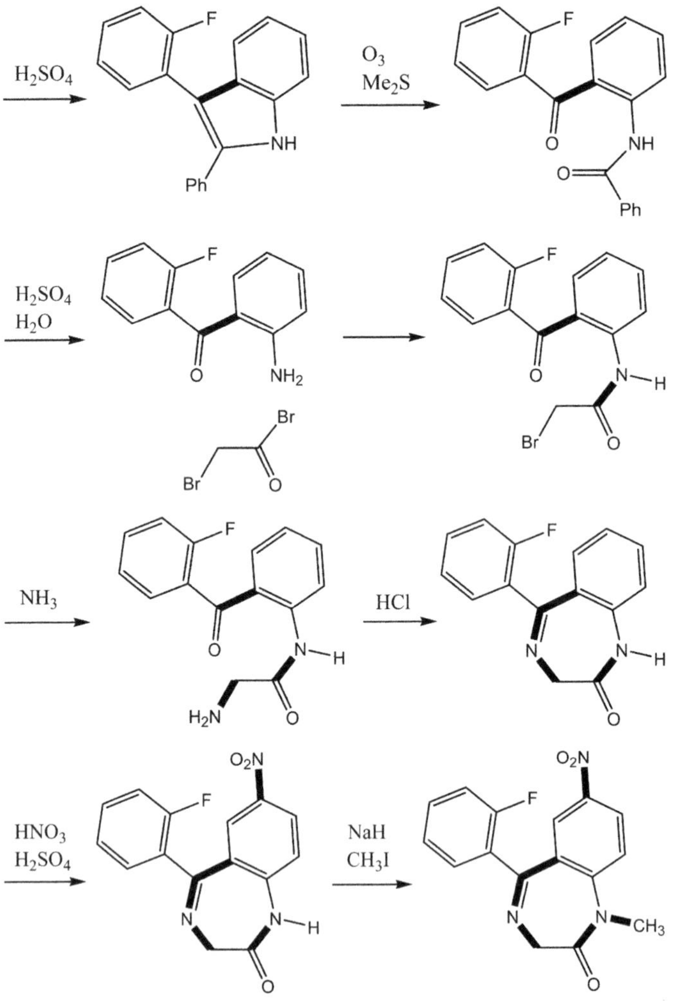

$\xrightarrow{H_2SO_4}$ $\xrightarrow{O_3,\ Me_2S}$

$\xrightarrow{H_2SO_4,\ H_2O}$ $\longrightarrow$

$\xrightarrow{NH_3}$ $\xrightarrow{HCl}$

$\xrightarrow{HNO_3,\ H_2SO_4}$ $\xrightarrow{NaH,\ CH_3I}$

Product structure target bond map:

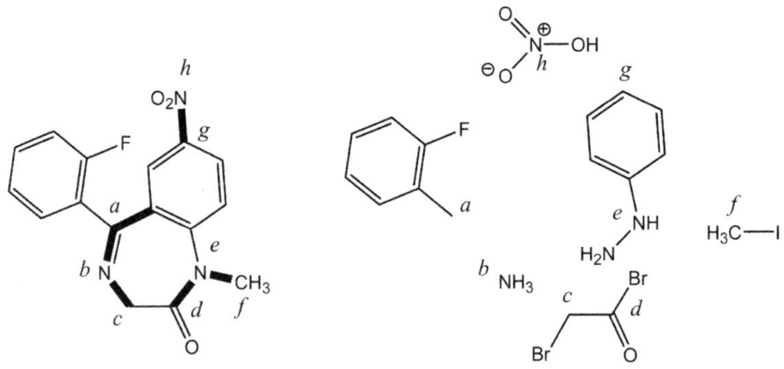

Target bond profile:

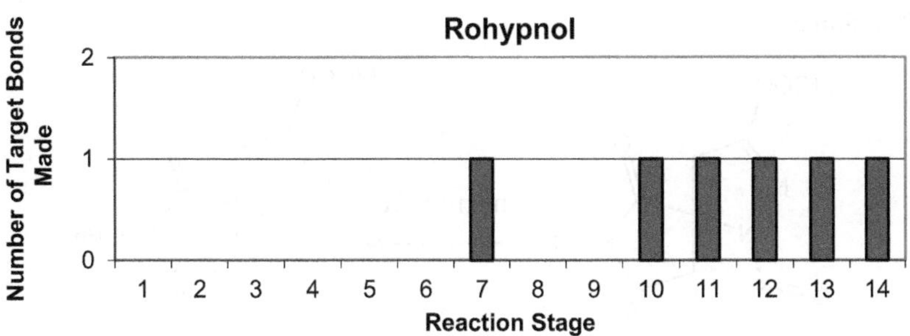

Number of construction steps = 6
Total number of steps = 14
Ideality = 6/14

5.3.1 Problems

PROBLEM 5.14 [68, 69]

Themes: balancing chemical equations, reaction mechanism, stoichiometry of reagents; synthesis strategy

When 1 equivalent of malononitrile is condensed with (*E*)-4-phenylbut-3-en-2-one in the presence of sodium methoxide in methanol, the product is 3-cyano-2-methoxy-6-methyl-4-phenylpyridine. When 2 equivalents of malononitrile are used under the same reaction conditions, the condensation product is 2,6-dicyano-3-methyl-5-phenylaniline.

Part 1

Write out mechanisms that account for both kinds of condensation products. Show the overall balanced chemical equations for each transformation.

Part 2

Draw out target bond maps for the product structure showing which bonds were made. For each case deduce the ring construction strategy.

PROBLEM 5.15 [70]

Themes: balancing chemical equations, reaction mechanism, ring construction strategy

Two microwave irradiation mediated reactions are shown below leading to fused oxazole products.

Reaction 1

O=N–N, N, EtOOC, O, O, N—Ph microwaves → O, N, EtOOC, O, O, N—Ph

Reaction 2

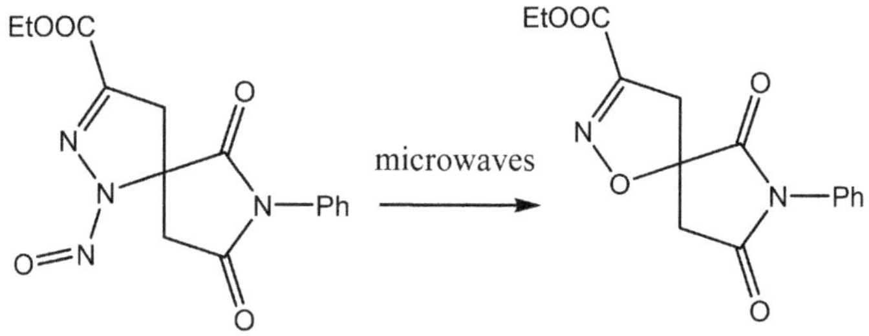
EtOOC
O
N
N
N—Ph
O=N
O
microwaves
EtOOC
O
N
O
N—Ph
O

Part 1

For each reaction write out balanced chemical equations showing all by-products.

Part 2

Suggest mechanisms for each reaction and highlight the target bonds made in the product structures.

PROBLEM 5.16 [71]

Themes: balancing chemical equations, multi-component reactions, target bond mapping

Two different product outcomes arise from multi-component reactions involving the same ingredients depending on the acidity of the reaction medium.

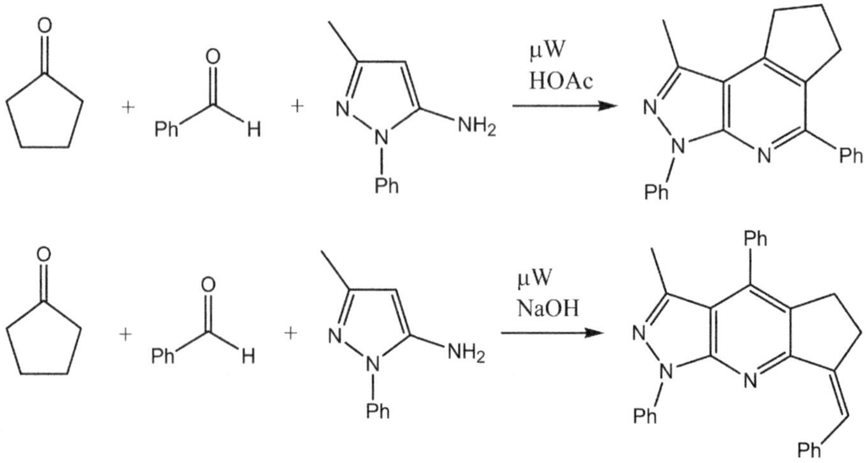
O
+
O
Ph
H
+
N
N
NH2
Ph
μW
HOAc
N
N
N
Ph
Ph
O
+
O
Ph
H
+
N
N
NH2
Ph
μW
NaOH
Ph
N
N
N
Ph
Ph

For each case provide balanced chemical equations, show the target bond mapping in the product structure.

PROBLEM 5.17 [72]

Themes: product structure target bond maps

Rimonabant is an anti-obesity wonder drug marketed by Sanofi-Synthélabo under the name Acomplia®. Its synthesis is shown below.

Working backwards in the synthesis plan, determine the product structure target bond map. Draw the target bond profile for the synthesis plan and determine its ideality.

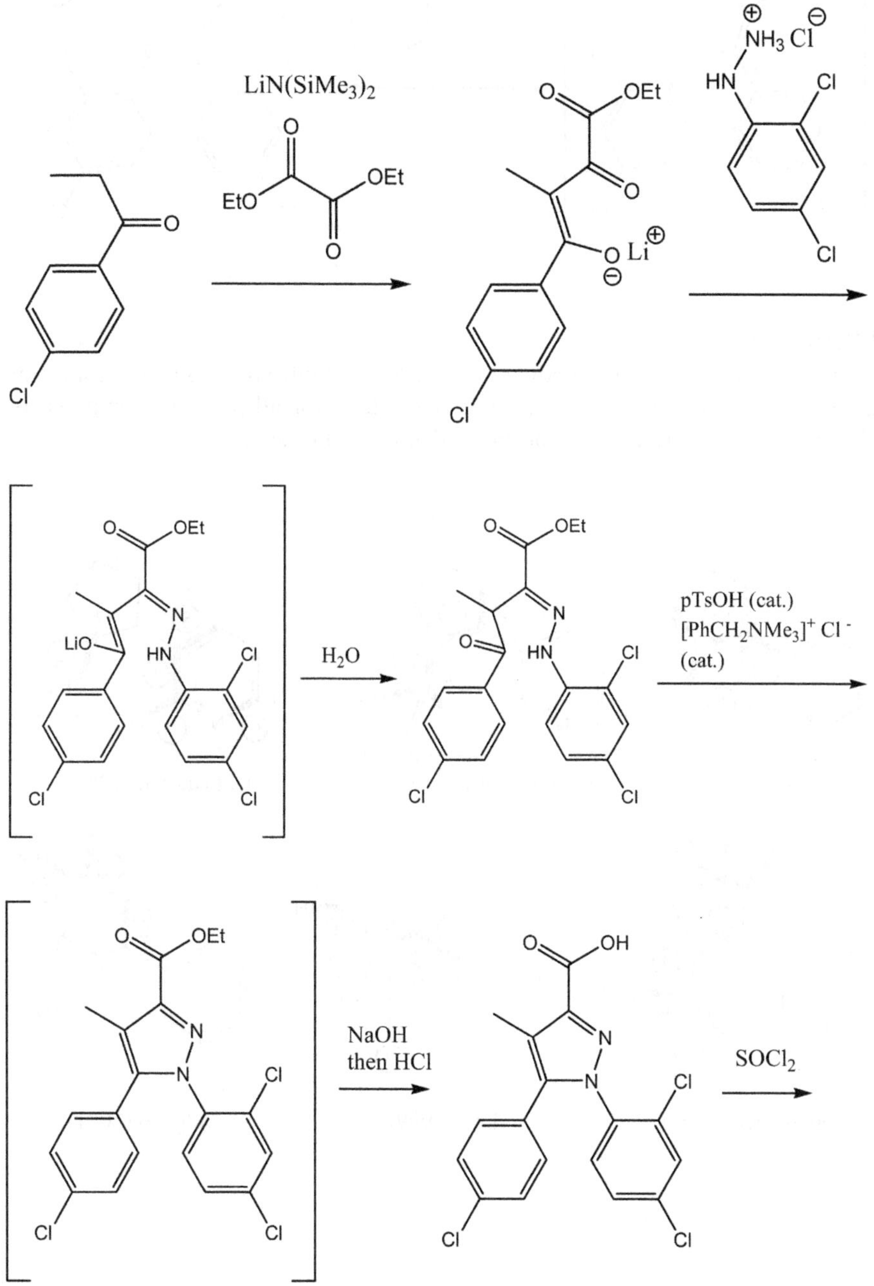

$LiN(SiMe_3)_2$

H_2O

pTsOH (cat.)
$[PhCH_2NMe_3]^+ Cl^-$
(cat.)

NaOH
then HCl

$SOCl_2$

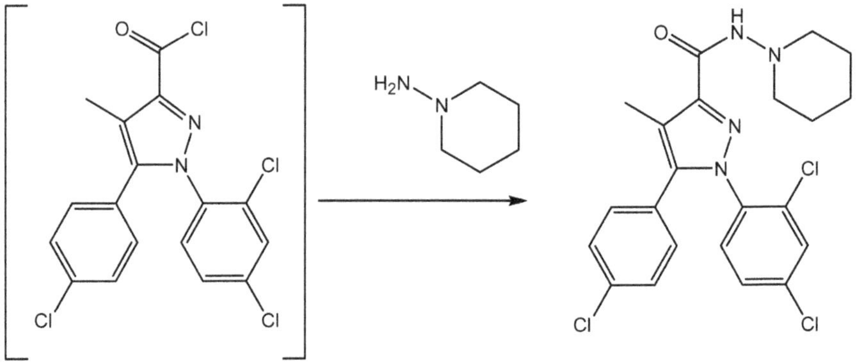

PROBLEM 5.18 [73–78]

Six target bond maps corresponding to six different synthesis plans to make nicotine are shown below. Determine the Tanimoto indices for all pairwise comparisons. Classify the route strategies according to degree of similarity.

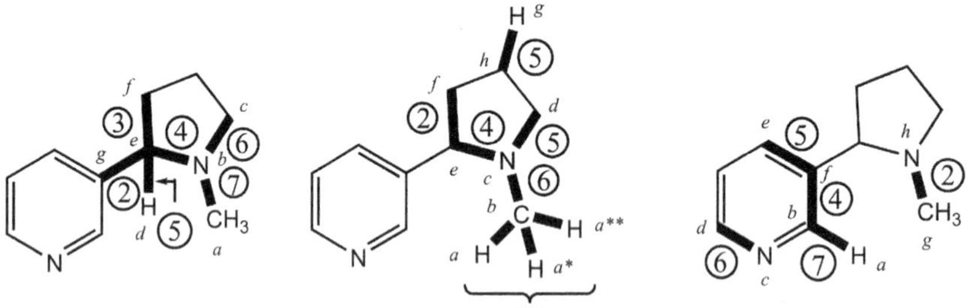

Craig (1933) Lebreton (2001) Chavdarian (1982)

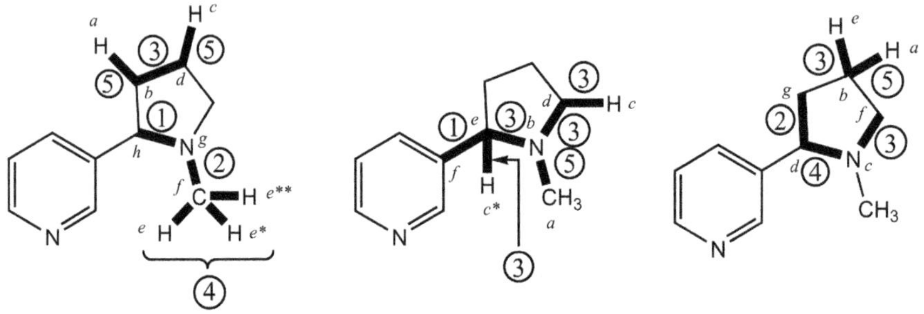

Helmchen (2005) Loh (1999) Delgado (2001)

Solution

From the 6 synthesis plans there are 0.5*(6*6 – 6) = 15 possible pairwise comparisons.

Pair	$b(A \cap B)$	$b(A)$	$b(B)$	T
Craig–Lebreton	4	6	8	0.4
Craig–Chavdarian	1	6	5	0.1
Craig–Helmchen	2	6	8	0.167
Craig–Loh	5	6	6	0.714
Craig–Delgado	3	6	5	0.375
Lebreton–Chavdarian	1	8	5	0.083
Lebreton–Helmchen	6	8	8	0.6
Lebreton–Loh	3	8	6	0.273
Lebreton–Delgado	4	8	5	0.444
Chavdarian–Helmchen	1	5	8	0.083
Chavdarian–Loh	1	5	6	0.1
Chavdarian–Delgado	0	5	5	0
Helmchen–Loh	2	8	6	0.167
Helmchen–Delgado	2	8	5	0.182
Loh–Delgado	2	6	5	0.222

Based on the target bond maps, the relative order of pairwise similarity is as follows:

Craig–Loh > Lebreton–Helmchen > Lebreton–Delgado > Craig–Lebreton > Craig–Delgado > Lebreton–Loh > Loh–Delgado > Helmchen–Delgado > Helmchen–Loh = Craig–Helmchen > Craig–Chavdarian = Chavdarian–Loh > Lebreton–Chavdarian = Chavdarian–Helmchen > Chavdarian–Delgado

The most similar plans are the Craig and Loh pair followed by the Lebreton and Helmchen pair. The least similar plans are the Chavdarian and Delgado pair with no bonds in common. Clearly, the Chavdarian plan is distinctly different from the others because it is the only plan that synthesizes the pyridine ring instead of the pyrrolidine ring.

5.4 TRACKING OXIDATION STATE CHANGES

Hypsicity [79]

(Gk: *hypsos*, level or height)

A term coined by James Hendrickson to track the changes in oxidation state, or hypsicity, of key atoms involved in bond-making and bond-breaking steps. He proposed that synthesis plans should be designed to aim for the *isohypsic* (same oxidation level) condition where a synthesis plan is characterized by a zero net change in oxidation state of all atoms of starting materials and intermediates involved until the target product is reached. This goal can be achieved by designing synthesis plans that eliminate redox reactions entirely. If these cannot be avoided due to practical considerations then the next best thing to achieve the isohypsic condition is to strategically sequence redox reactions in such a way that for every increase in oxidation level of an atom occurring in a step it is matched immediately by a decrease in oxidation level of equal magnitude in the next step, or vice versa. This cuts down on the accumulation of excess gains or losses in oxidation level of atoms, as the case may be, in starting materials and intermediates with respect to the oxidation levels of those atoms in the final target molecule over the course of the synthesis.

Hypsicity Index (HI) [80]

A metric that tracks the oxidation changes step-by-step in a synthesis plan given by Equation (5.3).

$$HI = \frac{\sum_{\text{stages},j}\left[\sum_{\text{atoms},i}\left[(Ox)_{\text{stage},j}^{\text{atom},i} - (Ox)_{\text{stage},N}^{\text{atom},i}\right]\right]}{N+1} = \frac{\sum_{\text{stages},j}\Delta_j}{N+1} \tag{5.3}$$

The following sequence of steps may be followed to determine *HI* for a synthesis:

1. Enumerate atoms in the target structure that are only involved in the building-up process from corresponding starting materials. This set of atoms defines those that are involved in bonding changes occurring in the relevant reaction steps.
2. Work backwards intermediate by intermediate to trace the oxidation numbers of the above set of atoms back to original starting materials, as appropriate, following the reaction stages back to the zeroth stage.
3. For each key atom, *i*, in each reaction stage, *j*, determine the difference in oxidation number of that atom with respect to what it is in the final target structure. Hence,

$$(Ox)_{\text{stage},j}^{\text{atom},i} - (Ox)_{\text{stage},N}^{\text{atom},i}.$$

4. Sum the differences determined in step 3 over all key atoms in stage *j*. This yields the term

$$\sum_{\text{atoms},i}\left[(Ox)_{\text{stage},j}^{\text{atom},i} - (Ox)_{\text{stage},N}^{\text{atom},i}\right] = \Delta_j.$$

5. Finally, take the sum $\sum_{\text{stage},j}\Delta_j$ over the number of stages and divide by $N+1$ accounting for the extra zeroth reaction stage.

If *HI* is zero, then the synthesis is *isohypsic* (same oxidation level). If *HI* is positive valued then to get to the target molecule, a net reduction is required over the course of the synthesis since an accumulated gain in oxidation level has resulted. Such a condition is termed *hyperhypsic* (higher oxidation level). Conversely, if *HI* is negative valued then to get to the target molecule a net oxidation is required over the course of the synthesis since an accumulated loss in oxidation level has resulted. Such a condition is termed *hypohypsic* (lower oxidation level).

The *calculator-hypsicity-index.xls* spreadsheet facilitates the computation of *HI* according to the five steps given above for any synthesis plan once all structures of starting materials, intermediates, and product are written out and the target bonds are identified.

Example 5.30

Themes: product structure target bond maps, hypsicity index

For the following generalized reactions, identify the target bonds made and determine the hypsicity index.

(a)

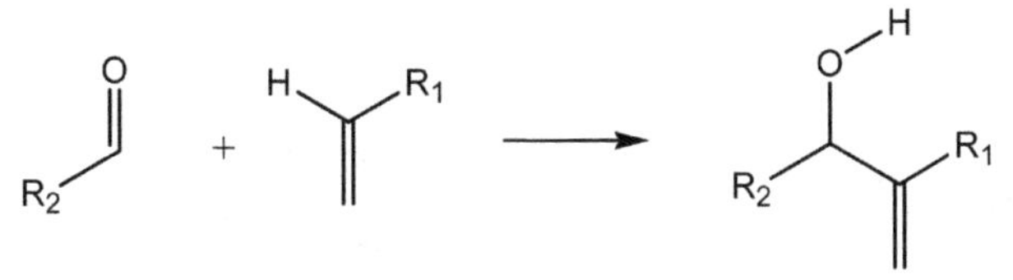

(b)

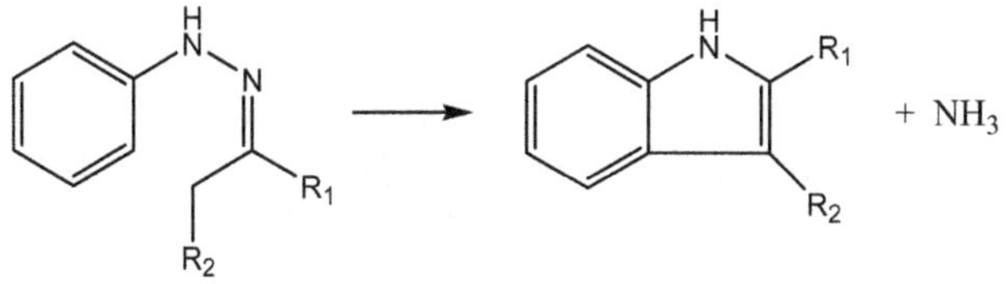

+ NH_3

(c)

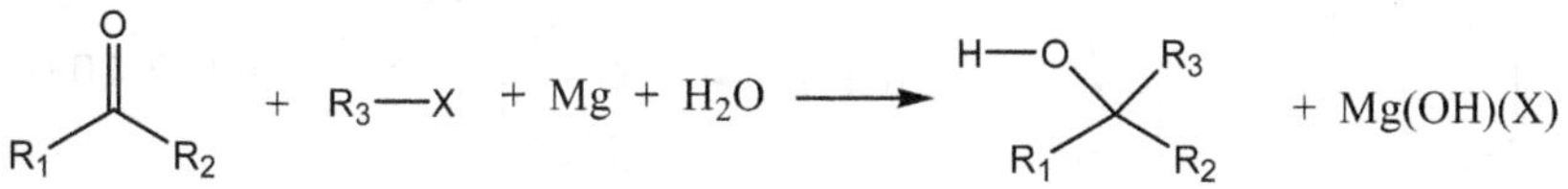

+ R_3—X + Mg + H_2O ⟶ + Mg(OH)(X)

(d)

+ R_4—X + Et_3N ⟶ + $Et_3NH^{\oplus}$ $X^{\ominus}$

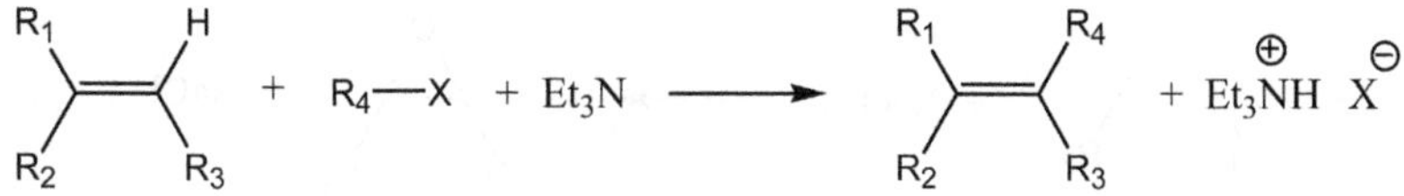

(e)

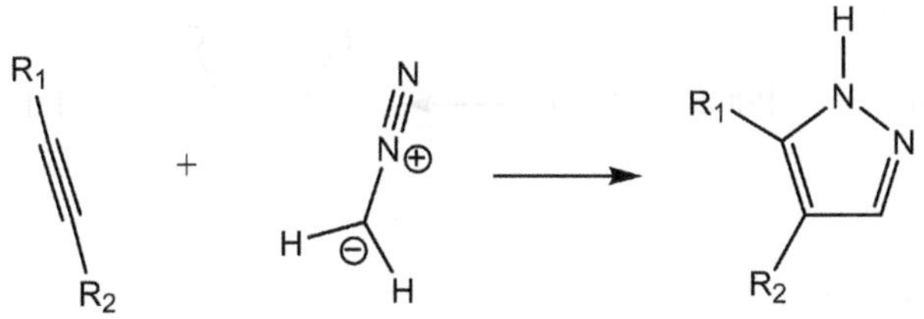

(f)

$R_1CH=CHR_2$ + ICH_2I + Zn ⟶ cyclopropane(R_1, R_2) + ZnI_2

(g)

$RCH=O$ + $(EtO)_2P(O)CH_2CN$ + NaH ⟶ $RCH=CHCN$ + H_2 + $(EtO)_2P(O)O^{\ominus}$ $Na^{\oplus}$

(h)

$R-C_6H_4-N_2^{\oplus}$ $BF_4^{\ominus}$ ⟶ $R-C_6H_4-F$ + N_2 + BF_3

(i)

$R_1CH_2R_2$ + $Hg(OAc)_2$ ⟶ $R_1CH(OAc)R_2$ + HOAc + Hg

(j)

$X-C_6H_4-NH_2$ + $NaNO_2$ + 2 HCl ⟶ $X-C_6H_4-N_2^{\oplus}$ $Cl^{\ominus}$ + NaCl + 2 H_2O

(k)

$R_1CH=CHR_2$ + NaOCl ⟶ epoxide(R_1, R_2) + NaCl

(l)

$H_2C=CH_2$ + $PdCl_2$ + H_2O ⟶ CH_3CHO + Pd + 2 HCl

(m)

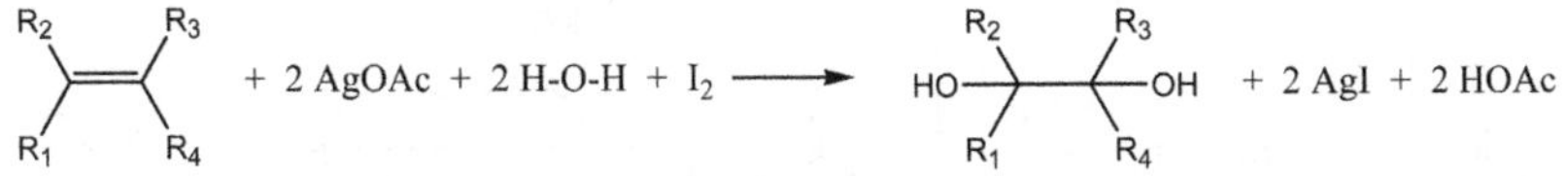

R_2 R_3 R_1 R_4 $+ 2\,AgOAc + 2\,H\text{-}O\text{-}H + I_2 \longrightarrow$ HO R_2 R_3 OH R_1 R_4 $+ 2\,AgI + 2\,HOAc$

(n)

R = EDG

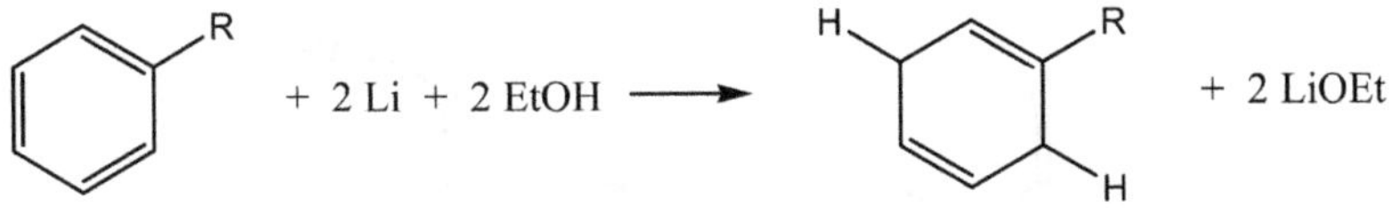

R $+ 2\,Li + 2\,EtOH \longrightarrow$ H R H $+ 2\,LiOEt$

(o)

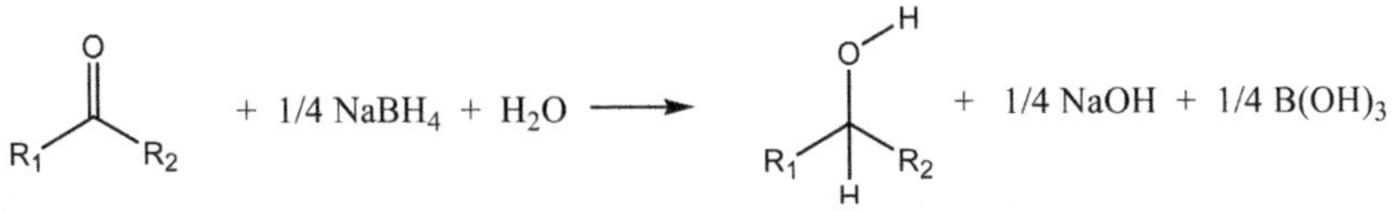

O R_1 R_2 $+ 1/4\,NaBH_4 + H_2O \longrightarrow$ H O R_1 R_2 H $+ 1/4\,NaOH + 1/4\,B(OH)_3$

(p)

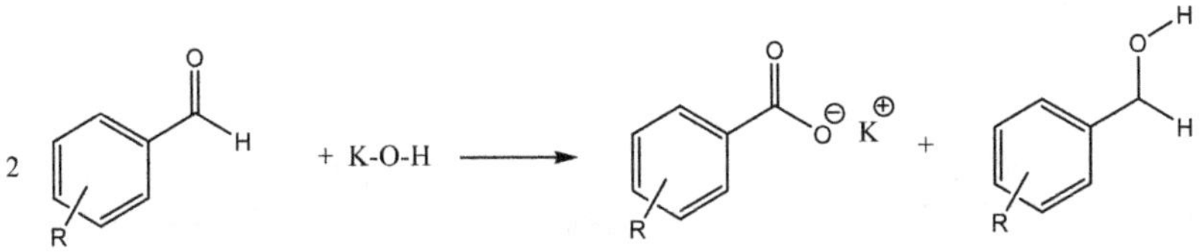

2 O H R $+$ K-O-H $\longrightarrow$ O $O^{\ominus}$ $K^{\oplus}$ R $+$ H O H R

(q)

O R_1 R_2 $+ 2\,Zn + 4\,HCl \longrightarrow$ H H R_1 R_2 $+ 2\,ZnCl_2 + H_2O$

(r)

$R_1CH{=}CHR_2 + H{-}H \longrightarrow$ R_1 R_2 H H

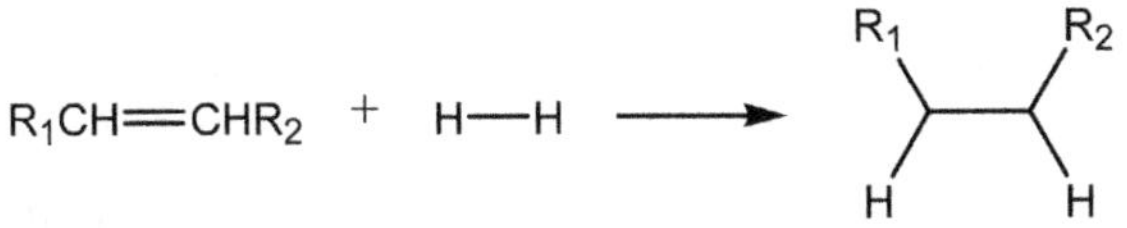

(s)

$O^{\ominus}$ $N^{\oplus}$ O R $+$ H S H $\longrightarrow$ H N H R $+ SO_2$

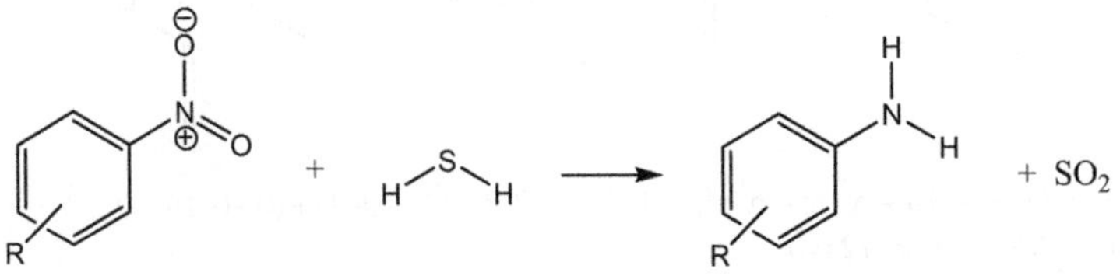

(t)

$R_1C(=O)R_2 + NH_2NH_2 \xrightarrow{-H_2O} [R_1C(=N-NH_2)R_2] \xrightarrow[-KOCH_2CH_2OK,\ -N_2,\ -2\ H_2O]{2\ KOH,\ HOCH_2CH_2OH} R_1CH_2R_2$

(u)

$RC(=O)CH_2SiMe_3 \rightarrow RC(OSiMe_3)=CH_2$

(v)

(w)

$H^{\oplus}$ (cat.) ; $-H_2O$; $-H^{\oplus}$

(x)

$-N_2$; R-O-H

Solution

(a)

$$\Delta = \Delta\,(a) + \Delta\,(b) + \Delta\,(c) + \Delta\,(d) = (1-1) + (-2-(-2)) + (0-1) + (0-(-1)) = 0$$

$$HI = -\Delta/(N+1) = 0/2 = 0$$

(b)

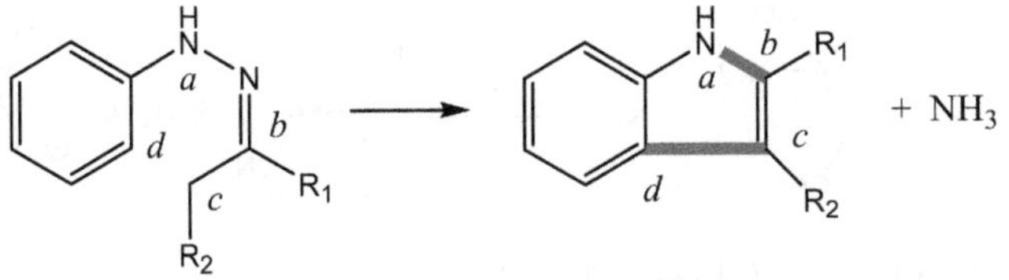

+ NH_3

$$\Delta = \Delta\,(a) + \Delta\,(b) + \Delta\,(c) + \Delta\,(d) = (-3-(-3)) + (1-2) + (0-(-2)) + (0-(-1)) = +2$$
$$HI = -\Delta/(N+1) = -2/2 = -1$$

(c)

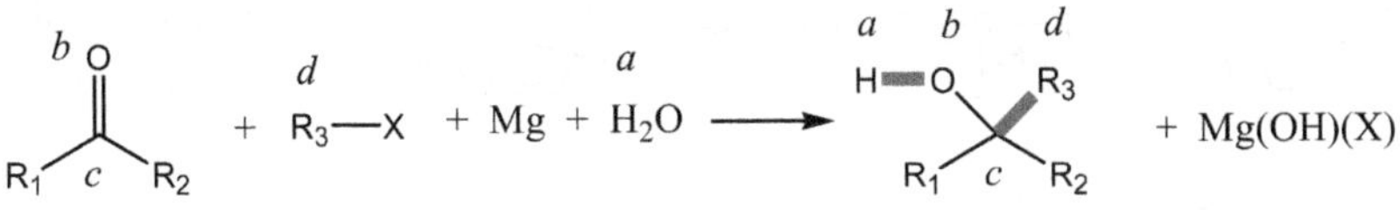

R_3—X + Mg + H_2O ⟶ + $Mg(OH)(X)$

$$\Delta = \Delta\,(a) + \Delta\,(b) + \Delta\,(c) + \Delta\,(d) = (1-1) + (-2-(-2)) + (1-2) + (-3-(-2)) = -2$$
$$HI = -\Delta/(N+1) = 2/2 = +1$$

(d)

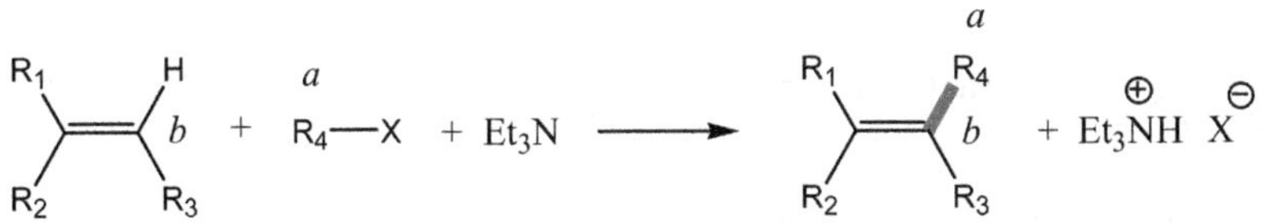

R_4—X + Et_3N ⟶ + $Et_3NH^{\oplus}$ $X^{\ominus}$

$$\Delta = \Delta\,(a) + \Delta\,(b) = (-3-(-2)) + (0-(-1)) = 0$$
$$HI = -\Delta/(N+1) = 0/2 = 0$$

(e)

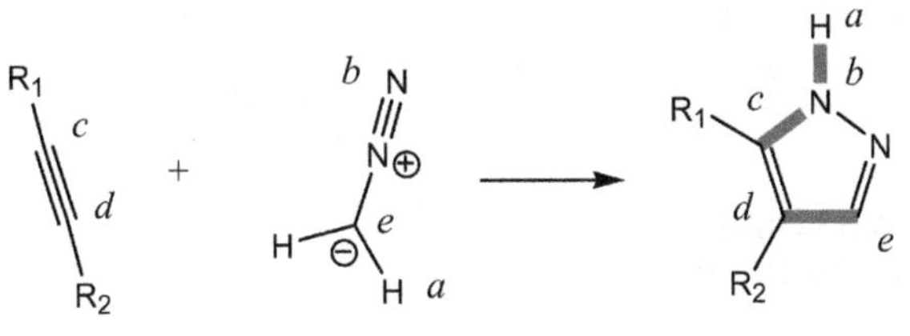

$$\Delta = \Delta\,(a) + \Delta\,(b) + \Delta\,(c) + \Delta\,(d) = (1-1) + (-2-0) + (1-0) + (0-0) + (1-(-2)) = 2$$
$$HI = -\Delta/(N+1) = -2/2 = -1$$

(f)

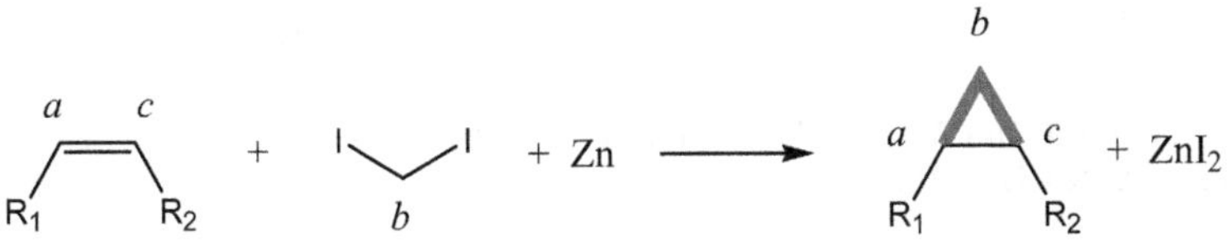

+ Zn ⟶ + ZnI_2

$$\Delta = \Delta\,(a) + \Delta\,(b) + \Delta\,(c) = (-1-(-1)) + (-2-0) + (-1-(-1)) = -2$$
$$HI = -\Delta/(N+1) = 2/2 = +1$$

(g)

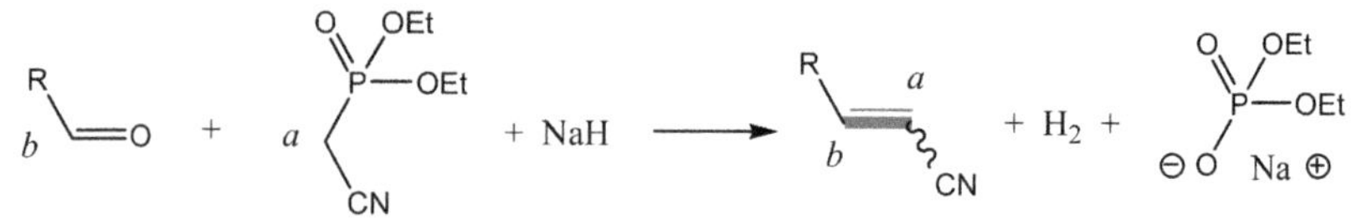

$$\Delta = \Delta\ (a) + \Delta\ (b) = (-1-(-1)) + (-1-1) = -2$$
$$HI = -\Delta/(N+1) = 2/2 = +1$$

(h)

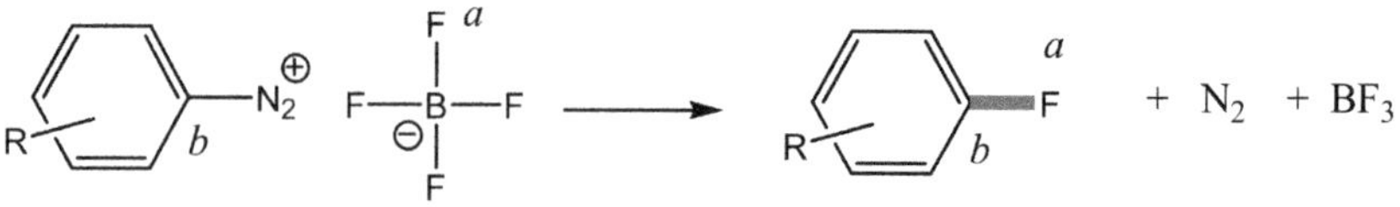

$$\Lambda = \Delta\ (a) + \Delta\ (b) = (-1-(-1)) + (1-1) = 0$$
$$HI = -\Delta/(N+1) = 0/2 = 0$$

(i)

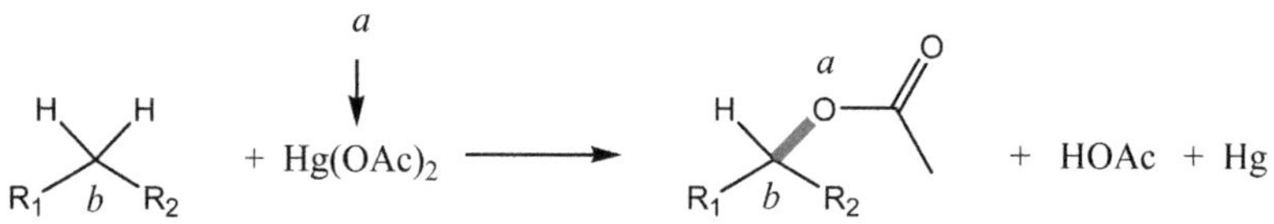

$$\Delta = \Delta\ (a) + \Delta\ (b) = (-2-(-2)) + (0-(-2)) = +2$$
$$HI = -\Delta/(N+1) = -2/2 = -1$$

(j)

$$\Delta = \Delta\ (a) + \Delta\ (b) = (0-3) + (0-(-3)) = 0$$
$$HI = -\Delta/(N+1) = 0/2 = 0$$

(k)

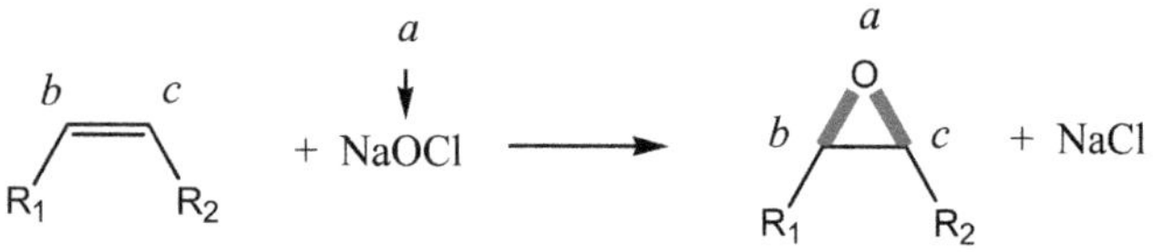

$$\Delta = \Delta\ (a) + \Delta\ (b) + \Delta\ (c) = (-2-(-2)) + (0-(-1)) + (0-(-1)) = +2$$
$$HI = -\Delta/(N+1) = -2/2 = -1$$

(l)

$$H_2C{=}CH_2 + PdCl_2 + H_2O \longrightarrow CH_3CHO + Pd + 2\,HCl$$

$\Delta = \Delta\,(a) + \Delta\,(b) + \Delta\,(c) = (-2-(-2)) + (1-(-2)) + (-3-(-1)) = +1$
$HI = -\Delta/(N+1) = -1/2 = -\frac{1}{2}$

(m)

$$R_1R_2C{=}CR_3R_4 + 2\,AgOAc + 2\,H\text{-}O\text{-}H + I_2 \longrightarrow R_1R_2C(OH){-}C(OH)R_3R_4 + 2\,AgI + 2\,HOAc$$

$\Delta = \Delta\,(a) + \Delta\,(b) + \Delta\,(c) + \Delta\,(d) = (-2-(-2)) + (1-0) + (1-0) + (-2-(-2)) = +2$
$HI = -\Delta/(N+1) = -2/2 = -1$

(n)

R = EDG

$$C_6H_5R + 2\,Li + 2\,EtOH \longrightarrow \text{1-R-cyclohexa-1,4-diene} + 2\,LiOEt$$

$\Delta = \Delta\,(a1) + \Delta\,(a2) + \Delta\,(b) + \Delta\,(c) = (1-1) + (1-1) + (-2-(-1)) + (-2-(-1)) = -2$
$HI = -\Delta/(N+1) = 2/2 = +1$

(o)

$$R_1C(O)R_2 + 1/4\,NaBH_4 + H_2O \longrightarrow R_1CH(OH)R_2 + 1/4\,NaOH + 1/4\,B(OH)_3$$

$\Delta = \Delta\,(a) + \Delta\,(b) + \Delta\,(c) + \Delta\,(d) = (1-1) + (-2-(-2)) + (0-2) + (1-(-1)) = 0$
$HI = -\Delta/(N+1) = 0/2 = 0$

(p)

$$2\,R\text{-}C_6H_4CHO + K\text{-}O\text{-}H \longrightarrow R\text{-}C_6H_4CO_2^{\ominus}K^{\oplus} + R\text{-}C_6H_4CH_2OH$$

Product is carboxylate salt:
$\Delta = \Delta\,(b) + \Delta\,(c) + \Delta\,(e) = (-2-(-2)) + (3-1) + (-2-(-2)) = +2$
$HI = -\Delta/(N+1) = -2/2 = -1$

Product is alcohol:
$\Delta = \Delta\ (a) + \Delta\ (b) + \Delta\ (c) + \Delta\ (d) = (1-1) + (-2-(-2)) + (-1-1) + (1-1) = -2$
$HI = -\Delta/(N+1) = 2/2 = +1$

(q)

$$R_1C(=O)R_2 + 2\,Zn + 4\,HCl \longrightarrow R_1CH_2R_2 + 2\,ZnCl_2 + H_2O$$

$\Delta = \Delta\ (a) + \Delta\ (a^*) + \Delta\ (b) = (1-1) + (1-1) + (-2-2) = -4$
$HI = -\Delta/(N+1) = 4/2 = +2$

(r)

$$R_1CH{=}CHR_2 + H{-}H \longrightarrow R_1CH_2{-}CH_2R_2$$

$\Delta = \Delta\ (a) + \Delta\ (a^*) + \Delta\ (b) + \Delta\ (c) = (1-0) + (1-0) + (-2-(-1)) + (-2-(-1)) = 0$
$HI = -\Delta/(N+1) = 0/2 = 0$

(s)

$$R{-}C_6H_4{-}NO_2 + H_2S \longrightarrow R{-}C_6H_4{-}NH_2 + SO_2$$

$\Delta = \Delta\ (a) + \Delta\ (b) + \Delta\ (c) = (1-1) + (-3-3) + (1-1) = -6$
$HI = -\Delta/(N+1) = 6/2 = +3$

(t)

$$R_1C(=O)R_2 + NH_2NH_2 \xrightarrow{-H_2O} [R_1C(=N{-}NH_2)R_2] \xrightarrow[-KOCH_2CH_2OK,\ -N_2,\ -2\,H_2O]{2\,KOH,\ HOCH_2CH_2OH} R_1CH_2R_2$$

$\Delta = \Delta\ (a) + \Delta\ (a^*) + \Delta\ (b) = (1-1) + (1-1) + (-2-2) = -4$
$HI = -\Delta/(N+1) = 4/2 = +2$

(u)

$$RC(=O)CH_2SiMe_3 \longrightarrow RC(OSiMe_3){=}CH_2$$

$\Delta = \Delta\ (a) + \Delta\ (b) = (-2-(-4)) + (-2-(-2)) = +2$
$HI = -\Delta/(N+1) = -2/2 = -1$

(v)

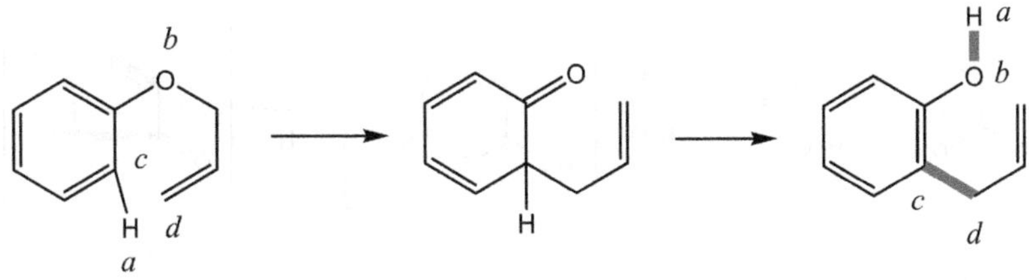

$\Delta = \Delta$ (a) + Δ (b) + Δ (c) + Δ (d) = (1–1) + (–2–(–2)) + (0–(–1)) + (–2–(–2)) = +1
HI = $-\Delta/(N+1) = -1/2 = -½$

(w)

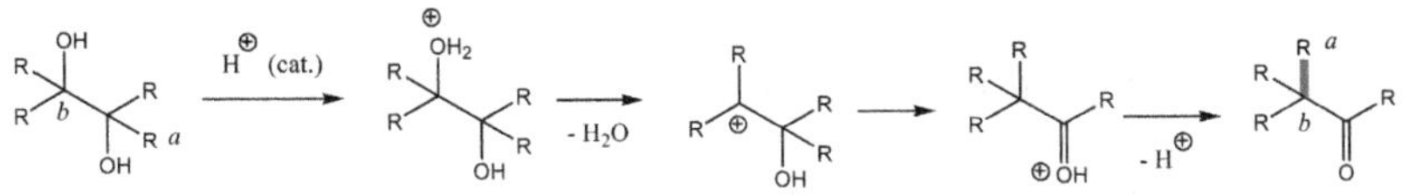

$\Delta = \Delta$ (a) + Δ (b) = (–3–(–3)) + (0–1) = –1
HI = $-\Delta/(N+1) = 1/2 = ½$

(x)

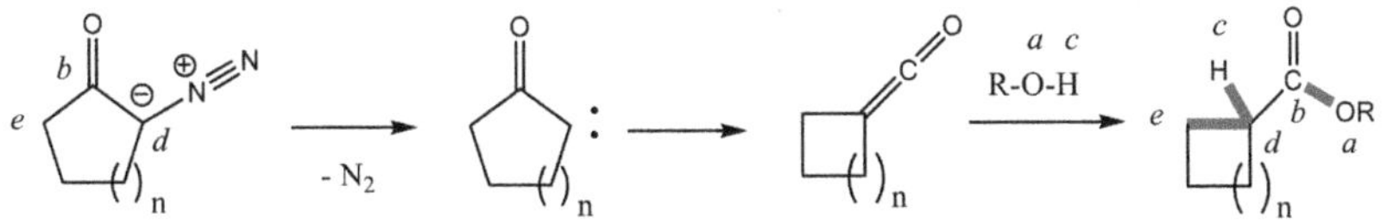

$\Delta = \Delta$ (a) + Δ (b) + Δ (c) + Δ (d) + Δ (e) = (–2–(–2)) + (3–2) + (1–1) + (–1–0) + (–2–(–2
)) = 0
HI = $-\Delta/(N+1) = 0/2 = 0$

Example 5.31 [81]

Themes: product structure target bond map, hypsicity

The Grimaux plan to synthesize allantoin is shown below.

Working backwards in the synthesis plan, determine the product structure target bond map. Determine the hypsicity index for the synthesis plan and display the hypsicity profile. Characterize the synthesis as isohypsic, hyperhypsic, or hypohypsic.

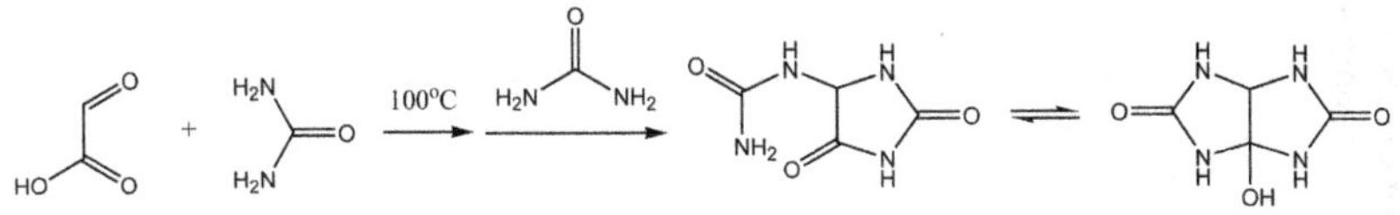

Solution

Product structure target bond map:

Hypsicity index:

Rxn stage	a	b	c	d	E	f	g	h	SUM	Δ (row)	
2	−3	−3	3	−2	1	−3	−3	1	−9	0	
1	−3	−3	3	−2	1	−3	−3	1	−9	0	
0	−3	−3	3	−2	1	−3	−3	1	−9	0	
										0	SUM
N	2										
HI	0										

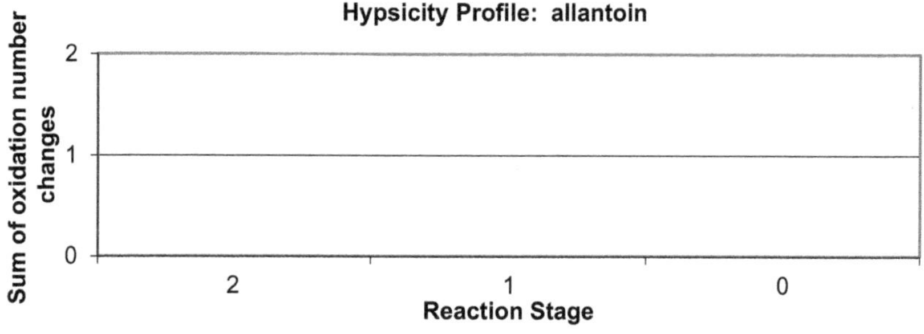

Since $HI = 0$, the synthesis plan is isohypsic.

Example 5.32

Themes: product structure target bond map, hypsicity

Captodiame is a sedative and tranquilizer that is produced by a convergent synthesis.

Working backwards in the synthesis plan, determine the product structure target bond map. Determine the hypsicity index for the synthesis plan and display the hypsicity profile. Characterize the synthesis as isohypsic, hyperhypsic, or hypohypsic.

OH SOCl$_2$ Cl HN N N

SH Cl S

O O OH Ph Cl Ph NaBH$_4$ Ph S S

NH$_2$ NH$_2$ Cl S NH$_2$ SOCl$_2$ Cl S NH$_2$ Ph Ph S S

SH Cl N S N NaOH Ph Ph S S

Solution

OH SOCl$_2$ Cl HN N N

SH Cl S

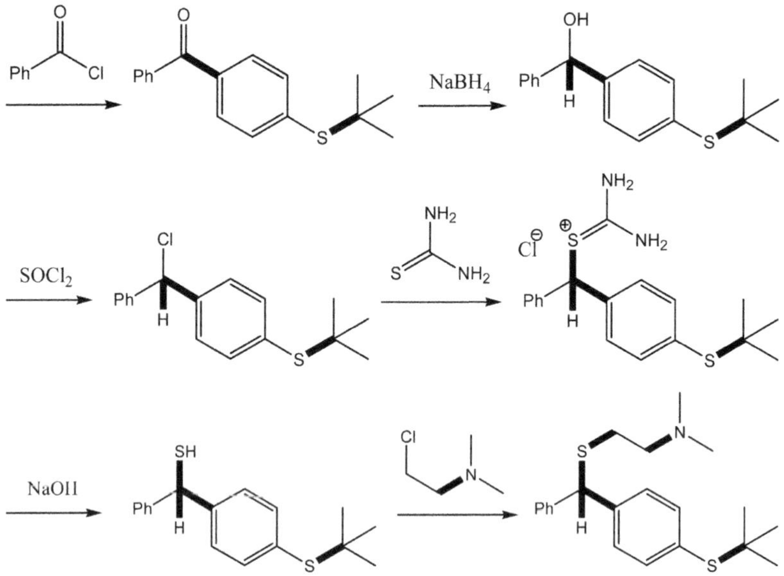

$NaBH_4$

$SOCl_2$

NaOH

Product structure target bond map:

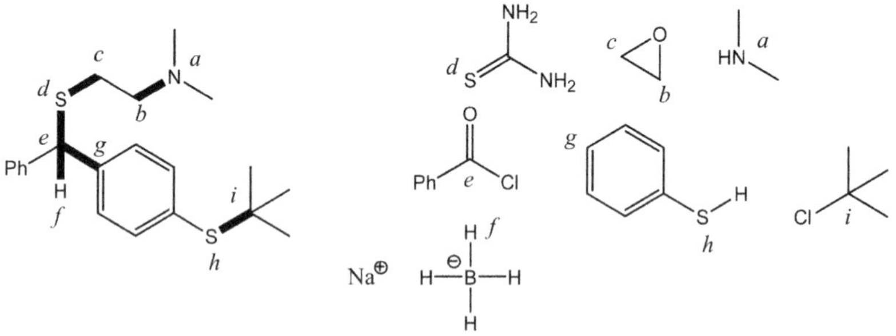

Hypsicity index:

Rxn stage	a	b	c	d	e	f	g	h	i	SUM	Δ (row)	
7	–3	–1	–1	–2	0	1	0	–2	1	–7	0	
6	–3	–1	–1	–2	0	1	0	–2	1	–7	0	
5	–3	–1	–1	–2	0	1	0	–2	1	–7	0	
4	–3	–1	–1	–2	0	1	0	–2	1	–7	0	
3	–3	–1	–1	–2	0	1	0	–2	1	–7	0	
2	–3	–1	–1	–2	2	–1	0	–2	1	–7	0	
1	–3	–1	–1	–2	3	1	–1	–2	1	–5	2	
0	–3	–1	–1	–2	0	1	–1	–2	1	–8	–1	
											1	SUM
N	7											
HI	0.125											

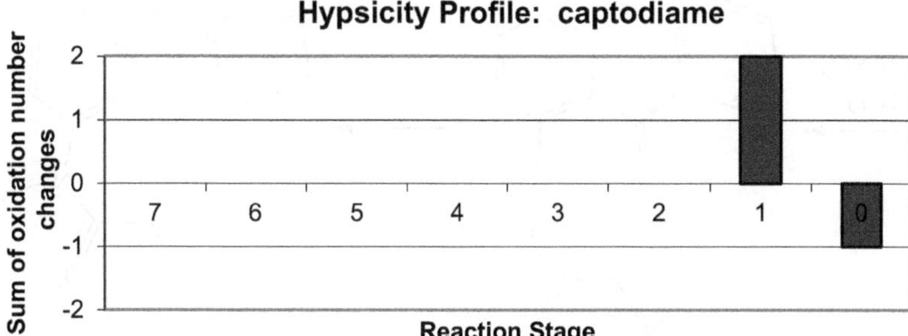

Since $HI > 0$, the synthesis plan is hyperhypsic.

Example 5.33 [64–67]

Themes: product structure target bond map, hypsicity

Rohypnol (flunitrazepam) is a controversial sedative because it has been misused to spike people's drinks and hence it is known as the "date rape" drug. Its synthesis is shown below.

Working backwards in the synthesis plan, determine the product structure target bond map. Determine the hypsicity index for the synthesis plan and display the hypsicity profile. Characterize the synthesis as isohypsic, hyperhypsic, or hypohypsic.

F Br_2 $h\nu$ F Br NaCN F N H_2SO_4 H_2O F O OH $SOCl_2$

F O Cl PhH $AlCl_3$ F O Ph HN NH_2 F Ph N NH

H_2SO_4 F NH Ph O_3 Me_2S F O NH O Ph

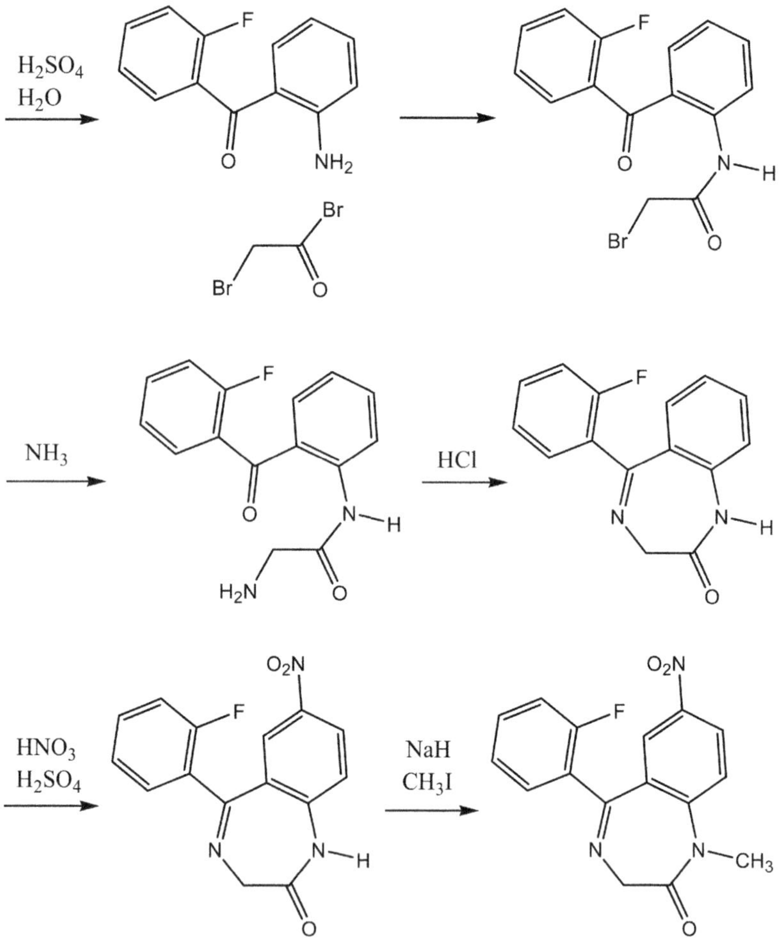

H_2SO_4
H_2O

F
O NH2
Br
Br O

F
O N H
Br O

NH_3

F
O N H
H2N O

HCl

F
N N H
O

HNO_3
H_2SO_4

O2N
F
N N H
O

NaH
CH_3I

O2N
F
N N CH3
O

Solution

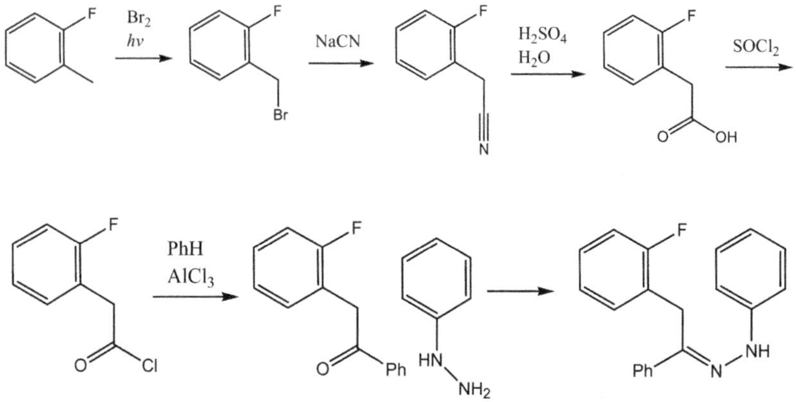

F
Br_2
$h\nu$
F
Br
NaCN
F
N
H_2SO_4
H_2O
F
O OH
$SOCl_2$

F
O Cl
PhH
$AlCl_3$
F
O Ph
HN NH2
F
Ph N NH

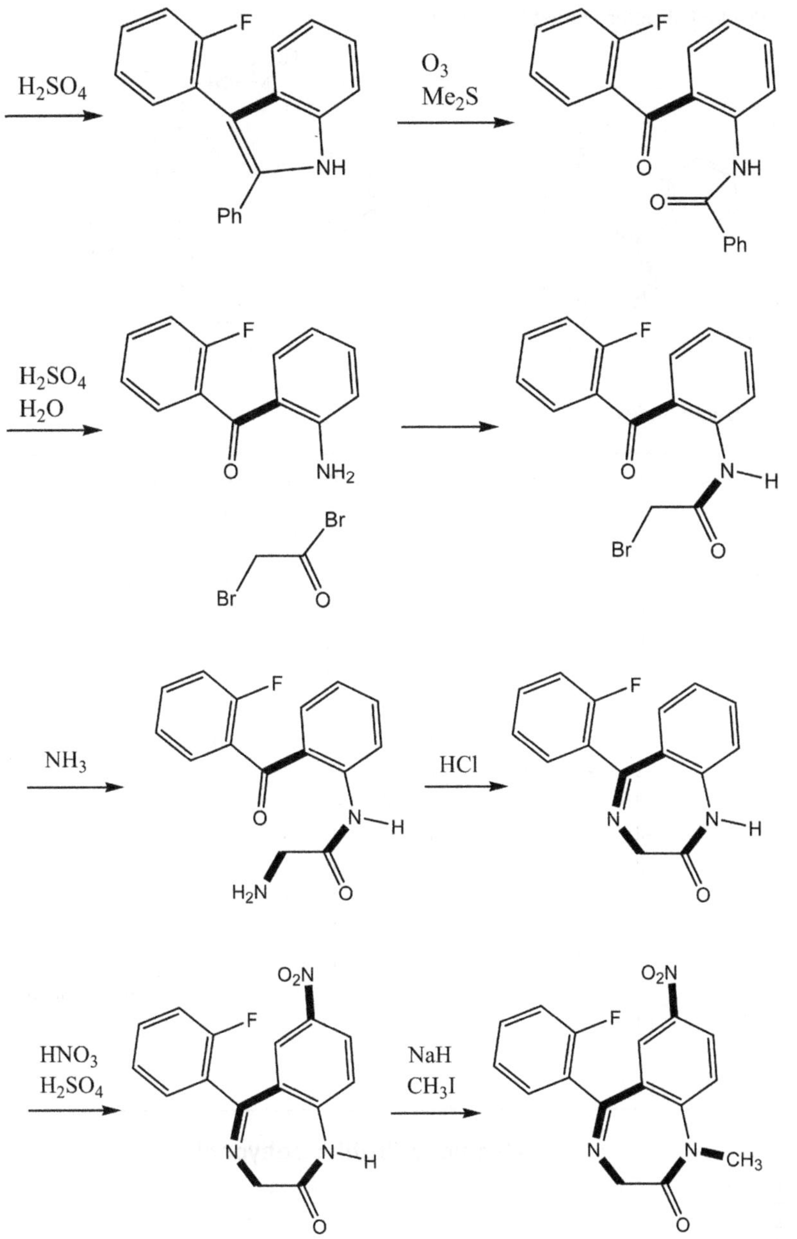

$\xrightarrow{H_2SO_4}$

$\xrightarrow{O_3,\ Me_2S}$

$\xrightarrow{H_2SO_4,\ H_2O}$

$\longrightarrow$

$\xrightarrow{NH_3}$

$\xrightarrow{HCl}$

$\xrightarrow{HNO_3,\ H_2SO_4}$

$\xrightarrow{NaH,\ CH_3I}$

Product structure target bond map:

Hypsicity index:

Rxn stage	a	b	c	d	e	f	g	h	SUM	Δ (row)	
14	2	−3	−1	3	−3	−2	1	3	0	0	
13	2	−3	−1	3	−3	−2	1	3	0	0	
12	2	−3	−1	3	−3	−2	−1	5	0	0	
11	2	−3	−1	3	−3	−2	−1	3	−2	−2	
10	2	−3	−1	3	−3	−2	−1	3	−2	−2	
9	2	−3	−1	3	−3	−2	−1	3	−2	−2	
8	2	−3	−1	3	−3	−2	−1	3	−2	−2	
7	0	−3	−1	3	−3	−2	−1	3	−4	−4	
6	−2	−3	−1	3	−1	−2	−1	3	−4	−4	
5	−2	−3	−1	3	−1	−2	−1	3	−4	−4	
4	−2	−3	−1	3	−3	−2	1	3	−4	−4	
3	−2	−3	−1	3	−3	−2	1	3	−4	−4	
2	−2	−3	−1	3	−3	−2	1	3	−4	−4	
1	−1	−3	−1	3	−3	−2	1	3	−3	−3	
0	−3	−3	−1	3	−3	−2	1	3	−5	−5	
										−40	SUM
N	14										
HI	−2.67										

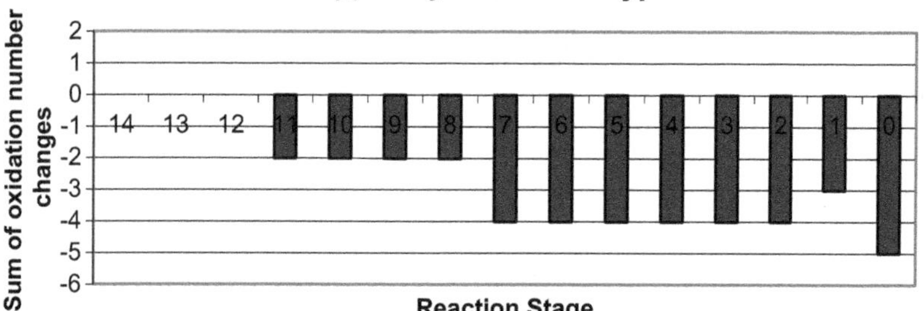

Since $HI < 0$, the synthesis plan is hypohypsic.

Oxidation Length, $|UD|$ [80]

A parameter that tracks the "oxidation length" traversed over the course of a synthesis plan equal to the sum of the absolute values of all the "ups" and "downs" in a hypsicity profile or bar graph beginning with the zeroth reaction stage. A hypsicity profile is a histogram or bar graph of the sum of oxidation number changes (see hypsicity) versus reaction stage. Some useful trends are:

UD, the algebraic sum of "ups" and "downs" in any hypsicity profile, is always zero.
|UD| is always an even number.

For monotonic increases (hypohypsic profile) or monotonic decreases (hyperhypsic profile), |UD| is twice the height of the first bar in the hypsicity profile.

Example 5.34

Determine the oxidation lengths for the synthesis plans for allantoin, captodiamine, and rohypnol.

Solution
For allantoin, $|UD|=0$.
For captodiamine, $|UD|=|-1|+|3|+|-2|=1+3+2=6$.
For rohypnol, $|UD|=|-5|+|2|+|-1|+|2|+|2|=5+2+1+2+2=12$.

Redox Economy [50, 51]

A term coined by Philip Baran at Scripps Institute to qualitatively describe efficient transformations that do not involve net increases or decreases in oxidation number for atoms involved in forming target bonds. The concept is to minimize the use of correction redox reactions in a synthesis plan.

5.4.1 Problems

PROBLEM 5.19 [82]

Themes: product structure target bond map, hypsicity

An industrial synthesis of vinyl acetate is shown below.

Working backwards in the synthesis plan, determine the product structure target bond map. Determine the hypsicity index for the synthesis plan and display the hypsicity profile. Characterize the synthesis as isohypsic, hyperhypsic, or hypohypsic.

CaC_2 —(H_2O)→ HC≡CH —(H_2O, H_2SO_4 (cat.), Hg_2SO_4 (cat.))→ [acetaldehyde: O, H] —(1/2 O_2, $Mn(OAc)_2$ (cat.))→ [acetic acid: O, OH]

—(HC≡CH)→ [vinyl acetate: O, O]

PROBLEM 5.20 [83, 84]

Themes: product structure target bond map, hypsicity

The Sherwin-Williams plan to synthesize the artificial sweetener saccharin is shown below.

Working backwards in the synthesis plan, determine the product structure target bond map. Determine the hypsicity index for the synthesis plan and display the hypsicity profile. Characterize the synthesis as isohypsic, hyperhypsic, or hypohypsic.

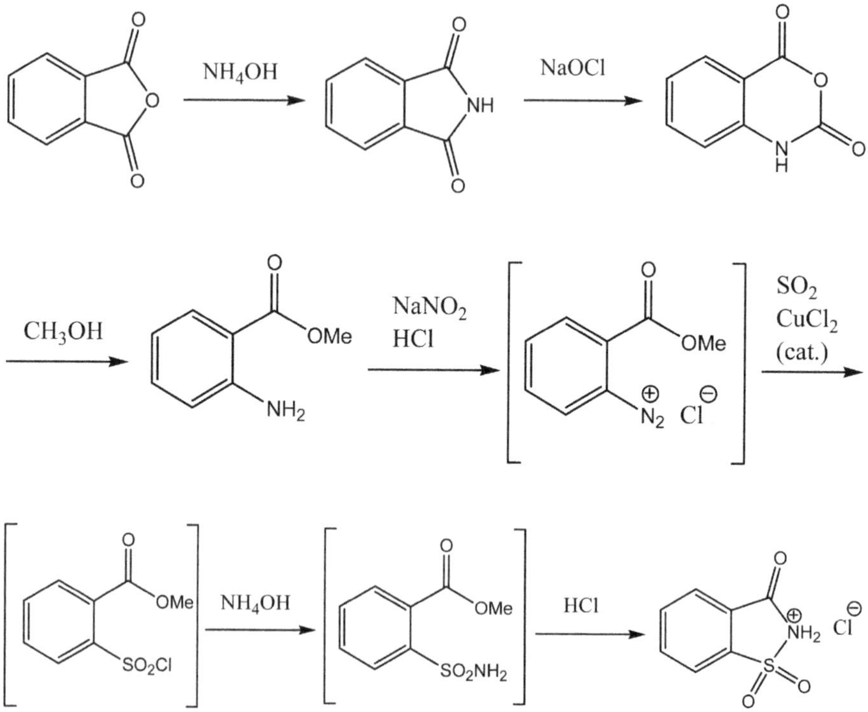

NH_4OH → NH; NaOCl → O, N H, O

CH_3OH → OMe, NH_2; $NaNO_2$, HCl → [OMe, $N_2^{\oplus}$ $Cl^{\ominus}$]; SO_2, $CuCl_2$ (cat.) →

[OMe, SO_2Cl] NH_4OH → [OMe, SO_2NH_2] HCl → O, $NH_2^{\oplus}$ $Cl^{\ominus}$, S, O, O

PROBLEM 5.21 [85]

Themes: balancing chemical equations, synthesis plan analysis, target bond mapping, hypsicity index, reaction mechanism, side product α-Herbertenol is a fungicidal sesquiterpene found in liverworts. Its synthesis plan is shown below.

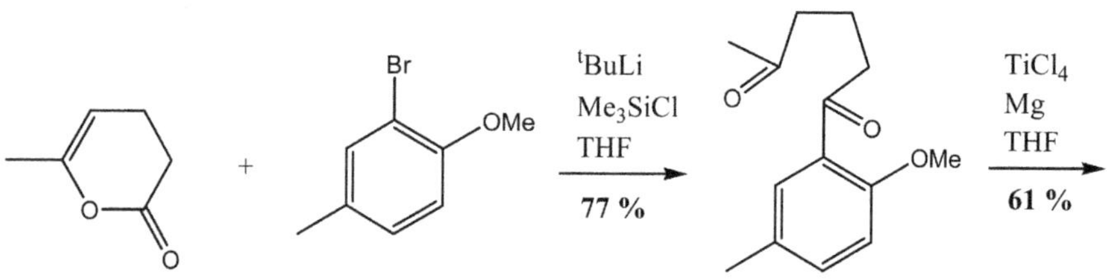

O, O + Br, OMe → ${}^tBuLi$, Me_3SiCl, THF, **77 %** → O, O, OMe → $TiCl_4$, Mg, THF, **61 %** →

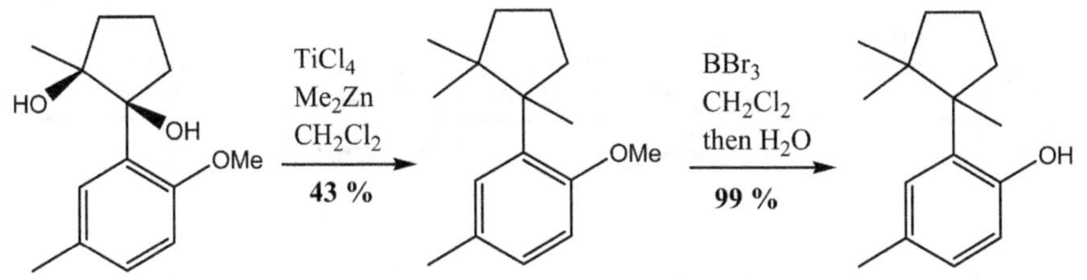

TiCl$_4$
Me$_2$Zn
CH$_2$Cl$_2$
43 %

BBr$_3$
CH$_2$Cl$_2$
then H$_2$O
99 %

α-herbertenol

A side product of step 3 formed in 38% yield is shown below.

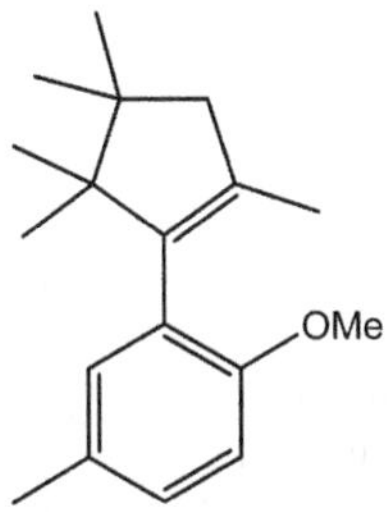

Part 1

Balance each chemical reaction in the plan showing all by-products. Determine the corresponding atom economies.

Part 2

Show the balanced equation that accounts for the side product observed in step 3. Suggest a mechanism for its formation. Compare with the mechanism for the formation of the target product intended for the synthesis plan.

Part 3

Rewrite the plan showing the target bonds made.

Part 4

From Part 3, track the oxidation numbers of atoms involved in target bond forming steps and determine the hypsicity index using the *calculator-hypsicity-index.xls* spreadsheet.

PROBLEM 5.22 [86]

Themes: product structure target bond map, hypsicity

The first synthesis of ascorbic acid (vitamin C) from a non-carbohydrate source was reported in 1998.

Working backwards in the synthesis plan, determine the product structure target bond map. Determine the hypsicity index for the synthesis plan and display the hypsicity profile. Characterize the synthesis as isohypsic, hyperhypsic, or hypohypsic.

O_2
H_2O
Pseudomonas putida

$Me_2C(OMe)_2$
pTsOH (cat.)

*m*CPBA
$NaHCO_3$

$PhCH_2OH$
CF_3SO_3H (cat.)

O_3
CH_3OH

- HCl

$NaBH_3CN$
then H_2O

Pd / C (cat.)

PhCHO
CF_3SO_3H (cat.)

$^tBuPh_2SiCl$
imidazole

Dess-Martin
periodinane

H_2O

H_2O
HOAc (cat.)

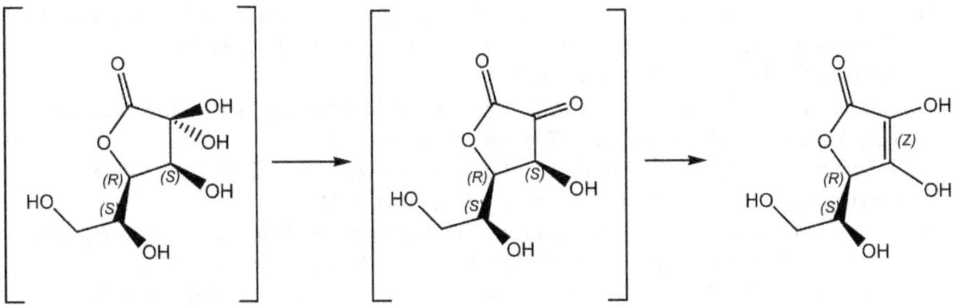

REFERENCES

1. Kidwai, M.; Mohan, R. *Found. Chem.* 2005, *7*, 269.
2. Solomons, T.W.G. *Organic Chemistry*, 6th ed., Wiley: Hoboken, 1996, p. 530.
3. Furniss, B.S.; Hannaford, A.J.; Rogers, V.; Smith, P.W.G.; Tatchell, A.R. *Vogel's Textbook of Practical Organic Chemistry*, 5th ed., Longman: London, 1989, pp. 1150–1151.
4. Vogel, A.I. *A Textbook of Practical Organic Chemistry*, 3rd ed., Longman: London, 1956, pp. 477–478, 998.
5. Zeng, F.; Alper, H. *Org. Lett.* 2010, *12*, 1188.
6. Liang, H.; Ciufolini, M.A. *Org. Lett.* 2010, *12*, 1760.
7. Lee, H.Y.; Jung, Y.; Yoon, Y.; Kim, B.G.; Kim, Y. *Org. Lett.* 2010, *12*, 2672.
8. Ramazani, A.; Rezaei, A. *Org. Lett.* 2010, *12*, 2852.
9. Burke, J.P.; Sabat, M.; Iovan, D.A.; Myers, W.H.; Chruma, J.J. *Org. Lett.* 2010, *12*, 3192.
10. Trofimov, B.A.; Shemyakina, O.A.; Malkina, A.G.; Ushakov, I.A.; Kazheva, O.N.; Alexandrov, G.G.; Dyachenko, O.A. *Org. Lett.* 2010, *12*, 3200.
11. Zeng, F.; Alper, H. *Org. Lett.* 2010, *12*, 3642.
12. Chen, W.; Hu, M.; Wu, J.; Zou, H.; Yu, Y. *Org. Lett.* 2010, *12*, 3863.
13. Miao, L.; Haque, I.; Manzoni, M.R.; Tham, W.S.; Chemler, S.R. *Org. Lett.* 2010, *12*, 4739.
14. Cleary, L.; Yoo, H.; Shea, K.J. *Org. Lett.* 2011, *13*, 1781.
15. Zheng, H.; Lejkowski, M.; Hall, D.G. *Tetrahedron Lett.* 2013, *54*, 91.
16. Keck, G. E.; Wager, T. T.; Rodriquez, J. F. D. *J. Am. Chem. Soc.*, 1999, *121*, 5176.
17. Kermack, W.O.; Robinson, R. *J. Chem. Soc.* 1922, *121*, 427.
18. Flynn, A.B.; Ogilvie, W.W. *J. Chem. Educ.* 2015, *92*, 803.
19. Nowak, D.M.; Lansbury, P.T. *Tetrahedron* 1998, *54*, 319.
20. Tolman, C.A.; Jesson, J.P. *Science* 1973, *181*, 501.
21. Tolman, C.A.; Faller, J.W. *Homogeneous Catalysis with Metal Phosphine Complexes.* In L.H. Pignolet (Ed.), Plenum Press: New York, 1983, pp. 13–136.
22. Volla, C.M.R.; Backväll, J.E. *Angew. Chem. Int. Ed.* 2013, 52, 14209.
23. Bart, H. DE 250264 (1910).
24. Bart, H. DE 254092 (1912).
25. Bart, H. DE 264924 (1913).
26. Bart, H. DE 268172 (1913).
27. Bart, H. *Ann. Chem.* 1922, *429*, 55.
28. Eghbali, N.; Eddy, J.; Anastas, P.T. *J. Am. Chem. Soc.* 2008, *73*, 6932.
29. Chen, G.; Gong, W.; Zhuang, Z.; Andrä, M.S.; Chen, Y.Q.; Hong, X.; Yang, Y.F.; Liu, T.; Houk, K.N.; Yu, J.Q. *Science* 2016, *353*, 1023.
30. Takaki, K.; Ohno, A.; Hino, M.; Shitaoka, T.; Komeyama, K.; Hiroto, Y. *Chem. Commun.* 2014, 12285.

31. O'Brien, C.J.; Tellez, J.L.; Nixon, Z.S.; Kang, L.J.; Carter, A.L.; Kunkel, S.R.; Przeworski, K.C.; Chass, G.A. *Angew. Chem. Int. Ed.* 2009, *48*, 6836.
32. Peplow, M. *ACS Cent. Sci.* 2017, *3*, 261.
33. Liu, X.; Conte, M.; Elias, D.; Lu, L.; Morgan, D.J.; Freakley, S.J.; Johnston, P.; Kiely, C.J.; Hutchings, G.J. *Catal. Sci. Technol.* 2016, *6*, 5144.
34. Johnston, P.; Carthey, N.; Hutchings, G.J. *J. Am. Chem. Soc.* 2015, *137*, 14548.
35. Davies, C. PhD thesis, University of Cardiff, 2013, p. 48.
36. Simon, M.; Jonk, P.; Wuhl-Couturier, G.; Halbach, S. *Ullmann's Encyclopedia of Industrial Chemistry*, Wiley-VCH: Weinheim: 2012, vol 22, pp. 559–594.
37. Aggarwal, V.K.; Harvey, J.N.; Richardson, J. *J. Am. Chem. Soc.* 2002, *124*, 5747.
38. Catellani, M.; Frignani, F.; Rangoni, A. *Angew. Chem. Int. Ed.* 1997, *36*, 119.
39. Habibi, D.; Pakravan, N.; Nematollahi, D. *Electrochem. Commun.* 2014, *49*, 65.
40. Nematollahi, D.; Karbasi, H. *J. Iran. Chem. Soc.* 2011, *8*, 48.
41. Ito, S.; Katayama, R.; Kunai, A.; Sasaki, K. *Tetrahedron Lett.* 1989, *30*, 205.
42. Zhang, L.; Zha, Z.; Wang, Z.; Fu, S. *Tetrahedron Lett.* 2010, *51*, 1426.
43. Pütter, H.; Hannebaum, H. DE 19618854 (BASF, 1997).
44. Gao, Y. US 2013134047 (2013).
45. Hilt, G. *Angew. Chem. Int. Ed.* 2003, *42*, 1720.
46. Chou, C.F.; Chou, T.C. *J. Appl. Electrochem.* 2003, *33*, 741.
47. Li, W.; Nonaka, T. *Electrochim. Acta* 1999, *44*, 2605.
48. Jow, J.J.; Lee, A.C.; Chou, T.C. *J. Appl. Electrochem.* 1987, *17*, 753.
49. Andraos, J. *Org. Process Res. Dev.* 2006, *10*, 212.
50. Burns, N.Z.; Baran, P.S.; Hoffmann, R.W. *Angew Chem. Int. Ed.* 2009, *48*, 2854.
51. Newhouse, T.; Baran, P.S.; Hoffmann, R.W. *Chem Soc Rev.* 2009, *38*, 3010.
52. Gaich, T.; Baran, P.S. *J. Org. Chem.* 2010, *75*, 4657.
53. Rogers, D.J.; Tanimoto, T.T. *Science* 1960, *132*, 1115.
54. Tanaka, A.; Kawai, T.; Takabatake, T.; Oka, N.; Okamoto, H.; Bersohn, M. *Tetrahedron* 2007, *63*, 10226.
55. Tanaka, A.; Kawai, T.; Takabatake, T.; Oka, N.; Okamoto, H.; Bersohn, M. *Tetrahedron Lett.* 2006, *47*, 6733.
56. Bersohn, M. *Bull. Chem. Soc. Jpn.* 1972, *45*, 1897.
57. Ruzicka, L. *Experentia* 1953, *9*, 357.
58. Burke, M.D.; Schreiber, S.L. *Angew. Chem. Int. Ed.* 2004, *43*, 46.
59. Dai, W.M.; Shi, J. *Combin. Chem. High Throughput Screening* 2007, 10, 837.
60. Harris, G.D. Jr.; Herr, R.J.; Weinreb, S.M. *J. Org. Chem.* 1992, *57*, 2528.
61. Heathcock, C.H.; Ruggeri, R.B.; McClure, K.F. *J. Org. Chem.* 1992, *57*, 2585.
62. Stevens, R.V.; Lee, A.W.M. *J. Am. Chem. Soc.* 1979, *101*, 7032.
63. Lindsley, C.W.; Wisnoski, D.D.; Wang, Y.; Leister, W.H.; Zhao, Z. *Tetrahedron Lett.* 2003, *44*, 4495.
64. Kariss, J.; Newmark, H.L. US 3116203 (Hoffmann-LaRoche, 1963).
65. Kariss, J.; Newmark, H.L. US 3123529 (Hoffmann-LaRoche, 1964).
66. Keller, O.; Steiger, N.; Sternbach, H. US 3203990 ((Hoffmann-LaRoche, 1965).
67. Sternbach, L.H.; Fryer, R.I.; Keller, O.; Metlesics, W.; Sach, G.; Steiger, N. *J. Med. Chem.* 1963, *6*, 261.
68. Victory, P.; Alvarez-Larena, A.; Germain, G.; Kessels, R.; Piniella, J.F.; Vidal-Ferran, A. *Tetrahedron* 1995, *51*, 235.
69. Victory, P.; Borrell, J.L.; Vidal-Ferran, A. *Heterocycles* 1993, *36*, 769.
70. Stepakov, A.V.; Galkin, I.A.; Kostikov, R.R.; Starova, G.L.; Starikova, Z.A.; Molchanov, A.P. *Synlett* 2007, 1235.
71. Wang, S.L.; Liu, Y.P.; Xu, B.H.; Wang, X.H.; Jiang, B.; Tu, S.J. *Tetrahedron* 2011, *67*, 9417.

72. Barth, F.; Casellas, P.; Cong, C.; Martinez, S.; Renaldi, M.; Anne-Archard, G. US 5624941 (Sanofi, 1997).
73. Craig, L.C. *J. Am. Chem. Soc.* 1933, *55*, 2854.
74. Felpin, F.X.; Girard, S.; Vo-Thanh, G.; Robins, R.J.; Villieras J.; Lebreton, J. *J. Org. Chem.* 2001, *66*, 6305.
75. Chavdarian, C.G.; Sanders, E.B.; Bassfield, R.L. *J. Org. Chem.* 1982, *47*, 1069.
76. Welter, C.; Moreno, R.M.; Streiff, S.; Helmchen, G. *Org. Biomol. Chem.* 2005, *3*, 3266.
77. Loh, T.P.; Zhou, J.R.; Li, X.R.; Sim, K.Y. *Tetrahedron Lett.* 1999, *40*, 7847.
78. Marquez, F.; Llebaria, A.; Delgado, A. *Tetrahedron Asymm.* 2001, *12*, 1625.
79. Hendrickson, J.B. *J. Am. Chem. Soc.* 1971, *93*, 6847.
80. Andraos, J. Application of green metrics analysis to chemical reactions and synthesis plans, in green chemistry metrics. In A. Lapkin; D.C. Constable (Eds.), *Green Chemistry Metrics*, Blackwell Scientific: Oxford, 2008, pp. 69–199.
81. Grimaux, E. *Ann. Chim. Phys.* 1877, *11(5)*, 356.
82. Faith, W.L.; Keyes, D.B.; Clark, R.L. *Industrial Chemicals*, 3rd ed., Wiley: Hoboken, 1966, p. 1, 12, 28, 800.
83. Riggin, R.M.; Kinzer, G.W. *Food Chem. Tox.* 1983, *21*, 1.
84. Tonne, P.; Jaedicke, H. US 4464537 (BASF, 1984).
85. Harrowven, D.C.; Hannam, J.C. *Tetrahedron* 1999, *55*, 9333.
86. Banwell, M.; Blakey, S.; Harfoot, G.; Longmore, R. *J. Chem. Soc. Perkin Trans I* 1998, 3141.

6 Waste Production and Input Material Consumption

After having gone through the first five chapters of this book mastering the preliminary skills of identifying the role of materials used in a chemical reaction, balancing a chemical reaction, linking that balanced chemical equation to a reaction mechanism, knowing how to read and evaluate experimental procedures published in the literature, knowing how to categorize a given reaction into one or more of the main classifications, knowing how to parse a product structure back to its reactant component structures and to link that to synthesis strategy and target bond mapping, knowing how to properly draw chemical structures following the principle of conservation of structural aspect, and tracking oxidation states of key atoms throughout a synthesis plan, we are now in a position to tackle the fundamental problem of quantifying how material efficient a given reaction is. In this chapter, we first examine the basic question of what constitutes waste. Next, we introduce successively various metrics based on the perspectives of waste production, input material consumption, and product formation. These form a sequential ladder of understanding toward the goal of unifying these metrics into a simplified hierarchy that succinctly describes how to quantify the material efficiency of any reaction.

At this stage, we must state that the development of green metrics over the last 25 years has focused primarily on the parameterization of reaction efficiency from two main perspectives, which can be classified according to the "glass half full" and "glass half empty" philosophies. The "glass half full" philosophy corresponds to the view of how much of the input materials used in carrying out a reaction end up in the desired product. The opposite "glass half full" philosophy corresponds to the view of how much waste material is produced. By common sense, these views are directly linked through the concept of balanced chemical equations and the law of conservation of mass [1] developed by Antoine Lavoisier in 1775. We will justify this statement *vide infra*. Despite this obvious realization to any intelligent person, chemists have taken nearly two and half centuries to figure this out. Figure 6.1 shows a historical timeline of key events in the development of our modern understanding of parameterizing material efficiency in chemical reactions. Lavoisier and Richter laid out the fundamental starting concepts of balancing and stoichiometry [2]. Wöhler's breakthrough urea synthesis [3, 4] launched the era of organic synthesis and introduced the concept of reaction yield as the first measure of efficiency. Between 1828 and 1991 academic synthetic chemists had no interest in worrying about synthesis efficiency since they were engaged in discovering various aspects of chemical structure such as functional groups and developing methods of identifying them chemically and

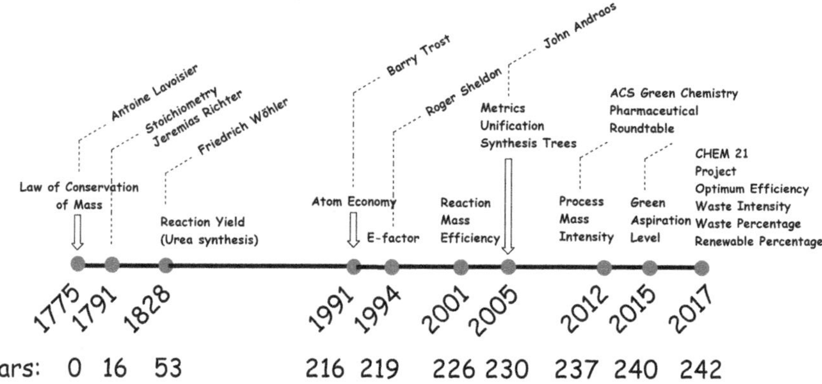

FIGURE 6.1 Timeline of introduction of key concepts in the chemistry literature pertaining to quantifying material efficiency of a chemical reaction.

spectroscopically, and in developing the vast library of chemical reactions that have become the working toolbox for any practicing chemist. Their main focus of research and discovery was therefore on determining whether their isolated substances were pure, determining the chemical structural identities of those substances, and in developing synthetic methods of making them from simpler molecular entities as a means to prove their structural identity. In 1991, after spending 163 years in this endeavor, the concept of atom economy (AE) was introduced as a *qualitative* concept to induce chemists to invent reactions that did not produce by-products. Though Barry Trost's seminal publication in the journal *Science* [5] has been the most cited reference in the field of green chemistry literature, few have actually read the paper. When one does, one finds the astounding feature that *all* of the unimolecular, bimolecular, and multi-component chemical reactions discussed in that paper did not produce any by-products and all were catalytic. Even more amazing was that there was no *quantitative* expression describing the atom economy concept in mathematical terms, which would represent the first introduction of a "green" metric in the literature. Chemical reactions with high atom economies were branded as "green" and those with low atom economies were not. However, no threshold value was used to benchmark the cut-off between "green" and "non-green" in the same way that no cut-off value exists between "good" reaction yields and "modest" or "poor" reaction yields. Nevertheless, atom economy was the first metric introduced from a "glass half full" perspective. Three years later, another metric called the *E*-factor [6] was introduced to measure efficiency from the "glass half empty" perspective, namely the amount of waste produced per unit mass of desired product. Unlike atom economy, the *E*-factor was introduced in quantitative terms. However, those who promulgated these two metrics in the literature did so in competition with one another and neglected to connect them via the concept of balancing chemical equations and the law of conservation of mass. An attempt to do so was made via the introduction in 2001 of another "glass half full" metric called reaction mass efficiency (RME) [7], which was originally defined from the point of view of mass of target product versus masses of reactants only. My entry into the field of green chemistry came about from the following sentences that appeared in papers in the journal *Green Chemistry* written by people working in the pharmaceutical industry: (a) "Atom economy may be useful as an organizing concept or in combination with other metrics, but at this time it is not considered to be useful as a stand-alone metric" [8] and (b) "The overall yield of the process is included as a conventional measure of efficiency (to include all arms of the synthesis if convergent) although this provides no information on the amount or type of waste generated" [9]. The first sentence demonstrated that the authors at the time did not understand that true reaction optimization begins first by *designing* reactions that minimize by-product formation. Instead, they were fixated with waste reduction by reducing or eliminating auxiliary materials used in carrying reactions such as solvents, work-up materials, and purification materials. This idea seems on the surface to make sense since the bulk of material used in chemical reactions comes from auxiliary material consumption. Implicit in this notion, however, was an acceptance that chemists should continue to use the existing chemical reactions in their working toolbox and "green" them up by addressing auxiliary material consumption by minimization or recycling methodologies. The second sentence implied that the authors believed that "overall yield" could be applied to convergent synthesis plans and was calculated by multiplying together all reaction yields from all branches. This second statement is mathematically incorrect as we will demonstrate in Chapter 1 of *Synthesis Green Metrics: Problems, Exercises, and Solutions*. Both of these statements triggered in me the idea that primarily synthetic organic chemists fundamentally did not understand or appreciate the meaning of balanced chemical equations. This came as no surprise since the subject is taught from the perspective of reagents operating on a substrate of interest that leads to a product of interest as is customarily done in a sequential series of reactions appearing in a synthesis plan. In 2005, I decided to examine the question of quantitative parameterization of reaction performance from the beginning, that is, from the fundamental Lavoisier principle [10]. I realized that the "glass half full" and "glass half empty" philosophies were describing both sides of the same coin, and that the competition between them in the literature was a false one and only benefited the promulgators of those philosophies. Figure 6.2 nicely illustrates the inverse relationship between input material maximization and waste material minimization via the ancient *yin-yang* symbol. The net result was the development of two master equations shown in Equations (6.1) and (6.2) that linked these ideas in a quantitative way.

$$\mathrm{AE} = \frac{1}{1+E_{mw}} \tag{6.1}$$

$$\mathrm{RME} = \frac{1}{1+E} = \frac{1}{PMI} \tag{6.2}$$

Equation (6.1) applicable to any balanced chemical equation links the atom economy concept with the *E*-factor concept on the basis of *molecular weight*. Equation (6.2), also applicable to any balanced chemical equation, takes into account the *masses* of all materials used in carrying out a chemical reaction and so links the concept of reaction mass efficiency (broadened to include all input materials as opposed to only reactants) to the concept of *E*-factor on the basis of *mass*. The parallel formalism of Equations (6.1) and (6.2) is obvious. Moreover, another "glass half full" metric now adopted by

FIGURE 6.2 Yin-yang relationship between waste material minimization and input material maximization.

process chemists in the pharmaceutical industry as the gold standard measure of efficiency, process mass intensity (formerly called mass intensity), was also linked to both *E*-factor and RME. Clearly, from these quantitative relationships we can see that the goal of minimizing waste (or maximizing input materials toward desired product) is directly linked to minimizing both *E*-factor and *PMI* and maximizing both AE and RME. Essentially, Equations (6.1) and (6.2) form the basis of a *complete* understanding of the key variables involved in reaction optimization from a materials consumption perspective. Furthermore, RME for a single chemical reaction could be decomposed into its constituent parameters as shown by the relationship in Equation (6.3) that includes reaction yield, Ɛ, which was the traditional metric used to gauge reaction efficiency.

$$\mathrm{RME} = (\mathrm{AE})(\varepsilon)\left(\frac{1}{\mathrm{SF}}\right)\left(\frac{1}{1+(\mathrm{AE})(\varepsilon)\left(\frac{1}{\mathrm{SF}}\right)\left(\frac{m_{\mathrm{aux}}}{m_{\mathrm{prod}}}\right)}\right) \quad (6.3)$$

where:

SF is the stoichiometric factor taking into account excess reagent consumption relative to the limiting reagent

m_{aux} is the sum of masses of all auxiliary materials

m_{prod} is the mass of desired product collected

As far as estimating the material efficiency of any chemical reaction the parameters shown in Equations (6.1) through (6.3) are all that is needed to solve the problem. However, since this development in 2005, the green chemistry literature has ballooned with derivative "me-too" material efficiency metrics that serve no other purpose than to re-brand existing ideas with "new" names and to carve personal niches in the literature for the promulgators. This cottage industry resulted in a confusion of nomenclature and falsely increased the complexity of solving a very simple problem.

The net backlash for the green chemistry community is that it hampered interested newcomers from engaging in the field and adopting green chemistry practice in their work, and worse, it fueled the skeptics in viewing the field as undergoing an inertia in which there existed an unresolved "disagreement and debate" about which metric to adopt. Intertwined in this embroil is the well-known deficiency of numeracy skills in the education of synthetic organic chemists who have been taught to believe that quantitative analysis and reasoning have no place in that subject. These problems were recently discussed in a special review article on the state of affairs in green chemistry research [11].

The main goals of this chapter are first and foremost to describe in simple language the parameterization of material efficiency metrics, to streamline the thinking behind reaction optimization and to illustrate the power of visual representation as a means to quickly and correctly interpret the meaning of the numerical results. In this exposition, we have ignored all superfluous, repetitive, and redundant metrics. Each section of this chapter (Sections 6.1 through 6.4) begins with a list of associated terminologies fully defined followed by illustrative examples. A hierarchy and connection between metrics are shown in Section 6.5. Then, a section showing fully worked examples (Section 6.6) based on current research from the literature illustrate how the methods can be used and decisions made about material "greenness." We conclude the chapter with a section on problems for the reader to demonstrate their understanding (Section 6.7).

6.1 DEFINITION OF WHAT CONSTITUTES WASTE

6.1.1 Terms, Definitions, and Examples

Kernel Mass of Waste

Kernel mass of waste refers to the total mass of waste originating from reaction by-products and stoichiometric unreacted starting materials in a given chemical reaction or a synthesis plan.

Example 6.1 [12]

A chemist reacts benzyl alcohol (10.81 g, 0.10 mol) with *p*-toluene sulfonyl chloride (21.9 g, 0.115 mol) in toluene (500 g) and triethylamine (15 g) to give 23.6 g of the sulfonate ester. Determine the kernel mass of waste produced in this reaction.

Solution

The balanced chemical equation for this transformation is given below.

PhCH$_2$OH + p-CH$_3$C$_6$H$_4$SO$_2$Cl + Et_3N → p-CH$_3$C$_6$H$_4$SO$_2$OCH$_2$Ph + $[Et_3NH]Cl$

108	190.45	101	262	137.45

MW Reactants: $108 + 190.45 + 101 = 399.45$
MW Products: $262 + 137.45 = 399.45$

Reactant	MW (g/mol)	Mass (g)	Obs. moles	Stoich. moles	Moles reacted	Stoich. moles unreacted
Benzyl alcohol	108	10.81	0.1001	0.1001	0.0901	0.0990
pTsCl	190.45	21.9	0.1150	0.1001	0.0901	0.0990
Et3N	101	15	0.1485	0.1001	0.0901	0.0990

The limiting reagent is benzyl alcohol.

The by-product of the reaction is triethylammonium chloride.

The observed number of moles of sulfonate ester product is 23.6/262 = 0.0901 moles.

Since the balanced chemical equation shows that 1 mole of by-product is produced for every mole of product, then the observed mass of triethylammonium chloride is 0.0901*137.45 = 12.38 g.

Assuming that no other side reactions have occurred, the number of moles and masses of unreacted stoichiometric reagents is as follows.

Reactant	MW (g/mol)	Mass (g)	Input moles	Stoich. moles	Moles reacted	Stoich. moles unreacted	Stoich. mass unreacted (g)
Benzyl alcohol	108	10.81	0.1001	0.1001	0.0901	0.0099	1.09
pTsCl	190.45	21.9	0.1150	0.1001	0.0901	0.0099	1.91
Et3N	101	15	0.1485	0.1001	0.0901	0.0099	1.01

Therefore, the total mass of stoichiometric unreacted reagents is 1.09 + 1.91 + 1.01 = 4.01 g.

Therefore, the kernel mass of waste for this reaction is 12.38 + 4.01 = 16.39 g.

Mass of Waste

Mass of waste refers to the difference between the mass of all input materials and the mass of desired product collected in a single chemical reaction. For a synthesis plan it refers to the sum of all scaled masses of input materials for the entire plan minus the mass of the final desired target product collected at the end of the synthesis.

Example 6.2

From the reaction shown in Example 6.1, determine the mass of waste produced. Itemize the masses of all waste constituents and verify that the sum of masses of inputs balances with the sum of masses of outputs.

Solution

The masses of input materials are given below.

Input material	Mass (g)
Benzyl alcohol	10.81
p-TsCl	21.9
Toluene	500
Et3N	15
TOTAL	547.71

The total mass of input materials is 547.71 g.

The mass of sulfonate ester product collected is 23.6 g.

Therefore, the mass of waste produced in this reaction is 547.71 − 23.6 = 524.11 g.

The sources of waste in this reaction are the reaction solvent, unreacted starting materials, and by-product. From Example 6.2 we determined that the mass of by-product is 12.38 g. The mass of unreacted starting materials is found as follows.

Reactant	MW (g/mol)	Mass (g)	Input moles	Moles reacted	Moles unreacted	Mass unreacted (g)
Benzyl alcohol	108	10.81	0.1001	0.0901	0.01	1.08
pTsCl	190.45	21.9	0.1150	0.0901	0.0249	4.74
Et3N	101	15	0.1485	0.0901	0.0584	5.90

The total mass of unreacted reagents is 1.08 + 4.74 + 5.90 = 11.72 g.

If we add the mass of toluene solvent, we obtain a total mass of waste of 12.38 + 11.72 + 500 = 524.10 g. This value agrees with the mass of waste determined by taking the difference of masses of total input and product given above.

6.2 METRICS ASSOCIATED WITH WASTE PRODUCTION

6.2.1 Terms, Definitions, and Examples

E-aux

The contribution to the total *E*-factor of a reaction from auxiliary materials is given by Equation (6.4).

$$E_{\text{aux}} = \frac{m_{\text{solvent}} + m_{\text{catalyst}} + m_{\text{workup}} + m_{\text{purification}}}{m_{\text{product}}} \tag{6.4}$$

where the m parameters refer to the masses of reaction solvent, catalyst, work-up materials, purification materials, and target product.

Example 6.3

Determine *E*-aux for the reaction given in Example 6.1.

Solution

The only auxiliary input material used in the reaction is toluene reaction solvent. Hence, E-aux = 500/23.6 = 21.19

E-excess

The contribution to the total E-factor of a reaction from excess reagents is given by Equation (6.5).

$$E_{\text{excess}} = \frac{m_{\text{excess reagents}}}{m_{\text{product}}} \tag{6.5}$$

where the m parameters refer to the masses of excess reagents and target product.

Example 6.4

Determine E-excess for the reaction given in Example 6.1.

Solution

Reactant	MW (g/mol)	Mass (g)	Input moles	Moles excess	Mass excess reagents (g)
Benzyl alcohol	108	10.81	0.1001	0	0
pTsCl	190.45	21.9	0.1150	0.0149	2.84
Et3N	101	15	0.1485	0.0484	4.89

Total mass of excess reagents is 0 + 2.84 + 4.89 = 7.73 g.
Hence, E-excess = 7.73/23.6 = 0.33.

E-factor [6]

The E-factor metric measures the mass ratio of total waste produced in a chemical reaction to the mass of target product collected and is given by Equation (6.6). In an ideal reaction producing no waste of any kind $E=0$. The total E-factor takes into account waste from all sources including reaction by-products and side products, reaction solvent, catalysts and other additives, work-up materials, and purification materials. Process chemists sometimes, but not always, include any solvents used for cleaning equipment. The concept was introduced in 1994 by Roger Sheldon. There have been corrupted definitions used for this metric in the literature which omit aqueous waste sources, particularly from work-up procedures, in order to artificially reduce its magnitude. The strict definition includes waste from *all* sources.

$$E_{\text{factor}} = \frac{m_{\text{waste}}}{m_{\text{product}}} \tag{6.6}$$

Example 6.5

Determine the E-factor for the reaction given in Example 6.1.

Solution

From Example 6.2 we found that the total mass of waste produced is 524.11 g. Hence, *E*-factor is 524.11/23.6 = 22.21.

E-flash chrom. [13]

The contribution to the total *E*-factor of a reaction from materials used in flash chromatography such as mass of silica gel and mass of eluent solvents is given by Equation (6.7).

$$E_{\text{flash chrom.}} = \frac{m_{\text{silica gel}} + m_{\text{eluent}}}{m_{\text{product}}} \tag{6.7}$$

E-flash chrom. may be estimated from the following expression given by Equation (6.8)

$$E_{\text{flash chrom.}} = A\left(\frac{m_{\text{crude product}}}{m_{\text{pure product}}}\right)\left[1+\rho_{\text{eluent}}\left(\frac{\varepsilon_{\text{silica}}}{\rho_{\text{silica}}}\right)\left[\frac{0.64}{R_f}\left(1+\frac{2}{\sqrt{N}}\right)+1\right]\right] \tag{6.8}$$

where:

$m_{\text{crude product}}$	is the mass of product before chromatography
$m_{\text{pure product}}$	is the mass of product after chromatography
A	is a degree of difficulty separation factor (facile separations are assigned $A = 10$; difficult separations are assigned $A = 152$)
ρ_{eluent}	is the density of the eluent solvent
$\varepsilon_{\text{silica}}$	is the porosity of silica (0.9)
ρ_{silica}	is the density of silica (0.5 g/cm^3)
R_f	is the retention factor for the product
N	is the number of theoretical plates in the column

The number of theoretical plates, *N*, is estimated using the relations given by Equations (6.9a) and (6.9b).

$$N = \begin{cases} 33.64\left(\dfrac{m_{\text{crude product}}}{m_{\text{pure product}}}\right)^{0.44}, & \text{facile separation} \\ 51.70\left(\dfrac{m_{\text{crude product}}}{m_{\text{pure product}}}\right)^{0.44}, & \text{difficult separation} \end{cases} \tag{6.9a,b}$$

If an experimental procedure provides only the masses of crude and pure products as well as the identity of the eluent solvent, then minimum and maximum estimates of *E*-flash chrom. can be made according to Equations (6.10a) and (6.10b).

$$E\text{-flash chrom.(min)} = 10\left(\frac{m_{\text{crude product}}}{m_{\text{pure product}}}\right)\left[1+6.20\rho_{\text{eluent}}\right]$$

$$E\text{-flash chrom.(max)} = 152\left(\frac{m_{\text{crude product}}}{m_{\text{pure product}}}\right)\left[1+6.20\rho_{\text{eluent}}\right] \tag{6.10a,b}$$

where $R_f = 0.35$ and $N = 35$ were used, and the weighted average density of eluent is given by Equation (6.11)

$$\rho_{\text{eluent}} = \frac{\sum_j \rho_j V_{f,j}}{\sum_j V_{f,j}} \tag{6.11}$$

where:

ρ_j is the density of the jth eluent solvent
$V_{f,j}$ is the volume fraction of the jth eluent solvent

The *flash-chromatography-mass intensity-template.xls* spreadsheet was created to automate the calculations described by Equations (6.7) through (6.11) inclusive. The spreadsheet determines *E*-flash chrom. under four conditions: (a) minimum and maximum exact calculations based on known quantities of all parameters; and (b) minimum and maximum approximate calculations based on default valucs described above. The minimum and maximum values in each case refer to facile and difficult separations.

Example 6.6

An experimental procedure calls for purification of 10 g of crude product using EtOAc:hexane (1:10) by volume. Two liters of eluent are required to obtain 8 g of pure product. Use the *flash-chromatography-mass intensity-template.xls* spreadsheet to determine the minimum and maximum *E*-factor contribution for this operation.

Solution

Density of EtOAc = 0.901 g/mL
Density of hexane = 0.684 g/mL
Volume of EtOAc = (1/11)*2000 = 181.8 mL
Volume of hexane = (10/11)*2000 = 1818.2 mL
Density of combined eluent = (0.901*181.8 + 0.684*1818.2)/2000 = 0.704 g/mL
$R_f = 0.35$ (assumed)
N = 35 (assumed)
Minimum *E*-factor calculation:
Mass of silica gel = 10*mass crude product = 10*10 = 100 g
Void volume = (9/5)*mass of silica gel = (9/5)*100 = 180 cm^3
Mass of eluent
= (void volume)*(density of combined eluent)*((0.64/R_f)(1 + (2/sqrt(N))) + 1)
= 180*0.704*((0.64/0.35)(1 + (2/sqrt(35))) + 1)
= 436.8 g
E-factor = (mass of silica gel + mass of eluent)/mass pure product
= (100 + 436.8)/8 = 67.1
Maximum *E*-factor calculation:

Mass of silica gel = 152*mass crude product = 152*10 = 1520 g
Void volume = (9/5)*mass of silica gel = (9/5)*1520 = 2736 cm^3
Mass of eluent
= (void volume)*(density of combined eluent)*((0.64/R_f)(1 + (2/sqrt(N))) + 1)
= 2736*0.704*((0.64/0.35)(1 + (2/sqrt(35))) + 1)
= 6636.3 g
E-factor = (mass of silica gel + mass of eluent)/mass pure product
= (1520 + 6636.3)/8 = 1019.5

E-kernel

The contribution to the total *E*-factor of a reaction from reaction by-products and stoichiometric unreacted reagents is given by Equation (6.12)

$$E_{\text{kernel}} = \frac{m_{\text{byproducts}} + m_{\text{stoichiometric unreacted reagents}}}{m_{\text{product}}} \tag{6.12}$$

where the m parameters refer to the masses of by-products, stoichiometric unreacted reagents, and target product.

Example 6.7

Determine *E*-kernel from the reaction given in Example 6.1.

Solution

The kernel mass of waste is 16.39 g. Hence, *E*-kernel = 16.39/23.6 = 0.69.

E-solvent

The contribution to the total *E*-factor of a reaction from *reaction* solvents given by Equation (6.13)

$$E_{\text{solvent}} = \frac{m_{\text{reaction solvent}}}{m_{\text{product}}} \tag{6.13}$$

where the m parameters refer to the masses of reaction solvent and target product. Note that this component does not include solvents used in the work-up or purification operations. These other solvent components are included in the *E*-work-up, *E*-purif, and *E*-aux *E*-factors.

Example 6.8

Determine *E*-solvent from the reaction given in Example 6.1.

Solution

Since 500 g of toluene is used as reaction solvent, *E*-solvent = 500/23.6 = 21.19.

E-total

This metric corresponds to the global *E*-factor and is given by Equation (6.14)

$$E_{\text{total}} = \frac{m_{\text{byproducts}} + m_{\text{stoich. unreacted reagents}} + m_{\text{excess reagents}} + m_{\text{reaction solvent}} + m_{\text{catalyst}} + m_{\text{workup}} + m_{\text{purification}}}{m_{\text{product}}} \tag{6.14}$$

where the *m* parameters refer to the masses of by-products, stoichiometric unreacted reagents, reaction solvent, excess reagents, catalysts, work-up materials, purification materials, and target product. This is the explicit and unambiguous definition for *E*-factor pertaining to a single chemical reaction. It should be noted that *E*-total is the sum of the contributing *E*-factors shown in Equation (6.15) where each component *E*-factor is determined with respect to the mass of product of a given reaction.

$$E_{\text{total}} = E_{\text{kernel}} + E_{\text{excess}} + E_{\text{solvent}} + E_{\text{cat}} + E_{\text{workup}} + E_{\text{purification}} = E_{\text{kernel}} + E_{\text{excess}} + E_{\text{aux}} \tag{6.15}$$

Example 6.9

Verify the sum relationship for the component *E*-factors given in Equation (6.15) for the reaction given in Example 6.1.

Solution

From Example 6.4 we found that *E*-excess = 0.33.
From Example 6.7 we found that *E*-kernel = 0.69.
From Example 6.8 we found that *E*-solvent = 21.19.
Since there were no catalysts, work-up, or purification materials reported, then *E*-cat = 0, *E*-work-up = 0, and *E*-purification = 0.
Therefore, *E*-total = 0.69 + 0.33 + 21.19 + 0 + 0 + 0 = 22.21. This value agrees with the *E*-factor obtained in Example 6.5 using Equation (6.6).

E_{mw} [10]

For a balanced chemical equation this metric measures the molecular weight ratio of reaction by-products to target product. It is the corollary metric to atom economy. The connecting relationship between these molecular weight-based metrics is given by Equation (6.16).

$$AE = \frac{1}{1 + E_{mw}} \tag{6.16}$$

An E_{mw} equal to 0 corresponds to a chemical reaction producing no by-products with an $AE = 1$, or 100 %. An example is the [4 + 2] Diels–Alder cycloaddition of a dienophile and a diene.

Example 6.10

Determine E_{mw} for the reaction given in Example 6.1.

Solution

The molecular weight of triethylammonium chloride by-product is 137.45 g/mol.
The molecular weight of sulfonate ester product is 262 g/mol.
Therefore, $E_{mw} = 137.45/262 = 0.525$.
The corresponding atom economy according to Equation (6.16) is $1/(1 + 0.525) = 0.656$, or 65.6%.

Global *E*-factor (Overall *E*-factor)

This metric synonymous to overall *E*-factor is applicable to single balanced chemical reactions and to synthesis plans (linear or convergent) composed of sets of appropriately scaled balanced chemical equations. The generalized mathematical representation of this metric is given by Equation (6.17).

$$gE = E_{\text{overall}} = \frac{\sum_j m_{\text{waste},j}}{m_{\text{product}}} = \frac{\sum_j m_{\text{input},j} - m_{\text{product}}}{m_{\text{product}}} = gPMI - 1 \qquad (6.17)$$

where the *m* terms refer to the masses of final target product and all input materials including reagents, catalysts, reaction solvents, work-up materials, and purification materials. Masses of intermediates formed as products in step 1 to step *N*–1 are not included since they are made and consumed in consecutive steps in an appropriately scaled synthesis plan. The parameter *gPMI* is the global process mass intensity.

Kernel *E*-factor

An *E*-factor based only on the mass of waste due to reaction by-products and stoichiometric unreacted reagents in a given chemical reaction or a synthesis plan. For a single chemical reaction, it is given by Equation (6.18) which is identical to Equation (6.12)

$$E_{\text{kernel}} = \frac{m_{\text{reagents}} - m_{\text{excess reagents}} - m_{\text{product}}}{m_{\text{product}}} = \frac{m_{\text{stoichiometric reagents}} - m_{\text{product}}}{m_{\text{product}}}$$
$$= \frac{m_{\text{byproducts}} + m_{\text{stoichiometric unreacted reagents}}}{m_{\text{product}}} \qquad (6.18)$$

where the *m* terms refer to the respective masses specified.

Example 6.11

Verify numerically that the numerators shown in Equation (6.18) are identical using the reaction described in Example 6.1.

Solution

Total mass of reagents = 10.81 + 21.9 + 15 = 47.71 g
Total mass of excess reagents = 2.84 + 4.89 = 7.73 g
Mass of product = 23.6 g

First numerator = (47.71 − 7.73) − 23.6 = 16.38 g
Total mass of stoichiometric reagents = 10.81 + 19.06 + 10.11 = 39.98 g
Second numerator = 39.98 − 23.6 = 16.38 g
Total mass of by-products = 12.38 g
Total mass of stoichiometric unreacted reagents = 1.08 + 1.91 + 1.01 = 4.00 g
Third numerator = 12.38 + 4.00 = 16.38 g

Solvent Intensity

Solvent intensity is the ratio of total mass of solvents used (including water) to mass of product made. This metric is identical to *E*-solvent.

6.3 METRICS ASSOCIATED WITH INPUT CONSUMPTION

6.3.1 Terms, Definitions, and Examples

Actual Atom Economy [14]

The multiplicative product of the atom economy and the reaction yield. Each of these quantities is expressed as a fraction between 0 and 1 so that the resulting actual atom economy is also a fraction between 0 and 1. Alternatively, each of these quantities can be expressed as a percentage.

Example 6.12

Determine the actual atom economy for the reaction given in Example 6.1.

Solution

From Example 6.9 we determined that the AE = 0.656.

We found that the limiting reagent is benzyl alcohol in which 0.1001 moles were used.

The number of moles of sulfonate ester product is 0.0901 moles. Since the mole ratio of limiting reagent and product is 1:1, the reaction yield is 0.0901/0.1001 = 0.90, or 90%.

The actual atom economy is therefore 0.656*0.90 = 0.590, or 59.0%.

Atom Economy (AE) [5, 15]

In a balanced chemical reaction, atom economy is the molecular weight ratio of the target product to the sum of all reactants as given in Equation (6.19). Essentially, it measures the molecular weight fraction of reactants that end up in the desired product. The concept was introduced in 1991 by Barry Trost at Stanford University.

$$AE = \frac{MW_{product}}{\sum_j v_j (MW)_{reactant,j}} \tag{6.19}$$

where:

V_j is the stoichiometric coefficient of the *j*th reactant
MW parameters refer to the respective molecular weights

Example 6.13

Determine the atom economy of the reaction in Example 6.1 according to the definition given in Equation (6.16).

Solution

Based on the molecular weights of all chemical species appearing the balanced chemical equation, we have

AE = 262/[(1)*108 + (1)*190.45 + (1)*101] = 0.656, or 65.6%.

Experimental Atom Economy [16–19]

Experimental atom economy is identical to actual atom economy and kernel reaction mass efficiency which is the multiplicative product of atom economy and reaction yield as shown in Equation (6.20).

$$AE_{\text{experimental}} = RME_{\text{kernel}} = (AE)(\varepsilon) \tag{6.20}$$

Kernel Process Mass Intensity

A process mass intensity based only on the stoichiometric masses of input reactants in a given chemical reaction or a synthesis plan.

Kernel Reaction Mass Efficiency [10]

Kernel reaction mass efficiency is a reaction mass efficiency based only on the stoichiometric masses of input reactants in a given chemical reaction or a synthesis plan.

The mathematical expressions for a single reaction and a linear synthesis plan are given in Equations (6.21) and (6.22), respectively.

single reaction

$$RME_{\text{kernel}} = \frac{m_{\text{product}}}{\sum_j m_{\text{stoichiometric reactant},j}} = (AE)(\varepsilon) \tag{6.21}$$

linear synthesis plan

$$\begin{aligned} RME_{\text{kernel}} &= \frac{m_{\text{product}}}{\sum_j m_{\text{reactant},j}} \\ &= \frac{MW_{product}}{\dfrac{\sum_{j,stepN} MW_{\text{reactant},j}}{\varepsilon_N} + \dfrac{\sum_{j,stepN-1} MW_{\text{reactant},j}}{\varepsilon_N \varepsilon_{N-1}} + \ldots + \dfrac{\sum_{j,step1} MW_{\text{reactant},j}}{\varepsilon_N \varepsilon_{N-1} \ldots \varepsilon_1}} \end{aligned} \tag{6.22}$$

Note that Equation (6.18) is identical to the definition of *actual atom economy.*

Example 6.14

Determine the kernel reaction mass efficiency for the reaction given in Example 6.1 using the definition shown in Equation (6.18).

Solution

The mass of sulfonate ester product collected is 23.6 g.
The sum of stoichiometric masses of benzyl alcohol, *p*-toluenesulfonyl chloride, and triethylamine reactants is 10.81 + 19.06 + 10.11 = 39.98. Hence, the RME_{kernel} = 23.6/39.98 = 0.590, or 59.0%. Note that this value is identical to that found for the actual or experimental atom economy found by multiplying AE by the reaction yield as shown in Example 6.11.

Lavoisier Number [11]

The Lavoisier number is defined as the reciprocal of atom economy and is related to the *E*-factor based on molecular weight according to Equation (6.23).

$$\mathrm{LN} = \frac{1}{\mathrm{AE}} = E_{mw} + 1. \tag{6.23}$$

This expression is analogous to the connecting relationship between process mass intensity and reaction mass efficiency based on mass given by Equation (6.24).

$$PMI = \frac{1}{\mathrm{RME}} = E + 1. \tag{6.24}$$

An ideally designed reaction produces no by-products at the kernel level and so its atom economy and Lavoisier numbers are both equal to 1, or 100%. Lavoisier numbers for reactions suggest a convenient scale that measures how far away from ideality they are at the kernel level. For example, a reaction producing a target molecule with a Lavoisier number of 2 (i.e., with an atom economy equal to 0.5 or 50%) means that it is two times further away from ideality than another reaction producing the same target molecule with no by-products. Reactions having high Lavoisier numbers much larger than 1 produce significant by-products and have low atom economies. Figure 6.3 shows a series of histograms of Lavoisier numbers for various named organic reactions categorized according to addition (carbon-carbon forming), addition (non-carbon-carbon forming), condensation, multi-component, elimination, rearrangement, substitution, sequence, reduction, and oxidation type reactions. It is clear that from the database of known named organic reactions available that oxidation and reduction reactions have the worst Lavoisier number profiles corresponding to poor atom economies compared with the other reaction categories. This is directly associated with significant by-product formation in these types of reactions. On the other hand, condensation, multi-component, and rearrangement reactions have narrowly distributed Lavoisier numbers close to one, indicating their very high atom economies and low by-product formation.

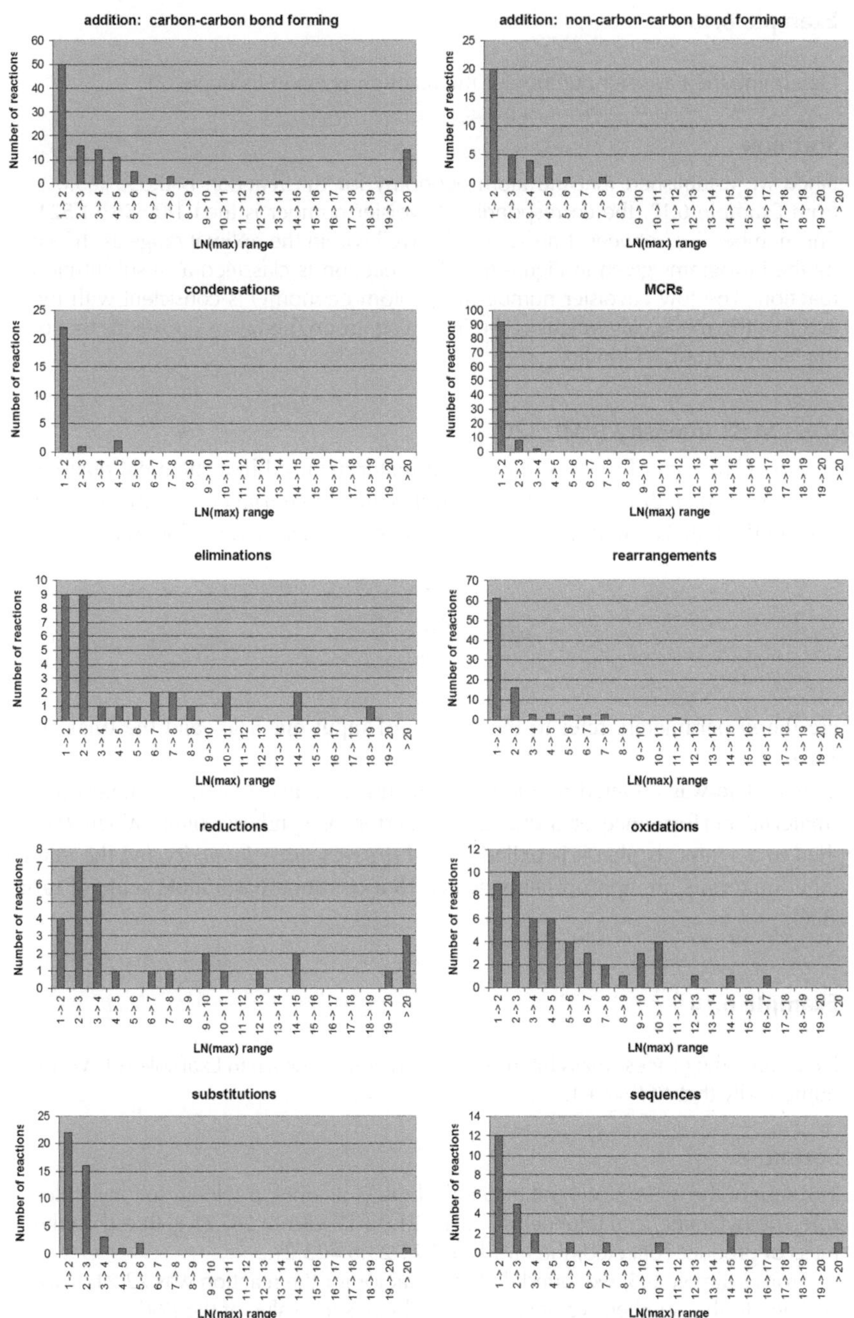

FIGURE 6.3 Histograms showing distribution of Lavoisier numbers for major categories of named organic reactions.

Example 6.15

Determine the Lavoisier number for the reaction given in Example 6.1.

Solution

Since we established that the atom economy for the reaction is 0.656 (65.6%) from Example 6.12, the corresponding Lavoisier number is then 1/0.656 = 1.524. This number lies between 1 and 2 and fits well within the optimal range as shown by the histograms given in Figure 6.3. The reaction is classified as a substitution reaction. The low Lavoisier number (high atom economy) is consistent with the fact that the molecular weight of the incoming group (benzyloxy) exceeds that of the leaving group (chloride).

Process Mass Intensity (*PMI*) [20]

The ratio of the sum of masses of all input materials used in a chemical reaction or synthesis plan including reactants and all auxiliary materials to the mass of target product collected. The mathematical expression is given by Equation (6.25).

$$PMI = \frac{\sum_j m_{\text{input},j}}{m_{\text{product}}} \tag{6.25}$$

where the m parameters refer to the masses of input materials and the final target product.

This metric was adopted by the pharmaceutical industry as the standard metric for material performance of a chemical reaction or synthesis plan. When *PMI* is applied to a synthesis plan it is called *global process mass intensity* and the masses of input materials are appropriately scaled to a common basis mole scale of target product.

Example 6.16

Determine the process mass intensity of the reaction shown in Example 6.1. Verify numerically that $PMI = E + 1$.

Solution

The sum of the masses of input materials (benzyl alcohol, *p*-toluenesulfonyl chloride, triethylamine, and toluene) is 10.81 + 21.9 + 15 + 500 = 547.71 g. Based on the mass of product collected, the *PMI* is 547.71/23.6 = 23.21.

From Example 6.5, we found that the *E*-factor for the reaction was 22.21. If we add one to this number, we obtain 23.21 which is the value of the *PMI*.

Raw Material Use

Raw material use is the ratio of mass of total raw materials used to mass of product made. This quantity is identical to process mass intensity.

Reaction Mass Efficiency (Andraos Definition) [10]

For a single chemical reaction or a synthesis plan, the reaction mass efficiency is given by Equation (6.26).

$$\mathrm{RME}_{\mathrm{Andraos}} = \frac{m_{\mathrm{product}}}{\sum_j m_{\mathrm{input},j}} = \frac{1}{PMI} = \frac{1}{E+1} \tag{6.26}$$

where the m parameters refer to the masses of target product and all input materials (reagents and auxiliaries). RME is the reciprocal of the process mass intensity. For a synthesis plan, the masses of all input materials are appropriately scaled to a common basis scale for the final target product in the plan. For a single reaction, RME can be decomposed into its contributing metrics as shown in Equation (6.27).

$$\mathrm{RME}_{\mathrm{Andraos}} = (\mathrm{AE})(\varepsilon)\left(\frac{1}{\mathrm{SF}}\right)(\mathrm{MRP}) \tag{6.27}$$

where:

- AE is atom economy
- ε is reaction yield
- SF is stoichiometric factor
- MRP is the material recovery parameter

Example 6.17

Determine the Andraos reaction mass efficiency for the reaction given in Example 6.1 according to Equation (6.26).

Solution

From Example 6.15, the total mass of input materials is 547.71 g and the mass of product collected is 23.6 g. Hence, RME(Andraos) = 23.6/547.71 = 0.043, or 4.3%.

Reaction Mass Efficiency (Curzons Definition) [7–9]

For a single reaction or synthesis plan, the Curzons reaction mass efficiency is given by Equation (6.28).

$$\mathrm{RME}_{\mathrm{Curzons}} = \frac{m_{\mathrm{product}}}{\sum_j m_{\mathrm{reagent},j}} \tag{6.28}$$

where the m parameters refer to the masses of target product and all reagents. Masses of all other auxiliaries are excluded. For a synthesis plan, the masses of all reagents are appropriately scaled to a common basis scale for the final target product in the plan. For a single reaction, RME can be decomposed into its contributing metrics as shown in Equation (6.29)

$$\mathrm{RME}_{\mathrm{Curzons}} = (\mathrm{AE})(\varepsilon)\left(\frac{1}{\mathrm{SF}}\right) \tag{6.29}$$

where:

AE is atom economy
ε is reaction yield
SF is stoichiometric factor

Example 6.18

Determine the Curzons RME for the reaction given in Example 6.1 according to Equation (6.28).

Solution

Mass of sulfonate ester product = 23.6 g.

Sum of masses of reagents (benzyl alcohol, p-toluenesulfonyl chloride, and triethylamine) is 10.81 + 21.9 + 15 = 47.71 g.

Therefore, RME(Curzons) = 23.6/47.71 = 0.495, or 49.5%.

Reaction Yield

The reaction yield is defined as shown in Equation (6.30) for a balanced chemical equation where ν refers to the respective stoichiometric coefficients.

$$Y = \varepsilon = \left(\frac{\text{moles}_{\text{product}}}{\text{moles}_{\text{limiting reagent}}}\right)\left(\frac{\nu_{\text{limiting reagent}}}{\nu_{\text{product}}}\right) \tag{6.30}$$

It is often expressed as a percentage, however for the purposes of calculating other material efficiency metrics for synthesis plans it is more convenient to use the conventional fractional decimal notation. Reaction yield can be regarded as a measure of the probability that a given reaction will occur between a set of reagents under a given set of conditions that leads to a desired product. A reaction with a high yield is considered energetically favorable and hence the probability that it will produce its reaction products is also high. The converse is true for reactions with low reaction yields. The factors that govern the probability of reaction producing a desired product include the following physical parameters: reaction time, reaction temperature, reaction pressure, choice of solvent, choice of catalyst, choice of ligand, relative concentration of reagents, mixing characteristics, and method of activation (conventional heating, microwave heating, continuous flow, etc.); and the following intrinsic molecular parameters pertaining to the respective nucleophilic and electrophilic structural moieties of the reagents: electronic effects (bond forming and bond breaking energies, σ-bond inductive, π-bond mesomeric, acidity-basicity) and steric effects (bulkiness or size, conformations).

Example 6.19

Determine the reaction yield for the reaction shown in Example 6.1 according to Equation (6.30).

Solution

Number of moles of product = 0.0901 moles.
Number of moles of limiting reagent = 0.1001 moles.
Stoichiometric coefficient of limiting reagent = 1
Stoichiometric coefficient of product = 1
Therefore, reaction yield is (0.0901/0.1001)*(1/1) = 0.900, or 90.0%.

6.4 METRICS ASSOCIATED WITH PRODUCT FORMATION

6.4.1 Terms, Definitions, and Examples

Biocatalysis Yield [21]

The biocatalysis yield is given by Equation (6.31).

$$\text{Biocatalyst yield} = \frac{\text{Mass product}}{\text{Mass biocatalyst}}. \tag{6.31}$$

The units are g/g.

Conversion

This metric, often expressed as a percentage, is the fractional amount of starting material that gets transformed or converted to all products in a chemical reaction given by $(m_{\text{initial}} - m_{\text{final}})/m_{\text{initial}}$ where the m terms refer to the masses of starting material at the beginning and end of a reaction. The starting material is taken as the limiting reagent in the reaction.

Example 6.20

Determine the % conversion of each reagent used in the reaction given in Example 6.1.

Solution

Since 0.0901 moles of sulfonate ester product is produced in the reaction, we can conclude that 0.0901 moles of benzyl alcohol reacted, 0.0901 moles of *p*-toluenesulfonyl chloride reacted, and 0.0901 moles of triethylamine reacted. Initially, the reaction began with 0.1001 moles of benzyl alcohol, 0.1150 moles of *p*-toluenesulfonyl chloride, and 0.1485 moles of triethylamine. Therefore, we have the following % conversions:

1. benzyl alcohol, 0.0901/0.1001 = 0.900, 90.0%
2. *p*-toluenesulfonyl chloride, 0.0901/0.1150 = 0.783, or 78.3%
3. triethylamine, 0.0901/0.1485 = 0.607, or 60.7%.

Note that for the limiting reagent (benzyl alcohol), the % conversion is identical to the % reaction yield.

Yield Based on Recovered Starting Materials

A calculation of reaction yield that includes both the intended target product and unreacted starting material in a chemical reaction as the desired products; this is usually reported in papers when the true reaction yield to the intended target compound is lower than 50%.

Example 6.21

Suppose for the reaction shown in Example 6.1 that any unreacted benzyl alcohol was recovered. Determine the yield of product based on recovered starting materials.

Solution

Number of moles of sulfonate ester = 0.0901 moles.
Number of moles of starting benzyl alcohol = 0.1001 moles.
Number of moles of unreacted benzyl alcohol that is recovered = 0.1001 − 0.0901 = 0.01 moles.
MW sulfonate ester = 262 g/mol
MW benzyl alcohol = 108 g/mol
Mass of sulfonate ester = 262*0.0901 = 23.6 g
Mass of recovered benzyl alcohol = 108*0.01 = 1.08 g
Theoretical number of moles of sulfonate ester expected = 0.1001 moles.
Theoretical mass of sulfonate ester expected = 262*0.1001 = 26.23 g
Yield based on recovered starting material = (23.6 + 1.08)/26.23 = 0.941, or 94.1%.
(Cf. actual reaction yield of product of 90.0%)

Example 6.22

Repeat the determination of yield based on recovered starting material if unreacted *p*-toluenesulfonyl chloride is also recovered in addition to benzyl alcohol.

Solution

Number of moles of starting *p*-toluenesulfonyl chloride = 0.1150 moles.
Number of moles of unreacted *p*-toluenesulfonyl chloride that is recovered = 0.1150 − 0.0901 = 0.0249 moles.
MW *p*-toluenesulfonyl chloride = 190.45 g/mol
Mass of recovered *p*-toluenesulfonyl chloride = 190.45*0.0249 = 4.74 g.
Yield based on recovered starting material = (23.6 + 1.08 + 4.74)/26.23 = 1.122, or 112.2%.

Example 6.23

Themes: percent conversion, percent yield, percent mole selectivity

Part 1

For the generalized unimolecular reaction shown that produces two isomeric products, determine the percent conversion of starting material, percent mole selectivity for each product, and percent yield for each product. The reaction begins with x moles of A and ends up with y moles of P_1 and z moles of P_2. What is the connecting relationship between these three metrics?

$$\underset{x\ mol}{A} \longrightarrow \underset{y\ mol}{P_1} + \underset{z\ mol}{P_2}$$

Part 2

An example of such a reaction is the Fries rearrangement that yields *ortho*- and *para*-substituted products [22].

$AlCl_3$, 150°C

A: 35 %

B: 50 %

Interpreting the percent values as reaction yields, determine the percent conversion of starting material and the percent mole selectivities to each product. What percent of the starting material remains unreacted?

Part 3

For the generalized bimolecular reaction shown that produces two isomeric products P_1 and P_2, determine the percent conversion of starting materials A and B, percent mole selectivity for each product, and percent yield for each product. The reaction begins with a moles of A and b moles of B and produces p_1 moles of P_1 and p_2 moles of P_2. What is the connecting relationship between these three metrics?

$$\underset{a\ mol}{A} + \underset{b\ mol}{B} \longrightarrow \underset{p_1\ mol}{P_1} + \underset{p_2\ mol}{P_2} + \underset{q_1\ mol}{Q_1} + \underset{q_2\ mol}{Q_2}$$

Assume that A is the limiting reagent, Q_1 is the by-product formed as a consequence of producing P_1, and Q_2 is the by-product formed as a consequence of producing P_2.

Part 4

An example of a bimolecular reaction producing two structural isomers is shown below [23].

$$CH_3CH_2—C(=O)—CH{=}CHCl \xrightarrow[-\,H_2O\ \ -\,HCl]{NH_2OH} \mathbf{A} + \mathbf{B}$$

A 60 % **B** 40 %

Interpreting the percent values as a product ratio, determine the percent conversion of each starting material and the percent yields and percent mole selectivities to each isoxazole product. What percent of each starting material remains unreacted? Assume that hydroxylamine is used in 10% excess.

Solution

Part 1

Percent conversion:

Number of moles of A converted to products P1 and P2 = y + z
Number of moles of A remaining = x − (y + z)
Fraction of A that is converted to products P1 and P2 = $(y+z)/x$
Percent conversion = $100*(y+z)/x$

Percent yield:

Percent yield of product P1 = 100*(y/x)
Percent yield of product P2 = 100*(z/x)

Percent selectivity:

Products P1 and P2 are produced in the ratio y to z.
Percent mole selectivity to product P1 = $100*y/(y+z)$
Percent mole selectivity to product P2 = $100*z/(y+z)$

Connecting relationship:

conversion fraction X mole selectivity fraction = yield fraction

For product P1 we have $\left(\frac{y+z}{x}\right)\left(\frac{y}{y+z}\right)=\frac{y}{x}$.

For product P2 we have $\left(\frac{y+z}{x}\right)\left(\frac{z}{y+z}\right)=\frac{z}{x}$.

Part 2

$$\underset{x\ mol}{\text{phenyl propanoate}} \xrightarrow[150^{\circ}C]{AlCl_3} \underset{y\ mol}{\mathbf{A}} + \underset{z\ mol}{\mathbf{B}}$$

yield of *ortho*-product $= y/x = 0.35$
yield of *para*-product $= z/x = 0.5$
mole selectivity for *ortho*-product =

$$\frac{y}{y+z} = \frac{1}{1+\frac{z}{y}} = \frac{1}{1+\left(\frac{z}{x}\right)\left(\frac{x}{y}\right)} = \frac{1}{1+(0.5)\left(\frac{1}{0.35}\right)} = \frac{1}{1+\frac{10}{7}} = \frac{7}{17} = 0.412\text{, or } 41.2\%$$

mole selectivity for *para*-product $= 1-(7/17) = 10/17 = 0.588$, or 58.8%

Since $\left(\frac{y+z}{x}\right)\left(\frac{y}{y+z}\right) = \frac{y}{x}$, we have

$$\left(\frac{y+z}{x}\right)\left(\frac{7}{17}\right) = 0.35$$

$$\left(\frac{y+z}{x}\right) = \left(\frac{7}{20}\right)\left(\frac{17}{7}\right) = \frac{17}{20} = 0.85$$

Therefore, the percent conversion of starting material to both products is 85%.

Similarly, from $\left(\frac{y+z}{x}\right)\left(\frac{z}{y+z}\right) = \frac{z}{x}$, we have

$$\left(\frac{y+z}{x}\right)\left(\frac{10}{17}\right) = 0.5$$

$$\left(\frac{y+z}{x}\right) = \left(\frac{1}{2}\right)\left(\frac{17}{10}\right) = \frac{17}{20} = 0.85$$

Therefore $100 - 85 = 15\%$ of the starting material remains unreacted.

Part 3

Since P_1 and Q_1 are mechanistically connected, $p_1 = q_1$.
Since P_2 and Q_2 are mechanistically connected, $p_2 = q_2$.

Therefore, the revised scheme is as shown below.

$$\begin{array}{ccccccccccc} A & + & B & \longrightarrow & P_1 & + & P_2 & + & Q_1 & + & Q_2 \\ a\ mol & & b\ mol & & p_1\ mol & & p_2\ mol & & p_1\ mol & & p_2\ mol \end{array}$$

Since A is the limiting reagent the yields of products P_1 and P_2 are pegged against A.

Yield of $P_1 = p_1/a$

Yield of $P_2 = p_2/a$

Mole selectivity for $P_1 = p_1/(p_1 + p_2)$

Mole selectivity for $P_2 = p_2/(p_1 + p_2)$

Since the chemical structures of P_1 and P_2 contain part of the structure of A, we interpret that a portion of A is converted to P_1 and P_2. We also assume that the structures of by-products Q_1 and Q_2 also contain a portion of A.

Number of moles of A converted to products P_1 and $P_2 = p_1 + p_2$

Number of moles of A remaining $= a-(p_1 + p_2)$

Fraction of A that is converted to products P_1 and $P_2 = (p_1 + p_2)/a$

Therefore, for product P_1: $\left(\dfrac{p_1 + p_2}{a}\right)\left(\dfrac{p_1}{p_1 + p_2}\right) = \dfrac{p_1}{a}$.

Therefore, for product P_2: $\left(\dfrac{p_1 + p_2}{a}\right)\left(\dfrac{p_2}{p_1 + p_2}\right) = \dfrac{p_2}{a}$.

Hence, the same connecting relationship is true as found in Part 2:

conversion fraction × mole selectivity fraction = yield fraction

As before we assume that the chemical structures of P_1, P_2, Q_1, and Q_2 contain a part of structure B.

Number of moles of B converted to products P_1 and $P_2 = p_1 + p_2$

Number of moles of B remaining $= b - (p_1 + p_2)$

Fraction of B that is converted to products P_1 and $P_2 = (p_1 + p_2)/b$

If we peg the reaction yields of P_1 and P_2 against the excess reagent B, we obtain

Yield of $P_1 = p_1/b$

Yield of $P_2 = p_2/b$

Note that since $b > a$, these yield values are smaller than those determined with respect to limiting reagent A.

The mole selectivity fractions for products P_1 and P_2 remain the same as before.

Therefore, for product P_1: $\left(\dfrac{p_1 + p_2}{b}\right)\left(\dfrac{p_1}{p_1 + p_2}\right) = \dfrac{p_1}{b}$.

Therefore, for product P_2: $\left(\dfrac{p_1 + p_2}{b}\right)\left(\dfrac{p_2}{p_1 + p_2}\right) = \dfrac{p_2}{b}$.

Again, the connecting relationship **conversion fraction × mole selectivity fraction = yield fraction** holds true.

Part 4

$$CH_3CH_2-C(=O)-CH=CHCl + NH_2OH \rightarrow$$

a mol   *1.1 a mol*

A [3-ethylisoxazole] $+ H_2O + HCl$

p_1 *mol* p_1 *mol* p_1 *mol*

B [5-ethylisoxazole] $+ H_2O + HCl$

p_2 *mol* p_2 *mol* p_2 *mol*

Product ratio, A to B $= p_1/p_2 = 60/40 = 1.5$
Yield of A with respect to 1-chloro-pent-1-en-3-one $= p_1/a$
Yield of B with respect to 1-chloro-pent-1-en-3-one $= p_2/a$
Yield of A with respect to hydroxylamine $= p_1/b$
Yield of B with respect to hydroxylamine $= p_2/b$
Mole selectivity for A $= p_1/(p_1+p_2) = 1.5/(1.5+1) = 0.6$
Mole selectivity for B $= p_2/(p_1+p_2) = 1/(1.5+1) = 0.4$
conversion of 1-chloro-pent-1-en-3-one to products A and B $= (p_1/a)/0.6 = (p_2/a)/0.4$
conversion of hydroxylamine to products A and B
$= (p_1/b)/0.6 = (p_2/b)/0.4$
$= (p_1/1.1\ a)/0.6 = (p_2/1.1\ a)/0.4$
fraction of unreacted 1-chloro-pent-1-en-3-one $= 1-(p_1/a)/0.6$, or $1-(p_2/a)/0.4$
fraction of unreacted hydroxylamine $= 1-(p_1/1.1\ a)/0.6 = 1-(p_2/1.1\ a)/0.4$

Note that if the yield of A with respect to 1-chloro-pent-1-en-3-one is 60%, then $p_1/a = 0.6$ and the conversion of 1-chloro-pent-1-en-3-one to both isoxazole products is $(p_1/a)/0.6 = 0.6/0.6 = 1$, or 100%. Similarly, the conversion of hydroxylamine to both isoxazole products is $(p_1/1.1\ a)/0.6 = (0.6/1.1)/0.6 = 1/1.1 = 0.909$, or 90.9%. The yield of product A with respect to hydroxylamine would be $p_1/b = p_1/1.1\ a = 0.6/1.1 = 0.545$, or 54.5%. Similarly, the yield of product B with respect to hydroxylamine would be $p_2/b = p_2/1.1\ a = 0.4/1.1 = 0.364$, or 36.4%. If 100% of 1-chloro-pent-1-en-3-one is converted to both isoxazole products, then 0% of it remains unreacted. If 90.9% hydroxylamine is converted to both isoxazole products, then 9.1% of it remains unreacted.

6.5 CONNECTIONS AND HIERARCHY

KERNEL

A descriptor that refers to the core material efficiency performance of a reaction based only on its by-product formation. All auxiliary materials such as excess reagents, catalysts, reaction solvents, work-up materials, and purification materials are ignored so that the focus is entirely on the inherent or intrinsic material performance of the reaction based on the design aspect of constructing the target molecule from a set of reactants. It represents the first stage of optimization with respect to material efficiency for any given chemical reaction.

Kernel Mass of Reagents

Kernel mass of reagents refers to the total mass of stoichiometric reagents in a given chemical reaction or a synthesis plan.

Mass Balance

The difference in mass between the total mass of input materials used in a chemical reaction or synthesis plan and the mass of target product collected.

Mass Index [24]

The mass index is identical to the reciprocal of the global reaction mass efficiency.

Mass Intensity

This term is identical in meaning to the process mass intensity.

Mass of Target Product

The mass of target product collected from a chemical reaction or a synthesis plan. This quantity is used in the determination of reaction yield, *E*-factor, reaction mass efficiency, and process mass intensity.

Mass of Waste

The difference in mass between the total mass of input materials used in a chemical reaction or synthesis plan and the mass of target product collected.

Material Recovery Parameter (MRP) [10]

This metric takes into account auxiliary materials used in the reaction and post-reaction phases (work-up and purification) such as reaction solvents, catalysts, solvents and washings for extractions, and solvents for chromatography and recrystallization. The mathematical expression for MRP for a single chemical reaction is given by Equation (6.32).

$$\mathrm{MRP} = \frac{1}{1+\dfrac{(\varepsilon)(\mathrm{AE})\left(m_{\mathrm{solvent}}+m_{\mathrm{catalyst}}+m_{\mathrm{workup}}+m_{\mathrm{purification}}\right)}{(\mathrm{SF})\,m_{\mathrm{product}}}} \tag{6.32}$$

$$= \frac{1}{1+\mathrm{RME}_{\mathrm{Curzons}}E_{\mathrm{aux}}}$$

The notion of recovery is used in the name since reaction and extraction solvents constitute the bulk of material used in a given chemical reaction and are consequently the first materials that are easily recoverable for re-use or recycling.

Example 6.24

Determine the MRP for the reaction given in Example 6.1 based on Equation (6.32).

Solution

We know the following quantities:
Mass of reaction solvent = 500 g
Mass of catalyst = 0 g
Mass of work-up materials = 0 g
Mass of purification materials = 0 g
Mass of product = 23.6 g
Reaction yield = 0.90 (90%)
AE = 0.656 (65.6%).
We determined in Example 6.18 that the RME(Curzons) = 0.495.
We determined in Example 6.3 that *E*-aux = 21.19.
Therefore, MRP = 1/(1 + 0.495*21.19) = 0.0870.

Mol%

This metric is usually used to describe quantitatively catalyst loading and is given by $(n_{catalyst}/n_{limiting\ reagent}) \times 100\%$, where n refers to the number of moles of each species.

Optimum Efficiency [25]

Optimum efficiency is defined as the ratio of global reaction mass efficiency and atom economy according to Equation (6.33).

$$OE = \frac{RME}{AE} = \frac{\varepsilon(AE)\left(\frac{1}{SF}\right)(MRP)}{AE} = \varepsilon\left(\frac{1}{SF}\right)(MRP) \qquad (6.33)$$

Example 6.25

Determine the optimum efficiency for the reaction given in Example 6.1 according to Equation (6.30).

Solution

From Example 6.17 we found that RME = 0.043.
From Example 6.13 we found that AE = 0.656.
Hence, *OE* = 0.043/0.656 = 0.0656, or 6.56%.

Pot Economy or One-Pot Economy

A strategy employed in organic synthesis that carries out more than one chemical transformation in a sequential fashion in the same reaction vessel without isolating the intermediate products generated after each transformation. In process chemistry this technique is called telescoping. The great benefit of this technique is that there is a great saving in solvent consumption since work-up and purification are carried only out after the last chemical transformation has taken place. A major provision for successfully carrying out this technique is that the sequenced reactions must be compatible in the same reaction solvent and that the by-products of prior reactions do not interfere adversely in subsequent reactions by jeopardizing reaction yield performance or causing the production of unwanted side reactions. If there are reaction solvent incompatibilities with reagents in subsequent steps due to solubility issues, then solvent switching is done where a solvent from a prior reaction is evaporated before the next one is added to the crude reaction residue. Though the technique has been used for a long time, it was recently demonstrated using this terminology for the synthesis of the anti-influenza neuramidase inhibitor oseltamivir phosphate using three one-pot operations [26] and for the synthesis of dipeptidyl peptidase IV (DPP4) inhibitor ABT-341 using a single one-pot operation involving six consecutive reaction steps [27]. The topic was nicely reviewed by Hayashi in 2016 [28].

Pot-Atom-Step Economy (PASE) [29, 30]

A phrase coined by Paul A. Clarke of the University of York (UK) describing:

> a technique in organic synthesis that combines as many transformations as possible into a single reaction vessel, without the need for work-up and isolation of intermediate compounds. Ideally, all the reagents should also be incorporated into the final product. This should lead to a reduction in the amount of waste generated by a synthetic route due to the minimization of by-products, reaction and extraction solvents, silica gel for chromatographic purification of intermediate compounds and alike.

The technique was first demonstrated using this terminology for the synthesis of tetrahydropyran-4-ones.

Radial Diagram (or Spider Diagram)

A diagram of a regular *n*-sided polygon whose axes emanating from the center pertain to *n* metrics. Often each axis has the same length. The outer perimeter of the polygon pertains to metrics values corresponding to ideal greenness (usually 1) and the center point pertains to metrics values that correspond to complete anti-greenness (usually 0).

Radial Pentagon [31]

A radial diagram with five axes pertaining to the following metrics: atom economy; reaction yield; stoichiometric factor; material recovery parameter; and global reaction mass efficiency. This diagram is a useful visual aid to gauge any bottlenecks and strengths in the material efficiency performances of individual reactions. Since all parameters used to construct the radial pentagon are numbers between 0 and 1, it is very easy to see which parameters are high performing and which are poor for any given reaction. The ideal limiting situation is designated by a green colored perimeter where all parameters are equal to 1. The more distorted is the pentagon toward the center, the worse is the overall material efficiency performance of the reaction.

Example 6.26

Draw the radian pentagon for the reaction given in Example 6.1.

Solution

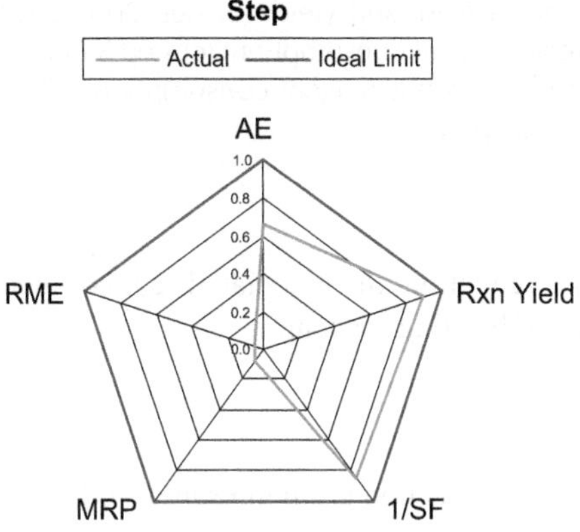

Example 6.27

The radial pentagons for each reaction in a synthesis plan for an unknown pharmaceutical are shown below. Based only on the shapes of the pentagons without knowledge of the chemistries involved, point out the bottlenecks in the plan and suggest what needs to be addressed to improve the overall material performance of the plan.

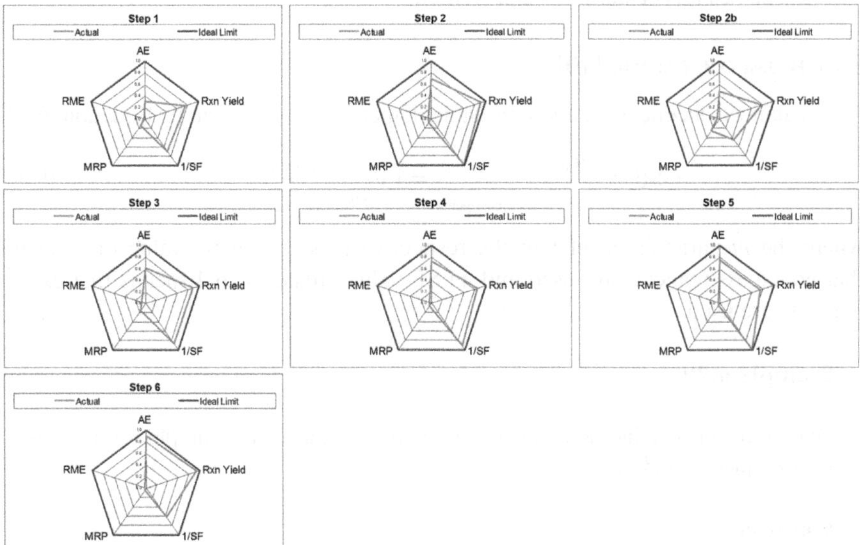

Solution

AE liabilities: steps 1 > 2b > 3
Reaction yield liabilities: 1 > 2b > 5
Excess reagent consumption: steps 2b > 6 > 1
Auxiliary material consumption: steps 6, 5, 4 > 2 > 3 > 1 > 2b

The main improvements for further material consumption performance include: (a) step 1 needs to increase AE and yield; (b) step 2b needs to increase AE and yield and reduce excess reagent consumption; (c) step 3 needs to increase AE; and (d) step 6 needs to reduce excess reagent consumption. All steps need to reduce auxiliary material consumption.

Scale of Reaction

For a balanced chemical equation, the associated scale of that reaction corresponds to the number of moles of limiting reagent.

Example 6.28

Determine the scale of the reaction given in Example 6.1

Solution

Number of moles of limiting benzyl alcohol reagent = 0.1001 moles.

Therefore, the scale of the reaction is 0.1001 moles.

Stoichiometric Coefficient

In a balanced chemical equation, stoichiometric coefficients are the integer coefficients appearing before the chemical species. If the stoichiometric coefficient is 1 for a given species it is customarily not written.

Stoichiometric Factor [10]

For a balanced chemical equation, the stoichiometric factor is given by Equation (6.34)

$$SF = \frac{m_{\text{actual reagents used}}}{m_{\text{stoichiometric reagents}}} = 1 + \frac{m_{\text{excess reagents}}}{m_{\text{stoichiometric reagents}}} \tag{6.34}$$

where the m parameters refer to the respective masses. An SF value of 1 means that no excess reagents are used and an SF value greater than 1 means that excess reagents are used.

Example 6.29

Determine the stoichiometric factor for the reaction given in Example 6.1 according to Equation (6.34).

Solution

Total mass of reagents used = 10.81 + 21.9 + 15 = 47.71 g
Total mass of stoichiometric reagents = 10.81 + 19.06 + 10.11 = 39.98 g

Therefore, SF = 47.71/39.98 = 1.1933.
Total mass of excess reagents = 0 + 2.84 + 4.89 = 7.73 g (see Example 6.4).
Therefore, SF = 1 + (7.73/39.98) = 1.1933.

Example 6.30

Verify the relationship given by Equation (6.24) for the Andraos RME using the data for the reaction given in Example 6.1.

Solution

We know the following quantities:
AE = 0.656 (see Example 6.13)
Y = 0.900 (see Example 6.19)
SF = 1.1933 (see Example 6.29)
MRP = 0.0870 (see Example 6.24)

Implementing Equation (6.24) yields 0.656*0.9*(1/1.1933)*0.0870 = 0.043, or 4.3%. This value is in agreement with that found in Example 6.17 for the Andraos RME.

Turnover Frequency (TOF)

In a chemical reaction employing a catalyst, the TOF is given by Equation (6.35).

$$\text{TOF} = \left(\frac{\text{moles}_{\text{product}}}{\text{moles}_{\text{catalyst}}}\right)\left(\frac{1}{\text{reaction time}, h}\right) \tag{6.35}$$

Units: h^{-1}.

Turnover Number (TON)

In a chemical reaction employing a catalyst, the TON is given by Equation (6.36).

$$\text{TON} = \frac{\text{moles}_{\text{product}}}{\text{moles}_{\text{catalyst}}}. \tag{6.36}$$

Units: dimensionless

Example 6.31

For each of the following examples, determine the following parameters based on the data given: catalyst loading, turnover number, and turnover frequency. Balance each reaction showing by-products.

(a) [32]

OH + $PhCH_2NH_2$ → (0.02 g $B(OH)_3$ (cat.), 16 hours) N H Ph

4.92 g 3.32 g 6.9 g

(b) [33]

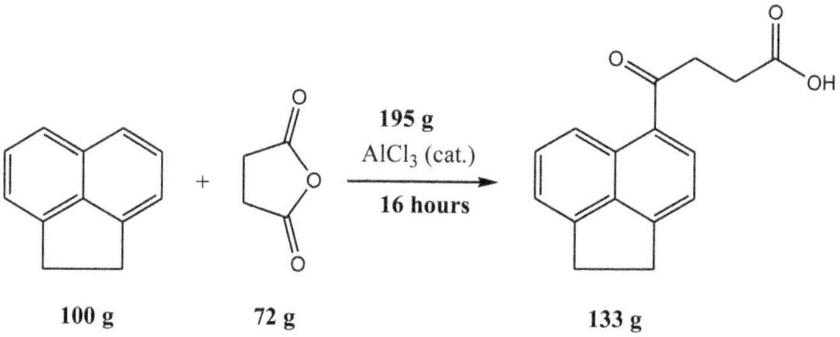

195 g $AlCl_3$ (cat.) 16 hours

100 g + 72 g → 133 g

(c) [34]

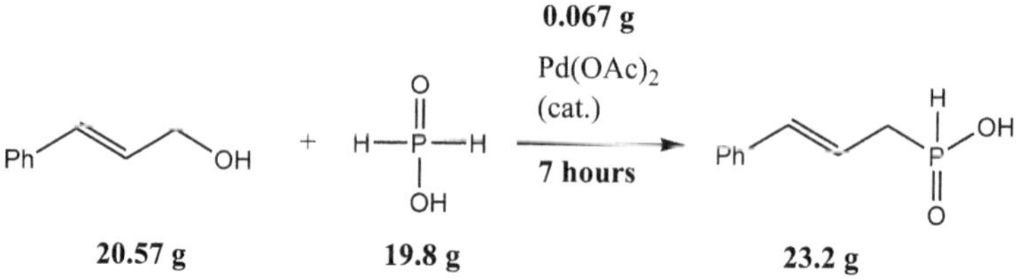

0.067 g $Pd(OAc)_2$ (cat.) 7 hours

20.57 g + 19.8 g → 23.2 g

(d) [35]

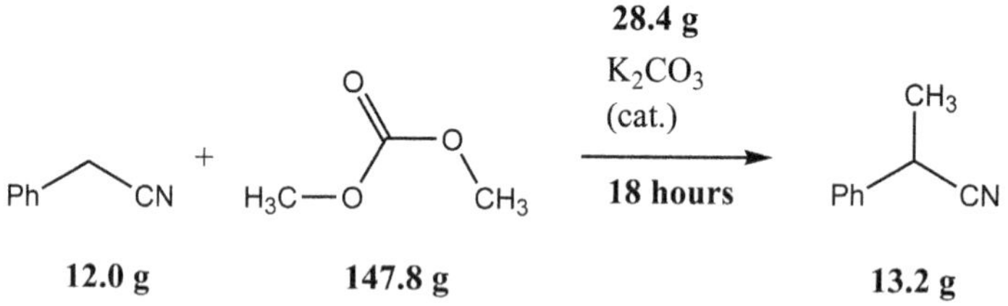

28.4 g K_2CO_3 (cat.) 18 hours

12.0 g + 147.8 g → 13.2 g

(e) [36]

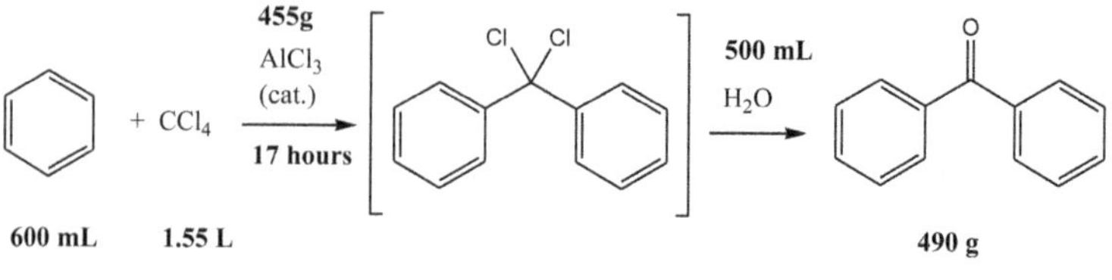

455g $AlCl_3$ (cat.) 17 hours; 500 mL H_2O

600 mL + CCl_4 1.55 L → 490 g

(f) [37]

HCl (cat.) 16 mL 37 wt%
$LiClO_4$-($MeOCH_2CH_2OMe$) (cat.) 57.3 g
1.5 hours

27.2 g furan
58.1 g acetone

10.2 g

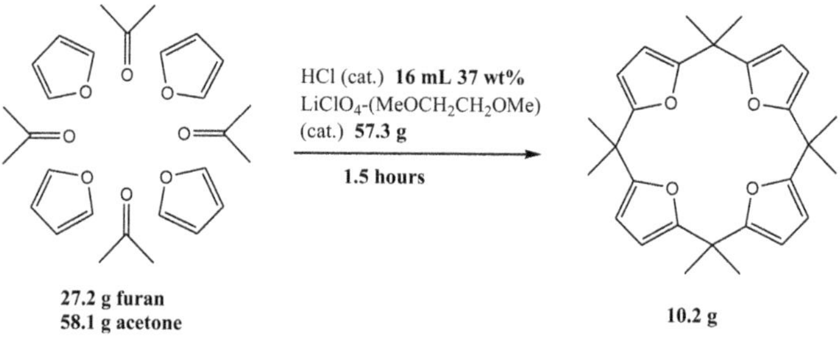

(g) [38]

NaOH **3.1 g**
K_2CO_3 **6.04 g**
$[Bu_4N][HSO_4]$ **1.48 g**
(cat.)

2 hours

4.0 g **3.9 g** **5.0 g**

(h) [39]

65 mg
(cat.)
Et_3N **5.56 g**
3 hours

(±)

17.91 g **9.14 g** **14.25 g** **6.60 g**

(i) [40]

OsO_4 (cat.) **1 g**
dihydroquinidine
4-chlorobenzoate
(chiral aux.) **23.25 g**

33 hours

180.25 g

260 mL of
60 wt% aq. soln.

155 g

Solution

$$\text{Catalyst loading: } Mol\% = \left(\frac{\text{Moles catalyst}}{\text{Moles limiting reagent}}\right)100$$

$$\text{Turnover number: TON} = \frac{\text{Moles product}}{\text{Moles catalyst}}$$

$$\text{Turnover frequency: TOF} = \frac{\text{TON}}{\text{Reaction time in hours}}$$

(a)

0.02 g (0.00032 moles)
62
$B(OH)_3$ (cat.)
+ $PhCH_2NH_2$
- H_2O

164 107 253

4.92 g (0.03 moles) **3.32 g (0.031 moles)** **6.9 g (0.0273 moles)**

Limiting reagent is 4-phenyl-butyric acid.
catalyst loading: mol % = (0.00032/0.03)*100 = 1.1%
turnover number (TON): TON = 0.0273/0.00032 = 85.3
turnover frequency (TOF): TOF = 85.3/16 = 5.3 h^{-1}

(b)

133.35
$AlCl_3$ (cat.)
195 g
(1.46 moles)

154 | 100 | 254

100 g (0.65 moles) | 72 g (0.72 moles) | 133 g (0.524 moles)

Limiting reagent is acenaphthene.
catalyst loading: mol% = (1.46/0.65)*100 = 225%
turnover number (TON): TON = 0.524/1.46 = 0.359
turnover frequency (TOF): TOF = 0.359/16 = 0.0224 h^{-1}

(c)

0.067 g (0.000298 moles)
$Pd(OAc)_2$ 224.51
(cat.)
7 hours
+ H_2O

134 | 66 | 182

20.57 g (0.154 moles) | 19.8 g (0.3 moles) | 23.2 g (0.127 moles)

Limiting reagent is 3-phenyl-prop-2-en-1-ol.
catalyst loading: mol % = (0.000298/0.154)*100 = 0.194%
turnover number (TON): TON = 0.127/0.000298 = 426.2
turnover frequency (TOF): TOF = 426.2/7 = 60.9 h^{-1}

(d)

28.4 g (0.206 moles)
K_2CO_3 138
(cat.)
18 hours
+ CO_2 + CH_3OH

117 | 90 | 131

12.0 g (0.103 moles) | 147.8 g (1.642 moles) | 13.2 g (0.101 moles)

Limiting reagent is benzylcyanide.
catalyst loading: mol % = (0.206/0.103)*100 = 200%
turnover number (TON): TON = 0.101/0.206 = 0.490
turnover frequency (TOF): TOF = 0.490/18 = 0.027 h^{-1}

(e)

2 benzene (78) + CCl_4 (153.8) → [455g (3.412 moles), 133.35 $AlCl_3$ (cat.), - 2 HCl] → [dichlorodiphenylmethane] → [500 mL H_2O 18, - 2 HCl] → benzophenone (182)

600 mL
d = 0.879 g/mL
527.4 g
(6.762 moles)

1.55 L
d = 1.594 g/mL
2470.7 g
(16.064 moles)

490 g
(2.692 moles)

Limiting reagent is benzene.
catalyst loading: mol % = (3.412/6.762)*100 = 50.5%
turnover number (TON): TON = 2.692/3.412 = 0.79
turnover frequency (TOF): TOF = 0.79/17 = 0.046 h^{-1}

(f)

furan (68), acetone (58) → [36.45 HCl (cat.) 16 mL 37 wt% d = 1.1837 g/mL, 7.01 g 0.192 moles; 197.45 $LiClO_4$-($MeOCH_2CH_2OMe$) (cat.) 57.3 g (0.290 moles); 1.5 hours] → product (432)

27.2 g furan (0.4 moles)
58.1 g acetone (1.002 moles)

10.2 g
(0.0236 moles)

Limiting reagent is furan.
HCl

catalyst loading: mol % = (0.192/0.4)*100 = 48%
turnover number (TON): TON = 0.0236/0.192 = 0.12
turnover frequency (TOF): TOF = 0.12/1.5 = 0.08 h^{-1}

$LiClO_4$-dimethoxyethane

catalyst loading: mol % = (0.290/0.4)*100 = 72.5%
turnover number (TON): TON = 0.0236/0.290 = 0.081
turnover frequency (TOF): TOF = 0.081/1.5 = 0.054 h^{-1}

(g)

Limiting reagent is furan-2-yl-carbamic acid *tert*-butyl ester.
K_2CO_3

catalyst loading: mol % = (0.0438/0.0219)*100 = 200%
turnover number (TON): TON = 0.0199/0.0438 = 0.45
turnover frequency (TOF): TOF = 0.45/2 = 0.23 h^{-1}

$[Bu_4N][HSO_4]$

catalyst loading: mol % = (0.00437/0.0219)*100 = 20%
turnover number (TON): TON = 0.0199/0.00437 = 4.6
turnover frequency (TOF): TOF = 4.6/2 = 2.3 h^{-1}

(h)

Limiting reagent is benzoyl chloride.
catalyst loading: mol % = (0.000301/0.065)*100 = 0.46%
turnover number (TON): TON = 0.0504/0.000301 = 167.4
turnover frequency (TOF): TOF = 167.4/3 = 55.8 h^{-1}

(i)

Limiting reagent is *trans*-stilbene.
OsO_4

catalyst loading: mol % = (0.00393/1.001)*100 = 0.39%
turnover number (TON): TON = 0.724/0.00393 = 184.2
turnover frequency (TOF): TOF = 184.2/33 = 5.58 h^{-1}

dihydroquinidine 4-chlorobenzoate

catalyst loading: mol % = (0.0501/1.001)*100 = 5.01%
turnover number (TON): TON = 0.724/0.0501 = 14.5
turnover frequency (TOF): TOF = 14.5/33 = 0.44 h^{-1}

Vector Magnitude Ratio (VMR) [10]

Given a set of metrics each having a value ranging between 0 (complete anti-greenness condition) and 1 (complete greenness condition) used to describe the degree of greenness of a chemical reaction or synthesis plan, VMR is a global parameter that measures the overall degree of greenness with respect to the set of metrics chosen. It is a convenient quantity that allows for unbiased ranking of synthesis plans based on a given set of metrics.

The mathematical expression for VMR is given by Equation (6.37).

$$\mathrm{VMR} = \frac{\sqrt{\sum_{j=1}^{s}\left(\mathrm{metric}_j\right)^2}}{\sqrt{s}} \tag{6.37}$$

where s is the number of metrics considered. In principle, an unlimited number of metrics can be used for determining VMR so long as the same set of contributing metrics are used when ranking more than one synthesis plan to the same target molecule.

VMR is an unbiased quantity unlike an area determination from a radial diagram which depends on the order of metrics used to create the diagram. VMR depends on the vector length of the set of metrics and is independent of the order of the metrics selected.

Example 6.32

Determine the VMR for the five metrics comprising the radial pentagon shown in Example 6.26 for the reaction given in Example 6.1.

Solution

Number of metrics = 5
$AE = 0.656$
$Y = 0.900$
$1/SF = 1/1.1933 = 0.838$
$MRP = 0.087$
$RME = 0.043$

Therefore, $\mathrm{VMR} = \sqrt{\frac{0.656^2 + 0.9^2 + 0.838^2 + 0.087^2 + 0.043^2}{5}} = 0.6248.$

Waste Intensity (WI) [25]

WI is defined as the ratio of mass of total waste produced in a reaction to the total mass of input materials used as shown in Equation 6.38.

$$\begin{aligned} \mathrm{WI} &= \frac{m_{\text{waste}}}{m_{\text{input}}} = \frac{m_{\text{input}} - m_{\text{product}}}{m_{\text{input}}} = 1 - \frac{m_{\text{product}}}{m_{\text{input}}} \\ &= 1 - \mathrm{RME} = 1 - \frac{1}{PMI} = \frac{E}{PMI} = \frac{E}{E+1} \end{aligned} \tag{6.38}$$

Example 6.33

Determine WI for the reaction given in Example 6.1 according to Equation (6.38).

Solution

Mass of waste = 524.11 g
Total mass of input materials = 547.71 g
Therefore, *WI* = 524.11/547.71 = 0.957.
We can verify that *WI* = 1–RME as follows: *WI* = 1 – 0.043 = 0.957.
We can verify that *WI* = *E*/(*E* + 1) as follows: *WI* = 22.21/(1 + 22.21) = 0.957.

Waste Percentage (WP) [25]

WP is defined as the ratio of waste intensity to process mass intensity according to Equation (6.39).

$$\begin{aligned} \mathrm{WP} &= \left(\frac{\mathrm{WI}}{PMI}\right)100 = \left(\frac{E / PMI}{PMI}\right)100 \\ &= \left(\frac{E}{(PMI)^2}\right)100 = \left(\frac{E}{(E+1)^2}\right)100 \end{aligned} \tag{6.39}$$

Example 6.34

Determine the waste percentage for the reaction given in Example 6.1 according to Equation (6.39).

Solution

WI = 0.957
PMI = 23.21
Therefore, *WP* = 100*(0.957/23.21) = 4.12%.
We can verify that *WP* = 100*(*E*/(*E* + 1)^2) as follows: *WP* = 100*(22.21/(22.21 + 1)^2) = 4.12%.

Figure 6.4 shows a flowchart that summarizes the essential material efficiency metrics introduced in this chapter shown in the shadowed box covering

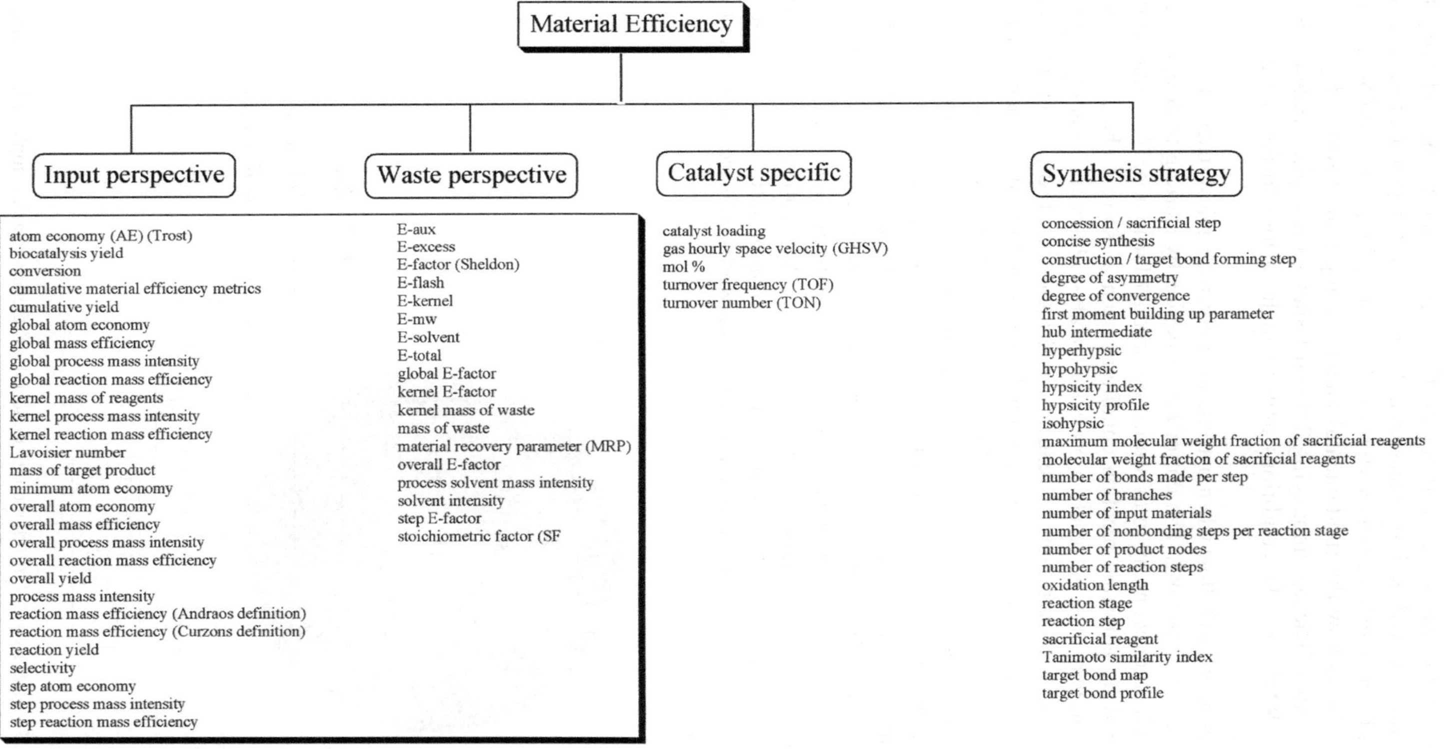

FIGURE 6.4 Flowchart showing hierarchy of material efficiency quantification metrics. Metrics listed in shadowed box are the essential metrics to consider in evaluating material efficiency for a single chemical reaction or a synthesis plan.

input material and waste material perspectives. The other metrics referring to synthesis strategy will be introduced and discussed in Chapter 1 of the sequel companion book *Synthesis Green Metrics: Problems, Exercises, and Solutions*. Figure 6.5 shows how the fundamental metrics comprising various degrees of RME are related. The accompanying Table 6.1 shows the explicit relationships for various RME and the conditions under which various forms of waste are considered.

The main recipe to follow when optimizing a given chemical reaction for material efficiency is as follows. The optimization begins by starting to design reactions for high atom economy to minimize by-product formation. Next, maximize reaction yield by adjusting reaction conditions such as solvent choice, reaction time, reaction temperature, reaction pressure, catalyst choice and loading, ligand choice, relative concentrations of reagents, and inclusion of any other additives as necessary. Use of excess reagents should be justified based on chemical reasoning such as increasing reaction rate or shifting equilibria to the product side; otherwise, stoichiometric quantities should be used as prescribed by the balanced chemical equation. The common practice of material efficiency optimization is to take existing reactions regardless of how atom economical they are and to reduce MRP as far as possible in an effort to reduce auxiliary material consumption. This approach is limited by physical parameters such as solubility and heat dissipation when it comes to solvent choice and quantities used. Application of green chemistry principles suggest that the direction of optimization follows the trajectory indicated by Figure 6.5 that starts at the center and works outward rather than the other way around.

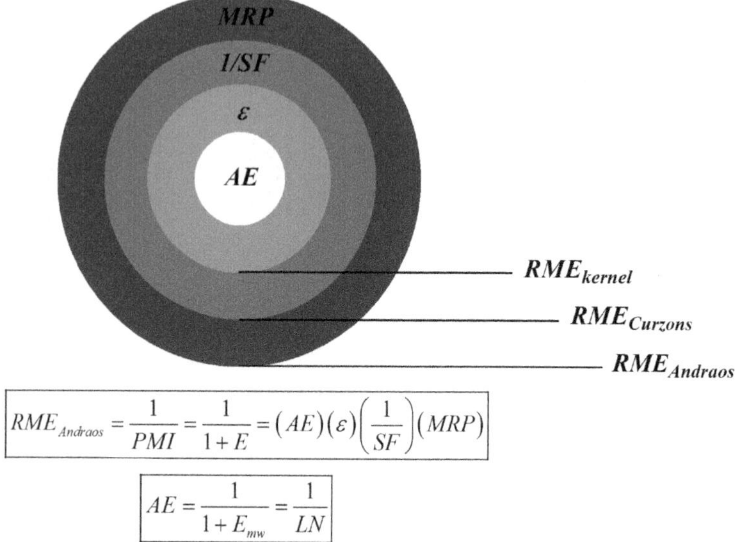

FIGURE 6.5 Hierarchy of fundamental material efficiency metrics applicable to single chemical reactions.

TABLE 6.1
Summary of Reaction Mass Efficiency Expressions under Various Reaction Conditions

Expression	By-products waste	Excess reagents waste	Auxiliary waste
$RME_{Andraos} = (AE)(\varepsilon)\left(\frac{1}{SF}\right)(MRP)$	Yes	Yes	Yes
$RME_{Curzons} = (AE)(\varepsilon)\left(\frac{1}{SF}\right)$	Yes	Yes	No
$RME_{stoich} = (AE)(\varepsilon)(MRP)$	Yes	No	Yes
$RME_{kernel} = (AE)(\varepsilon)$	Yes	No	No

6.6 WORKED EXAMPLES

Example 6.35 [41]

Part 1

In order to produce pentyl fluoride 0.90 g of 1-pentanol is combined with 1.55 g of NaF, and 4 g of sulfuric acid is added. What is the theoretical yield? If the actual yield is 0.80 g of pentyl fluoride, what is the percentage yield?

Solution

The balanced chemical equation for the process is given below.

$$\text{1-pentanol (OH)} + NaF + H_2SO_4 \longrightarrow \text{pentyl fluoride (F)} + NaHSO_4 + H_2O$$

88 42 98 90 120 18

Reactants: 88 + 42 + 98 = 228
Products: 90 + 120 + 18 = 228

Reactant	MW (g/mol)	Mass (g)	Moles
1-Pentanol	88	0.9	0.0102
NaF	42	1.55	0.0369
H_2SO_4	98	4	0.0408

The limiting reagent is 1-pentanol.

Since 1 mole of 1-pentanol produces 1 mole of pentyl fluoride, then 0.0102 moles of 1-pentanol are expected to produce 0.0102 moles of pentyl fluoride. The theoretical yield of pentyl fluoride is 0.0102*90 = 0.918 g.

The observed % yield of pentyl fluoride is 100*(0.8/0.918) = 87.1%.

Part 2

What is the % atom economy for the reaction?

Solution

MW pentyl fluoride = 90
Sum of MW of reactants = 228
% AE = 100*(90/228) = 39.5%.

Part 3

What is the *E*-factor for the reaction?

Solution

Mass of product collected = 0.8 g
Mass of input materials = 0.9 + 1.55 + 4 = 6.45 g
Mass of waste = 6.45 – 0.8 = 5.65 g
E-factor is 5.65/0.8 = 7.06

Part 4

What is the process mass intensity for the reaction?

Solution

Mass of input materials = 6.45 g
Mass of product collected = 0.8 g
PMI is 6.45/0.8 = 8.06

Example 6.36 [42]

A zirconium oxide catalytic process to produce 3-pentanone from propionic acid at 380°C also yields carbon dioxide and water as by-products. Determine the atom economy for the reaction. If the percent yield of 3-pentanone is 98.8% and 99.5% of propionic acid is converted, determine the mass of carbon dioxide generated and the mass of propionic acid that is needed if 1000 kg of 3-pentanone of 99.6% purity is to be made. How much propionic acid is wasted in the process? What is the *PMI* for the process? Do a mass balance for the reaction and determine if there is a mass deficit based on the information given.

Solution

The balanced chemical equation is given below.

74 **74** **86** **44** **18**

Reactants: 2*74 = 148
Products: 86 + 44 + 18 = 148
% AE = 100*(86/148) = 58.1%

1000 kg of 3-pentanone of 99.6% purity = 0.996*1000*1000/86 = 11581.4 moles
Mass of propionic acid required = 2*74*11581.4/0.988 = 1734866 g = 1734.866 kg
Moles of carbon dioxide generated = moles of 3-pentanone = 11581.4
Mass of carbon dioxide generated = 11581.4*44 = 509582 g = 509.582 kg
If 99.5% of propionic acid is converted, then 0.5% remains. This represents 0.005*1734.866 = 8.67 kg.
PMI = 1734.866/1000 = 1.73
Input materials:
1734.866 kg propionic acid
Output materials:
996 kg 3-pentanone
4 kg impurity
509.582 kg carbon dioxide
208.465 kg water
8.67 kg unreacted propionic acid
Adding the output materials = 996 + 4 + 509.582 + 208.465 + 8.67 = 1726.717 kg
Unaccounted output materials that are side products = 1734.866–1726.717 = 8.149 kg

Example 6.37 [43]

A new heterogeneous alkylation zeolite catalyst is discovered and a chemist desires to test it on a simple system. p-Cresol (15.5 g, 0.141 mol) is combined with methyl *tert*-butyl ether (MTBE, 12.4 g, 0.141 mol) and the zeolite catalyst (2 wt%) and the mixture is heated at 125°C for 2 h. After cooling, the products were identified by gas chromatography. The main product was 2-*tert*-butyl-*p*-cresol (10.5 g), with 7.95 g of *p*-cresol remaining unreacted. From the information given, determine: the reaction yield of 2-*tert*-butyl-*p*-cresol, atom economy, *E*-factor, *PMI*, % conversion of *p*-cresol, and selectivity of this particular catalyst.

Solution

The balanced chemical equation for the reaction is given below.

OH + OMe —zeolite catalyst→ OH + MeOH

108 88 164 32

Reactants: 108 + 88 = 196
Products: 164 + 32 = 196

Reactant	MW (g/mol)	Mass (g)	Moles
p-Cresol	108	15.5	0.1435
MTBE	88	12.4	0.1409

The limiting reagent is MTBE.

Since 1 mole of MTBE produces 1 mole of 2-*tert*-butyl-*p*-cresol, then 0.1409 moles of MTBE are expected to produce 0.1409 moles of 2-*tert*-butyl-*p*-cresol.
The actual yield of 2-*tert*-butyl-*p*-cresol is 10.5 g or 10.5/164 = 0.0640 moles.
The observed % yield is 100*(0.0640/0.1409) = 45.4%
MW of 2-*tert*-butyl-*p*-cresol product = 164
MW of reactants = 196
% AE = 100*(164/196) = 83.7%
Mass of 2-*tert*-butyl-*p*-cresol product = 10.5 g
Mass of input materials = 15.5 + 12.4 + 2%*12.4 = 28.15 g
[Note: The mass of zeolite is determined with respect to the limiting reagent.]
Mass of waste = 28.15 − 10.5 = 17.65 g
E-factor is 17.65/10.5 = 1.68
PMI is 28.15/10.5 = 2.68
15.5 g *p*-cresol at start
7.95 g *p*-cresol unreacted
Therefore, 15.5 − 7.95 = 7.55 g p-cresol reacted.
Hence, the % conversion of *p*-cresol is 100*(7.55/15.5) = 48.7%.
% selectivity for 2-*tert*-butyl-*p*-cresol product = 100*(fractional yield of product)/(fraction of substrate converted)
% selectivity for 2-*tert*-butyl-*p*-cresol product = 100*(0.454/0.487) = 93.2%

Example 6.38

Given the following three-step synthesis plan:

$$A + B \xrightarrow{\varepsilon_1} I_1 + Q_1$$

$$I_1 + C \xrightarrow{\varepsilon_2} I_2 + Q_2$$

$$I_2 + D \xrightarrow{\varepsilon_3} P + Q_3$$

where I_1 and I_2 are intermediate products of steps 1 and 2; Q_1, Q_2, and Q_3 are by-products of steps 1, 2, and 3; ε_1, ε_2, and ε_3 are reaction yields of steps 1, 2, and 3.

Part 1

Write out expressions for the step atom economies.

Part 2

Write out an expression for the global atom economy.

Part 3

Rewrite the expression for the global atom economy in terms of the step atom economies.

Generalize the result to an *N*-step linear plan.

Part 4

Based on the expressions found in Parts 1 and 2, under what condition is it true that the product of the step atom economies is equal to the global atom economy?

Solution

Part 1

AE of step 1: $(\text{AE})_1 = \dfrac{i_1}{a+b}$

AE of step 2: $(\text{AE})_2 = \dfrac{i_2}{i_1+c}$

AE of step 3: $(\text{AE})_3 = \dfrac{p}{i_2+d}$

Part 2

Global AE: $(\text{AE})_T = \dfrac{p}{a+b+c+d}$

Part 3

$$(\text{AE})_T = \frac{p}{a+b+c+d}$$

$$= \frac{p}{\dfrac{i_1}{(\text{AE})_1} + \dfrac{i_2}{(\text{AE})_2} - i_1 + \dfrac{p}{(\text{AE})_3} - i_2}$$

$$= \frac{p}{i_1\left(\dfrac{1}{(\text{AE})_1} - 1\right) + i_2\left(\dfrac{1}{(\text{AE})_2} - 1\right) + \dfrac{p}{(\text{AE})_3}}$$

$$(\text{AE})_T = \frac{p}{i_1\left(\dfrac{1}{(\text{AE})_1} - 1\right) + i_2\left(\dfrac{1}{(\text{AE})_2} - 1\right) + \cdots + i_{N-1}\left(\dfrac{1}{(\text{AE})_{N-1}} - 1\right) + \dfrac{p}{(\text{AE})_N}}$$

$$\boxed{(\text{AE})_T = \frac{p}{\left[\displaystyle\sum_{j=1}^{N-1} i_j\left(\dfrac{1}{(\text{AE})_j} - 1\right)\right] + \dfrac{p}{(\text{AE})_N}}}$$

$(\text{AE})_j$ is the atom economy of step j
i_j is the molecular weight of intermediate I_j
p is the molecular weight of the final target product P

Part 4

The product of the step economies is given by

$$(\text{AE})_1(\text{AE})_2(\text{AE})_3 = \left(\frac{i_1}{a+b}\right)\left(\frac{i_2}{i_1+c}\right)\left(\frac{p}{i_2+d}\right)$$

The global atom economy is given by

$$(\text{AE})_T = \frac{p}{a+b+c+d}$$

Equating these quantities suggests that

$$\left(\frac{i_1}{a+b}\right)\left(\frac{i_2}{i_1+c}\right)\left(\frac{p}{i_2+d}\right)=\frac{p}{a+b+c+d}$$

$$\left(\frac{i_1}{a+b}\right)\left(\frac{i_2}{i_1+c}\right)\left(\frac{1}{i_2+d}\right)=\frac{1}{a+b+c+d}$$

$$i_1 i_2(a+b+c+d)=(a+b)(i_1 i_2+ci_2+i_1 d+cd)$$

$$i_1 i_2(c+d)=(a+b)(ci_2+i_1 d+cd)$$

Case 1:

$c+d=a+b$ and $i_1 i_2=ci_2+i_1 d+cd$

This is a fortuitous scenario that has no practical significance.

Case 2:

$c=0$ and $d=0$

This scenario implies that the first step is any kind of reaction and the next two steps must be either rearrangements or eliminations.

Example 6.39

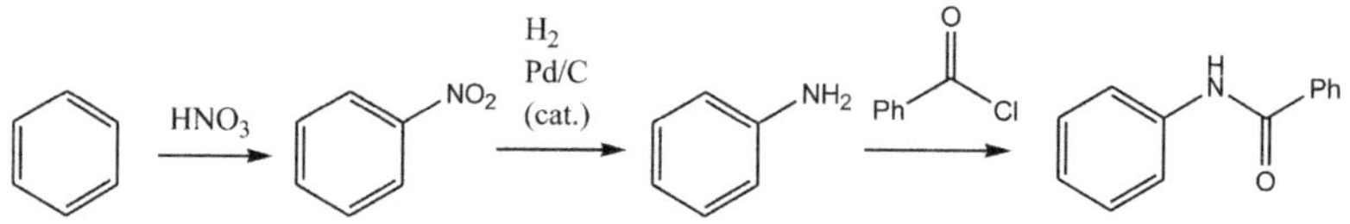

Balance each reaction step and determine the step AEs and global AE. Verify that the formula found in Part 3 of Example 6.38 is true.

Solution

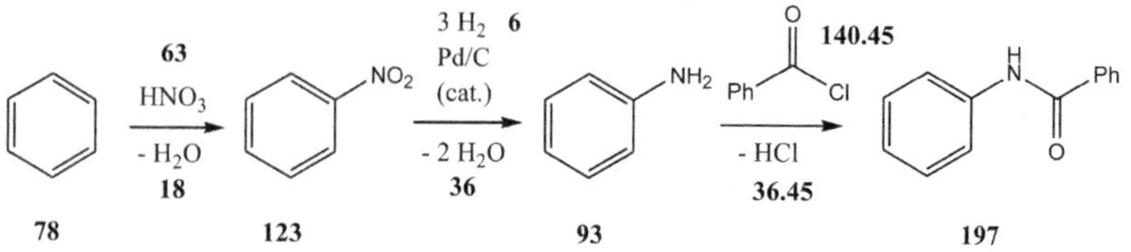

Step 1:

Reactants: $78+63=141$

Products: $123+18=141$

AE(1) = 123/141 = 0.872, or 87.2%

Step 2:

Reactants: 123 + 3*2 = 129
Products: 93 + 2*18 = 129
AE(2) = 93/129 = 0.721, or 72.1%

Step 3:

Reactants: 93 + 140.45 = 233.45
Products: 197 + 36.45 = 233.45
AE(3) = 197/233.45 = 0.844, or 84.4%
AE(global) = 197/(78 + 63 + 3*2 + 140.45) = 0.685, or 68.5%

$$(\mathrm{AE})_T = \frac{p}{i_1\left(\frac{1}{(\mathrm{AE})_1}-1\right)+i_2\left(\frac{1}{(\mathrm{AE})_2}-1\right)+\frac{p}{(\mathrm{AE})_3}}$$

$$= \frac{197}{123\left(\frac{141}{123}-1\right)+93\left(\frac{129}{93}-1\right)+197\left(\frac{233.45}{197}\right)}$$

$$= \frac{197}{141-123+129-93+233.45}$$

$$= \frac{197}{287.45}$$

$$= 0.685$$

or 68.5%
This result agrees with the one found for AE(global).

Example 6.40 [44]

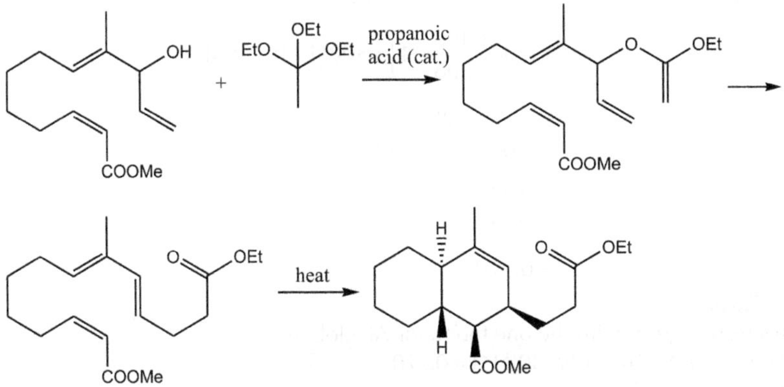

Balance each reaction step and determine the step AEs and global AE. Verify that the formula found in Part 3 of Example 6.38 is true. Show that the product of the step AEs is equal to the global AE.

Solution

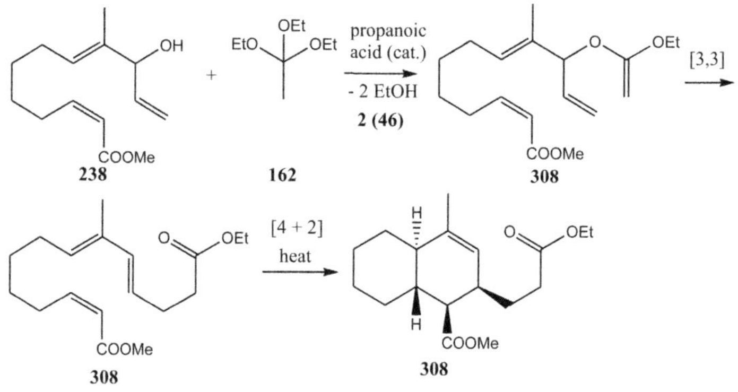

Step 1:

Reactants: 238 + 162 = 400
Products: 308 + 2*46 = 400
AE(1) = 308/400 = 0.77, or 77.0%

Step 2:

AE(2) = 1, or 100%

Step 3:

AE(3) = 1, or 100%

$$(AE)_T = \frac{p}{i_1\left(\frac{1}{(AE)_1} - 1\right) + i_2\left(\frac{1}{(AE)_2} - 1\right) + \frac{p}{(AE)_3}}$$

$$= \frac{308}{308\left(\frac{400}{308} - 1\right) + 308(1-1) + 308(1)}$$

$$= \frac{308}{400 - 308 + 308}$$

$$= \frac{308}{400}$$

$$= 0.770$$

or 77.0%

This result agrees with the one found for AE(global).
AE(1)*AE(2)*AE(3) = (308/400)*1*1 = 0.770
AE(global) = 0.77

Example 6.41

Given the following three-step synthesis plan:

$$A + B \xrightarrow{\varepsilon_1} I_1 + Q_1$$

$$I_1 + C \xrightarrow{\varepsilon_2} I_2 + Q_2$$

$$I_2 + D \xrightarrow{\varepsilon_3} P + Q_3$$

where I_1 and I_2 are intermediate products of steps 1 and 2; Q_1, Q_2, and Q_3 are by-products of steps 1, 2, and 3; ε_1, ε_2, and ε_3 are reaction yields of steps 1, 2, and 3.

Part 1

Write out expressions for the step RMEs.

Part 2

Write out an expression for the global RME.

Part 3

Rewrite the expression for the global RME in terms of the step RMEs.

Generalize the result to an N-step linear plan.

SOLUTION

Part 1

Kernel RME of step 1: $(\mathrm{RME})_1 = (\mathrm{AE})_1\,\varepsilon_1 = \left(\dfrac{i_1}{a+b}\right)\varepsilon_1$

Kernel RME of step 2: $(\mathrm{RME})_2 = (\mathrm{AE})_2\,\varepsilon_2 = \left(\dfrac{i_2}{i_1+c}\right)\varepsilon_2$

Kernel RME of step 3: $(\mathrm{RME})_3 = (\mathrm{AE})_3\,\varepsilon_3 = \left(\dfrac{p}{i_2+d}\right)\varepsilon_3$

Part 2

Global kernel RME: $(\mathrm{RME})_T = \dfrac{p}{\dfrac{d}{\varepsilon_3} + \dfrac{c}{\varepsilon_2\varepsilon_3} + \dfrac{a+b}{\varepsilon_1\varepsilon_2\varepsilon_3}}$

Part 3

$$(\mathrm{RME})_T = \frac{p}{\dfrac{d}{\varepsilon_3} + \dfrac{c}{\varepsilon_2\varepsilon_3} + \dfrac{a+b}{\varepsilon_1\varepsilon_2\varepsilon_3}}$$

$$= \frac{p}{\dfrac{1}{\varepsilon_3}\left[\dfrac{p\varepsilon_3}{(\mathrm{RME})_3} - i_2\right] + \dfrac{1}{\varepsilon_2\varepsilon_3}\left[\dfrac{i_2\varepsilon_2}{(\mathrm{RME})_2} - i_1\right] + \dfrac{1}{\varepsilon_1\varepsilon_2\varepsilon_3}\left[\dfrac{i_1\varepsilon_1}{(\mathrm{RME})_1}\right]}$$

$$= \frac{m_p}{\left[\dfrac{m_p}{(\mathrm{RME})_3} - m_{I_2}\right] + \left[\dfrac{m_{I_2}}{(\mathrm{RME})_2} - m_{I_1}\right] + \left[\dfrac{m_{I_1}}{(\mathrm{RME})_1}\right]}$$

$$= \frac{m_p}{\dfrac{m_p}{(\mathrm{RME})_3} + \left[\dfrac{m_{I_2}}{(\mathrm{RME})_2} - m_{I_2}\right] + \left[\dfrac{m_{I_1}}{(\mathrm{RME})_1} - m_{I_1}\right]}$$

$$(\text{RME})_T = \frac{m_p}{\dfrac{m_p}{(\text{RME})_N} + \sum_{j=1}^{N-1}\left[\dfrac{m_{I_j}}{(\text{RME})_j} - m_{I_j}\right]}$$

$$\boxed{(\text{RME})_T = \frac{m_p}{\dfrac{m_p}{(\text{RME})_N} + \sum_{j=1}^{N-1} m_{I_j}\left[\dfrac{1}{(\text{RME})_j} - 1\right]}}$$

$(\text{RME})_j$ is the kernel RME of step j
m_{Ij} is the mass of intermediate I_j
m_p is the mass of the final target product P

Example 6.42

If the connecting relationship between step RME and global RME for a linear synthesis plan is given by

$$\boxed{(\text{RME})_T = \frac{m_p}{\dfrac{m_p}{(\text{RME})_N} + \sum_{j=1}^{N-1} m_{I_j}\left[\dfrac{1}{(\text{RME})_j} - 1\right]}}$$

where:
$(\text{RME})_j$ is the RME of step j
m_{Ij} is the mass of intermediate I_j
m_p is the mass of the final target product P
derive analogous expressions connecting step *PMI* with global *PMI*, and step *E*-factor with global *E*-factor.

Solution

Since RME = 1/*PMI*, then

$$(\text{RME})_T = \frac{m_p}{\dfrac{m_p}{(\text{RME})_N} + \sum_{j=1}^{N-1} m_{I_j}\left[\dfrac{1}{(\text{RME})_j} - 1\right]}$$

$$\frac{1}{(PMI)_T} = \frac{m_p}{m_p (PMI)_N + \sum_{j=1}^{N-1} m_{I_j}\left[(\text{PMI})_j - 1\right]}$$

$$\boxed{(PMI)_T = \frac{m_p (PMI)_N + \sum_{j=1}^{N-1} m_{I_j}\left[(\text{PMI})_j - 1\right]}{m_p}}$$

Since $PMI = E + 1$, then

$$(PMI)_T = \frac{m_p (PMI)_N + \sum_{j=1}^{N-1} m_{I_j} \left[(PMI)_j - 1 \right]}{m_p}$$

$$E_T + 1 = \frac{m_p (E_N + 1) + \sum_{j=1}^{N-1} m_{I_j} \left[(E_j + 1) - 1 \right]}{m_p}$$

$$E_T = \frac{m_p (E_N + 1) + \sum_{j=1}^{N-1} m_{I_j} E_j}{m_p} - 1$$

$$\boxed{E_T = \frac{m_p (E_N) + \sum_{j=1}^{N-1} m_{I_j} E_j}{m_p}}$$

Example 6.43 [45, 46]

In 1914, Emil Fischer devised an ingenious experiment that proved that a switching of groups about an asymmetric carbon center results in a concomitant switching of optical rotation. He did this by the following sequence of reactions starting from (S)-2-(aminocarbonyl)-3-methylbutanoic acid. Note that the switching of the positions of the COOH and $CONH_2$ groups is achieved without breaking the C-C bonds linking these groups to the asymmetric center!

$CONH_2$ / (S) / COOH $\xrightarrow{CH_2N_2}$ $CONH_2$ / (S) / $COOCH_3$ $\xrightarrow{HNO_2}$ COOH / (S) / $COOCH_3$

$\xrightarrow{NH_2NH_2}$ COOH / (R) / $CONHNH_2$ $\xrightarrow{NaNO_2,\ H_2SO_4}$ COOH / (R) / CON_3 $\xrightarrow{NH_3}$ COOH / (R) / $CONH_2$

Part 1

Balance each reaction step accounting for by-products.

Part 2

Suggest mechanisms for steps 2 and 4.

Part 3

Determine the overall atom economy for each step and for the entire synthesis.

Part 4

How does the overall atom economy compare with the product of the step atom economies? How about the inverse sum of the reciprocals of the step atom economies?

Part 5

Using the AE values found in Part 3, verify numerically that the following relation connecting the overall AE with the step AEs is true for linear sequences.

$$(AE)_T = \frac{(MW)_P}{\left(\sum_{j=1}^{N-1} \frac{(MW)_{Y_j}}{(AE)_j} - (MW)_{Y_j}\right) + \frac{(MW)_P}{(AE)_N}}$$

where:

- $(AE)_T$ is the overall atom economy
- $(MW)_{Yj}$ is the molecular weight of intermediate Y_j
- $(MW)_p$ is the molecular weight of final product

Solution

Part 1

$C_6H_{11}NO_3$ (Mol. Wt.: 145) $\xrightarrow[-N_2]{CH_2N_2}$ $C_7H_{13}NO_3$ (Mol. Wt.: 159) $\xrightarrow[-H_2O,\ -N_2]{HNO_2}$ $C_7H_{12}O_4$ (Mol. Wt.: 160) $\xrightarrow[-CH_3OH]{NH_2NH_2}$ $C_6H_{12}N_2O_3$ (Mol. Wt.: 160) $\xrightarrow[-NaHSO_4,\ -2H_2O]{NaNO_2,\ H_2SO_4}$ $C_6H_9N_3O_3$ (Mol. Wt.: 171) $\xrightarrow[-HN_3]{NH_3}$ $C_6H_{11}NO_3$ (Mol. Wt.: 145)

Step 1:

Reactants: 145 + 42 = 187
Products: 159 + 28 = 187

Step 2:

Reactants: 159 + 47 = 206
Products: 160 + 28 + 18 = 206

Step 3:

Reactants: 160 + 32 = 192
Products: 160 + 32 = 192

Step 4:

Reactants: 160 + 69 + 98 = 327
Products: 171 + 120 + 2*18 = 327

Step 5:

Reactants: 171 + 17 = 188
Products: 145 + 43 = 188

Part 2

Step 2

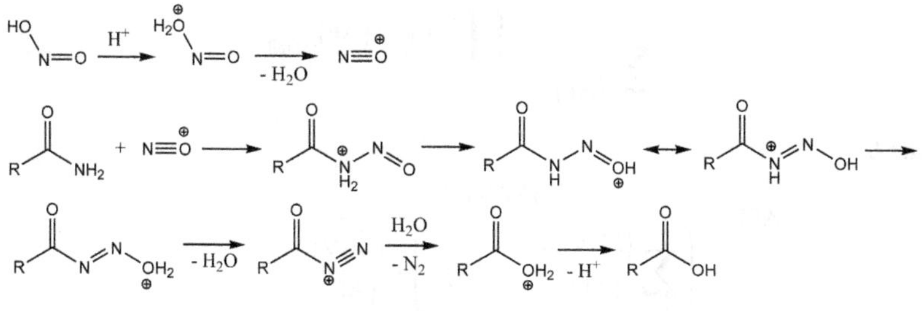

Step 4

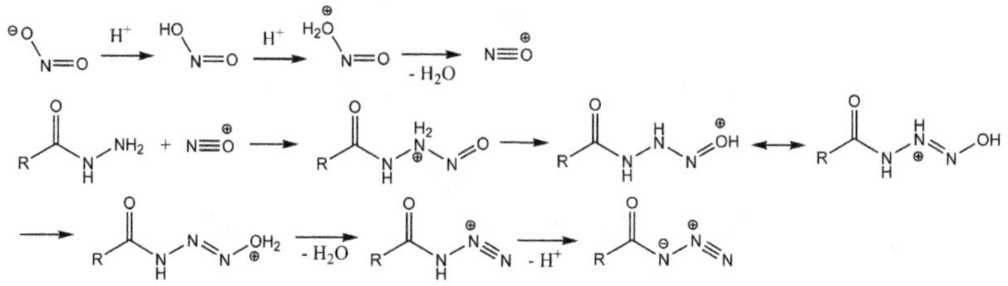

Part 3

Step 1

AE = 159/187 = 0.850, or 85.0%

Step 2

AE = 160/206 = 0.777, or 77.7%

Step 3

AE = 160/192 = 0.833, 83.3%

Step 4

AE = 171/327 = 0.523, or 52.3%

Step 5

AE = 145/188 = 0.771, or 77.1%

Overall synthesis:

Reactants: 145 + 42 + 47 + 32 + 69 + 98 + 17 = 450
Product: 145
AE = 145/450 = 0.322, or 32.2%

Part 4

Product of step atom economies:
0.850*0.777*0.833*0.523*0.771 = 0.222, or 22.2%

Inverse of sum of reciprocals of step atom economies:
1/((1/0.850) + (1/0.777) + (1/0.833) + (1/0.523) + (1/0.771)) = 0.145, or 14.5%

$$\frac{1}{\sum_j \frac{1}{(AE)_j}} < \prod_j (AE)_j < (AE)_{overall}$$

Part 5

$$(AE)_T = \frac{(MW)_P}{\left(\sum_{j=1}^{N-1} \frac{(MW)_{Y_j}}{(AE)_j} - (MW)_{Y_j}\right) + \frac{(MW)_P}{(AE)_N}}$$

$$(AE)_T = \frac{145}{\left(\frac{159}{0.850} - 159\right) + \left(\frac{160}{0.777} - 160\right) + \left(\frac{160}{0.833} - 160\right) + \left(\frac{171}{0.523} - 171\right) + \frac{145}{0.771}}$$

$$= \frac{145}{28.059 + 45.920 + 32.077 + 155.960 + 188.067}$$

$$= \frac{145}{450}$$

$$= 0.322$$

or 32.2%, which agrees with the value obtained in Part 3.

Example 6.44 [47]

The following sequential reaction produces both homocoupling and cross-coupling products.

$PdCl_2(PPh_3)_2$ (cat.)
$[Bu_4N]F$
THF / H_2O
O_2 (air)

A 49 % + B 34 % + C 17 %

Part 1

Provide balanced chemical equations for each product.

Part 2

Determine the atom economies and kernel reaction mass efficiencies for each product.

Part 3

Suggest a mechanism that accounts for the formation of all observed products.

Part 4

The experimental conditions called for 1 equivalent of naphthyl boronate, 0.95 equivalents of palladium chloride, and 0.95 equivalents of tetrabutylammonium fluoride for the first stage of the reaction. In the second stage 0.95 equivalents of *p*-tolyl boronate and 0.95 equivalents of tetrabutylammonium fluoride were used. What is the limiting reagent if product A is the target product? What is the limiting reagent if product B is the target product? What is the limiting reagent if product C is the target product?

Solution

Parts 1 and 2

We assume that both of the starting boronate esters are completely converted to products. Therefore, the percent values below each product structure are interpreted as reaction yields.

PRODUCT A

$$2 \text{ (naphthyl boronate)} + O_2 + H_2O + 2\,[Bu_4N]F \longrightarrow \text{(binaphthyl)} + 2 \text{ (F–B(O)(O) ester)} + [Bu_4N]OH + [Bu_4N]OOH$$

2 (240) | 32 | 18 | 2 (261) | 254 | 2 (132) | 259 | 275

Reactants: 2*240 + 32 + 18 + 2*261 = 1052
Products: 254 + 2*132 + 259 + 275 = 1052
AE = 254/1052 = 0.241, or 24.1%
Kernel RME = AE*yield = 0.241*0.49 = 0.118, or 11.8%

PRODUCT B

$$2 \text{ (p-tolyl boronate)} + O_2 + H_2O + 2\,[Bu_4N]F \longrightarrow \text{(4,4'-dimethylbiphenyl)} + 2 \text{ (F–B(O)(O) ester)} + [Bu_4N]OH + [Bu_4N]OOH$$

2 (204) | 32 | 18 | 2 (261) | 182 | 2 (132) | 259 | 275

Reactants: 2*204 + 32 + 18 + 2*261 = 980
Products: 182 + 2*132 + 259 + 275 = 980
AE = 182/980 = 0.186, or 18.6%
Kernel RME = AE*yield = 0.186*0.34 = 0.063, or 6.3%

PRODUCT C

254 + 204 + O_2 + H_2O + 2 [Bu_4N]F ⟶ 232 + 2 (132) + [Bu_4N]OH + [Bu_4N]OOH

254 | 204 | 32 | 18 | 2 (261) | 232 | 2 (132) | 259 | 275

Reactants: 254 + 204 + 32 + 18 + 2*261 = 1030
Products: 232 + 2*132 + 259 + 275 = 1030
AE = 232/1030 = 0.225, or 22.5%
Kernel RME = AE*yield = 0.225*0.17 = 0.038, or 3.8%

Part 3

Product A
Product C
Product B

$Ph_3P-Pd^0-PPh_3$ ⟶ (O_2, H_2O) $Ph_3P-Pd^{II}(OOH)(OH)-PPh_3$ ⟶ (2 Bu_4NCl; Bu_4NOH, $Bu_4N(OOH)$) $Cl-Pd^{II}(PPh_3)_2-Cl$

Part 4

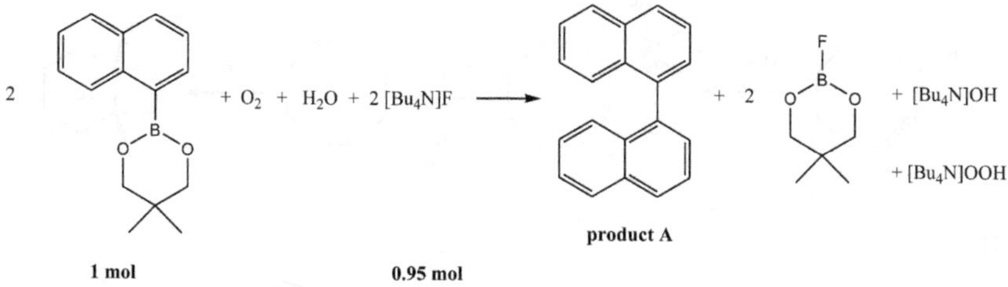

2 [naphthyl boronate] + O_2 + H_2O + 2 $[Bu_4N]F$ → product A + 2 [fluoroboronate] + $[Bu_4N]OH$ + $[Bu_4N]OOH$

1 mol 0.95 mol

Tetrabutylammonium fluoride is the limiting reagent.

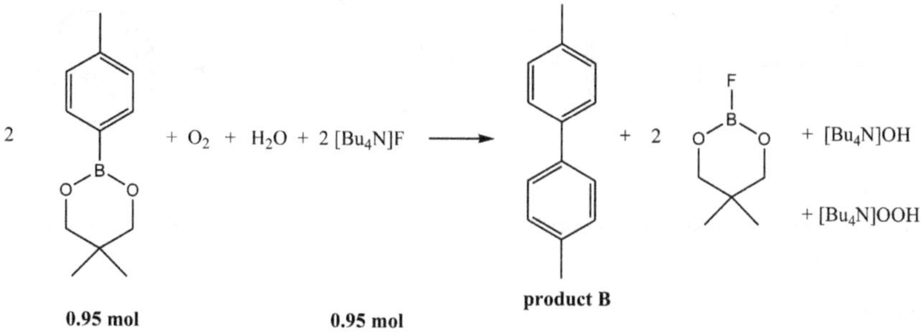

2 [p-tolyl boronate] + O_2 + H_2O + 2 $[Bu_4N]F$ → product B + 2 [fluoroboronate] + $[Bu_4N]OH$ + $[Bu_4N]OOH$

0.95 mol 0.95 mol

Both reagents are used in stoichiometric amounts, so no excess reagents are used.

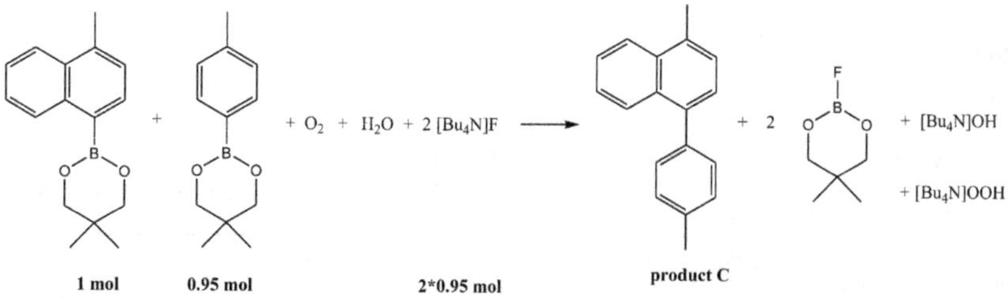

[naphthyl boronate] + [p-tolyl boronate] + O_2 + H_2O + 2 $[Bu_4N]F$ → product C + 2 [fluoroboronate] + $[Bu_4N]OH$ + $[Bu_4N]OOH$

1 mol 0.95 mol 2*0.95 mol

The naphthyl boronate starting material is used in excess. Hence, the *p*-tolyl boronate starting material and tetrabutylammonium fluoride are the limiting reagents.

Example 6.45 [48]

Part 1

For the following sequences, determine the step and overall atom economies.

(a)

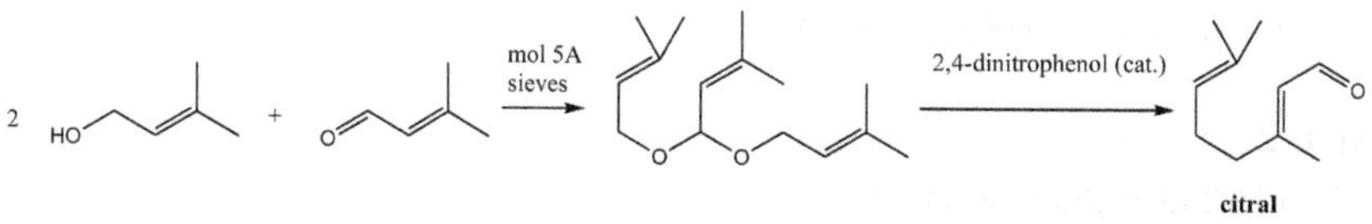

2 [prenol] + [prenal] —mol 5A sieves→ [acetal] —2,4-dinitrophenol (cat.)→ citral

(b)

H_2O_2 165°C *hv*

(c)

$Ni(CN)_2$ (cat.) 15 atm heat *hv*

Part 2

What is the relationship between the step atom economies and the overall atom economy?

Part 3

What is the relationship between the step kernel RME values and the overall kernel RME?

Solution

(a)

mol 5A seives - H_2O 2,4-dinitrophenol (cat.) 86 − HO

86 84 86 238

152

Step 1:

AE = 238/(84 + 172) = 0.930, or 93.0%

Step 2:

AE = 152/238 = 0.639, or 63.9%

Overall AE:

AE = 152/(172 + 84) = 0.594, or 59.4%

(b)

H_2O_2 / - H_2O ; 165°C / - Me_2NOH ; hv

153 169 108 108

Step 1:

AE = 169/(153 + 34) = 0.904, or 90.4%

Step 2:

AE = 108/169 = 0.639, or 63.9%

Step 3:

AE = 108/108 = 1, or 100%

Overall AE:

AE = 108/(153 + 34) = 0.578, or 57.8%

(c)

8 ||| —$Ni(CN)_2$, 15 atm→ 2 ; heat ; hv ; 78

8 (26) 2 (104) 208 130

Step 1:

AE = (2*104)/(8*26) = 1, or 100%

Step 2:

AE = 208/(2*104) = 1, or 100%

Step 3:

AE = 130/208 = 0.625, or 62.5%

Part 2

In the three cases shown, the second and subsequent reactions are either elimination or rearrangement reactions. The first reaction in the sequence can be any type of reaction. In examples (a) and (c) the first reactions were multi-component reactions. In example (b), the first reaction is an elimination reaction.

The general form of the synthesis tree diagram (See Chapter 1 of *Synthesis Green Metrics: Problems, Exercises, and Solutions*) for these special kinds of reaction sequences takes on the form shown below.

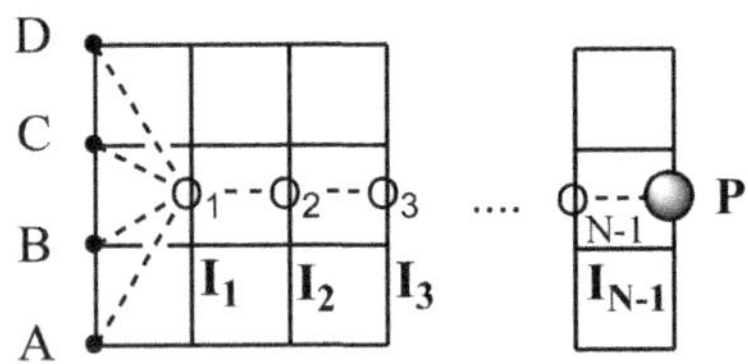

If the molecular weights of the starting materials, intermediates, and final product are a, b, c, d, i_1, i_2, i_3, ..., p, then the step atom economies are as follows:

$$(\text{AE})_1 = \frac{i_1}{a+b+c+d}$$

$$(\text{AE})_2 = \frac{i_2}{i_1}$$

$$(\text{AE})_3 = \frac{i_3}{i_2}$$

$$\vdots$$

$$(\text{AE})_j = \frac{i_j}{i_{j-1}}$$

$$(\text{AE})_N = \frac{p}{i_{N-1}}$$

The overall AE is given by $(\text{AE})_{\text{overall}} = \dfrac{p}{a+b+c+d}$.

For a three-step sequence, this relationship can be rewritten as follows.

$$\begin{aligned}(\text{AE})_{\text{overall}} &= \frac{p}{a+b+c+d}\\ &= \frac{(AE)_3\, i_2}{a+b+c+d}\\ &= \frac{(AE)_3 (AE)_2\, i_1}{a+b+c+d}\\ &= \frac{(AE)_3 (AE)_2 (AE)_1 (a+b+c+d)}{a+b+c+d}\\ &= (AE)_3 (AE)_2 (AE)_1\end{aligned}$$

Therefore, the overall AE is equal to the product of the step AE values.

For example (a), 0.93*0.639 = 0.594 = AE(overall).
For example (b), 0.904*0.639*1 = 0.578 = AE(overall).
For example (c), 1*1*0.625 = 0.625 = AE(overall).

Part 3

Modifying the generalized synthesis tree diagram to include reaction yields we have the following.

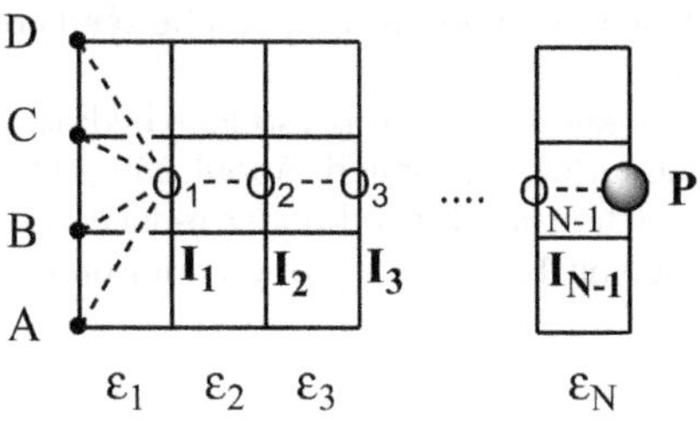

The step kernel RME expressions are as follows.

$$(\text{RME})_1 = \frac{i_1}{\frac{a+b+c+d}{\varepsilon_1}} = \frac{i_1\varepsilon_1}{a+b+c+d} = (\text{AE})_1\,\varepsilon_1$$

$$(\text{RME})_2 = (\text{AE})_2\,\varepsilon_2$$

$$(\text{RME})_3 = (\text{AE})_3\,\varepsilon_3$$

$$\vdots$$

$$(\text{RME})_N = (\text{AE})_N\,\varepsilon_N$$

The overall kernel RME is given by $(\text{RME})_{\text{overall}} = \frac{p}{\frac{a+b+c+d}{\varepsilon_1\varepsilon_2\ldots\varepsilon_N}} = (\text{AE})_{\text{overall}}\,\varepsilon_1\varepsilon_2\ldots\varepsilon_N$.

Since $(\text{AE})_{\text{overall}} = (\text{AE})_1(\text{AE})_2\ldots(\text{AE})_N$, we have

$$\begin{aligned}(\text{RME})_{\text{overall}} &= (\text{AE})_1(\text{AE})_2\ldots(\text{AE})_N\,\varepsilon_1\varepsilon_2\ldots\varepsilon_N \\ &= \left[(\text{AE})_1\,\varepsilon_1\right]\left[(\text{AE})_2\,\varepsilon_2\right]\ldots\left[(\text{AE})_2\,\varepsilon_2\right] \\ &= (\text{RME})_1(\text{RME})_2\ldots(\text{RME})_N\end{aligned}$$

Therefore, the overall kernel RME is equal to the product of the step RME values.

The relationships found in Parts 2 and 3 apply only to this special type of synthesis sequence where the first reaction can be any kind of reaction and all subsequent reactions are either rearrangement or elimination reactions.

Example 6.46 [49]

For the following procedure, determine the reaction mass efficiency for the production of furfural.

In a 12 L flask are placed 1.5 kg dry corn cobs, 5 L of 10% sulfuric acid (d = 1.0661 g/mL), and 2 kg sodium chloride. The flask is shaken in order to obtain a homogeneous mixture and is then connected with an upright tube, water condenser, and return tube. Heat is applied from a ring burner, the flame being adjusted so that the liquid distills at a rapid rate. The distillation process is continued until practically no more furfural can be seen collecting. The operation requires 5 to 10 h.

This distillate is then treated with enough sodium hydroxide so that the mixture is faintly acidic and the furfural separated. About 200 g of crude furfural is collected which is then fractionally distilled under reduced pressure with the temperature bath not exceeding 130°C. 165 g of pure furfural (b.p 90°C/65 mm Hg) s collected.

Solution

Input materials used:

1.5 kg dry corn cobs
1.0661*5000 = 5330.5 g 10 wt% sulphuric acid
2 kg sodium chloride
Total mass of input materials = 1500 + 5330.5 + 2000 = 8830.5 g
Mass of product collected = 165 g
Reaction mass efficiency = 165/8830.5 = 0.0187, or 1.87%
Process mass intensity = 8830.5/165 = 53.52
E-factor = (8830.5 − 165)/165 = 52.52

Note that since the mass of sodium hydroxide in the neutralization step was not specified the estimate for RME is an upper limit, and the estimates of *PMI* and *E*-factor are lower limits.

Example 6.47 [50]

The Cannizzaro reaction produces two useful products from the same starting material. Two equivalents of an aromatic aldehyde react via a hydride transfer mechanism to produce an aromatic alcohol and an aromatic carboxylate under basic conditions as shown in the example below. The alcohol is first extracted from the basic aqueous medium, then the filtrate is acidified to form the carboxylic acid which is then obtained by precipitation.

$$2\ \text{C}_6\text{H}_5\text{CHO} \xrightarrow[\text{then HCl}]{\text{KOH}} \text{C}_6\text{H}_5\text{COOH} + \text{C}_6\text{H}_5\text{CH}_2\text{OH}$$

Part 1

Based on the given masses of input materials used and products collected, determine the reaction yield, AE, *E*, RME, and *PMI* for each reaction product.
Reaction materials:

32 g benzaldehyde
29 g KOH in 27 mL water
80 mL of 37 wt% HCl
80 mL water
100 g crushed ice

Work-up materials:

105 mL water
80 mL diethyl ether
5 mL saturated sodium bisulfite soln.
10 mL 10 wt% sodium carbonate soln.
10 mL water
10 g magnesium sulfate

Product materials:

13 g benzyl alcohol
13.5 g benzoic acid

Part 2

Determine the global reaction yield, AE, E, RME, and *PMI* for obtaining both products.

Part 3

How are the individual metrics for each product related to the global metrics for both products?

Part 4

If both products arise from the same mechanism, then it is expected that the number of moles produced of each product should be the same, therefore both products should be made in equal percent yields. From your yield calculations, explain why the observed yields are different.

Solution

Part 1

Mass balance

Input materials	**Mass (g)**
Benzaldehyde	32
KOH	29
HCl (37 wt%)	94.696
Water	107
Crushed ice	100
Water	105
Et2O	56.64
Sat'd $NaHSO_3$ soln.	6
10 wt % $Na2CO_3$ soln.	11.029
Water	10
$MgSO_4$	10
Total	561.365
Products	**Mass (g)**
Benzyl alcohol	13
Benzoic acid	13.5

Balanced chemical equation:

$$2\ PhCHO + KOH + HCl \rightarrow PhCH_2OH + PhCOOH + KCl$$

$$2(106) \quad 56 \quad 36.45 \quad 108 \quad 122 \quad 74.45$$

Metrics

Product	Limiting reagent	AE	RY	E	RME	*PMI*
Benzyl alcohol	Benzaldehyde	0.355	0.797	42.2	0.02316	43.2
Benzoic acid	Benzaldehyde	0.401	0.733	40.6	0.02405	41.6

Part 2

Metrics

Product	AE	RY	E	RME	*PMI*
Benzyl alcohol + benzoic acid	0.755	1.531	20.2	0.04721	21.2

Note that the overall yield for both products exceeds 100%.

Part 3

The subscripts 1 and 2 in the formulas below refer to benzyl alcohol and benzoic acid products, respectively. The subscript T refers to the total value of the metric.

$$(AE)_T = (AE)_1 + (AE)_2$$

$$(RY)_T = (RY)_1 + (RY)_2$$

$$(RME)_T = (RME)_1 + (RME)_2$$

$$\frac{1}{(PMI)_T} = \frac{1}{(PMI)_1} + \frac{1}{(PMI)_2}$$

$$(PMI)_T = \frac{1}{\frac{1}{(PMI)_1} + \frac{1}{(PMI)_2}}$$

$$= \frac{(PMI)_1 (PMI)_2}{(PMI)_1 + (PMI)_2}$$

We note that the total *PMI* is the harmonic mean of the individual *PMI* values.

$$\frac{1}{1+E_T} = \frac{1}{1+E_1} + \frac{1}{1+E_2}$$

$$1+E_T = \frac{1}{\frac{1}{1+E_1} + \frac{1}{1+E_2}}$$

$$E_T = \frac{1}{\frac{1}{1+E_1} + \frac{1}{1+E_2}} - 1$$

$$= \frac{(1+E_1)(1+E_2)}{E_1 + E_2 + 2} - 1$$

$$= \frac{1+E_1+E_2+E_1E_2 - (E_1+E_2+2)}{E_1+E_2+2}$$

$$= \frac{E_1E_2 - 1}{E_1+E_2+2}$$

Part 4

The difference in reaction yields of both products arises due to losses encountered in the extraction process to obtain the alcohol and in the precipitation process to obtain the acid. Since the observed yield of acid product is less than that of the alcohol product, we may conclude that some of the carboxylate salt may have been lost during the first ether extraction to separate out the benzyl alcohol, or the carboxylic acid, formed after acidification of the saved aqueous layer from that extraction, may have been lost during precipitation and recrystallization.

Example 6.48 [48, 51]

Part 1

Given a general linear sequence given by the synthesis tree shown below, determine general expressions for the overall *E*-factor and overall *PMI* for the sequence in terms of the step *E*-factors and step *PMI*, respectively. Assume that all masses of reagents are properly scaled.

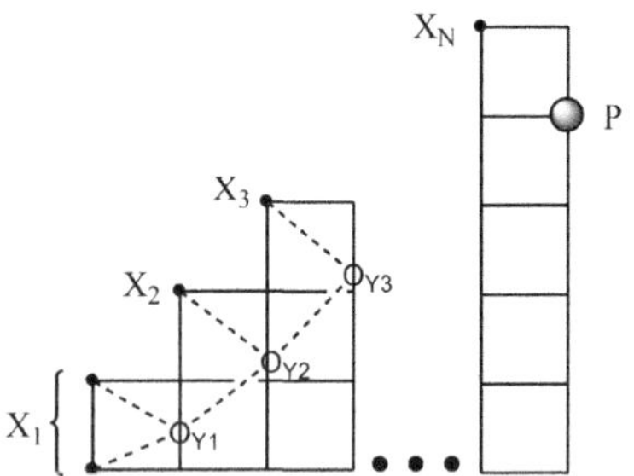

Part 2

From the synthesis tree, what are the step *PMI* and *E*-factor contributions to overall *PMI* and overall *E*-factor? Verify that the sum of the step *PMI* contributions is the overall *PMI* and that the sum of the step *E*-factor contributions is the overall *E*-factor.

Part 3

The following 5-step sequence is employed to make 1000 kg of an intermediate that is used to manufacture sildenafil (Viagra®). In each step, the product collected is committed as a reagent in the following step.

$C_6H_{10}O_4$ Mol. Wt.: 146

$C_5H_{10}O$ Mol. Wt.: 86

Na + EtOH, $- 1/2\ H_2$, [NaOEt] (cat.), - EtOH

A3 $C_9H_{14}O_4$ Mol. Wt.: 186

H_2NNH_2, H_2O, HOAc (cat.), $- 3\ H_2O$

A4 $C_9H_{14}N_2O_2$ Mol. Wt.: 182

$(MeO)_2SO_2$, $- (HO)SO_2(OMe)$

A5 $C_{10}H_{16}N_2O_2$ Mol. Wt.: 196

NaOH (cat.), EtOH, $- H_2O$

A6 $C_8H_{12}N_2O_2$ Mol. Wt.: 168

HNO_3, H_2SO_4 (cat.), $- H_2O$

P $C_8H_{11}N_3O_4$ Mol. Wt.: 213

Part (i)

Draw a synthesis tree diagram for this 5-step linear sequence.

Part (ii)

The inventory of materials used in this sequence is shown in Table 6.2. Determine the overall mass of input materials used and the overall waste produced. Determine the overall *E*-factor and overall *PMI*. Determine the step *E*-factors, step *PMI* values, step *E*-factor contributions, and step *PMI* contributions by completing Table 6.3. Verify that the sum of the step *E*-factor contributions is equal to the overall *E*-factor and verify that the sum of the step *PMI* contributions is equal to the overall *PMI*.

Solution

Part 1

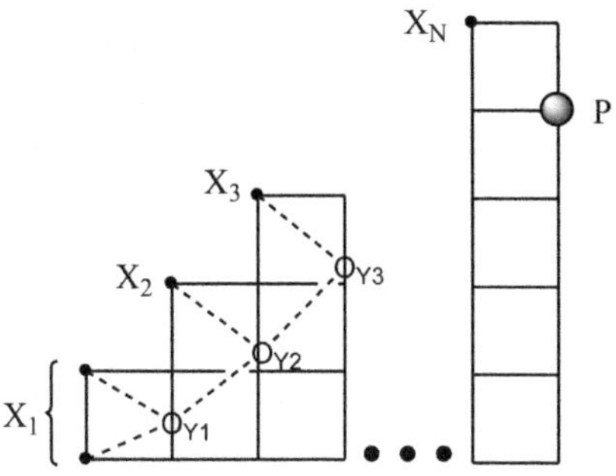

TABLE 6.2
Inventory of Masses of Input Materials

Step	Reagents	Solvents	Catalysts	Auxiliaries	Product
S1	1894 g diethyl oxalate; 1117 g 5-pentanone	18940 g ethanol	149 g sodium metal	None	1690 g product A3
S2	1690 g product A3; 499 g hydrazine hydrate	3540 g acetic acid	None	None	1587 g product A4
S3	1587 g product A4; 1107 dimethyl sulfate	11630 g dichloromethane	None	1088 g sodium bicarbonate; 8777 g water	1351 g product A5
S4	1351 g product A5; 6882 g water	None	826 g sodium hydroxide	2065 g 37 wt% hydrochloric acid	822 g product A6
S5	822 g product A6; 421 g nitric acid	7335 g sulfuric acid	None	17580 g water	1000 g product P

TABLE 6.3
***E*-factor and *PMI* Metrics**

Step	Inputs (g)	Waste (g)	Step E	Step *PMI*	Step E contribution	Step *PMI* contribution
S1	22100	20410	20410/1690 =12.077	22100/1690 =13.077	20410/1000 =20.410	22100/1000 =22.100
S2	5729	4142	4142/1587 =2.610	5729/1587 =3.610	4142/1000 =4.142	4039/1000 =4.039
S3	24189	22838	22838/1351 =16.905	24189/1351 =17.905	22838/1000 =22.838	22602/1000 =22.602
S4	11124	10302	10302/822 =12.533	11124/822 =13.533	10302/1000 =10.302	9773/1000 =9.773
S5	26158	25158	25158/1000 =25.158	26158/1000 =26.158	25158/1000 =25.158	25336/1000 =25.336

Step 1

$$PMI_1 = \frac{m_{X_1} + \Delta_1}{m_{Y_1}} \qquad E_1 = PMI_1 - 1 = \frac{m_{X_1} + \Delta_1}{m_{Y_1}} - 1$$

Step 2

$$PMI_2 = \frac{m_{Y_1} + m_{X_2} + \Delta_2}{m_{Y_2}} \qquad E_2 = PMI_2 - 1 = \frac{m_{Y_1} + m_{X_2} + \Delta_2}{m_{Y_2}} - 1$$

Step 3

$$PMI_3 = \frac{m_{Y_2} + m_{X_3} + \Delta_3}{m_{Y_3}} \qquad E_3 = PMI_3 - 1 = \frac{m_{Y_2} + m_{X_3} + \Delta_3}{m_{Y_3}} - 1$$

Step N

$$PMI_N = \frac{m_{Y_{N-1}} + m_{X_N} + \Delta_N}{m_P} \qquad E_N = PMI_N - 1 = \frac{m_{Y_{N-1}} + m_{X_N} + \Delta_N}{m_P} - 1$$

The Δ_j values represent auxiliary materials used in step j.

Overall *PMI* for the sequence is given by

$$PMI_T = \frac{\sum_j m_{X_j} + \sum_j \Delta_j}{m_P}$$

$$= \frac{(m_{X_1} + \Delta_1) + (m_{X_2} + \Delta_2) + \ldots + (m_{X_N} + \Delta_N)}{m_P}$$

$$= \frac{m_{Y_1} PMI_1 + (m_{Y_2} PMI_2 - m_{Y_1}) + (m_{Y_3} PMI_3 - m_{Y_2}) + \ldots (m_P PMI_N - m_{Y_{N-1}})}{m_P}$$

$$= \frac{m_{Y_1}(PMI_1 - 1) + m_{Y_2}(PMI_2 - 1) + m_{Y_3}(PMI_3 - 1) + \ldots + m_{Y_{N-1}}(PMI_{N-1} - 1) + m_P PMI_N}{m_P}$$

Therefore,

$$PMI_T = \frac{m_{pPMI_N +} \sum_j^{N-1} m_{Y_j}(PMI_j - 1)}{m_P}.$$

Since $PMI = E + 1$, overall E-factor for the sequence is given by

$$E_T + 1 = \frac{m_P(E_N + 1) + \sum_j^{N-1} m_{Y_j}(E_j)}{m_P}$$

$$E_T = \frac{m_P(E_N + 1) + \sum_j^{N-1} m_{Y_j}(E_j)}{m_P} - 1$$

Part 2

Solution

Step 1

$$PMI_{1,\text{contrib}} = \frac{m_{X_1} + \Delta_1}{m_P} \qquad E_{1,\text{contrib}} = \frac{m_{X_1} + \Delta_1 - m_{Y_1}}{m_P}$$

Step 2

$$PMI_{2,\text{contrib}} = \frac{m_{X_2} + \Delta_2}{m_P} \qquad E_{2,\text{contrib}} = \frac{m_{Y_1} + m_{X_2} + \Delta_2 - m_{Y_2}}{m_P}$$

Step 3

$$PMI_{3,\text{contrib}} = \frac{m_{X_3} + \Delta_3}{m_P} \qquad E_{3,\text{contrib}} = \frac{m_{Y_2} + m_{X_3} + \Delta_3 - m_{Y_3}}{m_P}$$

Step N

$$PMI_{N,\text{contrib}} = \frac{m_{X_N} + \Delta_N}{m_P} \qquad E_{N,\text{contrib}} = \frac{m_{Y_{N-1}} + m_{X_N} + \Delta_N - m_P}{m_P}$$

If we sum the step *PMI* contributions we obtain

$$\sum_j^N PMI_{j,\text{contrib}} = \frac{(m_{X_1} + \Delta_1) + (m_{X_2} + \Delta_2) + \ldots + (m_{X_N} + \Delta_N)}{m_P}$$

$$= \frac{\sum_j^N (m_{X_j} + \Delta_j)}{m_P}$$

$$= PMI_T$$

If we sum the step E contributions we obtain

$$\sum_j^N E_{j,\text{contrib}} = \frac{(m_{X_1} + \Delta_1 - m_{Y_1}) + (m_{Y_1} + m_{X_2} + \Delta_2 - m_{Y_2}) + (m_{Y_2} + m_{X_3} + \Delta_3 - m_{Y_3}) + \cdots + (m_{Y_{N-1}} + m_{X_N} + \Delta_N - m_P)}{m_P}$$

$$= \frac{(m_{X_1} + \Delta_1) + (m_{X_2} + \Delta_2) + (m_{X_3} + \Delta_3) + \cdots + (m_{X_N} + \Delta_N - m_P)}{m_P}$$

$$= \frac{(m_{X_1} + \Delta_1) + (m_{X_2} + \Delta_2) + (m_{X_3} + \Delta_3) + \cdots + (m_{X_N} + \Delta_N)}{m_P} - 1$$

$$= PMI_T - 1$$

$$= E_T$$

Part 3

Part (i)

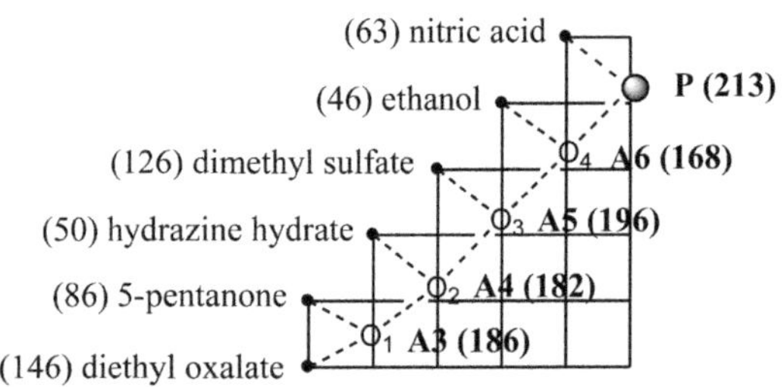

Part (ii)

Step S1

Total inputs = 1894 + 1117 + 18940 + 149 = 22100 g
Waste = 22100 − 1690 = 20410 g

Step S2

Total inputs = 1690 + 499 + 3540 = 5729 g
Waste = 5729 − 1587 = 4142 g

Step S3

Total inputs = 1587 + 1107 + 11630 + 1088 + 8777 = 24189 g
Waste = 24189 − 1351 = 22838 g

Step S4

Total inputs = 1351 + 6882 + 826 + 2065 = 11124 g
Waste = 11124 − 822 = 10302 g

Step S5

Total inputs = 822 + 421 + 7335 + 17580 = 26158 g
Waste = 26158 − 1000 = 25158 g

Overall

Total inputs = (22100) + (5729 − 1690) + (24189 − 1587) + (11124 − 1351) + (26158 − 822) = 22100 + 4039 + 22602 + 9773 + 25336 = 83850 g
Total waste = 20410 + 4142 + 22838 + 10302 + 25158 = 82850 g
Overall *E*-factor = 82850/1000 = 82.85
Overall *PMI* = 83850/1000 = 83.85

Verification:

Sum of step *E*-contributions = 20.410 + 4.142 + 22.838 + 10.302 + 25.158 = 82.85
Sum of step *PMI* contributions = 22.100 + 4.039 + 22.602 + 9.773 + 25.336 = 83.85

Note the following relationships:

step *PMI* = 1 + step E
(These are calculated based on the mass of product collected at each step.)
step *PMI* contribution ≠ 1 + step E contribution
(These are calculated based on the mass of final product in the sequence.)

Example 6.49 [52]

Derive an expression for the overall *PMI* for a linear synthesis plan containing *N* steps in terms of the step *PMIs*.

Solution

For a linear sequence of N steps involving the production of intermediate products P_1, P_2, …, and P_N in steps 1, 2, …, N where all of the preceding intermediate in any given step is committed as a reagent in the next step, we can write the step *PMI* for step j as shown in Equation (1).

$$(\text{PMI})_j = \frac{\phi_j}{m_{P_j}} \tag{1}$$

where:

ϕ_j is the mass of input materials used in step j
m_{P_j} is the mass of intermediate product collected in step j

For the entire synthesis plan covering all N steps, the overall *PMI* is given by Equation (2) where we have substituted the step *PMI* expression given in Equation (2).

$$\begin{aligned}(PMI)_{1\to N} &= \frac{\phi_1 + \phi_2 + \ldots + \phi_N - \left(m_{P_1} + m_{P_2} + \ldots + m_{P_{N-1}}\right)}{m_{P_N}} \\ &= \frac{1}{m_{P_N}}\left[\sum_{j=1}^{N}\phi_j - \sum_{j=1}^{N-1} m_{P_j}\right] \\ &= \frac{1}{m_{P_N}}\left[\sum_{j=1}^{N} m_{P_j}(PMI)_j - \sum_{j=1}^{N-1} m_{P_j}\right] \\ &= \frac{1}{m_{P_N}}\left[m_{P_N}(PMI)_N + \sum_{j=1}^{N-1} m_{P_j}\left((PMI)_j - 1\right)\right]\end{aligned} \tag{2}$$

Example 6.50

From the results found above, derive an analogous expression for the overall E-factor for a linear synthesis plan in terms of the step E-factors. Compare your result with the following statement given in the paper: "For a three-step process, the overall E factor is $E(\text{total}) = E_1 + E_2 + E_3$; whereas the overall *PMI* is $PMI(\text{total}) = PMI_1 + E_2 + E_3$." Does your result agree with this statement? Comment on the result.

Solution

Transforming the expression given in Equation (2) using the substitution $PMI = E + 1$ yields Equation (3).

$$E_{1\to N}+1=\frac{1}{m_{P_N}}\left[m_{P_N}\left(E_N+1\right)+\sum_{j=1}^{N-1}m_{P_j}\left(\left(E_j+1\right)-1\right)\right]$$

$$=\frac{1}{m_{P_N}}\left[m_{P_N}\left(E_N+1\right)+\sum_{j=1}^{N-1}m_{P_j}E_j\right] \tag{3}$$

$$E_{1\to N}=E_N+\frac{1}{m_{P_N}}\sum_{j=1}^{N-1}m_{P_j}E_j$$

The above results show that the overall E-factor for a linear synthesis plan is not the sum of the step E-factors. Therefore, the Sheldon statement is formally incorrect. Based on the formula given in Equation (3), $E_{1\to N}=\sum_{j=1}^{N}E_j$ under the very special condition when the masses of the intermediate products are each identical to the final product of the linear synthesis, that is, $m_{P_N}=m_{P_j}$ for $j=1, 2, 3, \ldots, N$. Such a scenario implies a linear sequence comprising consecutive rearrangement reactions where the molecular weight of each intermediate product remains the same and each reaction yield is 100% with no loss of mass arising from by-products or unreacted starting materials along the way.

Example 6.51

For a linear synthesis plan prove the recursive relation shown for the cumulative *PMI*.

$$\left(cPMI\right)_{1\to i}=\frac{m_{P_{i-1}}}{m_{P_i}}\left[\left(\mathrm{cPMI}\right)_{1\to i-1}-1\right]+\left(PMI\right)_i$$

Solution

For a three-step plan, the step *PMIs* for steps 1, 2, and 3 are given by Equations (4) through (6).

$$\left(PMI\right)_1=\frac{\phi_1}{m_{P_1}}$$

$$\left(PMI\right)_2=\frac{\phi_2}{m_{P_2}} \tag{4–6}$$

$$\left(PMI\right)_3=\frac{\phi_3}{m_{P_3}}$$

The cumulative *PMI* for steps 1 and 2 is given by Equation (7).

$$
\begin{aligned}
(cPMI)_{1\to 2} &= \frac{\phi_1+\phi_2-m_{P_1}}{m_{P_2}} = \frac{\phi_1}{m_{P_2}}+\frac{\phi_2}{m_{P_2}}-\frac{m_{P_1}}{m_{P_2}} \\
&= \frac{\phi_1}{m_{P_1}}\frac{m_{P_1}}{m_{P_2}}+\frac{\phi_2}{m_{P_2}}-\frac{m_{P_1}}{m_{P_2}} = \frac{m_{P_1}}{m_{P_2}}\left[\frac{\phi_1}{m_{P_1}}-1\right]+\frac{\phi_2}{m_{P_2}} \\
&= \frac{m_{P_1}}{m_{P_2}}\left[(PMI)_1-1\right]+(PMI)_2 \\
&= \frac{m_{P_1}}{m_{P_2}}\left[(cPMI)_{1\to 1}-1\right]+(PMI)_2
\end{aligned}
\tag{7}
$$

where it is understood that the step *PMI* for step 1 is identical to the cumulative *PMI* for step 1. Similarly, the cumulative *PMI* for steps 1, 2, and 3 is given by Equation (8).

$$
\begin{aligned}
(cPMI)_{1\to 3} &= \frac{\phi_1+\phi_2+\phi_3-m_{P_1}-m_{P_2}}{m_{P_3}} = \frac{m_{P_2}(cPMI)_{1\to 2}+\phi_3-m_{P_2}}{m_{P_3}} \\
&= \frac{m_{P_2}}{m_{P_3}}\left[(cPMI)_{1\to 2}-1\right]+(PMI)_3
\end{aligned}
\tag{8}
$$

It is readily apparent from the emerging pattern of Equations (7) and (8) that the stated recursive relation is verified for general step *i*.

Example 6.52

What is the analogous recursive expression for cumulative *E*-factor?

Solution

Substituting *PMI* = E + 1 for the step *PMIs* and c*PMI* = cE + 1 for the cumulative *PMIs* yields Equation (9).

$$
\begin{aligned}
(cE)_{1\to i}+1 &= \frac{m_{P_{i-1}}}{m_{P_i}}\left[(cE)_{1\to i-1}+1-1\right]+(E)_i+1 \\
(cE)_{1\to i} &= \frac{m_{P_{i-1}}}{m_{P_i}}(cE)_{1\to i-1}+(E)_i
\end{aligned}
\tag{9}
$$

Example 6.53

A four-step transformation is shown below which converts a 5-oxo starting material to a 4-oxo product. Determine how material efficient this sequence is by calculating its overall kernel RME.

NaBH$_4$ EtOH 98 %

amberlyst-15 toluene 93 %

*m*CPBA $CHCl_3$ 97 %

BF_3 OEt_2 benzene 79 %

Solution

We first write the scheme showing balanced chemical equations for each step:

98 % NaBH$_4$ 4 EtOH - NaOEt - $B(OEt)_3$ - 3 H_2

93 % - H_2O

277 **279** **261**

172.45

97 %

79 %

277 **277**

The overall kernel RME is given by

$$RME_{kernel} = \frac{277}{\frac{172.45}{0.97*0.79} + \frac{277+38+4*46}{0.98*0.93*0.97*0.79}}$$

$$= \frac{277}{225.042+714.484}$$

$$= \frac{277}{939.527}$$

$$= 0.295$$

or 29.5%.

Example 6.54 [53]

Themes: balancing chemical equations, catalytic cycles, redox reactions, atom economy, process mass intensity

The oxidation of 4-methylbenzyl alcohol to 4-methylbenzaldehyde was carried under catalytic conditions using TEMPO functionalized Co/C nanoparticles. Sodium hypochlorite solution was used to regenerate the catalyst. The catalyst is separated from the reaction solution by magnetic decantation.

R O N O• (cat.)

OH NaOCl / $NaHCO_3$ O

Part 1

Write out a catalytic cycle for the reaction that illustrates the mechanism.

Part 2

Determine the overall balanced chemical equation for the reaction and calculate its atom economy.

Part 3

The Co/C TEMPO functionalized catalyst was prepared according to the scheme shown below. Balance all chemical equations.

HO NH_2 + Co $NaNO_2$ HCl HO Co EtOOC—N=N—COOEt HN_3 PPh_3 N≡N N Co

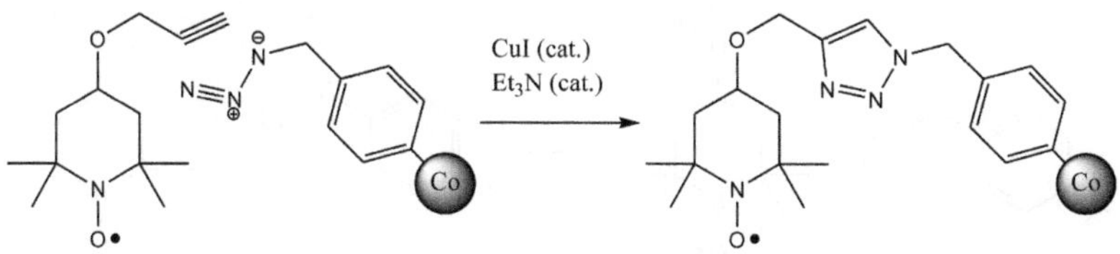

CuI (cat.)
Et_3N (cat.)

Part 4

The experimental procedure for the 4-methylbenzyl alcohol oxidation calls for the following materials.

Reaction:

336 mg 4-methylbenzyl alcohol
6 mL dichloromethane ($d = 1.336$ g/mL)
120 mg potassium bromide
750 mg Co/C-TEMPO nanoparticles (0.1 mmol/g; 2.5 mol% TEMPO)
2.4 mL of 10 wt% NaOCl solution ($d = 1.206$ g/mL)
120 mg sodium bicarbonate

Work-up:

3 × 10 mL dichloromethane
1 g magnesium sulfate (estimated)
Catalyst washing for re-use:
3×5 mL water ($d = 1$ g/mL)
6×5 mL acetone ($d = 0.791$ g/mL)
Yield of 4-methylbenzaldehyde = 89%

Based on this information determine the *PMI* for the reaction using the REACTION spreadsheet. Determine the *E*-factor profile.

SOLUTION

Part 1

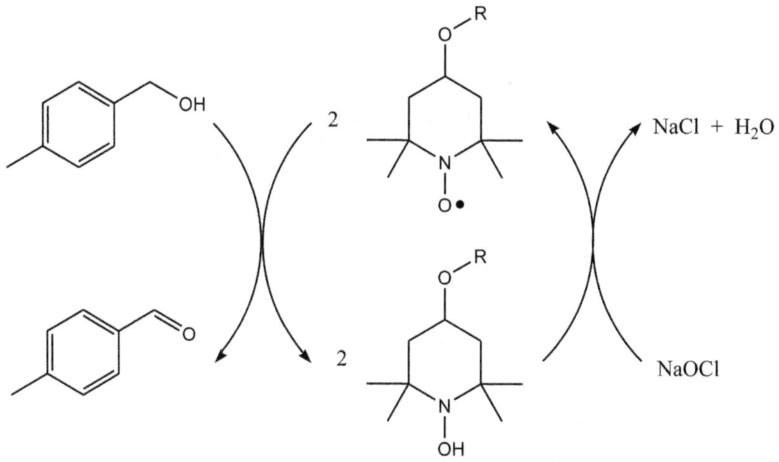

2

$NaCl + H_2O$

2

NaOCl

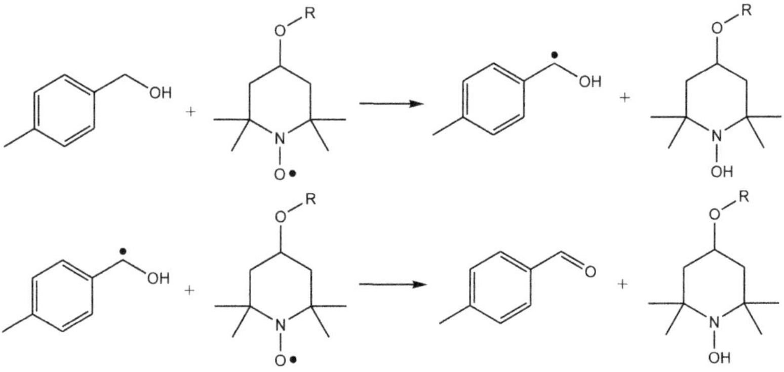

Net reaction:

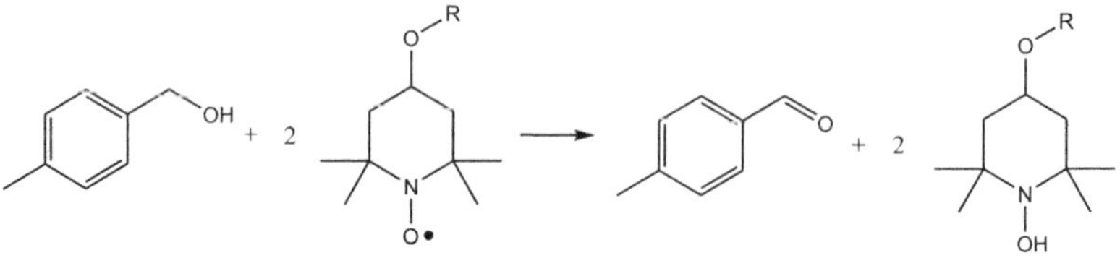

Catalyst recycling:

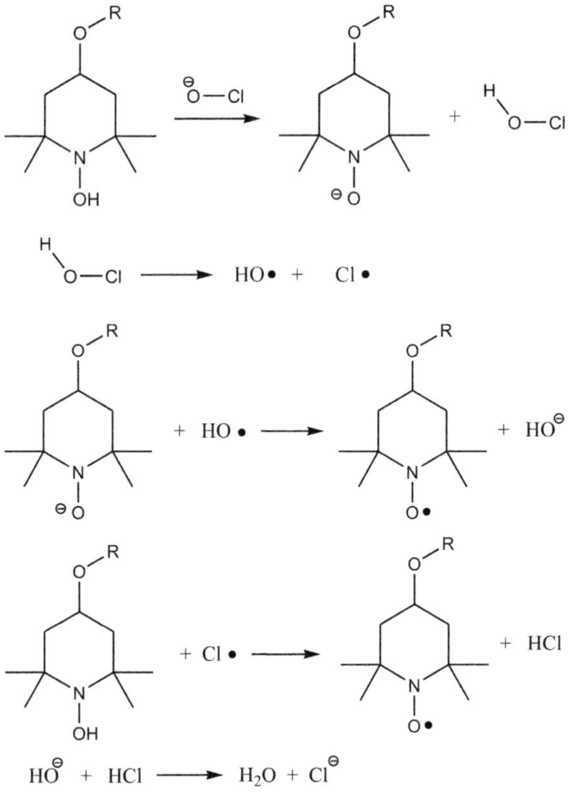

$HO^{\ominus}$ + HCl ⟶ H_2O + $Cl^{\ominus}$

Net reaction:

$$2\ \text{(4-RO-TEMPO-H, N–OH)} + {}^{\ominus}OCl \longrightarrow 2\ \text{(4-RO-TEMPO, N–O}\bullet) + H_2O + Cl^{\ominus}$$

Part 2

Since the TEMPO catalyst is regenerated, the net consumption of oxidant is coming from sodium hypochlorite. Therefore, the balanced chemical equation is given by

$$\text{4-methylbenzyl alcohol} + NaOCl \longrightarrow \text{4-methylbenzaldehyde} + NaCl + H_2O$$

122	**74.45**	**120**	**58.45**	**18**

AE = 120/(122 + 74.45) = 0.611, or 61.1%.

Part 3

$$\text{HO–CH}_2\text{–C}_6\text{H}_4\text{–NH}_2 + \text{Co} \xrightarrow[\text{- NaCl, - 2 }H_2O\text{, - }N_2]{NaNO_2\text{, 2 HCl}} \text{HO–CH}_2\text{–C}_6\text{H}_4\text{–Co} \xrightarrow[\text{- EtOOC–NH–NH–COOEt, - O=}PPh_3]{\text{EtOOC–N=N–COOEt, }HN_3\text{, }PPh_3} N_3\text{–CH}_2\text{–C}_6\text{H}_4\text{–Co}$$

$$\text{4-(propargyloxy)-TEMPO} + N_3\text{–CH}_2\text{–C}_6\text{H}_4\text{–Co} \xrightarrow{\text{CuI (cat.), }Et_3N\text{ (cat.)}} \text{TEMPO–O–CH}_2\text{–triazole–CH}_2\text{–C}_6\text{H}_4\text{–Co}$$

Part 4

The *PMI* for the reaction is 313.49.

The *E*-factor profile is given below.

Raw *E*-factor profile		
E-kernel		0.84
E-excess		0.33
E-rxn solvent		36.49
E-catalyst		3.37
E-work-up		271.46
E-purification		0
	E-aux	311.32
E-total		312.49

Example 6.55 [54]

Themes: process chemistry, material efficiency metrics, waste stream profile

The details of a process flow sheet for the preparation of an intermediate is shown below.

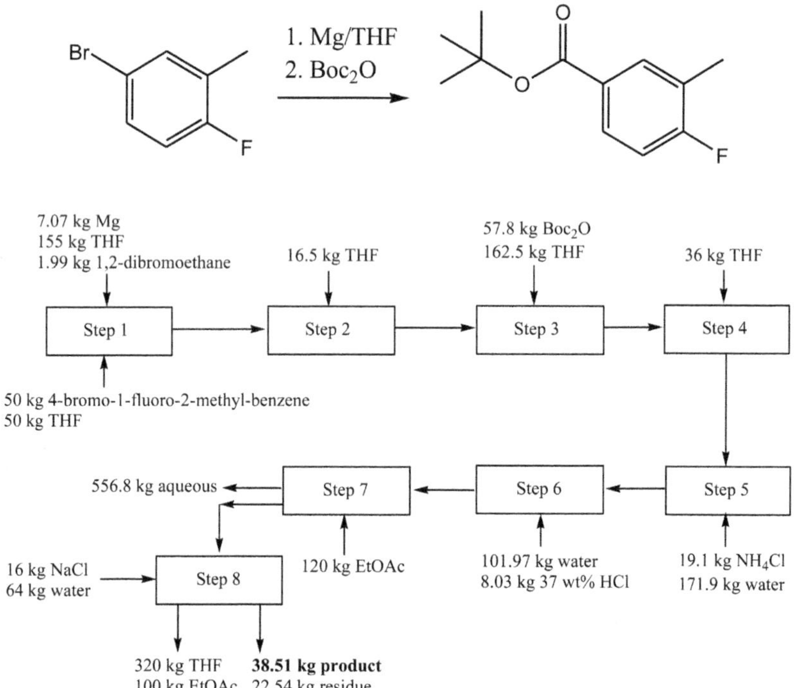

Based on the data given determine the % yield, % AE, *E*-factor, and *PMI*. Is all of the waste accounted for in the waste stream? What is the composition of the waste stream?

SOLUTION

The balanced chemical equation for the reaction is given below.

188.9 —(Mg 24.3)→ [BrMg intermediate] —(218; then NH_4Cl 53.45; $- CO_2$ 44; $- {}^tBuOH$ 90; $- Mg(Br)(Cl)$ 139.65; $- NH_3$ 17)→ 210

Reactants: 188.9 + 24.3 + 218 + 53.45 = 484.65
Products: 210 + 44 + 74 + 139.65 + 17 = 484.65
% AE = 100*(210/484.65) = 43.3%

Reactant	MW (g/mol)	Mass (g)	Exp. moles	Stoich. moles	Excess moles
4-bromo-1-fluoro-2-methyl-benzene	188.9	50000	265	265	0
Mg	24.3	7070	291	265	26
(Boc)2O	218	57800	265	265	0
NH4Cl	53.45	19100	357	265	93

Limiting reagent is 4-bromo-1-fluoro-2-methyl-benzene.

Yield of product is 38.51 kg or 38510/210 = 183.38 moles.
% Yield of product = 100*(183.38/265) = 69.2%.
Mass of reactants = 50 + 7.07 + 57.8 + 19.1 = 133.97 kg
Mass of input THF = 155 + 50 + 16.5 + 162.5 + 36 = 420 kg
Mass of input EtOAc = 120 kg
Mass of input water = 171.9 + 101.97 + 64 = 337.87 kg
Mass of other input auxiliaries = 1.99 + 8.03 + 16 = 26.02 kg
Total mass of input materials = 133.97 + 420 + 120 + 337.87 + 26.02 = 1037.86 kg
Mass of waste = 1037.86 − 38.51 = 999.35 kg
E-factor = 999.35/38.51 = 25.95
PMI = 1037.86/38.51 = 26.95
Mass of recovered THF = 320 kg (100*320/420 = 76.2% recovery)
Mass of recovered EtOAc = 100 kg (100*100/120 = 83.3% recovery)
Mass of aqueous waste = 556.8 kg
Mass of residue = 22.54 kg
Total mass of waste recovered = 320 + 100 + 556.8 + 22.54 = 999.34 kg
Mass of waste lost in process = 999.35 − 999.34 = 0.01 kg

Almost all of the waste material is accounted for. About 100*556.8/999.35 = 55.7% of this waste is aqueous; 100*(420/999.35) = 42.0% is organic; and 100*22.54/999.35 = 2.3% is non-recoverable residue.

Example 6.56

Themes: balancing chemical equations, ring construction strategies, minimum metrics thresholds

Two multi-component reactions are shown below for the synthesis of 3,4-dihydro-1 *H*-benzo[e][1,4]diazepine-2,5-dione derivatives.

(a) [55]

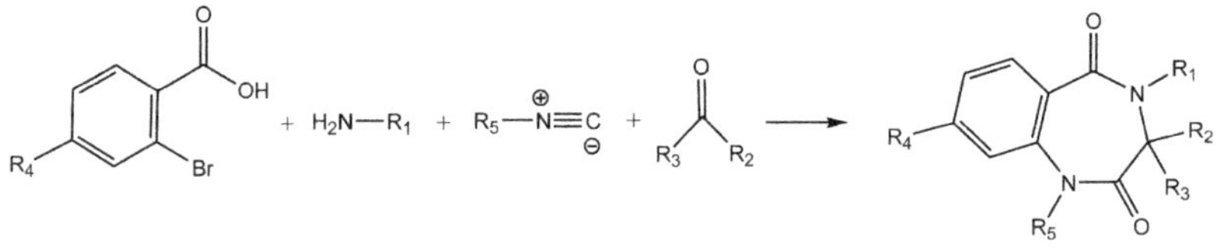

(b) [56]

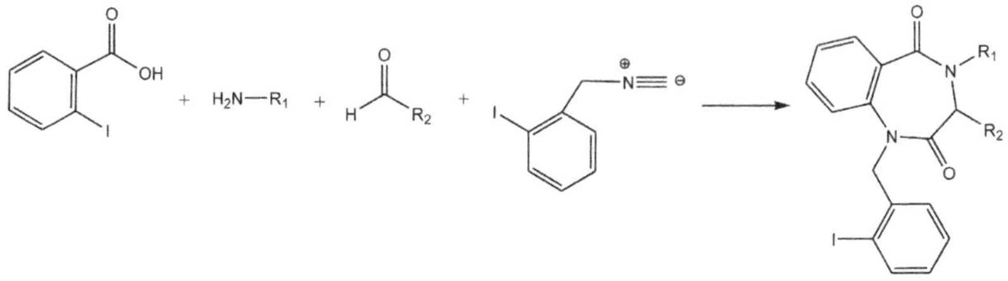

Part 1

Provide a balanced chemical equation for each transformation and determine its minimum atom economy.

Part 2

Determine the ring construction strategy employed in each reaction.

Part 3

If the kernel RME is 60% for each reaction, determine the corresponding minimum reaction yield that meets this threshold?

Part 4

If the global RME is 20% and the minimum reaction yield is as found in Part 3, determine the minimum *E*-factor for auxiliary materials. Assume that both reactions are carried out under stoichiometric conditions.

Part 5

Repeat Part 4 with the additional restriction that the benzoic acid substrate is the limiting reagent and that each of the other reagents is used in 5% excess.

Solution

Part 1

(a)

$R_4 + 199.9$ $R_5 + 26$ $R_1 + 16$ $R_2 + R_3 + 28$ → (- HBr, - H_2O) $R_1 + R_2 + R_3 + R_4 + R_5 + 171$

Reactants: R1 + R2 + R3 + R4 + R5 + 199.9 + 16 + 28 + 26 = R1 + R2 + R3 + R4 + R5 + 269.9

Products: R1 + R2 + R3 + R4 + R5 + 171 + 80.9 + 18 = R1 + R2 + R3 + R4 + R5 + 269.9

AE = (R1 + R2 + R3 + R4 + R5 + 171)/(R1 + R2 + R3 + R4 + R5 + 269.9)

Let R1 = Me, R2 = Me, R3 = Me, R4 = H, and R5 = Me.

AE(min) = (15 + 15 + 15 + 1 + 15 + 171)/(15 + 15 + 15 + 1 + 15 + 269.9) = 0.701, or 70.1%

(b)

248 $R_1 + 16$ $R_2 + 29$ 243 → (- H_2O, - HI) $R_1 + R_2 + 390$

Reactants: R1 + R2 + 248 + 16 + 29 + 243 = R1 + R2 + 536

Products: R1 + R2 + 390 + 18 + 128 = R1 + R2 + 536

AE = (R1 + R2 + 390)/(R1 + R2 + 536)

Let R1 = Me and R2 = Me.

AE(min) = (15 + 15 + 390)/(15 + 15 + 536) = 0.742, or 74.2%

Part 2

Both reactions employ a [3 + 2 + 1 + 1] strategy to construct the 7-membered ring.

Part 3

(Kernel RME)min = (AE)min*(Y)min

For reaction (a):

0.6 = 0.701*(Y)min

Therefore, (Y)min = 0.6/0.701 = 0.86, or 86%.

For reaction (b):

0.6 = 0.742 * (Y)min

Therefore, (Y)min = 0.6/0.742 = 0.81, or 81%.

Part 4

(Global RME)min = (AE)min * Y(min) * (1/SF) * (MRP)
= (Kernel RME)min * (1/SF) * (MRP)
MRP = 1/(1 + (AE)min*(Y)min*(1/SF)*(mass aux/mass product))
= 1/(1 + (AE)min*(Y)min*(1/SF)*(*E*-aux)) = 1/(1 + (Kernel RME)min*(1/SF)*(*E*-aux))

Therefore,

E-aux = ((1/MRP) − 1)/((Kernel RME)min * (1/SF))

For reactions (a) and (b), (AE)min * Y(min) = (Kernel RME)min = 0.6.

Therefore,

0.2 = 0.6 * (1/1) * MRP
MRP = 0.2/0.6 = 0.333
E-aux = ((1/0.333) − 1)/(0.6 * (1/1)) = (3−1)/0.6 = 2/0.6 = 3.33

Part 5

SF = 1 + (mass excess reagents/mass stoichiometric reagents)
= 1 + (AE/MW(product))*sum (mole excess * MW(reagent))

For reaction (a):

AE(min) = 0.701
MW(product)(min) = 15 + 15 + 15 + 1 + 15 + 171 = 232
Mole excess = 0.05
Minimum MW amine = 15 + 16 = 31
Minimum MW isocyanide = 15 + 26 = 41
Minimum MW ketone = 15 + 15 + 28 = 58
SF = 1 + (0.701/232)*0.05*(31 + 41 + 58) = 1.0196
MRP = (Global RME)min/((Kernel RME)min * (1/SF))
MRP = 0.2/(0.6*(1/1.0196)) = 0.34
E-aux = ((1/MRP) − 1)/((Kernel RME)min * (1/SF))
E-aux = ((1/0.34) − 1)/(0.6 * (1/1.0196)) = 3.3

For reaction (b):

AE(min) = 0.742
MW(product)(min) = 15 + 15 + 390 = 420
Mole excess = 0.05
Minimum MW amine = 15 + 16 = 31
Minimum MW isocyanide = 243
Minimum MW aldehyde = 15 + 29 = 44
SF = 1 + (0.742/420)*0.05*(31 + 243 + 44) = 1.0281
MRP = (Global RME)min/((Kernel RME)min * (1/SF))
MRP = 0.2/(0.6*(1/1.0281)) = 0.343
E-aux = ((1/MRP) – 1)/((Kernel RME)min * (1/SF))
E-aux = ((1/0.343) – 1)/(0.6 * (1/1.0281)) = 3.282

Example 6.57 [57]

Themes: balancing chemical equations, linear synthesis plan, step versus total materials metrics

A four-step linear plan to make 2,6-dihydroxyacetophenone from resorcinol is shown below.

HO OH + EtO O O → H_2SO_4 (cat.) 82 % → HO O O → Ac_2O 90 % → AcO O O

→ $AlCl_3$ (cat.) 73 % → HO O O (with acetyl) → NaOH then HCl 73 % → HO OH (with acetyl)

Part 1

Rewrite the scheme showing balanced chemical equations for all steps.

Part 2

Look up the experimental procedures for each step and use the *REACTION-template.xls* and *SYNTHESIS-template.xls* spreadsheets to determine the step and overall values for the following metrics: AE, RME, *PMI*, and *E*-factor. State any assumptions made in the calculations.

Part 3

Verify the following connecting relationships between the step and overall metrics.

$$(AE)_T = \frac{p}{\left[\sum_{j=1}^{N-1} l_j \left(\frac{1}{(AE)_j} - 1\right)\right] + \frac{p}{(AE)_N}}$$

$$(RME)_T = \frac{m_p}{\frac{m_p}{(RME)_N} + \sum_{j=1}^{N-1} m_{l_j}\left[\frac{1}{(RME)_j} - 1\right]}$$

$$(PMI)_T = \frac{m_p (PMI)_N + \sum_{j=1}^{N-1} m_{l_j}\left[(PMI)_j - 1\right]}{m_p}$$

$$E_T = \frac{\sum_{j=1}^{N-1} m_{l_j} E_j + m_P E_N}{m_P}$$

Compare the sum of the step *E*-factors with the total *E*-factor.
Compare the sum of the step *PMIs* with the total *PMI*.

Part 4

Determine the step *E*-factor contributions to the total *E*-factor.
Determine the step *PMI* contributions to the total *PMI*.
What is the relationship between the sum of the step metric contributions and the total metric value in each case?

Solution

Part 1

HO OH + EtO O O — H_2SO_4 (cat.), - EtOH, - H_2O → HO O O — Ac_2O, -HOAc → AcO O O

110 130 176 218

$AlCl_3$ (cat.) → HO O O O — 2 NaOH then 2 HCl, - 2 NaCl → HO OH O + HO O O

218 152 102

Part 2

Step	Assumptions
1	None
2	None
3	None
4	None

Step	AE (%)	RME (%)	*PMI*	E	MW (I_j), g/mol	Mass (I_j), g
1	73.3	1.54	64.98	63.98	176	2433
2	78.4	3.51	28.46	27.46	218	2722
3	100	1.87	53.41	52.41	218	1973
4	41.0	1.74	57.38	56.38	152	1000
Overall	30.7	0.256	391.24	390.24		

Part 3

$$\text{Verification of}\,(\text{AE})_T = \frac{p}{\left[\sum_{j=1}^{N-1} I_j\left(\frac{1}{(\text{AE})_j}-1\right)\right]+\frac{p}{(\text{AE})_N}}$$

$N=4$, $p=152$, $(\text{AE})_1=0.733$, $(\text{AE})_2=0.784$, $(\text{AE})_3=1$, $(\text{AE})_N=0.41$
$I_1=176$, $I_2=218$, $I_3=218$
RHS = 152/(176*((1/0.733) − 1) + 218*((1/0.784) − 1) + 218*((1/1) − 1) + (152/0.410)) = 0.307

From the SYNTHESIS spreadsheet we obtain $(\text{AE})_T=0.307$

$$\text{Verification of}\,(\text{RME})_T = \frac{m_p}{\frac{m_p}{(\text{RME})_N}+\sum_{j=1}^{N-1} m_{I_j}\left[\frac{1}{(\text{RME})_j}-1\right]}$$

$N=4$, $m_p=1000$, $(\text{RME})_1=0.0154$, $(\text{RME})_2=0.0351$, $(\text{RME})_3=0.0187$, $(\text{RME})_N=0.0174$
$m(I_1)=2433$, $m(I_2)=2722$, $m(I_3)=1973$
RHS = 1000/(2433*((1/0.0154) − 1) + 2722*((1/0.0351) − 1) + 1973*((1/0.0187) − 1) + (1000/0.0174)) = 0.00256

From the SYNTHESIS spreadsheet $(\mathrm{RME})_T = 0.00256$

$$\text{Verification of } (PMI)_T = \frac{m_p (PMI)_N + \sum_{j=1}^{N-1} m_{I_j} \left[(PMI)_j - 1 \right]}{m_p}$$

$N = 4$, $m_p = 1000$, $(PMI)_1 = 64.98$, $(PMI)_2 = 28.46$, $(PMI)_3 = 53.41$, $(PMI)_N = 57.38$
$m(I_1) = 2433$, $m(I_2) = 2722$, $m(I_3) = 1973$
RHS = (1000*57.38 + 2433*(64.98–1) + 2722*(28.46–1) + 1973*(53.41–1))/1000 = 391.194

From the SYNTHESIS spreadsheet $(PMI)_T = 391.24$

$$\text{Verification of } E_T = \frac{\sum_{j=1}^{N-1} m_{I_j} E_j + m_P E_N}{m_P}$$

$N = 4$, $m_p = 1000$, $(E)_1 = 63.98$, $(E)_2 = 27.46$, $(E)_3 = 52.41$, $(E)_N = 56.38$
$m(I_1) = 2433$, $m(I_2) = 2722$, $m(I_3) = 1973$
RHS = (2433*63.98 + 2722*27.46 + 1973*52.41 + 1000*56.38)/1000 = 390.194
From the SYNTHESIS spreadsheet $(E)_T = 390.24$
Therefore, all four formulas are verified.
Sum of step *E*-factors = 63.98 + 27.46 + 52.41 + 56.38 = 200.23
The sum of the step *E*-factors is NOT equal to the overall *E*-factor.
Sum of step *PMIs* = 64.98 + 28.46 + 53.41 + 57.38 = 204.23
The sum of the step *PMIs* is NOT equal to the overall *PMI*.

Part 4

Step	E	Mass (I_j), g	Mass waste (g)	Mass input materials (g)
1	63.98	2433	63.98*2433 = 155663	2433 + 155663 = 158096
2	27.46	2722	27.46*2722 = 74746	2722 + 74746 – 2433 = 75035
3	52.41	1973	52.41*1973 = 103405	1973 + 103405 – 2722 = 102656
4	56.38	1000	56.38*1000 = 56380	1000 + 56380 – 1973 = 55407
Overall	390.24	1000	390194	391194

Mass of waste for step $j = (E)_j$* Mass(I_j)
Mass of input materials for step 1 = Mass of waste for step 1 + Mass (I_1)
Mass of input materials for step j = Mass of waste for step j + Mass(I_j)–Mass(I_{j-1}), $j > 1$

Step	% Contribution of waste	Step *E*-factor contribution	% Contribution to input materials	Step *PMI* contribution
1	39.9	155.66	40.4	158.10
2	19.2	74.75	19.2	75.04
3	26.5	103.40	26.2	102.66
4	14.4	56.38	14.2	55.41
Overall	100	390.19	100	391.19

% contribution of waste in step j = (mass of waste in step j/mass of total waste)*100
step *E*-factor contribution of step j = (% contribution of waste for step j)*overall *E*-factor
% contribution to input materials for step j = (mass of input materials for step j/mass of total input materials)*100
step *PMI* contribution of step j = (% contribution to input materials for step j)*overall *PMI*

Sum of step *E*-factor contributions is equal to overall *E*-factor.
Sum of step *PMI* contributions is equal to overall *PMI*.

The *E*-factor contribution of step j to the overall *E*-factor is not the same as the step *E*-factor for step j. Similarly, the *PMI* contribution of step j to the overall *PMI* is not the same as the step *PMI* for step *j*.

Supplementary Question

Verify the formula $E_T = \frac{\sum_{j=1}^{N-1} m_{I_j} E_j + m_P E_N}{m_P}$ at the kernel and Curzons levels using the *E*-factor data obtained from the REACTION and SYNTHESIS spreadsheets.

Step	*E*-kernel	*E*-excess	*E*-Curzons
1	0.66	0	0.66
2	0.41	1.27	1.68
3	0.38	0	0.38
4	2.36	5.44	7.80
Overall	5.82	8.90	14.72

E-Curzons = *E*-kernel + *E*-excess

Solution

Kernel level:

$N = 4$, $m_p = 1000$, $(E)_1 = 0.66$, $(E)_2 = 0.41$, $(E)_3 = 0.38$, $(E)_N = 2.36$
$m(I_1) = 2433$, $m(I_2) = 2722$, $m(I_3) = 1973$
RHS = (2433*0.66 + 2722*0.41 + 1973*0.38 + 1000*2.36)/1000 = 5.83
Note that the sum of the *E*-kernel factors is 0.66 + 0.41 + 0.38 + 2.36 = 3.81.

Curzons level:

$N = 4$, $m_p = 1000$, $(E)_1 = 0.66$, $(E)_2 = 1.68$, $(E)_3 = 0.38$, $(E)_N = 7.80$
$m(I_1) = 2433$, $m(I_2) = 2722$, $m(I_3) = 1973$
RHS = (2433*0.66 + 2722*1.68 + 1973*0.38 + 1000*7.80)/1000 = 14.73
Note that the sum of the *E*-Curzons factors is 0.66 + 1.68 + 0.38 + 7.80 = 10.52.

We conclude that the sum of *E*-factors at any level is not equal to the true overall *E*-factor at that level.

Example 6.58 [58–60]

Themes: balancing chemical equations, material efficiency metrics, reaction comparison

Since benzofused sultams containing the benzoxazepine-1,1-dioxide motif were identified as important probes for various biological processes they were selected as targets to make compound libraries for further study. Microwave-assisted continuous flow organic synthesis (MACOS) was implemented as a strategy to synthesize these compounds using a "numbering up" or "scaling out" approach. In such a strategy, reactions are carried out in several parallel micro-reactors at a given small scale that when added together produce a desired large scale of final product. This is in contrast to running a single scaled-up reaction at the intended larger scale. According to the authors "splitting reactant streams and sending reactions through our bundled (multi)capillary reactor system into a common collection device will multiply output again by the number of capillaries through which any one reaction is sent."

The diversity organic synthesis reaction used to generate the compound library is shown below. The first reaction is carried out under conventional conditions. The second reaction is run under the MACOS numbering up strategy.

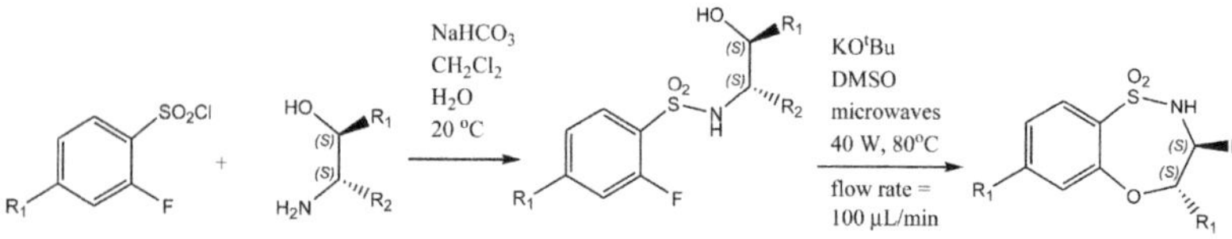

In order to make 2.07 g of the benzosultam shown below, the following reaction conditions were applied.

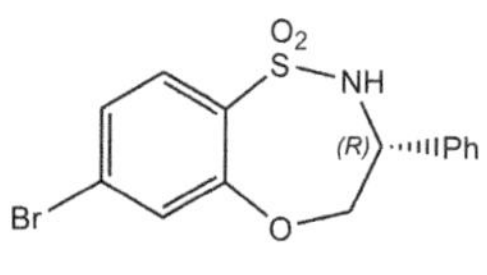

A stock solution of DMSO containing KO^tBu and sulfonamide is made up with a combined concentration of reagents of 0.8 M. The mole equivalents of KO^tBu to sulfonamide is 3:1. This solution is divided into two syringes arranged in parallel where each syringe contains 20 mL of liquid. Both syringes are connected via capillary tubing primed with DMSO to a syringe pump that draws out liquid at a flow rate of 100 μL/min to a microwave reactor set at 40 W.

Yield of product = 73%

Part 1

Write out balanced chemical equations for both steps in order to make the final (R)-7-bromo-3-phenyl-3,4-dihydro-2 *H*-benzo[b][1,4,5]oxathiazepine 1,1-dioxide product. Determine their respective atom economies.

Solution

Step 1

SO_2Cl / Br / F (273.35) + H_2N / (R) / Ph / OH (137) $\xrightarrow{NaHCO_3}$ - NaCl, - H_2O, - CO_2 → O_2S–NH / (R) / Ph / OH / Br / F (373.9)

Reactants: 273.35 + 137 + 84 = 494.35
Products: 373.9 + 58.45 + 18 + 44 = 494.35
AE = 373.9/494.35 = 0.756, or 75.6%

Step 2

O_2S–NH / (R) / Ph / OH / Br / F (373.9) $\xrightarrow{KO^tBu}$ - HO^tBu, - KF → O_2S–NH / (R) / Ph / O / Br (353.9)

Reactants: 373.9 + 112 = 485.9
Products: 353.9 + 74 + 58 = 485.9
AE = 353.9/485.9 = 0.728, or 72.8%

Part 2

Using the *REACTION-template.xls* spreadsheet and the details of the MACOS experiment, determine the material efficiency for the second step. What is the expected processing time to carry out the reaction? State any assumptions made in the calculations.

Solution

2.07 g of product represents 2.07/353.9 = 0.00585 moles
The limiting reagent is the sulfonamide.
If the reaction yield is 73%, then the number of moles of sulfonamide used is 0.00585/0.73 = 0.00801.
Therefore, the number of moles of potassium *t*-butoxide used is 3*0.00801 = 0.0240.
The combined concentration of sulfonamide and potassium *t*-butoxide is 0.8 M.
Hence, the volume of DMSO required for the stock solution containing both reagents is (0.00801 + 0.024)*1000/0.8 = 40 mL.
If the flow rate from each syringe is 100 μL/min, then the total processing time based on the volume of each syringe (20 mL) is 20e−3/100e−6 = 200 min.

E-kernel		0.88
E-excess		0.87
E-rxn solvent		21.28
E-catalyst		0
E-work-up		0
E-purification		0
	E-aux	21.28
E-total		23.02
Yield		73.0%
AE		72.8%
PMI		24.02
RME (global)		4.2%

The waste metrics are lower limit estimates because they exclude the volume of DMSO solvent used to prime the entire system.

Example 6.59 [61]

Themes: balancing chemical equations, redox reactions, material efficiency metrics

A bioreduction of aromatic aldehydes to alcohols using extract from *Aloe vera* under microwave irradiation conditions was claimed as a green alternative to standard metal or metal hydride reducing agents. An example procedure is shown below.

$$\text{PhCHO} \xrightarrow[\textbf{65 \%}]{\textit{A. vera} \text{ extract, microwaves}} \text{PhCH}_2\text{OH}$$

Experimental Procedure:
An amount (63 g) of the pulp of *A. vera* was extracted, cut into small pieces ($0.5\ cm^3$) and blended for 60 s. Then, aromatic aldehyde (0.6 g) was added and deionized water was also added to the reaction mixture for a total volume of

60 mL. The reaction mixture was placed in the microwave reactor at 100 watts potency (100°C at 300 psi) for 70 min. The reaction was monitored by TLC. After reaction time, the hot mixture was allowed to cool to room temperature. The suspension was extracted with 50 mL of ethyl acetate three times. The organic phase was filtrated under vacuum to remove any plant solid, dried over anhydrous sodium sulfate and concentrated in a rotary evaporator. The residue was passed through a short silica gel column using chloroform as eluent. The solvent was dried and removed to yield a clear liquid.

Part 1

Based on the experimental procedure, determine the *PMI* for the reaction. State any assumptions made in the calculation.

Solution

Assumptions: 60 mL deionized water used; 5 g sodium sulfate drying agent

Input material	Volume (mL)	Density (g/mL)	Mass (g)
A. vera pulp			63
Benzaldehyde			0.6
Water	60	1	60
EtOAc	150	0.902	135.3
Na_2SO_4			5
Total			263.9

0.6 g benzaldehyde (MW 106 g/mol) represents 0.6/106 = 0.00566 moles

A yield of 65% of benzyl alcohol (MW 108 g/mol) represents 0.65*0.00566 = 0.00368 moles or 108*0.00368 = 0.4 g.

PMI = 263.9/0.4 = 659.75

Part 2

The authors did not specify what reducing agents are found in *A. vera* extract that would affect the reduction reaction. Suggest how the reduction takes place.

Solution

Possible reducing agents found in biological systems are $NADH_2$ (nicotinamide adenosine dinucleotide) and $NADPH_2$ (nicotinamide adenosine dinucleotide phosphate). The reduction could take place via the same transfer hydrogenation mechanism that is operative in Hantzsch dihydropyridine reagents.

Another possible biological reducing agent is a reductase enzyme.

Without a precise knowledge of the reducing agent immediately implies that a fully balanced chemical equation cannot be written for the transformation. This, in turn, means that an atom economy cannot be determined.

Example 6.60

Themes: balancing chemical equations, step versus global material efficiency metrics

EXPERIMENTAL PROCEDURES:

Step 1 [62]

500 g chloroacetic acid in 700 mL water
290 g sodium carbonate
294 g sodium cyanide in 750 mL water
200 mL cold water addition
694 g conc. hydrochloric acid (d = 1.156 g/mL)
600 mL 95% ethanol washing
500 mL 95% ethanol washing
600 + 300 mL absolute ethanol
10 + 4 mL conc. sulfuric acid
Yield of ethyl cyanoacetate = 474 g

Step 2 [63]

400 g ethyl cyanoacetate
500 mL conc. aqueous ammonia (d = 0.90 g/mL)
2 × 50 mL cold ethanol washings
350 + 100 mL 95% ethanol recrystallization solvent
Yield of cyanoacetamide = 255 g

Part 1

Write out balanced chemical equations for both steps.

Part 2

Using the *REACTION-template.xls* and *SYNTHESIS-template.xls* spreadsheets and the details of the experimental procedure, determine the step and global material efficiency metrics. State any assumptions made in the calculations.

Step	Yield	AE	1/SF	MRP	RME	*PMI*	E
1							
2							
Overall							

Part 3

From Part 2 summarize the following for each step:

kernel level → RME, *PMI, E*
Curzons level → RME, *PMI, E*
Global level → RME, *PMI, E*

Step	RME (kernel)	*PMI* (kernel)	E (kernel)
1			
2			
Overall			

Step	RME (Curzons)	*PMI* (Curzons)	E (Curzons)
1			
2			
Overall			

Step	RME (global)	*PMI* (global)	E (global)
1			
2			
Overall			

Verify numerically the following relationships for each step:

$$(\mathrm{RME})_{\mathrm{kernel}} = (\mathrm{AE})\varepsilon$$

$$(\mathrm{RME})_{\mathrm{Curzons}} = (\mathrm{AE})\varepsilon\left(\frac{1}{\mathrm{SF}}\right)$$

$$(\mathrm{RME})_{\mathrm{global}} = (\mathrm{AE})\varepsilon\left(\frac{1}{\mathrm{SF}}\right)(\mathrm{MRP})$$

Part 4

For the two-step plan investigate the connection between the product of the step RMEs and the overall RME at the kernel, Curzons, and global levels. Repeat for the connection between the sum of the step *E*-factors and the overall *E*-factor at the three levels. Repeat for the connection between the sum of the step *PMIs* and the overall *PMI* at the three levels.

Part 5

Verify numerically the following connecting relationships between step and overall metrics at the kernel, Curzons, and global levels:

$$(\mathrm{RME})_T = \frac{m_p}{\dfrac{m_p}{(\mathrm{RME})_N} + \sum_{j=1}^{N-1} m_{l_j}\left[\dfrac{1}{(\mathrm{RME})_j} - 1\right]}$$

$$(PMI)_T = \frac{m_p (PMI)_N + \sum_{j=1}^{N-1} m_{l_j}\left[(PMI)_j - 1\right]}{m_p}$$

$$E_T = \frac{m_p\left(E_N + 1\right) + \sum_{j=1}^{N-1} m_{I_j} E_j}{m_p} - 1$$

Solution

Part 1

Step 1

$$\text{ClCH}_2\text{COOH} \; (94.45) \xrightarrow[-\,0.5\,H_2O,\; -\,0.5\,CO_2]{0.5\,Na_2CO_3} \text{ClCH}_2\text{COONa} \xrightarrow[-\,2\,NaCl]{NaCN,\; HCl} \text{NCCH}_2\text{COOH} \xrightarrow[-\,H_2O]{H_2SO_4\,(cat.),\; EtOH} \text{NCCH}_2\text{COOEt} \; (113)$$

Reactants: 94.45 + 0.5*106 + 49 + 36.45 + 46 = 278.9
Products: 113 + 0.5*18 + 0.5*44 + 2*58.45 + 18 = 278.9

Step 2

$$\text{NCCH}_2\text{COOEt} \; (113) \xrightarrow[-\,H_2O,\; -\,EtOH]{NH_4OH} \text{NCCH}_2\text{CONH}_2 \; (84)$$

Reactants: 113 + 35 = 148
Products: 84 + 18 + 46 = 148

Part 2

Step 1

Assumptions: conc. hydrochloric acid aqueous solution is 37 wt%.

Step 2

Assumptions: conc. ammonium hydroxide aqueous solution is 28 wt%.

Step	Yield	AE	1/SF	MRP	RME	*PMI*	E
1	0.792	0.405	0.720	0.405	0.094	10.68	9.68
2	0.858	0.568	0.996	0.406	0.197	5.08	4.08
Overall	0.680	0.268	0.738	0.368	0.049	20.26	19.26

Part 3

Step	RME (kernel)	*PMI* (kernel)	E (kernel)
1	0.322	3.11	2.11
2	0.488	2.05	1.05
Overall	0.186	5.37	4.37

Step	RME(Curzons)	*PMI*(Curzons)	E(Curzons)
1	0.231	4.32	3.32
2	0.485	2.06	1.06
Overall	0.137	7.28	6.28

Step	RME(Global)	*PMI*(Global)	E(Global)
1	0.094	10.68	9.68
2	0.197	5.08	4.08
Overall	0.049	20.26	19.26

$$(\text{RME})_{\text{kernel}} = (\text{AE})\varepsilon$$

Step 1: (AE)*Yield = 0.405*0.792 = 0.321
Step 2: (AE)*Yield = 0.568*0.858 = 0.487
Overall: (AE)*Yield = 0.268*0.680 = 0.182

$$(\text{RME})_{\text{Curzons}} = (\text{AE})\varepsilon\left(\frac{1}{\text{SF}}\right)$$

Step 1: (AE)*Yield*(1/SF) = 0.405*0.792*0.720 = 0.231
Step 2: (AE)*Yield*(1/SF) = 0.568*0.858*0.996 = 0.485
Overall: (AE)*Yield*(1/SF) = 0.268*0.680*0.738 = 0.134

$$(\text{RME})_{\text{global}} = (\text{AE})\varepsilon\left(\frac{1}{\text{SF}}\right)(\text{MRP})$$

Step 1: (AE)*Yield*(1/SF)*(MRP) = 0.405*0.792*0.720*0.405 = 0.094
Step 2: (AE)*Yield*(1/SF)*(MRP) = 0.568*0.858*0.996*0.406 = 0.197
Overall: (AE)*Yield*(1/SF)*(MRP) = 0.268*0.680*0.738*0.368 = 0.049

Note: The MRP factor appearing in the global RME expression is a fudge factor, not equal to a true MRP as defined for a single reaction, since the full expression for the global RME cannot be factored into the four factors as indicated. Also, the kernel, Curzons, and global RME expressions for the overall case (taking into account both reaction steps) do not give a numerical value that is exactly the same as the true kernel, Curzons, and global RME values. For example, the actual overall kernel RME is 0.186, whereas the value obtained by multiplying overall AE times overall yield is 0.182. This is due to the fact that the overall kernel RME expression cannot be factored into overall AE and overall yield as shown below.

$$(RME)_{kernel,overall} = \frac{84}{\frac{35}{0.86} + \frac{94.45+53+49+36.45+46}{0.86*0.79}}$$

$$= \frac{84}{\frac{35}{0.86} + \frac{278.9}{0.86*0.79}}$$

$$= 0.86*0.79*\frac{84}{35*0.79+278.9}$$

$$= Y_{overall}\frac{84}{35*0.79+278.9}$$

$$(AE)_{overall} = \frac{84}{35+278.9}$$

The only way that $(RME)_{overall} = (AE)_{overall} Y_{overall}$ for the two-step plan is when the yield of step 1 is 100%. In general, at the kernel level for any linear synthesis plan, $(RME)_{overall} = (AE)_{overall} Y_{overall}$ is true if all reaction yields up to the penultimate step are 100%. The last step can have any yield value that will appear as a common factor in the expression for overall kernel RME. This means that $(RME)_{overall} = (AE)_{overall} Y_{overall}$ reduces to $(RME)_{overall} = (AE)_{overall} \varepsilon_N$, where ε_N is the yield of the last step.

Part 4

Kernel level

RME(1)*RME(2) = 0.322*0.488 = 0.157 vs RME(overall) = 0.186

Curzons level

RME(1)*RME(2) = 0.231*0.485 = 0.112 vs RME(overall) = 0.137

Global level

RME(1)*RME(2) = 0.094*0.197 = 0.019 vs RME(overall) = 0.049

In each case, RME(1)*RME(2) < RME(overall).

Kernel level

$E(1) + E(2) = 2.11 + 1.05 = 3.16$ vs $E(\text{overall}) = 4.37$

Curzons level

$E(1) + E(2) = 3.32 + 1.06 = 4.38$ vs $E(\text{overall}) = 6.28$

Global level

$E(1) + E(2) = 9.68 + 4.08 = 13.76$ vs $E(\text{overall}) = 19.26$

In each case, *E*(1) + *E*(2) < *E*(overall).
Kernel level

$$PMI(1)+PMI(2)=3.11+2.05=5.16 \text{ vs } PMI(\text{overall})=5.37$$

Curzons level

$$PMI(1)+PMI(2)=4.32+2.06=6.38 \text{ vs } PMI(\text{overall})=7.28$$

Global level

$$PMI(1)+PMI(2)=10.68+5.08=15.76 \text{ vs } PMI(\text{overall})=20.26$$

In each case, *PMI*(1) + *PMI*(2) < *PMI*(overall).

Part 5

$$\boxed{(\text{RME})_T=\frac{m_p}{\dfrac{m_p}{(\text{RME})_N}+\displaystyle\sum_{j=1}^{N-1} m_{I_j}\left[\dfrac{1}{(\text{RME})_j}-1\right]}}$$

At the kernel level:

$$(\text{RME})_T=\frac{84}{\dfrac{84}{0.488}+\dfrac{113}{0.86}\left[\dfrac{1}{0.322}-1\right]}=\frac{84}{448.796}=0.187$$

At the Curzons level:

$$(\text{RME})_T=\frac{84}{\dfrac{84}{0.485}+\dfrac{113}{0.86}\left[\dfrac{1}{0.231}-1\right]}=\frac{84}{610.612}=0.138$$

At the global level:

$$(\text{RME})_T=\frac{84}{\dfrac{84}{0.197}+\dfrac{113}{0.86}\left[\dfrac{1}{0.094}-1\right]}=\frac{84}{1692.823}=0.0496$$

$$\boxed{(PMI)_T=\frac{m_p(PMI)_N+\displaystyle\sum_{j=1}^{N-1} m_{I_j}\left[(PMI)_j-1\right]}{m_p}}$$

At the kernel level:

$$(PMI)_T=\frac{84*2.05+\dfrac{113}{0.86}[3.11-1]}{84}$$

$$=2.05+1.564(2.11)=5.35$$

At the Curzons level:

$$(PMI)_T = \frac{84*2.06+\frac{113}{0.86}[4.32-1]}{84}$$

$$= 2.06+1.564(3.32) = 7.25$$

At the global level:

$$(PMI)_T = \frac{84*5.08+\frac{113}{0.86}[10.68-1]}{84}$$

$$= 5.08+1.564(9.68) = 20.22$$

$$\boxed{E_T = \frac{m_p(E_N+1)+\sum_{j=1}^{N-1} m_{l_j}E_j}{m_p} - 1}$$

At the kernel level:

$$E_T = \frac{84(1.05+1)+\frac{113}{0.86}(2.11)}{84} - 1$$

$$= 2.05+1.564*2.11-1 = 4.35$$

At the Curzons level:

$$E_T = \frac{84(1.06+1)+\frac{113}{0.86}(3.32)}{84} - 1$$

$$= 2.06+1.564*3.32-1 = 6.25$$

At the global level:

$$E_T = \frac{84(4.08+1)+\frac{113}{0.86}(9.68)}{84} - 1$$

$$= 5.08+1.564*9.68-1 = 19.22$$

Example 6.61 [64]

Themes: balancing chemical equations, complete materials metrics analysis, one substrate leading to more than one product, side reactions

Reaction of phenol with nitric acid or related derivatives leads to major products that arise from aromatic nitration and minor benzoquinone products that arise from aromatic oxidation. The nitration products and the benzoquinone product arise from different reaction mechanisms.

An example reaction conducted under reaction solvent-free conditions using a mechanochemical strategy has been reported as shown below.

oxalic acid
$NaNO_3$
grinding

28 % 42 % 4 %

Reaction materials:

1 mmol phenol
1.2 mmol oxalic acid
1.2 mmol sodium nitrate

Work-up materials:

20 mL dichloromethane

Product materials:

28% *o*-nitrophenol
42% *p*-nitrophenol
4% *p*-benzoquinone

Part 1

Write a balanced chemical equation for the main nitration reaction showing all associated by-products.

Part 2

Write a balanced chemical equation for the side oxidation reaction showing all associated by-products.

Part 3

For the main reaction, determine the reaction yield, AE, *E*, RME, and *PMI* for the production of *o*-nitrophenol and for the production of *p*-nitrophenol.

Part 4

For the side reaction, determine the reaction yield, AE, *E*, RME, and *PMI* for the production of *p*-benzoquinone.

Part 5

Determine the global reaction yield, AE, *E*, RME, and *PMI* for both nitrophenols produced in this reaction. Verify that the connecting relationships found in Part 3 of the Cannizzaro reaction problem are true for the individual nitrophenols and for both nitrophenols.

Part 6

Determine the global *E*, RME, and *PMI* for all products produced in this reaction.

Part 7

Explain why a global AE and reaction yield for all products cannot be determined.

Solution

Part 1

Mass balance	
Input materials	**Mass (g)**
Phenol	0.094
Oxalic acid	0.108
Sodium nitrate	0.102
Dichloromethane	26.72
TOTAL	27.024
Product materials	**Mass (g)**
o-Nitrophenol	0.039
p-Nitrophenol	0.058
p-Benzoquinone	0.0043

Main nitration reaction:

$$C_6H_5OH + NaNO_3 + HOOC{-}COOH \longrightarrow HOC_6H_4NO_2 + HOOC{-}COONa + H_2O$$

94 **85** **90** **139** **112** **18**

Check balancing:

Reactants: 94 + 85 + 90 = 269
Products: 139 + 112 + 18 = 269

Part 2

Side oxidation reaction:

$$C_6H_5OH + 2\,NaNO_3 \longrightarrow C_6H_4O_2 + H_2O + 2\,NaNO_2$$

94 **2(85)** **108** **18** **2(69)**

Check balancing:

Reactants: 94 + 2*85 = 264
Products: 108 + 18 + 2*69 = 264

Part 3

Metrics

Product	**Limiting reagent**	**AE**	**RY**	**E**	**RME**	***PMI***
o-Nitrophenol	Phenol	0.517	0.28	691.9	0.00144	692.9
p-Nitrophenol	Phenol	0.517	0.42	464.9	0.00215	465.9

Part 4

Metrics

Product	**Limiting reagent**	**AE**	**RY**	**E**	**RME**	***PMI***
p-Benzoqui-none	NaNO3	0.409	0.066	6283.7	0.00016	6284.7

Part 5

Metrics

Product	**Limiting reagent**	**AE**	**RY**	**E**	**RME**	***PMI***
Both nitrophenols	Phenol	0.517	0.70	277.6	0.00359	278.6

The atom economy for producing o-nitrophenol is the same as that for producing *p*-nitrophenol.

RY(*o*-nitrophenol) = 0.28
RY(*p*-nitrophenol) = 0.42
RY(both nitrophenols) = 0.28 + 0.42 = 0.70
E(*o*-nitrophenol) = 691.9
E(*p*-nitrophenol) = 464.9
E(both nitrophenols) = (691.9*464.9 – 1)/(691.9 + 464.9 + 2) = 277.6
RME(*o*-nitrophenol) = 0.00144
RME(*p*-nitrophenol) = 0.00215
RME(both nitrophenols) = 0.00144 + 0.00215 = 0.00359
PMI(*o*-nitrophenol) = 692.9
PMI(*p*-nitrophenol) = 465.9
PMI(both nitrophenols) = 692.9*465.9/(692.9 + 465.9) = 278.6

Part 6

Metrics

Product	**E**	**RME**	***PMI***
All products	265.8	0.00375	266.8

$PMI = 27.024/(0.039 + 0.058 + 0.0043) = 266.8$
$\mathrm{RME} = 1/PMI = 1/266.8 = 0.00375$
$\mathrm{E} = PMI - 1 = 266.8 - 1 = 265.8$

Recall that determination of global *E*, global *PMI*, and global RME does not depend on knowledge of any chemical reaction.

Part 7

The nitration reaction and the oxidation reaction are different reactions occurring by different mechanisms. Therefore, they cannot be added to obtain global AE and RY values.

Example 6.62 [65, 66]

Themes: balancing chemical equations, recycling, process mass intensity

Part 1

Using the *REACTION-template.xls* and *SYNTHESIS-template.xls* spreadsheets, determine the *PMI* for the Wittig reaction shown below given the experimental procedure details and if 1000 g of methylenecyclohexane are to be made.

$$Ph_3P + CH_3Br \longrightarrow Ph_3\overset{\oplus}{P}{-}CH_3\ \ \overset{\ominus}{Br} \xrightarrow{\text{nBuLi}}$$

Step 1

55 g triphenylphosphine
28 g methylbromide
45 mL ($d = 0.879$ g/mL) benzene
74 g of methyl triphenylphosphonium bromide collected which was recrystallized from
500 mL benzene

Step 2

35.7 g methyl triphenylphosphonium bromide
10.8 g cyclohexanone
62.5 mL of 1.6 M nBuLi in hexane solution ($d = 0.68$ g/mL)
200 mL ($d = 0.708$ g/mL) diethyl ether
Workup: 100 mL diethyl ether, 100 mL water
3.8 of methylenecyclohexane was collected

Part 2

If 1 kg of methylenecyclohexane is produced by this method, what is the expected mass of triphenylphosphine oxide that is co-produced?

Part 3

The triphenylphosphine oxide by-product from the Wittig reaction can be recycled back to triphenylphosphine using the reaction shown below.

$$Ph_3P{=}O + Cl_2C{=}O \longrightarrow [Ph_3PCl_2] \xrightarrow{Al} Ph_3P$$

The experimental details are as follows.

880 g/h of a 12.5 wt% triphenylphosphine oxide solution in chlorobenzene
9200 g phosgene
1519 g aluminum
21000 g triphenylphosphine collected after 211 h

Determine the *PMI* for the recycling reaction.

Part 4

Suppose the collected mass of triphenylphosphine oxide in Part 2 is recycled back to triphenylphosphine according to the procedure given in Part 3. What is the expected mass of triphenylphosphine that is recovered? What fraction of the initial triphenylphosphine input is recovered?

Part 5

Determine the *PMI* for the combined Wittig and recycling reaction if 1 kg of methylenecyclohexane is produced. Is it worth carrying out the recycling reaction based on *PMI* considerations?

SOLUTION

Part 1

Target product: 1000 g methylenecyclohexane

Input materials:

		Mass (g)
Step 1	Triphenylphosphine	6982.58
	Methyl bromide	3554.77
	Benzene solvent	5021.11
	Benzene recrystallization solvent	55797.16
Step 2	Cyclohexanone	2842.11
	nBuLi	1684.21
	Et_2O	37263.17
	Hexane	9500
	Total	122645.1

PMI for Wittig reaction = 122645.1/1000 = 122.65

Part 2

Based on experimental procedure for step 2:

Number of moles of methylenecyclohexane collected = 0.03958
Number of moles of triphenylphosphine oxide = 0.03958
Mass of triphenylphosphine oxide = 0.03958*278 = 11.00 g
11 g triphenylphosphine oxide is produced along with 3.8 g methylenecyclohexane
Therefore, if 1000 g methylenecyclohexane are collected, then 1000*(11/3.8) = 2894.74 g triphenylphosphine oxide is expected.

Part 3

PMI for recycling reaction = 9.35

Part 4

From experimental procedure of recycling reaction:

23210 g triphenylphosphine oxide is converted to 21000 g triphenylphosphine
Therefore, 2894.74 g triphenylphosphine oxide from Part 2 can be converted to 2894.74*(21000/23210) = 2619.11 g triphenylphosphine.
To produce 1000 g methylenecyclohexane, 6982.58 g of triphenylphosphine are required.
Fraction of triphenylphosphine recovered = 2619.11/6982.58 = 0.375, or 37.5%

Part 5

Target product: 1000 g methylenecyclohexane
Recovered material: 2619.11 g triphenylphosphine

Input materials used:

		Mass (g)
Step 1	Triphenylphosphine	6982.58
	Methyl bromide	3554.77
	Benzene solvent	5021.11
	Benzene recrystallization solvent	55797.16
Step 2	Cyclohexanone	2842.11
	^{n}BuLi	1684.21
	Et2O	37263.17
	Hexane	9500
Recycling reaction	Phosgene	1147.42
	Aluminum	189.45
	Chlorobenzene	20263.18
	TOTAL	144245.2

Mass of material needed for recycling reaction $= M_R = 1147.42 + 189.45 + 20263.18 = 21600.1$ g
Mass of recovered material $= M_r = 2619.11$ g
Mass of product (methylenecyclohexane) $= M_P = 1000$ g
Mass of input materials used in original plan $= M_0 = 122645.1$ g
PMI for combined Wittig and recycling reactions = (144245.2 – 2619.11)/1000 = 141.63
Cf. *PMI* for Wittig reaction = 122.65

Since the *PMI* for the combined reactions is larger than the *PMI* for the original Wittig reaction, it appears that recycling triphenylphosphine oxide back to triphenylphosphine is not worth it. This conclusion is consistent with the observation that the mass of recovered material (M_r) is much less than the mass of material needed for carrying out the recycling reaction (M_R). The ratio of M_R to M_r is 21600.1/2619.11 = 8.25.

Since $M_R > M_r$, we can numerically show that the *PMI* of the combined Wittig and recycling reactions will always be larger than the *PMI* of the Wittig reaction alone regardless of the yield of the Wittig reaction.

Let the yield of the Wittig reaction be *x*.

1000 g of methylenecyclohexane = 1000/96 = 10.417 moles

Mass of reagents required:

Cyclohexanone: (10.417/*x*)*98 g
nBuLi: (10.417/*x*)*64 g
MeBr: (10.417/(*x**0.99))*94.9 g
Ph_3P: (10.417/(*x**0.99))*262 g
Scaling factor for step 2: 1000/(*x**0.1*96)
Scaled mass of limiting reagent in step 2 (ylide): (1000/(*x**0.1*96))*35.7 g
Scaling factor for step 1: (1000/(*x**0.1*96))*35.7/74

Mass of auxiliary materials required:

Benzene (step 1): (39.55 + 439.5)*((1000/(x*0.1*96))*35.7/74) g
Ether (step 2): 141.6*(1000/(*x**0.1*96)) g
Hexane (step 2): 36.1*(1000/(*x**0.1*96)) g

Expected mass of triphenylphosphine oxide produced in Wittig reaction

= (*x**0.1*278)*(1000/(*x**0.1*96)) = 2895.83 g

If 1000 g (10.417 moles) methylenecyclohexane are collected, then 10.417 moles (10.417*278 = 2895.93 g) of triphenylphosphine oxide are also produced as a co-product.

If 23210 g triphenylphosphine oxide can be recycled to 21000 g triphenylphosphine, then (*x**0.1*278)*(1000/(*x**0.1*96)) g triphenylphosphine oxide can be recycled to (*x**0.1*278)*(1000/(*x**0.1*96))*(21000/23210) g triphenylphosphine

Therefore, mass of recovered material = M_r = **2620.1 g**

Scaling factor for recycling reaction:

(*x**0.1*278)*(1000/(*x**0.1*96))*(21000/23210)/21000

Mass of materials required for recycling reaction:

Phosgene: 9200*(*x**0.1*278)*(1000/(*x**0.1*96))*(21000/23210)/21000 = 1147.85 g
Aluminum: 1519*(*x**0.1*278)*(1000/(*x**0.1*96))*(21000/23210)/21000 = 189.52 g
Chlorobenzene: 162470*(*x**0.1*278)*(1000/(*x**0.1*96))*(21000/23210)/21000 = 20270.83 g
Mass of material needed for recycling reaction = M_R = **21608.21 g**
Mass of input materials used in original plan = M_0
Cyclohexanone: (10.417/*x*)*98 = 1020.87/*x* g
nBuLi: (10.417/*x*)*64 = 666.69/*x* g
MeBr: (10.417/(*x**0.99))*94.9 = 998.56/*x* g
Ph_3P: (10.417/(*x**0.99))*262 = 2756.82/*x* g
Benzene (step 1): (39.55 + 439.5)*((1000/(*x**0.1*96))*35.7/74) = 24073.88/*x* g
Ether (step 2): 141.6*(1000/(*x**0.1*96)) = 14750/*x* g
Hexane (step 2): 36.1*(1000/(*x**0.1*96)) = 3760.42/*x* g
M_0 = 48027.23/x g
PMI for Wittig reaction = (48027.23/*x*)/1000 = 48.03/*x*
PMI for combined Wittig and recycling reactions = ((48027.23/*x*) + 21608.21 – 2620.1)/1000

x	*PMI* (Wittig)	*PMI* (Wittig + recycling)
0.1	480.3	499.3
0.2	240.1	259.1
0.3	160.1	179.1
0.4	120.1	139.1
0.5	96.1	115.0
0.6	80.0	99.0
0.7	68.6	87.6
0.8	60.0	79.0
0.9	53.4	72.4
1	48.0	67.0

Under all circumstances, *PMI* (Wittig + recycling) > *PMI* (Wittig), no matter what the yield for the Wittig reaction is. The difference between the *PMI* values is constant at 19.

The initial mass of triphenylphosphine used is (10.417/(*x**0.99))*262 = 2756.82/*x* g.

The recovered mass of triphenylphosphine = 2620.1 g.
The optimal % recovery of triphenylphosphine occurs when $x = 1$, that is, at 2620.1/2756.82 = 0.95, or 95%.

Example 6.63 [48]

Themes: synthesis tree diagrams, atom economy scenario analysis

Suppose in a linear plan consisting of five steps an oxidation reaction of a hydroxyl group to a carbonyl group occurs. Suppose that the AE for the oxidation reaction is 30% and the molecular weight of the product of the reaction is 150. If the oxidizing agent is replaced by another that results in an increase of the atom economy for the reaction by 20%, explore what impact this will have on the overall AE for the entire synthesis?

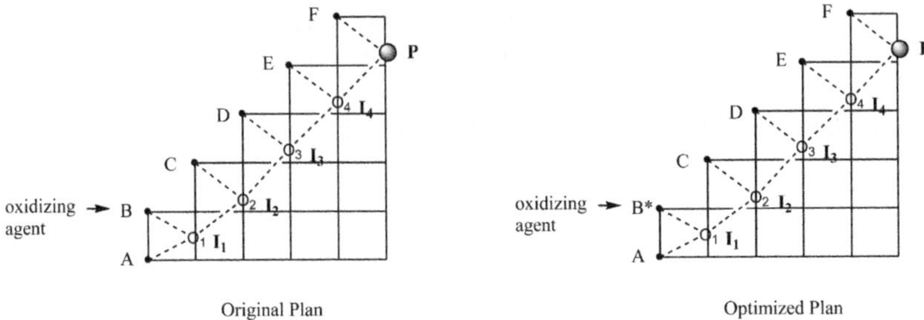

Solution

Oxidation reaction in original plan:

$$(\mathrm{AE})_1 = \frac{i_1}{a+b} = 0.3$$

Atom economy for the entire plan is

$$(\mathrm{AE})_{\mathrm{overall}} = \frac{p}{a+b+X} = \frac{p}{\frac{i_1}{0.3}+X}$$

where $X = c+d+e+f$.
Oxidation reaction in optimized plan:
Increasing the atom economy of step 1 by 20% means that the new atom economy for the same reaction is

0.3*(1 + 0.2) = 0.36 (36%).

Hence,

$$(\mathrm{AE})_1^* = \frac{i_1}{a+b^*} = 0.36$$

Atom economy for the entire plan is
The resulting change in overall atom economy is

$$\Delta(\mathrm{AE})_{\text{overall}} = \frac{p}{\dfrac{i_1}{0.36}+X} - \frac{p}{\dfrac{i_1}{0.3}+X}$$

$$= p\left[\frac{0.36}{i_1+0.36X} - \frac{0.3}{i_1+0.3X}\right]$$

The percent change in overall atom economy is

$$\%\Delta(\mathrm{AE})_{\text{overall}} = \frac{p\left[\dfrac{0.36}{i_1+0.36X} - \dfrac{0.3}{i_1+0.3X}\right]}{p\left[\dfrac{0.3}{i_1+0.3X}\right]} * 100$$

$$= \left[\frac{0.36}{0.3} * \frac{i_1+0.3X}{i_1+0.36X} - 1\right] * 100$$

If the molecular weight of the product of the oxidation reaction is 150 ($i_1 = 150$), then

$$\%\Delta(\mathrm{AE})_{\text{overall}} = \left[1.2 * \frac{150+0.3X}{150+0.36X} - 1\right] * 100$$

X	% Change in AE
0	20
50	17.9
100	16.1
150	14.7
200	13.5
250	12.5
300	11.6
350	10.9
400	10.2
450	9.6
500	9.1
550	8.6
600	8.2
650	7.8
700	7.5
750	7.1
800	6.8
850	6.6
900	6.3
950	6.1
1000	5.9
1050	5.7
1100	5.5

X	% Change in AE
1150	5.3
1200	5.2
1250	5.0

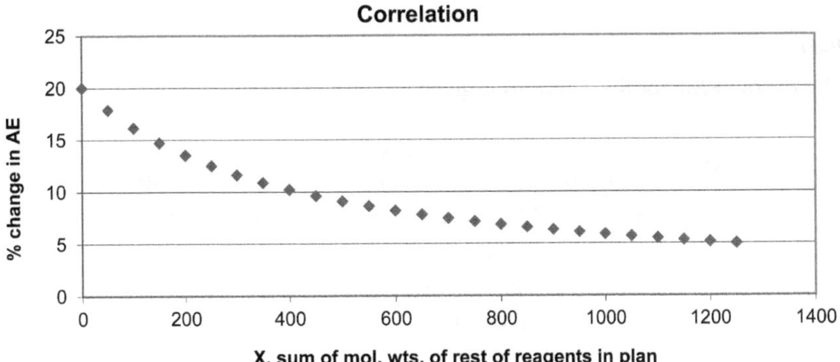

The % change in overall AE diminishes as X increases. The maximum impact is felt if the oxidation reaction were the only reaction in the plan (i.e., 20% increase in overall AE).

Example 6.64 [48]

Themes: synthesis tree diagrams, kernel *E*-factor, kernel reaction mass efficiency, scenario analysis

A linear synthesis plan has the following synthesis tree diagram with reaction yields and molecular weights of input reagents shown.

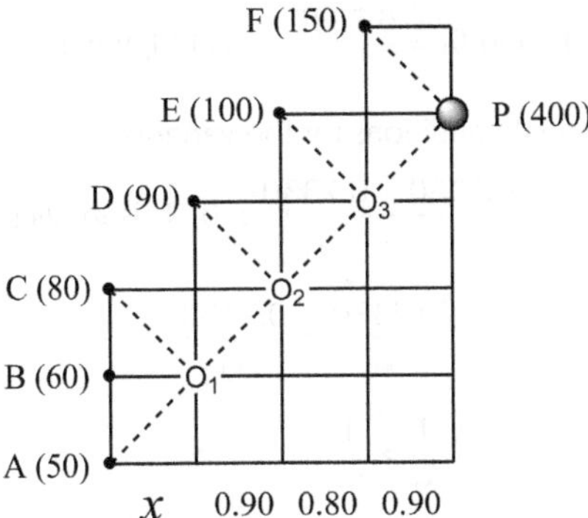

Part 1

If it is desired to reduce the overall *E*-factor for the plan by *y* percent, derive an expression that relates the dependence of the percent increase in reaction yield for step 1 with y? Assume that no excess reagents are used and all auxiliary materials are neglected.

Part 2

What is the maximum percent decrease in E-factor that is possible if the starting reaction yield for step 1 is 70%?

Solution

Part 1

The overall RME for the synthesis is given by

$$\mathrm{RME} = \frac{400}{\dfrac{50+60+80}{x*0.9*0.8*0.9}+\dfrac{90}{0.9*0.8*0.9}+\dfrac{100}{0.8*0.9}+\dfrac{150}{0.9}}$$

$$= \frac{400}{\dfrac{190}{0.648x}+\dfrac{90}{0.648}+\dfrac{100}{0.72}+\dfrac{150}{0.9}}$$

$$= \frac{400}{\dfrac{293.21}{x}+444.44}$$

The E-factor is given by

$$E = \frac{1}{\mathrm{RME}}-1 = \frac{\dfrac{293.21}{x}+444.44}{400}-1$$

$$= \frac{0.7330}{x}+1.1111-1 = \frac{0.7330}{x}+0.1111.$$

Therefore,

$E_1 = \dfrac{0.7330}{x_1}+0.1111$ and $E_2 = \dfrac{0.7330}{x_2}+0.1111$, where the subscripts 1 and 2 refer to the initial and final conditions for the variables.

If $E_2 < E_1$, then $E_1 - E_2 = \dfrac{0.7330}{x_1} - \dfrac{0.7330}{x_2} > 0$. This implies that

$$\frac{0.7330}{x_1} > \frac{0.7330}{x_2}$$

$$\frac{1}{x_1} > \frac{1}{x_2}$$

$$x_1 < x_2$$

This is consistent with the result of optimizing (increasing) the reaction yield for step 1 from x_1 to x_2.

If E is reduced by y percent, the final E-factor value is $E_2 = \left(1-\dfrac{y}{100}\right)E_1$.

Let $z = 1 - \frac{y}{100}$.

$$E_1 - E_2 = \frac{0.7330}{x_1} - \frac{0.7330}{x_2}$$

$$E_1 - zE_1 = 0.7330\left(\frac{1}{x_1} - \frac{1}{x_2}\right)$$

Let the reaction yield be increased by w percent. Then $x_2 = \left(1 + \frac{w}{100}\right)x_1$. Let $v = 1 + \frac{w}{100}$

Therefore,

$$E_1(1-z) = 0.7330\left(\frac{1}{x_1} - \frac{1}{vx_1}\right) = \frac{0.7330}{x_1}\left(1 - \frac{1}{v}\right)$$

$$\frac{E_1(1-z)x_1}{0.7330} = 1 - \frac{1}{v}$$

$$\frac{1}{v} = 1 - \frac{E_1(1-z)x_1}{0.7330}$$

$$v = \frac{1}{1 - \frac{E_1(1-z)x_1}{0.7330}}$$

Substituting $E_1 = \frac{0.7330}{x_1} + 0.1111$ we have

$$1 + \frac{w}{100} = \frac{1}{1 - (1-z)\left(1 + \frac{0.1111x_1}{0.7330}\right)}$$

$$= \frac{1}{1 - (1-z)(1 + 0.1516x_1)}$$

$$= \frac{1}{1 - \left(\frac{y}{100}\right)(1 + 0.1516x_1)}$$

$$w = \left[\frac{1}{1 - \left(\frac{y}{100}\right)(1 + 0.1516x_1)} - 1\right] * 100$$

Part 2

As an example, if the original yield of the first step in the plan is 70% and we wish to reduce the E-factor by 20%, then substituting $x_1 = 0.7$ and $y = 20$ in the above formula leads to $w = 28.4$. This means that the reaction yield for the first step needs to be increased by 28.4%. Increasing a yield of 70% by 28.4% leads to a final optimized yield of 89.9%.

The maximum yield for x_2 is 100%, which means that x_1 has a maximum possible increase of 42.86% ($w = 42.86$). The maximum percent decrease in *E*-factor is therefore 27.1% ($y = 27.1$).

A plot of *y* (% change in *E*-factor) versus *w* (% change in step 1 reaction yield) is shown below for a starting yield for the first step of 70% ($x_1 = 0.7$).

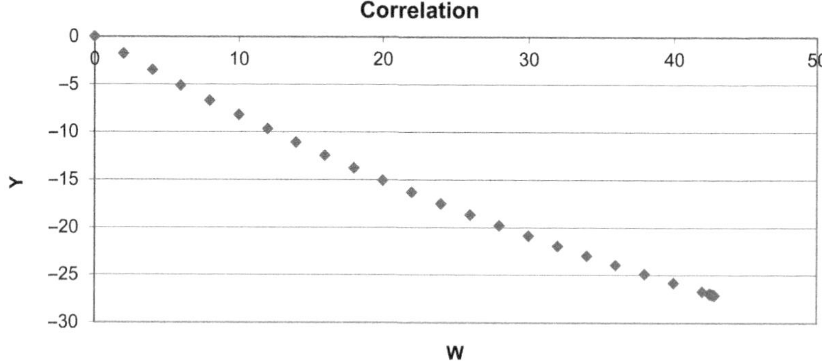

Example 6.65 [48]

Themes: synthesis tree diagrams, mass of reaction solvent scenario analysis

Part 1

Suppose in a 5-step linear plan every reaction yield is 80%. The experimental procedure for step 1 of the synthesis calls for 100 mL of reaction solvent at a 0.05 mole scale of limiting reagent. Determine the volume and mass of reaction solvent that is required for this step if it is desired to make 10 moles of final product P (MW = 400) in the synthesis plan (density of solvent = 1.5 g/mL). Assume that the molecular weight of all reagents is the same.

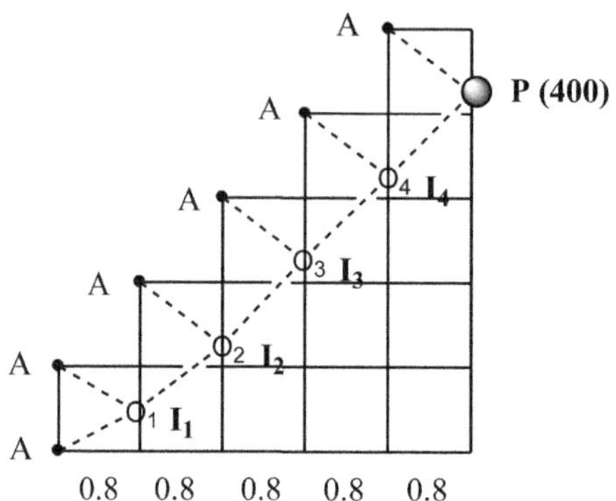

Part 2

Suppose it is desired to reduce the solvent usage in step 1 by 30%. How will this impact the component of overall *E*-factor for solvent usage for the entire plan given the data presented in the table below?

Scale of P	10	Moles				
Step	Yield	Mole scale (synthesis tree)	Mole scale (experimental procedure)	Volume of solvents used (mL)	Solvent density (g/mL)	Mass of solvent used (g)
5	0.8		0.05	100	1.5	
4	0.8		0.05	100	1.5	
3	0.8		0.05	100	1.5	
2	0.8		0.05	100	1.5	
1	0.8		0.05	100	1.5	

Part 3

Repeat the above analysis if the solvent usage in step 5 is reduced by 30%.

Solution

Part 1

10 moles of product P requires

$$10*\left[\frac{1}{(0.8)^5}\right] = 30.518 \text{ moles of reagent A in step 1.}$$

If 100 mL of reaction solvent is required when 0.05 moles of reagent A are used in step 1, then 100*(30.518/0.05) = 61035.16 mL of solvent are required when 30.518 moles of reagent A are used. Mass of solvent = 61035.16*1.5 = 91552.74 g.

Part 2

Original Plan:

Scale of P	10	Moles					
Step	Yield	Mole scale (synthesis tree)	Mole scale (experimental procedure)	Volume of solvents used (mL)	Solvent density (g/mL)	Mass of solvent used (g)	
5	0.8	12.500	0.05	100	1.5	37500	
4	0.8	15.625	0.05	100	1.5	46875	
3	0.8	19.531	0.05	100	1.5	58594	
2	0.8	24.414	0.05	100	1.5	73242	
1	0.8	30.518	0.05	100	1.5	91553	
						307764	SUM
MW of P	400	g/mol					
Mass of P	4000	g					
E-solvent	769.41						

E-solvent = 307764/4000 = 769.41

Optimized Plan (reduction of solvent usage in step 1 by 30%):

Scale of P	10	Moles					
Step	**Yield**	**Mole scale (synthesis tree)**	**Mole scale (experimental procedure)**	**Volume of solvents used (mL)**	**Solvent density (g/mL)**	**Mass of solvent used (g)**	
5	0.8	12.500	0.05	100	1.5	37500	
4	0.8	15.625	0.05	100	1.5	46875	
3	0.8	19.531	0.05	100	1.5	58594	
2	0.8	24.414	0.05	100	1.5	73242	
1	0.8	30.518	0.05	100	1.5	64087	***Reduced by 30%**
						280298	SUM
MW of P	400	g/mol					
Mass of P	4000	g					
E-solvent	700.74						
	−8.92	% change in *E*-solvent					

E-solvent = 280298/4000 = 700.74
% change in *E*-solvent = 100*(700.74 – 769.41)/769.41 = −8.92%

Part 3

Optimized Plan (reduction of solvent usage in step 5 by 30%):

Scale of P	10	Moles					
Step	**Yield**	**Mole scale (synthesis tree)**	**Mole scale (experimental procedure)**	**Volume of solvents used (mL)**	**Solvent density (g/mL)**	**Mass of solvent used (g)**	
5	0.8	12.500	0.05	100	1.5	26250	***Reduced by 30%**
4	0.8	15.625	0.05	100	1.5	46875	
3	0.8	19.531	0.05	100	1.5	58594	
2	0.8	24.414	0.05	100	1.5	73242	
1	0.8	30.518	0.05	100	1.5	91553	
						296514	SUM
MW of P	400	g/mol					
Mass of P	4000	g					
E-solvent	741.28						
	−3.66	% change in *E*-solvent					

E-solvent = 296514/4000 = 741.28
% change in *E*-solvent = 100*(741.28 – 769.41)/769.41 = −3.66%

We observe that a 30% reduction in solvent usage in step 1 has greater impact on *E*-solvent than when it is applied to step 5 since the mole scale for that step as prescribed by the synthesis tree diagram is more than twice as large; that is, 30.5 versus 12.5.

Example 6.66 [48]

Themes: synthesis tree diagrams, reaction yield scenario analysis

Part 1

Suppose in a 5-step linear plan every reaction yield is 80%. If the reaction yield of step 1 is increased by 10% by means of a catalyst, what impact will this have on the kernel RME, kernel *PMI*, kernel *E*-factor, and overall yield? Assume that the molecular weight of all reagents is the same.

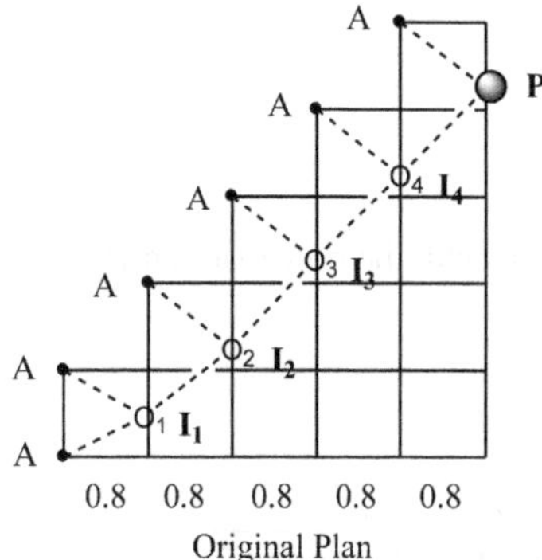

Original Plan

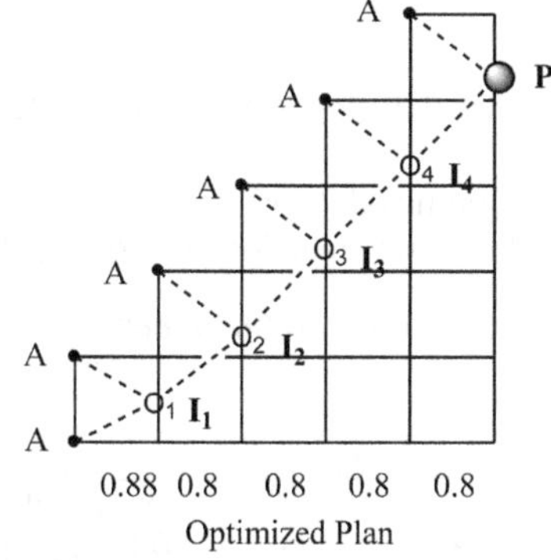

Optimized Plan

Part 2

Repeat the above analysis if the 10% increase was applied to the last step in the plan.

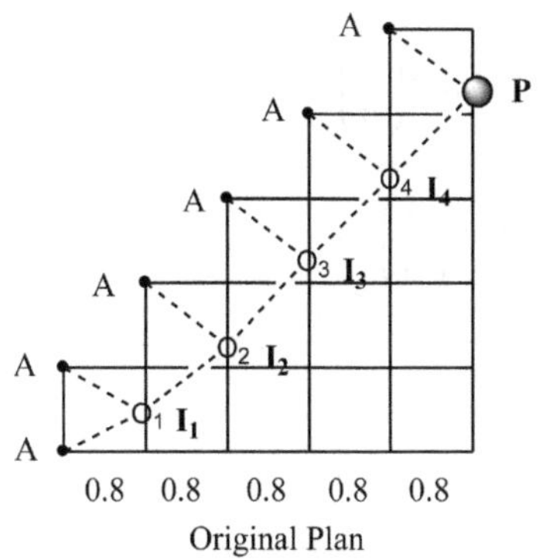

Original Plan

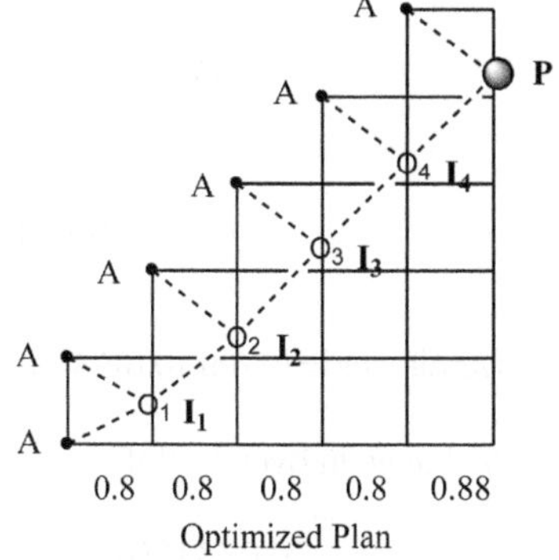

Optimized Plan

Solution

Part 1

Original Plan:

$$(\text{RME})_{\text{overall}} = \frac{p}{\frac{a}{0.8}+\frac{a}{(0.8)^2}+\frac{a}{(0.8)^3}+\frac{a}{(0.8)^4}+\frac{2a}{(0.8)^5}}$$

$$= \left(\frac{p}{a}\right)*\left(\frac{1}{13.311}\right) = 0.0751*\left(\frac{p}{a}\right)$$

$$(PMI)_{\text{overall}} = \frac{1}{(\text{RME})_{\text{overall}}} = 13.311*\left(\frac{a}{p}\right)$$

$$E_{\text{overall}} = (PMI)_{\text{overall}} - 1 = 13.311*\left(\frac{a}{p}\right) - 1$$

overall yield = (0.8)^5 = 0.328 (32.8%)

Optimized Plan:
Increasing reaction yield of step 1 by 10% means that the new reaction yield is

0.8*(1 + 0.1) = 0.88 (or 88%)

$$(\text{RME})_{\text{overall}} = \frac{p}{\frac{a}{0.8}+\frac{a}{(0.8)^2}+\frac{a}{(0.8)^3}+\frac{a}{(0.8)^4}+\frac{2a}{0.88*(0.8)^4}}$$

$$= \left(\frac{p}{a}\right)*\left(\frac{1}{12.756}\right)$$

$$= 0.0784*\left(\frac{p}{a}\right)$$

$$(PMI)_{\text{overall}} = \frac{1}{(\text{RME})_{\text{overall}}} = 12.756*\left(\frac{a}{p}\right)$$

$$E_{\text{overall}} = (PMI)_{\text{overall}} - 1 = 12.756*\left(\frac{a}{p}\right) - 1$$

overall yield = (0.88)*(0.8)^4 = 0.360 (36.0%)

% change in overall RME:

$$\frac{0.0784*(p/a) - 0.0751*(p/a)}{0.0751*(p/a)}*100 = 4.35\%$$

% change in overall *PMI*:

$$\frac{12.756*(a/p)-13.311*(a/p)}{13.311*(a/p)}*100=-4.17\%$$

% change in overall *E*:

$$\frac{12.756*(a/p)-1-\left[13.311*(a/p)-1\right]}{13.311*(a/p)-1}*100=\frac{-0.555*(a/p)}{13.311*(a/p)-1}*100$$

% change in overall yield:

$$\frac{0.360-0.328}{0.328}*100=10\%$$

Part 2

Optimized Plan:

$$(\text{RME})_{\text{overall}}=\frac{p}{\frac{a}{0.88}+\frac{a}{0.88*(0.8)}+\frac{a}{0.88*(0.8)^2}+\frac{a}{0.88*(0.8)^3}+\frac{2a}{0.88*(0.8)^4}}$$

$$=\left(\frac{p}{a}\right)*\left(\frac{1}{12.101}\right)=0.0826*\left(\frac{p}{a}\right)$$

$$(PMI)_{\text{overall}}=\frac{1}{(\text{RME})_{\text{overall}}}=12.101*\left(\frac{a}{p}\right)$$

$$E_{\text{overall}}=(PMI)_{\text{overall}}-1=12.101*\left(\frac{a}{p}\right)-1$$

overall yield = (0.88)*(0.8)^4 = 0.360 (36.0%)

% change in overall RME:

$$\frac{0.0826*(p/a)-0.0751*(p/a)}{0.0751*(p/a)}*100=10\%$$

% change in overall *PMI*:

$$\frac{12.101*(a/p)-13.311*(a/p)}{13.311*(a/p)}*100=-9.1\%$$

% change in overall E:

$$\frac{12.101*(a/p)-1-\left[13.311*(a/p)-1\right]}{13.311*(a/p)-1}*100=\frac{-1.210*(a/p)}{13.311*(a/p)-1}*100$$

% change in overall yield:

$$\frac{0.360-0.328}{0.328}*100=10\%$$

We observe that a yield increment of 10% applied to the last step has a greater impact on reducing the overall kernel *PMI* (increasing the kernel RME) than one applied to the first step. We also note that percent changes in RME and *PMI* may be determined without knowledge of the molecular weights of A and P, unlike for percent changes in *E*. The presence of a minus one in the denominator in the expression for percent change in overall *E* spoils the possibility of canceling the factor *a/p*.

If we conduct the same analysis to the case where a 10% increase in yield is applied at steps 2, 3 and 4, we will obtain the following results:

10% Yield increase Applied to step no.	**$RME_{optimized}$**	**$PMI_{optimized}$**
1	0.07840 (p/a)	12.75600 (*a/p*)
2	0.07978 (p/a)	12.53374 (*a/p*)
3	0.08093 (p/a)	12.35618 (*a/p*)
4	0.08187 (p/a)	12.21413 (*a/p*)
5	0.08260 (p/a)	12.10100 (*a/p*)

Recognizing that the RME and *PMI* of the original non-optimized plan were 0.0751 (*p/a*) and 13.311 (*a/p*), we can calculate the changes in RME and *PMI* for the optimized plans as compared with the original:

10% Yield increase applied to step no.	**% Change in RME**	**% Change in *PMI***
1	4.39	–4.17
2	6.23	–5.84
3	7.76	–7.17
4	9.01	–8.24
5	10.0	–9.10

6.7 PROBLEMS

PROBLEM 6.1 [67]

Themes: balancing chemical equations, catalytic cycles, atom economy, kernel reaction mass efficiency

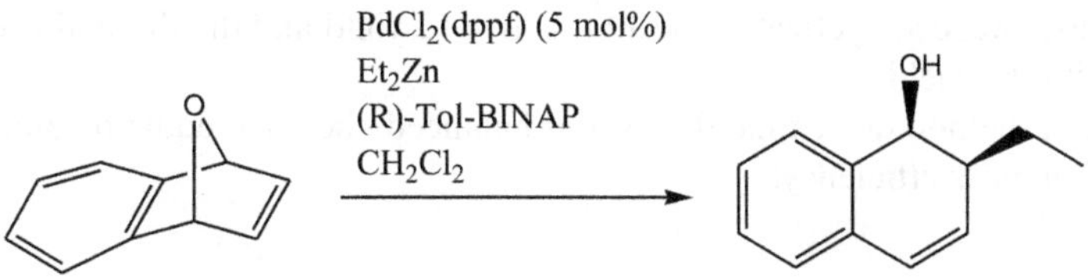

$PdCl_2(dppf)$ (5 mol%)
Et_2Zn
(R)-Tol-BINAP
CH_2Cl_2

OH

98 %, 96 % ee

dppf = 1,1′-bis(diphenylphosphino)ferrocene

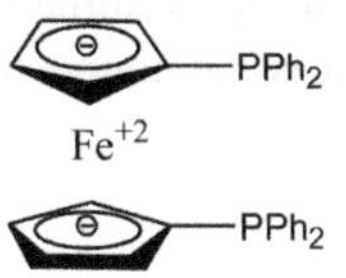

PPh2
Fe^{+2}
PPh2

dppf = 1,1'-bis(diphenylphosphino)ferrocene

(R) PAr_2 PAr_2

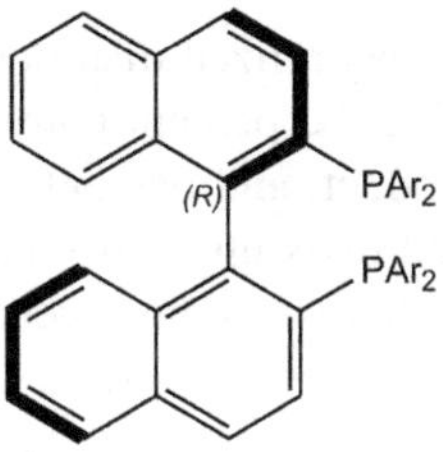

Ar = tolyl
(R)-Tol-BINAP

Part 1

Write out a catalytic cycle for the reaction.

Part 2

Write out a balanced chemical equation for the reaction and determine its atom economy.

Part 3

Determine the kernel reaction mass efficiency.

PROBLEM 6.2 [68]

Themes: balancing chemical equations, kernel reaction mass efficiency

Two methods for preparing methyl 1-hydroxy-naphthalene-2-carboxylate are shown below.

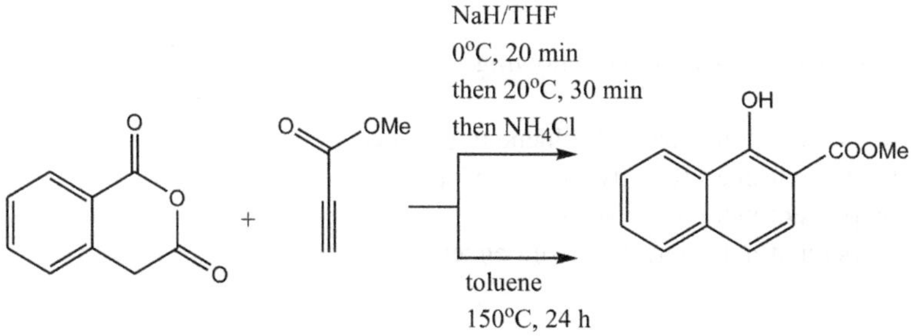

NaH/THF
0°C, 20 min
then 20°C, 30 min
then NH_4Cl

OMe + OH COOMe

toluene
150°C, 24 h

The base induced cyclization occurs with 49% yield and the thermal cyclization occurs with 9% yield.

For each method determine the overall balanced chemical equation and the kernel reaction mass efficiency.

PROBLEM 6.3 [69]

Themes: percent conversion, percent yield, percent mole selectivity

Part 1

For the generalized disproportionation reaction shown that produces two products, determine the percent conversion of starting material, percent mole selectivity for each product, and percent yield for each product. The reaction begins with x moles of A and ends up with y moles of P_1 and z moles of P_2. Assume that P_2 is made as a mechanistic consequence of producing P_1. What is the connecting relationship between these three metrics?

$$\underset{x\ mol}{2\,A} \longrightarrow \underset{y\ mol}{P_1} + \underset{z\ mol}{P_2}$$

Part 2

The Mobil toluene disproportionation process (MTDP) to produce xylenes and benzene from toluene is shown below.

toluene + toluene → {o-xylene + m-xylene + p-xylene} + benzene

After carrying out the reaction for an hour at 900°F, 400 psig, and a weight hourly space velocity (WHSV) of 9.8 g toluene per hour per gram of zeolite catalyst (ZSM-5), a sample of the reaction mixture was analyzed and found to contain 80 wt% toluene, 8.3 wt% benzene, and 11.4 wt% mixed xylenes; 81% of the xylenes was the *para* isomer; 1.4 g of catalyst were used.

Part A

From these data determine the following:

1. the percent conversion of toluene to products;
2. the percent mass selectivity of *p*-xylene;
3. the percent yield of *p*-xylene;
4. the percent mass selectivity of benzene;

5. the percent yield of benzene;
6. the masses of benzene and *p*-xylene produced; and
7. the mass of toluene remaining and the mass of toluene that reacted.

Part B

Check to see if the connecting relation found in Part 1 holds for the results found above. Explain.

PROBLEM 6.4 [70]

Themes: cumulative *PMI*

Use the *calculator-cumulative-PMI.xls* spreadsheet to determine the cumulative *PMIs* for the synthesis plan depicted in the synthesis tree shown below.

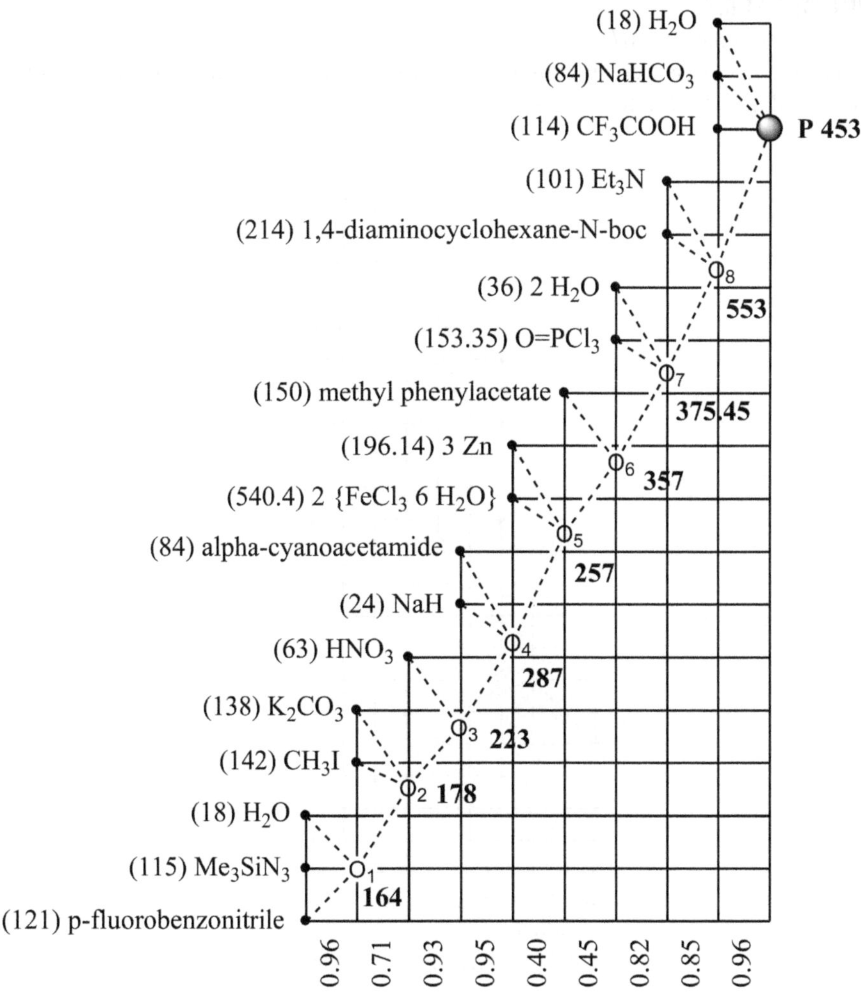

The step *PMIs* are as follows:

Step *PMI*
118.70
85.3
191.12
48.13
1210.09
57.49
1337.22
300.26
519.33

PROBLEM 6.5

Themes: cumulative *PMI*

For the synthesis tree diagram shown below determine recursive expressions for the cumulative *PMI* in terms of the step *PMIs*.

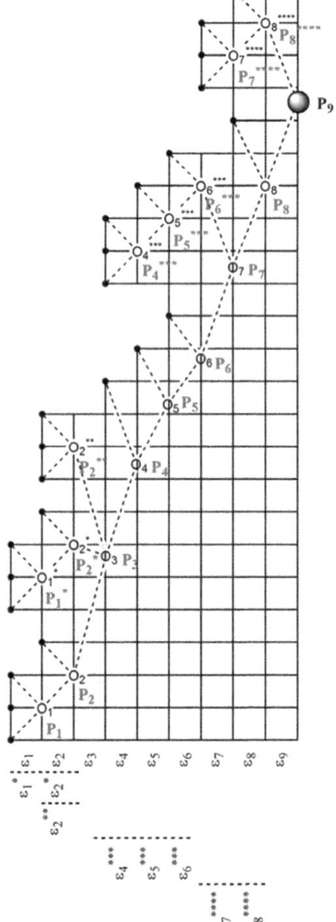

PROBLEM 6.6 [71]

Themes: cumulative *PMI*

The anti-coagulant pharmaceutical apixaban developed by Bristol-Myers Squibb has been synthesized using the following convergent strategy.

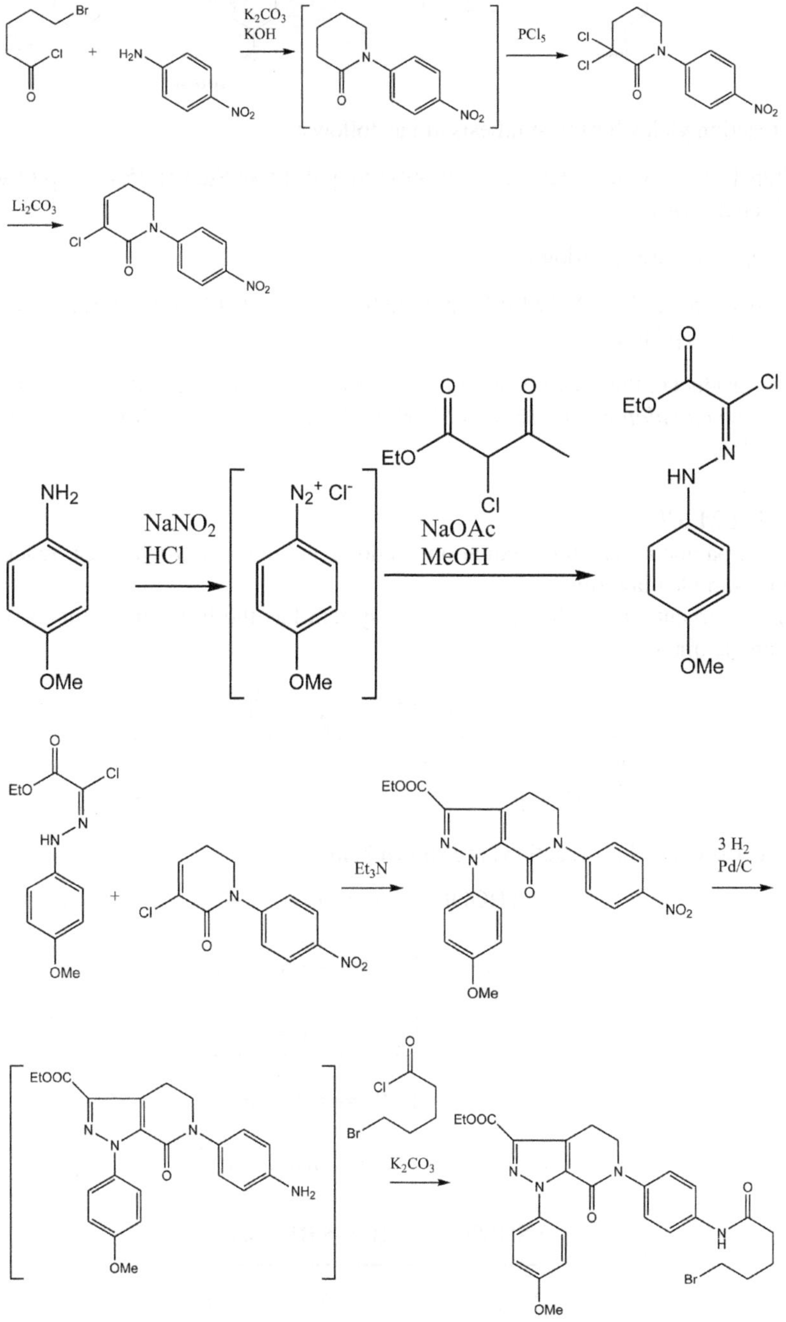

K_2CO_3
KOH
PCl_5
Li_2CO_3

$NaNO_2$
HCl
NaOAc
MeOH

Et_3N
3 H_2
Pd/C

K_2CO_3

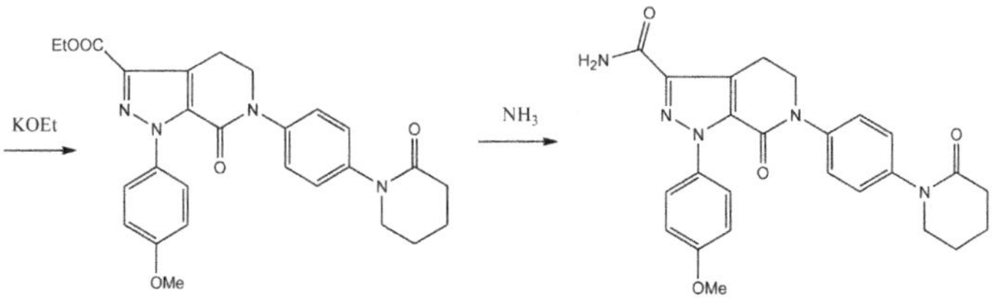

KOEt → NH$_3$ →

apixaban

The reaction yields for the synthesis are as follows:

Step 1: 76%; Step 2: 92%; Step 3: 89%; Step 4: 85%; Step 5: 75%; Step 6: 95%; and Step 2b: 78%.

The step *PMIs* are as follows:

Step 1: 45.92; Step 2: 19.19; Step 3: 13.92; Step 4: 25.61; Step 5: 36.03; Step 6: 39.85; and Step 2b: 24.28.

The product of the convergent branch is used in 6.6% excess in the convergent step. Use the *calculator-cumulative-PMI.xls* to determine the overall *PMI* for the synthesis plan.

PROBLEM 6.7

Themes: balancing chemical equations, redox reactions, electrochemical reactions, cathode–anode diagrams

A convergent electrochemical reaction system has the following template cathode–anode diagram.

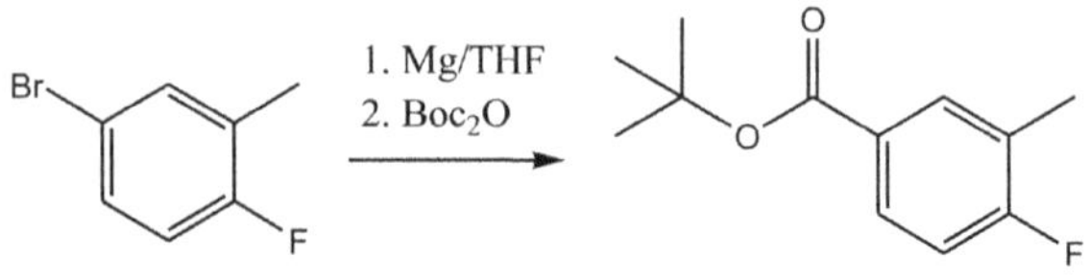

1. Mg/THF
2. Boc$_2$O

The associated half reactions are shown below.

CATHODE $A + n\,e \Longrightarrow A^*$

$A^* \Longrightarrow A + n\,e$

$S_1 + n\,e \Longrightarrow P$

$S_2 \Longrightarrow P + n\,e$

$B^* + n\,e \Longrightarrow B$

ANODE $B \Longrightarrow B^* + n\,e$

Net: $S_1 + S_2 \Longrightarrow 2\,P$

Such a reaction is useful for converting two starting materials at different oxidation levels to the same product. The species A and B are sacrificial mediators.

An example is the electrochemical conversion of glyoxal and oxalic acid to glyoxylic acid in aqueous HCl solution [72].

CATHODE $2\,H^+ + 2\,e \Rightarrow H_2$

$H_2 \Rightarrow 2\,H^+ + 2\,e$

HOOC-COOH $+ 2\,H^+ + 2\,e \Rightarrow$ OHC-COOH $+ H_2O$

OHC-CHO $+ H_2O \Rightarrow$ OHC-COOH $+ 2\,H^+ + 2\,e$

$Cl_2 + 2\,e \Rightarrow 2\,Cl^-$

ANODE $2\,Cl^- \Rightarrow Cl_2 + 2\,e$

The net reaction is

$$\text{HOOC-COOH} + \text{OHC-CHO} \longrightarrow 2\ \text{OHC-COOH}$$

Part 1

Apply the same reasoning to conjecture an analogous electrochemical system for the reverse Cannizzaro reaction that converts an organic alcohol and its related organic acid to a common organic aldehyde. Write out the associated half reactions.

Part 2

Determine the minimum atom economy for the production of aldehyde.

PROBLEM 6.8 [73]

Themes: balancing chemical equations, reaction mechanism, Curzons reaction mass efficiency

33 %

F_3C–CF_2–CF_2–C(=O)–Cl + H–C(=O)–NMe_2 $\xrightarrow{Et_3N}$ F_3C–CF_2–CF_2–C(=O)–C(=O)–NMe_2

α-ketoperfluorovaleric acid dimethylamide

Part 1

Write a mechanism that accounts for how HCl is produced as a by-product which is then neutralized by triethylamine.

Part 2

Write a balanced chemical equation for the reaction.

Part 3

Experimental Procedure:

In a glass tube are placed 24 g (0.33 mol) of anhydrous dimethylformamide and 6.5 g (0.064 mol) of dry triethylamine. The tube is cooled to –10°C to –20°C, then 15 g (0.064 mol) of perfluorobutyryl chloride is added. The tube is sealed and allowed to stand at 0°C for 15 h. Then the tube is cooled, its contents poured into ice-water, the organic layer is separated and the aqueous one extracted several times with ether. The organic layer and the ether extract are combined, washed with water, dried over $MgSO_4$, then maintained over P_2O_5 and distilled in vacuum. The yield of alpha-keto-perfluorovaleric acid dimethylamide is 5.7 g (33%); bp 62°–66°C/7 mm Hg.

Based on the information given determine the Curzons reaction mass efficiency.

PROBLEM 6.9 [74]

Themes: balancing chemical equations, biotransformation, material efficiency metrics

Chlorobenzene $\xrightarrow[O_2 \text{ (air)}]{\textit{Pseudomonas putida} \text{ 39/D}}$ 3-chloro-cyclohexa-3,5-diene-1,2-diol (Cl, OH, OH)

Pre-culture broth:

50 mL mineral salt broth
0.2% L-arginine hydrochloride
P. putida colony loop

Shake flask:

500 mL mineral salt broth
0.2% L-arginine hydrochloride
50 mL pre-culture broth
10 mL chlorobenzene

Work-up:

4 × 100 mL ethyl acetate
sodium sulfate drying agent
Yield of dihydroxylated product: 190 mg

Mineral salt broth is a blend of three solutions and distilled water in a volume ratio of A: B: C: distilled water=4: 2: 1.5: 92.5.
For 1 L of solution A:

136 g KH_2PO_4
268.1 g $Na_2HPO_4{\cdot}7H_2O$

For 1 L of solution B:

10 g nitrilotriacetic acid
7.5 g KOH
29.6 g $MgSO_4{\cdot}7H_2O$
3.3 g $CaCl_2{\cdot}2H_2O$
9.3 mg $(NH_4)_6Mo_7O_{24}{\cdot}4H_2O$
99 mg $FeSO_4{\cdot}7H_2O$
50 mL Metals 44 solution

For 100 mL Metals 44 solution:

250 mg ethylenediaminetetraacetic acid
1.095 g $ZnSO_4{\cdot}7H_2O$
500 mg $FeSO_4{\cdot}7H_2O$
154 mg $MnSO_4{\cdot}H_2O$
39.2 mg $CuSO_4{\cdot}5H_2O$
24.8 mg $Co(NO_3)_2{\cdot}6H_2O$
17.7 mg $Na_2B_4O_7{\cdot}10H_2O$

For 1 L Solution C:

200 mg $(NH_4)_2SO_4$

Part 1

Write a balanced chemical equation for the reaction and determine its atom economy.

Part 2

Suggest mechanisms that can account for the *cis*-dihydroxylation.

Part 3

Using the *REACTION-template.xls* spreadsheet and the experimental details determine the material efficiency metrics for the transformation. State any assumptions made in the calculations.

Part 4

Repeat the analysis of Part 3 excluding the consumption of water in the mineral broths. What is the percent decrease in the *PMI* when water is not counted?

PROBLEM 6.10 [51, 75, 76]

Themes: balancing chemical equations, synthesis tree diagrams, and step and overall *E*-factors

In a recent review, Roger Sheldon expounded the superiority of the *E*-factor metric over *PMI* and RME to describe synthesis efficiency. He stated that:

> None of these alternative metrics offer any particular advantage over the E factor for describing how wasteful a process is. The ideal *PMI* is 1, whereas the ideal E factor is 0, which more clearly reflects the ultimate goal of zero waste. The E factor also has the advantage that, in evaluating a multi-step process, E factors of individual steps are *additive* but *PMIs* are not because *PMI* doesn't discount step products from the mass balance.

Corma applied this additive strategy to evaluate the overall *E*-factors for both linear and convergent synthesis plans. Two reaction networks are shown below showing various routes to 4-(4-methoxyphenyl)-but-3-en-2-one and 4-(6,6-dimethylbicyclo[3,1,1]hept-2-en-2-yl)-2-butanone. *E*-factors for individual reactions are shown below the reaction arrows.

For 4-(4-methoxyphenyl)-but-3-en-2-one, three routes can be traced with the following overall *E*-factors as noted by Corma:

Route 1 (linear): $0.78+0.04+9.69=10.51$
Route 2 (linear): $2.77+9.69=12.46$
Route 3 (convergent): $0.30+0.60+1.96=2.86$

Similarly, for 4-(6,6-dimethylbicyclo[3,1,1]hept-2-en-2-yl)-2-butanone, two routes can be traced with the following overall *E*-factors as noted by Corma:

Route 4 (convergent): $1.18+2.78+0.86+0.63=5.45$
Route 5 (convergent): $0.06+1.30+0.05=1.41$

Network #1

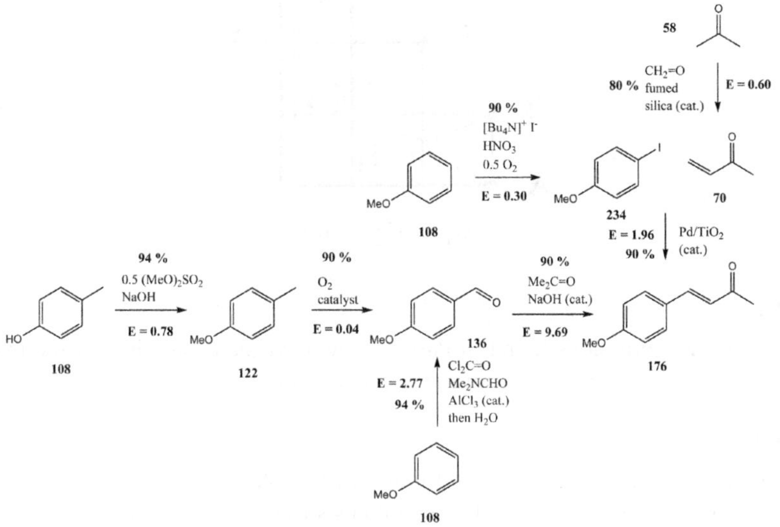

Network #2

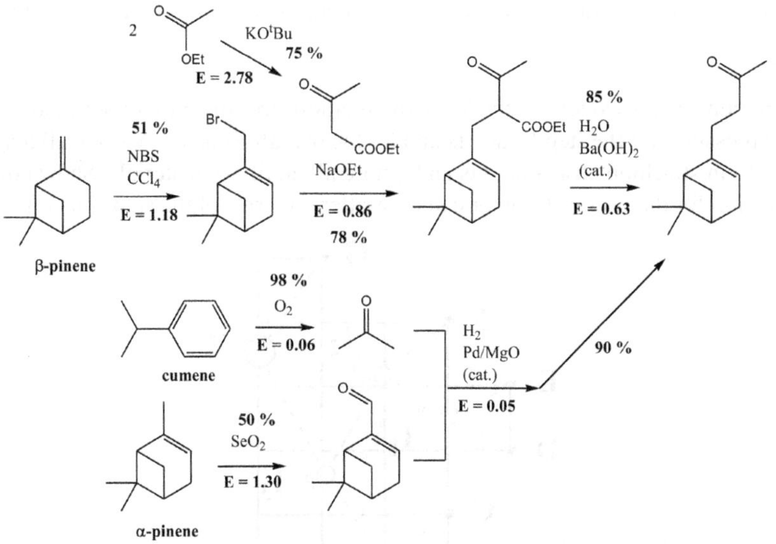

Part 1

For the linear sequence depicted in the synthesis tree diagram shown below, write out expressions for the step *E*-factors and for the overall *E*-factor. Assume all reagents are used in stoichiometric amounts and ignore all auxiliary materials.

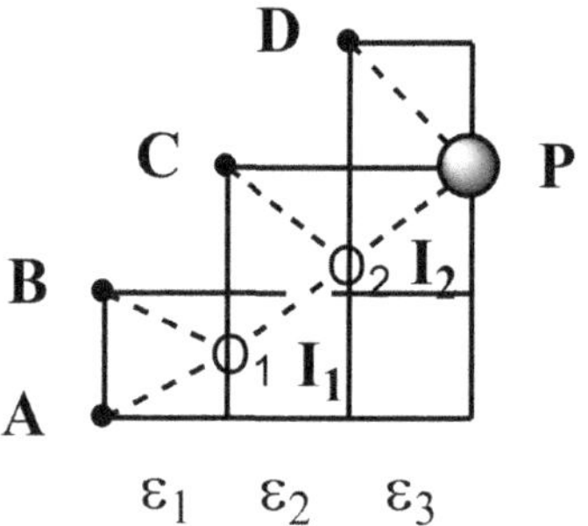

Part 2

Based on the expressions found in Part 1, show how the expression for the overall *E*-factor can be written in terms of the step *E*-factors.

Part 3

Based on the form of the expression found in Part 2, are Sheldon's statement and Corma's analysis correct for the linear synthesis routes shown in networks #1? Use synthesis tree diagrams for each route to support your answer.

Part 4

Is Sheldon's claim valid that step *E*-factors are additive while step *PMIs* are not?

Part 5

For the convergent sequence depicted in the synthesis tree diagram shown below, write out expressions for the step *E*-factors and for the overall *E*-factor. Assume all reagents are used in stoichiometric amounts and ignore all auxiliary materials. Show how the expression for the overall *E*-factor can be written in terms of the step *E*-factors.

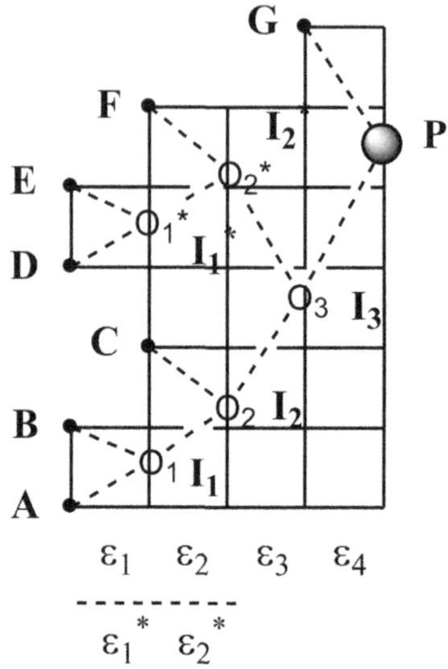

Part 6

Based on Part 5, determine the overall *E*-factors for the convergent synthesis routes shown in networks #1 and #2. Use synthesis tree diagrams for each route to support your answer. Compare the results with those reported by Corma.

PROBLEM 6.11 [77]

Themes: balancing chemical equations, catalysts, atom economy, oxidation number changes, process mass intensity

Magnetic decantation is a novel technique that has been used to facilitate separation of catalysts from reaction media without the need of filtration. Filtration of catalysts by conventional methods leads to significant loss of material in filter papers or frits of funnels. The new method involves placing a magnet outside the flask, which results in the catalyst made up of carbon nanoparticles with cobalt or iron cores to migrate toward the magnet leaving behind a clear reaction solution. The catalysts can be reused several times without loss of activity. An example reaction in which this technique has been used is the hydroxycarbonylation reaction shown below.

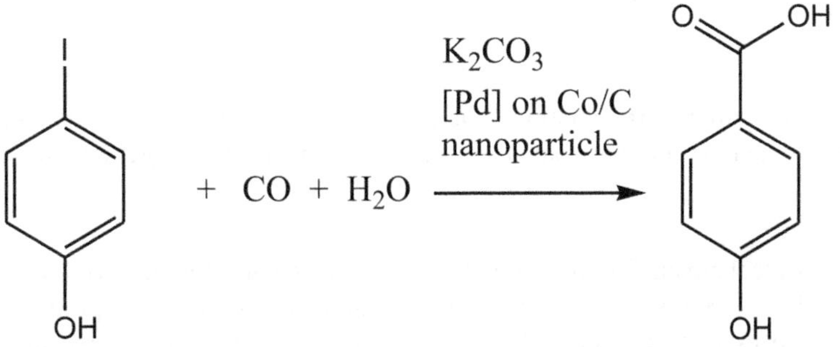

I, OH + CO + H_2O —(K_2CO_3, [Pd] on Co/C nanoparticle)→ O, OH, OH

The synthesis of the catalyst is shown in the following scheme.

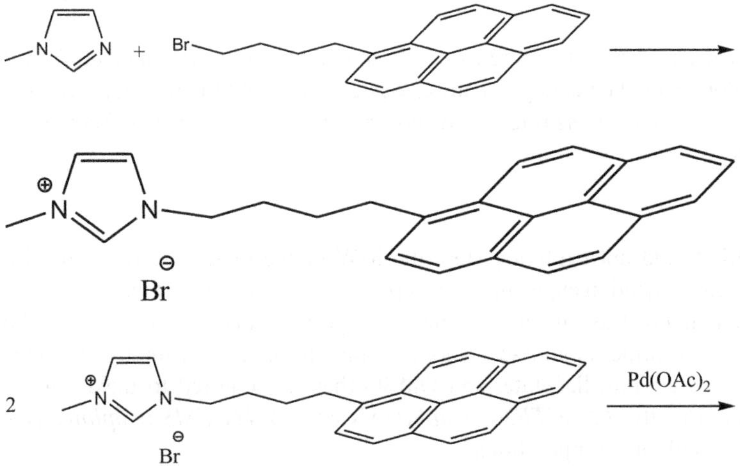

N, N + Br →

⊕ N, N, ⊖ Br

2 ⊕ N, N, ⊖ Br —$Pd(OAc)_2$→

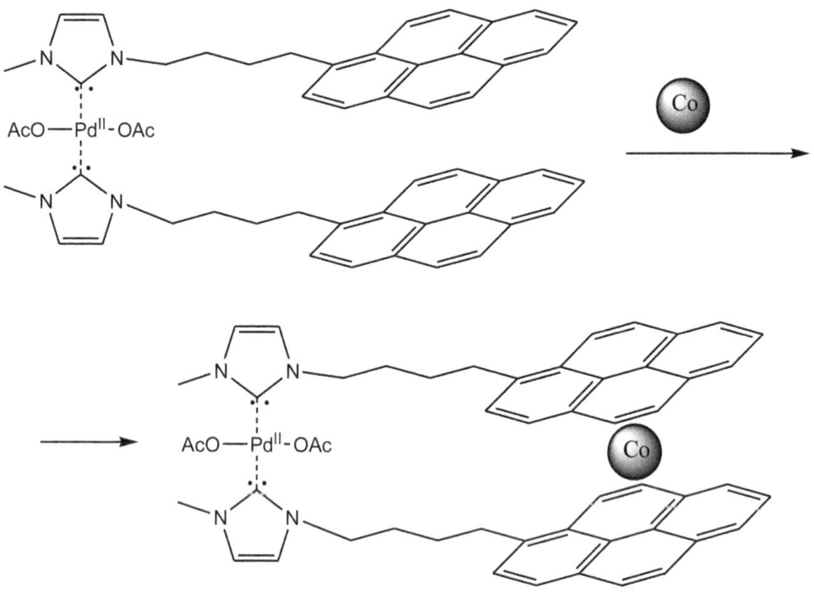

Part 1

Balance the hydroxycarbonylation equation, calculate the atom economy, and determine any oxidation number changes in key atoms involved in target forming bonds.

Part 2

Suggest a mechanism for the second step in the scheme for the synthesis of the catalyst. In the Supporting Information the authors described the product of the second step as a "palladium(VI) bromide" complex. Does this make any sense mechanistically?

Part 3

In the hydroxycarbonylation reaction, 66 mg of product were collected from 110 mg of 4-iodophenol. The catalyst loading was 2 mol % and the reaction time was 10 h. From these data, determine the turnover number and turnover frequency of the catalyst.

Part 4

The authors did not explicitly state the MW or the mass of catalyst used in their hydroxycarbonylation experiment, except to say that the catalyst loading was 2 mol%.

Based only on this information and the experimental details to make the Pd-NHC complex determine the masses of 4-(4-bromobutyl)pyrene, 1-methylimidazole, acetonitrile, palladium diacetate, and DMSO that are required to make the Pd-NHC complex. Use the *REACTION-template.xls* and *SYNTHESIS-template.xls* spreadsheets to facilitate computations.

Determine the *PMI* for the process. Ignore materials used in chromatographic purification for both steps.

Part 5

In making the Co/C Pd-NHC catalyst, the final step calls for ten times the mass of Co nanoparticles compared with the Pd-NHC complex. Determine the mass of Co nanoparticles that is required in the final step to make the catalyst in order to satisfy the 2 mol% catalyst loading in the hydroxycarbonylation reaction. Determine the total mass of catalyst material that was used in the hydroxycarbonylation reaction.

Part 6

Using the data in the experimental procedure to make *p*-hydroxybenzoic acid and the mass of catalyst determined in Part 5, determine the *PMI* for the hydroxycarbonylation reaction. Assume ten equivalents of carbon monoxide were used. Determine the overall *PMI* for making the catalyst and the hydroxycarbonylation reaction.

PROBLEM 6.12 [75–81]

Themes: balancing chemical equations, atom economy

Various methods to generate isocyanides via dehydration of N-formamides are given below. Determine which dehydrating agent should be chosen so that the highest atom economy can be attained for the reaction.

Method 1

R–NH–CHO + Cl–C(=O)–Cl $\xrightarrow{Et_3N}$ $R–N^{\oplus}{\equiv}C^{\ominus}$

Method 2

R–NH–CHO + $POCl_3$ $\xrightarrow{Et_3N}$ $R–N^{\oplus}{\equiv}C^{\ominus}$

Method 3

R–NH–CHO + $Cl_2P(=O)OPh$ $\xrightarrow{Et_3N}$ $R–N^{\oplus}{\equiv}C^{\ominus}$

Method 4

$RNHCHO + Cl_2P(O)OEt \xrightarrow{Et_3N} R-N^{\oplus}\equiv C^{\ominus}$

Method 5

$RNHCHO + PPh_3 \xrightarrow{I_2,\ Et_3N} R-N^{\oplus}\equiv C^{\ominus}$

Method 6

$RNHCHO + Et_2N-S^{\oplus}F_2\ BF_4^{\ominus} \xrightarrow{Et_3N} R-N^{\oplus}\equiv C^{\ominus}$

Method 7

$RNHCHO + (\text{polymer})-SO_2Cl \xrightarrow{Et_3N,\ CH_2Cl_2,\ \mu W} R-N^{\oplus}\equiv C^{\ominus}$

Method 8

$RNHCHO + \text{2,4,6-trichloro-1,3,5-triazine} \xrightarrow{Et_3N} R-N^{\oplus}\equiv C^{\ominus}$

PROBLEM 6.13 [5]

Themes: balancing chemical equations, atom economy, ring construction mapping

Barry Trost coined the phrase "atom economy" in his seminal 1991 paper, which has received and will continue to receive the highest number of citations of any paper published in the field of green chemistry. Yet, this paper gives no numerical examples, formulae, method of calculation of atom economy, or compares the performances of various reaction classes encountered in organic chemistry. In 1998, Trost was awarded a US Presidential Green Chemistry Challenge Award in the Academic Awards category. It is the only such award given to a concept or idea. Look up this paper and write out balanced chemical equations for all reactions cited. Determine the atom economies of each reaction. What do the results have in common? If ring-containing compounds are produced highlight the target bonds made and describe the ring construction strategy employed.

PROBLEM 6.14 [82]

Themes: balancing chemical equations, stoichiometry, atom economy, kernel reaction mass efficiency

Supramolecular structures with polygon shapes were synthesized according to the reaction shown below. Products A3, B4, C5, D6, E7, and F8 are triangle, square, pentagon, hexagon, heptagon, and octagon macrocycles. The structure of the triangle product is shown as an example.

AcO, AcO (binaphthyl diyne) + $Et_3P-Pt^{II}(Cl)_2-PEt_3$ $\xrightarrow{\text{CuCl (cat.), } Et_2NH, CH_2Cl_2}$ A_3 + B_4 + C_5 + D_6 + E_7 + F_8

5 % 18 % 16 % 10 % 5 % 4 %

A_3 = [structure: OAc OAc; Et_3P, Pt^{II}, PEt_3; AcO; OAc] = [triangle: binaphthyl moiety; Pt complex]

Write out balanced chemical equations for each product. Determine the corresponding atom economy and kernel RME values. How does the AE vary with polygon size?

PROBLEM 6.15 [83]

Themes: balancing chemical equations, reaction yield, atom economy, *E*-factor, process mass intensity, stoichiometric factor, energy intensity

Dimethyl malonate is produced according to the following scheme.

$$\mathrm{ClCH_2COOH} \xrightarrow{\mathrm{NaOH}} \mathrm{ClCH_2COONa} \xrightarrow{\mathrm{NaCN}} \mathrm{NCCH_2COONa} \xrightarrow[\mathrm{H_2SO_4}]{\mathrm{MeOH}} \mathrm{MeOC(O)CH_2C(O)OMe}$$

To produce 1000 kg/h of 97% pure dimethyl malonate, 3.5 MJ/h of electricity, 101 MJ/h of steam, and 57 MJ/h of refrigeration are needed, and it is required to dissipate 3165 MJ/h of heat using cooling water.

Monochloroacetic acid in water is mixed with cracked ice. Sodium hydroxide is added until the solution is made alkaline. Subsequently, sodium chloroacetate is formed. Sodium cyanide in water is added carefully to form a solution of sodium cyanoacetate. This solution is evaporated under reduced pressure to form a crude sodium cyanoacetate cake. The cake is hydrolyzed and esterified in the presence of methanol and sulfuric acid. Three extractions are performed with toluene. The dried product is distilled, at first under atmospheric pressure, and finally under reduced pressure, to remove any remaining toluene.

The list and amounts of chemicals used in the process are given below.

Chemical	Amount (kg/h)
Chloroacetic acid	715
Methanol	500
Sodium cyanide	372
Sodium hydroxide (50 wt%)	688
Sulfuric acid (98 wt%)	1835
Toluene	55
Water	3254

From these data, determine the % yield of dimethyl malonate, % AE, *E*-factor, *PMI*, stoichiometric factor, and energy intensity for the process.

PROBLEM 6.16 [84]

Themes: balancing chemical equations, material efficiency metrics, biosynthetic route

Citric acid monohydrate is produced biosynthetically from starch under aerobic fermentation conditions with *Aspergillus niger.* In order to produce 1 kg of product the following ingredients are needed: 1.27 kg starch, 0.51 kg oxygen, 0.01 kg ammonium nitrate, 0.02 kg α-amylase, 0.01 kg hydrogen chloride, 0.01 kg magnesium sulfate, 0.01 kg potassium dihydrogen phosphate, 0.01 kg sodium hydroxide, 0.01 kg fats, and 14.98 kg water.

The waste stream consists of 0.01 kg ammonium nitrate, 0.16 kg biomass, 0.41 kg carbon dioxide, 0.01 kg glucose, 0.04 kg dissolved salts, 0.07 kg citric acid (loss), and 15.12 kg water.

The authors report that for every 100 kg of starch the bioreaction yield of citric acid monohydrate is 84% and that 12,630 tons of product can be produced over 330 days in 550 batches.

Part 1

Write out a balanced chemical equation for the process. Assume that starch is a polymer of glucose. Determine the atom economy based on this equation.

Part 2

Perform a mass balance analysis. Do the masses of all inputs and all outputs balance out?

Part 3

Verify the yield claim for the process.

Part 4

Determine the *PMI* for the bioreaction.

Part 5

Determine the mass of carbon dioxide produced over a one-year period from this bioreaction. What is the source of carbon dioxide waste?

PROBLEM 6.17 [85, 86]

Themes: balancing chemical equations, material efficiency metrics

The mechanochemical technique called ball milling is advertised as a green method for carrying out organic transformations under the condition that such reactions are carried out without reaction solvent. An example transformation is the organocatalyzed enantioselective aldol reaction shown below using ball milling and conventional stirring conditions.

anti syn

(S)-proline (10 mol%)

ball milling	99 % yield; anti:syn = 89 : 11; 94 % ee
conventional	98 % yield; anti:syn = 87 : 13; 94 % ee

The relevant experimental procedures are given below.

Ball milling procedure:

A zirconium oxide bowl containing 60 grinding balls was charged with solid aldehyde (2.0 mmol, 1.0 equiv), ketone (2.2 mmol, 1.1 equiv) and (*S*)-proline (23 mg, 0.2 mmol, 0.1 equiv). The mixture was milled in the ball mill at a speed of 250–400 rpm. A milling cycle, which includes several minutes of milling followed by a cooling pause, was selected (15 min at 400 rpm + 5 min pause; total time 5.5 h). The crude product was washed off the vessel and the balls with ether (4×40 mL). The combined fractions were filtered and concentrated *in vacuo*. Purification by flash chromatography on silica gel (pentane/EtOAc 100:0 80:20) afforded the desired pure aldol products.

Conventional procedure:

A 5 mL round bottom flask containing a magnetic stirring bar was charged with solid aldehyde (2.0 mmol, 1.0 equiv), ketone (2.2 mmol, 1.1 equiv), and (*S*)-proline (23 mg, 0.2 mmol, 0.1 equiv). The mixture was stirred at room temperature for 96 h. The crude product was washed off the flask with ether (10×4 mL). The combined fractions were filtered and concentrated *in vacuo*. The residue was subjected to flash chromatography on silica gel (pentane/EtOAc 100:0 → 80:20) to afford the desired pure aldol products.

Using the *REACTION-template.xls* spreadsheet and the details of both procedures, determine the material efficiency metrics for each method. State any assumptions made in the calculations. Comment on the results. Have the authors made their point convincingly?

PROBLEM 6.18 [87]

Themes: balancing chemical equations, multi-component reactions, material efficiency metrics

Monastrol is a mitotic Eg5 kinesin inhibitor having a dihydropyrimidine structure that can be built using the three-component Biginelli reaction. Kappe and coworkers undertook a statistical design of experiments (DoE) optimization of reaction parameters in order to determine the best protocol to make this compound. Reaction parameters included reaction solvent, catalyst type, catalyst loading, reaction time, and reaction temperature. The final optimized experimental procedure selected involved a microwave irradiation protocol and is given below.

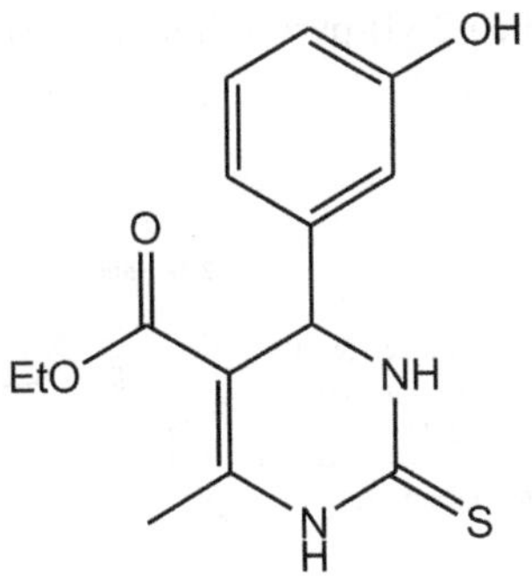

monastrol

Optimized experimental procedure:

In a 10 mL Pyrex microwave vial, 3-hydroxybenzaldehyde (100 mg, 0.82 mmol, 1 equiv), ethyl acetoacetate (157 mL, 1.23 mmol, 1.5 equiv), thiourea (60.7 mg, 0.82 mmol, 1 equiv), and $LaCl_3$ (24.5 mg, 0.082 mmol, 0.10 equiv) were added to 1 mL anhydrous ethanol. After magnetic stirring for 2 min, the vial was closed, placed in the microwave cavity, and irradiated at 140°C for 30 min (fixed hold time). After the reaction was finished, the solvent was removed under reduced pressure and the residue subjected to silica gel chromatography (chloroform/acetone = 5/1), resulting in the isolation of the pure product in 82% yield.

Part 1

Write out a balanced chemical equation for the transformation and determine its atom economy. Show the target bonds made in the product structure and determine the ring construction strategy.

Part 2

Using the *REACTION-template.xls* spreadsheet and the details of the optimized experimental procedure, determine the materials efficiency metrics for the reaction. State any assumptions made in the calculations.

Part 3

Determine the catalyst loading and the stoichiometric factor.

PROBLEM 6.19 [88]

Themes: balancing chemical equations, reaction mechanism, material efficiency metrics

Heck coupling of 5-bromopyrazine and 2-phenylthiophene under microwave irradiation conditions results in two products: 5-(5-phenyl-thiophen-2-yl)-pyrimidine (major) and 5-(5-phenyl-thiophen-3-yl)-pyrimidine (minor). By contrast, in the presence of potassium ferricyanide the same heterocycles couple differently to give

5-bro mo-4-(5-phenyl-thiophen-2-yl)-pyrimidine (major) and 4-(5-phenyl-thiophen-2-yl)-pyrimidine (minor).

Reaction 1

52 % isolated

$Pd(OAc)_2$ (cat.)
PCy_3 (ligand)
K_2CO_3
microwaves

39 : 0.2

Reaction 2

CF_3COOH
$K_3Fe(CN)_6$
KOH
room temperature

76 % isolated

82 : 7.5

Part 1

For each reaction write out balanced chemical equations for each product outcome and determine the respective atom economies.

Part 2

Based on the information given, for each reaction determine the expected yields of the minor products.

Part 3

Suggest a mechanism for Reaction 2.

PROBLEM 6.20 [89, 90]

Themes: balancing chemical equations, atom economy, synthesis plan analysis

Hydrogen peroxide is now considered a green oxidant mainly because it produces water as a reaction by-product in oxidation reactions. In the chemical industry, it has replaced chlorine as a bleaching agent in paper manufacture, and is used in epoxidations of olefins such as propylene and in the ammoximation of cyclohexanone for the manufacture of ε-caprolactam. There are five known ways of generating hydrogen peroxide as shown in the following series of examples. The anthraquinone process is the one used most in the chemical industry for hydrogen peroxide production.

Thenard method (1818) [91]

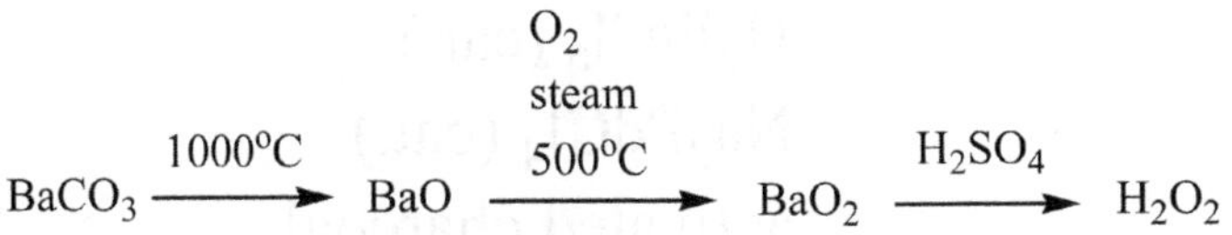

$$BaCO_3 \xrightarrow{1000^\circ C} BaO \xrightarrow[500^\circ C]{O_2,\ \text{steam}} BaO_2 \xrightarrow{H_2SO_4} H_2O_2$$

Meidinger–Berthelot method (1878) [92, 93]

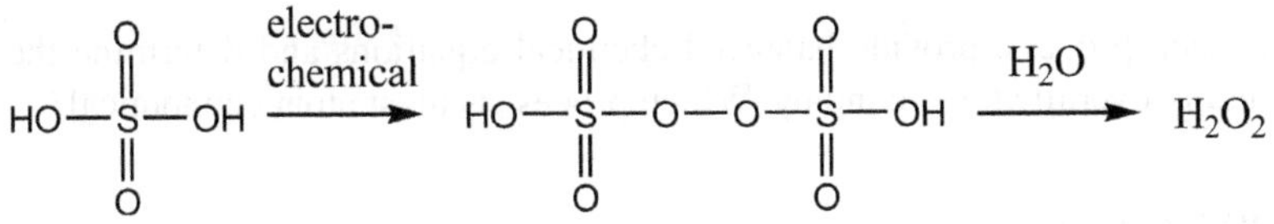

$$HO{-}S(=O)_2{-}OH \xrightarrow{\text{electro-chemical}} HO{-}S(=O)_2{-}O{-}O{-}S(=O)_2{-}OH \xrightarrow{H_2O} H_2O_2$$

Azo-hydrazobenzene method (1932) [94]

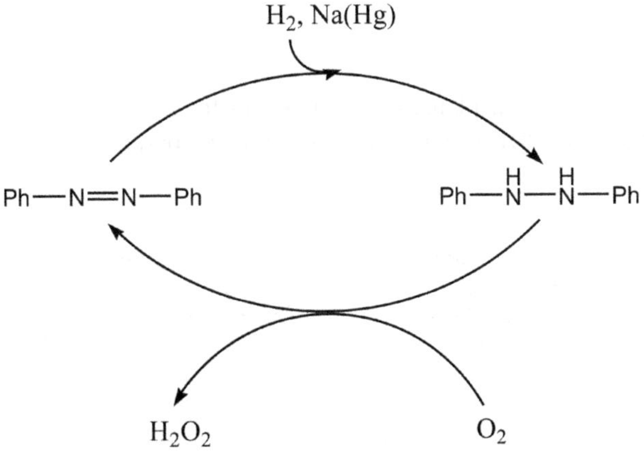

H_2, Na(Hg)

Ph—N=N—Ph → Ph—NH—NH—Ph

H_2O_2 O_2

Anthaquinone process (1940 s) [95–97]

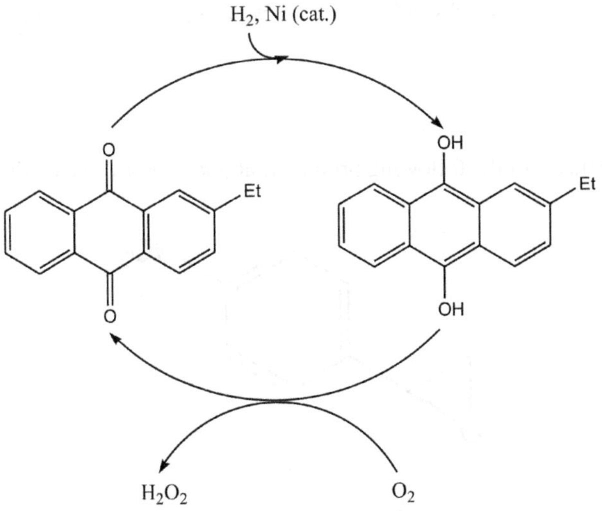

H_2, Ni (cat.)

Et

OH

Et

OH

H_2O_2 O_2

Direct Synthesis (2000 s) [95–100]

$$H_2 + O_2 \xrightarrow[\text{}]{H_2PtCl_6 \text{ (cat.)},\ Na_2PdCl_4 \text{ (cat.)},\ \text{activated charcoal}} H_2O_2$$

For each process, provide balanced chemical equations and determine the corresponding overall atom economy. Which process is most atom economical?

PROBLEM 6.21

Themes: balancing chemical equations, scenario analysis, synthesis planning, metrics constraints

Part 1

For the following synthesis plans, suggest reagents for each step that satisfy the constraint AE>60%. Balance each reaction step showing by-products.

Plan 1

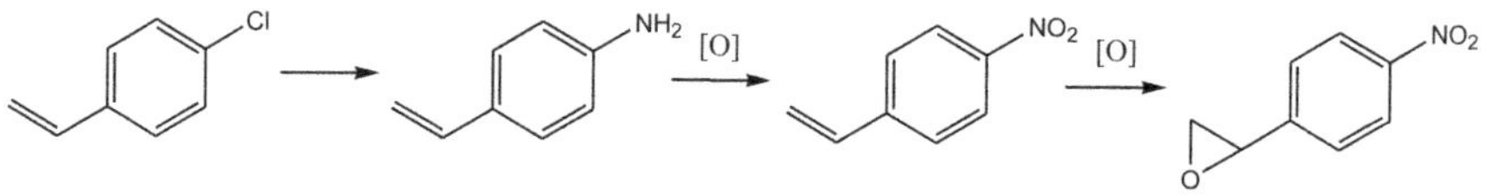

Plan 2

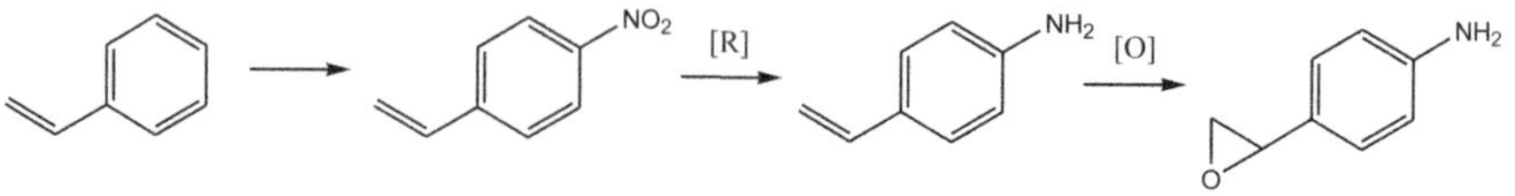

Part 2

Suggest a synthesis of the following product that does not involve oxidation or reduction steps.

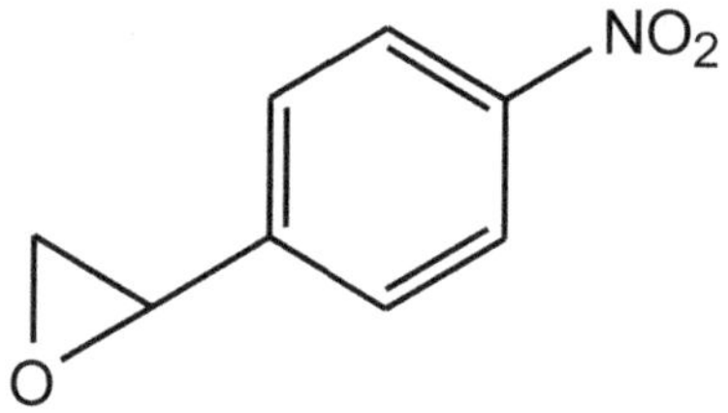

PROBLEM 6.22 [101]

Themes: balancing chemical equations, side products, kernel reaction mass efficiency

For the reactions shown, determine the kernel reaction mass efficiencies for each of the products obtained.

(a)

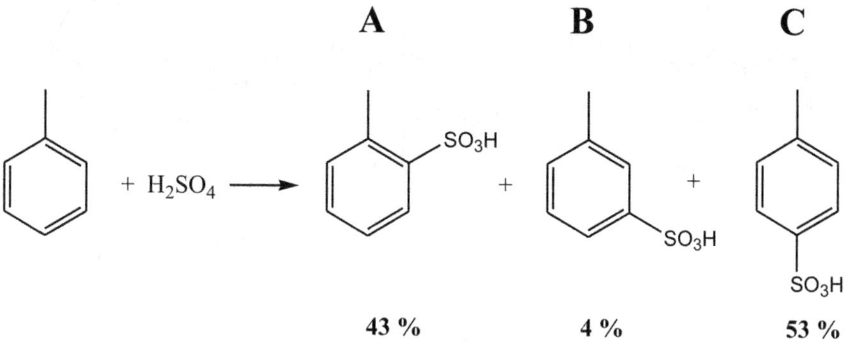

toluene + H_2SO_4 → A + B + C

A (ortho, SO_3H) 43 %

B (meta, SO_3H) 4 %

C (para, SO_3H) 53 %

(b)

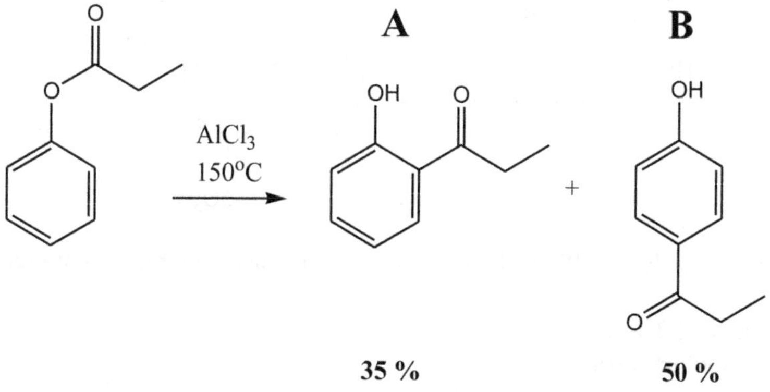

phenyl propanoate, $AlCl_3$, 150°C → A + B

A (OH, ortho-propanoyl) 35 %

B (OH, para-propanoyl) 50 %

(c)

$CH_3CH_2—C(=O)—CH=CHCl$ —(NH_2OH, EtOH, heat)→ A + B

A (3-ethylisoxazole; N, O) 60 %

B (5-ethylisoxazole; N, O) 40 %

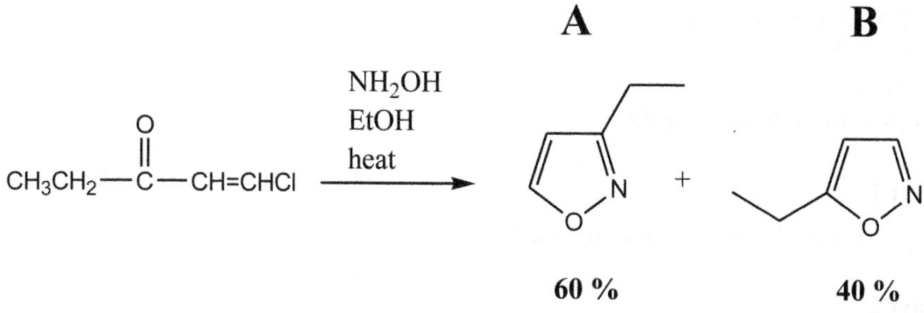

PROBLEM 6.23 [102]

Themes: balancing chemical equations, atom economy, stoichiometric factors

Chrome Blue Black R (C.I. 15705) is an azo dye made according to the following plan.

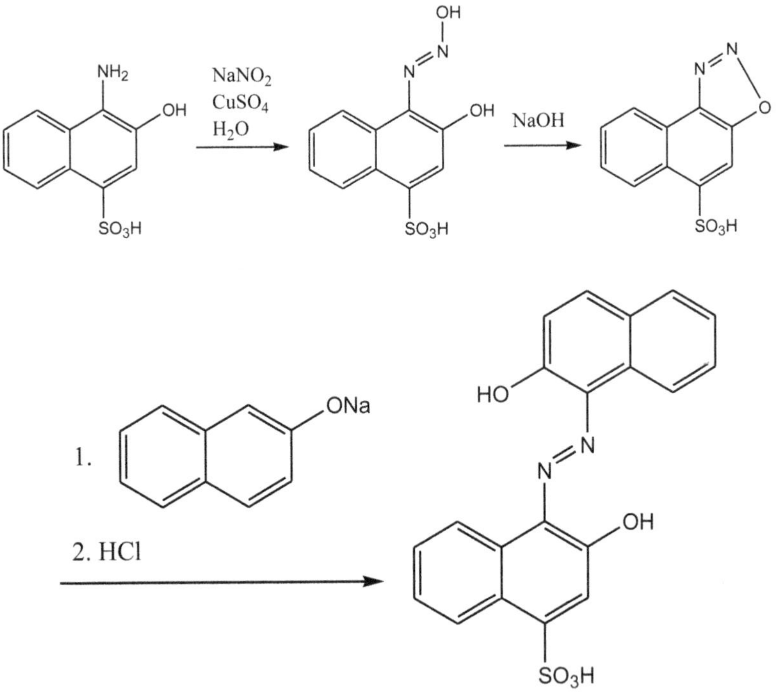

NaNO$_2$
CuSO$_4$
H$_2$O

NaOH

1.

2. HCl

To make this dye in 80% yield the following amounts of ingredients are required:

1200 lb. of 1,2,4-acid
365 lb. sodium nitrite
15 lb. copper(II)sulfate pentahydrate
600 lb. caustic soda
730 lb. sodium β-naphtholate
500 lb. hydrochloric acid
8250 lb. water
2700 lb. ice
1900 lb. sodium chloride

Part 1

Suggest a mechanism for the first step.

Part 2

Write balanced chemical equations for each step.

Part 3

For each step, determine the atom economy.

Part 4

For the entire plan, determine the overall atom economy, overall yield, overall stoichiometric factor, process mass intensity, and overall reaction mass efficiency.

$$\text{Verify that RME} = \varepsilon(\text{AE})\left(\frac{1}{\text{SF}}\right)(\text{MRP}) = \varepsilon(\text{AE})\left(\frac{1}{\text{SF}}\right)\left(\frac{1}{1+\frac{(\text{AE})\varepsilon(\text{mass})_{\text{aux}}}{(\text{SF})(\text{mass})_p}}\right).$$

PROBLEM 6.24 [103]

Themes: balancing chemical equations, catalytic cycles, recycling, material efficiency metrics

$Pd(OAc)_2$ (10 mol%)
$B_2(pin)_2$ (3 eq.)
H_2O

95 %

Experimental Procedure:

To a Schenk tube equipped with a stir bar, 152.4 mg of B_2pin_2 (0.6 mmol), 4.5 mg $Pd(OAc)_2$ (0.02 mmol) were added. The Schenk tube was capped with a septum, degassed and backfilled with N_2 for at least three times. Then, water (2.0 ml) and 25.8 mg quinoline (0.2 mmol) were added via syringe. The mixture was stirred at room temperature for about 10 h. Then, the mixture was extracted with ethyl acetate, repeated three times. The combined organic layer was evaporated under reduced pressure, and the product was purified by flash chromatography using petroleum ether and ethyl acetate as eluent. Yield of 1,2,3,4-tetrahydroquinoline as a pale yellow oil: 25.3 mg, 95%.

Part 1

Write out a catalytic cycle for the transformation.

Part 2

Based on the catalytic cycle, write out the overall balanced chemical equation for the transformation and determine its atom economy.

Part 3

Explain the logic of using three equivalents of bis(pinacolato)diborane.

Part 4

What opportunities exist for by-product recycling?

Part 5

Using the *REACTION-template.xls* spreadsheet and the details of the experimental procedure, determine the material efficiency metrics for the reaction. State any assumptions made in the calculations.

PROBLEM 6.25 [104]

Themes: balancing chemical equations, reaction mechanism, material efficiency metrics

Experimental Procedure:

The reaction mixture obtained by heating 2-chloro-3-nitropyridine (0.079 g, 0.5 mmol), alcohol phenyl-pyridin-2-yl-methanol (0.324 g, 1.75 mmol) and methyl acrylate (0.180 mL, 2 mmol) in acetonitrile (1 mL) at 110°C for 96 h, was resolved by flash chromatography (CH_2Cl_2/EtOAc 100:3), leading to phenyl-pyridin-2-yl-methanone (0.244 g, 89% referred to 2-chloro-3-nitropyridine), 3-amino-2-chloropyridine (0.005 g, 8%), and 3-(2-chloro-pyridin-3-ylamino)-propionic acid methyl ester (0.073 g, 68%) as a dark yellow oil.

Part 1

For each pyridinoamine target product, determine the balanced chemical equation that leads to it.

Part 2

Show the redox couple for the production of 3-(2-chloro-pyridin-3-ylamino)-propionic acid methyl ester. Show the redox couple for the production of 3-amino-2-chloropyridine.

Part 3

Write out a mechanism that accounts for production of 3-(2-chloro-pyridin-3-ylamino)-propionic acid methyl ester.

Part 4

Using the *REACTION-template.xls* spreadsheet and the details of the experimental procedure, determine the material efficiency metrics for each reaction. State any assumptions made in the calculations.

Part 5

One of the advantages of the hydrogenation transfer reagent phenyl-pyridin-2-yl-methanol is that its oxidized ketone by-product can be easily recycled back to it via sodium borohydride reduction. The experimental procedure for this recycling reaction is shown below [105]. This scenario may be contrasted with the one in which such a recycling reaction has been demonstrated when a Hantzsch dihydropyridine is used as a transfer hydrogenation agent.

20.65 g (0.113 mol) of phenyl-pyridin-2-yl-methanone in 120 ml of methanol at 0°C were treated in small portions with 5.00 g (0.132 mol) of solid $NaBH_4$. After stirring for 2.5 h at room temperature, 250 ml of water was added slowly, and the mixture stirred for another 0.5 h. Then it was extracted with five 75 ml portions of CH_2Cl_2 and the organic extracts dried over $MgSO_4$. After removal of the solvent *in vacuo* the solid residue was recrystallized from *n*-heptane/benzene (1:1), yielding 16.58 g (79%) of colorless phenyl-pyridin-2-yl-methanol, m.p. 132°C.

Based on this reaction, determine the percent phenyl-pyridin-2-yl-methanol that is recovered from the amount of phenyl-pyridin-2-yl-methanone collected in the transfer hydrogenation reaction. Is it worth carrying out the recycling reaction?

REFERENCES

1. Lavoisier, A. Mémoire sur la nature du principe qui se combine avec les métaux pendant leur calcination, et qui en augmente le poids. *Mèmoires de l'Académie Royale des Sciences* 1778, *1775*, 520.
2. Richter, J.B. *Anfangsgründe der Stöchiometrie oder Messkunst chemischer Elemente*, J.F. Korn: Breslau and Hirschberg, 1792–1794.
3. Wöhler, F. *Ann. Chim. Phys.* 1828, *37*(1), 330.
4. Wöhler, F. *Ann. Phys. Chem.* 1828, *12*(2), 253.
5. Trost, B.M. *Science* 1991, *254*, 1471.
6. Sheldon, R.A. *ChemTech* 1994, *24*(3), 38.
7. Curzons, A.D.; Constable, D.J.C.; Mortimer, D.N.; Cunningham, V.L. *Green Chem.* 2001, *3*, 1.
8. Constable, D.J.C.; Curzons, A.D.; Cunningham, V.L. *Green Chem.* 2002, *4*, 521.
9. Constable, D.J.C.; Curzons, A.D.; dos Santos, L.M.F.; Geen, G.R.; Hannah, R.E.; Hayler, J.D.; Kitteringham, J.; McGuire, M.A.; Richardson, J.E.; Smith, P.; Webb, R.L.; Yu, M. *Green Chem.* 2001, *3*, 7.
10. Andraos, J. *Org. Process Res. Dev.* 2005, *9*, 149.
11. Andraos, J. *ACS Sust. Chem. Eng.* 2018, *6*, 3206.
12. Jimenez-Gonzalez, C.; Constable, D.J.C. *Green Chemistry and Engineering: A Practical Design Approach*, Wiley: Hoboken, NJ, 2011, p. 86.
13. Pessel, F.; Augé, J.; Billault, I.; Scherrmann, M.C. *Beilstein J. Org. Chem.* 2016, *12*, 2351.
14. Ruiz-Mercado, G.J.; Smith, R.L.; Gonzalez, M.A. *Ind. Eng. Chem. Res.* 2012, *51*, 2309.
15. Trost, B.M. *Angew. Chem. Int. Ed.* 1995, *34*, 259.

16. Cann, M.C.; Connelly, M. *Real World Cases in Green Chemistry*, American Chemical Society, Washington, DC, 1998.
17. Cann, M.C. The University of Scranton. (http://academic.scranton.edu/faculty/CANNM1/organicmodule.html)
18. Cheung, L.L.W.; Styler, S.A.; Dicks, A.P. *J. Chem. Educ.* 2010, *87*, 628.
19. Aktoudianakis, E.; Chan, E.; Edward, A.R.; Jarosz, I.; Lee, V.; Mui, L.; Thatipamala, S.S.; Dicks, A.P. *J. Chem. Educ.* 2009, *86*, 730.
20. Jiménez-González, C.; Ponder, C. S.; Broxterman, Q. B.; Manley, J. B. *Org. Process Res. Dev.* 2011, *15*, 912.
21. Whittall, J.; Sutton, P.W.; Kroutil, W. *Practical Methods for Biocatalysis and Biotransformations 3*, Wiley, Chichester, 2016, p. 32.
22. Streitweiser, A. Jr.; Heathcock, C.H. *Introductory Organic Chemistry*, 3rd ed., Macmillan Publishing Company: New York, 1985, p. 824.
23. Streitweiser, A. Jr.; Heathcock, C.H. *Introductory Organic Chemistry*, 3rd ed., Macmillan Publishing Company: New York, 1985, p. 1046.
24. Eissen, M.; Metzger, J.O. *Chem. Eur. J.* 2002, *8*, 3580.
25. McElroy, C.R.; Constantinou, A.; Jones, L.C.; Summerton, L.; Clark, J.H. *Green Chem.* 2015, *17*, 3111.
26. Ishikawa, H.; Suzuki, T.; Hayashi, Y. *Angew. Chem. Int. Ed.* 2009, *48*, 1304.
27. Ishikawa, H.; Honma, M.; Hayashi, Y. *Angew. Chem. Int. Ed.* 2011, *50*, 2824.
28. Hayashi, Y. *Chem. Sci.* 2016, *7*, 866.
29. Clarke, P.A.; Santos, S.; Martin, W.H.C. *Green Chem.* 2007, *9*, 438.
30. Clarke, P.A.; Zaytzev, A.V.; Whitwood, A.C. *Tetrahedron Lett.* 2007, *48*, 5209.
31. Andraos, J.; Sayed, M. *J. Chem. Educ.* 2007, *84*, 1004.
32. Tang, P. *Org. Synth.* 2005, *81*, 262.
33. Fieser, L.F. *Org. Synth. Coll.* 1955, *3*, 6.
34. Bravo-Altamirano, K.; Montchamp, J.L. *Org. Synth.* 2008, *85*, 96.
35. Tundo, P.; Selva, M.; Bomben, A. *Org. Synth.* 1999, *76*, 169.
36. Marvel, C.S.; Sperry, W.M. *Org. Synth. Coll.* 1941, *1*, 95.
37. Chastrette, M.; Chastrette, F.; Sabadie, J. *Org. Synth. Coll.* 1988, *6*, 856.
38. Padwa, A.; Brodney, M.A.; Lynch, S.M. *Org. Synth. Coll.* 2001, *78*, 202.
39. Terakado, D.; Oriyama, T. *Org. Synth.* 2006, *83*, 70.
40. McKee, B.H.; Gilheany, D.G.; Sharpless, K.B. *Org. Synth. Coll.* 1998, *9*, 383.
41. Jimenez-Gonzalez, C.; Constable, D.J.C. *Green Chemistry and Engineering: A Practical Design Approach*, Wiley: Hoboken, NJ, 2011, p. 79.
42. Jimenez-Gonzalez, C.; Constable, D.J.C. *Green Chemistry and Engineering: A Practical Design Approach*, Wiley: New Jersey, 2011, pp. 243–244.
43. Jimenez-Gonzalez, C.; Constable, D.J.C. *Green Chemistry and Engineering: A Practical Design Approach*, Wiley: Hoboken, NJ, 2011, p. 82.
44. Mulzer, J.; Bock, H.; Eck, W.; Buschmann, J.; Luger, P. *Angew. Chem. Int. Ed.* 1991, *30*, 414.
45. Fischer, E.; Brauns, F. *Chem. Ber.* 1914, *47*, 3181.
46. Fischer, E. *Ann. Chem.* 1914, *402*, 364.
47. Punna, S.; Diaz, D.D.; Finn, M.G. *Synlett* 2004, 2351.
48. Andraos, J.; Hent, A. *J. Chem. Educ.* 2015, *92*, 1831.
49. Adams, R.; Voorhees, V. *Org. Synth. Coll. Vol.* 1941, *1*, 280.
50. Furniss, B.S.; Hannaford, A.J.; Smith, P.W.G.; Tatchell, A.R. *Vogel's Textbook of Practical Organic Chemistry*, 5th ed., Longman, Harlow, 1989, p. 1029.
51. Roschangar, F.; Sheldon, R.; Senanake, C.H. *Green Chem.* 2015, *17*, 752.
52. Sheldon, R.A. *ACS Sust. Chem. Eng.* 2018, *6*, 32.
53. Schätz, A.; Grass, R.N.; Stark, W.J.; Reiser, O. *Chem. Eur. J.* 2008, *14*, 8262.

54. Jimenez-Gonzalez, C.; Constable, D.J.C. *Green Chemistry and Engineering: A Practical Design Approach*, Wiley, Hoboken, NJ, 2011, pp. 126–127.
55. Kalinski, C.; Umkehrer, M.; Ross, G.; Kolb, J.; Burdack, C.; Hiller, W. *Tetrahedron Lett.* 2006, *47*, 3423.
56. Salcedo, A.; Neuville, L.; Rondot, C.; Retailleau, P.; Zhu, J. *Org. Lett.* 2008, *10*, 857.
57. Russell, A.; Frye, J.R. *Org. Synth. Coll. Vol.* 1955, *3*, 281.
58. Ullah, F.; Samarakoon, T.; Rolfe, A.; Kurtz, R.D.; Hanson, P.R.; Organ, M.G. *Chem. Eur. J.* 2006, *16*, 10959.
59. Bremner, W.S.; Organ, M.G. *J. Comb. Chem.* 2007, *9*, 14.
60. Watts, P.; Haswell, S.J. *Chem. Eng. Technol.* 2005, *28*, 290.
61. Leyva, E.; Moctezuma, E.; Santos-Diaz, M.S.; Loredo-Carrillo, S.E.; Hernández-González, O. *Revista Latinoamer. Quim.* 2012, *40*, 140.
62. Inglis, J.K.H. *Org. Synth. Coll. Vol.* 1941, *1*, 254.
63. Corson, B.B.; Scott, R.W.; Vose, C.E. *Org. Synth. Coll. Vol.* 1941, *1*, 179.
64. Beheshti, S.; Kianmehr, E.; Yahyaee, M.; Tabatabai, K. *Bull. Korean Chem. Soc.* 2006, *27*, 1056.
65. Wittig, G.; Schoellkopf, U. *Org. Synth. Coll.* 1973, *5*, 751.
66. Hermeling, D.; Bassler, P.; Hammes, P.; Hugo, R.; Lechtken, P.; Siegel, H. US 5527966 (BASF, 1996).
67. Lautens, M.; Renaud, J.L.; Hiebert, S. *J. Am. Chem. Soc.* 2000, *122*, 1804.
68. Tamura, Y.; Sasho, M.; Nakagawa, K.; Tsugoshi, T.; Kita, Y. *J. Org. Chem.* 1984, *49*, 473.
69. Haag, W.O.; Olson, D.H. US4097543 (Mobil Corp., 1978).
70. Sauvageau, G.; Gareau, Y. US2015011543 (U. Montreal, 2015).
71. Shapiro, R.; Rossano, L.T.; Mudryk, B.M.; Cuniere, N.; Obelholzer, M.; Zhang, H.; Chen, B.C. Process for Preparing 4,5-Dihydro-pyrazolo-[3,4c]-pyrid-2-ones. US2006069258 (Bristol-Myers Squibb Company, 2006).
72. Jalbout, A.F.; Zhang, S. *Acta Chim. Slov.* 2002, *49*, 917.
73. Knunyants, I.L.; Yakobson, G.G. *Syntheses of Fluoroorganic Compounds*, Springer-Verlag: Berlin, 1985, p. 70.
74. Hudlicky, T.; Stabile, M.R.; Gibson, D.T.; Whited, G.M. *Org. Synth. Coll.* 2004, *10*, 217.
75. Climent, M.J.; Corma, A.; Iborra, S.; Mifsud, M.; Velty, A. *Green Chem.* 2010, *12*, 99.
76. Sheldon, R.A. *Green Chem.* 2017, *19*, 18.
77. Wittmann, S.; Schätz, A.; Grass, R.N.; Stark, W.J.; Reiser, O. *Angew. Chem. Int. Ed.* 2010, *49*, 1867.
78. Crosignani, S.; Launay, D.; Linclau, B.; Bradley, M. *Mol. Divers.* 2003, *7*, 203.
79. Kobayashi, G.; Saito, T.; Kitano, Y. *Synthesis* **2011**, 3225.
80. Wang, X.; Wang, Q.G.; Luo, Q.L. *Synthesis* 2015, 49.
81. Keita, M.; Vandamme, M.; Mahé, O.; Paquin, J.F. *Tetrahedron Lett.* 2015, *56*, 461.
82. Jiang, H.; Lin, W. *J. Am. Chem. Soc.* 2003, *125*, 8084.
83. Jimenez-Gonzalez, C.; Constable, D.J.C. *Green Chemistry and Engineering: A Practical Design Approach*, Wiley, Hoboken, NJ, 2011, pp. 102–103.
84. Jimenez-Gonzalez, C.; Constable, D.J.C. *Green Chemistry and Engineering: A Practical Design Approach*, Wiley, Hoboken, NJ, 2011, pp. 228–230.
85. Rodriguez, B.; Rantanen, T.; Bolm, C. *Angew. Chem. Int. Ed.* 2006, *45*, 6924.
86. Rodriguez, B.; Bruckmann, A.; Bolm, C. *Chem. Eur. J.* 2007, *13*, 4710.
87. Glasnov, T.N.; Tye, H.; Kappe, C.O. *Tetrahedron* 2008, *64*, 2035.
88. Verbitskyiy, E.V.; Cheprakova, E.M.; Zhilina, E.F.; Kodess, M.I.; Ezhikova, M.A.; Pervova, M.G.; Slepukhin, P.A.; Subbotina, J.O.; Schepochkin, A.V.; Rusinov, G.L.; Chupakhin, O.N.; Charushin, V.N. *Tetrahedron* 2013, *69*, 5164.

89. Ciriminna, R.; Albanese, L.; Meneguzzo, F.; Pagliaro, M. *ChemSusChem.* 2016, *9*, 3374.
90. Strukul, G. (ed.) *Catalytic Oxidations with Hydrogen Peroxide*, Kluwer Academic Publishers: Dordrecht, 1992.
91. Thenard, L.J. *Ann. Chim. Phys.* 1818, *8*, 306.
92. Berthelot, H. *Comptes Rend.* 1878, *186*, 71.
93. Meidinger, H. *Ann. Chem.* 1853, *88*, 57.
94. Walton, J.H.; Filson, G.W. *J. Am. Chem. Soc.* 1932, *54*, 3228.
95. Riedl, H.J.; Pfleiderer, G. US 2158525 (I.G. Farben, 1939).
96. Riedl, H.J.; Pfleiderer, G. US 2215883 (I.G. Farben, 1940).
97. Pfleiderer, G.; Riedl, H.J. US 2369912 (I.G. Farben, 1945).
98. Fischer, M.; Kaibel, G.; Stammer, A.; Flick, K.; Quaiser, S.; Harder, W.; Massone, K. US 2001003578 (2001).
99. Paparatto, G.; Rivetti, F.; Andrigo, P.; De Albertini, G. US 2002025293 (ENI S.p.A., 2002).
100. Haas, T.; Stochniol, G.; Rollmann, J. US 2004223904 (2004).
101. Streitweiser, A. Jr.; Heathcock, C.H. *Introductory Organic Chemistry*, 3rd ed., Macmillan Publishing Company, New York, 1985, p. 769, 824, 1046.
102. Shreve, Joseph, Brink, A. Jr.; Shreve, R.N. *Chemical Process Industries*, McGraw-Hill, New York, 1977, p. 740–741.
103. Xuan, Q.; Song, Q. *Org. Lett.* 2016, *18*, 4250.
104. Giomi, D.; Alfini, R.; Brandi, A. *Tetrahedron* 2011, *67*, 167.
105. Sudbrake, C.; Vahrenkamp, H. *Inorg. Chim. Acta* 2001, *318*, 23.

7 Intrinsic Greenness

7.1 MINIMUM ATOM ECONOMY

7.1.1 Terms, Definitions, and Examples

Minimum Atom Economy, $(AE)_{min}$ [1]

From a Markush type balanced chemical equation, general expressions for AE and E_{mw} as functions of R group size are written under the assumption of 100% reaction yield and stoichiometric operating conditions. No auxiliary materials are considered. It is then possible to estimate the minimum atom economy, $(AE)_{min}$, or the maximum unit waste expected per unit target product, E_{max}, according to the dependence of these metrics on R group size. The minimum AE is obtained by setting the R group as small as possible in the generalized expression for AE for a given reaction. Usually it is either R=H (MW=1) or R=methyl (MW=15).

Example 7.1 [2]

Themes: balancing chemical equations, by-product identification, roles of reagents, and reaction mechanism

Amides may be dehydrated to nitriles under mild acidic conditions according to the following transformation:

$$\mathrm{RC(=O)NH_2} \xrightarrow[\mathrm{CH_2{=}O}]{\mathrm{CH_3CN,\ HCOOH}} \mathrm{R{-}C{\equiv}N}$$

Part 1

Balance the chemical equation showing all by-products and propose a mechanism that accounts for the roles of acetonitrile, formic acid, and formaldehyde in this reaction.

Hint: If the amide oxygen atom is labeled as ^{18}O, then this label ends up as ^{18}O labeled formaldehyde.

Part 2

Based on the balanced chemical equation, what is the expected minimum atom economy for this transformation?

Part 3

Compare the minimum atom economy performance of the transformation with the one shown below where phosphorus oxychloride is used as the dehydrating agent. Assume an aqueous workup.

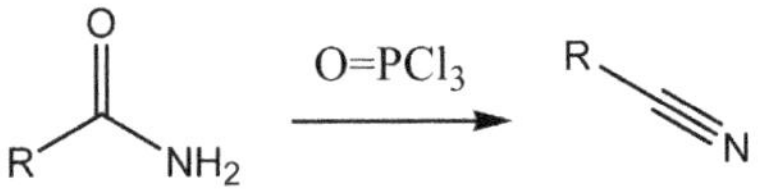

Solution

Part 1

Balanced chemical equation:

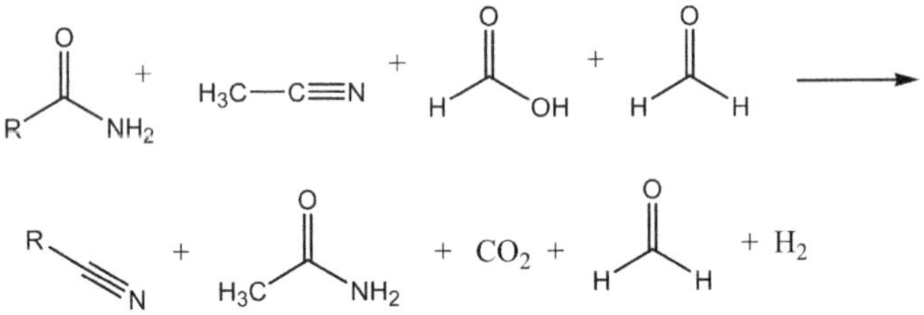

Mechanism:

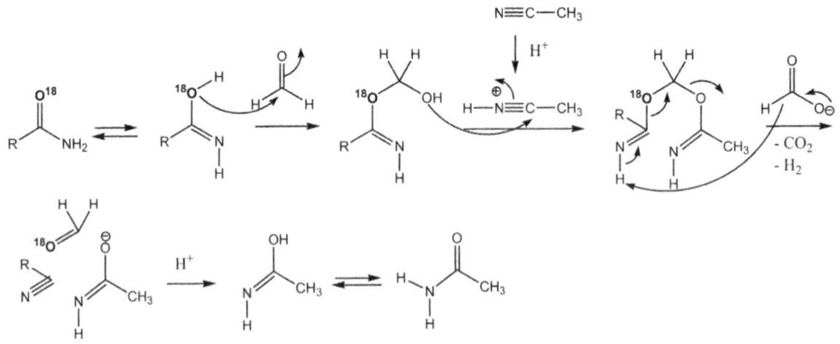

Acetonitrile has a dual role in this reaction. It is a solvent for the reaction and is converted to acetamide by-product where the oxygen atom of formaldehyde is transferred to acetamide.

Formic acid acts as an acid as well as a hydride donor, thereby resulting in its conversion to carbon dioxide.

Formaldehyde is a sacrificial reagent that transfers its original oxygen atom to acetonitrile giving acetamide and then acquires the ^{18}O label from the amide substrate to produce ^{18}O labeled formaldehyde.

Part 2

The balanced chemical showing molecular weights of all reactants and products is shown below.

$RC(=O)NH_2$ + $H_3C{-}C{\equiv}N$ + $HC(=O)OH$ + $HC(=O)H$ ⟶

R + 44 **41** **46** **30**

$R{-}C{\equiv}N$ + $H_3C{-}C(=O)NH_2$ + CO_2 + $HC(=O)H$ + H_2

R + 26 **59** **44** **30** **2**

The sum of the molecular weights of the reactants is

$$R + 44 + 41 + 46 + 30 = R + 161.$$

The sum of the molecular weights of the products is

$$R + 26 + 59 + 44 + 30 + 2 = R + 161.$$

Therefore, both sides of the chemical equation balance.

The expression for the atom economy of the reaction is given by

$$AE = \frac{R + 26}{R + 161}.$$

The minimum atom economy is found for the smallest sized R group. For practical purposes, the smallest R group is taken as methyl with a molecular weight of 15 g/mol.

Hence, the minimum atom economy for the transformation is given by

$$AE_{min} = \frac{15 + 26}{15 + 161} = \frac{41}{176} = 0.233, \text{ or } 23.3\%.$$

Part 3

The balanced chemical equation is

$RC(=O)NH_2$ + $Cl{-}P(=O)Cl_2$ + $2\,H_2O$ ⟶ $R{-}C{\equiv}N$ + $HO{-}P(=O)(OH){-}OH$ + $3\,HCl$

R + 44 **153.35** **2 (18)** **R + 26** **98** **3 (36.45)**

Check for equation balancing:

Reactants: $R + 44 + 153.35 + 2(18) = R + 233.35$
Products: $R + 26 + 98 + 3(36.45) = R + 233.35$

Minimum atom economy when $R = 15$:

$$AE_{min} = \left.\frac{R + 26}{R + 233.35}\right|_{R=15} = \frac{15 + 26}{15 + 233.35} = 0.165, \text{ or } 16.5\%.$$

At the level of atom economy considerations, the milder dehydrating method is about 7% more material efficient.

Example 7.2

Themes: balancing chemical equations, Markush structures, minimum atom economy

For the following named C=C bond forming reactions determine the minimum atom economies and rank the reactions.

(a) Horner–Emmons–Wadsworth reaction

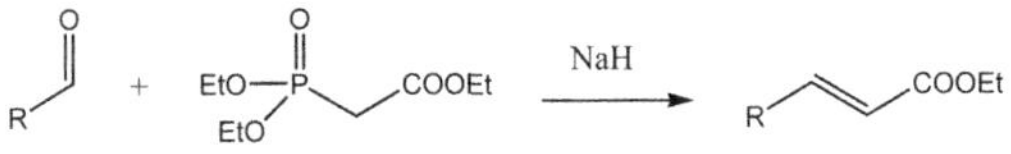

NaH

(b) Julia–Kocienski olefination

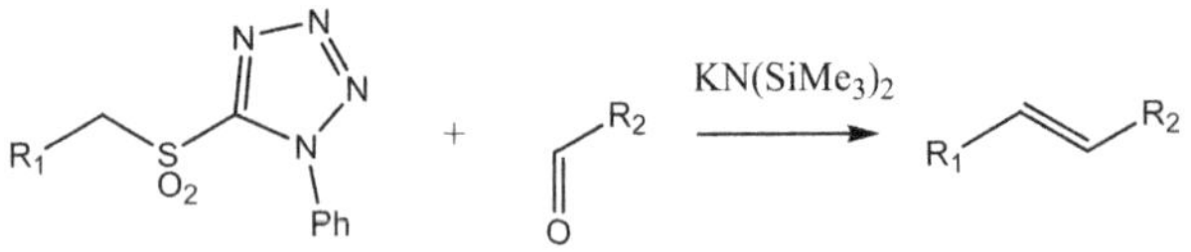

$KN(SiMe_3)_2$

(c) Julia–Lythgoe olefination

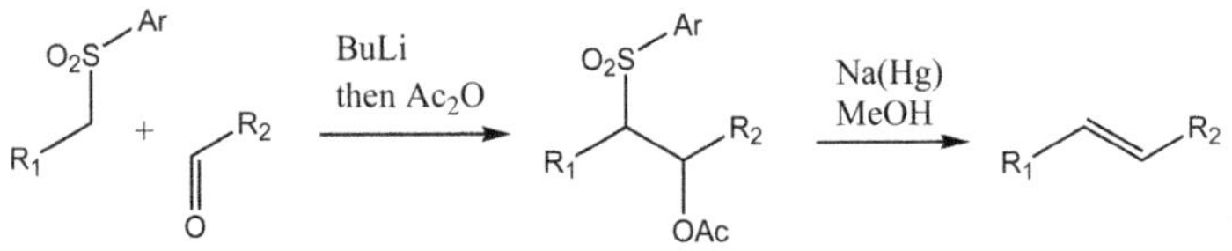

BuLi then Ac_2O; Na(Hg) MeOH

(d) McMurry coupling

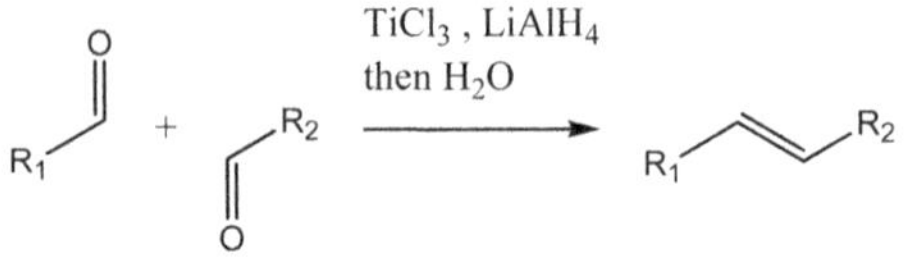

$TiCl_3$, $LiAlH_4$ then H_2O

(e) Perkin reaction

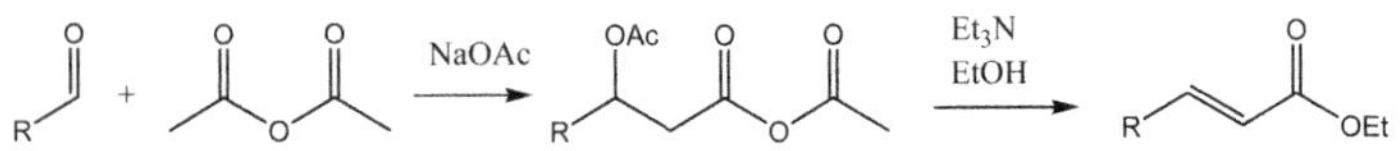

NaOAc; Et_3N EtOH

(f) Peterson olefination

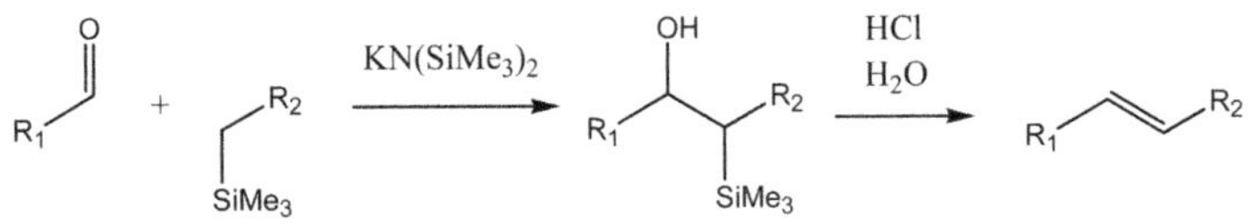

$KN(SiMe_3)_2$; HCl H_2O

(g) Schlosser–Wittig modification

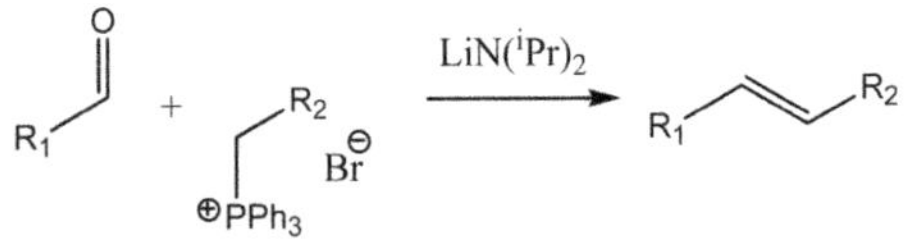

$LiN(^iPr)_2$

Solution

(a) Horner–Emmons–Wadsworth reaction

$$\text{RCHO} + (\text{EtO})_2\text{P(O)CH}_2\text{COOEt} \xrightarrow[-\text{H}_2\ (2)]{\text{NaH}\ (24)} \text{RCH=CHCOOEt} + (\text{EtO})_2\text{P(O)ONa}$$

R + 29   224   R + 99   176

Reactants: R + 29 + 224 + 24 = R + 277
Products: R + 99 + 176 + 2 = R + 277
AE = (R + 99)/(R + 277)
When R = Me, AE(min) = (15 + 99)/(15 + 277) = 0.390, or 39.0%

(b) Julia–Kocienski olefination

$$\text{R}_1\text{CH}_2\text{SO}_2\text{-(1-phenyltetrazol-5-yl)} + \text{R}_2\text{CHO} \xrightarrow[-\text{HN(SiMe}_3)_2\ (161)]{\text{KN(SiMe}_3)_2\ (199)} \text{R}_1\text{CH=CHR}_2 + \text{KOSO}_2\text{-(1-phenyltetrazol-5-yl)}$$

R_1 + 223 R_2 + 29 R_1 + R_2 + 26   264

Reactants: R1 + R2 + 223 + 29 + 199 = R1 + R2 + 451
Products: R1 + R2 + 26 + 264 + 161 = R1 + R2 + 451
AE = (R1 + R2 + 26)/(R1 + R2 + 451)
When R1 = R2 = Me, AE(min) = (15 + 15 + 26)/(15 + 15 + 451) = 0.116, or 11.6%.

(c) Julia–Lythgoe olefination

$$\text{R}_1\text{CH}_2\text{SO}_2\text{Ar} + \text{R}_2\text{CHO} \xrightarrow[-\text{BuH},\ -\text{LiOAc}]{\text{BuLi, then Ac}_2\text{O}} \text{R}_1\text{CH(SO}_2\text{Ar)CH(OAc)R}_2 \xrightarrow[-\text{HOAc},\ -\text{NaOMe}]{2\ \text{Na(Hg), MeOH}} \text{R}_1\text{CH=CHR}_2 + \text{ArS(O)ONa}$$

R_1 + Ar + 78 R_2 + 29 R_1 + R_2 + 26   Ar + 87

Reactants: R1 + R2 + Ar + 78 + 29 + 64 + 102 + 2*23 + 32 = R1 + R2 + Ar + 351
Products: R1 + R2 + Ar + 26 + 87 + 58 + 66 + 60 + 54 = R1 + R2 + Ar + 351
AE = (R1 + R2 + 26)/(R1 + R2 + Ar + 351)
When R1 = R2 = Me and Ar = Ph, AE(min) = (15 + 15 + 26)/(15 + 15 + 77 + 351) = 0.122, or 12.2%.

(d) McMurry coupling

$$\text{R}_1\text{CHO} + \text{R}_2\text{CHO} \xrightarrow[-5/2\ \text{H}_2]{\text{TiCl}_3\ (154.25),\ \text{LiAlH}_4\ (38)\ \text{then}\ 3\ \text{H}_2\text{O}\ (54)} \text{R}_1\text{CH=CHR}_2 + \text{TiO}_2 + \text{LiCl} + \text{Al(OH)}_3 + 2\ \text{HCl}$$

R_1 + 29 R_2 + 29 R_1 + R_2 + 26   79.9   42.45   78   72.9

Reactants: R1 + R2 + 29 + 29 + 154.25 + 38 + 3*18 = R1 + R2 + 304.25
Products: R1 + R2 + 26 + 79.9 + 42.45 + 78 + 72.9 + (5/2)*2 = R1 + R2 + 304.25

AE = (R1 + R2 + 26)/(R1 + R2 + 304.25)
When R1 = R2 = Me, AE(min) = (15 + 15 + 26)/(15 + 15 + 304.25) = 0.168, or 16.8%.

(e) Perkin reaction

NaOAc (cat.), Ac_2O; - HOAc 60; 2 Et_3N 101, then EtOH 46; - 2 $[Et_3NH][OAc]$ 2(161)

R + 29 102 R + 99

Reactants: R + 29 + 2*102 + 2*101 + 46 = R + 481
Products: R + 99 + 60 + 2*161 = R + 481
AE = (R + 99)/(R + 481)
When R = Me, AE(min) = (15 + 99)/(15 + 481) = 0.230, or 23.0%.

(f) Peterson olefination

199 $KN(SiMe_3)_2$; - $HN(SiMe_3)_2$ 161; HCl, H_2O; - KCl 74.45; - $HOSiMe_3$ 90; - H_2O 18

$R_1 + 29$ $R_2 + 87$ $R_1 + R_2 + 26$

Reactants: R1 + R2 + 29 + 87 + 199 + 36.45 = R1 + R2 + 351.45
Products: R1 + R2 + 26 + 161 + 74.45 + 90 = R1 + R2 + 351.45
AE = (R1 + R2 + 26)/(R1 + R2 + 351.45)
When R1 = R2 = Me, AE(min) = (15 + 15 + 26)/(15 + 15 + 351.45) = 0.147, or 14.7%.

(g) Schlosser–Wittig modification

107 $LiN(^iPr)_2$; - LiBr 86.9; - $HN(^iPr)_2$ 101; - $O{=}PPh_3$ 278

$R_1 + 29$ $R_2 + 355.9$ $R_1 + R_2 + 26$

Reactants: R1 + R2 + 29 + 355.9 + 107 = R1 + R2 + 491.9
Products: R1 + R2 + 26 + 86.9 + 101 + 278 = R1 + R2 + 491.9
AE = (R1 + R2 + 26)/(R1 + R2 + 491.9)
When R1 = R2 = Me, AE(min) = (15 + 15 + 26)/(15 + 15 + 491.9) = 0.107, or 10.7%.

The following group of reactions can be ranked head-to-head since they lead to the same product structure given below.

R_1 R_2

Reaction	AE(min)
McMurry coupling	0.168
Peterson olefination	0.147
Julia–Lythgoe olefination	0.122
Julia–Kocienski olefination	0.116
Schlosser–Wittig modification	0.107

The following group of reactions can be ranked head-to-head since they lead to the same product structure given below.

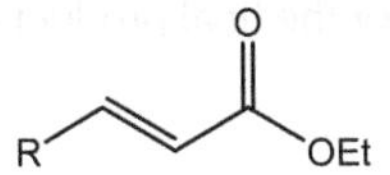

Reaction	AE(min)
Horner–Emmons–Wadsworth reaction	0.390
Perkin reaction	0.230

Example 7.3 [3]

Themes: balancing chemical equations, multi-component reactions, ring construction strategy, target bond mapping

A three-component coupling reaction to make cyclopropane adducts is shown below. The chiral auxiliary allows four stereogenic centers to be set from achiral starting materials.

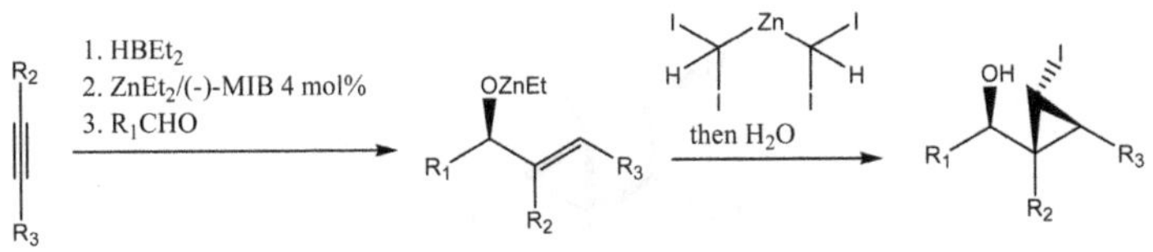

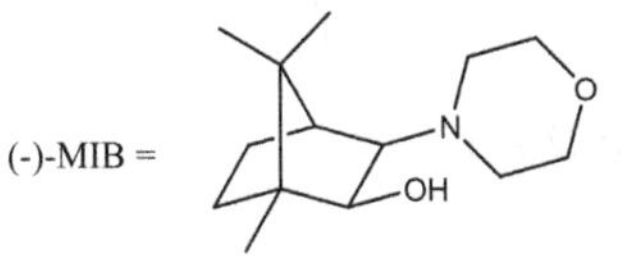

1. $HBEt_2$
2. $ZnEt_2$/(-)-MIB 4 mol%
3. R_1CHO

then H_2O

(-)-MIB =

(-)-3-morpholinoborneol

Part 1

Show a complete balanced chemical equation for the transformation and determine its minimum atom economy.

Part 2

Draw a target bond map for the final product structure and describe the ring construction strategy.

Part 3

The same product can be assembled via a different strategy as shown below.

$ZnEt_2$ + CF_3CH_2OH + CH_2I_2

Zn(R$_1$)$_2$/(-)-MIB 4 mol%

then H_2O

(a) Show a complete balanced chemical equation for the transformation and determine its minimum atom economy.

(b) Draw a target bond map for the final product structure and describe the ring construction strategy.

Solution

Part 1

70 HBEt$_2$

R_1 + 29

$ZnEt_2$ **123.38**

- BEt_3

0.5 (599.38)

1/2

then 2 H_2O

- 1/2 ZnI_2

- $Zn(OH)_2$

- EtH

R_2 + R_3 + 24

R_1 + R_2 + R_3 + 195

Reactants: R1 + R2 + R3 + 24 + 70 + 29 + 123.38 + 0.5*599.38 + 2*18 = R1 + R2 + R3 + 582.07

Products: R1 + R2 + R3 + 195 + 98 + 0.5*319.38 + 99.38 + 30 = R1 + R2 + R3 + 582.07

AE = (R1 + R2 + R3 + 195)/(R1 + R2 + R3 + 582.07)

Let R1 = R2 = R3 = Me.

Minimum AE = (15 + 15 + 15 + 195)/(15 + 15 + 15 + 582.07) = 0.383, or 38.3%

Part 2

Ring construction strategy: [2 + 1]

Part 3

(a)

123.38 **100** **268**

$ZnEt_2$ + CF_3CH_2OH + CH_2I_2

- 2 EtH

65.38 + 2 R_1

Zn(R$_1$)$_2$/(-)-MIB 4 mol%

then 2 H_2O

- R_1H

- $Zn(OH)_2$

- $Zn(I)(OCH_2CF_3)$

R_2 + R_3 + 54

R_1 + R_2 + R_3 + 195

Reactants: 2*R1 + R2 + R3 + 54 + 65.38 + 123.38 + 100 + 268 + 2*18
= 2*R1 + R2 + R3 + 646.76
Products: 2*R1 + R2 + R3 + 195 + 2*30 + 1 + 99.38 + 291.38 = 2*R1 + R2 + R3 + 646.76
AE = (R1 + R2 + R3 + 195)/(2*R1 + R2 + R3 + 646.76)
Let R1 = R2 = R3 = Me.
Minimum AE = (15 + 15 + 15 + 195)/(2*15 + 15 + 15 + 646.76) = 0.340, or 34.0%

(b)

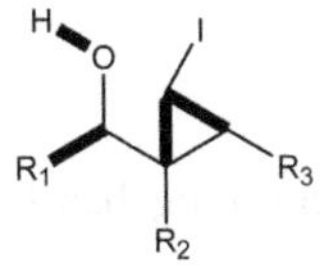
H
O
I
R1
R3
R2

Ring construction strategy: [2 + 1]

7.1.2 Problems

PROBLEM 7.1

For the following sequences involving inversion of configuration determine balanced chemical equations and minimum atom economies for each step.

(a)

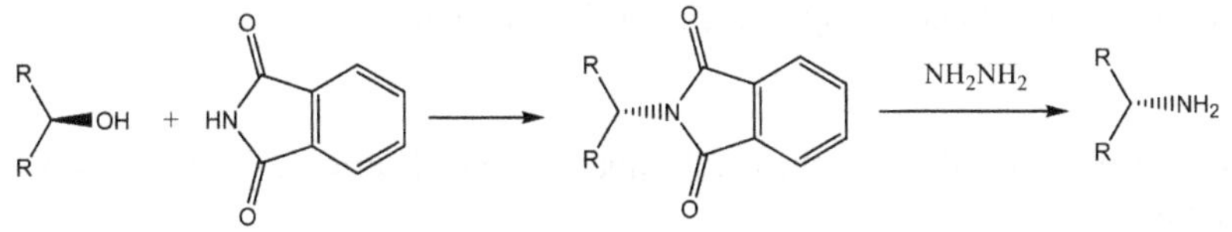
R
OH
R
HN
O
O
NH2NH2
NH2

(b)

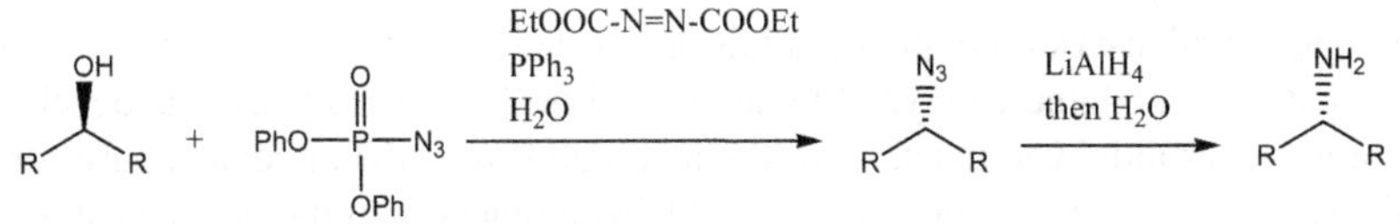
OH
PhO—P—N3
OPh
EtOOC-N=N-COOEt
PPh3
H2O
N3
LiAlH4
then H2O
NH2
R

(c)

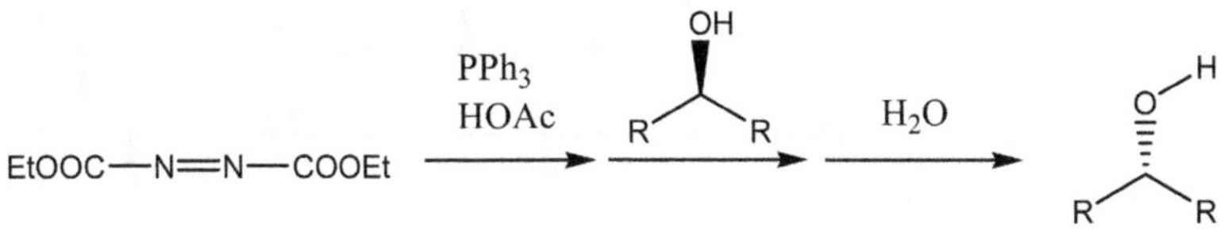
EtOOC—N=N—COOEt
PPh3
HOAc
OH
R
H2O
O
H

PROBLEM 7.2 [4]

For the four-component Krohnke pyridine synthesis reaction shown below, highlight the target bonds made in the product structure and determine the ring construction strategy. Write out a balanced chemical equation for the reaction showing all by-products and determine its minimum atom economy.

NH_4OAc
HOAc
microwaves

PROBLEM 7.3 [5]

A combined diazotization Heck coupling has been achieved to produce styrenes from anilines and olefins.

$Pd(OAc)_2$
$^{t}BuO\text{-}N{=}O$
$MeSO_3H$
MeOH

Part 1

Write a mechanism and catalytic cycle for the reaction.

Part 2

Write the overall balanced chemical equation and determine its minimum atom economy.

PROBLEM 7.4 [6]

A one-pot condensation-conjugate addition cascade reaction involving two aldehydes and nitromethane catalyzed by acids and proline derivatives was developed. The more polar aldehyde reacts with the acid catalyst in the aqueous phase for the condensation reaction and the less polar aldehyde reacts with the proline catalyst in the organic phase for the conjugate addition reaction.

$+ CH_3NO_2$

aqueous phosphate buffer

Part 1

Write a balanced chemical equation for the transformation and determine its minimum atom economy.

Part 2

Write out a catalytic cycle for the reaction that includes the roles of both catalysts.

PROBLEM 7.5 [7]

A library of substituted pyrimidines were synthesized using a catch and release strategy via solid-supported reagents under microwave conditions. The first step is a 3-component coupling between a piperazine bound resin, an imidazole acetal, and a β-ketoester. The resultant product is bound to the resin ("catch"). In the second step, a guanidine salt is reacted to release the pyrimidine product from the piperazine resin.

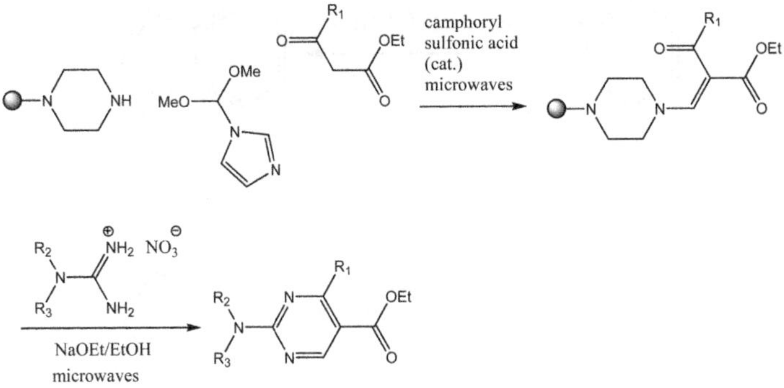

Part 1

Show balanced chemical equations for both steps. Determine the overall minimum atom economy.

Part 2

Highlight the target bonds made in the pyrimidine product and describe the ring construction strategy.

PROBLEM 7.6

Part 1

For the following named indole forming reactions, determine the minimum atom economies and rank the reactions.

Baeyer

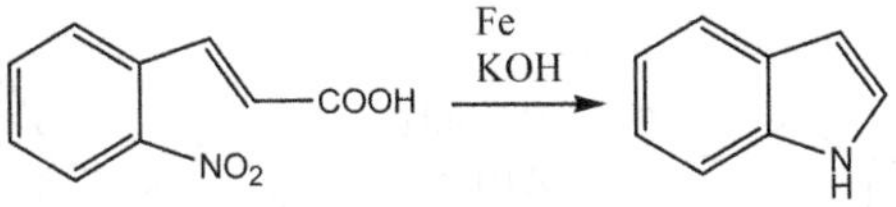

Bartoli

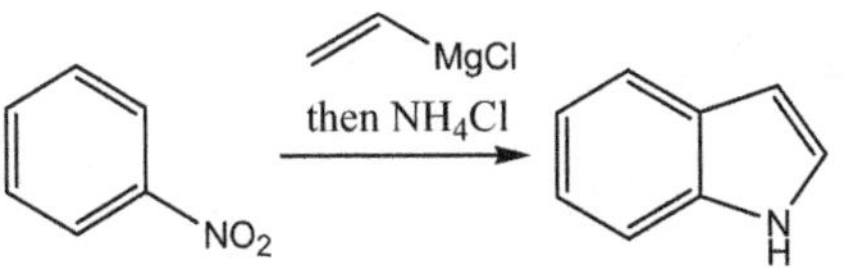

Batcho–Leimgruber

OMe
+ Me_2N
OMe
NO_2
DABCO
DMF
NMe_2
NO_2
H_2
Raney Ni
N
H

Cadogan–Sundberg

NO_2
$P(OEt)_3$
N
H

Castro

NH_2
CuI
DMF
N
H

Fischer

NH_2
N
H
+
O
HCl
N
H

Fukuyama G1

R_2
N
C
Bu$_3$SnH
AIBN
then H_2O
R_2
N
H

Fukuyama G2

R_2
NH
S
R_1
Bu$_3$SnH
AIBN
R_2
R_1
N
H

Fürstner

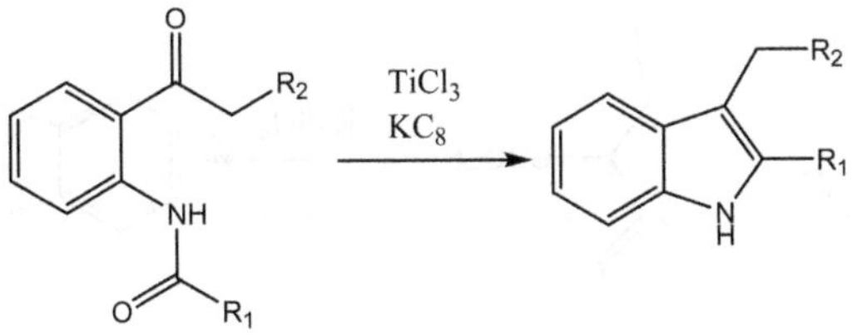
O
R2
NH
O
R1
TiCl3
KC8
R2
R1
N
H

Gassman

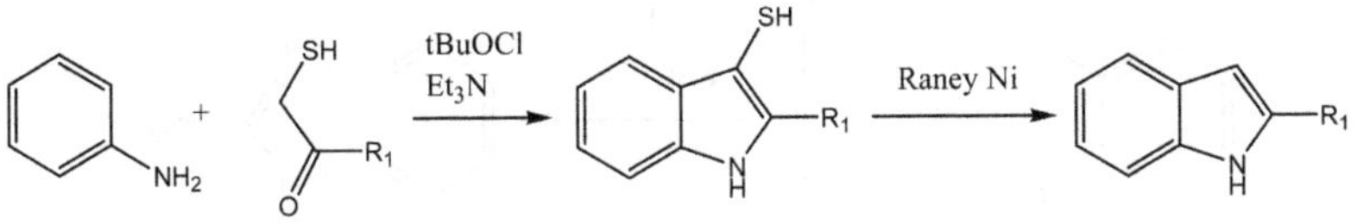
NH2
+
SH
O
R1
tBuOCl
Et3N
SH
R1
N
H
Raney Ni
R1
N
H

Hegedus G1

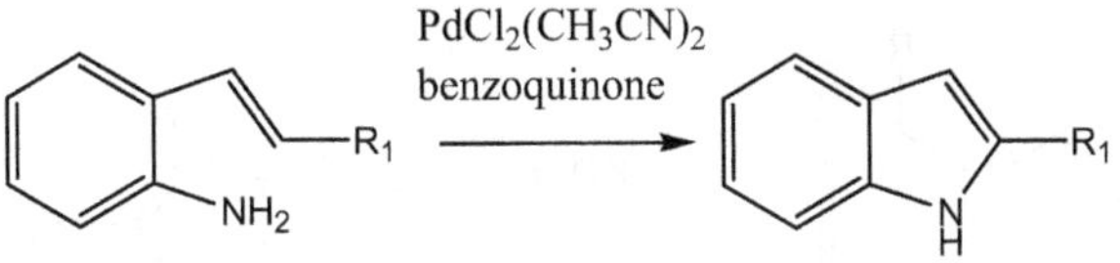
R1
NH2
PdCl2(CH3CN)2
benzoquinone
R1
N
H

Hegedus G2

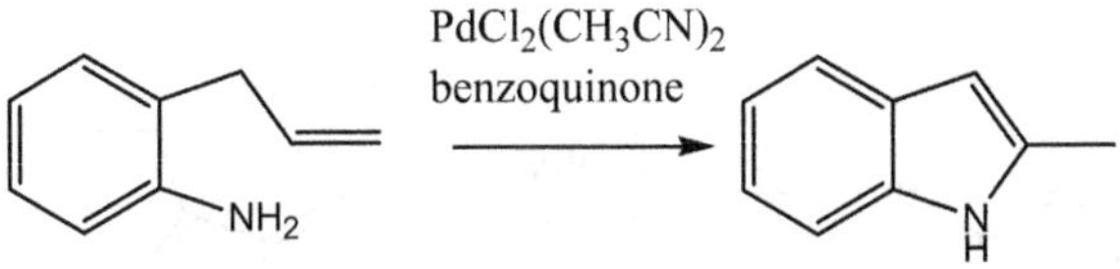
NH2
PdCl2(CH3CN)2
benzoquinone
N
H

Hegedus G3

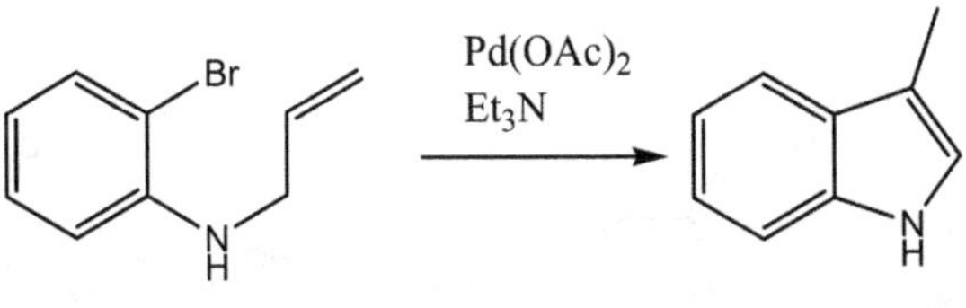
Br
N
H
Pd(OAc)2
Et3N
N
H

Hemetsberger

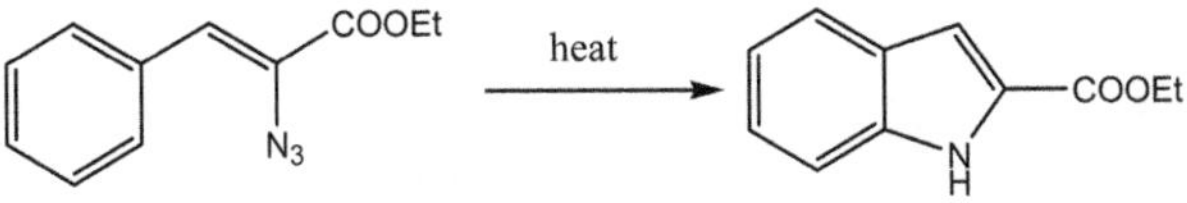
COOEt
N3
heat
COOEt
N
H

Japp–Klingemann

H_2SO_4

Larock

$Pd(OAc)_2$
LiCl
K_2CO_3

Madelung

KOEt

Mori–Ban

$Pd(PPh_3)_4$
Ag_2CO_3

Nenitzescu

HOAc

Reissert

1. NaOMe
2. $SnCl_2$
3. Cu_2O/quinoline

Sugasawa

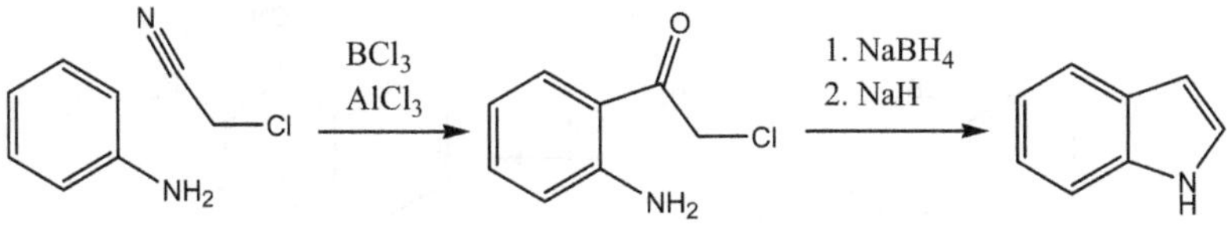

BCl_3, $AlCl_3$; 1. $NaBH_4$ 2. NaH

Sundberg

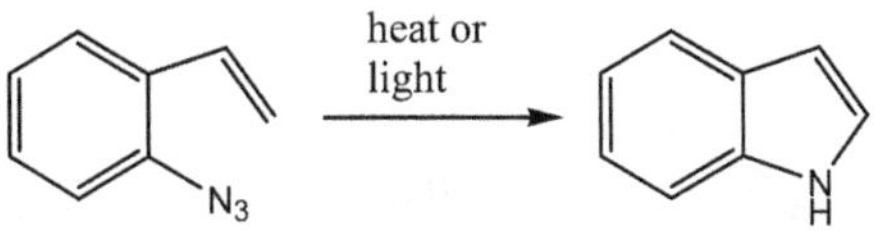

heat or light

Wender

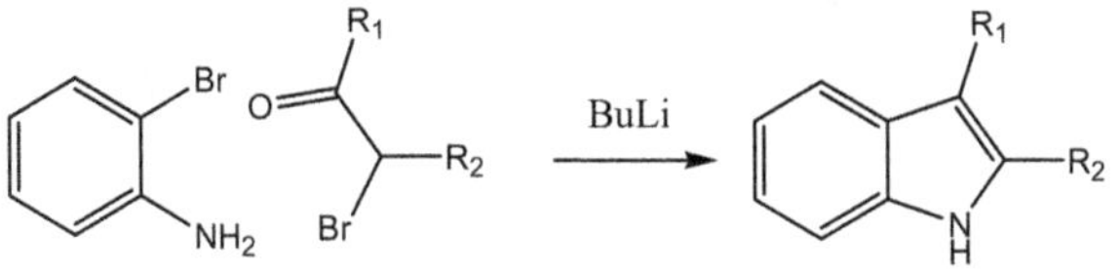

BuLi

Part 2

Categorize the ring construction strategies employed by the named reactions.

Part 3

Enumerate all possible one- and two-partitions of the indole ring that leaves the benzene ring intact. From this list indicate which combinations have NOT been employed by any of the named reactions.

Part 4

For the combinations found in Part 3 that are not employed by any of the named reactions, suggest viable reactions that could lead to those partition patterns. Justify with mechanisms.

PROBLEM 7.7 [8–10]

Various methods to prepare the isoquinoline heterocyclic ring are shown below.

For each reaction provide a balanced chemical equation showing all by-products and determine the minimum atom economy. Also, show the ring construction strategy by highlighting which bonds are made in the target product structure.

(1)

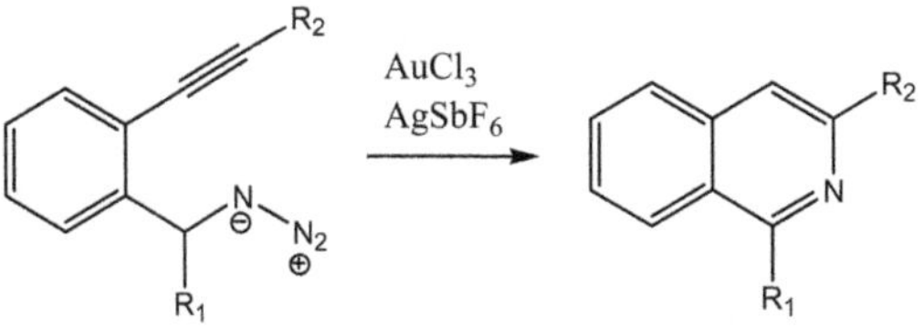

$AuCl_3$, $AgSbF_6$

(2)

CN, Br, R_2, N, OAc, R_1 — CuI → CN, R_2, N, R_1

(3)

R_2, N, R_1 — CF_3SO_3H microwaves → R_2, N, R_1

(4)

S, O, R_1 + R_2 — $[Rh(nbd)_2][BF_4]$ xantphos → [R_2, O, S, R_1] — NH_4OAc → R_2, N, R_1

(5)

R_2, R_3, O, R_1, H_2N–OH HCl — $[CpRhCl_2]_2$ KOAc → R_2, R_3, N, R_1

(6)

R_2, R_3, N, OH, R_1 — $[RuCl_2(p\text{-cymene})_2]_2$ NaOAc → R_2, R_3, N, R_1

(7)

COOEt
+
AgOTf
DBU
COOEt
N
R
R

(8)

Br
R
NH_4OAc
O
$Pd(OAc)_2$
PPh_3
KOAc
microwaves
R
N

(9)

R_2
N
N_2
R_1
1. $AgSbF_6$
2. $NaHCO_3$
R_2
N
R_1

(10)

R_1
R_2
OMs
COOEt
NaN_3
K_2CO_3
O_2
O
N
N
R_1
R_2
N
COOEt

(11)

R_1
N
O
R_2
AgOTf
TfOH
R
N

(12)

OTf, SiMe$_3$ + R$_2$, NH, O, R$_1$ — $[Bu_4N]F$ → R$_2$, N, R$_1$

(13)

COOEt, $N^{\ominus}$, $N_2^{\oplus}$ + O, R$_2$, R$_1$, N$_2$ — PPh_3 → COOEt, N, R$_1$, R$_2$

(14)

HN, O — $P(OPh)_3$, Br_2, Et_3N → N

(15)

Br, NH + COOEt, R, O — CuI, K_2CO_3, air → COOEt, R, N

(16)

HN, O, R — Tf_2O, $O{=}PPh_3$, then MnO_2 → N, R

(17)

R$_3$, R$_2$, N, Ph, R$_1$ — $Rh(PPh_3)_3Cl$ → R$_3$, R$_2$, N, R$_1$

(18)

$MnBr(CO)_5$

(19)

$[Ru(p\text{-cymene})Cl_2]_2$
KPF_6
$Cu(OAc)_2$ H_2O

(20)

$[Cp^*RhCl_2]_2$
NaOAc

(21)

$[Cp^*RhCl_2]_2$
$Cu(OAc)_2$
HOAc

(22)

P_2O_5

PROBLEM 7.8

Part 1

Determine the minimum atom economies for the pyrrole forming named reactions shown below.

Barton–Zard reaction

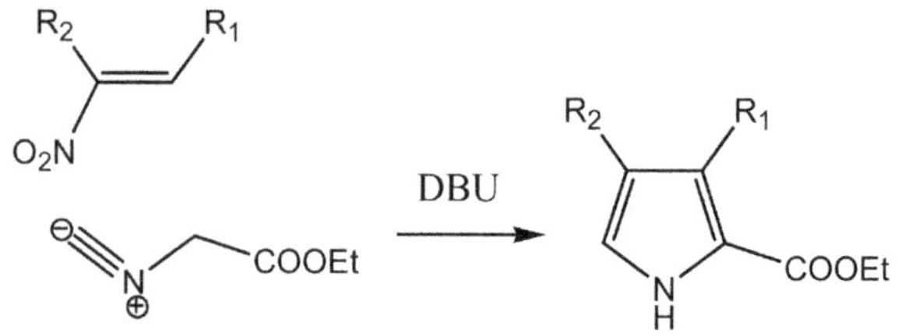

DBU

Hanzsch pyrrole synthesis

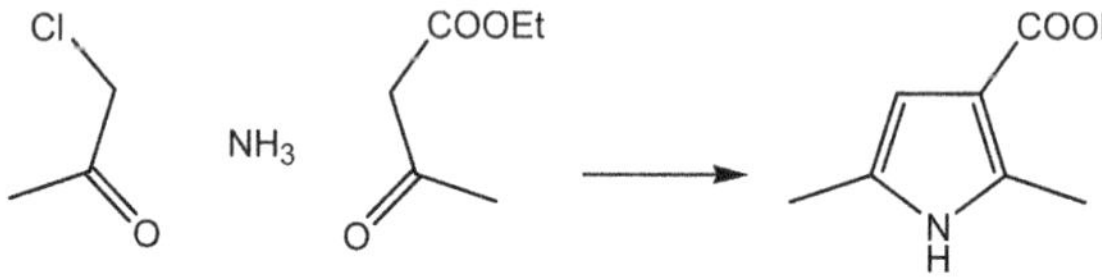

Paal–Knorr pyrrole synthesis

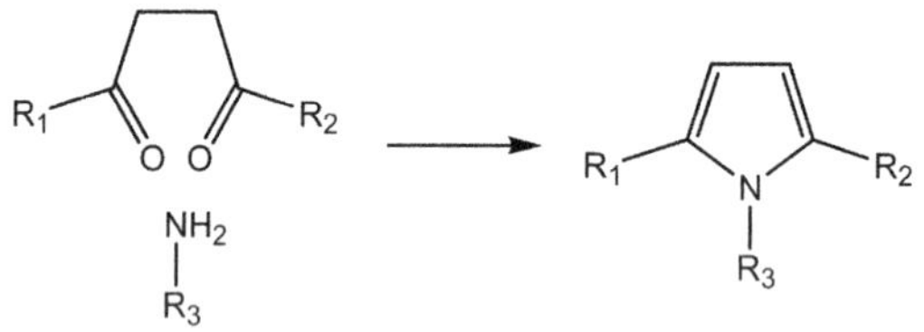

Part 2

Show all possible two-partitions of the pyrrole ring and determine which partitions corresponds to the named methodologies shown above. Repeat for all possible three-partitions.

PROBLEM 7.9 [11]

Instead of using conventional oxidants to convert secondary alcohols to ketones, a new anaerobic oxidation reaction using aromatic halides as oxidants was achieved. This reaction is carried out using an N-heterocyclic carbene palladium(II) catalyst in basic solution under microwave irradiation conditions.

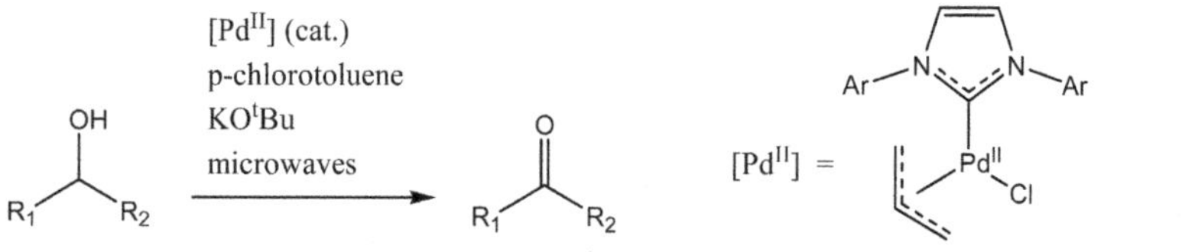

$[Pd^{II}]$ (cat.)
p-chlorotoluene
KO^tBu
microwaves

Ar = 2,6-diisopropylbenzene

Part 1

Show a balanced chemical equation for the transformation showing all reaction by-products.

Part 2

Compare its atom economy performance versus the Jones oxidation.

Part 3

The authors of the paper did not show a mechanism for the reaction. Suggest one.

Part 4

The authors also extended the method to a one-pot transformation as shown below.

$[Pd^{II}]$ (cat.)
p-chlorotoluene
KO^tBu (2 equiv.)
R_3-Cl

OH R_1 R_2 → O R_1 R_2 R_3

Parse this reaction showing all steps with balanced chemical equations.

PROBLEM 7.10

Various named cyclopropanation reactions are listed below.

(a) [12–14]

R R + MeO–P(=O)(OMe)–CH=N₂ —Cu (cat.)→ R R cyclopropyl–P(=O)(OMe)–OMe

(b) [15, 16]

R R + Me_3Si–CH=N_2 → R R cyclopropyl–$SiMe_3$

(c) [17]

R R + I–CH₂–I —Zn/Cu→ R R cyclopropane

(d) [18, 19]

R R + Ph–Hg–CF_3 —NaI (cat.)→ R R cyclopropane F F

(e) [20–22]

Sm

(f) [23, 24]

(g) [25]

CH_3OTf

(h) [26, 27]

KO^tBu

(i) [28]

NaOH

(j) [29]

$Ti(O^iPr)_4$ (cat.)

H_2O

(k) [30–35]

KO^tBu

(l) [36–39]

(m) [40, 41]

Cp_2ZrCl_2
CH_3CH_2MgBr
then $AlCl_3$
H_2O

(n) [42–44]

Zn

Part 1

For each reaction, write out balanced chemical equations and determine the minimum atom economy. Classify the reactions according to the ring construction strategy employed and whether or not they involve carbene insertion.

Part 2

Determine if any of the reactions either are redox reactions that can be decomposed into redox couples, are internal redox reactions, or are non-redox reactions.

7.2 PROBABILITY OF DETERMINING INTRINSIC GREENNESS FOR ANY CHEMICAL REACTION

7.2.1 Terms, Definitions, and Examples

Golden Atom Economy Threshold [1]

Green reactions are characterized as having atom economy values greater than or equal to 0.618, or 61.8%. This number corresponds to the golden ratio and arises directly from the connecting relationship between atom economy and E_{mw} given by Equation (7.1).

$$\mathrm{AE} = \frac{1}{1 + E_{mw}} \tag{7.1}$$

A plot of AE versus E_{mw} reveals that it can be divided into two regions: one in which the magnitude of AE exceeds E_{mw} and the other in which the opposite is true. By setting AE equal to E_{mw}, one finds that the magnitudes of the atom economy and environmental impact factor based on molecular weight match at a value given by 0.61803. This means that a true "green" reaction is one which produces at least 0.618 mass units of target product per mass unit of all reactants used (AE > 0.618), or conversely, one that produces at most 0.618 mass units of waste by-products per mass unit of target product ($E_{mw} < 0.618$). It should be emphasized that this threshold arises directly from the inverse relationship between AE and E_{mw} that in turn arises directly from a simple analysis of a balanced chemical equation and therefore is not an arbitrary designation.

In deciding a kernel RME threshold value, α, the universal connecting relationship between RME and *E*-factor given by Equation (7.2) may be used, which resembles Equation (7.1) in form.

$$\mathrm{RME} = \frac{1}{1+E} \tag{7.2}$$

Setting RME = $E = x$, we obtain a quadratic $x^2 + x - 1 = 0$ whose positive real solution is given by Equation (7.3).

$$x = \frac{1}{2}\left(\sqrt{5} - 1\right) = 0.618 \tag{7.3}$$

Imposing such a threshold value guarantees that the magnitude of kernel RME exceeds that of the corresponding kernel *E*-factor. Put another way, the kernel process mass intensity (i.e., molar mass of reagents required per molar mass of target product) is guaranteed to be less than the inverse of the kernel *E*-factor (molar mass of target product per molar mass of by-product and unreacted reagent waste). The magnitude of the threshold value given in Equation (7.3) is recognizable as the "golden ratio," which suggests that intrinsically green reactions are inherently the gold standard of chemical transformations.

Intrinsic Greenness

The assessment of greenness of a chemical reaction or synthesis plan based on kernel metrics determined solely from balanced chemical equations under stoichiometric conditions such as atom economy, reaction yield, kernel RME, kernel *PMI*, and kernel *E*-factor. All other masses of excess reagents and auxiliary materials are not counted.

See *probability of intrinsic greenness.*

Probability of Achieving Intrinsic Greenness [45]

The probability of achieving intrinsic greenness refers to the probability of satisfying a threshold target value of kernel reaction mass efficiency based on boundary conditions for both atom economy and reaction yield for a given chemical reaction. The boundary conditions are minimum atom economy and minimum reaction yield. The selected threshold target value of kernel RME is 0.618 corresponding to the golden number.

See intrinsic greenness; golden atom economy threshold; minimum atom economy.

We begin with the general expression for the global reaction mass efficiency (RME) for any single chemical reaction given by Equation (7.4).

$$\mathrm{RME}_{\mathrm{global}} = (\mathrm{AE})(\varepsilon)\left(\frac{1}{\mathrm{SF}}\right)(\mathrm{MRP}) \tag{7.4}$$

where:

AE is atom economy

ε is reaction yield

1/SF is the inverse of the stoichiometric factor which takes into account excess reagent consumption

MRP is the material recovery parameter which takes into account all auxiliary material consumption (reaction solvent, work-up, and purification materials)

All of these parameters are fractions between 0 and 1. Note that the inverse of global RME is identical to the process mass intensity (*PMI*) which has now been adopted by the pharmaceutical industry as a key green metric. When dealing with synthesis strategy performance Equation (7.4) reduces to the simplified expression given by Equation (7.5), for the kernel RME, which involves only the variables that govern intrinsic chemical performance.

$$\mathrm{RME}_{\mathrm{kernel}} = (\mathrm{AE})(\varepsilon) \tag{7.5}$$

Under these conditions, Equation (7.5) implies that all auxiliary materials are eliminated or recovered and the reaction is run under stoichiometric conditions. We now impose a threshold value, α, for the kernel RME such that the condition given by Equation (7.6) is satisfied.

$$\mathrm{RME}_{\mathrm{kernel}} = (\mathrm{AE})(\varepsilon) \geq \alpha \tag{7.6}$$

Equation (7.6) may be rewritten as shown in Equation (7.7) so that a plot can be made of AE versus reaction yield as given in Figure 7.1 for various values of α.

$$\mathrm{AE} \geq \alpha/\varepsilon \tag{7.7}$$

We note from Figure 7.1 that as the full spectrum of values of AE and reaction yield are considered for a given reaction, any combination of these two variables must lie somewhere in the domain of the square bounded by the points (0,0), (0,1), (1,0), and (1,1). In order to satisfy the inequalities given by Equation (7.6) or Equation (7.7) such a combination of AE and reaction yield values must fall above the line given by the equality relationship. Also, we note that as the threshold value decreases, the area of the domain above the line increases. Since the parameters AE and reaction yield are independent variables and they each range in value between 0 and 1, we may directly interpret the area of the domain above the equality relationship given by Equation (7.7) as the probability that the combination of AE and reaction yield for a given reaction will yield a kernel RME value that exceeds the threshold value, α.

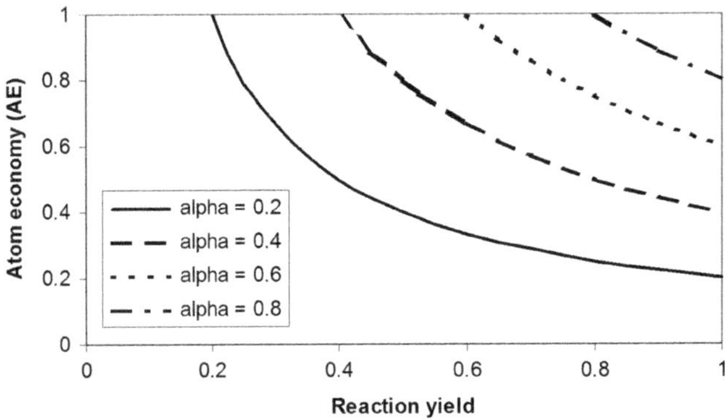

FIGURE 7.1 Plot of AE versus reaction yield for various values of α.

For any given threshold, α, and without imposing any restriction on AE and reaction yield other than $0 \leq \varepsilon \leq 1$ and $0 \leq AE \leq 1$, such a probability is given by Equation (7.8).

$$p = \frac{\text{Area-above-hyperbola}}{\text{Area-of-square}} = 1 - \alpha + \alpha \ln \alpha \tag{7.8}$$

When a low threshold value is set for the kernel RME, the probability of achieving it is high. Conversely, when a high threshold is set the likelihood of achieving it diminishes. In other words, if we aim for mediocrity we are more likely to achieve it, whereas, if we aim for excellence we are less likely to achieve it. A plot of Equation (7.8) is given in Figure 7.2. When dealing with a specific reaction where a minimum atom economy may be determined for a generalized structure of the intended target product, and a minimum reaction yield is imposed, the probability of achieving a kernel RME exceeding α is given by expressions listed in Table 7.1. There are seven cases to consider depending on where the AE(min) and Ɛ (min) values fall in the square in relation to α.

The derivations of Equations (7.9) through (7.15) follow based on the graphs shown in Figures 7.3 through 7.9.

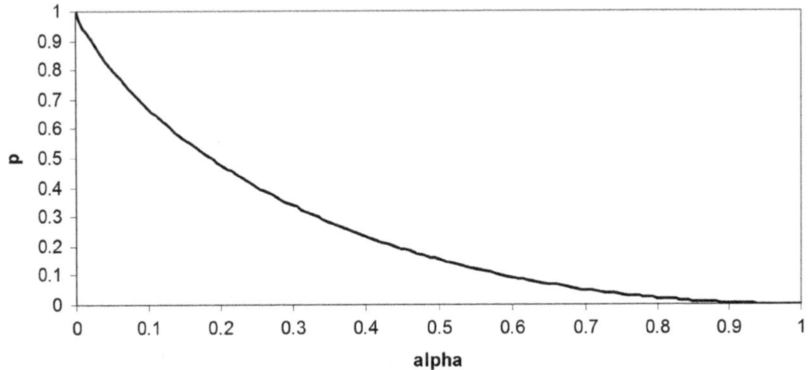

FIGURE 7.2 Plot of probability versus kernel RME threshold according to Equation (7.8).

Case I: (AE)min > α and 0 < ε (min) < α

probability = area above dashed curve that falls inside black rectangle/area of black rectangle

Area black rectangle $= \left[1 - \mathrm{AE}(\min)\right]\left[1 - \varepsilon(\min)\right]$

Black rectangle intersects dashed curve at $\varepsilon(\min) = \alpha/\mathrm{AE}(\min)$.

Area above dashed curve that falls inside black rectangle

$$= \left[1 - \frac{\alpha}{\mathrm{AE}(\min)}\right]\left[1 - \mathrm{AE}(\min)\right] + \left[\frac{\alpha}{\mathrm{AE}(\min)} - \alpha\right] - \int_{\alpha}^{\alpha/\mathrm{AE}(\min)} \frac{\alpha}{\varepsilon}\, d\varepsilon$$

$$= 1 - \frac{\alpha}{\mathrm{AE}(\min)} - \mathrm{AE}(\min) + \alpha + \frac{\alpha}{\mathrm{AE}(\min)} - \alpha - \alpha\left[\ln \varepsilon\right]_{\alpha}^{\alpha/\mathrm{AE}(\min)}$$

$$= 1 - \mathrm{AE}(\min) - \alpha\left[\ln\left(\frac{\mathrm{a}}{\mathrm{AE}(\min)}\right) - \ln \alpha\right]$$

$$= 1 - \mathrm{AE}(\min) - \alpha\left[\ln \alpha - \ln\left(\mathrm{AE}(\min)\right) - \ln \alpha\right]$$

$$= 1 - \mathrm{AE}(\min) + \alpha \ln \mathrm{AE}(\min)$$

TABLE 7.1

Summary of Probability Relationships Based on AE(min) and Minimum Reaction Yield Thresholds

Equation	Probability	Threshold conditions
7.9	$p = \dfrac{1 - \mathrm{AE(min)} + \alpha \ln(\mathrm{AE})\min}{[1 - \mathrm{AE(min)}][1 - \varepsilon(\min)]}$	(AE)min ≥ α and 0 ≤ ε (min) ≤ α
7.10	$p = \dfrac{1 - \mathrm{AE(min)} + \alpha - \varepsilon(\min) - \alpha \ln\left[\dfrac{\alpha}{(\mathrm{AE})\min \varepsilon(\min)}\right]}{[1 - \mathrm{AE(min)}][1 - \varepsilon(\min)]}$	(AE)min ≥ α and α ≤ ε (min) ≤ α/AE(min)
7.11	$p = 1$	(AE)min ≥ α and α/AE(min) ≤ ε (min) ≤ 1
7.12	$p = \dfrac{1 - \alpha + \alpha \ln \alpha}{[1 - \alpha][1 - \varepsilon(\min)]}$	(AE)min = α and 0 < ε (min) < α
7.13	$p = \dfrac{1 - \varepsilon(\min) + \alpha \ln \varepsilon(\min)}{[1 - \alpha][1 - \varepsilon(\min)]}$	(AE)min = α and α ≤ ε (min) ≤ 1
7.14	$p = \dfrac{1 - \alpha + \alpha \ln \alpha}{[1 - AE(\min)][1 - \varepsilon(\min)]}$	(AE)min ≤ α and 0 ≤ ε (min) ≤ α
7.15	$p = \dfrac{1 - \varepsilon(\min) + \alpha \ln \varepsilon(\min)}{[1 - \mathrm{AE(min)}][1 - \varepsilon(\min)]}$	(AE)min ≤ α and α ≤ ε (min) ≤ 1

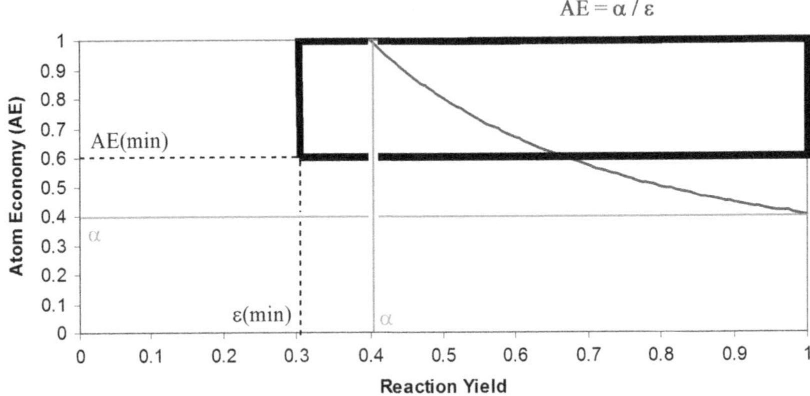

FIGURE 7.3 Graph for case I with domains (AE)min > α and 0 < ε (min) < α.

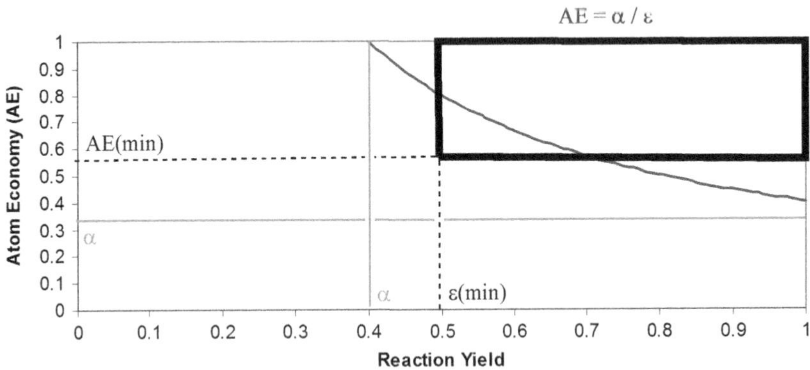

FIGURE 7.4 Graph for case II with domains (AE)min > α and α < ε (min) < α/AE(min). Probability = area above dashed curve that falls inside black rectangle/area of black rectangle.

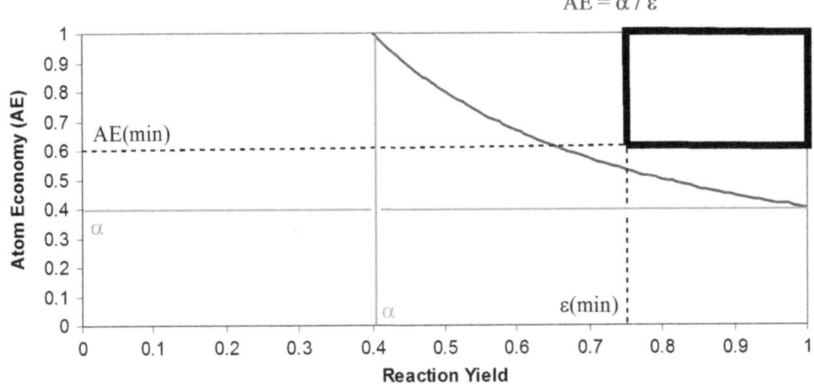

FIGURE 7.5 Graph for case III with domains (AE)min > α and α/AE(min) < ε (min) < 1. Probability = area above dashed curve that falls inside black rectangle/area of black rectangle.

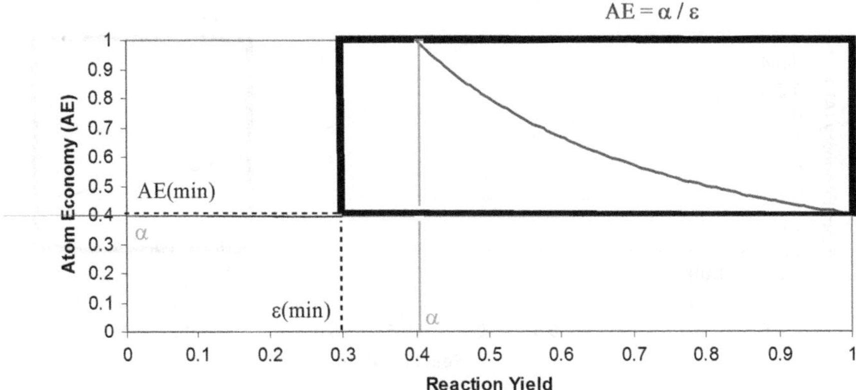

FIGURE 7.6 Graph for case IV with domains (AE)min = α and 0 < ε (min) < α.

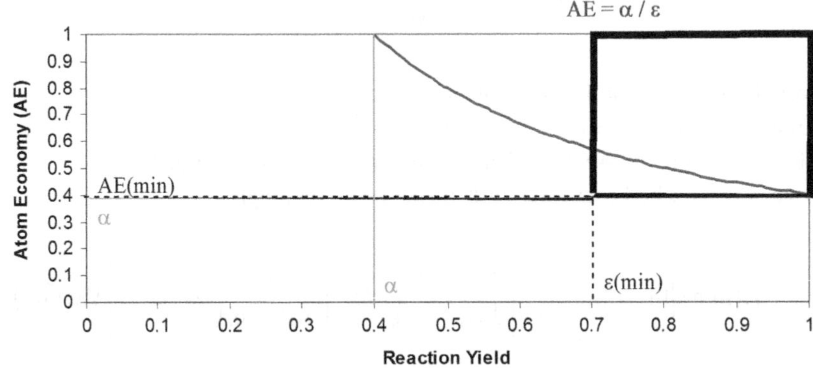

FIGURE 7.7 Graph for case V with domains (AE)min = α and α < ε (min) < 1.

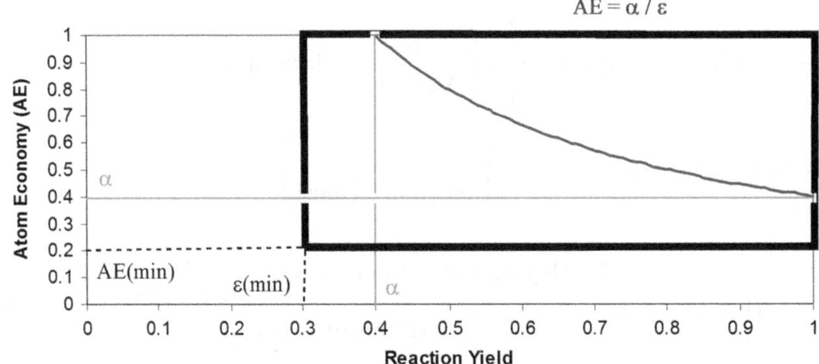

FIGURE 7.8 Graph for case VI with domains (AE)min < α and 0 < ε (min) < α.

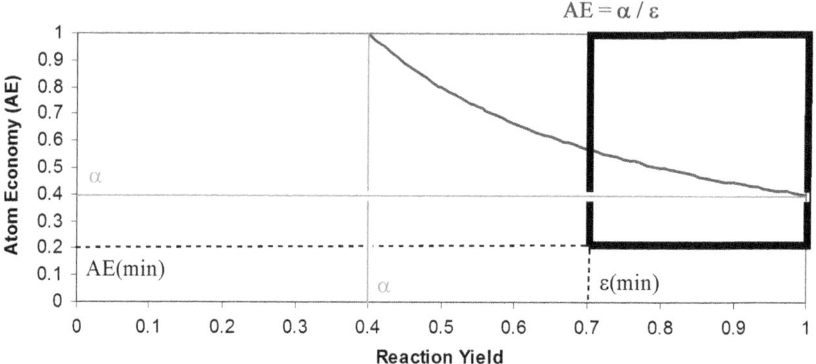

FIGURE 7.9 Graph for case VII with domains (AE)min<α and α<Ɛ (min)<1.

$$\text{Therefore, } p = \frac{1-\text{AE(min)}+\alpha\ln(\text{AE})\min}{[1-\text{AE(min)}][1-\varepsilon(\min)]}. \tag{7.9}$$

Case II: (AE)min>α and α<Ɛ (min)<α/AE(min)

Area black rectangle $=[1-\text{AE}(\min)][1-\varepsilon(\min)]$

Black rectangle intersects dashed curve at $\varepsilon(\min)=\alpha/\text{AE}(\min)$.

Area above dashed curve that falls inside black rectangle

$$=[1-\text{AE}(\min)][1-\varepsilon(\min)]-\left[\int_{\varepsilon(\min)}^{\alpha/\text{AE}(\min)}\frac{\alpha}{\varepsilon}d\varepsilon-\left[\frac{\alpha}{\text{AE}(\min)}-\varepsilon(\min)\right]\text{AE}(\min)\right]$$

$$=1-\text{AE}(\min)-\varepsilon(\min)+\text{AE}(\min)\varepsilon(\min)+\alpha-\text{AE}(\min)\varepsilon(\min)-\int_{\varepsilon(\min)}^{\alpha/\text{AE}(\min)}\frac{\alpha}{\varepsilon}d\varepsilon$$

$$=1-\text{AE}(\min)-\varepsilon(\min)+\alpha-\alpha[\ln\varepsilon]_{\varepsilon(\min)}^{\alpha/\text{AE}(\min)}$$

$$=1-\text{AE}(\min)-\varepsilon(\min)+\alpha-\alpha\left[\ln\left(\frac{\alpha}{\text{AE}(\min)}\right)-\ln\varepsilon(\min)\right]$$

$$=1-\text{AE}(\min)-\varepsilon(\min)+\alpha-\alpha\ln\left(\frac{\alpha}{\text{AE}(\min)\varepsilon(\min)}\right)$$

$$\text{Therefore, } p = \frac{1-\text{AE(min)}+\alpha-\varepsilon(\min)-\alpha\ln\left[\dfrac{\alpha}{(\text{AE})\min\varepsilon(\min)}\right]}{[1-\text{AE(min)}][1-\varepsilon(\min)]}. \tag{7.10}$$

Case III: (AE)min > α and α/AE(min) < ε (min) < 1

$$\text{Since the entire black rectangle falls inside the region above the blue curve, } p = 1 \tag{7.11}$$

Case IV: (AE)min = α and 0 < ε (min) < α

probability = area above dashed curve that falls inside black rectangle/area of black rectangle

Area black rectangle $= \left[1 - \text{AE}(\min)\right]\left[1 - \varepsilon(\min)\right]$

Area above dashed curve that falls inside black rectangle

$$= 1 - \alpha - \int_{\alpha}^{1} \frac{\alpha}{\varepsilon} d\varepsilon$$

$$= 1 - \alpha - \alpha\left[\ln \varepsilon\right]_{\alpha}^{1}$$

$$= 1 - \alpha - \alpha\left[0 - \ln \alpha\right]$$

$$= 1 - \alpha + \alpha \ln \alpha$$

$$\text{Therefore, } p = \frac{1 - \alpha + \alpha \ln \alpha}{\left[1 - \alpha\right]\left[1 - \varepsilon(\min)\right]}. \tag{7.12}$$

Case V: (AE)min = α and α < ε (min) < 1

probability = area above dashed curve that falls inside black rectangle/area of black rectangle

Area black rectangle $= \left[1 - \text{AE}(\min)\right]\left[1 - \varepsilon(\min)\right] = \left[1 - \alpha\right]\left[1 - \varepsilon(\min)\right]$

Area above dashed curve that falls inside black rectangle

$$= \left[1 - \alpha\right]\left[1 - \varepsilon(\min)\right] - \left[\int_{\varepsilon(\min)}^{1} \frac{\alpha}{\varepsilon} d\varepsilon - \left(1 - \varepsilon(\min)\right)\alpha\right]$$

$$= 1 - \alpha - \varepsilon(\min) + \alpha\varepsilon(\min) + \alpha - \alpha\varepsilon(\min) - \int_{\varepsilon(\min)}^{1} \frac{\alpha}{\varepsilon} d\varepsilon$$

$$= 1 - \varepsilon(\min) - \alpha\left[\ln \varepsilon\right]_{\varepsilon(\min)}^{1}$$

$$= 1 - \varepsilon(\min) - \alpha\left[0 - \ln \varepsilon(\min)\right]$$

$$= 1 - \varepsilon(\min) + \alpha \ln \varepsilon(\min)$$

$$\text{Therefore, } p = \frac{1 - \varepsilon(\min) + \alpha \ln \varepsilon(\min)}{\left[1 - \alpha\right]\left[1 - \varepsilon(\min)\right]} \tag{7.13}$$

Case VI: (AE)min < α and 0 < ε (min) < α

probability = area above dashed curve that falls inside black rectangle/area of black rectangle

Area black rectangle $=\left[1-\mathrm{AE}(\min)\right]\left[1-\varepsilon(\min)\right]$

Area above dashed curve that falls inside black rectangle

$$=1-\alpha-\int_{\alpha}^{1}\frac{\alpha}{\varepsilon}d\varepsilon$$

$$=1-\alpha-\alpha\left[\ln\varepsilon\right]_{\alpha}^{1}$$

$$=1-\alpha-\alpha\left[0-\ln\alpha\right]$$

$$=1-\alpha+\alpha\ln\alpha$$

$$\text{Therefore, } p=\frac{1-\alpha+\alpha\ln\alpha}{\left[1-AE(\min)\right]\left[1-\varepsilon(\min)\right]}. \tag{7.14}$$

Case VII: (AE)min < α and α < ε (min) < 1

probability = area above dashed curve that falls inside black rectangle/area of black rectangle

Area black rectangle $=\left[1-\mathrm{AE}(\min)\right]\left[1-\varepsilon(\min)\right]$

Area above dashed curve that falls inside black rectangle

$$=\left[1-\mathrm{AE}(\min)\right]\left[1-\varepsilon(\min)\right]-\left[\int_{\varepsilon(\min)}^{1}\frac{\alpha}{\varepsilon}d\varepsilon-\mathrm{AE}(\min)\left[1-\varepsilon(\min)\right]\right]$$

$$=1-\mathrm{AE}(\min)-\varepsilon(\min)+\mathrm{AE}(\min)\varepsilon(\min)+\mathrm{AE}(\min)$$

$$-\mathrm{AE}(\min)\varepsilon(\min)-\int_{\varepsilon(\min)}^{1}\frac{\alpha}{\varepsilon}d\varepsilon$$

$$=1-\varepsilon(\min)-\alpha\left[\ln\varepsilon\right]_{\varepsilon(\min)}^{1}$$

$$=1-\varepsilon(\min)-\alpha\left[0-\ln\varepsilon(\min)\right]$$

$$=1-\varepsilon(\min)+\alpha\ln\varepsilon(\min)$$

$$\text{Therefore, } p=\frac{1-\varepsilon(\min)+\alpha\ln\varepsilon(\min)}{\left[1-AE(\min)\right]\left[1-\varepsilon(\min)\right]}. \tag{7.15}$$

Setting $\alpha=0.618$ and $\varepsilon(\min)=0.8$ as the desired threshold parameters for intrinsic greenness, so that α < ε (min) < 1 is satisfied, yields expressions for the corresponding probability of intrinsic greenness as a function of AE(min) only as given by Equations (7.16) through (7.18). These may be obtained by direct substitution in Equations (7.15), (7.13), (7.10), and (7.11), respectively, found in Table 7.1. Each equation is associated with its own valid domain range along the AE(min) axis. This means that a plot of p versus AE(min) is a smooth continuous piecewise function comprising four regions as shown in Figure 7.10.

For region 1, AE(min) < 0.618 and Equation (7.15) reduces to Equation (7.16).

$$p = \frac{0.31}{1 - \text{AE}(\text{min})} \tag{7.16}$$

For region 2, AE(min) = 0.618 and Equation (7.13) reduces to Equation (7.17).

$$p = 0.813 \tag{7.17}$$

For region 3, 0.618 < AE(min) < 0.773 and Equation (7.10) reduces to Equation (7.18).

$$p = \frac{0.818 - \text{AE}(\text{min}) - 0.618 \ln\left[\frac{0.773}{\text{AE}(\text{min})}\right]}{0.2\left[1 - \text{AE}(\text{min})\right]} \tag{7.18}$$

For region 4, AE(min) > 0.773 and $p = 1$.

Example 7.4 [1, 45, 46]

In Example 7.2, it was found that the relative minimum AE values for various named olefin forming reactions were as follows.

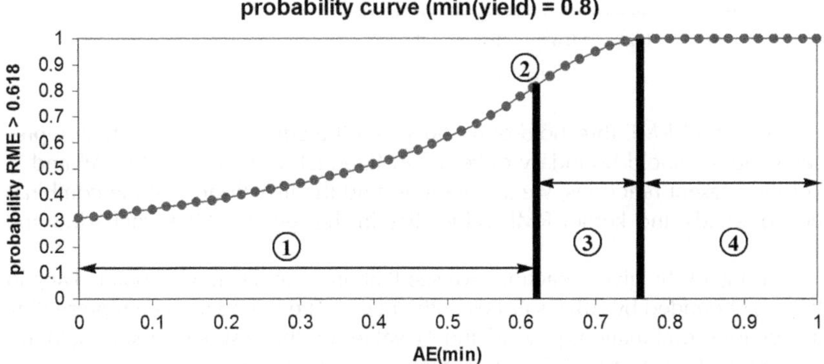

FIGURE 7.10 Relationship between probability of intrinsic greenness and AE(min) given a kernel RME threshold of $\alpha = 0.618$, and a reaction yield threshold of ϵ (min) = 0.8. The four regions are defined by Equations (7.16) through (7.18).

Reaction	AE(min)
McMurry coupling	0.168
Peterson olefination	0.147
Julia–Lythgoe olefination	0.122
Julia–Kocienski olefination	0.116
Schlosser–Wittig modification	0.107

Since yield (Y) and atom economy (AE) are core metrics that define material efficiency at the kernel level, we associate them with intrinsic greenness. Suppose we set a threshold yield value of 80% and a threshold kernel RME value of $\alpha = 60\%$ for these reactions as a definition of intrinsic greenness where kernel RME = AE*Y. For the given reactions, what would be the corresponding probabilities of achieving the intrinsic greenness criteria?

Solution

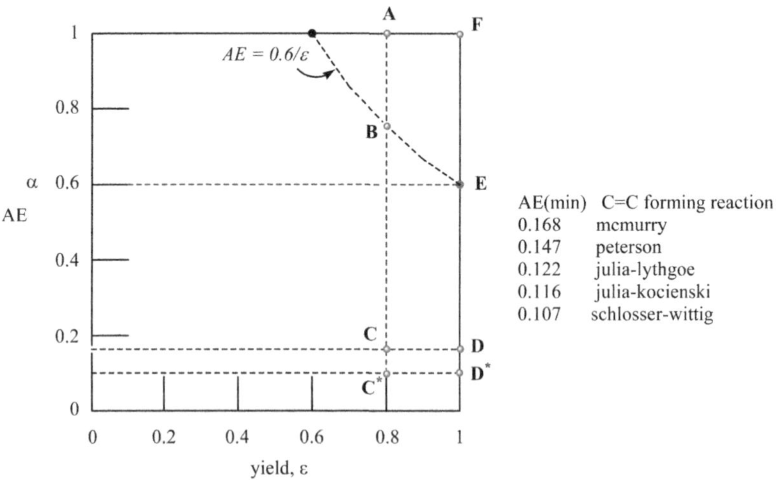

The kernel RME threshold of $\alpha = 60\%$ sets the curve along BE, which represents the threshold boundary of kernel RME equal to the product of AE and Y. For the present reactions, we are asked to find the likelihood that the combination of yields and kernel RME values lies in the region RME(kernel) > 0.6 and Y > 0.8.

For any of the given reactions we see that any combination of points lying in the area bounded by ABEF satisfies both criteria. The boundary area representing all possible combinations of yield and AE values for the best performing McMurry reaction that satisfies Y > 0.8 and AE(min) > 0.168 is given by ACDF. Therefore, the required probability of intrinsic greenness is given by the ratio of area ABEF to area ACDF. Similarly, the boundary area representing all possible combinations of yield and AE values for the worst performing Schlosser–Wittig reaction that satisfies Y > 0.8 and AE(min) > 0.107 is given by AC*D*F.

In general, we have

$$p = \frac{\text{Area}(\text{ABEF})}{(1-0.8)(1-\text{AE}_{\text{min}})}$$

Area ABEF is given by

$$\begin{aligned}\text{Area}(\text{ABEF}) &= (1-0.8)(1) - \int_{0.8}^{1} \frac{0.6}{\varepsilon} d\varepsilon \\ &= 0.2 - 0.6\left[\ln\varepsilon\right]_{0.8}^{1} \\ &= 0.2 - 0.6\left[\ln(1) - \ln(0.8)\right] \\ &= 0.2 + 0.6\ln(0.8) \\ &= 0.06611\end{aligned}$$

Reaction	**AE(min)**	***p***
McMurry coupling	0.168	0.397
Peterson olefination	0.147	0.388
Julia–Lythgoe olefination	0.122	0.377
Julia–Kocienski olefination	0.116	0.374
Schlosser–Wittig modification	0.107	0.370

Supplementary Questions

Suppose we lower the yield threshold to 60%. How do the probabilities of intrinsic greenness change?

Solution

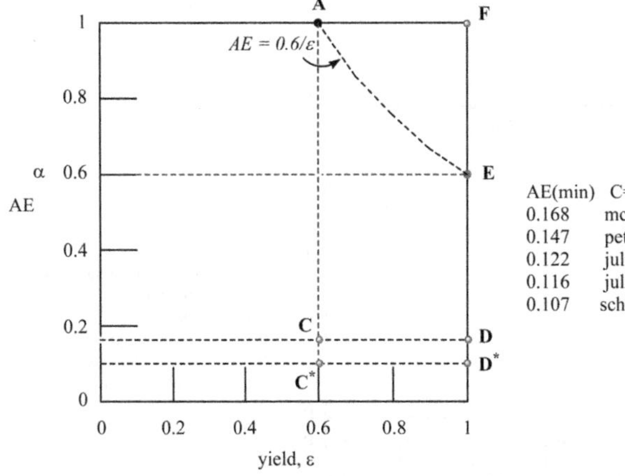

The region of possible Y and AE scenarios that satisfy Y>0.6 and RME(kernel)>0.6 is given by area AFE. For the given reactions, the region of possible Y and AE scenarios that satisfy Y>0.6 and AE>AE(min) is given by area ACDF for the McMurry reaction (best case) and area AC*D*F for the Schlosser–Wittig reaction (worst case).

The required probability is given by

$$p = \frac{\text{Area}(\text{AFE})}{(1-0.6)(1-\text{AE}_{\text{min}})}$$

Area AFE is given by

$$\begin{aligned}\text{Area}(\text{AFE}) &= (1-0.6)(1) - \int_{0.6}^{1} \frac{0.6}{\varepsilon} d\varepsilon \\ &= 0.4 - 0.6\left[\ln\varepsilon\right]_{0.6}^{1} \\ &= 0.4 - 0.6\left[\ln(1) - \ln(0.6)\right] \\ &= 0.4 + 0.6\ln(0.6) \\ &= 0.09351\end{aligned}$$

Reaction	**AE(min)**	***p***
McMurry coupling	0.168	0.281
Peterson olefination	0.147	0.274
Julia–Lythgoe olefination	0.122	0.266
Julia–Kocienski olefination	0.116	0.264
Schlosser–Wittig modification	0.107	0.262

How do the probabilities change if we do not impose a yield threshold?

Solution

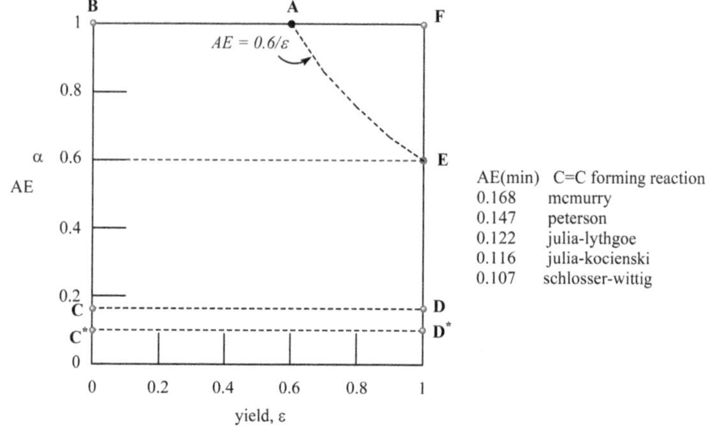

The target area AFE remains the same. The area corresponding to possible Y and AE scenarios for the reactions is given by the region BCDF for the McMurry reaction (best case) and BC*D*F for the Schlosser–Wittig reaction (worst case).

The required probability is given by

$$p = \frac{\text{Area}(\text{AFE})}{(1 - \text{AE}_{min})}$$

Reaction	AE(min)	*p*
McMurry coupling	0.168	0.112
Peterson olefination	0.147	0.110
Julia–Lythgoe olefination	0.122	0.106
Julia–Kocienski olefination	0.116	0.106
Schlosser–Wittig modification	0.107	0.105

Example 7.5 [45, 47]

The Biginelli reaction is a one-pot three-component coupling reaction used to synthesize substituted 3,4-dihydro-1 H-pyrimidin-2-ones, as shown below.

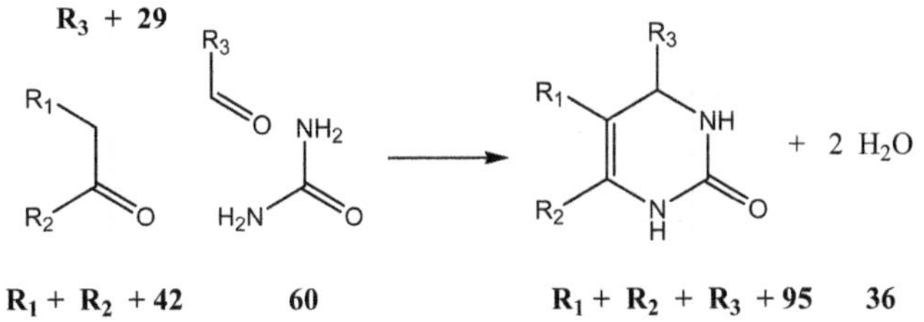

$R_3 + 29$

$R_1 + R_2 + 42$ 60   → $R_1 + R_2 + R_3 + 95$ $+ 2\ H_2O$ 36

Part 1

Determine the minimum AE and maximum *E*(mw) (*E*-factor based on molecular weight) for this reaction.

Part 2

Suppose a minimum target kernel RME of 60% is imposed on this reaction. What is the probability that this target will be met given the minimum AE found in Part 1?

Part 3

Suppose we add another constraint that the reaction yield should have a minimum value of 70%. What is the probability that the kernel RME will be at least 60% given both constraints of a minimum AE found in Part 1 and a minimum reaction yield of 70%?

Solution

Part 1

$$AE = \frac{R_1 + R_2 + R_3 + 95}{R_1 + R_2 + 42 + R_3 + 29 + 60} = \frac{R_1 + R_2 + R_3 + 95}{R_1 + R_2 + R_3 + 131}$$

The minimum AE is found when the R groups are set to hydrogen atoms. Hence R1 = R2 = R3 = 1.

AE(min) = (1 + 1 + 1 + 95)/(1 + 1 + 1 + 131) = 0.731, or 73.1%

Since $AE_{min} = \frac{1}{1 + E_{mw}}$, then $E_{mw} = \frac{1}{AE_{min}} - 1 = \frac{1}{0.731} - 1 = 0.367$.

Part 2

Let RME(kernel) threshold be $\alpha = 0.6$.

$$RME_{kernel} = \varepsilon(AE) \geq 0.6$$

$$AE \geq \frac{0.6}{\varepsilon}$$

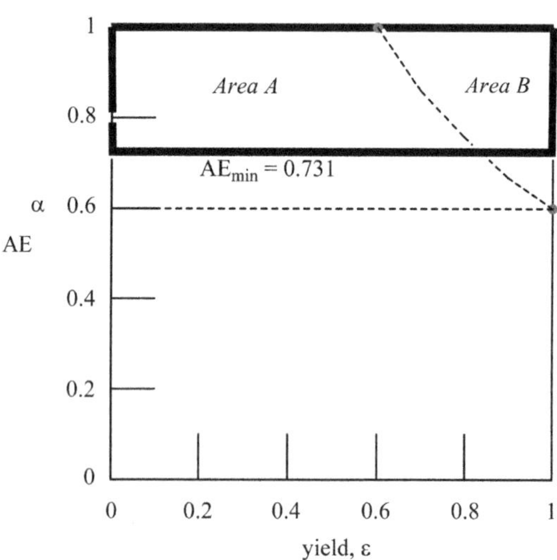

Probability:

$$p = \frac{\text{Area B}}{\text{Area A+Area B}}$$

$$= \frac{1 - AE_{min} + \alpha \ln AE_{min}}{1 - AE_{min}}$$

$$= \frac{1 - 0.731 + 0.6 * \ln(0.731)}{1 - 0.731}$$

$$= \frac{0.0809}{0.269}$$

$$= 0.301$$

or 30.1%

Part 3

Let RME(kernel) threshold be $\alpha = 0.6$.

AE(min) = 0.731

ε (min) = 0.7

$$AE_{min} \geq \alpha \text{ and } \alpha \leq \varepsilon_{min} \leq \frac{\alpha}{AE_{min}}$$

$$AE_{min} \geq 0.6 \text{ and } 0.6 \leq \varepsilon_{min} \leq \frac{0.6}{0.731} = 0.820$$

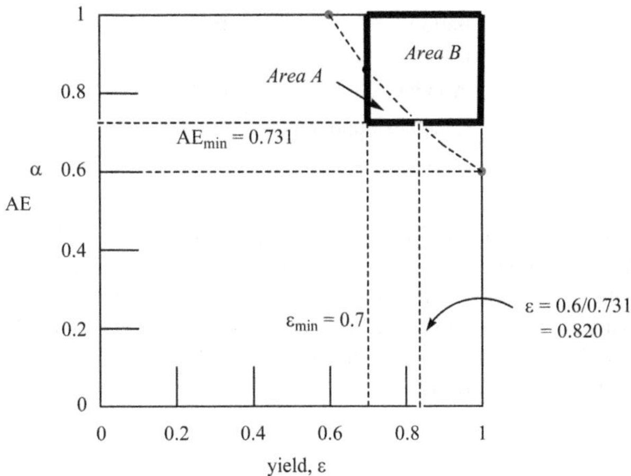

Probability:

$$p = \frac{\text{Area B}}{\text{Area A+Area B}}$$

$$= \frac{1 - AE_{min} + \alpha - \varepsilon_{min} - \alpha \ln\left(\frac{\alpha}{AE_{min}\varepsilon_{min}}\right)}{(1 - AE_{min})(1 - \varepsilon_{min})}$$

$$= \frac{1 - 0.731 + 0.6 - 0.7 - 0.6 * \ln\left(\frac{0.6}{0.731 * 0.7}\right)}{(1 - 0.731)(1 - 0.7)}$$

$$= \frac{0.0734}{0.0805}$$

$$= 0.911$$

or 91.1%.

Example 7.6 [1, 45, 46]

The relative minimum AE values for various named indole forming reactions are as follows.

Method	AE(min)
Sundberg	0.838
Hemetsberger	0.838
Fischer	0.670
Madelung	0.587
Hegedus G1	0.569
Japp–Klingemann	0.561
Baeyer	0.393
Bartoli	0.357
Larock	0.353
Wender	0.322
Cadogan–Sundberg	0.285

Part 1

Since yield (Y) and atom economy (AE) are core metrics that define material efficiency at the kernel level, we associate them with intrinsic greenness. Suppose we set a threshold yield value of 80% and a threshold kernel RME value of $\alpha = 60\%$ for these reactions as a definition of intrinsic greenness where kernel RME = AE*Y. For the given reactions, what would be the corresponding probabilities of achieving the intrinsic greenness criteria?

Solution

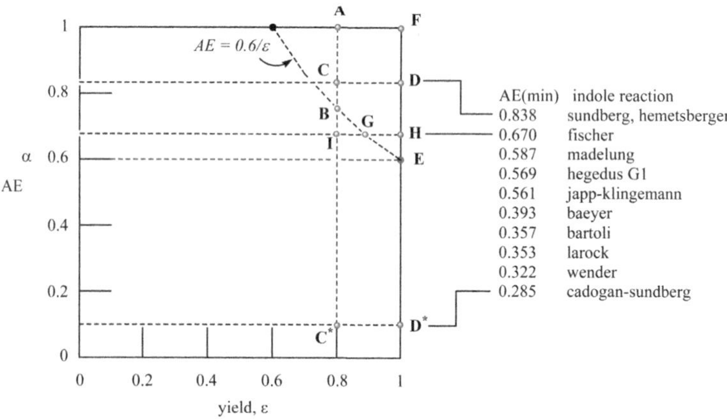

The kernel RME threshold of 60% sets the curve along BGE, which represents the threshold boundary of kernel RME equal to the product of AE and Y. For the present reactions, we are asked to find the likelihood that the combination of yields and AE values lies above RME(kernel) > 0.6 and Y > 0.8.

The list of reactions can be classified into three groups depending on where the value of AE(min) intersects the boundary curve defined by RME(kernel) = 0.6.

CASE I (AE(min) > 0.6/0.8 = 0.75 and Y > 0.8; region ACDF)

For the Sundberg and Hemetsberger reactions, AE(min) = 0.838. The boundary area representing all possible combinations of yield and AE values for these reactions that satisfies Y>0.8 and AE(min)>0.838 is given by ACDF. Since this region lies in the target region ABEF, the associated probability is 1.

CASE II (0.6<AE(min)<0.75 and Y>0.8; region ABGHF)

For the Fischer reaction, the region AIHF gives the area representing all possible combinations of yield and AE values that satisfy the thresholds. The target area is given by region ABGHF. Therefore, the required probability of intrinsic greenness is given by the ratio of area ABGHF to area AIHF.

Area(AIHF) = (1 – 0.67)*(1 – 0.8) = 0.066

Area(ABGHF) = Area(AIHF) – Area(BIG)

The required probability is given by

$$p = \frac{\text{Area}(\text{AIHF}) - \text{Area}(\text{BIG})}{\text{Area}(\text{AIHF})}$$

$$= \frac{0.066 - \left[\int_{0.8}^{0.6/0.67} \frac{0.6}{\varepsilon} d\varepsilon - (0.67)\left(\frac{0.6}{0.67} - 0.8\right)\right]}{0.066}$$

$$= \frac{0.066 - \left[0.6 \ln[\varepsilon]_{0.8}^{0.6/0.67} - 0.6 + (0.67)(0.8)\right]}{0.066}$$

$$= \frac{0.066 - \left[0.6\left\{\ln\left(\frac{0.6}{0.67}\right) - \ln(0.8)\right\} - 0.6 + (0.67)(0.8)\right]}{0.066}$$

$$= 0.9443$$

CASE III (AE(min)<0.6 and Y>0.8; region AC*D*F)

For the remaining reactions, the boundary area representing all possible combinations of yield and AE values that satisfies Y>0.8 and AE>AE(min) is given by AC*D*F. The target region is given by area ABEF.

In general, we have

$$p = \frac{\text{Area}(\text{ABEF})}{(1 - 0.8)(1 - \text{AE}_{min})}$$

Area ABEF is given by

$$\text{Area}(\text{ABEF}) = (1 - 0.8)(1) - \int_{0.8}^{1} \frac{0.6}{\varepsilon} d\varepsilon$$

$$= 0.2 - 0.6[\ln \varepsilon]_{0.8}^{1}$$

$$= 0.2 - 0.6[\ln(1) - \ln(0.8)]$$

$$= 0.2 + 0.6 \ln(0.8)$$

$$= 0.06611$$

Method	AE(min)	p
Madelung	0.587	0.800
Hegedus G1	0.569	0.767
Japp–Klingemann	0.561	0.753
Baeyer	0.393	0.545
Bartoli	0.357	0.514
Larock	0.353	0.511
Wender	0.322	0.488
Cadogan–Sundberg	0.285	0.462

Part 2

Suppose we lower the yield threshold to 60%. How do the probabilities of intrinsic greenness change?

Solution

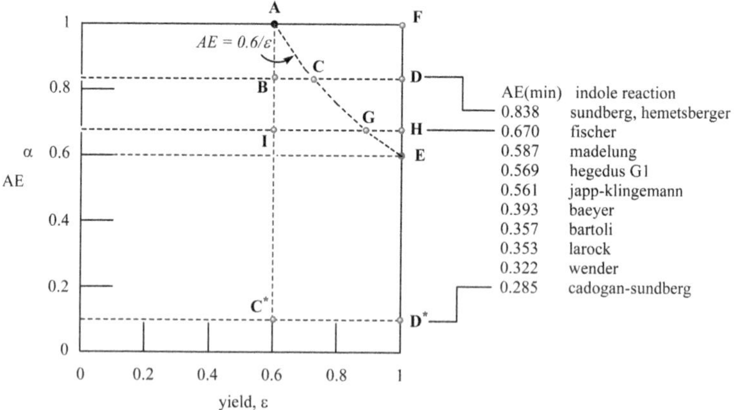

CASE I (AE(min) > 0.6 and Y > 0.6; regions ACDF and AGHF)
This applies to the Sundberg, Hemetsberger, and Fischer indole reactions.
For the Sundberg and Hemetsberger reactions, the required probability is given by

$$p = \frac{\text{Area}(ACDF)}{\text{Area}(ABDF)} = \frac{\text{Area}(ABDF) - \text{Area}(ABC)}{\text{Area}(ABDF)} = 1 - \frac{\text{Area}(ABC)}{\text{Area}(ABDF)}$$

$$= 1 - \frac{\int_{0.6}^{0.6/0.838} \frac{0.6}{\varepsilon} d\varepsilon - (0.838)\left(\frac{0.6}{0.838} - 0.6\right)}{(1-0.6)(1-0.838)}$$

$$= 1 - \frac{0.6\left[\ln\left(\frac{0.6}{0.838}\right) - \ln(0.6)\right] - 0.6 + (0.838)(0.6)}{(1-0.6)(1-0.838)}$$

$$= 0.8635$$

For the Fischer reaction, the required probability is given by

$$p = \frac{\text{Area}(AGHF)}{\text{Area}(AIHF)} = \frac{\text{Area}(AIHF) - \text{Area}(AIG)}{\text{Area}(AIHF)} = 1 - \frac{\text{Area}(AIG)}{\text{Area}(AIHF)}$$

$$= 1 - \frac{\displaystyle\int_{0.6}^{0.6/0.67} \frac{0.6}{\varepsilon}\, d\varepsilon - (0.67)\left(\frac{0.6}{0.67} - 0.6\right)}{(1-0.6)(1-0.67)}$$

$$= 1 - \frac{0.6\left[\ln\left(\frac{0.6}{0.67}\right) - \ln(0.6)\right] - 0.6 + (0.67)(0.6)}{(1-0.6)(1-0.67)}$$

$$= 0.6796$$

CASE II (AE(min) < 0.6 and Y > 0.6; region AEF)

For the remaining reactions, the required probability is given by

$$p = \frac{\text{Area}(AEF)}{(1-0.6)(1-AE_{min})} = \frac{(1-0.6)(1) - \displaystyle\int_{0.6}^{1} \frac{0.6}{\varepsilon}\, d\varepsilon}{0.4(1-AE_{min})}$$

$$= \frac{0.4 - 0.6\left[\ln(1) - \ln(0.6)\right]}{0.4(1-AE_{min})}$$

$$= \frac{0.4 + 0.6\left[\ln(0.6)\right]}{0.4(1-AE_{min})}$$

Method	**AE(min)**	***p***
Madelung	0.587	0.566
Hegedus G1	0.569	0.542
Japp–Klingemann	0.561	0.532
Baeyer	0.393	0.385
Bartoli	0.357	0.364
Larock	0.353	0.361
Wender	0.322	0.345
Cadogan–Sundberg	0.285	0.327

Part 3

How do the probabilities change if we do not impose a yield threshold?

Solution

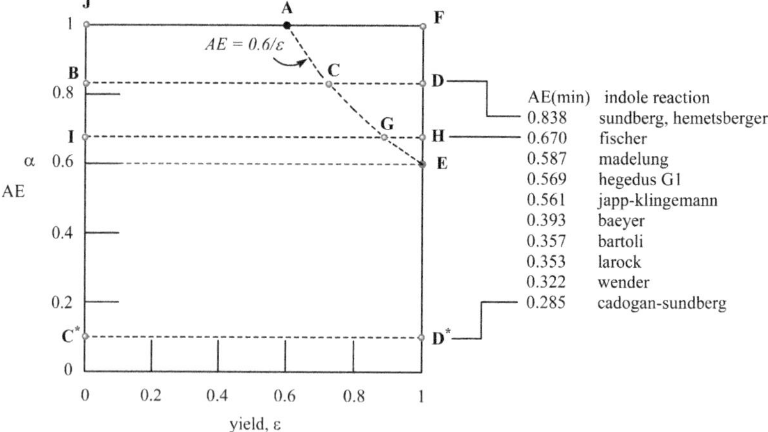

CASE I (AE(min) > 0.6; regions ACDF and AGHF)
This applies to the Sundberg, Hemetsberger, and Fischer indole reactions.
For the Sundberg and Hemetsberger reactions, the required probability is given by

$$p = \frac{\text{Area}(\text{ACDF})}{\text{Area}(\text{JBDF})}$$

$$= \frac{(1-0.6)(1-0.838) - \left[\int_{0.6}^{0.6/0.838} \frac{0.6}{\varepsilon} d\varepsilon - (0.838)\left(\frac{0.6}{0.838} - 0.6\right)\right]}{(1-0.838)(1)}$$

$$= \frac{0.0648 - \left[0.6\left\{\ln\left[\frac{0.6}{0.838}\right] - \ln[0.6]\right\} - 0.6 + (0.838)(0.6)\right]}{0.162}$$

$$= 0.3454$$

For the Fischer reaction, the required probability is given by

$$p = \frac{\text{Area}(\text{AGHF})}{\text{Area}(\text{JIHF})}$$

$$= \frac{(1-0.6)(1-0.67) - \left[\int_{0.6}^{0.6/0.67} \frac{0.6}{\varepsilon} d\varepsilon - (0.67)\left(\frac{0.6}{0.67} - 0.6\right)\right]}{(1-0.67)(1)}$$

$$= \frac{0.132 - \left[0.6\left\{\ln\left[\frac{0.6}{0.67}\right] - \ln[0.6]\right\} - 0.6 + (0.67)(0.6)\right]}{0.33}$$

$$= 0.2719$$

CASE II (AE(min) < 0.6; region AEF)

For the remaining reactions, the required probability is given by

$$p = \frac{\text{Area}(AEF)}{(1 - AE_{min})} = \frac{(1-0.6)(1) - \int_{0.6}^{1} \frac{0.6}{\varepsilon} d\varepsilon}{(1 - AE_{min})}$$

$$= \frac{0.4 - 0.6[\ln(1) - \ln(0.6)]}{(1 - AE_{min})}$$

$$= \frac{0.4 + 0.6[\ln(0.6)]}{(1 - AE_{min})}$$

Method	AE(min)	*p*
Madelung	0.587	0.226
Hegedus G1	0.569	0.217
Japp–Klingemann	0.561	0.213
Baeyer	0.393	0.154
Bartoli	0.357	0.145
Larock	0.353	0.145
Wender	0.322	0.138
Cadogan–Sundberg	0.285	0.131

Example 7.7 [1, 45, 46]

The relative minimum AE values for various esterification reactions were as follows.

Method	AE(min)
Fischer	0.804
Cyanohydrin-oxidation-substitution sequence	0.394
Saponification-substitution sequence	0.306
Diazomethane method	0.236

Part 1

Since yield (Y) and atom economy (AE) are core metrics that define material efficiency at the kernel level, we associate them with intrinsic greenness. Suppose we set a threshold yield value of 80% and a threshold kernel RME value of $\alpha = 60\%$ for these reactions as a definition of intrinsic greenness where kernel RME = AE*Y. For the given reactions, what would be the corresponding probabilities of achieving the intrinsic greenness criteria?

Solution

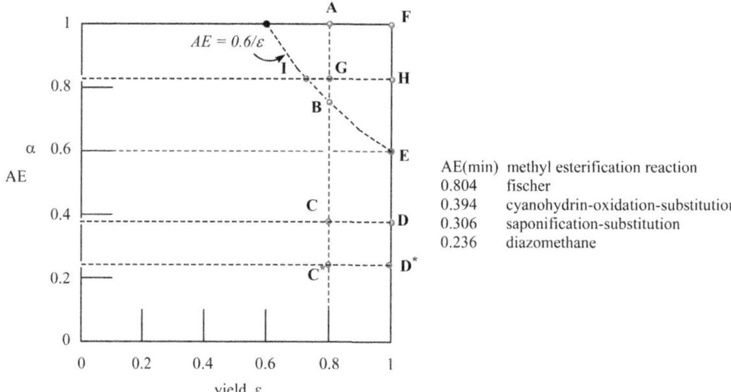

The kernel RME threshold of $\alpha = 60\%$ sets the curve along BE, which represents the threshold boundary of kernel RME equal to the product of AE and Y. For the present reactions, we are asked to find the likelihood that the combination of yields and kernel RME values lies in the region RME(kernel) > 0.6 and Y > 0.8.

CASE I: AE(min) > 0.6 and Y > 0.8; region AGHF

For the Fischer esterification reaction, the required probability is $p = 1$ since the region AGHF lies in the target region ABEF.

CASE II: AE(min) < 0.6 and Y > 0.8; region ABEF

For the remaining esterification reactions, the required probability is given by

$$p = \frac{\text{Area}(\text{ABEF})}{(1-0.8)(1-\text{AE}_{\text{min}})}$$

Area ABEF is given by

$$\text{Area}(\text{ABEF}) = (1-0.8)(1) - \int_{0.8}^{1} \frac{0.6}{\varepsilon} d\varepsilon$$

$$= 0.2 - 0.6[\ln \varepsilon]_{0.8}^{1}$$

$$= 0.2 - 0.6[\ln(1) - \ln(0.8)]$$

$$= 0.2 + 0.6 \ln(0.8)$$

$$= 0.06611$$

Method	AE(min)	*p*
Cyanohydrin-oxidation-substitution sequence	0.394	0.545
Saponification-substitution sequence	0.306	0.476
Diazomethane method	0.236	0.433

Part 2

Suppose we lower the yield threshold to 60%. How do the probabilities of intrinsic greenness change?

Solution

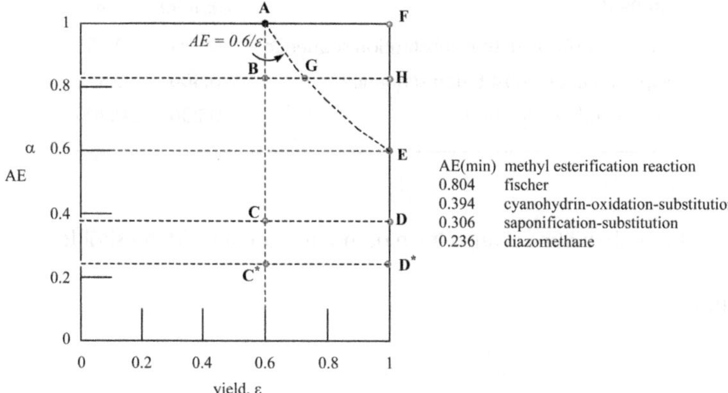

CASE I: AE(min) > 0.6 and Y > 0.6; region AGHF
For the Fischer esterification reaction, the required probability is given by

$$p = \frac{\text{Area}(\text{AGHF})}{\text{Area(ABHF)}} = \frac{\text{Area}(\text{ABHF}) - \text{Area}(\text{ABG})}{\text{Area(ABHF)}}$$

$$= 1 - \frac{\int_{0.6}^{0.6/0.804} \frac{0.6}{\varepsilon} d\varepsilon - (0.804)\left(\frac{0.6}{0.804} - 0.6\right)}{(1-0.6)(1-0.804)}$$

$$= 1 - \frac{0.6\left[\ln\left(\frac{0.6}{0.804}\right) - \ln(0.6)\right] - 0.6 + (0.804)(0.6)}{0.0784}$$

$$= 0.8304$$

CASE II: AE(min) < 0.6 and Y > 0.6; region AFE
For the remaining oxidation reactions, the required probability is given by

$$p = \frac{\text{Area}(\text{AFE})}{(1-0.6)(1-\text{AE}_{min})}$$

Area AFE is given by

$$\text{Area}(\text{AFE}) = (1-0.6)(1) - \int_{0.6}^{1} \frac{0.6}{\varepsilon} d\varepsilon$$

$$= 0.4 - 0.6[\ln\varepsilon]_{0.6}^{1}$$

$$= 0.4 - 0.6[\ln(1) - \ln(0.6)]$$

$$= 0.4 + 0.6\ln(0.6)$$

$$= 0.09351$$

Method	AE(min)	p
Cyanohydrin-oxidation-substitution sequence	0.394	0.386
Saponification-substitution sequence	0.306	0.337
Diazomethane method	0.236	0.306

Part 3

How do the probabilities change if we do not impose a yield threshold?

Solution

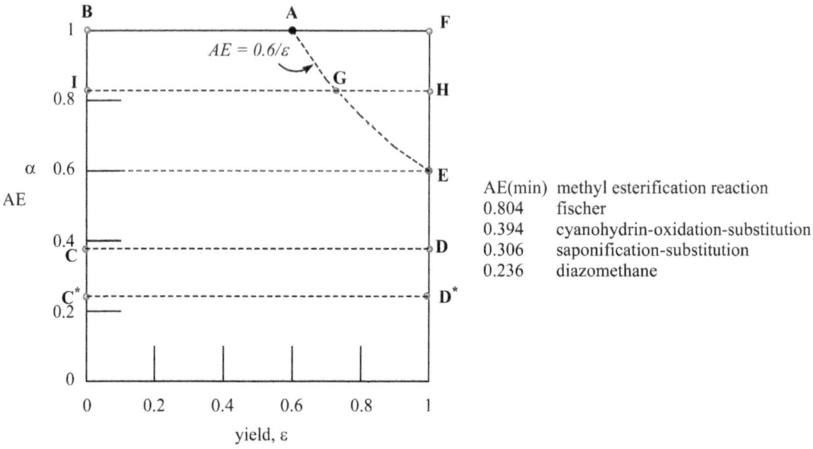

CASE I: AE(min) > 0.6; region AGHF
For the Fischer esterification, the required probability is given by

$$p = \frac{\text{Area}(\text{AGHF})}{\text{Area}(\text{BIHF})}$$

$$= \frac{(1-0.6)(1-0.804) - \left[\int_{0.6}^{0.6/0.804} \frac{0.6}{\varepsilon}\, d\varepsilon - (0.804)\left(\frac{0.6}{0.804} - 0.6\right)\right]}{(1-0.804)(1)}$$

$$= \frac{0.0784 - \left[0.6\left\{\ln\left[\frac{0.6}{0.804}\right] - \ln[0.6]\right\} - 0.6 + (0.804)(0.6)\right]}{0.196}$$

$$= 0.3322$$

CASE II: AE(min) < 0.6; region AEF
For the remaining esterification reactions, the required probability is given by

$$p = \frac{\text{Area}(\text{AEF})}{(1-\text{AE}_{\min})} = \frac{(1-0.6)(1) - \int_{0.6}^{1} \frac{0.6}{\varepsilon} d\varepsilon}{(1-\text{AE}_{\min})}$$

$$= \frac{0.4 - 0.6[\ln(1) - \ln(0.6)]}{(1-\text{AE}_{\min})}$$

$$= \frac{0.4 + 0.6[\ln(0.6)]}{(1-\text{AE}_{\min})}$$

Method	AE(min)	*p*
Cyanohydrin-oxidation-substitution sequence	0.394	0.154
Saponification-substitution sequence	0.306	0.135
Diazomethane method	0.236	0.122

7.2.2 Problems

PROBLEM 7.11 [45, 47]

Themes: balancing chemical equations, minimum atom economy, kernel reaction mass efficiency, probability of intrinsic greenness

Four different routes to a generalized γ-butyrolactone are considered.

Route #1 (one pot)

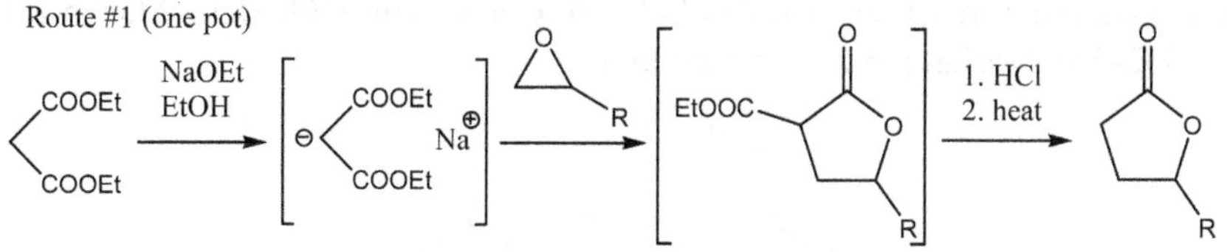

NaOEt EtOH; 1. HCl 2. heat

Route #2

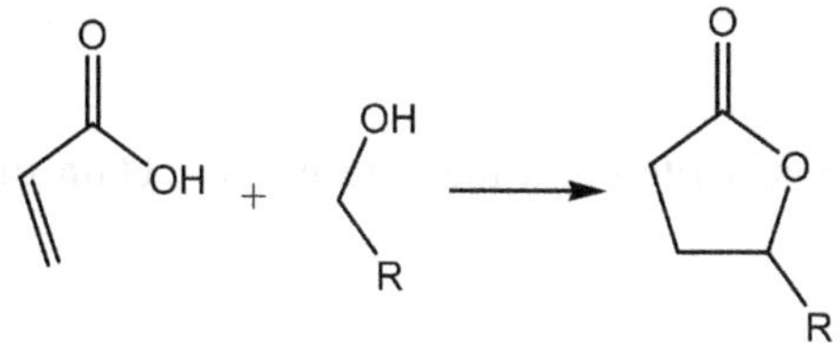

Route #3

$+ KMnO_4 \longrightarrow$ $+ Mn(OAc)_2$

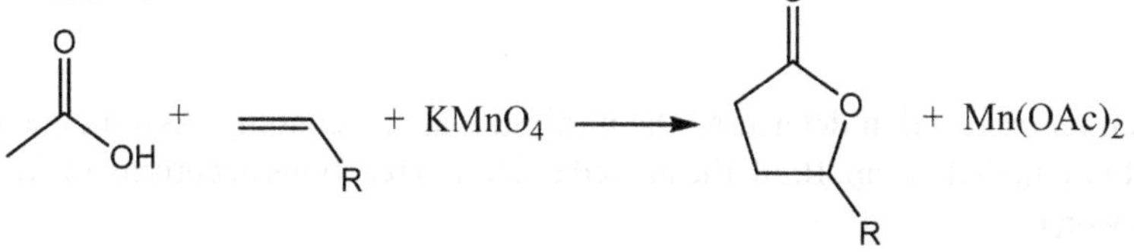

Route #4

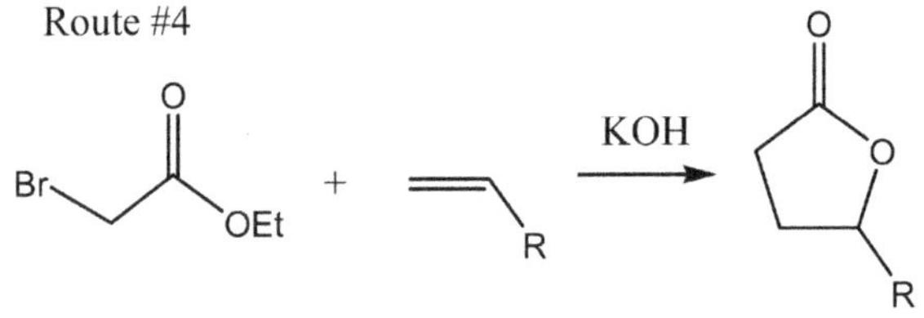

Br–CH₂–C(=O)OEt + CH₂=CH–R —KOH→ γ-lactone (5-R-dihydrofuran-2-one)

Part 1

For each route, provide balanced chemical equations.

Part 2

For each balanced equation, determine the minimum AE.

Part 3

If the kernel RME for each route is to be above a threshold value of 0.5, determine the range of permissible reaction yield values when AE = AE(min), and the probability of achieving such a threshold RME value in each case.

PROBLEM 7.12

Themes: balancing chemical equations, redox reactions, minimum atom economy, reaction mechanism, redox couples, probability of intrinsic greenness

References:

Alcohols may be oxidized to ketones according to the following six known named organic oxidation reactions: Collins [48, 49], Corey–Kim [50], Dess–Martin [51], Jones [52–55], Moffatt [56], and Swern [57].

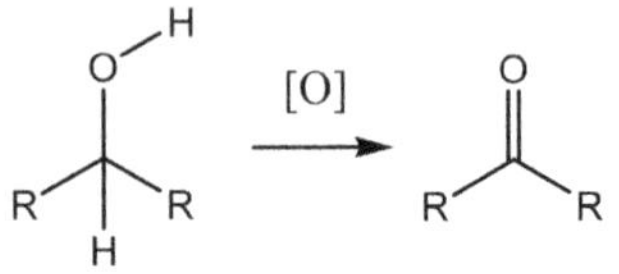

R₂CH–OH —[O]→ R₂C=O

Part 1

For each redox reaction, write out a mechanism and obtain the overall balanced chemical equation.

Part 2

For each redox reaction, write out the accompanying redox couple.

Part 3

For each redox reaction determine the minimum atom economy. Assume the minimum R is a methyl group. Rank the named oxidation reactions according to AE from best to worst.

Part 4

If each redox reaction were to have a kernel RME of at least 60%, what is the associated probability that such a criterion would be met? Rank the named oxidation reactions according to probability of intrinsic greenness.

PROBLEM 7.13

Themes: balancing chemical equations, redox reactions, redox couples, minimum atom economy, probability of intrinsic greenness

Part 1

For each epoxidation reaction shown below, determine the minimum atom economy. Assume the minimum R is a methyl group. Rank the named epoxidation reactions according to AE from best to worst.

Part 2

If each epoxidation reaction were to have a kernel RME of at least 60%, what is the associated probability that such a criterion would be met? Rank the named epoxidation reactions according to probability of intrinsic greenness.

Corey–Chaykovsky epoxidation [58, 59]
Jacobsen epoxidation [60, 61]
Prilezhaev reaction (epoxidation of olefins) [62]
Sharpless epoxidation [63]
Shi asymmetric epoxidation [64, 65]

PROBLEM 7.14 [1, 45, 46]

Themes: probability of intrinsic greenness

The following relative minimum AE values apply to various alcohol oxidation reactions.

Reaction	AE(min)
Hydrogen peroxide	0.617
Permanganate	0.351
Jones	0.258
Swern	0.120

Part 1

Since yield (Y) and atom economy (AE) are core metrics that define material efficiency at the kernel level, we associate them with intrinsic greenness. Suppose we set a threshold yield value of 80% and a threshold kernel RME value of $\alpha = 60\%$ for these reactions as a definition of intrinsic greenness where kernel RME = AE*Y. For the given reactions, what would be the corresponding probabilities of achieving the intrinsic greenness criteria?

Part 2

Suppose we lower the yield threshold to 60%. How do the probabilities of intrinsic greenness change?

Part 3

How do the probabilities change if we do not impose a yield threshold?

PROBLEM 7.15 [1, 45, 46]

Themes: probability of intrinsic greenness

The relative minimum AE values for various named amide forming reactions are as follows.

Method	AE(min)
Hydrolysis of nitriles	1
Oxidation of nitrlles	0.330
Acid chloride-Schotten–Baumann sequence	0.277
Carbonyldiimidazole (CDI) method	0.247
Carbodiimide method	0.208
Umpolung method	0.0654

Part 1

Since yield (Y) and atom economy (AE) are core metrics that define material efficiency at the kernel level, we associate them with intrinsic greenness. Suppose we set a threshold yield value of 80% and a threshold kernel RME value of $\alpha=60\%$ for these reactions as a definition of intrinsic greenness where kernel RME=AE*Y. For the given reactions, what would be the corresponding probabilities of achieving the intrinsic greenness criteria?

Part 2

Suppose we lower the yield threshold to 60%. How do the probabilities of intrinsic greenness change?

Part 3

How do the probabilities change if we do not impose a yield threshold?

7.3 AUTOMATED DETERMINATION OF AE(MIN), *p* INTRINSIC GREENNESS, AND HYPSICITY FOR ANY CHEMICAL REACTION

7.3.1 *CALCULATOR-PROBABILITY-INTRINSIC-GREENNESS.XLS* SPREADSHEET

The ideas presented in this chapter emphasize the importance of selecting reactions that have a high potential of meeting well-defined "green" standards in a systematic way. When such reactions are selected in designing a synthesis plan to a complex target molecule, the overall probability of achieving greenness for the entire plan will therefore be high. The powerful analysis described in the previous sections is based on the following economies of synthesis on a per-step basis: atom economy and yield economy. We can also add redox economy to this list if we incorporate the quantification of hypsicity described in Chapter 5. All of these aspects are incorporated in the *calculator-probability-intrinsic-greenness.xls* spreadsheet that determines for any reaction written out as a generalized Markush representation the following key design green metrics: AE(min); probability of intrinsic greenness; and hypsicity index (HI). This is a highly useful tool that facilitates such computations that can be tedious. Readers who attempt to go through the examples and posed problems in Sections 7.1 and 7.2 will appreciate the difference between carrying out pencil and paper computations versus the calculator spreadsheet.

The input data required for the *calculator-probability-intrinsic-greenness.xls* spreadsheet includes the following: (a) literature reference for chemical reaction to be analyzed; (b) designation of variable R groups usually set as methyl groups; (c) list of reaction by-products along with their molecular weights; (d) list of sacrificial reagents that do not become incorporated in the target product structure along with their molecular weights; (e) molecular weights of input reagents added in the reaction balancing checking section; (f) the oxidation numbers of key atoms in target construction bonds appearing in the product structure and the oxidation numbers of those same atoms in the reagent structures; and (g) the molecular weight of the constant part of the product structure as shown in the Markush representation of the chemical equation. The spreadsheet determines whether the chemical equation is properly balanced and calculates the HI, the minimum atom economy, the maximum *E*-factor based on molecular weight, the maximum fraction of sacrificial reagents for the reaction, and the probability of intrinsic greenness based on a benchmark threshold kernel RME of 61.8% and a threshold of reaction yield of 80%. Both threshold values can be changed if desired and the resulting probabilities will be determined accordingly. The correct value of a given probability corresponds to the entry where the conditional statement based on the domain inequalities listed in column 3 of Table 7.1 appears as "TRUE." An accompanying plot of AE versus reaction yield is shown similar to Figures 7.3 through 7.9, which gives a visual representation of the kernel RME threshold and AE(min) for the given entered chemical reaction. For reactions producing ring constructions, entry cells are provided for specifying the ring size created and the ring construction strategy according to the notation described in Chapter 4.

7.3.2 Trends for Named Organic Reactions

We present results of applying the *calculator-probability-intrinsic-greenness.xl s* spreadsheet to all the well-known named organic reactions that form the basic toolbox from which an organic chemist draws upon to design synthesis plans to any desired chemical structure. Figure 7.11 shows probability curves similar to Figure 7.10 for each reaction class: carbon–carbon (C-C) bond forming reactions, condensations, eliminations, multi-component reactions (MCR), non-carbon–carbon bond forming reactions, oxidations, rearrangements, reductions, sequences, and substitutions. From these figures by looking at the spreading of the points along

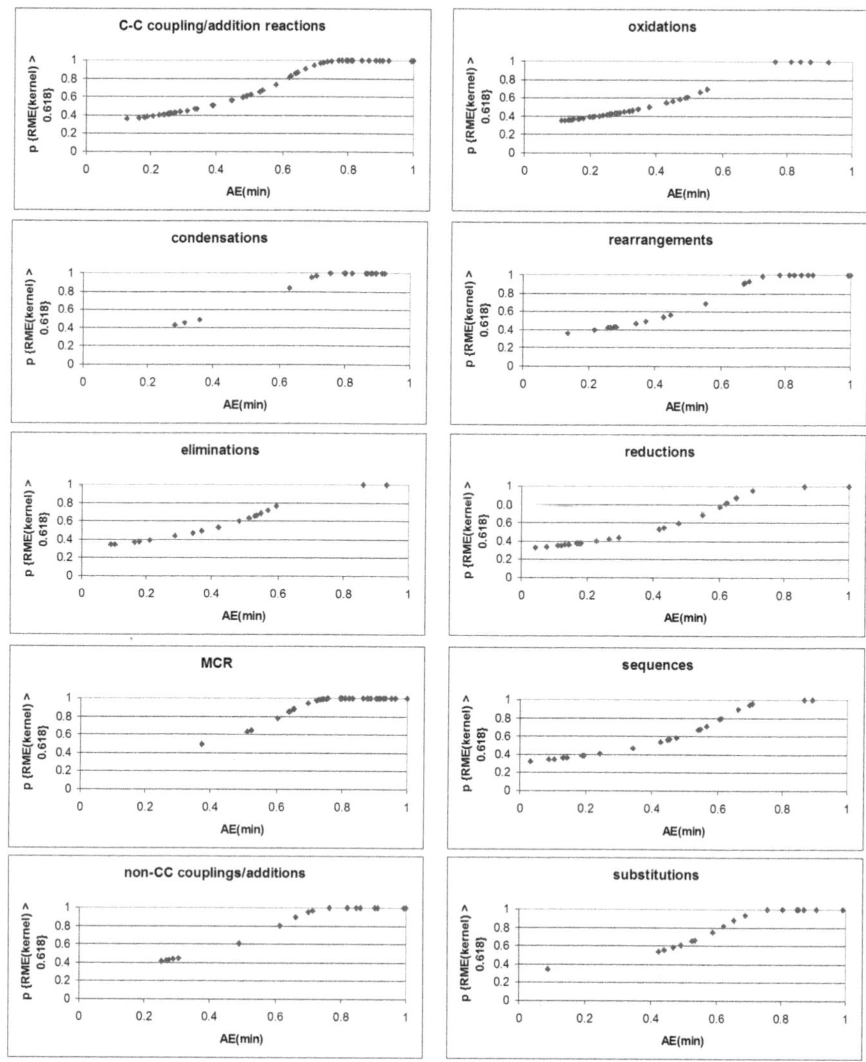

FIGURE 7.11 Probability curves showing AE(min) and intrinsic probability values for reactions in various organic reaction classes that satisfy a kernel RME threshold of $\alpha=0.618$ and a reaction yield threshold of ε (min)=0.8.

the curve, we note that condensations and MCRs tend to have high probabilities of intrinsic greenness and that eliminations, oxidations, and reductions tend to have low such values. If we plot the values for the average AE(min) and average intrinsic greenness probability for each reaction class, we obtain a graph as shown in Figure 7.12. We observe readily that those reaction classes at the top right corner are the ones that should be selected for optimizing "greenness" in designing synthesis plans; whereas, those appearing in the lower left corner should be avoided or at least kept to a minimum. Essentially, the probability analysis presented in

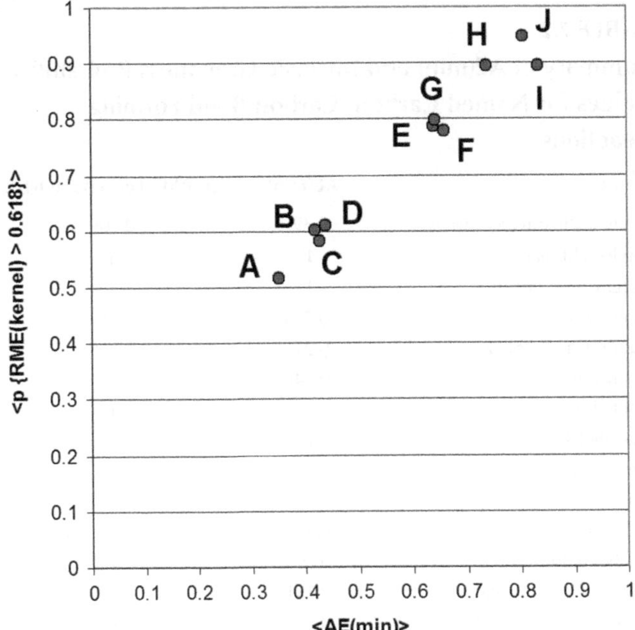

FIGURE 7.12 Plot of average values of AE(min) and intrinsic greenness probability for each reaction class: oxidations (A), sequences (B), eliminations (C), reductions (D), non-carbon–carbon forming reactions (E), carbon–carbon forming reactions (F), substitutions (G), condensations (H), rearrangements (I), and MCRs (J).

this chapter supplies quantitative evidence behind general statements in the literature about choosing the former kinds of reactions for designing "ideal syntheses." Tables 7.2 through 7.11 summarize the values of AE(min) and intrinsic greenness probability values for each reaction in each reaction class. The accompanying web-based material to this book contains calculator Excel files for each reaction included in the named organic reaction database.

TABLE 7.2
Summary of AE(min) and Intrinsic Greenness Probability Values for Named Carbon–Carbon Bond Forming Reactions

Name	AE(min)	p {RME(kernel) > 0.618}
Acetoacetic ester synthesis	0.39	0.509
Baylis–Hillman	1	1
Bergmann cyclization	1	1
Blanc ($n = 1$)	0.339	0.469
Cadiot-Chodkiewitz	0.27	0.425
Ciamician	0.446	0.56
Danheiser	1	1
Danishefsky	1	1
Diels–Alder	1	1
Doetz	0.923	1
Eglinton	0.813	1
Fischer indole	0.895	1
Friedel–Crafts acylation	0.781	1
Friedel–Crafts alkylation	0.737	0.987
Friedlander	0.799	1
Gatterman–Koch	1	1
Glaser	0.224	0.4
Gomberg–Bachman	0.625	0.827
Grignard	0.696	0.946
Hammick	0.75	0.995
Heck	0.506	0.628
Henry (weak acid)	0.579	0.737
Henry (strong acid)	0.479	0.596
Hiyama	0.238	0.408
Horner–Emmons–Wadsworth	0.273	0.427
Hosomi–Sakurai	0.241	0.409
Houben–Hoesch	0.884	1
Jacobs	0.904	1
Kiliani–Fischer	0.841	1
Koch–Haaf	1	1
Kolbe	1	1
Kulinkovich	0.482	0.6
Kumada	0.809	1
Liebeskind–Srogl	0.257	0.417
Malonic ester synthesis	0.64	0.857
Marshalk	0.715	0.968
McMurry	0.189	0.383
Meerwein	0.862	1

(Continued)

TABLE 7.2 (CONTINUED)
Summary of AE(min) and Intrinsic Greenness Probability Values for Named Carbon–Carbon Bond Forming Reactions

Name	AE(min)	p {RME(kernel) > 0.618}
Michael	1	1
Mukaiyama	0.334	0.466
Mukaiyama–Michael	0.288	0.436
Nazarov	1	1
Negishi	0.504	0.626
Nieuwland	1	1
Nozaki	0.62	0.816
Organocuprate	0.207	0.391
Paterno–Buchi	1	1
Pechmann	1	1
Perkin	0.725	0.978
Peterson	0.54	0.674
Pinacol	0.669	0.908
Prins	1	1
Reformatskii	0.492	0.611
Reimer–Tiemann	0.309	0.449
Richter	0.449	0.563
Robinson	0.53	0.66
Sakurai	0.262	0.421
Simmons–Smith	0.18	0.378
Skraup	0.253	0.416
Sonogashira	0.994	1
Stetter	1	1
Stille	0.387	0.507
Suzuki	0.77	1
Tebbe	0.164	0.371
Thorpe	1	1
Tishchenko	1	1
Trost	1	1
Ullmann	0.795	1
Vilsmeier–Haack–Arnold	0.388	0.507
Vorbrueggen	0.646	0.867
Wender–Trost	1	1
Wurtz–Fittig	0.127	0.356
Zincke–Suhl	0.861	1
Average	0.659	0.780

TABLE 7.3
Summary of AE(min) and Intrinsic Greenness Probability Values for Named Condensation Reactions

Name	AE(min)	p {RME(kernel) > 0.618}
Acyloin	0.282	0.432
Aldol	0.822	1
Bamberger–Goldschmidt	0.897	1
Bamberger–Goldschmidt triazenes	0.699	0.95
Benzoin	0.922	1
Claisen	0.884	1
Darzens	0.63	0.837
Dieckmann	0.894	1
Hinsberg	0.864	1
Knoevenagel	0.912	1
Knorr	0.869	1
Mukaiyama aldol	0.313	0.452
Nenitzescu	0.359	0.484
Pechmann	0.714	0.968
Pellizzari	0.755	0.997
Pictet–Spengler	0.88	1
Schiff	0.798	1
Stobbe	0.802	1
Average	0.739	0.896

TABLE 7.4
Summary of AE(min) and Intrinsic Greenness Probability Values for Named Elimination Reactions

Name	AE(min)	p {RME(kernel) > 0.618}
Azetidine	0.42	0.535
Borodin–Hunsdiecker	0.21	0.393
Burgess	0.179	0.378
Chugaev	0.341	0.471
Cope	0.534	0.667
Dakin	0.547	0.686
Edman	0.932	1
Grob	0.369	0.492
Haloform	0.529	0.659
Hofmann	0.104	0.346
Kochi	0.092	0.342
Lossen	0.595	0.766
McFadyen–Stevens	0.286	0.435
Norrish Type II	0.859	1
Pyrolysis of sulfoxides	0.511	0.635
RCM	0.481	0.599
Ruff–Fenton	0.163	0.371
Tiffeneau–Demjanov	0.568	0.718
Average	0.429	0.583

TABLE 7.5
Summary of AE(min) and Intrinsic Greenness Probability Values for Named Multi-Component Reactions

Name	AE(min)	p {RME(kernel) > 0.618}
Alper	1	1
Asinger	0.799	1
Asymmetric Mannich	0.742	0.99
Aza-Diels–Alder	0.934	1
Betti	0.951	1
Biginelli	0.795	1
Bucherer	0.864	1
Chichibabin	0.652	0.879
Doebner	0.822	1
Dornow–Wiehler	0.64	0.857
Feist–Benary	0.697	0.948
Gewald	0.911	1
Grieco	0.928	1
Guareschi–Thorpe	0.755	0.997
Hantzsch dihydropyridines	0.832	1
Hantzsch pyrroles	0.739	0.989
Heck–Diels–Alder	0.636	0.85
Knoevenagel–hetero-Diels–Alder	0.963	1
Mannich-1,6	1	1
Mannich	0.878	1
Michael-aldol-HWE	0.51	0.633
Michael–Michael–Wittig	0.373	0.494
Nenitzescu–Praill	0.722	0.975
Passerini	1	1
Pauson–Khand	1	1
Petasis	0.654	0.882
Petrenko–Kritschenko	0.888	1
Pinner	0.735	0.986
Radziszewski	0.697	0.947
Robinson–Schoepf	0.876	1
Reppe	1	1
Riehm	0.751	0.995
Roelen	1	1
Rothemund	0.744	0.991
Strecker	0.795	1
Tandem Asinger–Ugi	0.639	0.855
Tandem Passerini–Wittig	0.523	0.651

(*Continued*)

TABLE 7.5 (CONTINUED)
Summary of AE(min) and Intrinsic Greenness Probability Values for Named Multi-Component Reactions

Name	AE(min)	p {RME(kernel) > 0.618}
Tandem Petasis–Ugi	0.729	0.981
Thiele	0.808	1
Trost g,d-unsaturated ketones	1	1
Trost 1,5-diketones	1	1
Ugi	0.905	1
Weiss	0.917	1
Wulff	0.604	0.784
Average	0.805	0.947

TABLE 7.6
Summary of AE(min) and Intrinsic Greenness Probability Values for Named Non-Carbon–Carbon Bond Forming Reactions

Name	AE(min)	p {RME(kernel) > 0.618}
Arbusov–Michaelis	0.914	1
Aziridine	0.253	0.415
Azo coupling	0.847	1
Buchwald–Hartwig	0.994	1
Eschweiler–Clarke	0.488	0.606
Griess	0.614	0.805
Hoch–Campbell	0.289	0.436
Hofmann–Loffler–Freytag ($n=1$)	0.7	0.951
Leuckart	0.488	0.606
Menshutkin	1	1
Paal–Knorr (X=O)	0.714	0.968
Polonovski	0.765	0.999
Ritter	0.662	0.896
Stahl	0.86	1
Staudinger	0.819	1
Stoltz	0.905	1
Wenker	0.268	0.424
Williamson	0.277	0.429
Wohler	0.305	0.446
Average	0.640	0.788

TABLE 7.7
Summary of AE(min) and Intrinsic Greenness Probability Values for Named Oxidation Reactions

Name	AE(min)	p {RME(kernel) > 0.618}
Baeyer–Villiger	0.84	1
Bamford-Stevens	0.227	0.402
Boyland–Sims	0.316	0.454
Corey g-lactone	0.279	0.43
Corey–Chaykovsky	0.238	0.408
Corey–Kim	0.163	0.371
Criegee	0.207	0.391
Dess–Martin	0.213	0.395
Elbs	0.289	0.436
Etard	0.282	0.432
Fehling	0.251	0.414
Fenton	0.327	0.461
Fleming	0.163	0.371
Forster	0.377	0.498
Graham	0.53	0.661
Harries (oxidative)	0.87	1
Harries (reductive)	0.53	0.661
Hooker	0.319	0.456
Jacobsen	0.552	0.693
Jones	0.258	0.418
Lemieux–Johnson 1	0.448	0.562
Lemieux–Johnson 2	0.289	0.437
Malaprade	0.429	0.544
Oppenauer	0.345	0.474
Oxymercuration	0.148	0.364
Permanganate	0.486	0.604
Pfitzner–Moffatt	0.134	0.359
Prevost	0.112	0.35
Prilezhaev	0.926	1
Riley	0.47	0.586
Rubottom	0.263	0.421
Sarett	0.145	0.363
Sharpless–Jacobenw dihydroxylation	0.178	0.378
Sharpless epoxidation	0.493	0.613
Sharpless oxyamination	0.256	0.417
Shi epoxidation	0.273	0.427
Swern	0.124	0.354

(Continued)

TABLE 7.7 (CONTINUED)
Summary of AE(min) and Intrinsic Greenness Probability Values for Named Oxidation Reactions

Name	AE(min)	p {RME(kernel) > 0.618}
Tollens	0.137	0.36
Uemura	0.763	0.999
Wacker–Tsugi	0.197	0.387
Wessely	0.301	0.444
Willgerodt	0.811	1
Woodward	0.167	0.373
Average	0.352	0.516

TABLE 7.8
Summary of AE(min) and Intrinsic Greenness Probability Values for Named Rearrangement Reactions

Name	AE(min)	p {RME(kernel) > 0.618}
Acyl	0.848	1
Allylic	0.992	1
Aza-Cope	1	1
Bamberger	1	1
Bamberger–Goldschmidt	0.446	0.561
Barton	1	1
Beckmann	1	1
Benzidine	1	1
Benzylic	1	1
Brook	1	1
Camphene	1	1
Chapman	1	1
Ciamician	1	1
Claisen	1	1
Claisen–Ireland	0.342	0.472
Cope	1	1
Cornforth	1	1
Criegee	1	1
Curtius	0.256	0.418
Dieckmann–Thorpe	1	1
dienone–phenol	1	1
Di-pi-methane	1	1
Epoxide	0.371	0.494
Favorskii	0.551	0.691

(Continued)

TABLE 7.8 (CONTINUED)
Summary of AE(min) and Intrinsic Greenness Probability Values for Named Rearrangement Reactions

Name	AE(min)	p {RME(kernel) > 0.618}
Fischer–Hepp	1	1
Fries	1	1
Fritsch–Buttenberg–Wiechell	0.283	0.433
Hofmann–Matius	0.136	0.359
Hydroboration-borane	0.424	0.539
Hydroperoxide	0.811	1
Kemp	1	1
Lossen	0.674	0.916
Martynoff	1	1
Meisenheimer	1	1
Meyer–Schuster	1	1
Nametkin	0.993	1
Neber	0.266	0.423
Norrish Type I	1	1
Orton	1	1
Oxy-Cope	1	1
Payne	1	1
Pentazadiene	1	1
Perkin	0.669	0.907
Pinacol	0.825	1
Polonovski	1	1
Pummerer	1	1
Pyridine N-oxide-pyridone	1	1
Ramberg–Bäckland	0.278	0.43
Retro Sommelet–Hauser	1	1
Rupe	1	1
Schmidt	0.217	0.396
Sigmatropic	1	1
Smiles	1	1
Stevens	0.729	0.981
Sulfenate-sulfoxide	1	1
Tiemann	0.276	0.429
Vinyl ether	1	1
Vinylogous Wolff	0.868	1
Wagner–Meerwein	0.883	1
Wallach	1	1
Wawzonek–Yeakey	1	1
Wittig	0.685	0.931
Wolff	0.781	1
Average	0.835	0.895

TABLE 7.9
Summary of AE(min) and Intrinsic Greenness Probability Values for Named Reduction Reactions

Name	AE(min)	p {RME(kernel) > 0.618}
Benkeser	0.6	0.776
Birch	0.475	0.591
Borch	0.619	0.814
Borohydride	0.702	0.954
Bouveault-Blanc	0.145	0.363
Cannizzaro (product is alcohol)	0.432	0.547
Cannizzaro (product is K-acid)	0.623	0.823
Clemmensen	0.132	0.357
Corey–Bakshi–Shibata	0.476	0.593
Corey–Winter	0.143	0.362
Diimide	0.228	0.402
Gribble	0.266	0.423
Hydrogenation	1	1
Hydrogenolysis of benzyl ethers	0.863	1
LiAlH4	0.649	0.873
Meerwein–Pondorff–Verley	0.417	0.532
Midland	0.18	0.378
Noyori	1	1
Radical dehalogenation	0.043	0.325
Reduction of Nitroaromatics-H2S	0.617	0.81
Reduction of Nitroaromatics-Fe	0.077	0.336
Rosenmund	0.547	0.685
Shapiro	0.175	0.376
Stephen	0.295	0.44
Thioketal desulfurization	0.111	0.349
Tosylhydrazone	0.122	0.354
Wharton	0.652	0.878
Wolff–Kishner	0.167	0.373
Zincke	1	1
Average	0.440	0.611

TABLE 7.10
Summary of AE(min) and Intrinsic Greenness Probability Values for Named Sequences Reactions

Name	AE(min)	p {RME(kernel) > 0.618}
Arndt–Eisert	0.19	0.383
Barton decarboxylation	0.032	0.321
Bischler–Napieralski	0.445	0.56
Claisen–Ireland	0.342	0.472
Doering–LaFlamme	0.194	0.385
Japp–Klingemann	0.452	0.567
Julia olefination	0.131	0.357
Martinet dioxindole	0.545	0.683
Mitsunobu	0.105	0.347
Murahashi	0.142	0.362
Nef	0.241	0.409
Nenitzescu–Praill	0.424	0.539
Pfitzinger	0.662	0.897
Pictet–Spengler	0.891	1
Pinner	0.866	1
Pomeranz–Fritsch	0.54	0.675
Schlittler–Mueller	0.565	0.714
Reissert	0.609	0.795
Schlosser modification	0.088	0.34
Stolle	0.706	0.959
Stork	0.698	0.949
Urech	0.606	0.789
von Richter	0.473	0.589
Wittig	0.129	0.357
Average	0.420	0.602

TABLE 7.11
Summary of AE(min) and Intrinsic Greenness Probability Values for Named Substitution Reactions

Name	AE(min)	p {RME(kernel) > 0.618}
Bucherer	0.468	0.584
Chichibabin	0.691	0.94
Dakin–West	0.44	0.555
Delepine	0.87	1
Finkelstein	0.85	1
Fischer esterification	0.804	1
Gabriel	0.086	0.34
Gattermann	0.491	0.61
Helferich	0.854	1
Hell–Volhard–Zelinsky	0.654	0.883
Koenigs–Knorr	0.623	0.824
Lapworth	0.991	1
Sandmeyer	0.757	0.997
Sanger	0.908	1
Schiemann	0.525	0.653
Schotten–Baumann	0.532	0.664
Wohl–Ziegler	0.588	0.753
Zeisel	0.423	0.538
Average	0.642	0.797

REFERENCES

1. Andraos, J. *Org. Process Res. Dev.* 2005, *9*, 404.
2. Heck, M.P.; Wagner, A.; Mioskowski, C. *J. Org. Chem.* 1996, *61*, 6486.
3. Kim, H.Y.; Salvi, L.; Carroll, P.J.; Walsh, P.J. *J. Am. Chem. Soc.* 2009, *131*, 954.
4. Yan, C.G.; Cai, X.M.; Wang, Q.F.; Wang, T.Y.; Zheng, M. *Org. Biomolec. Chem.* 2007, *5*, 945.
5. Le Callonnec, F.; Fouquet, E.; Felpin, F.X. *Org. Lett.* 2011, *13*, 2646.
6. Scroggins, S.T.; Chi, Y.; Fréchet, J.M.J. *Angew. Chem. Int. Ed.* 2010, *49*, 2393.
7. Porcheddu, A.; Giacomelli, G.; De Luca, L.; Ruda, A.M. *J. Comb. Chem.* 2004, *6*, 105.
8. Jacob, J.; Varghese, N.; Rasheed, S.P.; Agnihotri, S.; Sharma, V.; Wakode, S. *World Pharm. Pharmaceut. Sci.* 2016, *5*, 1821.
9. He, R.; Huang, Z.T.; Zheng, Q.Y.; Wang, C. *Tetrahedron Lett.* 2014, *55*, 5705.
10. Wu, M.; Wang, S. *Synthesis* 2010, 587.
11. Landers, B.; Berini, C.; Wang, C.; Navarro, O. *J. Org. Chem.* 2011, *76*, 1390.
12. Seyferth, D.; Hilbert, P.; Marmor, R.S. *J. Am. Chem. Soc.* 1967, *89*, 4811.
13. Gilbert, J.C.; Weerasooriya, U. *J. Am. Chem. Soc.* 1979, *44*, 4997.
14. Gilbert, J.C.; Weerasooriya, U. *J. Am. Chem. Soc.* 1982, *47*, 1837.
15. Pfau, A.S.; Plattner, P.A. *Helv. Chim. Acta* 1939, *22*, 202.
16. Plattner, P.A.; Roniger, H. *Helv. Chim. Acta* 1942, *25*, 590.
17. Simmons, H.E.; Smith, R.D. *J. Am. Chem. Soc.* 1958, *80*, 5323.
18. Seyferth, D.; Mui, J.Y.P.; Gordon, M.E.; Burlitch, J.M. *J. Am. Chem. Soc.* 1965, *87*, 681.
19. Seyferth, D.; Burlitch, J.M.; Minasz, R.J.; Mui, J.Y.P.; Simmons, H.D. Jr.; Treiber, A.J.H.; Dowd, S.R. *J. Am. Chem. Soc.* 1965, *87*, 4259.

20. Girard, P.; Namy, J.L.; Kagan, H.B. *J. Am. Chem. Soc.* 1980, *102*, 2693.
21. Molander, G.A.; La Belle, B.E.; Hahn, G. *J. Org. Chem.* 1986, *51*, 5259.
22. Molander, G.A.; Harring, L.S. *J. Org. Chem.* 1989, *54*, 3525.
23. Fischer, E.O.; Maasböl, F. *Chem. Ber.* 1967, *100*, 2445.
24. Harvey, D.F.; Lund, K.P. *J. Am. Chem. Soc.* 1991, *113*, 8916.
25. Lavallo, V.; Mafhouz, J.; Canac, Y.; Donnadieu, B.; Schoeller, W.W.; Bertrand, G. *J. Am. Chem. Soc.* 2004, *126*, 8670.
26. von Doering, W.E.; Hoffmann, A.K. *J. Am. Chem. Soc.* 1954, *76*, 6162.
27. von Doering, W.E.; LaFlamme, P.M. *Tetrahedron* 1958, *2*, 75.
28. Vogel, E. *Angew. Chem.* 1960, *72*, 4.
29. Kulinkovich, O.G.; Sviridov, S.V.; Vasilevski, D.A. *Synthesis* 1991, 234.
30. Fraisse, J. *Bull. Soc. Chim. Fr.* 1957, 986.
31. McCoy, L.L. *J. Am. Chem. Soc.* 1958, *80*, 6568.
32. Mousseron, M. *Compt. Rend.* 1959, *248*, 887.
33. Mousseron, M. *Compt. Rend.* 1959, *248*, 465.
34. Mousseron, M. *Compt. Rend.* 1959, *248*, 2840.
35. Bojilova, A.; Trendafilova, A.; Ivanov, C.; Rodios, N.A. *Tetrahedron* 1993, *49*, 2275.
36. Ramberg, L.; Widequist, S. *Arkiv Kemi Mineral. Geolog.* 1937, *12A*, 12.
37. Widequist, S. *Arkiv Kemi Mineral. Geolog.* 1945, *20B*, 8.
38. Hart, H.; Freeman, F. *J. Am. Chem. Soc.* 1963, *85*, 1161.
39. Lee, J.Y.; Padias, A.B.; Hall, H.K. Jr. *Macromolecules* 1991, *24*, 17.
40. Al-Quntar, A.A.A.; Srebnik, M. *J. Organometallic Chem.* 2005, *690*, 2504.
41. Al-Quntar, A.A.A.; Dembitsky, D.M.; Srebnik, M. *Org. Prep. Proced. Int.* 2008, *40*, 505.
42. Freund, A. *Monatsh. Chem.* 1882, *3*, 625.
43. Gustavson, G. *J. Prakt. Chem.* 1887, *36*, 300.
44. Bartleson, J.D.; Burk, R.E.; Lankelma, H.P. *J. Am. Chem. Soc.* 1946, *68*, 2513.
45. Andraos, J. *ACS Sust. Chem. Eng.* 2013, *1*, 496.
46. Andraos, J. Application of green metrics analysis to chemical reactions and synthesis plans, In Lapkin, A.; Constable, D.C. (Eds.), *Green Chemistry Metrics in Green Chemistry Metrics.* Blackwell Scientific, Oxford, 2008, pp. 69–199.
47. Lapkin, A., Constable, D. J. C. *Green Chemistry Metrics: Measuring and Monitoring Sustainable Processes.* Wiley, Chichester, 2008, pp. 91–96.
48. Collins, J.C.; Hess, W.W.; Frank, F.J., *Tetrahedron Lett.* 1968, *9*, 3363.
49. Collins, J.C.; Hess, W.W., *Org. Synth. Coll. Vol. VI* 1988, *6*, 644.
50. Corey, E.J.; Kim, C.U. *J. Am. Chem. Soc.* 1972, *94*, 7586.
51. Dess, D.B.; Martin, J.C. *J. Org. Chem.* 1983, *48*, 4155.
52. Bowden, K.; Heilbron, I.M.; Jones, E.R.H.; Weedon, B.C.L. *J. Chem. Soc.* 1946, 39.
53. Curtis, R.G.; Heilbron, I.M.; Jones, E.R.H.; Woods, G.F. *J. Chem. Soc.* 1953, 457.
54. Halsall, T.G.; Hodges, R.; Jones, E.R.H. *J. Chem. Soc.* 1953, 3019.
55. Bowers, A.; Halsall, T.G.; Jones, E.R.H.; Lemin, A.J. *J. Chem. Soc.* 1953, 2548.
56. Pfitzner, K.E.; Moffatt, J.G. *J. Am. Chem. Soc.* 1963, *85*, 3027.
57. Omura, K.; Swern, D. *Tetrahedron* 1978, *34*, 1651.
58. Corey, E.J.; Chaykovsky, M. *J. Am. Chem. Soc.* 1962, *84*, 867.
59. Corey, E.J.; Chaykovsky, M. *J. Am. Chem. Soc.* 1965, *87*, 1353.
60. Zhang, W.; Loebach, J.L.; Wilson, S.R.; Jacobsen, E.N. *J. Am. Chem. Soc.* 1990, *112*, 2801.
61. Jacobsen, E.N.; Zhang, W.; Muci, A.R.; Ecker, J.R.; Deng, L. *J. Am. Chem. Soc.* 1991, *113*, 7063.
62. Prilezhaev, N. *Chem. Ber.* 1909, *42*, 4811.
63. Katsuki, T.; Sharpless, K.B., *J. Am. Chem. Soc.* 1980, *102*, 5974.
64. Wang, Z.X.; Tu, Y.; Frohn, M.; Zhang, J.R.; Shi, Y. *J. Am. Chem. Soc.* 1997, *119*, 11224.
65. Wang, Z.X.; Shi, Y. *J. Org. Chem.* 1997, *62*, 8622.

8 Automated Computation for a Single Reaction

8.1 EXCEL SPREADSHEET TOOLS: PART 1

In this section we describe seven Excel spreadsheets that facilitate the evaluation of material efficiency for a single reaction in an effortless automated fashion. Raw experimental data for the *DENSITIES.xls*, *densities-of-acid-solutions.xls*, *densities-of-base-solutions.xls*, and *densities-of-carbohydrate-solutions.xls* spreadsheets were taken from references [1, 2].

8.1.1 DENSITIES.XLS

The *densities.xls* spreadsheet contains a table of densities (in units of g/mL) for commonly used solvents in organic and inorganic reactions and another table of densities for acid, base, and salt solutions used in work-up procedures.

8.1.2 DENSITIES-OF-ACID-SOLUTIONS.XLS

The *densities-of-acid-solutions.xls* spreadsheet contains nine sheets corresponding to density data for the following acids in aqueous solution: hydrochloric, sulfuric, nitric, phosphoric, chromic, acetic, citric, formic, and tartaric acids. Specifically, polynomial correlation functions are determined from experimental data for the behavior of density of these acid solutions with respect to weight percent (wt%) and molar concentration (M, mol/L). In addition, polynomial correlation functions for the variation of wt% versus M and the variation of M versus wt% are given. All data were sufficiently fit to fourth-order polynomials except for aqueous acetic acid solutions where data were fit to a fifth order polynomial. For each acid, calculators are included for the determination of densities of acid solutions for any known concentrations in either wt% or M units within the concentration limits of the data fits. The correlation coefficients for all fits exceeded 99%. Tables 8.1 and 8.2 summarize the calculated coefficients according to the fitting Equations (8.1) and (8.2).

$$d_{g/mL} = A_1 w^4 + B_1 w^3 + C_1 w^2 + D_1 w + E_1 \tag{8.1}$$

$$d_{g/mL} = A_2 M^4 + B_2 M^3 + C_2 M^2 + D_2 M + E_2 \tag{8.2}$$

where:

- w is weight %
- M is molar concentration (mol/L)

TABLE 8.1
Summary of Coefficients for Data Fits of Density (g/mL) with Respect to wt% for Various Aqueous Solutions of Acids

Acid	A_1	B_1	C_1	D_1	E_1	F_1
HCl	−9.000E−09	4.000E−07	3.000E−06	4.800E−03	9.985E−01	
H_2SO_4	−2.000E−08	4.000E−06	−2.000E−04	1.000E−02	9.844E−01	
HNO_3	4.000E−09	−1.000E−06	9.000E−05	4.400E−03	1.002E+00	
H_3PO_4	−1.000E−09	3.000E−07	2.000E−05	5.300E−03	9.985E−01	
CrO_3	5.000E−09	2.000E−07	3.000E−05	7.500E−03	9.988E−01	
HOAc[a]	−4.000E−11	9.000E−09	−7.000E−07	2.000E−05	1.300E−03	9.987E−01
Citric	4.000E−10	8.000E−09	1.000E−05	3.800E−03	9.997E−01	
Formic	−6.000E−10	8.000E−08	−5.000E−06	2.600E−03	9.994E−01	
Tartaric	1.000E−09	−7.000E−08	2.000E−05	4.500E−03	9.999E−01	

[a] Fit according to a fifth order polynomial: $d_{g/mL} = A_1w^5 + B_1w^4 + C_1w^3 + D_1w^2 + E_1w + F_1$

TABLE 8.2
Summary of Coefficients for Data Fits of Density (g/mL) with Respect to mol/L for Various Aqueous Solutions of Acids

Acid	A_2	B_2	C_2	D_2	E_2	F_2
HCl	−7.000E−08	−1.000E−06	−1.000E−04	1.174E−02	9.986E−01	
H_2SO_4	−9.000E−06	3.000E−04	−3.400E−03	6.920E−02	9.952E−01	
HNO_3	1.000E−06	−5.000E−05	5.000E−05	3.330E−02	9.985E−01	
H_3PO_4	3.000E−07	−2.000E−05	−6.000E−05	5.050E−02	9.990E−01	
CrO_3	−9.000E−06	2.000E−04	−2.100E−03	7.430E−02	9.989E−01	
HOAc	−3.000E−07	1.000E−05	−2.000E−04	7.000E−04	7.200E−03	9.990E−01
Citric	−3.000E−06	1.000E−04	−1.500E−03	7.320E−02	9.997E−01	
Formic	−9.000E−08	5.000E−06	−2.000E−04	1.160E−02	9.995E−01	
Tartaric	7.000E−06	1.000E−06	−1.100E−03	6.780E−02	9.998E−01	

[a] Fit according to a fifth order polynomial: $d_{g/mL} = A_2w^5 + B_2w^4 + C_2w^3 + D_2w^2 + E_2w + F_2$

8.1.3 DENSITIES-OF-BASE-SOLUTIONS.XLS

The *densities-of-base-solutions.xls* spreadsheet contains three sheets corresponding to density data for the following bases in aqueous solution: ammonia, potassium hydroxide, and sodium hydroxide. The same kinds of polynomial fits were made as for aqueous acid solutions. Tables 8.3 and 8.4 summarize the calculated coefficients according to the fitting Equations (8.3) and (8.4).

$$d_{g/mL} = A_3w^4 + B_3w^3 + C_3w^2 + D_3w + E_3 \tag{8.3}$$

TABLE 8.3
Summary of Coefficients for Data Fits of Density (g/mL) with Respect to wt% for Various Aqueous Solutions of Bases

Base	A_3	B_3	C_3	D_3	E_3
NH_3	4.000E–09	−1.000E–06	7.000E–05	4.900E–03	9.995E–01
KOH	−3.000E–09	4.000E–07	9.000E–06	9.100E–03	9.992E–01
NaOH	3.000E–09	−8.000E–07	3.000E–05	1.080E–02	9.992E–01

TABLE 8.4
Summary of Coefficients for Data Fits of Density (g/mL) with Respect to mol/L for Various Aqueous Solutions of Bases

Base	A_4	B_4	C_4	D_4	E_4
NH_3	−6.000E–07	3.000E–05	−4.000E–04	−4.700E–03	9.946E–01
KOH	−2.000E–06	8.000E–05	−1.600E–03	5.010E–02	9.997E–01
NaOH	−5.000E–07	3.000E–05	−1.300E–03	4.260E–02	9.995E–01

$$d_{g/mL} = A_4M^4 + B_4M^3 + C_4M^2 + D_4M + E_4 \tag{8.4}$$

where:

w is weight %
M is molar concentration (mol/L)

8.1.4 DENSITIES-OF-CARBOHYDRATE-SOLUTIONS.XLS

The *densities-of-carbohydrate-solutions.xls* spreadsheet contains five sheets corresponding to density data for the following carbohydrates in aqueous solution: sucrose, D-fructose, D-glucose, lactose, and maltose. The same kinds of polynomial fits were made as for aqueous acid solutions. Tables 8.5 and 8.6 summarize the calculated coefficients according to the fitting Equations (8.5) and (8.6).

$$d_{g/mL} = A_5w^4 + B_5w^3 + C_5w^2 + D_5w + E_5 \tag{8.5}$$

$$d_{g/mL} = A_6M^4 + B_6M^3 + C_6M^2 + D_6M + E_6 \tag{8.6}$$

where:

w is weight %
M is molar concentration (mol/L)

TABLE 8.5
Summary of Coefficients for Data Fits of Density (g/mL) with Respect to wt% for Various Aqueous Solutions of Carbohydrates

Carbohydrate	A_5	B_5	C_5	D_5	E_5
Sucrose	−3.000E−10	7.000E−08	1.000E−05	3.900E−03	9.982E−01
D-fructose	−1.000E−10	5.000E−08	1.000E−05	3.900E−03	9.982E−01
D-glucose	2.000E−09	−2.000E−07	2.000E−05	3.700E−03	9.983E−01
Lactose	2.000E−07	−7.000E−06	7.000E−05	3.800E−03	9.982E−01
Maltose	−6.000E−09	6.000E−07	−4.000E−06	4.000E−03	9.985E−01

TABLE 8.6
Summary of Coefficients for Data Fits of Density (g/mL) with Respect to mol/L for Various Aqueous Solutions of Carbohydrates

Carbohydrate	A_6	B_6	C_6	D_6	E_6
Sucrose	2.000E−08	−2.000E−06	−4.000E−05	2.310E−02	9.983E−01
D-fructose	−1.000E−05	9.000E−05	−1.000E−03	7.050E−02	9.982E−01
D-glucose	6.000E−05	−4.000E−04	1.000E−04	6.820E−02	9.982E−01
Lactose	2.269E−01	−2.068E−01	5.030E−02	1.319E−01	9.982E−01
Maltose	−1.800E−03	6.400E−03	−7.400E−03	1.324E−01	9.987E−01

8.1.5 Flash-Chromatography-Mass Intensity-Template.xls

The *flash-chromatography-mass intensity-template.xls* spreadsheet determines the *E*-factor contribution from solvents used in flash chromatographic procedures during the purification phase of a chemical reaction according to a method described in 2016 [3]. Two kinds of calculations are worked out: (a) exact minimum and maximum estimates and (b) approximate minimum and maximum estimates. The minimum and maximum estimates refer to "easy" and "hard" separations, respectively where minimum and maximum volumes of solvent are used. The exact calculation pertains to the situation where the Rf value of the product to be separated based on thin layer chromatography (TLC) is given in an experimental procedure, and the approximate calculation pertains to the default situation where a Rf value of 0.35 and 35 theoretical plates are used in the analysis when no such data are given in an experimental procedure.

The contribution to the total *E*-factor of a reaction from materials used in flash chromatography such as mass of silica gel and mass of eluent solvents is given generally by Equation (8.7).

$$E - \text{flash chrom.} = \frac{m_{\text{silica gel}} + m_{\text{eluent}}}{m_{\text{product}}} \tag{8.7}$$

E-flash chrom. may be estimated from the following expression given by Equation (8.8).

$$E\text{-flash chrom.} = A\left(\frac{m_{\text{crude product}}}{m_{\text{pure product}}}\right)\left[1+\rho_{eluent}\left(\frac{\varepsilon_{\text{silica}}}{\rho_{\text{silica}}}\right)\left[\frac{0.64}{R_f}\left(1+\frac{2}{\sqrt{N}}\right)+1\right]\right] \quad (8.8)$$

where:

$m_{\text{crude product}}$	is the mass of product before chromatography
$m_{\text{pure product}}$	is the mass of product after chromatography
A	is a degree of difficulty separation factor (facile separations are assigned $A=10$; difficult separations are assigned $A=152$)
ρ_{eluent}	is the density of the eluent solvent
$\varepsilon_{\text{silica}}$	is the porosity of silica (0.9)
ρ_{silica}	is the density of silica (0.5 g/cm^3)
Rf	is the retention factor for the product
N	is the number of theoretical plates in the column

The number of theoretical plates, N, is estimated using the relations given in Equations (8.9a) and (8.9b).

$$N = \begin{cases} 33.64\left(\dfrac{m_{\text{crude product}}}{m_{\text{pure product}}}\right)^{0.44}, & \text{facile separation} \\ 51.70\left(\dfrac{m_{\text{crude product}}}{m_{\text{pure product}}}\right)^{0.44}, & \text{difficult separation} \end{cases} \quad (8.9\text{a,b})$$

If an experimental procedure provides only the masses of crude and pure products as well as the identity of the eluent solvent, then minimum and maximum estimates of E-flash chrom. can be made according to the following expressions given in Equations (8.10a) and (8.10b).

$$E\text{-flash chrom.(min)} = 10\left(\frac{m_{\text{crude product}}}{m_{\text{pure product}}}\right)\left[1+6.20\rho_{\text{eluent}}\right]$$

$$E\text{-flash chrom.(max)} = 152\left(\frac{m_{\text{crude product}}}{m_{\text{pure product}}}\right)\left[1+6.20\rho_{\text{eluent}}\right] \quad (8.10\text{a,b})$$

where Rf$=0.35$ and $N=35$ were used, and the density of eluent composed of multiple solvents is given by Equation (8.11).

$$\rho_{\text{eluent}} = \frac{\sum_j \rho_j V_{f,j}}{\sum_j V_{f,j}} \quad (8.11)$$

where:

ρ_j is the density of the jth eluent solvent
$V_{f,j}$ is the volume fraction of the jth eluent solvent

Example 8.1

An experimental procedure calls for purification of 10 g of crude product using EtOAc:hexane (1:10) by volume. Two liters of eluent are required to obtain 8 g of pure product. Use the *flash-chromatography-mass intensity-template.xls* spreadsheet to determine the minimum and maximum *E*-factor contribution for this operation.

Solution

Density of EtOAc = 0.901 g/mL
Density of hexane = 0.684 g/mL
Volume of EtOAc = (1/11)*2000 = 181.8 mL
Volume of hexane = (10/11)*2000 = 1818.2 mL
Density of combined eluent = (0.901*181.8 + 0.684*1818.2)/2000 = 0.704 g/mL
Rf = 0.35 (assumed)
N = 35 (assumed)

Minimum *E*-factor calculation:

Mass of silica gel = 10*mass crude product = 10*10 = 100 g
Void volume = (9/5)*mass of silica gel = (9/5)*100 = 180 cm^3

Mass of eluent

= (void volume)*(density of combined eluent)*((0.64/Rf)(1 + (2/sqrt(*N*))) + 1)
= 180*0.704*((0.64/0.35)(1 + (2/sqrt(35))) + 1)
= 436.8 g

E-factor = (mass of silica gel + mass of eluent)/mass pure product

= (100 + 436.8)/8 = 67.1

Maximum *E*-factor calculation:

Mass of silica gel = 152*mass crude product = 152*10 = 1520 g
Void volume = (9/5)*mass of silica gel = (9/5)*1520 = 2736 cm^3

Mass of eluent

= (void volume)*(density of combined eluent)*((0.64/Rf)(1 + (2/sqrt(*N*))) + 1)
= 2736*0.704*((0.64/0.35)(1 + (2/sqrt(35))) + 1)
= 6636.3 g

E-factor = (mass of silica gel + mass of eluent)/mass pure product

= (1520 + 6636.3)/8 = 1019.5

8.1.6 Template-Reaction.xls

Once a balanced chemical equation is written down for a given transformation and a literature experimental procedure is obtained for the synthesis of a given molecule based on a single step reaction, the *template-REACTION.xls* spreadsheet may be used to determine the following material efficiency metrics: reaction yield (Y), atom economy (AE), stoichiometric factor (SF), materials recovery parameter (MRP), process mass intensity (*PMI*), *E*-factor, and the following *E*-factor contributors according to by-product formation (*E*-kernel), excess reagent consumption (*E*-excess), reaction solvent (*E*-rxn solvent), catalyst (*E*-cat), work-up (*E*-workup), and purification (*E*-purif) material usage.

The spreadsheet also provides the following two visual aids for a user to view graphically the meaning of these parameters: (a) a radial pentagon displaying the 5 variables Y, AE, 1/SF, MRP, and RME (reaction mass efficiency) each varying between 0 and 1; and (b) a pie chart showing the percent contributions of the 6 contributing *E*-factors to the overall *E*-factor of a reaction. In addition, a vector magnitude ratio is determined according to Equation (8.12) for describing the overall material efficiency of the reaction with respect to an ideal situation where all 5 metric variables in the pentagon are each equal to 1.

$$\mathrm{VMR} = \frac{1}{\sqrt{5}}\left[\sqrt{(\mathrm{AE})^2 + (\mathrm{RY})^2 + \left(\frac{1}{\mathrm{SF}}\right)^2 + (\mathrm{MRP})^2 + (\mathrm{RME})^2}\right] \tag{8.12}$$

The data inputs required in the spreadsheet include: (a) the molecular weights, stoichiometric coefficients, and masses of reagents; (b) the molecular weights and stoichiometric coefficients of all reaction products according to the balanced chemical equation; (c) the densities and volumes of all auxiliary materials used such as solvents, catalysts, work-up extraction solvents, and purification solvents from which corresponding masses are automatically calculated; and (d) the mass of collected target product. The spreadsheet has a check calculation built in to verify whether or not the entered chemical equation is correctly balanced.

8.1.7 Template-Reaction-Input-Costs.xls

The *template-REACTION-input-costs.xls* spreadsheet is identical in layout to the *template-REACTION.xls* spreadsheet and contains all of the features and functions as described in Section 8.1.6. The only additional information is the cost of each input material used in the chemical reaction in units of \$/g from which the total input cost of raw materials may be determined for carrying out the specified reaction.

8.2 EVALUATING MATERIAL EFFICIENCY FOR A SINGLE CHEMICAL REACTION

Example 8.2 [4]

An alternative reaction to making biaryl compounds by the Suzuki coupling reaction is to couple aryl iodonium ions to tetraphenylborate. This reaction avoids the use of metal catalysts, ligands, and bases which are typically required in the Suzuki coupling protocol.

An example reaction is shown below along with its experimental procedure.

MeO–C6H4–I⊕(Ph) Br⊖ + Ph4B⊖ Na⊕ → (H_2O, 100°C, microwaves, 77 %) MeO–C6H4–Ph

EXPERIMENTAL PROCEDURE:

Sodium tetraphenylborate (513 mg, 1.5 mmol), *p*-methoxyphenyl phenyliodonium bromide (0.75 mmol), and water (5 mL) were placed in a 50 mL flask fitted with a condenser. The vessel was placed in the center of the microwave synthesizer and was then exposed to microwave irradiation (250 W) to heat it at reflux for 1–5 min at 100°C. After irradiation, the reaction mixture was allowed to cool to room temperature and extracted with diethyl ether (3 × 20 mL). The organic layer was dried with anhydrous

$MgSO_4$ and the solvents were removed *in vacuo*. The crude product was separated on a silica gel plate with hexane as eluent and the pure product was afforded in 77% yield.

Part 1

Using the *template-REACTION.xls* spreadsheet and the details of the experimental procedure determine the material efficiency metrics for the reaction. State any assumptions made in the calculations. Show the balanced chemical equation.

Solution

MeO–C6H4–I⊕(Ph) Br⊖ (390.9) + Ph4B⊖ Na⊕ (342) → MeO–C6H4–Ph (184) − BPh_3 − NaBr − PhI

Assumptions: 0.5 magnesium sulfate drying agent.

E-kernel	4.17
E-excess	2.41
E-rxn solvent	47.04
E-catalyst	0.00
E-workup	444.97
E-purification	0.00
E-aux	492.00
E-total	498.59
Yield	77.0%
AE	25.1%
PMI	499.59
RME (global)	0.2%

Part 2

Compare the above reaction performance with the Suzuki coupling methods for the same target product by microwave heating and by conventional heating and continuous flow [5]. Show the balanced chemical equations in each case.

Solution

Microwave heating:

Phenylboronic acid (122) + 4-bromoanisole (186.9) → 4-methoxybiphenyl (184)

Reagents: Pd (cat.), Na_2CO_3, 3 EtOH; by-products: - NaBr, - $B(OEt)_3$, - CO_2, - NaOH, - 2 H_2O

Assumptions: 5 g magnesium sulfate drying agent.

E-kernel	2.09
E-excess	5.37
E-rxn solvent	4.41
E-catalyst	0.00
E-workup	84.95
E-purification	0.00
E-aux	89.35
E-total	96.81
Yield	97.2%
AE	33.3%
PMI	97.81
RME (global)	1.0%

Conventional heating and continuous flow:

Phenylboronic acid (122) + 4-bromoanisole (186.9) → 4-methoxybiphenyl (184)

Reagents: Pd (cat.), 2 NaOH; by-products: - NaBr, - $NaB(OH)_4$

Assumptions: None.

E-kernel	1.37
E-excess	0.15
E-rxn solvent	10.91
E-catalyst	0.00
E-workup	12.17
E-purification	0.00
E-aux	23.08
E-total	24.60
Yield	89.1%
AE	47.3%
PMI	25.60
RME (global)	3.9%

Summary of materials metrics performances:

Method	% Yield	% AE	*PMI*
Iodonium method	77	25.1	499.6
Suzuki + microwave irradiation	97	33.3	97.8
Suzuki + conventional heating + continuous flow	89	47.3	25.6

Unfortunately, the authors' stated attributes of the iodonium method are wiped out by the excessive use of work-up ethyl acetate solvent (60 mL) used compared with the scale of the reaction (0.75 mmol) carried out in 5 mL of water. This results in the highest *PMI* value. The conventional heating and continuous flow Suzuki method has by far the lowest *PMI*.

8.3 EVALUATING INPUT COST FOR A SINGLE CHEMICAL REACTION

Example 8.3 [6, 7]

A 4-step procedure to prepare the sulfa drug sulfanilamide is given below.

REACTION 1

9.3 g of aniline, 12.24 g of acetic anhydride, 9.8 g sodium acetate, and 8 mL 98 wt% sulfuric acid (d = 1.8361 g/mL) are reacted in 250 mL water. The product collected, 12 g, was used directly in the next step without purification.

REACTION 2

12 g of product #1 and 52.4 g chlorosulfonic acid are reacted; 300 mL of water was used to extract 15 g of product #2 from the reaction mixture.

REACTION 3

All of product #2 was treated with 60 mL of concentrated (28 wt%) ammonium hydroxide aqueous solution (d=0.898 g/mL). After the reaction took place the mixture was neutralized with 10 mL 6 M (44 wt%) sulfuric acid solution (d=1.3384 g/mL) which afforded 11 g of product #3.

REACTION 4

All of product #3 was hydrolyzed in 60 mL water and 26 mL 3 M (10 wt%) aqueous HCl (d=1.047 g/mL) solution. After the reaction the mixture was neutralized with 15 mL 6 M (20 wt%) sodium hydroxide solution (d=1.219 g/mL) and 40 mL saturated sodium carbonate solution (d=1.1463 g/mL); 8.5 g of sulfanilamide were collected.

Costs of materials:

Aniline	$402.80/18 kg
Acetic anhydride	$364.30/18 kg
Sodium acetate	$153.70/3 kg
98 wt% sulfuric acid	$51.50/2.5 L
Chlorosulfonic acid	$57.10/kg
Conc. NH_4OH soln	$52.20/2 L
37 wt% HCl soln.	$268.40/16 kg
Sodium hydroxide	$212.20/12 kg
Sodium carbonate	$333.50/12 kg
Water	$0/kg

Part 1

Write out balanced chemical equations for each transformation.

Part 2

Use the Excel *template-REACTION-input-costs.xls* spreadsheet to determine AE, yield, RME, *PMI*, and RMC for each step.

Part 3

Determine the overall AE, overall yield, overall *PMI*, overall *PMI*, and overall RMC for the synthesis plan.

Part 4

Represent the plan as a synthesis tree diagram and determine the kernel RME.

Part 5

Determine the kernel RMC for the plan.

What is the most costly reagent? What is the least costly reagent?

Part 6

How much of the overall RMC spent on all input materials is destined for waste in this synthesis plan?

Solution

Part 1

Step 1

$$\text{PhNH}_2 + (\text{CH}_3\text{CO})_2\text{O} \xrightarrow[\text{(cat.)}]{\text{NaOAc}} \text{PhNHCOCH}_3 + \text{CH}_3\text{COOH}$$

93 102 135 60

Step 2

$$\text{PhNHCOCH}_3 + \text{Cl–SO}_2\text{–OH} \longrightarrow \text{HO–SO}_2\text{–C}_6\text{H}_4\text{–NHCOCH}_3 + \text{HCl}$$

135 116.45 215 36.45

Step 3

$$\text{HO–SO}_2\text{–C}_6\text{H}_4\text{–NHCOCH}_3 + \text{NH}_4\text{OH} \longrightarrow \text{H}_2\text{N–SO}_2\text{–C}_6\text{H}_4\text{–NHCOCH}_3 + 2\,\text{H}_2\text{O}$$

215 35 214 36

Step 4

$$\text{H}_2\text{N–SO}_2\text{–C}_6\text{H}_4\text{–NHCOCH}_3 + \text{H}_2\text{O} \longrightarrow \text{H}_2\text{N–SO}_2\text{–C}_6\text{H}_4\text{–NH}_2 + \text{CH}_3\text{COOH}$$

214 18 172 60

Part 2

Step	Yield (%)	AE (%)	RME (%)	*PMI*	RMC ($)
1	89	69	4.1	24.67	1.13
2	79	86	4.1	24.29	2.99
3	74	86	13.4	7.48	1.66
4	96	74	5.2	19.10	0.37

Part 3

	Yield (%)	AE (%)	RME (%)	*PMI*	RMC ($)
Overall	49	55	0.98	102.01	6.62

Part 4

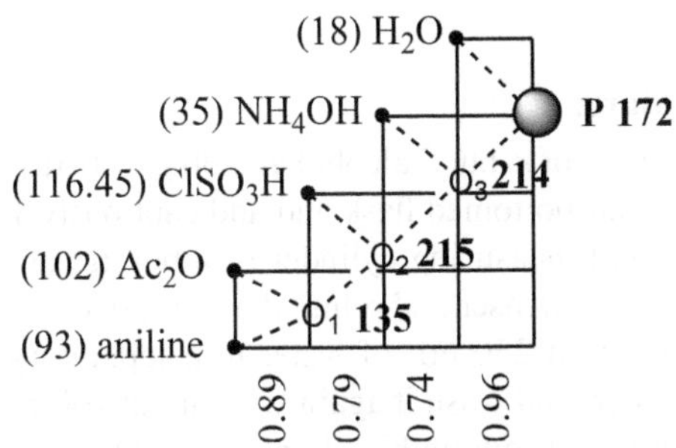

$$\text{RME}_{kernel} = \frac{172}{\frac{18}{0.96} + \frac{35}{0.74*0.96} + \frac{116.45}{0.79*0.74*0.96} + \frac{102+93}{0.89*0.79*0.74*0.96}}$$

$$= \frac{172}{18.750 + 49.268 + 207.496 + 390.404}$$

$$= \frac{172}{665.918}$$

$$= 0.258$$

or 25.8%.

Part 5

The kernel RMC is based on reagent materials only.

The *template-REACTION-input-costs.xls* spreadsheet gives the cost of reagents as $3.89 for the production of 8.5 g of sulfanilamide. This amounts to 59% of the total input material cost.

The most costly reagent is chlorosulfonic acid, which contributes \$2.99 or 77% out of the \$3.89 for total cost of input reagents.

The least costly reagent, other than water, is aniline which contributes \$0.21 or 5% out of the \$3.89 for total cost of input reagents.

Part 6

The total cost to produce 8.5 g sulfanilamide is \$6.62.

The overall RME for the 4-step plan is 0.0098.

Therefore, the amount of money spent on input materials that actually is directed to making the product is 0.0098*6.62 = \$0.065 = 6.5 cents. The balance, \$6.56, ends up producing waste materials.

8.4 PROBLEMS

PROBLEM 8.1 [8]

Themes: balancing chemical equations, experimental procedures, dual role solvent and reagent

An experimental procedure for a reaction using *n*-butanol as both solvent and reagent is given below.

Experimental Procedure:

Mix together 37 g (46 mL) of *n*-butyl alcohol and 60 g (60 mL) of glacial acetic acid in a 250 or 500 mL round-bottomed flask and add cautiously 1 mL of concentrated sulfuric acid (use a small measuring cylinder or a burette or a calibrated dropper pipette). Attach a reflux condenser and reflux the mixture on a wire gauze for 3-6 h. Pour the mixture into about 250 mL of water in a separatory funnel, remove the upper layer of crude ester, and wash it again with about 100 mL of water, followed by about 25 mL of saturated sodium bicarbonate solution and 50 mL of water. The ester must, of course, be separated between each washing. Dry the crude ester with 5 to 6 g of anhydrous sodium or magnesium sulfate. Filter through a small funnel containing a fluted filter paper or a small plug of cotton or glass wool into a dry 100 mL distilling flask. Add 2–3 fragments of porous porcelain and distill on a wire gauze or from an air bath. Collect the pure *n*-butyl acetate at 124 to 125°C. The yield is 40 g.

Part 1

Write out a balanced chemical equation for the transformation and determine its atom economy.

Part 2

Using the *template-REACTION.xls* spreadsheet and the details of the experimental procedure determine the material efficiency metrics for the reaction. State any assumptions made in the calculations.

Part 3

What mass of water is produced as a by-product?

Part 4

What fraction of the acetic acid participated as a solvent and what fraction participated as a reagent?

Part 5

What is the percent conversion of *n*-butanol?

Part 6

What is the catalyst loading in mol %?

PROBLEM 8.2 [9]

Themes: material efficiency metrics without a balanced chemical equation

n-Butanol can be obtained by fermentation of molasses with *Clostridium saccharobutyl acetonicum liquefaciens*. In the process acetone, ethanol, carbon dioxide, and hydrogen gas are also produced as side products.

Starch → glucose → acetone + *n*-butanol + ethanol + carbon dioxide + hydrogen

In order to produce 1000 kg of *n*-butanol, 5425 L of molasses, 80 cubic meters of water, and 3.5 kg of nutrients are required. 1 L of molasses (0.8 kg) yields 174 to 200 g *n*-butanol, 48 to 78 g acetone, and 8 to 12 g ethanol.

Based on these data, determine the process mass intensities for the production of *n*-butanol, acetone, and ethanol.

PROBLEM 8.3 [10]

Themes: balancing chemical equations, material efficiency metrics, experimental procedures

Me_3Si—CN

O, HN(Ph)–OH

$LiClO_4$
Et_2O
97 %

CN, N(Ph)–OH

Experimental Procedure:

To a mixture of aldehyde (2 mmol) in 5 M LPDE (lithium perchlorate/diethylether) (4 ml) was added hydroxylamine (2.2 mmol) at room temperature. The mixture was stirred for 2 min and trimethylsilylcyanide (2.2 mmol) was added. The mixture was stirred for 15 min then water was added and the product was extracted with CH_2Cl_2. The organic phase was collected, dried (Na_2SO_4), and evaporated to afford the crude product. The product was purified by flash chromatography (hexane–ethyl acetate). Yield: 97%.

Part 1

Write out the balanced chemical equation and determine its atom economy. Show the target bonds made in the product structure.

Part 2

Using the *template-REACTION.xls* spreadsheet and the experimental details determine the material efficiency metrics for the reaction. State any assumptions made in the calculations.

PROBLEM 8.4 [11]

Themes: balancing chemical equations, ring construction strategy, redox

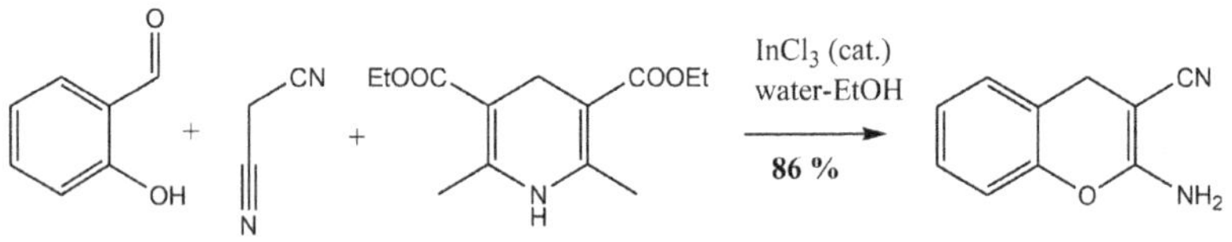

InCl$_3$ (cat.) water-EtOH 86 %

Experimental Procedure:

To a stirred mixture of appropriate salicylaldehyde (1 mmol), malononitrile (1 mmol) and Hantzsch dihydropyridine ester (1 mmol) in water:ethanol (1:1) (10 mL), a catalytic amount of indium(III) chloride (20 mol %) was added and the mixture was stirred at room temperature for 25 min. After complete conversion as indicated by TLC, the product was extracted with ethyl acetate (2×15 mL). The combined extracts were dried over anhydrous Na_2SO_4 and concentrated *in vacuo*. The resulting product was purified by column chromatography on silica gel (Merck, 60–120 mesh, ethyl acetate–hexane, 2:8) to afford pure product. Yield: 86%.

Part 1

Write a balanced chemical equation for the reaction. Determine the corresponding atom economy and state the ring construction strategy.

Part 2

Using the *template-REACTION.xls* spreadsheet and the details of the experimental procedure determine the material efficiency metrics for the reaction. State any assumptions made in the calculations.

PROBLEM 8.5 [12]

Themes: balancing chemical equations, experimental procedures, overall process mass intensity, raw material costs

A two-step synthesis to prepare modafinil, an anti-narcoleptic drug, in an undergraduate lab is carried out according to the experimental procedures below.

Step 1

Thiourea (5 g), potassium iodide (0.17 g), and water (15 mL) are added to a 250 mL round-bottomed flask. After the suspension is heated to 70°C, benzhydryl chloride (11 g) is added over 30 min. The mixture is heated to 95°C under a water condenser for 90 min.

The mixture is cooled at room temperature. Water (22.5 mL) followed by 30% sodium hydroxide aqueous solution (7.3 mL, 1.3279 g/mL) is added to the flask. When the mixture is heated to 70°C, triethylamine (8.3 mL, d=0.726 g/mL) is added. A hot solution of 2-chloroacetamide (5.5 g) in water (16.5 mL) is added in 1 h, separating an oil from the mixture. The mixture is stirred for another 15 min. While standing at room temperature the oil solidifies. The mixture can be cooled in an ice bath for 15 min to induce further crystallization.

The crude solid is isolated by vacuum filtration and dried for 10 min. Recrystallization from toluene (20 mL, d=0.865 g/mL) affords 9.78 g of an off-white solid.

Step 2

The product from step 1%, 30% hydrogen peroxide aqueous solution (3.1 mL, d=1.18 g/mL), and glacial acetic acid (41 mL, d=1.049 g/mL) are added to a 250 mL Erlenmeyer flask. The mixture is heated to 50°C while stirring vigorously for 2 h. The reaction mixture is allowed to cool to room temperature and 160 mL water is added to yield a white precipitate. The flask is placed in an ice bath to promote further precipitation. The crude solid is collected by vacuum filtration and dried for 5 min. The solid is recrystallized from methanol (20 mL, d=0.792 g/mL) and collected by vacuum filtration. 5.2 g of modafinil is collected.

The costs of all the input materials are given below.

Thiourea	\$47.30 per kg
Potassium iodide	\$477.80 per 2500 g
Benzhydryl chloride	\$185.20 per 500 g
30% NaOH	\$212.20 per 12 kg
Triethylamine	\$472.80 per 18 L
2-Chloroacetamide	\$59.90 per kg
Toluene	\$171.70 per 20 L
30% H_2O_2	\$252.00 per 4 L
Glacial HOAc	\$237.60 per 10 kg
Methanol	\$262.40 per 16 L
Water	\$0

Part 1

Write balanced chemical equations for each step.

Part 2

Write out a mechanism for step 1.

Part 3

Draw a synthesis tree diagram for the two-step plan. Using the *template-REACTION-input-costs.xls* spreadsheet determine the overall *PMI* for the synthesis and the overall RMC under conditions where all materials other than the target product are committed to waste.

PROBLEM 8.6 [13]

Themes: balancing chemical equations, material efficiency metrics

The asymmetric cyclopropanation reaction of styrene with ethyl diazoacetate was carried out using a solid-supported ruthenium(II) Pybox catalyst in a continuous flow reactor with supercritical carbon dioxide as solvent. The results of an experiment are shown below.

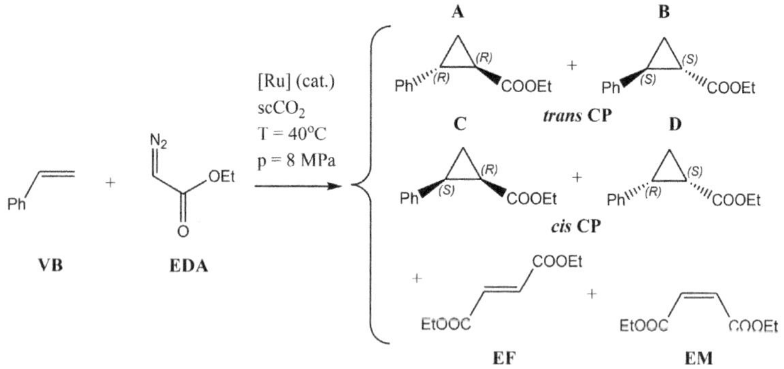

VB = vinyl benzene
EDA = ethyl diazoacetate
CP = cyclopropane
EF = ethyl fumarate
EM = ethyl maleate

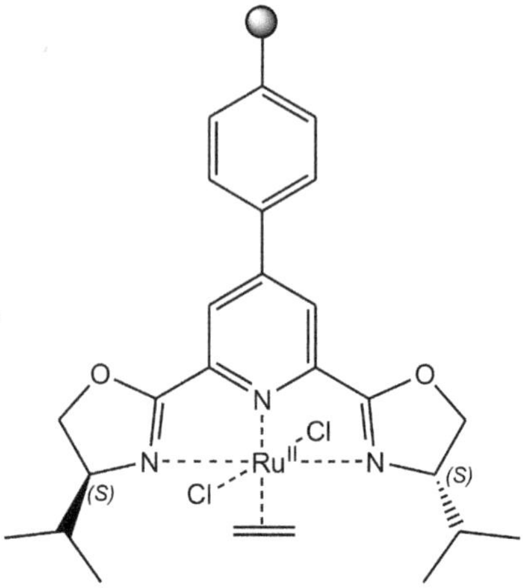

Reactor volume = 700 μL
0.00107 equivalents Ru per gram catalyst
0.4 g solid-supported catalyst
concentration of EDA in VB = 1.73 mol/L
mol ratio of VB to EDA = 4: 1
reaction time = 90 min
flow rate of organic substrates = 0.015 μL/min

flow rate of supercritical CO_2 = 0.15 μL/min
% conversion EDA = 100*(CP + EF + EM)/EDA = 91
% yield CP = 100*(CP/EDA) = 63
% chemoselectivity = 100*CP/(CP + EF + EM) = 82
trans CP/*cis* CP = (A + B)/(C + D) = 87/13
% *ee cis* CP = 100*(C – D)/(C + D) = 47
% *ee trans* CP = 100*(A – B)/(A + B) = 77

Part 1

From the authors' reported data, determine the % yields of each of the four cyclopropane stereoisomeric products.

Part 2

From the reported % chemoselectivity toward cyclopropane products, determine the number of moles of fumarate and maleate. Obtain the same quantity from the reported % conversion of EDA. Do the results agree?

Part 3

Using the *template-REACTION.xls* spreadsheet determine the material efficiency metrics for the cyclopropanation reaction. State any assumptions made in the calculations.

Part 4

Compare the material efficiency metrics with those found for the following catalytic reaction run under conventional non-flow reaction conditions [14].

Ph + N_2 OEt O [Cu] (cat.) - N_2 Ph COOEt [Cu] = MeOOC N H_2 Cl Cl Cu^{II} N N H_2 COOMe

250 mg ethyl diazoacetate dissolved in 250 mg styrene
10 mg copper catalyst dissolved in 250 mg styrene

225 mg of product is obtained in a *cis:trans* ratio of 1:2.2. The enantiomeric excess of each isomer was 20%. The reference stereoisomers on which the *ee* figures were obtained were not specified.

Part 5

Compare the material efficiency metrics with those found for the following non-catalytic thermal reaction run under conventional non-flow reaction conditions [15].

Ph + N_2 OEt O heat xylenes - N_2 Ph COOEt

179 g ethyl diazoacetate
163 g styrene
500 mL xylenes
155 g of product is obtained in a *cis:trans* ratio of 40: 65

Since no chiral catalyst is used we assume that both the *cis* and *trans* isomers are racemic.

Part 6 [16]

Ph–CH=CH$_2$ + H–C(=N$_2$)–C(=O)–OEt —[Rh] (cat.)→ *cis*: Ph (S), (R) COOEt; Ph (R), (S) COOEt — *trans*: Ph (S), (S) COOEt; Ph (R), (R) COOEt

Based on the following results, determine the % yield of each stereoisomer.

Yield of four stereoisomers with respect to ethyl diazoacetate = 39%
cis: trans ratio = 90: 10
% *ee cis* = 91 with respect to 1 S,2R isomer
% *ee trans* = 7 with respect to 1 S,2S isomer

PROBLEM 8.7 [17, 18]

Themes: balancing chemical equations, material efficiency metrics, chirality economy

Asymmetric hydrogenations of olefins are carried out with a rhodium(I) catalyst and a chiral monodendate BINOL-based monophosphonite ligand. In the example given below, when ligand L1 was used a product enantiomeric excess of 91.8% was obtained; whereas ligand L2 yielded a product *ee* of 93.3%. When both ligands were used simultaneously in a heterocombinatorial sense, the product *ee* increased further to 97.8%.

143 (MeOOC–C(=CH$_2$)–NH–C(=O)CH$_3$) —H_2, $[Rh^I(cod)_2]BF_4$ (cat.), L_1, L_2→ **145** (MeOOC–(S)CH(CH$_3$)–NH–C(=O)CH$_3$)

L_1: (R)-BINOL–P–CH$_3$ L_2: (R)-BINOL–P–C(CH$_3$)$_3$

cod = 1,5-cyclooctadiene
Reaction vessel = 50 mL
Homocombinations for ligands:

2 × 0.6 mL of 1.7 mM solution of L1 in dichloromethane (when L1 is used only)
2 × 0.6 mL of 1.7 mM solution of L2 in dichloromethane (when L2 is used only)

Heterocombination for ligands:

0.6 mL of 1.7 mM solution of L1 in dichloromethane
0.6 mL of 1.7 mM solution of L2 in dichloromethane

Catalyst:

0.5 mL of 2 mM solution of $[Rh(cod)_2]BF_4$ in dichloromethane
9 mL of 0.112 M solution of olefin substrate in dichloromethane
hydrogen: $T = 20°C$, $p = 1$ bar
% conversion of olefin substrate = 100%

Using the *template-REACTION.xls* spreadsheet and the given experimental conditions, determine the associated material efficiency metrics. In addition, determine the chirality economy (CE) given by

$$CE = \left(\frac{\%ee_{product}}{\%ee_{catalyst}}\right)\left(\frac{\%yield_{product}}{mol\,\%_{catalyst}}\right)$$

and the turnover number given by

$$TON = \frac{\text{Moles product}}{\text{Moles catalyst}}$$

PROBLEM 8.8 [19]

Themes: balancing chemical equations, non-unity stoichiometric equations

1000 kg of hexamethylenetetramine requires 3575 kg 37 wt% formaldehyde and 550 kg ammonia.

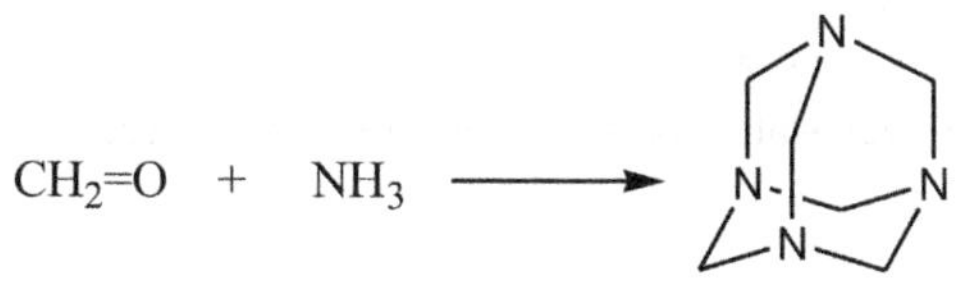

$$CH_2{=}O \;+\; NH_3 \longrightarrow$$

Part 1

Show a balanced chemical equation for the transformation and determine its atom economy.

Part 2

Which reagent is the limiting reagent in this reaction? From this determine the reaction yield.

Part 3

Using the *template-REACTION.xls* spreadsheet and the details of the experimental procedure, determine the material efficiency metrics for the reaction. State any assumptions made in the calculation.

PROBLEM 8.9 [20]

Themes: balancing chemical equations, catalyst recycling, side products

1 + 2 —(silica gel, toluene)→ 3 (66 %) + 4 (11 %)

Experimental Procedure:

A 100 mL three-necked flask was charged with silica gel (1 g) and heated at 180°C for 2 h under reduced pressure to remove traces of water. To the dried silica gel was added a solution of β,β'-dichloro sulfides **1** (0.5 mmol) in toluene (5 mL) and amine **2** (1.5 mmol). The mixture was then stirred at 100°C for 3 h. The reaction mixture was filtered on a Buchner funnel and the silica gel was washed with chloroform containing triethylamine (2 vol % $NEt_3/CHCl_3$, 20 mL). The combined filtrate was washed with an aqueous solution of NaOH (1 N, 20 mL) and the crude reaction products were extracted from the aqueous layer with chloroform (20 mL). The combined organic layer was dried over anhydrous sodium carbonate, the drying agent was removed by filtration, and the solvent was removed by evaporation. After purification by flash chromatography on silica gel eluting with ethyl acetate/hexane, **3** was obtained as a mixture of diastereomers.

Yield: 66% (**3**), 11% (**4**)

Catalyst recycling:

A 100 mL three-necked flask was charged with silica gel (1 g) that had been recovered from the former run and heated at 180°C for 2 h under reduced pressure to remove traces of water.

Yield of **3** after 1 cycle: 66%

Yield of **3** after 4 cycles at same scale and under same reaction conditions: 71%

Part 1

Write balanced chemical equations for the production of both products. Determine the respective atom economies.

Part 2

Using the *template-REACTION.xls* spreadsheet and the details of the experimental procedure determine the material efficiency metrics for producing product **3**. Repeat for product **4**. State any assumptions made in the calculations.

Part 3

Repeat Part 2 for product **3** using the yield results of the experiment for the 4th cycle.

PROBLEM 8.10 [21]

Themes: balancing chemical equations, side products, by-products, material efficiency metrics, synthesis plan analysis

A solvent-free Cannizzaro reaction was recently reported.

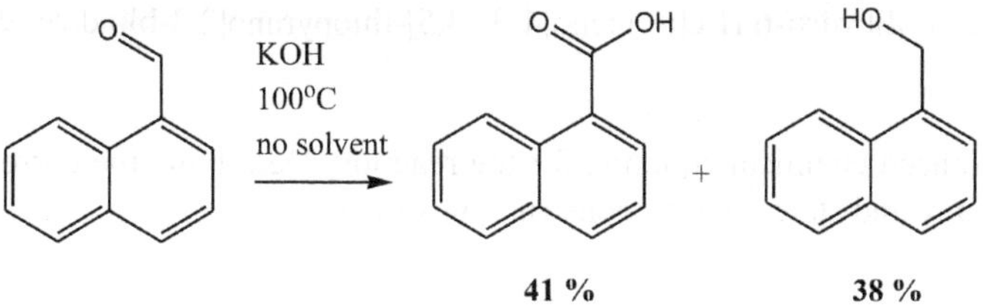

KOH
100°C
no solvent

41 % 38 %

Experimental Procedure:

After a mixture of powdered 1-formylnaphthalene (4.68 g, 30 mmol) and powdered KOH (2.52 g, 45 mmol) was kept at 100°C for 5 min, water was added to the reaction mixture, and was filtered to give 1-(hydroxymethyl) naphthalene, after Kugelrohr distillation, (0.485 g, 38% yield). The filtrate was acidified by conc. HCl and filtered to give 1-naphthoic acid (0.57 g, 41% yield).

Part 1

Write out balanced chemical equations for each product and determine the respective atom economies.

Part 2

Using the *template-REACTION.xls* spreadsheet and the details of the experimental procedure, determine the material efficiency metrics for producing the alcohol product, the acid product, and both products. State any assumptions made in the calculations.

PROBLEM 8.11 [22]

Themes: balancing chemical equations, ring construction strategies, mechanism, material efficiency metrics

ZnO (cat.)
CH_3CN
90 %

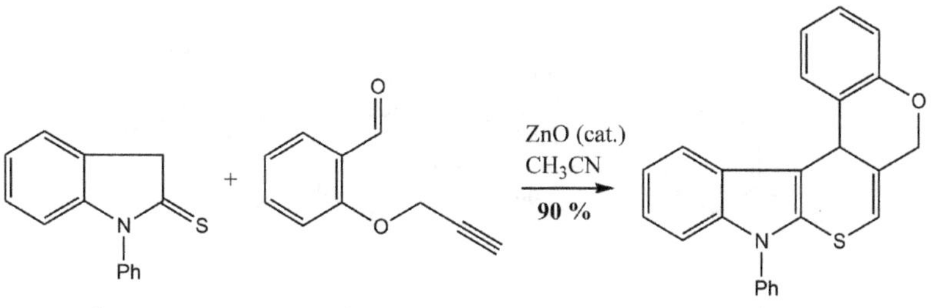

1 **2** **3**

Experimental Procedure:

To a stirred suspension of ZnO (10 mol %) in CH_3CN (10 ml) were added indolin-2-thione **1** (0.5 mmol) and O-propargylated salicylaldehyde derivative **2** (0.5 mmol). The reaction mixture was stirred at reflux for 3 h and the progress of the reaction was monitored by TLC. After completion of the reaction, the mixture was poured onto ice-cold water and stirred for 10 min, which resulted in the precipitation of the product **3**. The solid precipitate was filtered, dried, washed with petroleum ether to remove any residual starting material and after drying, recrystallized from ethanol. Yield of 9-Phenyl-9,13c-dihydro-6 H-chromeno[4′,3 ′:4,5]-thiopyrano[2,3-b]indole: 90%

Part 1

Write a balanced chemical equation for the reaction. Determine the corresponding atom economy and state the ring construction strategy.

Part 2

Write out a mechanism for the transformation.

Part 3

Using the *template-REACTION.xls* spreadsheet and the details of the experimental procedure, determine the material efficiency metrics for the reaction. State any assumptions made in the calculations.

PROBLEM 8.12 [23]

Themes: balancing chemical equations, redox reactions, material efficiency metrics

A microwave-assisted mono- and dibromination reaction of substituted toluenes using N-bromosuccinimide (NBS) was advertised as green based on the following criteria: shorter reaction times required compared with conventional heating, higher yields of products obtained, and recyclability of diethyl carbonate (DEC) solvent and succinimide by-product back to NBS. In the case of monobromination, two equivalents of NBS were used and for dibromination, three equivalents of NBS were used.

NBS (2 equiv.)
benzoyl peroxide
DMC
95 %

NBS (3 equiv.)
benzoyl peroxide
DMC
90 %

Experimental Procedures:

Monobromination

To an ethyl acetate (15 ml) or diethyl carbonate (10 ml) solution of 4-bromotoluene (1 g, 5.9 mmol), was added benzoyl peroxide (5 mg; 0.02 mmol, 0.34 mol %) and *N*-bromosuccinimide (1.89 g, 10.6 mmol, 1.8 equivalents). The reaction mixture was refluxed (30 min for ethyl acetate and 20 min for diethyl carbonate) under microwave heating (magnetron power of 400 W for ethyl acetate and 450 W diethyl carbonate). The clear reaction mixture was cooled at 5°C for 4–5 h. Formed white precipitate was separated by filtration and dried to give 990 mg (94%) for ethyl acetate and 1.0 g (95%) for diethyl carbonate of succinimide. The filtrate was then evaporated under vacuum to obtain a powdery material, which was purified by silica gel column chromatography with hexane–ethyl acetate (1: 4) as an eluent. The isolated yield of 1-bromo-4-bromomethylbenzene is 1.33 g (90%) for ethyl acetate and 1.40 g (95%) for diethyl carbonate as reaction media

Dibromination

The ethyl acetate (15 ml) or diethyl carbonate (10 ml) suspension of 4-bromotoluene (1 g, 5.9 mmol), benzoyl peroxide (5 mg; 0.02 mmol; 0.34 mol %) and *N*-bromosuccinimide (3.16 g, 17.7 mmol, 3 equivalents) was refluxed (30 min for ethyl acetate and 20 min for diethyl carbonate) under microwave heating (microwave magnetron power of 500 W for ethyl acetate and 600 W for diethyl carbonate). Clear reaction mixture was cooled at 5°C for 4–5 h. The formed white precipitate was separated by filtration and dried to give 1.61 g (92%) for ethyl acetate and 1.7 g (97%) for diethyl carbonate of succinimide. The filtrate was then evaporated under vacuum to a powdery material which was subjected to silica gel column chromatography with hexane–ethyl acetate (1:4) as an eluate. The isolated yield of ~96% pure product (HPLC) is 1.36 g (70%) for ethyl acetate and 1.75 g (90%) for diethyl carbonate as reaction media.

Part 1

For each transformation, write out a balanced chemical equation showing all by-products and determine its atom economy.

Part 2

The authors did not explain why they used an extra equivalent of NBS for each reaction unlike what the stoichiometry prescribes. Offer an explanation.

Part 3

Based on the details of the experimental procedures and using the *template-REACTION.xls* spreadsheet, determine the material efficiency metrics for the mono- and dibromination reactions. State any assumptions used in the calculations.

PROBLEM 8.13 [24]

Themes: balancing chemical equations, redox reactions, catalytic cycles

Benzil is prepared from diphenylacetylene via oxidation under microwave irradiation conditions in dimethylsulfoxide with iodine as catalyst. Experimental data for the yield of benzil with catalyst loading is shown in the table below.

Catalyst loading (mol %)	Yield of benzil (%)
10	19
20	21
30	64
40	74
50	95

Experimental Procedure:

To a 5 mL heat-resistant reaction vessel were added alkyne (0.2 mmol), I_2 (25 mg, 0.1 mmol), in DMSO (1 mL). The reaction vessel was then exposed to microwave irradiation until complete conversion. The reaction was cooled to room temperature. The crude product was extracted with ethyl acetate (3 × 10 mL). Organic layers were washed with 10% $Na_2S_2O_3$ solution (2 × 10 mL) and then saturated brine (1 × 10 mL), dried over $MgSO_4$, and filtered and concentrated under the reduced pressure. The crude mixture was then purified by column chromatography on silica gel.

Part 1

Show a balanced chemical equation for the transformation and determine its atom economy.

Part 2

Suggest a catalytic cycle and mechanism for the reaction.

Part 3

Using the *template-REACTION.xls* spreadsheet and the experimental data given, determine the mass of benzil, mass of iodine, and *PMI* as functions of catalyst loading. State any assumptions made in the calculations.

PROBLEM 8.14 [25, 26]

Themes: balancing chemical equations, redox reactions, material efficiency metrics
Asymmetrization of *meso* diol:

$KHSO_5$
$Na_2B_4O_7$
K_2CO_3
[$^nBu_4N$][HSO_4]
Shi catalyst

Ph, OH, (S), (R), Ph, OH
meso diol

Ph, O, (R), Ph, OH
80 % yield
87 % ee

Kinetic resolution of *rac*-diol:

$KHSO_5$
$Na_2B_4O_7$
K_2CO_3
$[^nBu_4N][HSO_4]$
Shi catalyst

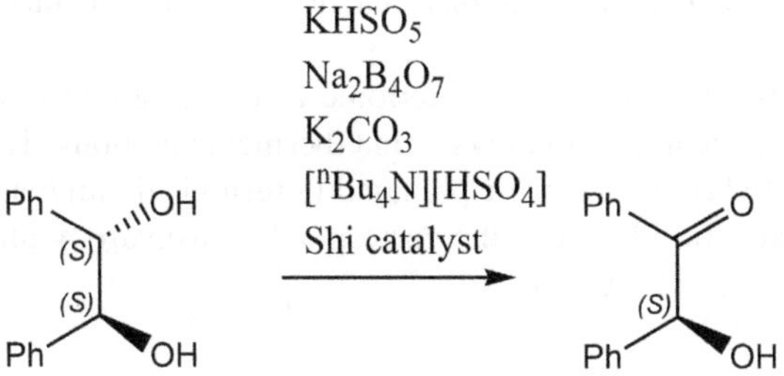

rac diol

48 % yield
87 % ee

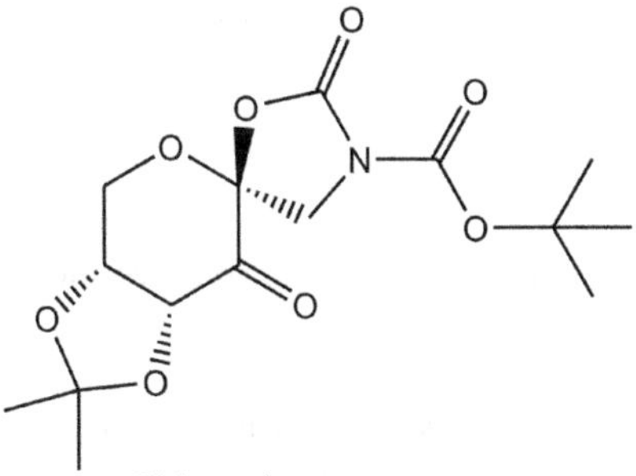

Shi catalyst

Experimental Procedure (asymmetrization of *m–eso* diols):

Diol (0.10 mmol), Shi catalyst (0.05 mmol, 50 mol %) and Bu_4NHSO_4 (4 mmol) in CH_3CN (1.5 mL) and $Na_2B_4O_7$ (0.5 mL)/K_2CO_3 buffer at 0–5°C.

Oxone (0.15 mmol, in 1.0 mL of 4×10^{-4} M aqueous solution of Na_2EDTA) and K_2CO_3 (0.63 mmol) were used.

Experimental Procedure (kinetic resolution of *rac* diols):

Diol (0.10 mmol), Shi catalyst (0.05 mmol, 50 mol %) and Bu_4NHSO_4 (4 mmol) in CH_3CN (1.5 mL) and $Na_2B_4O_7$ (0.5 mL)/K_2CO_3 buffer at 0–5°C.

Oxone (0.12 mmol, in 1.0 mL of 4×10^{-4} M aqueous solution of Na_2EDTA) and K_2CO_3 (0.58 mmol) were used.

Part 1

Write out balanced chemical equations for the asymmetrization of *meso*-diol and kinetic resolution of *rac*-diol.

Part 2

Using the *template-REACTION.xls* spreadsheet and the experimental details given, determine the material efficiency metrics for each process. State any assumptions made in the calculations.

PROBLEM 8.15 [27]

Themes: balancing chemical equations, reaction mechanism, materials efficiency metrics

Allyl tetronates were converted to tetronic acids via a domino Claisen–Conia ring-opening sequence under microwave and thermal conditions. The authors stated that the starting allyl tetronates were prepared in turn via domino addition/intramolecular Wittig alkenation between the corresponding α-hydroxy allyl ester and the phosphorus ylide $Ph_3P=C=C=O$.

microwaves
CH_3CN
150°C

MeOH

1 2 3 4

The experimental procedures for both reactions are shown below.

Reaction 1

Tetronate **1** (550 mg, 1.73 mmol) was placed in a glass vial with a magnetic stirrer and acetonitrile (5 mL). The vial was sealed and placed inside a CEM Discover single-mode microwave synthesizer where it was exposed to microwaves at 150°C for 1 h. The cap was then removed and the solvent evaporated *in vacuo*. The residue was purified by column chromatography (silica gel 60; diethyl ether/hexane, 1:1, v/v) to give 1-(2-Chloro-phenyl)-2-methyl-11-oxa-dispiro[2.1.5.2]dodecane-4,12-dione **3** (320 mg, 58%) and 3-[1-(2-Chloro-phenyl)-allyl]-4-hydroxy-1-oxa-spiro[4.5]dec-3-en-2-one **2** (230 mg, 42%).

Reaction 2

Spiro compound **3** (400 mg, 1.26 mmol) was dissolved in dry chloroform (20 mL) and methanol (5 mL). The solution was heated to reflux for 24 h after which the solvent was removed *in vacuo*. The residue was purified by column chromatography (silica gel 60; diethyl ether/hexane: 1:1, v/v) to give 3-[*syn*-2-(2-Chloro-phenyl)-2-methoxy-1-methyl-ethyl]-4-hydroxy-1-oxa-spiro[4.5]dec-3-en-2-one **4** as a white solid (278 mg, 63%).

Part 1

Show by means of a mechanism how the starting allyl tetronate **1** could be made via domino addition/intramolecular Wittig alkenation.

Part 2

Show how the products **2** and **3** arise from the starting allyl tetronate **1**.

Part 3

Using the *template-REACTION.xls* spreadsheet and the details of the experimental procedures determine the material metrics efficiencies for the production of compounds **2**, **3**, and **4**. State any assumptions made in the calculations.

PROBLEM 8.16 [28]

Themes: balancing chemical equations, material efficiency metrics

A demonstration of catalyst recycling and reuse using an example aldol reaction catalyzed by L-proline adsorbed onto an ionic liquid immobilized on silica gel was recently published. The synthesis of the silica bound ionic liquid is given below along with the details of the aldol reaction between acetone and benzaldehyde. The results show that the percent yield of aldol product and % *ee* are fairly constant over four cycles.

L-proline (30 mol%)
ionic liquid phase catalyst

cycle	% yield	% *ee*
1	57	74
2	53	74
3	55	74
4	56	74

Experimental Procedure:

A solution of aldehyde (0.5 mmol) in acetone (2 mL) was added to the ionic liquid silica bound system (520 mg, L-proline 30% mol). The mixture was stirred at room temperature for 24 h. After this time the modified silica gel was taken up with diethyl ether and filtered under reduced pressure. The solution was evaporated, checked by NMR and finally purified by chromatography (light petroleum/ethyl acetate) to give the aldol product. The catalytic system was dried for a few minutes and then re-used.

1. 300°C
2. MeO-Si(OMe)(OMe)-CH₂CH₂CH₂-SH
toluene

silica gel

NaBF$_4$
acetone

AIBN/CH_2Cl_2

Part 1

Using the *template-REACTION.xls* spreadsheet and the details of the experimental procedure, determine the material efficiency metrics for production of (R)-aldol product for each cycle. State any assumptions made in the calculations.

Part 2

The authors did not disclose reaction yields for the preparation of the immobilized ionic liquid. Based on the synthesis given determine its overall atom economy.

PROBLEM 8.17 [29]

Themes: balancing chemical equations, catalytic cycle, mechanism, organocatalysis

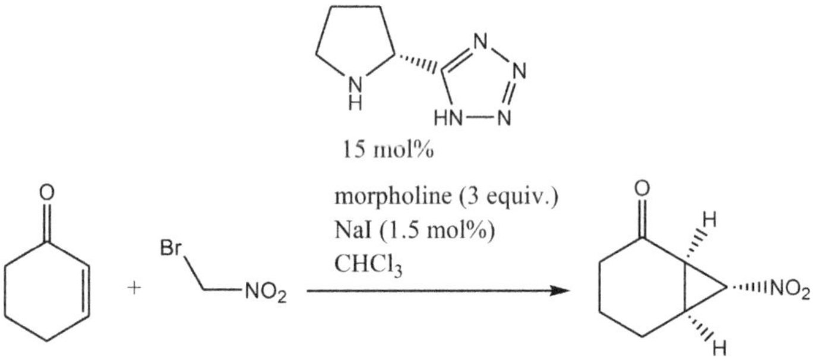

15 mol%

morpholine (3 equiv.)
NaI (1.5 mol%)
$CHCl_3$

Experimental Procedure:

To a stirred suspension of the enone starting material (0.5 mmol) and pyrrolidinyl tetrazole (15 mol%) in $CHCl_3$ (2 mL) were added NaI (1.5 mol%), morpholine (1.5 mmol) and bromonitromethane (1.0 mmol) at room temperature during the indicated time. The mixture was diluted with CH_2Cl_2 (10 mL) and washed with H_2O (10 mL). The aqueous phase was extracted with CH_2Cl_2 (3 × 10 mL). The combined organic phases were dried ($MgSO_4$), concentrated *in vacuo*, and the residue was purified by flash chromatography. Yield: 87%; 90% ee.

Part 1

Show a balanced chemical equation and determine the atom economy.

Part 2

Show a catalytic cycle and mechanism for the reaction.

Part 3

Using the *template-REACTION.xls* spreadsheet and the details of the experimental procedure, determine the material efficiency metrics for the reaction. State any assumptions made in the calculations.

PROBLEM 8.18 [30]

Themes: balancing chemical equations, material efficiency metrics

Microwave irradiation was demonstrated to effect Newman–Kwart rearrangements of O-thiocarbamates to S-thiocarbamates for substrates that are either electronically or sterically difficult to convert. Typically, such substrates require harsh conditions or specialized equipment such as heating oil baths, salt baths, autoclaves, or flash vacuum pyrolysis. The advantages of microwave irradiation in this case include rapid heating, a safer method to access high reaction temperatures, and a cleaner product profile.

DABCO, NMP, 77 % → microwaves, 300 C, DMA, 78 %

DABCO = 1,4-diazabicyclo[2.2.2]octane
NMP = N-methylpyrrolidone
DMA = dimethylacetamide

Experimental Procedure:

Reaction 1

2,6-Dimethoxyphenol (11.76 g, 75 mmol) and DABCO (11.30 g, 97.5 mmol, 1.3 equiv) were heated in NMP (60 mL) to 50°C with mechanical stirring to give a dark brown solution. Dimethyl thiocarbamoyl chloride (10.75 g, 82.5 mmol, 1.1 equiv) was dissolved in NMP (15 mL) and added dropwise to the previous solution over 18 min. Some fine precipitate formed in the dark red solution during this addition. The reaction was monitored by LC and was complete within 90 min at 50°C. Water (140 mL) was added over 15 min at 50°C. The original solid dissolved readily, but a yellow precipitate formed later in the addition, which persisted to the end. The reaction mixture was cooled smoothly to 20°C and the precipitate isolated by filtration. The product cake was slurry washed twice with water (24 mL each) and dried *in vacuo* at 50°C to yield the title compound as a fine, off-white crystalline solid (14.25 g, 77% yield).

Reaction 2

2,6-Dimethoxyphenyl-O-thiocarbamate (500 mg) was dissolved in DMA (2.0 mL) and heated at 300 W in a microwave tube to 300°C for 20 min. After compressed air cooling to room temperature, the dark reaction mixture was diluted with water (6 mL), from which a precipitate formed after about 5 min. The solid was isolated by filtration, washed with water (2×6 mL, then 2×3 mL) and dried *in vacuo* at 50°C to give the title compound as a white solid (390 mg, 78%).

Part 1

Show balanced chemical equations for both steps and determine the respective atom economies.

Part 2

Using the *template-REACTION.xls* spreadsheet and the details of the experimental procedures determine the material efficiency metrics for each step and for the overall synthesis. State any assumptions used in the calculations.

Part 3

Suggest a mechanism for the Newman–Kwart rearrangement.

PROBLEM 8.19 [31]

Themes: balancing chemical equations, catalytic cycles, mechanism, material efficiency metrics

(±) 2 → HCOOH, $Pd(OAc)_2$ (cat.), (S)-tBu-PHOX (ligand) → 1 (S)

1
88 % yield
94 % ee

(S)-^{t}Bu-PHOX

Experimental Procedure:

One hundred and eighty mg of oven-dried 4 Å molecular sieves (4 ÅMS) were placed in a 1 dram glass vial equipped with a magnetic stir bar, a screw cap, and a septum. The vial and 4 ÅMS were thoroughly flame dried under vacuum and backfilled with dry argon gas. The flame drying procedure was carried out twice more, and then the vial cooled to ambient temperature (20°C). To the cooled vial was added $Pd(OAc)_2$ (2.2 mg, 0.010 mmol, 0.10 equiv, 10 mol%), (*S*)-*t*-Bu-PHOX (4.8 mg, 0.0125 mmol, 0.125 equiv, 12.5 mol%), and freshly distilled *p*-dioxane (1.5 mL). The mixture was heated to 40°C for 30 min, at which point, neat HCO_2H was added (6 equiv.), followed immediately by addition of a solution of (±)-**2** (24.4 mg, 0.10 mmol, 1.0 equiv) in *p*-dioxane (1.5 mL). The reaction mixture was stirred at 40°C until TLC indicated complete consumption of (±)-**2** (about 10 h). After cooling to ambient temperature, the reaction mixture was filtered through a pad of celite. The filtrate was concentrated by rotary evaporation and the residue purified by flash chromatography on SiO_2 using 10% Et_2O in pentane as eluent. The ee of the product was determined by chiral HPLC with a Chiracel OD-H column using 1% 2-propanol in hexanes as eluent.

Part 1

Write out a balanced chemical equation for the reaction. Determine the atom economy, reaction yield for the (S) enantiomeric product, and the kernel reaction mass efficiency.

Part 2

Write out a catalytic cycle for the reaction mechanism.

Part 3

Using the *template-REACTION.xls* spreadsheet and the details of the experimental procedure determine the mass efficiency metrics for the reaction. State any assumptions made in the calculations.

PROBLEM 8.20

Themes: balancing chemical equations, process mass intensity, "solvent-free" claim

Two procedures for the synthesis of a benzodiazepine are advertised as "solvent-free." Use the *template-REACTION.xls* spreadsheet to evaluate the overall *PMI* and to determine the E-factor profile with respect to by-products, reaction solvent, catalyst, excess reagent, work-up, and purification.

Procedure #1: [32]

97 %

NH2 NH2 + 2 O; HOAc (cat.) microwaves solvent-free; - 2 H_2O; H N N

Materials used: 1 mmol *o*-phenylenediamine, 2.1 mmol acetone, 10 mol % acetic acid, undisclosed amount of n-hexane in chromatography

Procedure #2: [33]

93 %

NH2 NH2 + 2 O; $RuCl_3$ x H_2O (cat.); - 2 H_2O; H N N

Materials used: 1 mmol *o*-phenylenediamine, 2.5 mmol acetone, 5 mol % $RuCl_3 \times H_2O$, 20 mL EtOAc (d=0.901 g/mL), 45 mL saturated sodium bicarbonate solution (d=1.0581 g/mL), 45 mL saturated sodium chloride solution (d=1.1804 g/mL), undisclosed amount of sodium sulfate, and undisclosed amount of ethyl acetate:hexane for chromatography.

PROBLEM 8.21 [34]

Themes: balancing chemical equations, material efficiency metrics, ring construction strategies

OH + NC CN NO2 F; O N— EtOH; NH2 O CN NO2 F

Experimental Procedure:

A stirred mixture of 1-naphthol (1.44 g, 10 mmol) and 4-fluoro-3-nitrobenzylidenemalonitirle (2.17 g, 10 mmol) in ethanol (30 mL) was treated with 4-methylmorpholine (1.01 g, 10 mmol). A slight red color appeared. The mixture was warmed to reflux, yielding a red solution. Copious quantities of a pink solid precipitated out within 5 min of attaining reflux. The mixture was cooled to room temperature and the solid filtered off, washed with ethanol and ether, and dried *in vacuo* providing the product as a flesh-colored powder (2.49 g, 69%).

Part 1

Write out a mechanism for the reaction.

Part 2

Write a balanced chemical equation for the reaction and determine its atom economy.

Part 3

Using the *template-REACTION.xls* spreadsheet and details of the experimental procedure determine the material efficiency for the reaction. State any assumptions made in the calculations.

Determine the mol% of N-methylmorpholine used.

Part 4

Describe the ring construction strategy used by highlighting the target bonds in the product structure.

PROBLEM 8.22 [35]

Themes: balancing chemical equations, redox reactions, oxidation number analysis, mechanism

OMe
EtOOC
COOEt
N
H
1

$NaNO_2$
wet silica/
sulfuric acid
CH_2Cl_2

OMe
EtOOC
COOEt
N
2

Experimental Procedure:

A suspension of compound **1** (0.358 g, 1 mmol), silica modified sulfuric acid (0.3 g), wet SiO_2 (50% *w/w*, 0.2 g) and $NaNO_2$ (0.207 g, 3 mmol) in CH_2Cl_2 (6 mL) was stirred at room temperature for 20 min (the progress of the reaction was monitored by TLC) and then filtered. The residue was washed with CH_2Cl_2 (20 mL). Anhydrous

Na_2SO_4 (5 g) was added to the filtrate and filtered after off 20 min. Dichloromethane was removed. The yield was 0.352 g, (99%) of crystalline pale yellow solid (**2**), melting point 56–58°C.

Part 1

Write a mechanism for the reaction.

Part 2

Write a balanced chemical equation for the reaction.

Part 3

Write out the redox couple for the reaction. Indicate which atoms undergo a change in oxidation number that is consistent with the redox couple.

Part 4

Using the *template-REACTION.xls* spreadsheet and details of the experimental procedure determine the material efficiency for the reaction. State any assumptions made in the calculations.

PROBLEM 8.23 [36]

Themes: balancing chemical equations, reaction mechanism, catalytic cycles, redox couples, experimental procedure, material efficiency metrics

O OH OH
+
OMe OMe
$Rh^{III}(acac)_3$ (cat.)
DABCO
KOH
toluene

A pressure tube was charged with a mixture of acetophenone (1 mmol, 0.5 M in toluene) and (4-methoxyphenyl)methanol (1 mmol, 0.5 M in toluene), DABCO (50 mol%), KOH (1 mmol) and tris(acetylacetonato)-rhodium(III) (1.0 mol%). The reactants were stirred in 2 mL of toluene at 110°C for 4 h. The progress of the reaction was monitored by TLC. After the completion of the reaction, it was quenched using brine solution and extracted with ethyl acetate (3 × 50 mL). The combined organic extracts were dried over anhydrous sodium sulfate and filtered. Then, the solvent was evaporated under reduced pressure to give the crude product which was subjected to silica gel column chromatography using hexanes/ethyl acetate mixture as eluent to afford 3-(4-methoxyphenyl)-1-phenylpropan-1-ol; yield: 98%.

Part 1

Write out a mechanism for the transformation showing catalytic cycles.

Part 2

Write out the overall balanced chemical equation for the transformation.

Part 3

Write out the redox couple.

Part 4

Using the *template-REACTION.xls* spreadsheet and the details of the experimental procedure, determine the overall material efficiency metrics. State any assumptions made in the calculations.

PROBLEM 8.24 [36]

Themes: balancing chemical equations, reaction mechanism, catalytic cycles, redox couples, experimental procedure, material efficiency metrics

OH OH + OMe

$Rh^{III}(acac)_3$ (cat.)
DABCO
KOH
toluene

OH OMe

A pressure tube was charged with a mixture of 1-phenyl-ethanol (1.0 mmol, 0.5 M in toluene), (4-methoxyphenyl)methanol (1.0 mmol, 0.5 M in toluene), DABCO (50 mol%), KOH (1.0 mmol), and tris(acetylacetonato)-rhodium(III) (1.0 mol%). The reactants were stirred in 2 mL of toluene at 105–110°C for 12–16 h. The progress of the reaction was monitored by TLC analysis. After the completion of reaction, it was quenched using brine solution, and extracted with ethyl acetate (3 × 50 mL). The combined organic extracts were dried over anhydrous sodium sulfate and filtered. Then, the solvent was evaporated under reduced pressure to give the crude product which was subjected to silica gel column chromatography using hexanes/ethyl acetate mixture as eluent to afford 3-(4-methoxyphenyl)-1-phenylpropan-1-ol; yield: 98%.

Part 1

Write out a mechanism for the transformation showing catalytic cycles.

Part 2

Write out the overall balanced chemical equation for the transformation.

Part 3

Write out the redox couple.

Part 4

Using the *template-REACTION.xls* spreadsheet and the details of the experimental procedure, determine the overall material efficiency metrics. State any assumptions made in the calculations.

PROBLEM 8.25 [37]

Themes: balancing chemical equations, experimental procedures, material efficiency metrics

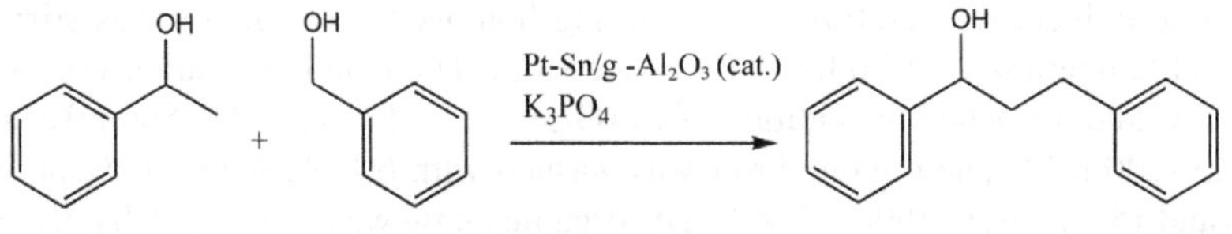

Under nitrogen atmosphere, to a 15-mL Pyrex glass screw-cap tube were added 1-phenylethanol (0.613 g, 5 mmol), benzyl alcohol (0.543 g, 5 mmol), K_3PO_4 (531 mg, 2.5 mmol), and Pt–Sn/γ-Al_2O_3 (catalyst: 147 mg, 0.075 mol% Pt). The resultant mixture was stirred in the sealed tube at 15°C for 48 h. After cooled to ambient temperature, the catalyst and base were removed by centrifugation and washed with Et_2O (2 X 5 mL). The combined supernatant was condensed under reduced pressure and subject to purification by silica gel column chromatography (eluent: petroleum ether (60–90°C)/EtOAc = 20:1, v/v), affording the product as a white solid (1.018 g, 96%).

Part 1

Write out the overall balanced chemical equation for the transformation.

Part 2

Write out the redox couple.

Part 3

Using the *template-REACTION.xls* spreadsheet and the details of the experimental procedure, determine the overall material efficiency metrics. State any assumptions made in the calculations.

PROBLEM 8.26 [38]

Themes: balancing chemical equations, organocatalysis, by-products, mechanism, material efficiency metrics

The following Michael reaction was carried out on an industrial scale at Merck as part of a synthesis plan to manufacture telcagepant, a CGRP (calcitonin gene-related peptide) receptor for antagonist for the treatment of migraine. The reaction employs a proline derived organocatalyst in order to carry out the stereoselective reaction.

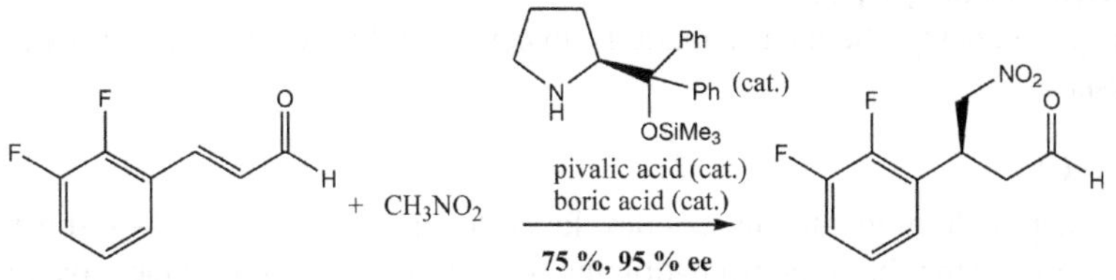

Experimental Procedure:

A concentrated solution of aldehyde **1** prepared as above (1.189 mol assay, 200 g) was diluted with THF to a volume of 2.8 L. Water (400 mL), pivalic acid (6.08 g, 0.0594 mol), boric acid (36.78 g, 0.594 mol), TMS-prolinol (19.36 g assay, 0.0594 mol), about 250 mL solution in THF/MTBE, and nitromethane (436 g, 7.143 mol) were added at ambient temperature. The resulting homogeneous solution was stirred at ambient temperature for 30 h (>94% conversion). The reaction solution was cooled to 2–7°C and quenched by addition of i-PrOAc (2 L), 15% aq NaCl (600 mL), and 1 N HCl (200 mL). The organic layer was washed with 6% $NaHCO_3$ in 5% aq NaCl (1 L) and 15% aq NaCl (600 mL × 2). The organic phase was azeotropically dried and solvent-switched to i-PrOAc while maintaining the internal temperature at < 15°C. The final solution (about 2 L, about 10 wt% of **2**) was directly used in the subsequent step or could be stored at 5°C; 73–75% assay yield of **2**; about 95% ee.

The following addition products were identified:

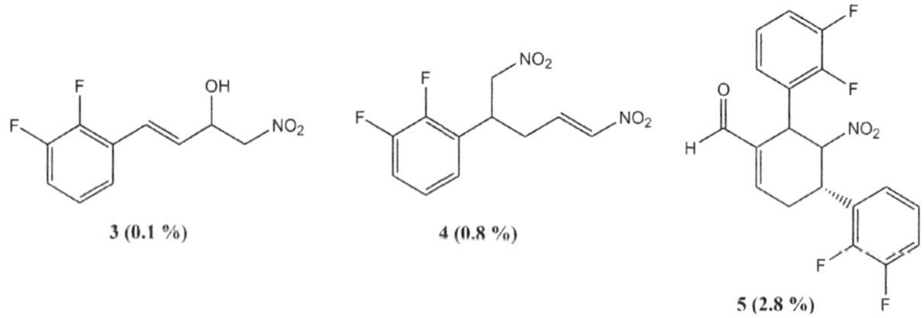
F
F
OH
NO2
F
F
NO2
NO2
F
F
O
H
NO2
F
F

3 (0.1 %) **4** (0.8 %) **5** (2.8 %)

Part 1

Write out the catalytic cycle that leads to the stereoselective product **2** and determine the resulting overall balanced chemical equation.

Part 2

Show by means of mechanisms how the additional products are formed. For each case write out balanced chemical equations for the reactions.

Part 3

Using the *template-REACTION.xls* spreadsheet and the details of the experimental section determine the material efficiency metrics for the production of each product formed in the reaction. State any assumptions made in the calculations.

PROBLEM 8.27 [39, 40]

Themes: balancing chemical equations, experimental procedure, materials metrics analysis

Method 1

A two-step undergraduate lab was developed to showcase the Wolff–Kishner reaction without using the standard hydrazine reagent in an effort to reduce toxicity and safety hazards associated with this reagent.

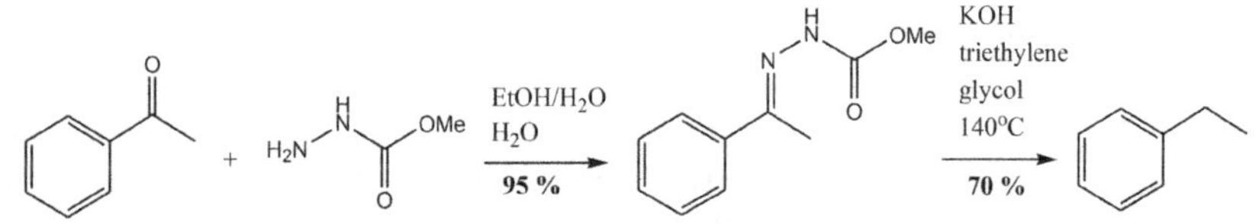
O
H2N
H
N
OMe
O
EtOH/H2O
H2O
95 %
H
N
N
OMe
O
KOH
triethylene
glycol
140°C
70 %

Step 1

Place acetophenone (8.00 g, 66.6 mmol) into a 250 mL round-bottomed flask with a large stir bar then add ethanol (25 mL). Next, add enough water such that the solution becomes cloudy (approximately 30 mL) and, once it is cloudy, add a few drops of ethanol until the solution becomes clear again. Then add methyl hydrazinocarboxylate (7.80 g, 86.6 mmol) followed by glacial acetic acid (1 mL). Place a reflux condenser on the reaction flask and heat the reaction under reflux for 15 min, then allow it to cool to room temperature. Once cool, place the flask on ice for 5 min then collect the product by Buchner filtration. Wash the filter cake with ice-cold 95% ethanol (20 mL) and allow to air dry for up to 30 min. Approximately 12.15 g, (95%) of hydrazone product is collected.

Step 2

Into a 250 mL round-bottomed flask, equipped with a large stir bar, place triethylene glycol (40 mL) followed by KOH (14.0 g, 250 mmol). Place a reflux condenser on the flask and heat the reaction, with vigorous stirring, to approximately 100°C to dissolve the KOH. Once dissolved, solution colors range from orange to almost black. Once the KOH has dissolved (takes up to 30 min), add the intermediate hydrazone (8.0 g, 41.6 mmol) using a powder funnel. The reaction effervesces strongly after initial addition. If it effervesces too much, stir the reaction mixture vigorously and remove it from the heat. Raise the temperature of the reaction to 140°C and continue to heat the reaction for 3 h.

Allow the reaction to cool to room temperature then add water (100 mL) and pour into a separating funnel. Add diethyl ether (20 mL) then extract the organic product into the ether layer. Re-extract the aqueous layer with a further portion of diethyl ether (20 mL) and then combine the organic layers. Dry the organic phase with $MgSO_4$ and then filter into a 250 mL round-bottomed flask. At this stage, add acetone (1 mL) to your flask then remove the majority of the solvent on the rotary evaporator. Leave approximately 7 mL of solvent in the flask. Pass the solvent through a small plug of silica gel encased in a syringe barrel and collect the washings into a pre-weighted 50 mL round-bottomed flask. Rinse the silica with pentane (20 mL). The excess solvent can be removed using a rotary evaporator without heating the water bath. A yellow oil is obtained in 70% yield. The authors report that about 5% of the ethyl benzene oil is an azine product, which is responsible for the yellow coloration.

Method 2

A second method to prepare ethyl benzene using neat hydrazine in propylene glycol and heated using a solar reflector on a sunny summer day was also reported.

KOH / NH_2NH_2
propylene
glycol
solar heating

20 %

A solution of propylene glycol (24.51 mL) and KOH (82.7 mmol (2.5 eq), 4.632 g) was heated to make the basic solvent solution needed for the reaction. After the solution had cooled, acetophenone (29.6 mmol (1 eq), 3.556 g) and neat hydrazine (127 mmol (4 eq), 4.070 g) were added and the solar reflector was moved back into position in which the focal point was directed at the bottom of the round bottom flask. Reflux was initiated for a period of 3 h at a temperature of 136–140°C. After acid work-up, pure ethyl benzene (5.83 mmol, 0.619 g, 19.7%) was collected through vacuum distillation at 4 in Hg at a temperature range of 71–73°C.

Part 1

For each method, write out balanced chemical equations.

Part 2

Show mechanistically how an azine product can arise in the second step of the undergraduate experiment.

Part 3

Using the *template-REACTION.xls* spreadsheet determine the material efficiencies of both methods and compare the results. State any assumptions made in the calculations.

PROBLEM 8.28

Themes: balancing chemical equations, experimental procedure, materials metrics analysis

Conventional Heating Method [41]

O, O, N, H; NH_2NH_2, 95 %; N—NH_2, O, N, H; NaOEt, EtOH, 73 %; O, N, H

Step 1

Isatin (25.0 g, 0.17 mol) is suspended in 200 mL of anhydrous methanol and 30.0 g of hydrazine hydrate (55% hydrazine content, 0.425 mol) is added in one portion. The solution is heated to a gentle reflux for 1 h. The solution is then cooled in an ice bath and the yellow crystals are filtered after half an hour. Yield: 26.0 g, 95% yield; melting point 215–219°C.

Step 2

While the above crystals are left on a watch glass to air dry, 2 g of sodium metal is dissolved in 50 mL of absolute ethanol in a 100 mL round-bottomed flask. When the sodium has completely dissolved, 5.0 g of the above hydrazone derivative is added in small portions, with shaking at 60°C to 70°C, over a 10 min interval. The solution is heated to reflux until the evolution of nitrogen gas has ceased (about 30 min).

The brown solution is then carefully poured on ice and acidified to pH = 1 with 10% HC1 solution. The mixture is extracted with 2 × 25 mL of ether, dried over calcium chloride, filtered, and concentrated to yield a crude, light-orange residue. The mass is recrystallized from 50 mL of water (a small amount of charcoal is added) to yield 3.0 g of oxindole (73% yield) as white needles (melting point 125–127°C).

Microwave Irradiation Method [42]

NH_2NH_2 ethylene glycol microwaves → 82 %; N—NH_2; KOH ethylene glycol microwaves → 32 %

Step 1

Isatin (0.25 g, 1.7 mmol), 55% hydrazine (0.30 g, 0.425 mmol) and ethylene glycol (1 mL) were added to a 50-mL beaker. The mixture was shaken gently to ensure proper mixing. The beaker was then covered with a watch glass and irradiated in the microwave oven at medium power for 30 s. After the beaker was removed from the oven and cooled to room temperature, the mixture was further cooled in an ice bath for 5 min. The yellow powders were collected in a suction flask, washed with cold ethanol (2 × 0.5 mL), and air-dried. Yield: 0.223 g (81.5%); mp: 219–221°C.

Step 2

A 50-mL beaker containing 0.5 mL of ethylene glycol and potassium hydroxide (62 mg, 1.1 mmol) was irradiated in the microwave oven for 10 s to dissolve the base. Isatin 3-hydrazone (58.5 mg, 0.36 mmol) was then added to the beaker and irradiated in the microwave oven for 10 s. The beaker was removed from the oven and cooled to room temperature. The brown solution was then diluted with 1 mL of deionized water, acidified with 6 M HCl until pH = 2, and extracted with diethyl ether (3 × 1.5 mL). The ether solution was dried with anhydrous sodium sulfate and evaporated in a hood to give a yellow solid. The solid was recrystallized from 0.7 mL of deionized water to yield 15.5 mg (32.4%) of oxindole as white needles, mp: 125–128°C.

Part 1

For each method write out balanced chemical equations.

Part 2

Using the *template-REACTION.xls* spreadsheet determine the material efficiencies of both methods and compare the results.

PROBLEM 8.29 [43]

Themes: balancing chemical equations, biocatalysis, catalytic cycles, redox reactions 1, ω-diols can be biocatalytically converted to 1, ω-diamines as shown by the example below. The process occurs sequentially starting from the diol that is oxidized to a monoaldehyde that reacts with ammonia to produce an imine which in turn gets reduced to the amine. This sequence is repeated on the other hydroxyl group. The authors described their work as the "first bioamination of primary alcohols employing a synthetic redox-self-sufficient cascade, that represents an environmentally benign concept yielding valuable building blocks for the polymer industry, otherwise only accessible through routes demanding harsh conditions."

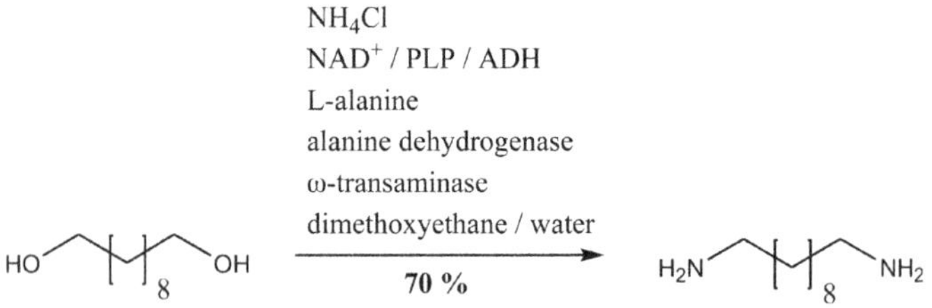

NH_4Cl
NAD^+ / PLP / ADH
L-alanine
alanine dehydrogenase
ω-transaminase
dimethoxyethane / water
70 %

HO–(…)$_8$–OH → H_2N–(…)$_8$–NH_2

ADH = alcohol dehydrogenase
NAD+ = nicotinamide adenine dinucleotide
PLP = pyridoxal 5'-phosphate

Experimental Procedure:

1,10-Decanediol (8.7 mg, 0.05 mmol) was dissolved in DME (0.1 mL, DME = 1,2-dimethoxyethane) in an Eppendorf tube (2 mL) and L-alanine (22.8 mg, 0.25 mmol) dissolved in 0.2 mL water was added as well as NH_4Cl (16 mg, 0.275 mmol in 0.075 mL water). The pH was adjusted to pH 8.5 by addition of 80 mL of a 6 M NaOH solution. NAD+ (0.5 mg, 0.0006 mmol) and PLP (0.1 mg, 0.0004 mmol; PLP = pyridoxal 5′-phosphate), each dissolved in 0.025 mL of water, were added as well as additional 0.585 mL of water, the ADH (alcohol dehydrogenase, 0.2 mL, 0.25 U), the ω-transaminase (0.2 mL, 0.2 U), and the alanine dehydrogenase (0.015 mL, 0.25 U). After shaking at 20°C, 300 rpm (orbital shaker, Infors Unitron) for 24 h the product was separated from the reaction mixture and purified through a weak acid cation exchanger. Finally, 6 mg (0.035 mmol, 70% yield) of 1,10-diaminodecane were obtained.

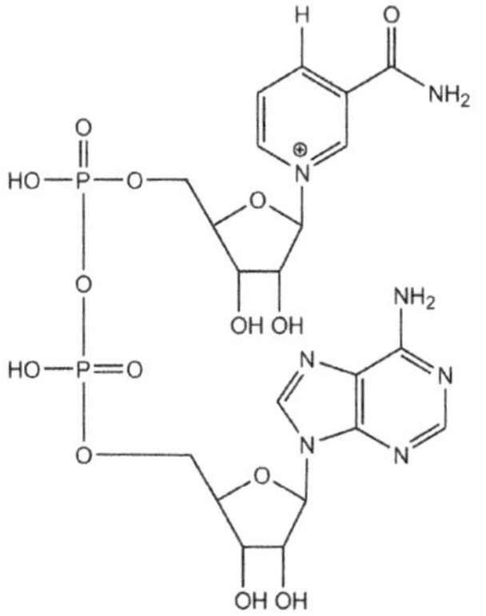

NAD^+
nicotinamide adenine dinucleotide

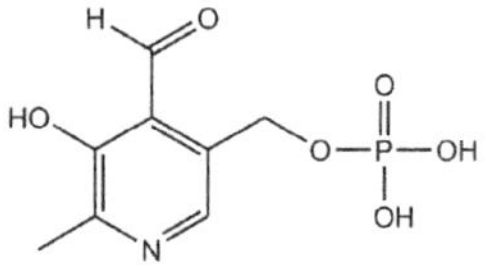

PLP
pyridoxal 5'-phosphate

Part 1

Write out the sequence of steps for the conversion of the diol to the diamine product.

Part 2

Write out a balanced chemical equation for the transformation and determine its atom economy.

Part 3

Using the *template-REACTION.xls* spreadsheet and the details of the experimental procedure determine material efficiency metrics for the transformation. State any assumptions made in the calculations.

PROBLEM 8.30 [44]

Themes: balancing chemical equations, catalytic cycles, mechanism, material efficiency metrics

Formylfurans and methylfurans are prepared via a coupling reaction between an alkyne and propargyl alcohol to give an enol ether intermediate which then cyclizes to the respective products depending on the catalyst used.

Reaction 1

COOEt EtOOC 1 HO 2 — DABCO (cat.) → [EtOOC EtOOC O] — O_2 CuI (cat.) → EtOOC EtOOC O O 3

Experimental Procedure:

Diethyl acetylenedicarboxylate (**1**, 0.5 mmol), prop-2-yn-1-ol (**2**, 0.5 mmol), and DABCO (0.05 mmol) in CH_2Cl_2 were stirred at room temperature for 10 min. The soln was evaporated to dryness under reduced pressure. Subsequently, 2.5% mmol CuI and DMF were added at 80°C. After completion of the reaction (TLC monitoring), the soln was evaporated to dryness under reduced pressure and then H_2O (8 mL) was added. The aq soln was extracted with Et_2O (3×8 mL) and the combined extracts were dried (anhyd $MgSO_4$). The solvent was removed and the crude product, diethyl 5-formylfuran-2,3-dicarboxylate, was separated by column chromatography to give a pure sample of **3**. Yield: 78%.

Reaction 2

COOEt EtOOC 170 HO 56 — DABCO (cat.) → [EtOOC EtOOC O] — PPh_3 (ligand) AgOAc (cat.) → EtOOC EtOOC O 226

Experimental Procedure:

Diethyl acetylenedicarboxylate (**1**, 0.5 mmol), prop-2-yn-1-ol (**2**, 0.5 mmol), and DABCO (0.05 mmol) in CH_2Cl_2 were stirred at room temperature for 10 min. The soln was evaporated to dryness under reduced pressure. Subsequently, AgOAc (5 mol%)/Ph_3P (10 mol%) and toluene were added at 50°C. After completion of the reaction (TLC monitoring), the soln was evaporated to dryness under reduced pressure and then H_2O (8 mL) was added. The aq soln was extracted with Et_2O (3×8 mL) and the combined extracts were dried (anhyd $MgSO_4$). The solvent was removed and the crude product, diethyl 5-methylfuran-2,3-dicarboxylate, was separated by column chromatography to give a pure sample of **4**. Yield: 79%.

Part 1

Write out balanced chemical equations for each reaction and determine the atom economies. Determine the ring construction strategy used.

Part 2

Write out catalytic cycles and mechanisms for each transformation.

Part 3

Using the *template-REACTION.xls* spreadsheet and the details of the experimental procedures, determine the material efficiency metrics for each reaction. State any assumptions used in the calculations.

PROBLEM 8.31 [45]

Themes: balancing chemical equations, ring construction strategies, catalytic cycles, material efficiency metrics

Experimental Procedure:

The aqueous solution of hydroxylamine hydrochloride (1.2 mmol) and aldehyde (1 mmol) were stirred at room temperature for 30 min. After complete conversion of aldehyde to oxime, N-chlorosuccinamide (1.3 mmol) was added to the reaction mixture and was allowed to stir for 3 h. The montmorillonite clay-Cu(II)/NaN_3 mixture (prepared by stirring 15 mol % clay-Cu catalyst and 7.5 mol % NaN_3 in water until the color changes from brown to black) and phenyl acetylene (1.3 mmol) was added and the reaction mixture was further stirred for another 3 h. After completion of reaction, the reaction mixture was filtered through Whatman filter paper, residue was washed with EtOAc. Organic layer was separated from filtrate and was dried over anhydrous sodium sulfate. Combined organic layers were concentrated *in vacuo* and crude reaction mixture was purified by silica gel (#100–200) column chromatography using EtOAc: hexane as eluting solvent to get corresponding 3,5-disubstituted isoxazoles. Yield: 68%; yellow solid; melting point 190–193°C.

Part 1

Write a balanced chemical equation for the reaction. Determine the corresponding atom economy and state the ring construction strategy.

Part 2

Write out a mechanism for the transformation showing a catalytic cycle. What is the purpose of sodium azide? Explain the regioselectivity of isoxazole product formed upon addition to phenylacetylene.

Part 3

Using the *template-REACTION.xls* spreadsheet and the details of the experimental procedure determine the material efficiency metrics for the reaction. State any assumptions made in the calculations.

PROBLEM 8.32 [46]

Themes: balancing chemical equations, redox reaction, sacrificial reagents, material efficiency metrics

Oxaziridines are mild aprotic oxidizing agents that are used to synthesize acid-sensitive species such as α-siloxyepoxides and sulfenic acids. A well-known example is the Davis reagent [47, 48]. They are made by oxidizing imines with urea-hydrogen peroxide (UHP), which is advertised as an inexpensive, easy to handle, and safe source of hydrogen peroxide. An example synthesis procedure is shown below involving the oxidation of *N*-benzyl-phenylimine (**1**) to 2-benzyl-3-phenyl oxaziridine (**2**) with the UHP/maleic anhydride system.

urea-H_2O_2
maleic anhydride
95 %

Ph, N, Ph **1** Ph, O, N, Ph **2** Ph, O, N, SO_2Ph Davis reagent

Experimental Procedure:

A suspension of compound **1** (0.390 g, 2 mmol), urea-hydrogen peroxide (0.188 g, 2 mmol), and maleic anhydride (0.176 g, 2 mmol) in methanol (8 mL) was stirred at 0°C. The progress of the reaction was monitored by TLC (eluent, EtOAc–*n*-hexane, 1:5). The reaction mixture was filtered and the filtrate was passed through a short pad of silica gel. Methanol was removed under reduced pressure. Highly pure oxaziridine **2** was obtained in 95% yield (0.354 g).

When the substrate is N-phenyl-phenylimine, a nitrone is produced instead.

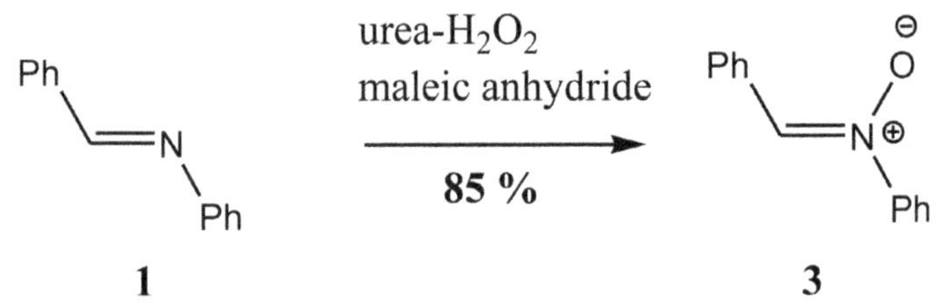

Ph–CH=N–Ph (1) → urea-H_2O_2, maleic anhydride, 85 % → Ph–CH=$N^{\oplus}$(–$O^{\ominus}$)–Ph (3)

Part 1

The authors did not address two points in their paper: the role of maleic anhydride in the oxidation reaction, and an explanation for why nitrones are formed as products instead of oxaziridines when N-phenyl substituents are present in the imine structure. Offer explanations for both observations.

Part 2

Write out a balanced chemical equation for the production of 2-benzyl-3-phenyl oxaziridine and determine the atom economy for the reaction.

Part 3

Using the *template-REACTION.xls* spreadsheet and the experimental details determine the material efficiency for producing 2-benzyl-3-phenyl oxaziridine. State any assumptions used in the calculations.

PROBLEM 8.33 [49]

Themes: balancing chemical equations, experimental procedures, materials efficiency metrics

An example Bargellini reaction promoted by a phase-transfer catalyst leads to two regioisomeric products depending on the orientation of coupling of the three components as shown below.

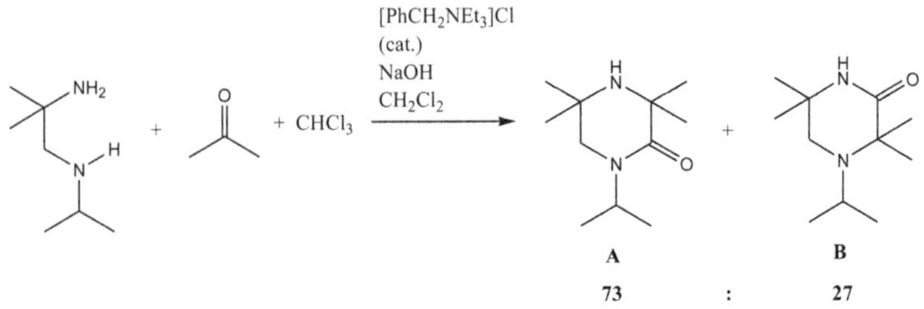

$[PhCH_2NEt_3]Cl$ (cat.), NaOH, CH_2Cl_2; + $CHCl_3$

A : B = 73 : 27

The product ratio was determined by gas chromatography. The isolated yield of product A is 52%.

Experimental Procedure:

N-Isopropyl-2-methyl-propane-1,2-diamine (50 mmol), chloroform (60 mmol), acetone (100 mmol), benzyltriethylammonium chloride (BTAC) (2 mmol), and 50 mL of CH_2Cl_2 are mixed and cooled while 50% NaOH (220 mmol) is added dropwise to keep the temperature below 5°C. The reaction is stirred at 5°C overnight and is worked up in the usual manner to obtain a clear oil which solidifies upon standing. Recrystallization from pentane affords a 52% yield of product A as colorless needles: melting point 82–84°C.

Part 1

Write out balanced chemical equations leading to each product.

Part 2

Using information given in the experimental procedure and the *template-REACTION.xls* spreadsheet, determine the material efficiency metrics for each reaction product. State any assumptions used in the calculations.

PROBLEM 8.34 [50]

Themes: balancing chemical equations, experimental procedures, materials efficiency metrics

An example solvent-free Beckmann rearrangement advertised in the literature is given below.

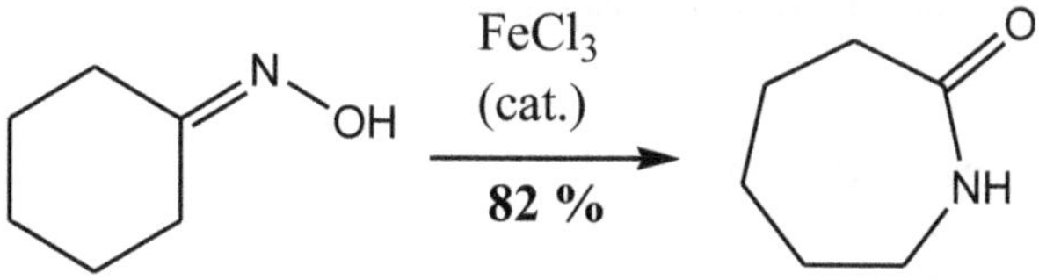

$FeCl_3$ (cat.)

82 %

10 mmol oxime
40 mmol ferric chloride
workup: 15 mL water, 3×20 mL chloroform

Part 1

Write out a balanced chemical equation for the reaction.

Part 2

Write a mechanism for the reaction.

Part 3

Using information given in the experimental procedure and the *template-REACTION.xls* spreadsheet, determine the material efficiency metrics for each reaction. State any assumptions used in the calculations.

PROBLEM 8.35 [51]

Themes: balancing chemical equations, experimental procedures, materials efficiency metrics

Two methods are reported for the preparation of 3-phenylindole via the Bischler indole synthesis under solvent-free and microwave irradiation conditions. Method A is a two-step process involving the isolation of N-phenacylaniline. In step 1, equimolar amounts of aniline and phenacylbromide are reacted in the presence of a base. In step 2, cyclization is carried out with anilinium bromide and DMF catalyst. Method B is a one-pot process involving two equivalents of aniline reacting with one equivalent of phenacylbromide with no added base.

Method A

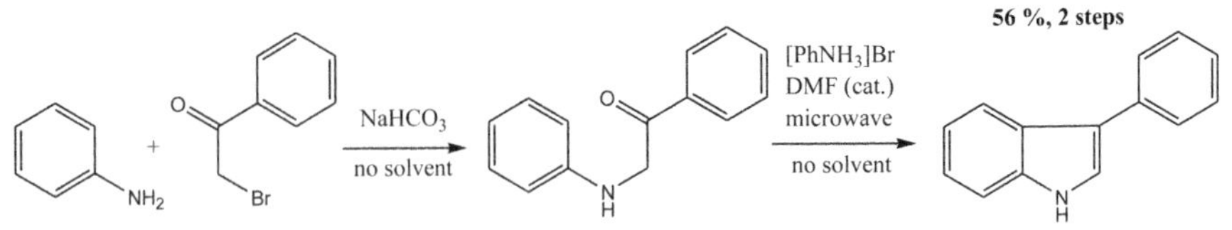

Step 1:

2 mmol aniline
2 mmol phenacylbromide
300 mg sodium bicarbonate
work-up: water for dilution and washing solid product

Step 2:

1 mmol of N-phenacylaniline
1.5 mmol anilinium bromide
3 to 4 drops DMF
purification: silica gel, 9:1 petroleum ether:EtOAc

Method B

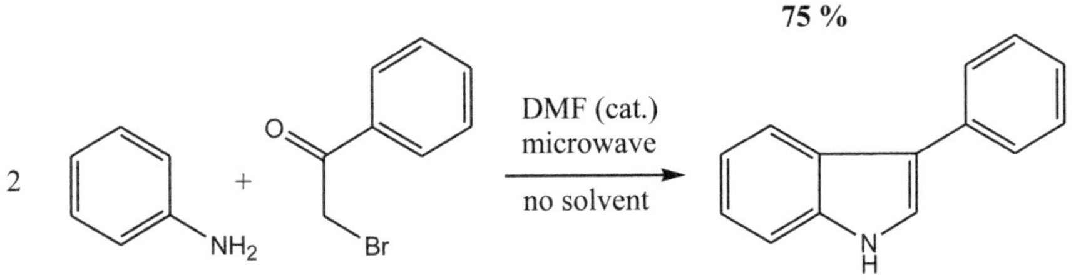

1 mmol phenacylbromide
2 mmol aniline
3 to 4 drops DMF
purification: silica gel, 9:1 petroleum ether:EtOAc

Use the above information given by the authors and the *template-REACTION.xls* spreadsheet to determine which method is more material efficient for the preparation of 1 mole of 3-phenylindole. State any assumptions made in the calculations.

PROBLEM 8.36 [52]

Themes: balancing chemical equations, experimental procedure, material efficiency metrics

An experimental procedure for a catalyst-free and reaction solvent-free Biginelli reaction is shown below.

A mixture of 530 g benzaldehyde (5 mol), 650 g ethyl acetoacetate (5 mol), and 360 g urea (6 mol) placed in a 2-L round-bottomed flask was stirred by a mechanical stirrer at room temperature for 2 min for uniform mixing, and then the temperature was raised to 100–105°C. No exothermic reaction was observed during addition and mixing. Stirring was continued for another hour at that temperature. During the progress of reaction (approximately during the first 20 min, all the urea dissolved and solids started to appear), the reaction mixture became a thick slurry, with the solids deposited.

At this stage, although no efficient stirring or agitation could be made, the reaction was not affected. After 1 h the reaction mixture was cooled in an ice water bath (0–5°C), and 200 mL water was added. The solid was carefully broken into pieces with a spatula and filtered. The yellow solid was then washed with 100 mL cold water followed by 50 mL cold rectified spirit (95% ethanol) to provide 1.025 kg colorless solid that was practically pure (79% yield), melting point 201–202°C.

Part 1

Write out the balanced chemical equation for the reaction.

Part 2

Using the *template-REACTION.xls* spreadsheet determine the material efficiency metrics for the reaction. State any assumptions made in the calculations.

Part 3

Draw the product structure target bond map and state the ring construction strategy used.

Part 4

Show all possible three-partitions of the Biginelli adduct that follow the strategy found in Part 3.

Part 5

From the list in Part 4 suggest a possible three-component coupling to the reaction product that results in an atom economy exceeding the value found in Part 2.

PROBLEM 8.37 [53]

Themes: balancing chemical equations, experimental procedures, material efficiency metrics

Hydrazobenzene (2.2 g) was obtained by reduction of 2.4 g azobenzene in 150 mL acetone with 12 g zinc powder and 15 mL saturated NH_4Cl solution. After being dried for 12 h over $CaCl_2$ and under vacuum, 1.5 g hydrazobenzene (8.15 mmol) was dissolved in a mixture of 306 mL freshly distilled 95% EtOH and 102 mL water. Another solution was prepared from 10.5 g LiCl (248 mmol), 41 mL 2.0 N HCl and 61 mL water. Each solution was cooled to 0°C under argon. The two solutions were mixed under a flow of argon and kept at 0°C for 30 min, after which 50 mL saturated KOH solution was added quickly. Air was bubbled through the alkaline solution for 24 h to oxidize unused hydrazobenzene to azobenzene. The solution was evaporated to a smaller volume at room temperature. Approximately 10 mg azobenzene distilled in the evaporation. The mixture that remained was extracted with ether (4 × 100 mL). The ether solution was washed with 2 N HCl (3 × 100 mL) and water (2 × 50 mL). The ether solution was dried over $MgSO_4$ and evaporated in a rotary evaporator to give 1015 mg azobenzene (5.58 mmol, 68.5%). This was sublimed to give 989 mg azobenzene (5.38 mmol, 66%). The combined aqueous washings were made alkaline with saturated KOH solution and were extracted with ether (3 × 100 mL). Workup gave 379 mg residue. This was dissolved in 8 mL CH_2Cl_2 and hexane was added slowly to the solution to precipitate 210 mg 4,4′-diaminobiphenyl, melting point 125–126°C. After filtration, the solution was evaporated to give 126 mg residue. This was chromatographed on silica gel with cyclohexane/absolute EtOH (5:1) to give 11 mg unknowns (two or more), 47.6 mg (0.26 mmol, 3.2%) 2,4'-diaminobiphenyl as an oil, and 77.6 mg 4,4'-diaminobiphenyl, melting point 125–126°C. The two portions of 4,4′-diaminobiphenyl were combined and crystallized from hot water to give 237 mg (1.29 mmol, 15.8%) 4,4′-diaminobiphenyl, melting point 125–126°C.

Part 1

Write out a balanced chemical equation for the reduction of azobenzene to hydrazobenzene using zinc. Using the *template-REACTION.xls* spreadsheet determine the material efficiency metrics for the reaction from the experimental details. State any assumptions used in the calculation.

Part 2

Write out a balanced chemical equation for the production of 4,4′-diaminobiphenyl from hydrazobenzene. Using the *template-REACTION.xls* spreadsheet determine the material efficiency metrics for the reaction from the experimental details. State any assumptions used in the calculation.

Part 3

Write out a balanced chemical equation for the production of 2,4′-diaminobiphenyl from hydrazobenzene. Using the *template-REACTION.xls* spreadsheet determine the material efficiency metrics for the reaction from the experimental details. State any assumptions used in the calculation.

Part 4

Determine the percent conversion of hydrazobenzene to biphenyl products and the percent selectivities for 4,4′-diaminobiphenyl and 2,4′-diaminobiphenyl.

For each biphenyl product, verify the relationship: conversion X selectivity = yield.

Part 5

Determine the percent of hydrazobenzene that does not react in both the benzidine rearrangement and the oxidation reaction.

PROBLEM 8.38 [54]

Themes: balancing chemical equations, experimental procedure, materials metrics analysis

Experimental Procedure:

Phenylacetaldehyde (120 g, 1 mol) and 0.5 mol of acetaldehyde were dissolved in 350 mL saturated solution of ammonia in 500 mL of absolute ethanol. The solution was sealed in a 1-L stainless-steel bomb and heated at 225–230°C for 6 h with constant rocking. The pressure was about 1000 psi. The bomb was then cooled and opened and the contents were transferred to a distilling flask. As much ethanol as possible was distilled on the steam bath at atmospheric pressure; 60.5 g of crude basic reaction product on standing overnight deposited 15.0 g 3,5-di-phenylpyridine, which was collected after diluting the mixture with 50 mL ligroin (boiling point 60–71°C). The material melted at 136–138°C. The remaining material was distilled to give three arbitrary fractions. The first fraction (boiling point up to 174°C, at 13 mmHg), weighed 3.9 g and was proved to be 2-methyl-5-phenylpyridine. The second fraction (boiling point 174–230°C, 13 mmHg), weighed 27.65 g and was proved to be 2-methyl-3,5-diphenylpyridine, with an actual of boiling point 226–228°C at 13 mmHg. The third fraction (boiling point 230–250°C, 13 mmHg), weighed 4.33 g and was proved to be 2-methyl-3,5-diphenylpyridine.

Part 1

Based on the experimental procedure given write out balanced chemical equations for each reaction product.

Part 2

Show the ring construction strategies by highlighting the target bonds made in the product structures.

Part 3

Using the *template-REACTION.xls* spreadsheet determine the material efficiency metrics for the production of each reaction product. State any assumptions used in the calculations.

PROBLEM 8.39 [55]

Themes: balancing chemical equations, experimental procedures, materials efficiency metrics

Two different reaction products arise when 1-cyclopropyl-2-phenylethanone reacts under Vilsmeier–Haack conditions depending on the reaction temperature as shown below.

O=PCl_3
Me_2CHO
100°C
71 %

A

O=PCl_3
Me_2CHO
120°C
53 %

B

Experimental Procedure:

The Vilsmeier–Haack reagent was prepared by adding $POCl_3$ (4.5 mmol) dropwise to ice-cold dry DMF (2 mL) under stirring. After 10 min, to the above Vilsmeier reagent was added 1-cyclopropyl-2-phenylethanone (48 mg, 0.3 mmol) as a solution in DMF (1.0 mL). The reaction mixture was stirred at 100°C (100°C/120°C) for 20 min. Then the mixture was poured into ice-cold water (20 mL) and extracted with dichloromethane (3×20 mL), and the combined organic phases were washed with water (3×20 mL), dried over $MgSO_4$, and filtered. The organic layer was removed under reduced pressure, and then the residue was purified by flash column chromatography.

Part 1

Write out a balanced chemical equation for each reaction.

Part 2

Write reaction mechanisms that account for the formation of each product.

Part 3

Highlight the target bonds made and describe the ring construction strategy for each product.

Part 4

Using information given in the experimental procedure and the *template-REACTION.xls* spreadsheet, determine the material efficiency metrics for each reaction. State any assumptions used in the calculations.

PROBLEM 8.40 [56, 57]

Themes: balancing chemical equations, catalytic cycles, experimental procedures, material efficiency metrics

COOMe PdCl₂, CuCl₂ O₂, H₂O 97 % COOMe

Experimental Procedure:

To a solution of 0.3 g methyl 7-allyl-6-hydroxy-5-oxo-1,2,3,5-tetrahydroindolizine-8- carboxylate (1.2 mmol) in 6 mL DMF at 25°C was added a solution containing 0.21 g of copper(II) chloride dihydrate (1.2 mmol) in 4 mL water followed by 0.021 g $PdCl_2$ (0.11 mmol), and the reaction mixture was stirred at room temperature under an oxygen atmosphere for 12 h. The dark reaction mixture was poured into a separatory funnel containing 100 mL water and was extracted with 100 mL ethyl acetate. The organic layer was dried over $MgSO_4$ and filtered through Celite. Upon removal of solvent, the resulting white solid was purified by silica gel column chromatography to afford 0.29 g methyl 2-methyl-8- oxo-5,6,7,8-tetrahydro-1-oxa-7a-azaindacene-4-carboxylate (97%) as a white solid, which was further purified by recrystallization from dichloromethane, melting point 159–160°C.

Part 1

Write a balanced chemical equation for the transformation and determine its atom economy.

Part 2

Write out a catalytic cycle showing the roles of palladium(II)chloride and copper(II) chloride.

Part 3

Write out the key redox couples involved in the reaction.

Part 4

Using the *template-REACTION.xls* spreadsheet and the experimental details determine the material efficiency metrics for the reaction. State any assumptions made in the calculations.

PROBLEM 8.41 [58]

Themes: balancing chemical equations, experimental procedure, materials metrics analysis

CuI (cat.)
$Pd(OAc)_2$ (cat.)
PPh_3 (ligand)
Et_3N

Experimental Procedure:

A 500-mL round-bottomed flask equipped with a Teflon-coated magnetic stirring bar (3.8×0.9 cm) is charged successively with 2-iodobenzamide (19.8 g, 80 mmol), PPh_3 (419 mg, 1.60 mmol, 0.02 equiv), CuI (152 mg, 0.80 mmol, 0.01 equiv), $Pd(OAc)_2$ (180 mg, 0.80 mmol, 0.01 equiv), *N,N*-dimethylformamide (DMF, 40 mL, anhydrous), Et_3N (120 mL), and 1-hexyne (12.0 mL, 104 mmol, 1.3 equiv). A glass inlet adapter (24/40) is fitted on the flask, which is then purged by two cycles of evacuation (2–5 sec.) and back-filling with nitrogen. A slight positive pressure of nitrogen is maintained. The flask is then placed in a preheated oil bath (oil bath temperature 60°C) and allowed to stir for 16 h, during which time the color of the slurry darkens progressively to brown. Completion of the reaction is confirmed by TLC monitoring (hexanes:EtOAc, 2:1, Rf=0.35), the reaction mixture is diluted with 200 mL of EtOAc and transferred to a 1 L separatory funnel. The organic phase is sequentially washed with saturated NH_4Cl solution (2×120 mL) and saturated brine (1×100 mL), and the aqueous layer is back-extracted with EtOAc (1×100 mL). The combined organic phase is dried over anhydrous Na_2SO_4 (15 g, 10 min), filtered, and concentrated by rotary evaporation (15–20 mmHg, 23°C) to give a dark orange solid. The solid is purified by column chromatography on silica gel to afford 2-(1-hexynyl) benzamide as a yellowish powder (14.4 g, 90%).

Column chromatography was carried out using SILICYCLE SiliaFlash P60 silica gel. A glass column (5×40 cm) was slurry-packed with 200 g of silica gel in hexane. The compound was loaded on the column as a solution in a small amount of CH_2Cl_2, and the column was eluted first with hexane, then hexane:EtOAc, 50:1 (ca. 1 L), then hexane:EtOAc, 10:1 (ca. 0.5 L), then hexane:EtOAc, 5:1 (ca. 0.5 L), and then hexane:EtOAc, 2:1 until all the product had eluted. Fractions containing the desired product were combined and concentrated using a rotary evaporator at 30°C (15–20 mmHg) and dried under high vacuum at 23°C (5–10 mmHg).

Using the *template-REACTION.xls* spreadsheet determine the material efficiency metrics for the coupling reaction.

PROBLEM 8.42 [59]

Themes: balancing chemical equations, experimental procedure, materials metrics analysis

The following reaction is a tested example of the Sharpless asymmetric dihydroxylation. Such a reaction involves a cocktail of reagents acting on an olefinic substrate.

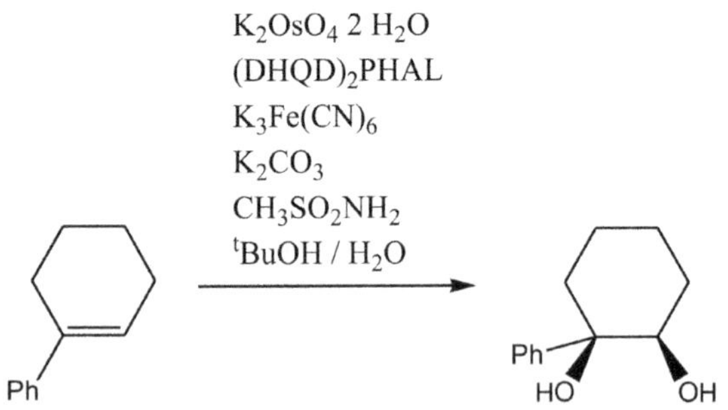

Part 1

Describe the role of each ingredient above the reaction arrow.

Part 2

Write out the redox couple for the reaction, a catalytic cycle, and a balanced chemical equation for the reaction.

Part 3

Using the *template-REACTION.xls* spreadsheet and the details of the following experimental procedure, determine the overall metrics for this reaction. State any assumptions used in the calculations.

A 3-L flask with a mechanical stirrer, thermometer, and an inlet port open to the atmosphere is charged with 375 mL of water. Stirring is started and the following reagents are added in the order indicated through a powder funnel: potassium ferricyanide (247 g, 0.75 mol, 3 equiv), anhydrous potassium carbonate (104 g, 0.75 mol, 3 equiv), methanesulfonamide (23.8 g, 0.25 mol, 1 equiv), potassium osmate dihydrate (46.1 mg, 0.125 mmol, 0.05 mol %), $(DHQD)_2PHAL$ [1,4-bis(9-*O*-dihydroquinidinyl)phthalazine, 486.9 mg, 0.625 mmol, 0.25 mol %], 1-phenylcyclohexene (39.55 g, 0.25 mol), and *tert*-butyl alcohol (250 mL). The slurry is stirred vigorously for 2 days at a rate of 500 rpm. During this time, the product crystallizes in the top organic phase, beginning after 4 h. Also, the appearance of the slurry gradually changes from a mixture containing red granules (ferricyanide) to yellow flakes, which are presumably a salt of iron (II).

After the reaction is complete the mixture is treated with ethyl acetate (250 mL) with stirring to dissolve the product. The resulting mixture is filtered through a 500-mL medium-fritted glass funnel and the flask and filter cake are washed with ethyl acetate (3 × 50 mL). The filtrate is transferred to a 2-L separatory funnel and the brown-colored aqueous phase is separated. The organic phase is washed with 2 M potassium hydroxide (KOH, 2 × 50 mL) with vigorous shaking to remove methanesulfonamide, then dried over magnesium sulfate ($MgSO_4$, 12.5 g). The solid is filtered, the cake is washed with ethyl acetate (2 × 37 mL) and the filtrate is evaporated, to afford a white solid. After the crude product is dried under reduced pressure overnight, it weighs 47.44 g (99%).

PROBLEM 8.43 [60]

Themes: balancing chemical equations, experimental procedure, materials metrics analysis

1. THF
2. pyridine
3. H_2N–NH–(2,4-dinitrophenyl)

O; O; O_2S; H; N; NH_2; NO_2; NO_2; +; O_2N; NO_2; H; N; N; O_2N; NO_2

Experimental Procedure:

To a solution of 0.576 g 2,4-dinitrophenylsulfonylhydrazine (2.2 mmol) in 20 mL THF cooled to −25°C was added 0.224 g 2,3-epoxy cyclohexan-1-one (2 mmol). The reaction mixture was kept at −25 to −30°C for 30 min and then at −10°C for 30 min. The mixture was allowed to warm to 0°C, dried at this temperature over anhydrous magnesium sulfate for 20 min, and filtered below 0°C. After being stirred for 1 min a drop of pyridine was added and the stirring was continued. The mixture was taken out of the ice bath, and after 2 min another drop of pyridine was added. This caused an extensive effervescence, and the mixture turned deep orange in color. After 2 min, more pyridine (a total of 0.16 g, 2 mmol) was added, and the stirring was continued for 5 min. The mixture was filtered through Celite 545, and the filtrate was stirred with 1 g powdered $CuSO_4 \cdot 5H_2O$ for 25 min. It was filtered, and the volatile substances from the filtrate were transferred under vacuum (0.03 mmHg) at ambient temperature into a receiver containing 0.792 g 2,4-dinitrophenylhydrazine (4 mmol) in 75 mL THF. The mixture was stored at 25–30°C for 48 h. The solvent was removed by rotary evaporation, and the residue was purified by preparative TLC on silica gel (methylene chloride, Rf=0.8) to give 342 mg of the orange crystalline 2,4-dinitrophenylhydrazone of 5-hexynal, in a yield of 62%, melting point 90–91°C.

Part 1

Write out a mechanism for the transformation.

Part 2

Based on the mechanism, write out a balanced chemical equation.

Part 3

Using the *template-REACTION.xls* spreadsheet, determine the material efficiency metrics for the reaction. State any assumptions used in the calculations.

PROBLEM 8.44 [61, 62]

Themes: balancing chemical equations, experimental procedures, materials efficiency metrics

Nifedipine is a coronary pharmaceutical made according to the Hantzsch dihydropyridine synthesis method. This transformation involves a four-component coupling of an aldehyde, ammonia, and two equivalents of a β-ketoester.

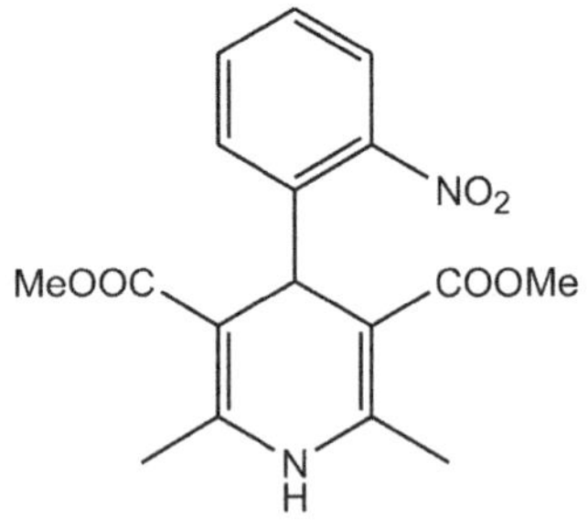

NO_2

MeOOC COOMe

N H

nifedipine

45 g 2-nitrobenzaldehyde
80 mL acetoacetic acid methyl ester
75 mL methanol
30 mL aqueous ammonia
Yield: 75 g nifedipine

Part 1

Write out a balanced chemical showing reactants and by-products that accounts for the synthesis of nifedipine according to the Hantzsch dihydropyridine strategy. Determine the associated atom economy.

Part 2

Use the *template-REACTION.xls* spreadsheet to determine the material efficiency metrics for the reaction.

Part 3

Highlight the target bonds made and describe the ring construction strategy.

Part 4

For a generally substituted symmetrical dihydropyridine, enumerate all possible three-partitions of the ring.

Part 5

From these possible partitions suggest an alternative strategy to synthesize the ring via a three-component coupling reaction. Support such a strategy with a reasonable reaction mechanism. Compare its expected minimum atom economy with that for the traditional four-component method.

Part 6

Suggest another [2 + 1 + 2 + 1] strategy to make the Hantzsch dihydropyridine ring. Compare its minimum atom economy with the traditional method.

PROBLEM 8.45 [63]

Themes: balancing chemical equations, experimental procedure, materials metrics analysis

An example of a Heck reaction conducted under microwave conditions is shown below.

Pd (cat.)
$[Bu_4N]Br$
K_2CO_3
water
microwaves

Br
MeO
+
MeO

Experimental Procedure:

In a 10-mL glass tube were placed 187 mg 4-bromoanisole (0.125 mL, 1.0 mmol), 208 mg styrene (0.230 mL, 2.0 mmol), 511 mg K_2CO_3 (3.7 mmol), 322 mg tetrabutylammonium bromide (1.0 mmol), 0.4 mL palladium stock solution (a 1000 ppm solution in water), and 1.6 mL water to give a total volume of water to 2 mL and a total palladium concentration of 200 ppm. The vessel was sealed with a septum, shaken, and placed into the microwave cavity. Initial microwave irradiation of 70 W was used and the temperature was increased from room temperature to the desired temperature of 170°C. Once this was reached, the reaction mixture was held at this temperature for 10 min. After allowing the mixture to cool to room temperature, the reaction vessel was opened, and the contents were poured into a separating funnel. Water (30 mL) and 30 mL EtOAc were added, and the organic material was extracted and removed. After further extraction of the aqueous layer with EtOAc, combining the organic washings and drying them over $MgSO_4$, the ethyl acetate was removed *in vacuo* to give a crude product that was purified via chromatography, in a yield of 90%.

Part 1

Write a balanced chemical equation for the reaction.

Part 2

Write out a mechanism for the reaction using a catalytic cycle.

Part 3

Based on the experimental procedure determine the material efficiencies of each reaction and the overall sequence. State any assumptions made in the calculations.

Determine the mass, moles, and mol% of Pd used as catalyst.

Determine the turnover number and turnover frequency.

PROBLEM 8.46 [64]

Themes: balancing chemical equations, experimental procedures, materials efficiency metrics

[Bu_4N]F 3 H_2O
polystyrene Pd
supported catalyst
(9.4 μmol Pd per g)

Si(OMe)3 + I

polystyrene Pd supported catalyst

Ph2P
Ph2P—Pd—PPh2
PPh2

The catalyst (43 mg, 0.1 mol % of supported palladium) was added to a solution of aryl iodide (0.40 mmol, 1.0 equiv), aryltrialkoxysilane (0.80 mmol, 2.0 equiv), and TBAF $3H_2O$ (252 mg, 0.8 mmol, 2.0 equiv) in a mixture of toluene (5 mL) and H_2O (50 mL). The reaction mixture was heated at 100°C for 20 h. After cooling to room temperature, 1a was filtered under vacuum on a 0.2 mm membrane. The catalyst was successively washed with toluene (10 mL) and Et_2O (10 mL). The combined organic phases were washed with H_2O (20 mL), dried with $MgSO_4$, filtered, and concentrated under vacuum. The residue was purified by flash chromatography on silica gel. The catalyst was dried under vacuum and can directly be used for another Hiyama coupling.

4-Methylbiphenyl: Elution with Et_2O/cyclohexane 1:99 afforded 66.5 mg (99% yield) of a white solid; melting point 47–48°C.

Part 1

Write out a balanced chemical equation for the transformation.

Part 2

Write out the catalytic cycle.

Part 3

Use the experimental information given and the *template-REACTION.xls* spreadsheet to determine the material efficiency metrics for the transformation. State any assumptions made in the calculation.

Part 4

Determine the catalyst loading, turnover number, and turnover frequency.

PROBLEM 8.47 [65]

Themes: balancing chemical equations, experimental procedures, materials efficiency metrics

The following Pschorr cyclization leads to two products B and A in the ratio 6 to 1.

86 % combined yield

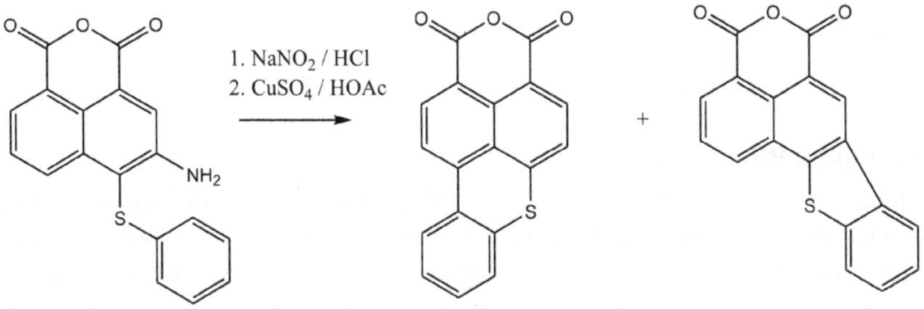

1. $NaNO_2$ / HCl
2. $CuSO_4$ / HOAc

NH_2 S O O O → B (6) : A (1)

Sodium nitrite (0.7 g) and glacial acid (2 mL) were added dropwise to concentrated sulfuric acid (10 mL). The mixture was cooled to 5°C and to it was added, portionwise over 1 h, 3-amino-4-phenylthio-1,8-naphthalic anhydride (3.2 g). After stirring for 1 h, the dark red viscous liquor was added over 90 min to a boiling solution of copper sulfate in water and glacial acetic acid. After the addition was complete, the liquor was refluxed for a further 30 min, cooled, filtered, dried and an orange solid (2.64 g, 86.3%) was collected. Recrystallization from DMF gave deep red needles, but TLC indicated it contained compound A (Rf=0.54) and B1 (Rf=0.68), which were separated on preparative thin layer chromatography using dichloromethane as eluent (three times) and gave two pure compounds. The compound A was a yellow–green solid, melting point 284–286°C. The compound B was an orange-red solid, melting point >300°C.

Part 1

Suggest mechanisms for the cyclization that lead to products A and B.

Part 2

Write out balanced chemical equations leading to each product.

Part 3

Using the experimental procedure details and the *template-REACTION.xls* spreadsheet determine the material efficiency metrics for producing each product. State any assumptions made in the calculations.

PROBLEM 8.48 [66]

Themes: balancing chemical equations, experimental procedures, materials efficiency metrics

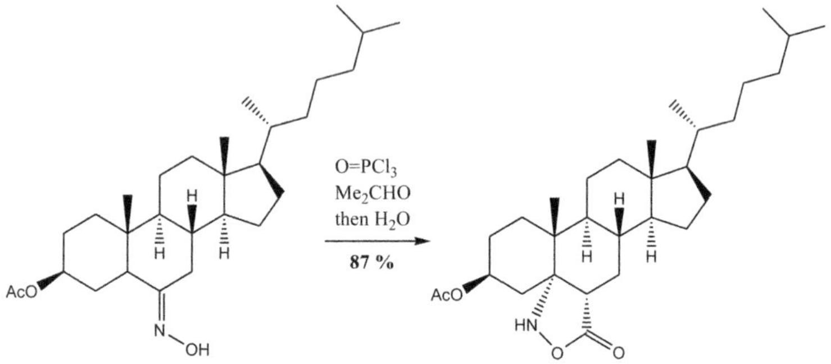

Experimental Procedure*:*

A solution of steroidal oxime (1 mmol) in DMF (5 mL) was added dropwise to an ice-cooled magnetically stirred mixture of Vilsmeier reagent prepared from DMF (3 equiv) and $POCl_3$ (3.5 equiv). The reaction mixture was kept at 0°C for 15 min and then it was gradually allowed to attain room temperature and maintained at 60–65°C for 30–50 min. After completion of the reaction (observed on TLC), it was allowed to cool at room temperature. The reaction mixture was poured into crushed ice and refrigerated overnight. The solid separated on neutralization with $NaHCO_3$ was filtered, washed with water (3×15 mL), and recrystallized from methanol to afford the product.

Part 1

Write out a balanced chemical equation for the reaction.

Part 2

Write a reaction mechanism that accounts for the formation of the product.

Part 3

Highlight the target bonds made and describe the ring construction strategy for the product.

Part 4

Using information given in the experimental procedure and the *template-REACTION.xls* spreadsheet, determine the material efficiency metrics for each reaction. State any assumptions used in the calculations.

PROBLEM 8.49 [67]

Themes: balancing chemical equations, material efficiency metrics, reaction mechanism, catalytic cycles

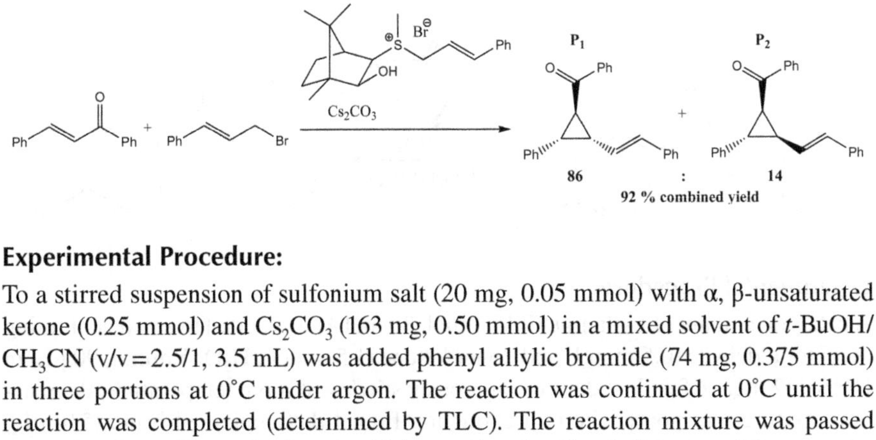

Cs_2CO_3

P_1 : P_2

86 : 14

92 % combined yield

Experimental Procedure:

To a stirred suspension of sulfonium salt (20 mg, 0.05 mmol) with α, β-unsaturated ketone (0.25 mmol) and Cs_2CO_3 (163 mg, 0.50 mmol) in a mixed solvent of *t*-BuOH/ CH_3CN (v/v = 2.5/1, 3.5 mL) was added phenyl allylic bromide (74 mg, 0.375 mmol) in three portions at 0°C under argon. The reaction was continued at 0°C until the reaction was completed (determined by TLC). The reaction mixture was passed through a short silica gel column, which was eluted with ethyl acetate. After concentration of the elution, the residue was purified by flash column chromatography to afford the cyclopropanes.

Part 1

Write out a balanced chemical equation for the transformation and determine its atom economy.

Part 2

Using the *template-REACTION.xls* spreadsheet and the details of the experimental procedure, determine the material efficiency metrics for the production of product P_1 and product P_2. State any assumptions made in the calculations.

Part 3

Write out a catalytic cycle for the reaction.

Part 4

The camphor-derived methyl sulfide catalyst was prepared as shown below [68].

1. LDA/THF
2. $PhSO_2SCH_3$/THF
3. $NaHSO_4$

1. DIBAL
2. NH_4Cl

(R)-(+)-camphor

Write out a fully balanced scheme showing all by-products for the synthesis. Determine the step atom economies and the overall atom economy.

PROBLEM 8.50 [69]

Themes: balancing chemical equations, catalytic cycle, mechanism, organocatalysis, material efficiency metrics

30 mol%

H_2O/DMSO

Experimental Procedure:

A catalytic amount of L-alanine (0.15 mmol, 30 mol%) was added to a vial containing acceptor aldehyde (0.5 mmol) and donor ketone (1.5 mmol), H_2O (5 mmol, 90 μL) in DMSO (2 mL). After 3–4 days of vigorously stirring at room temperature, the reaction mixture was poured into an extraction funnel that contained brine (5.0 mL), which was diluted with distilled H_2O (5.0 mL) and EtOAc (15 mL). The reaction vial was also washed with 2 mL of EtOAc, which was poured into the extraction funnel. The aqueous phase was extracted with EtOAc (2 × 15.0 mL). The combined organic phases were dried with Na_2SO_4 and the solvent removed under reduced pressure. The reaction can also be quenched by directly putting the reaction mixture on a silica-gel column. The crude aldol product was purified by silica-gel column chromatography (EtOAc:pentane-mixtures) to furnish the desired aldol product. Yield: 95%, d.r. (diastereomeric ratio) = 15:1.

Part 1

Show a balanced chemical equation and determine the atom economy.

Part 2

Show a catalytic cycle and mechanism for the reaction.

Part 3

Using the *template-REACTION.xls* spreadsheet and the details of the experimental procedure, determine the material efficiency metrics for the reaction that includes the synthesis of the catalyst. State any assumptions made in the calculations.

REFERENCES

1. Lange, N.A. *Handbook of Chemistry*, 10th ed., McGraw-Hill Book Company, Inc.: New York, 1961, pp. 1118–1171.
2. Lide, D.R. *CRC Handbook of Chemistry and Physics*, 85th ed., CRC Press: Boca Raton, 2005, Section 8, pp. 8–58, 8–84.
3. Pessel, F.; Augé, J.; Billault, I.; Scherrmann, M.C. *Beilstein J. Org. Chem.* 2016, *12*, 2351.
4. Yan, J.; Zhu, M.; Zhou, Z. *Eur. J. Org. Chem.* 2006, 2060.
5. Devine, W.G.; Leadbeater, N.E. *ARKIVOC* 2011, *5*, 127.
6. Galat, A. *Ind. Eng. Chem.* 1944, *36*, 192.
7. Vogel, A. I. *Textbook of Organic Chemistry*, Longman: London, 1956, pp. 1005–1007.
8. Vogel, A.I. *A Textbook of Practical Organic Chemistry*, 3rd ed., Longman: London, 1956, p. 382.
9. Lowenheim, F.A.; Moran, M.K. *Faith, Keyes, & Clark's Industrial Chemicals*, 4th ed., Wiley: Hoboken, 1975, p. 178.
10. Heydari, A.; Larijani, H.; Emani, J.; Karami, B. *Tetrahedron Lett.* 2000, *41*, 2471.
11. Shanthi, G.; Perumal, P.T. *Tetrahedron Lett.* 2007, *48*, 6785.
12. Aktoudianakis, E.; Lin, R.J.; Dicks, A.P. *J. Chem. Educ.* 2006, *83*, 1832.
13. Burguete, M.I.; Cornejo, A.; Verdugo, E.G.; Gil, M.J.; Luis, S.V.; Mayoral, J.A.; Martinez-Merino, V.; Sokolova, M. *J. Org. Chem.* 2007, *72*, 4344.
14. Basu, B.; Frejd, T. *Acta Chem. Scand.* 1996, *50*, 316.
15. Kaiser, C.; Weinstock, J.; Olmstead, M.P. *Org. Synth. Coll. Vol.* 1988, *6*, 913.
16. Estevan, F.; Lahuerta, P.; Lloret, J.; Sanau, M.; Ubeda, M.A.; Vila, J. *Chem. Commun.* 2004, 2408.
17. Reetz, M.T.; Sell, T.; Meiswinkel, A.; Mehler, G. *Angew. Chem. Int. Ed.* 2003, *42*, 790.
18. Patti, A. *Green Approaches to Asymmetric Catalytic Synthesis*, Springer: New York, 2011, p. 43.
19. Lowenheim, F.A.; Moran, M.K. *Faith, Keyes, & Clark's Industrial Chemicals*, 4th ed., Wiley, Hoboken, 1975, p. 445.
20. Mihara, M.; Ito, T.; Ishino, Y.; Oderaotoshi, Y.; Minikata, S.; Komatsu, M. *Tetrahedron Lett.* 2005, *46*, 8105.
21. Yoshizawa, K.; Toyota, S.; Toda, F. *Tetrahedron Lett.* 2001, *42*, 7983.
22. Kiamehr, M.; Moghaddam, F.M. *Tetrahedron Lett.* 2009, *50*, 6723.
23. Pingali, S.R.K.; Upadhyay, S.K.; Jursic, B.S. *Green Chem.* 2011, *11*, 928.
24. Chen, M.; Zhao, Q.; She, D.B.; Yang, M.Y.; Hui, H.H.; Huang, G.S. *J. Chem. Sci.* 2008, *120*, 347.
25. Jakka, K.; Zhao, C.G. *Org. Lett.* 2006, *8*, 3013.
26. Jakka, K.; Zhao, C.G. *Org. Lett.* 2007, *9*, 3485.
27. Schobert, R.; Gordon, G.J.; Mullen, G.; Stehle, R. *Tetrahedron Lett.* 2004, *45*, 1121.

28. Aprile, C.; Giacalone, F.; Gruttadauria, M.; Marculescu, A.M.; Noto, R.; Revell, J.D.; Wennemers, H. *Green Chem.* 2007, *9*, 1328.
29. Wascholowski, V.; Hansen, H.M.; Longbottom, D.A.; Ley, S.V. *Synthesis* 2008, 1269.
30. Moseley, J.D.; Lenden, P. *Tetrahedron* 2007, *63*, 4120.
31. Mohr, J.T.; Nishimata, T.; Behenna, D.C.; Stoltz, B.M. *J. Am. Chem. Soc.* 2006, *128*, 11348.
32. Pozarentzi, M., Stephanidou-Stephanatou, J., Tsoleridis, C. A. *Tetrahedron Lett.* 2002, *43*, 1755.
33. Saini, A., Sandhu, J. S. *Synth. Commun.* 2008, *38*, 3193.
34. Bloxham, J.; Dell, C.P.; Smith, C.W. *Heterocycles* 1994, *38*, 399.
35. Zolfigol, M.A.; Shirimi, F.; Choghamarani, A.G.; Mohammadpoor-Baltork, I. *Green Chem.* 2002, *4*, 562.
36. Satyanarayana, P.; Reddy, G.M.; Maheswaran, H.; Kantam, M.L. *Adv. Synth. Catal.* 2013, *355*, 1859.
37. Wu, K.; He, W.; Sun, C.; Yu, Z. *Tetrahedron Lett.* 2016, *57*, 4017.
38. Xu, F.; Zacuto, M.; Yoshikawa, N.; Desmond, R.; Hoermer, S.; Itoh, T.; Journet, M.; Humphrey, G.R.; Cowden, C.; Strotman, N.; Devine, P. *J. Org. Chem.* 2010, *75*, 7829.
39. Cranwell, P.B.; Russell, A.T. *J. Chem. Educ.* 2016, *93*, 949.
40. Agee, B.M.; Mullins, G.; Biernacki, J.J.; Swartling, D.J. *Green Chem. Lett. Rev.* 2014, *7*, 383.
41. Soriano, D.S. *J. Chem. Educ.* 1993, *70*, 332.
42. Parquet, E.; Lin, Q. *J. Chem. Educ.* 1997, *74*, 1225.
43. Sattler, J.H.; Fuchs, M.; Tauber, K.; Mutti, F.G.; Faber, K.; Pfeffer, J.; Haas, T.; Kroutil, W. *Angew. Chem. Int. Ed.* 2012, *51*, 9156.
44. Cao, H.; Jiang, H.; Huang, H. *Synthesis* 2011, 1019.
45. Bharate, S.B.; Padala, A.K.; Dar, B.A.; Yadav, R.R.; Singh, B.; Vishwakama, R.A. *Tetrahedron Lett.* 2013, *54*, 3558.
46. Damavandi, J.A.; Karami, B.; Zolfigol, M.A. *Synlett* 2002, 933.
47. Davis, F.A.; Nadir, J.U.; Kluger, E.W.; Sedergren, T.C.; Panunto, T.W.; Billmers, R.; Jenkins, R. Jr.; Tucchi, I.; Watson, W.H.; Chen, J.S.; Kimura, M. *J. Am. Chem. Soc.* 1980, *102*, 2000.
48. Davis, F.A.; Stringer, O.D. *J. Org. Chem.* 1982, *47*, 1774.
49. Lai, J. T. *J. Org. Chem.* 1980, *45*, 754.
50. Khodaei, M.M.; Meybodi, F.A.; Rezai, N.; Salehi, P. *Synth. Commun.* 2001, *31*, 2047.
51. Sridharan, V.; Perumal, S.; Avendano, C.; Menéndez, J.C. *Synlett* 2006, 91.
52. Ranu, B. C.; Hajra, A. and Dey, S. S., *Org. Proc. Res. Dev.* 2002, *6*, 817.
53. Subotkowski, W.; Kupczyk-Subotkowska, L. and Shine, H. J. *J. Am. Chem. Soc.* 1993, *115*, 5073.
54. Farley, C. P.; Eliel, E. L. *J. Am. Chem. Soc.* 1956, *78*, 3477.
55. Tang, X.Y.; Shi, M. *J. Org. Chem.* 2008, *73*, 8317.
56. Wang, Z. *Comprehensive Organic Name Reactions and Reagents*, Wiley: Hoboken, 2010, p. 2920.
57. Mmutlane, E. M.; Harris, J. M. and Padwa, A., *J. Org. Chem.*, 2005, *70*, 8055.
58. Okamoto, N.; Takeda, K.; Yanada, R. *Org. Synth.* 2014, *91*, 27.
59. Gonzalez, J.; Aurigemma, C.; Truesdale, L. *Org. Synth.* 2002, *79*, 93.
60. Corey, E. J., Sachdev, H. S., *J. Org. Chem.*, 1975, *40*, 579.
61. Bossert, F.; Vater, W. *Naturwissenschaften* 1971, *58*, 578.
62. Bossert, F.; Vater, W. US 3485847 (Bayer, 1969).
63. Arvela, R. K.; Leadbeater, N. E., *J. Org. Chem.*, 2005, *70*, 1786.
64. Diebold, C.; Derible, A.; Becht, J.M.; Le Drian, C. *Tetrahedron* 2013, *69*, 264.

65. Xu, Y.; Qian, X.; Yao, W.; Mao, P.; Cui, J. *Bioorg. Med. Chem.* 2003, *11*, 5247.
66. Shamzussaman; Khanam, H.; Mashrai, A.; Siddiqui, N. *Tetrahedron Lett.* 2013, *54*, 874.
67. Deng, X.M.; Cai, P.; Ye, S.; Sun, X.L.; Liao, W.W.; Li, K.; Tang, Y.; Wu, Y.D.; Dai, L.X. *J. Am. Chem. Soc.* 2006, *128*, 9730.
68. Li, A.H.; Dai, L.X.; Hou, X.L.; Huang, Y.Z.; Li, F.W. *J. Org. Chem.* 1996, *61*, 489.
69. Cordova, A.; Zou, W.; Ibrahem, I.; Reyes, E.; Engqvist, M.; Liao, W.W. *Chem. Commun.* 2005, 3586.

Appendix: Other Terminologies

Atom Utilization

This term is similar in meaning to atom economy except that it is based on the actual count of each kind of atom in a balanced chemical equation rather than on molecular weights of reactants and products. Hence, atom utilization compares the fraction of each kind of atom appearing in the reactants that end up in the target product. For example, if a reaction involves only carbon, oxygen, and hydrogen atoms then three atom utilization fractions are considered corresponding to the number of each kind of atom in the target product divided by the total number of atoms of that kind appearing in the reactants. The fraction corresponding to carbon atoms is called carbon efficiency. This term is no longer used and is superseded by atom economy, which includes all atoms in a chemical equation in a single metric.

Auxiliary Mass Factor *s* [1]

The mass ratio of all auxiliary materials used in carrying out a chemical reaction to the stoichiometric mass of reactants. For a single reaction this metric is related to the material recovery parameter (MRP) according to Equation (A.1).

$$\mathrm{MRP} = \frac{1+b}{1+b+s} \tag{A.1}$$

where b is the excess mass factor.

Balance Yield (Bilanzausbeute) [2]

This term is identical in meaning to the Andraos definition of reaction mass efficiency for a given chemical reaction that is balanced.

Carbon Efficiency

The number of carbon atoms appearing in the target product divided by the total number of carbon atoms appearing in the reactants of a balanced chemical equation. This metric focuses on carbon atoms since by frequency of their appearance in industrial commodity chemicals and natural products they form the fundamental building blocks of all organic molecules. It is no longer used and is superseded by atom economy.

Chirality Economy [3, 4]

The mathematical expression for chirality economy is given by Equation (A.2).

$$\mathrm{CE} = \left(\frac{\%ee_{\text{product}}}{\%ee_{\text{catalyst}}}\right)\left(\frac{\%\text{yield}_{\text{product}}}{\text{mol}\,\%_{\text{catalyst}}}\right) \tag{A.2}$$

Cycle Time

The total time to carry out a reaction or batch process involving all unit operations including: actual reaction time, filtration time, workup time, solvent evaporation time, recrystallization time, centrifugation time, drying time, and so on.

Diastereomeric Excess

If a reaction produces two diastereomers A and B, and A is the dominant diastereomer, then the diastereomeric excess is defined as shown in Equation (A.3).

$$de = \frac{m_A - m_B}{m_A + m_B} \tag{A.3}$$

where the m parameters refer to the masses of the respective diastereomers. This quantity is often expressed as a percentage.

Diastereomeric Ratio

If a reaction produces two diastereomers A and B, then the diastereomeric ratio of A to B is defined as shown in Equation (A.4).

$$dr = \frac{m_A}{m_B} \tag{A.4}$$

where the m parameters refer to the masses of the respective diastereomers. This quantity is related to diastereomeric excess according to Equation (A.5).

$$de = \frac{dr - 1}{dr + 1} \tag{A.5}$$

Effective Mass Yield [5]

Effective mass yield (EMY) is the ratio of mass of product to mass of all input materials excluding aqueous materials since they are considered to be benign. This quantity is a truncated form of the global reaction mass efficiency as shown in Equation (A.6).

$$\text{EMY} = \frac{\text{Mass product}}{\sum \text{Mass inputs} - \text{Mass aqueous inputs}} >> \frac{\text{Mass product}}{\sum \text{Mass inputs}} = \text{RME} \tag{A.6}$$

Effluent Load Factor [6]

A term used by chemical engineers that is identical to the E-factor.

Enantiomeric Excess

If a reaction produces two enantiomers A and B, and A is the dominant enantiomer, then the enantiomeric excess is defined as shown in Equation (A.7).

$$ee = \frac{m_A - m_B}{m_A + m_B} \tag{A.7}$$

where the m parameters refer to the masses of the respective enantiomers. This quantity is often expressed as a percentage.

Enantiomeric Ratio

If a reaction produces two enantiomers A and B, then the enantiomeric ratio of A to B is defined as shown in Equation (A.8).

$$er = \frac{m_A}{m_B} \tag{A.8}$$

where the m parameters refer to the masses of the respective enantiomers. This quantity is related to enantiomeric excess according to Equation (A.9).

$$ee = \frac{er - 1}{er + 1} \tag{A.9}$$

Excess Mass Factor *b* [1]

The ratio of excess mass of reactants used in carrying out a chemical reaction to the stoichiometric mass of reactants. For a single reaction this metric is related to the stoichiometric factor according to Equation (A.10).

$$\mathrm{SF} = 1 + b. \tag{A.10}$$

Gas Hourly Space Velocity

The gas hourly space velocity (GHSV) is applicable to catalytic gas phase chemical reactions and is given by Equation (A.11).

$$\mathrm{GHSV} = \frac{\left(\text{Total flow rate of gaseous reagents, cm}^3\,\text{min}^{-1}\right)}{\left(\text{Catalyst volume, cm}^3\right)} * 60\,\text{min}\,\text{h}^{-1} \tag{A.11}$$

Green Aspiration Level (GAL™) [7]

The green aspiration level (GAL) trademarked concept introduced by process chemists in the pharmaceutical industry adjusts calculated E-factors by correcting them for reaction complexity. The authors defined complexity as the difference between the total number of steps in a synthesis plan and the number of steps that are concession reactions; that is, reactions that are sacrificial and do not produce target bonds in the final product structure.

The authors used the terms "simplified E-factor" to refer to an E-factor that excludes solvent and water consumption (E_{min}), and "complete E-factor" to refer to an E-factor that includes them (E_{max}). The latter term is consistent with the traditional definition of E-factor. The mathematical relations for GAL_{min} and GAL_{max} are given by Equations (A.12) and (A.13).

$$\mathrm{GAL}_{min} = E_{min}\left(\frac{\text{Complexity}}{9}\right) = E_{min}\left(\frac{N_{total} - N_{concession}}{9}\right) \tag{A.12}$$

$$\text{GAL}_{\text{max}} = E_{\text{max}}\left(\frac{\text{Complexity}}{9}\right) = E_{\text{max}}\left(\frac{N_{\text{total}} - N_{\text{concession}}}{9}\right) \quad \text{(A.13)}$$

Both the minimum and maximum values of GAL are referenced to an arbitrarily chosen complexity value of 9. This number was determined under the assumption that the average number of synthesis steps (i.e., isolations) in a plan to a drug target is 7 and that each step involves on average of 1.3 chemical transformations excluding any sacrificial/concession transformations. Hence, the multiplication of 7 and 1.3 leads to 9.1.

Green Metric

A metric that is used to gauge the material efficiency, energy efficiency, environmental impact, safety-hazard impact, or synthesis strategy performance of a chemical reaction or synthesis plan against others for the same target molecule.

Metric

A quantitative measure of some property using some kind of scaled parameter.

Molar Efficiency (or Molar Yield) [8]

Molar efficiency is defined as a ratio of moles of substances used in a balanced chemical reaction according to Equation (A.14).

$$\text{Molar efficiency} = \left(\frac{\text{Moles product}}{\sum(\text{Moles reactants} + \text{moles additives}) + \text{moles catalyst}}\right) \times 100\% \quad \text{(A.14)}$$

Sometimes the denominator will include the moles of reaction solvent. This ratio may be converted to a reaction mass efficiency quantity based on mass by multiplying each type of number of moles by the molecular weight of the substance it refers to. Molar efficiency is relatable to reaction yield with respect to each reactant and to turnover number according to Equation (A.15).

$$\text{ME} = \frac{1}{\frac{1}{\varepsilon_A}\left(\frac{\nu_A}{\nu_P}\right) + \frac{1}{\varepsilon_B}\left(\frac{\nu_B}{\nu_P}\right) + \cdots + \frac{1}{\varepsilon_n}\left(\frac{\nu_n}{\nu_P}\right) + \frac{1}{\text{TON}}} \quad \text{(A.15)}$$

where:

- ε_A is the yield with respect to reactant *A*
- ε_B is the yield with respect to reactant *B*
- ε_n is the yield to reactant *n*
- V_A is the stoichiometric coefficient of reactant *A*
- V_B is the stoichiometric coefficient of reactant *B*
- V_n is the stoichiometric coefficient of reactant *n*
- TON is the turnover number equal to the number of moles of product divided by the number of moles of catalyst

If the reaction consists of n reactants with unit stoichiometric coefficients for all reactants and product and is run under stoichiometric conditions, then all the ε values are equal, all the stoichiometric coefficients are equal to 1, and the expression for ME reduces to Equation (A.16).

$$\mathrm{ME} = \frac{1}{\frac{n}{\varepsilon} + \frac{1}{\mathrm{TON}}} \approx \frac{\varepsilon}{n} \tag{A.16}$$

Inclusion of the number of moles of solvent in the computation of ME dramatically reduces its magnitude. The contribution from ligands and other additives is typically on the same order of magnitude as that from catalysts.

Process Excellence Index for Yield ($\mathrm{PEI}_{\mathrm{Yield}}$) [9]

A parameter used by process chemists to gauge the reproducibility and robustness of a reaction yield given by Equation (A.17).

$$\mathrm{PEI}_{\mathrm{yield}} = \frac{\mathrm{Yield}_{\mathrm{avg}}}{\mathrm{Yield}_{\mathrm{aspirational}}} \times 100\% \tag{A.17}$$

where:

$$\mathrm{Yield}_{\mathrm{avg}} = \frac{\sum_{j=1}^{N} \mathrm{Yield}_{\mathrm{run}_j}}{N}$$

N is the number of times a reaction is run (number of batches)

$$\mathrm{Yield}_{\mathrm{aspirational}} = \frac{\mathrm{Yield}_{\mathrm{best}} + \mathrm{Yield}_{\mathrm{median}}}{2}$$

The aim is to increase the reaction yield as far as possible.

Process Excellence Index for Cycle Time ($\mathrm{PEI}_{\mathrm{Cycle\ Time}}$) [9]

A parameter used by process chemists to gauge the reproducibility and robustness of the cycle time for a reaction or batch given by Equation (A.18).

$$\mathrm{PEI}_{\mathrm{cycle\ time}} = \frac{\mathrm{Cycle\ time}_{\mathrm{aspirational}}}{\mathrm{Cycle\ time}_{avg}} \times 100\% \tag{A.18}$$

where:

$$\mathrm{Cycle\ time}_{\mathrm{avg}} = \frac{\sum_{j=1}^{N} \mathrm{Cycle\ time}_{\mathrm{run}_j}}{N}$$

N is the number of times a reaction is run (number of batches)

$$\mathrm{Cycle\ time}_{\mathrm{aspirational}} = \frac{\mathrm{cycle\ time}_{\mathrm{best}} + \mathrm{Cycle\ time}_{\mathrm{median}}}{2}$$

The aim is to reduce the cycle time as far as possible.

Process Mass Intensity Complexity Model [10]

This metric estimates *PMI* for a synthesis of a given compound on the basis of its structural attributes without having to go through the tedious task of working through the experimental procedure for each reaction step-by-step. The expression is given by Equation (A.19).

$$PMI_{\text{est}} = 131 + 26C + 40H - 515A + 57(C - 1.5)(H - 8) \tag{A.19}$$

where:

- C is the number of chiral centers
- H is the number of heteroatoms
- A is the fraction of aromatic atoms given by Equation (A.20).

$$A = \frac{c_{\text{aromatic}} + H_{\text{aromatic}}}{c_{\text{total}} + H_{\text{total}}} \tag{A.20}$$

where:

- c(aromatic) is the number of carbon atoms in aromatic rings
- H(aromatic) is the number of heteroatoms (N, S, O) in aromatic rings
- c(total) is the total number of carbon atoms
- H(total) is the total number of heteroatoms

Quality Service Level (QSL) [9]

A parameter used in process chemistry that gauges the reliability and robustness of a given reaction in a production scale process. It is defined as shown in Equation (A.21).

$$\begin{aligned} \text{QSL} &= \frac{N - (n_{Q1}Q_1 + n_{Q2}Q_2 + n_{Q3}Q_3 + n_{Q4}Q_4)}{N} \\ &= \frac{N - (n_{Q_1}0 + n_{Q_2}0 + n_{Q_3}0.5 + n_{Q4}1)}{N} \end{aligned} \tag{A.21}$$

where:

- N is the number of times a reaction is repeated
- n_{Q_1} is the number of occurrences where full compliance is met with respect to the quality of the output product specifications as outlined in the master batch record (MBR)
- n_{Q_2} is the number of occurrences involving minor but acceptable deviations in the output product
- n_{Q_3} is the number of occurrences involving the output product being rejected for re-analysis but can be re-worked back into the process
- n_{Q4} is the number of occurrences involving the output product being rejected because it did not meet the standards in the MBR and is destined not to be used further in the overall production process

The corresponding Q values are penalty points.

For a synthesis plan involving M steps, the overall QSL value is given by Equation (A.22).

$$QSL_{overall} = \prod_{j=1}^{M} (QSL)_j \tag{A.22}$$

Relative Process Greenness [11]

The relative process greenness (RPG) metric is defined as the ratio of the green aspiration level (GAL) metric to the process mass intensity (*PMI*) metric. It was developed by a consortium of pharmaceutical companies to assess the synthesis performances of a given drug molecule in early, late, and commercial stage developments.

Renewability Index [12]

The renewability index is the ratio of total energy supplied from renewable resources to total energy supplied to a process. Essentially, it is the fraction of energy input to a process that originates from non-fossil fuel sources.

Renewables Intensity [13, 14]

Renewables intensity (RI) is defined as the ratio of mass of renewably derivable materials used in a reaction to mass of product collected.

Renewables Percentage [14]

Renewables percentage (RP) is defined as the ratio of renewables intensity and process mass intensity times 100 as shown in Equation (A.23).

$$RP = \left(\frac{RI}{PMI}\right)100 \tag{A.23}$$

See *renewables intensity.*

Selectivity

A term used to describe reactivity according to a specific region or group or structural motif in a given molecular structure. There are three types of selectivity: (a) *chemoselectivity*: a general term used to describe desired selectivity in carrying out a reaction according to some specific chemical group over all others in a given molecule; (b) *regioselectivity*: desired selectivity in carrying out a reaction according to a specific region or group over all others in a given molecule; (c) *stereoselectivity*: desired selectivity in carrying out a reaction according to a specific stereochemical group over all others in a given molecule.

Space Time Yield

A metric often used by process chemists to measure synthesis production efficiency is space-time-yield (STY) given by Equation (A.24).

$$STY = \frac{\text{Mass of product}}{(\text{Total process time})(\text{Total volume of input materials})} \tag{A.24}$$

It can be applied to single chemical reactions or synthesis plans. The units are kg/m^3/h, kg/m^3/s, and kg/L/h, kg/L/s. Sometimes the volume of the reactor is used instead of the volume of input materials.

Spider Diagram

This term is synonymous with the term "radial diagram."

Sustainability Metric

This term is synonymous with the term "green metric."

Volume-Time Output (VTO) [9]

A parameter used in process chemistry defined on a per-step basis as shown in Equation (A.25).

$$\text{VTO} = \frac{t \times \sum V_{\text{reactors}}}{m_{\text{product}}} \tag{A.25}$$

where:

t	is the time per batch (h)
$V_{reactors}$	is the nominal volume of all reactors (m^3)
$m_{product}$	is the mass of output product (kg)

REFERENCES

1. Augé, J. *Green Chem.* 2008, *10*, 225.
2. Steinbach, A.; Winkenbach, R. *Chem. Eng.* 2000, pp. 94–96, 98, 100, 102, 104.
3. Reetz, M.T.; Sell, T.; Meiswinkel, A.; Mehler, G. *Angew. Chem. Int. Ed.* 2003, *42*, 790.
4. Patti, A. *Green Approaches to Asymmetric Catalytic Synthesis*, Springer: New York, 2011, p. 43.
5. Hudlicky, T.; Frey, D.A.; Koroniak, C.D.; Claeboe, L.E.; Brammer, L.E. *Green Chem.* 1999, *2*, 57.
6. Lee, S.; Robinson, G. *Process Development: Fine Chemicals from Grams to Kilograms*, Oxford University Press: Oxford, 1995, p. 13.
7. Roschangar, F.; Sheldon, R.; Senanayake, C. H. *Green Chem.* 2015, *17*, 752.
8. McGonagle, F.I.; Sneddon, H.F.; Jamieson, C.; Watson, A.J.B. *ACS Sust. Chem. Eng.* 2014, *2*, 523.
9. Dach, R.; Song, J.J.; Roschangar, F.; Samstag, W.; Senanayake, C.H. *Org. Process Res. Dev.* 2012, *16*, 1697.
10. Kjell, D.P.; Watson, I.A.; Wolfe, C.N.; Spitler, J.T. *Org. Process Res. Dev.* 2013, *17*, 169.
11. Roschangar, F.; Colberg, J.; Dunn, P.J.; Gallou, F.; Hayler, J.D.; Koenig, S.G.; Kopach, M.E.; Leahy, D.K.; Mergelsberg, I.; Tucker, J.L.; Sheldon, R.A.; Senanayake, C.H. *Green Chem.* 2017, *19*, 281.
12. Ruiz-Mercado, G.J.; Smith, R.L.; Gonzalez, M.A. *Ind. Eng. Chem. Res.* 2012, *51*, 2309.
13. Jiménez-González, C.; Constable, D.J.C.; Ponder, C.S. *Chem. Soc. Rev.* 2012, *41*, 1485.
14. McElroy, C.R.; Constantinou, A.; Jones, L.C.; Summerton, L.; Clark, J.H. *Green Chem.* 2015, *17*, 3111.

Index

L

M

N

O

P

Q

R

S

T

V

W

Y

For Product Safety Concerns and Information please contact our EU representative GPSR@taylorandfrancis.com Taylor & Francis Verlag GmbH, Kaufingerstraße 24, 80331 München, Germany